Theory of Modeling and Simulation

Discrete Event and Iterative System Computational Foundations

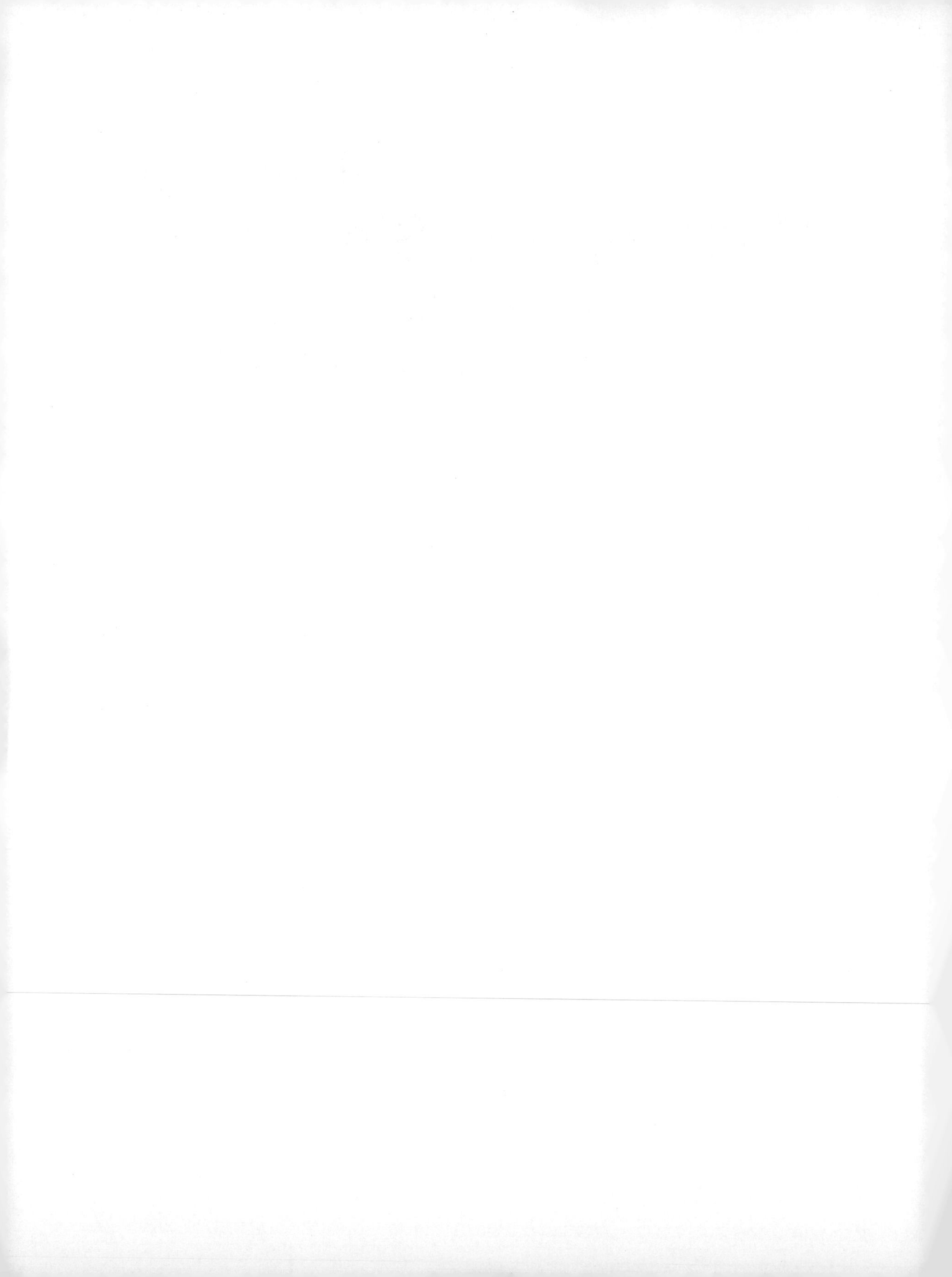

Theory of Modeling and Simulation

Discrete Event and Iterative System Computational Foundations

Third Edition

Bernard P. Zeigler
University of Arizona
Tucson, USA

Alexandre Muzy
CNRS, I3S Laboratory
Universté Côte d'Azur
Nice Sophia Antipolis, France

Ernesto Kofman
FCEIA – Universidad Nacional de Rosario
CIFASIS – CONICET
Rosario, Argentina

Academic Press is an imprint of Elsevier
125 London Wall, London EC2Y 5AS, United Kingdom
525 B Street, Suite 1650, San Diego, CA 92101, United States
50 Hampshire Street, 5th Floor, Cambridge, MA 02139, United States
The Boulevard, Langford Lane, Kidlington, Oxford OX5 1GB, United Kingdom

This is the third and revised edition of *Theory of Modeling and Simulation*, Bernard P. Zeigler, published by Wiley Interscience, 1976 with reissue by Krieger Pub. 1984

Library of Congress Cataloging-in-Publication Data
A catalog record for this book is available from the Library of Congress

British Library Cataloguing-in-Publication Data
A catalogue record for this book is available from the British Library

ISBN: 978-0-12-813370-5

For information on all Academic Press publications
visit our website at https://www.elsevier.com/books-and-journals

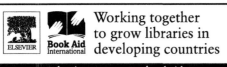

Publisher: Katey Birtcher
Acquisition Editor: Katey Birtcher
Editorial Project Manager: Karen Miller
Production Project Manager: Nilesh Kumar Shah
Designer: Mark Rogers

Typeset by VTeX

Contents

PART 2 ITERATIVE SYSTEM SPECIFICATION

PART 3 SYSTEM MORPHISMS: ABSTRACTION, REPRESENTATION, APPROXIMATION

Please find the companion website at http://textbooks.elsevier.com/web/Manuals.aspx?isbn=9780128133705

Contributions

The authors wish to acknowledge, with much thanks, important contributions of colleagues to this edition.

1. James J. Nutaro was the primary author of Chapter 14, "Parallel and distributed discrete event simulation."
2. Rodrigo Castro was the primary author of Chapter 24, "Open Research Problems: Systems Dynamics, Complex Systems."
3. Damian Vicino contributed to Section 4.7 on "Are DEVS state sets essentially discrete?"
4. Damien Foures, Romain Franceschini and Paul-Antoine Bisgambiglia contributed to Section 7.7 on "Multi-Component Parallel Discrete Event System Formalism."
5. Maria Julia Blas, Adelinde Uhrmacher, A. Hamri, contributed to Section 10.10 on "Closure under Coupling: Concept, Proofs, and Importance."
6. Jean-Fançois Santucci and Laurent Cappocci contributed to Section 17.7 "Handling Time Granularity Together with Abstraction."
7. Franck Grammont contributed to the description of Figure 23.1.
8. Ciro Barbageletta assisted with the LaTex formatting.
9. Daniel Foguelman, Pedro Rodríguez and Hernán Modrow. Students at the University of Buenos Aires, developed the XSMILE to DEVSML translation algorithm.

In addition we greatly appreciate and acknowledge the contributions of co-authors Herbert Praehorfer and Tag Fon Kim of the Second Edition that replicated in this edition.

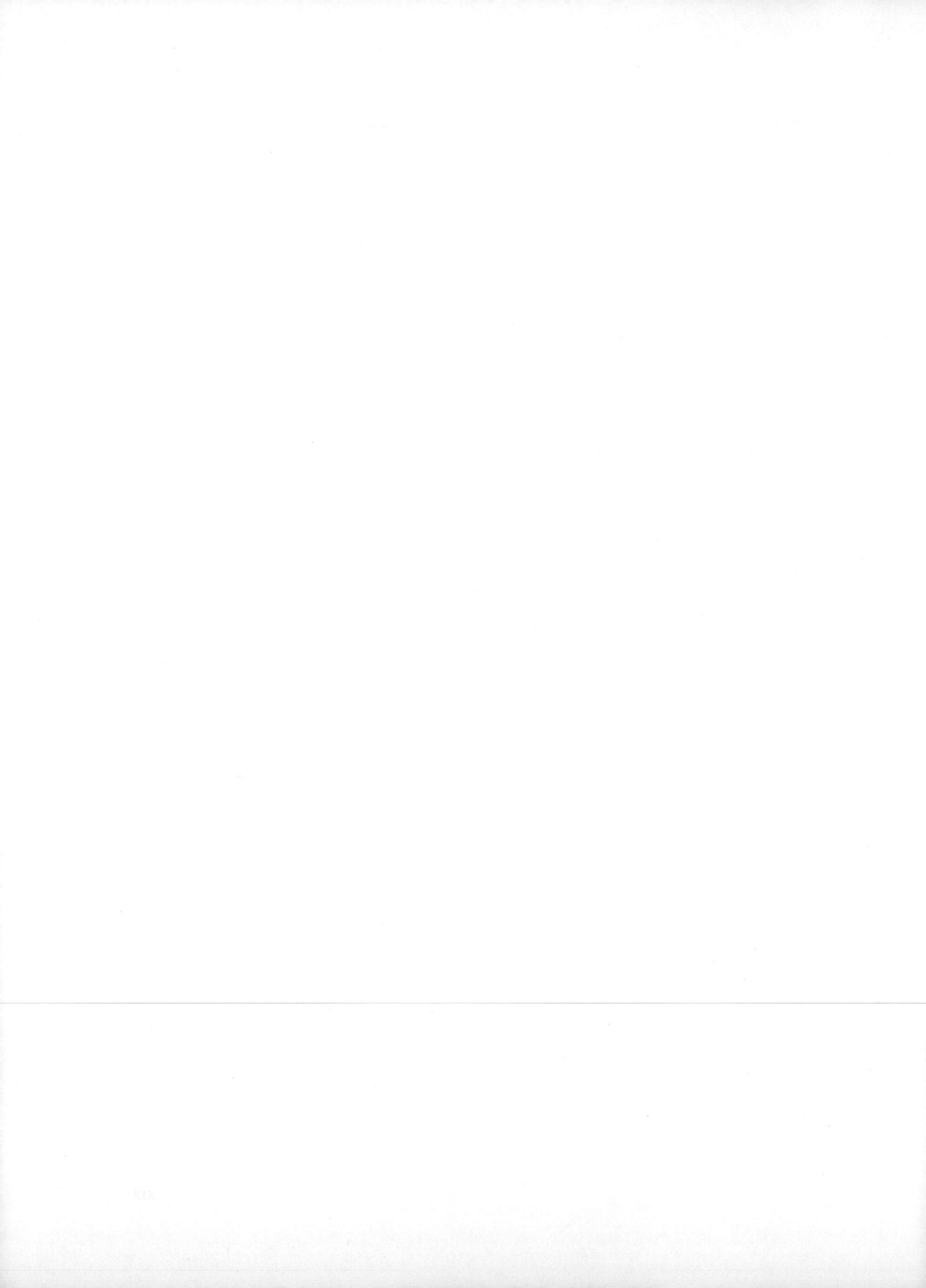

Preface to the Third Edition

A consensus on the fundamental status of theory of modeling and simulation is emerging – some recognize the need for a theoretical foundation for M&S as a science. Such a foundation is necessary to foster the development of M&S-specific methods and the use of such methods to solve real world problems faced by practitioners. "[Theory of Modeling and Simulation (1976)] gives a theory for simulation that is based on general system theory and this theory is considered the only major theory for simulation. This book showed that simulation has a solid foundation and is not just some ad hoc way of solving problems." (Sargent, 2017). "Theory of Modeling and Simulation is a major reference for modeling formalisms, particularly the Discrete Event Systems Specification (DEVS). ... We mention the System Entity Structures and Model Base (SES/MB) framework as breakthrough in this field [Model-base management]. It enables efficiency, reusability and interoperability." (Durak et al., 2017).

For others there is the acknowledgment that certain of the theory's basic distinctions such as the separation, and inter-relation, of models and simulators, are at least alternatives to be considered in addressing core M&S research challenges. Such challenges, and the opportunities to address them, are identified in areas including conceptual modeling, computational methods and algorithms for simulation, fidelity issues and uncertainty in M&S, and model reuse, composition, and adaptation (Fujimoto et al., 2017).

With the assertion that "an established body of knowledge is one of the pillars of an established discipline" (Durak et al., 2017), this third edition is dedicated to the inference that theory of M&S is an essential component, and organizing structure, for such a body of knowledge. A prime emphasis of this edition is on the central role of iterative specification of systems. The importance of iterative system specification is that it provides a solid foundation for the computational approach to complex systems manifested in modeling and simulation. While earlier editions introduced iterative specification as the common form of specification for unifying continuous and discrete systems, this edition employs it more fundamentally throughout the book. In addition to the new emphasis, throughout the book there are updates to earlier material outlining significant enhancements from a broad research community. To accommodate space for such additions some sections of the last edition have been omitted, not because of obsolescence – indeed, new editions may re-instate these parts.

This Third Edition coordinates with a second book. "Model Engineering for Simulation" (MES) to provide both a theoretical and application-oriented account of modeling and simulation. This makes sense as a coordinated "package", since most of the background theory material will be contained in this book and the application to model engineering will be contained in MES. This partitioning into theory and practice avoids unnecessary redundancy. The books will be published synchronously (or as closely timed as practical). The editor/leaders of the two books have coordinated closely to assure that a coherent whole emerges that is attractive to a large segment of the simulation community.

REFERENCES

Durak, U., Ören, T., Tolk, A., 2017. An Index to the Body of Knowledge of Simulation Systems Engineering. John Wiley & Sons, Inc., pp. 11–33.

Fujimoto, R., Bock, C., Chen, W., Page, E., Panchal, J.H., 2017. Research Challenges in Modeling and Simulation for Engineering Complex Systems. Springer.

Sargent, R.G., 2017. A perspective on fifty-five years of the evolution of scientific respect for simulation. In: Simulation Conference (WSC). 2017 Winter. IEEE, pp. 3–15.

Preface to the Second Edition

This is the second edition of Theory of Modeling and Simulation originally published by Wiley Interscience in 1976 and reissued by Krieger Publishers in 1984. The first edition made the case that a theory was necessary to help bring some coherence and unity to the ubiquitous field of modeling and simulation. Although nearly a quarter of a century later has seen many advances in the field, we believe that the need for a widely accepted framework and theoretical foundation is even more necessary today. Modeling and simulation lore is still fragmented across the disciplines making it difficult to share in the advances, reuse other discipline's ideas, and work collaboratively in multidisciplinary teams. As a consequence of the growing specialization of knowledge there is even more fragmentation in the field now then ever. The need for "knowledge workers" who can synthesize disciplinary fragments into cohesive wholes is increasingly recognized. Modeling and simulation – as a generic, non-discipline specific, set of activities – can provide a framework of concepts and tools for such knowledge work.

In the years since the first edition, there has been much significant progress in modeling and simulation but the progress has not been uniform across the board. Generally, model building and simulation execution have been made easier and faster by riding piggyback on the technology advances in software (e.g. object-oriented programming) and hardware (e.g., faster processors). However, hard, fundamental issues such as model credibility (e.g., validation, verification and model family consistency) and interoperation (e.g., repositories, reuse of components, and resolution matching) have received a lot less attention. But these issues are now moving to the front and center under the impetus of the High Level Architecture (HLA) standard mandated by the United States Department of Defense for all its contractors and agencies.

In this edition, two major contributors to the theory of modeling and simulation join with the original author to completely revise the original text. As suggested by its subtitle, the current book concentrates on the integration of the continuous and discrete paradigms for modeling and simulation. A second major theme is that of distributed simulation and its potential to support the co-existence of multiple formalisms in multiple model components.

Although the material is mostly new, the presentation format remains the same. There are three major sections. Part I introduces a framework for modeling and simulation and the primary continuous and discrete approaches to making models and simulating them on computers. This part offers a unified view of the field that most books lack and, written in an informal manner, it can be used as instructional material for undergraduate and graduate courses.

Part II revisits the introductory material but with a rigorous, multi-layered systems theoretic basis. It then goes on to provide an in-depth account of models as systems specifications, the major systems specification formalisms and their integration, and simulators for such formalisms, in sequential, parallel and distributed forms.

The fundamental role of systems morphisms is taken up in Part III: any claim relating systems, models and simulators to each other ultimately must be phrased with an equivalence or morphism of such kinds. Both perfect and approximate morphisms are discussed and applied to model abstraction and system representation. Especially, in the latter vein, we focus on the ability of the DEVS (Discrete Event System Specification) formalism to represent arbitrary systems including those specified in other discrete event and continuous formalisms. The importance of this discussion derives from two sources: the burgeoning use of discrete event approaches in high technology design (e.g., manufacturing control systems, communication, computers) and the HLA-stimulated growth of distributed simulation, for which discrete events match the discreteness of message exchange.

Part IV continues with the theme of DEVS-based modeling and simulation as a foundation for a high technology systems design methodology. We include integration with other formalisms for analysis and the system entity structure/model base concepts for investigating design alternatives and reusing good designs. Thoughts on future support of collaborative modeling and simulation close the book.

Although primarily intended as a reference, the structure of the book lends itself for use as a textbook in graduate courses on modeling and simulation. As a textbook, the book affords the advantage of providing an open systems view that mitigates the closed box trust-on-faith approach of many commercial domain-specific simulation packages. If nothing else, the student will have a more sophisticated skepticism about the model reliability and simulator correctness inside such boxes. For hands on experience, the book needs to be supplemented with an instructional modeling and simulation environment such as DEVSJAVA (available from the web site: https://acims.asu.edu/). Other books on statistical aspects of simulation and application to particular domains should be part of the background as well.

We suggest that Part IV might be a good place to start reading, or teaching, since most of the concepts developed earlier in the book are put into use in the last chapters. In this strategy, the learner soon realizes that new concepts are needed to achieve successful designs and is motivated to fill in the gaps by turning to the chapters that supply the requisite knowledge. More likely, a good teacher will guide the student back and forth between the later and earlier material.

Space limitations have prevented us from including all the material in the first edition. The decision on what to leave out was based on relevance to the current theme, whether significant progress had been made in an area, and whether this could be reduced to the requirements of a book. Thus, for example, a major omission is the discussion of structural inference in Chapters 14 and 15 of the original. We hope that a next revision would be able to include much more on developments in these important directions.

PART

1

BASICS: MODELING FORMALISMS AND SIMULATION ALGORITHMS

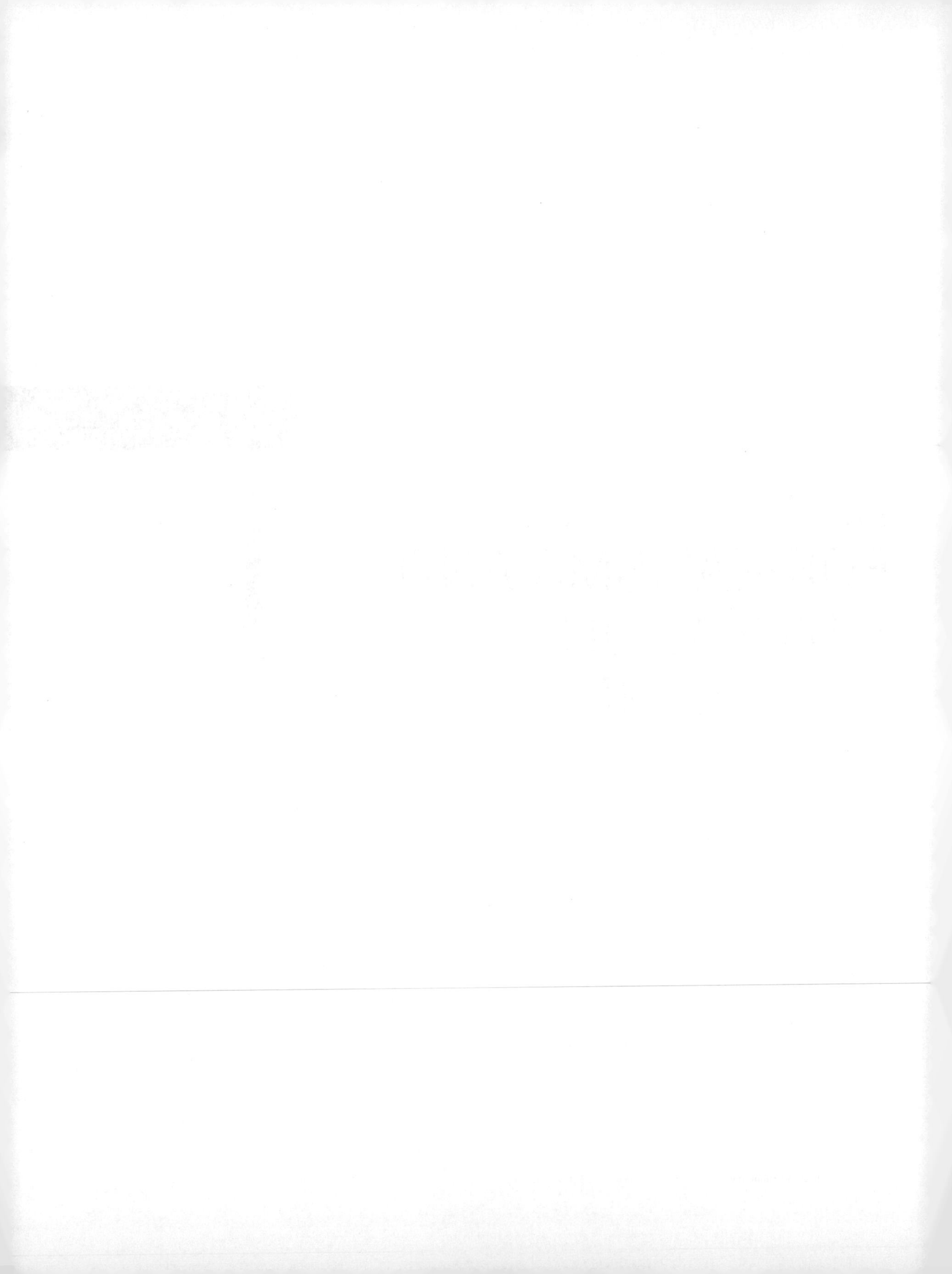

CHAPTER

INTRODUCTION TO SYSTEMS MODELING CONCEPTS 1

CONTENTS

This chapter introduces some key concepts that underlie the framework and methodology for modeling and simulation (M&S) originally presented in "Theory of Modeling and Simulation" published in 1976 – referred to hereafter as TMS76 to distinguish it from the current revised third edition TMS2018. Perhaps the most basic concept is that of mathematical systems theory. First developed in the nineteen sixties, this theory provides a fundamental, rigorous mathematical formalism for representing dynamical systems. There are two main, and orthogonal, aspects to the theory:

- *Levels of system specification*: these are the levels at which we can describe how systems behave and the mechanisms that make them work the way they do.
- *Systems specification formalisms*: these are the types of modeling styles, such continuous or discrete, that modelers can use to build system models.

Although the theory is quite intuitive, it does present an abstract way of thinking about the world that you will probably find unfamiliar. So we introduce the concepts in a spiral development consisting of easy-to-grasp stages – with each spiral revolution returning to a more faithful version of the full story.

Theory of Modeling and Simulation. https://doi.org/10.1016/B978-0-12-813370-5.00009-2

In this chapter we first introduce some basic systems concepts, then motivate the systems specification formalisms by describing their evolution over time. This also provides a way to point out the differences between earlier editions and this one (TMS2018). Finally, we discuss the levels of system specification, illustrating them with familiar examples. In this ground stage of our spiral development, the presentation is informal and prepares the way for the framework for M&S that comes in the next chapter. Later, in the second part of the book, we return to a more rigorous development of the concepts to lay a sound basis for the developments to come in the third part.

1.1 SYSTEMS SPECIFICATION FORMALISMS

System theory distinguishes between system structure (the inner constitution of a system) and behavior (its outer manifestation). Viewed as a black box (Fig. 1.1) the external behavior of a system is the relationship it imposes between its input time histories and output time histories. The system's input/output behavior consists of the pairs of data records (input time segments paired with output time segments) gathered from a real system or model. The internal structure of a system includes its state and state transition mechanism (dictating how inputs transform current states into successor states) as well as the state-to-output mapping. Knowing the system structure allows us to deduce (analyze, simulate) its behavior. Usually, the other direction (inferring structure from behavior) is not univalent – indeed, discovering a valid representation of an observed behavior is one of the key concerns of the M&S enterprise.

An important structure concept is that of *decomposition* namely, how a system may be broken down into component systems (Fig. 1.2). A second concept is that of *composition*, i.e., how component systems may be coupled together to form a larger system. Systems theory is *closed under composition* in that the structure and behavior of a composition of systems can be expressed in the original system theory terms. The ability to continue to compose larger and larger systems from previously constructed components leads to hierarchical construction. Closure under composition guarantees that such a composition results in a system, called its *resultant*, with well-defined structure and behavior. *Modular*

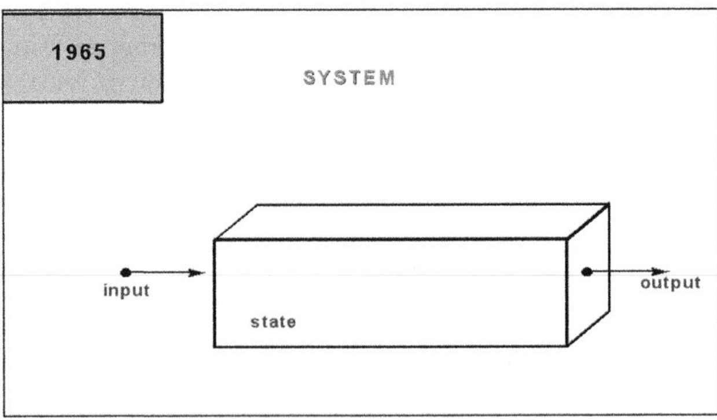

FIGURE 1.1

Basic System Concepts.

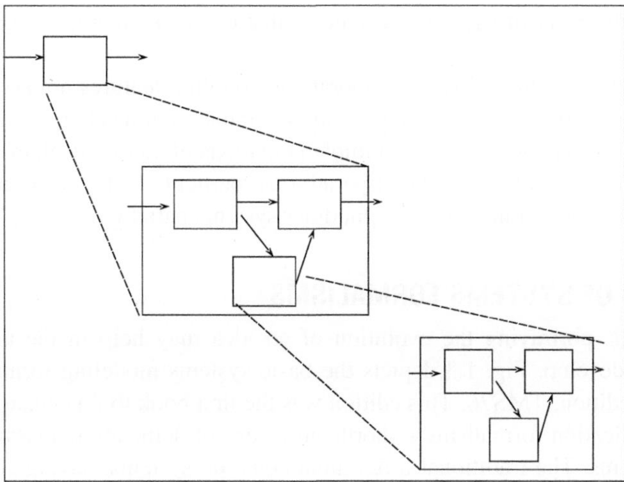

FIGURE 1.2

Hierarchical System Decomposition.

systems have recognized input and output ports through which all interaction with the environment occurs. They can be coupled together by coupling output ports to input ports and can have hierarchical structure in which component systems are coupled together to form larger ones.

The difference between a decomposed systems, as in Fig. 1.2, and undecomposed systems, as in Fig. 1.1, provides our first introduction to levels of systems specification. We'll say later that the former are at a higher level of specification than the latter since they provide more information about the structure of the system.

1.1.1 RELATION TO OBJECT ORIENTATION

Models developed in a system theory paradigm bear a resemblance to concepts of object-oriented programming. Both objects and system models share a concept of internal state. However, mathematical systems are formal structures that operate on a time base while programming objects typically do not have an associated temporal semantics. Objects in typical object oriented paradigms are not hierarchical or modular in the sense just described. The coupling concept in modular systems provides a level of delayed binding – a system model can place a value on one of its ports but the actual destination of this output is not determined until the model becomes a component in a larger system and a coupling scheme is specified. It can therefore: a) be developed and tested as a stand alone unit, b) be placed in a model repository and reactivated at will and c) reused in any applications context in which its behavior is appropriate and coupling to other components makes sense.

While coupling establishes output-to-input pathways, the systems modeler is completely free to specify how data flows along such channels. Information flow is one of many interactions that may be represented. Other interactions include physical forces and fields, material flows, monetary flows, and social transactions. The systems concept is broad enough to include the representation of any of these and supports the development of M&S environments that can make including many within the same large-scale model.

Although systems models have formal temporal and coupling features not shared by conventional objects, object-orientation does provide a supporting computational mechanism for system modeling. Indeed, there have been many object-oriented implementations of hierarchical, modular modeling systems These demonstrate that object-oriented paradigms, particularly for distributed computing, can serve as a strong foundation to implement the modular systems paradigm.

1.1.2 EVOLUTION OF SYSTEMS FORMALISMS

As in many situations, portraying the evolution of an idea may help in the understanding of the complexities as they develop. Fig. 1.3 depicts the basic systems modeling formalisms as they were presented in the first edition, TMS76. This edition was the first book to formulate approaches to modeling as system specification formalisms – shorthand means of delineating a particular system within a subclass of all systems. The traditional differential equation systems, having continuous states and continuous time, were formulated as the class of DESS (Differential Equation System Specifications). Also, systems that operated on a discrete time base such as automata were formulated as the class of DTSS (Discrete Time System Specifications). In each of these cases, mathematical representation had proceeded their computerized incarnations (it has been three hundred years since Newton-Leibnitz!).

However, the reverse was true for the third class, the Discrete Event System Specifications (DEVS). Discrete event models were largely prisoners of their simulation language implementations or algorithmic code expressions. Indeed, there was a prevalent belief that discrete event "world views" constituted new mutant forms of simulation, unrelated to the traditional mainstream paradigms. Fortunately, that situation has begun to change as the benefits of abstractions in control and design became clear. Witness the variety of discrete event dynamic system formalisms that have emerged (Ho, 1992). Examples are Petri Nets, Min-Max algebra, and GSMP (generalized semi-Markov processes). While each one has its

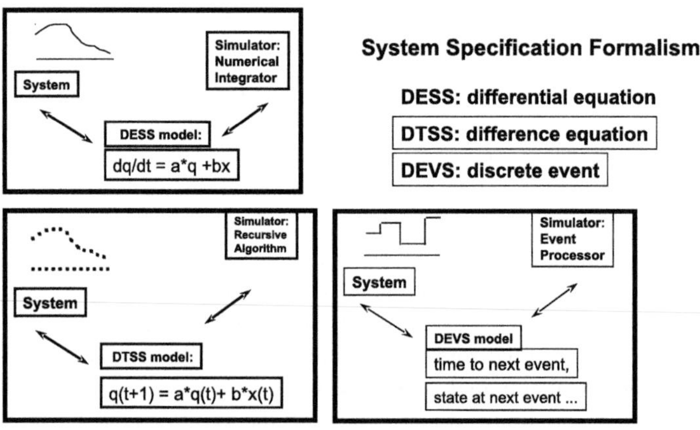

FIGURE 1.3

Basic Systems Specification Formalisms.

application area, none were developed deliberately as subclasses of the systems theory formalism. Thus to include such a formalism into an organized system-theory based framework requires "embedding" it into DEVS.

"Embedding." What could such a concept mean? The arrows in Fig. 1.4 indicate subclass relationships; for example, they suggest that DTSS is a "subclass of" DEVS. However, it is not literally true that any discrete time system is also discrete event system (their time bases are distinct, for example). So we need a concept of simulation that allows us to say when one system can do the essential work of another. One formalism can be embedded in another if any system in the first can be simulated by some system in the second. Actually, more than one such relationship, or morphism, may be useful, since, as already mentioned, there are various levels of structure and behavior at which equivalence of systems could be required. As a case in point, the TMS76 edition established that any DTSS could be simulated by a DEVS by constraining the time advance to be constant. However, this is not as useful as it could be until we can see how it applies to decomposed systems. Until that is true, we either must reconstitute a decomposed discrete time system to its resultant before representing it as a DEVS or we can represent each DTSS component as a DEVS but we can't network the DEVS together to simulate the resultant. TMS2000 established this stronger simulation relation and we discuss its application in Chapters 18 and 20 of this edition.

1.1.3 CONTINUOUS AND DISCRETE FORMALISMS

Skipping many years of accumulating developments, the next major advance in systems formalisms was the combination of discrete event and differential equation formalisms into one, the DEV&DESS. As shown in Fig. 1.5, this formalism subsumes both the DESS and the DEVS (hence also the DTSS) and thus supports the development of coupled systems whose components are expressed in any of the basic formalisms. Such *multi-formalism* modeling capability is important since the world does not

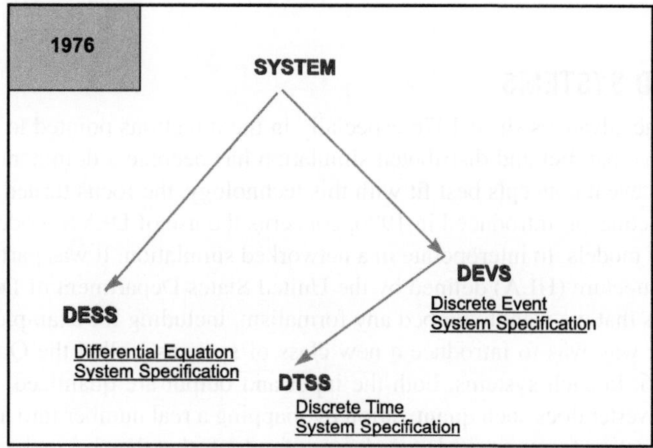

FIGURE 1.4

The Dynamics of Basic System Classes.

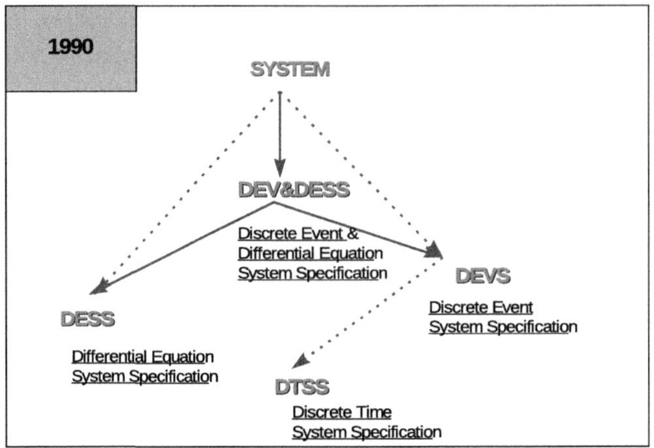

FIGURE 1.5

Introducing the DEV&DESS Formalism.

usually lend itself to using one form of abstraction at a time. For example, a chemical plant is usually modeled with differential equations while its control logic is best designed with discrete event formalisms. In 1990, Praehofer (Praehofer, 1991) showed that DEV&DESS was closed under coupling and in order to do so, had to deal with the pairs of input-output interfaces between the different types of systems. Closure under coupling also required that the DEV&DESS formalism provide a means to specify components with intermingled discrete and continuous expressions. Finally, simulator algorithms (so called *abstract simulator*) had to be provided to establish that the new formalism could be implemented in computational form (look ahead to Chapter 9 to see how this was all accomplished).

1.1.4 QUANTIZED SYSTEMS

TMS2000 built on the advances since 1976 especially in the directions pointed to by the introduction of DEV&DESS. Since parallel and distributed simulation has become a dominant form of model execution, and discrete event concepts best fit with this technology, the focus turned to a concept called the DEVS *bus*. This concept, introduced in 1996, concerns the use of DEVS models, as a "wrappers" to enable a variety of models, to interoperate in a networked simulation. It was particularly germane to the High Level Architecture (HLA) defined by the United States Department of Defense. One way of looking at this idea is that we want to embed any formalism, including for example, the DEV&DESS, into DEVS. Another way was to introduce a new class of systems, called the Quantized System, as illustrated in Fig. 1.6. In such systems, both the input and output are quantized. As an example, an analog-to-digital converter does such quantization by mapping a real number into a finite string of digits. In general, quantization forms equivalence classes of outputs that then become indistinguishable for downstream input receivers, requiring less data network bandwidth, but also possibly incurring error.

Quantization provides a process for representing and simulating continuous systems that is an alternative to the more conventional discretization of the time axis. While discretization leads to discrete

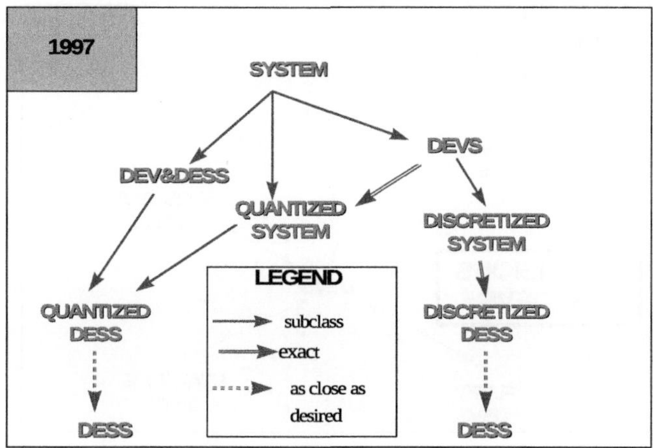

FIGURE 1.6

Introducing Quantized Systems.

time systems, quantization leads to discrete event systems. The theory of quantized state systems that has been developed since 2000 is presented in Chapter 20 of this edition.

When we restrict quantization to differential equation systems, we can express the resulting class, Quantized DESS, within DEV&DESS and study its properties, especially from the point of view of the DEVS bus. We can then study the approximation capability and simulation efficiency of DEVS in distributed simulation in comparison with classical time stepped integration and simulation approaches. Particularly with respect to reduction of message passing and network bandwidth (a major concern in distributed simulation) promising results are being obtained.

1.1.5 EXTENSIONS OF DEVS

Various extensions of DEVS have been developed as illustrated in Fig. 1.7. In the interest of space conservation, some of them are not discussed here while still available in TMS2000. Since our focus here is on Iterative System Specification as a modeling formalism, we present new types of such models in Chapter 12. These developments expand the classes of system models that can be represented and integrated within both DEVS and the parent systems theory formalism and UML can be used to classify newly developed variants and extensions (Blas and Zeigler, 2018).

These developments lend credence to the claim that DEVS is a promising computational basis for analysis and design of systems, particularly when simulation is the ultimate environment for development and testing (Fig. 1.8). The claim rests on the *universality* of the DEVS representation, namely the ability of DEVS bus to support the basic system formalisms. TMS2000 went some distance toward substantiating the claim that DEVS is the unique form of representation that underlies any system with discrete event behavior. In this edition, we expand the scope to consider Iterative Specification of Systems and the DEVS Bus support of co-simulation, the ability to correctly interoperate simulators embedding models from diverse formalisms within an integrated distributed simulation environment.

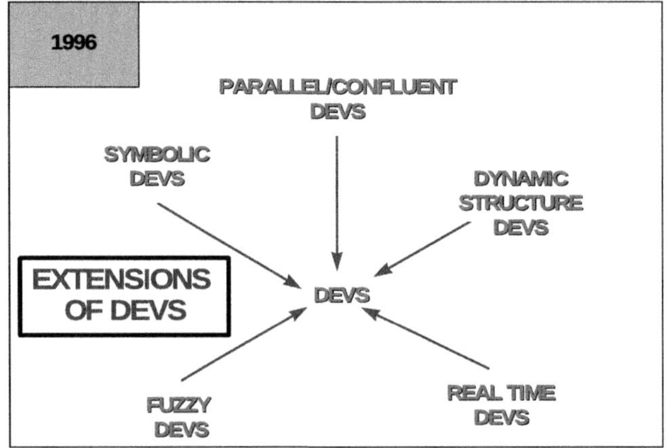

FIGURE 1.7

Extensions of the DEVS Formalism.

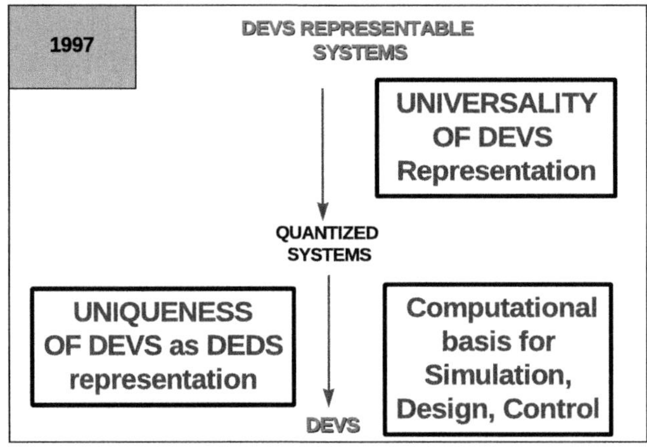

FIGURE 1.8

DEVS as a Computational Basis for Simulation, Design and Control.

1.2 LEVELS OF SYSTEM KNOWLEDGE

As already mentioned, the systems specification hierarchy is the basis for a framework for M&S which sets forth the fundamental entities and relationships in the M&S enterprise. The hierarchy is first presented in an informal manner and later in Chapter 5 in its full mathematical rigor. Our presentation starts with a review of George Klir's (Klir, 1985) systems framework.

Table 1.1 identifies four basic levels of knowledge about a system recognized by Klir. At each level we know some important things about a system that we didn't know at lower levels. At the lowest level,

Table 1.1 Levels of System Knowledge

Level	Name	What we know at this level
0	Source	what variables to measure and how to observe them
1	Data	data collected from a source system
2	Generative	means to generate data in a data system
3	Structure	components (at lower levels) coupled together to form a generative system

the *source level* identifies a portion of the real world that we wish to model and the means by which we are going to observe it. As the next level, the data level is a data base of measurements and observations made for the source system. When we get to Level 2, we have the ability to recreate this data using a more compact representation, such as a formula. Since typically, there are many formulas or other means to generate the same data, the generative level, or particular means or formula we have settled on, constitutes knowledge we didn't have at the data system level. When people talk about models in the context of simulation studies they are usually referring to the concepts identified at this level. That is, to them a model means a program to generate data. At the last level, the structure level, we have a very specific kind of generative system. In other words, we know how to generate the data observed at Level 1 in a more specific manner – in terms of component systems that are interconnected together and whose interaction accounts for the observations made. When people talk about systems, they are often referring to this level of knowledge. They think of reality as being made up of interacting parts – so that the whole is the sum (or a sometimes claimed, more, or less, than the sum) of its parts. Although some people use the term 'subsystems' for these parts, we call them **component** systems (and reserve the term subsystem for another meaning).

As we have suggested, Klir's terms are by no means universally known, understood, or accepted in the M&S community. However, his framework is a useful starting point since it provides a unified perspective on what are usually considered to be distinct concepts. From this perspective, there are only three basic kinds of problems dealing with systems and they involve moving between the levels of system knowledge (Table 1.2). In *systems analysis*, we are trying to understand the behavior of an existing or hypothetical system based on its known structure. *Systems inference* is done when we don't know what this structure is – so we try to guess this structure from observations that we can make. Finally, in *systems design*, we are investigating the alternative structures for a completely new system or the redesign of an existing one.

The central idea is that when we move to a lower level, we don't generate any really new knowledge – we are only making explicit what is implicit in the description we already have. One could argue that making something explicit can lead to insight, or understanding, which is a form of new knowledge, but Klir is not considering this kind of subjective (or modeler dependent) knowledge. In the M&S context, one major form of *systems analysis* is computer simulation which generates data under the instructions provided by a model. While no new knowledge (in Klir's sense) is generated, interesting properties may come to light of which we were not aware before the analysis. On the other hand, *systems inference* and *systems design* are problems that involve **climbing up** the levels. In both cases we have a low level system description and wish to come up with an equivalent higher level one. For *systems inference*, the lower level system is typically at the data system level, being data that we have observed from some existing source system. We are trying to find a generative system, or even a structure system, that can recreate the observed data. In the M&S context, this is usually called *model*

Table 1.2 Fundamental Systems Problems

Systems Problem	Does source of the data exist? What are we trying to learn about it?	Which level transition is involved?
systems analysis	The system being analyzed may exist or may be planned. In either case we are trying to understand its behavioral characteristics.	moving from higher to lower levels, e.g., using generative information to generate the data in a data system
systems inference	The system exists. We are trying to infer how it works from observations of its behavior.	moving from lower to higher levels, e.g., having data, finding a means to generate it
systems design	The system being designed does not yet exist in the form that is being contemplated.	We are trying to come up with a good design for it. moving from lower to higher levels, e.g. having a means to generate observed data, synthesizing it with components taken off the shelf.

construction. In the case of systems design, the source system typically does not yet exist and our objective is to build one that has a desired functionality. By functionality we mean what we want the system to do; typically, we want to come up with a structure system, whose components are technological, i.e., can be obtained off-the-shelf, or built from scratch from existing technologies. When these components are interconnected, as specified by a structure system's coupling relation, the result should be a real system that behaves as desired.

It is interesting to note that the process called *reverse engineering* has elements of both inference and design. To reverse engineer an existing system, such as was done in the case of the cloning of IBM compatible PCs, an extensive set of observations is first made. From these observations, the behavior of the system is inferred and an alternative structure to realize this behavior is designed thus bypassing patent rights to the original system design!

1.3 INTRODUCTION TO THE HIERARCHY OF SYSTEMS SPECIFICATIONS

At about the same time (in the early 1970's) that Klir introduced his epistemological (knowledge) levels, TMS76 formulated a similar hierarchy that is more oriented toward the M&S context. This framework employs a general concept of *dynamical system* and identifies useful ways in which such a system can be specified. These ways of describing a system can be ordered in levels as in Table 1.3. Just as in Klir's framework, at each level more information is provided in the specification that cannot be derived from lower levels. As can be seen in Table 1.3, these levels roughly correspond to those of Klir's framework.

The major difference between the two frameworks is that the System Specification Hierarchy recognize that simulation deals with **dynamics**, the way in which systems behave over time. Therefore, time is the base upon which all events are ordered. We also view systems as having *input* and *output* interfaces through which they can interact with other systems. As illustrated in Fig. 1.9, systems receive stimuli ordered in time through their *input ports*, and respond on their *output ports*. The term "port" sig-

Table 1.3 Relation between System Specification Hierarchy and Klir's levels

Level	Specification Name	Corresponds to Klir's	What we know at this level
0	Observation Frame	Source System	how to stimulate the system with inputs; what variables to measure and how to observe them over a time base;
1	I/O Behavior	Data System	time-indexed data collected from a source system; consists of input/output pairs.
2	I/O Function		Knowledge of initial state; given an initial state, every input stimulus produces a unique output.
3	State Transition	Generative System	How states are affected by inputs; given a state and an input what is the state after the input stimulus is over; what output event is generated by a state.
4	Coupled Component	Structure System	Components and how they are coupled together. The components can be specified at lower levels or can even be structure systems themselves – leading to hierarchical structure.

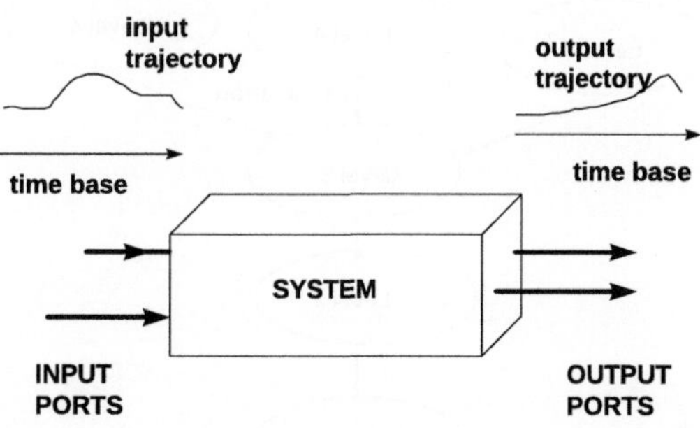

FIGURE 1.9

Input/Output System.

nifies a specific means of interacting with the system. Whether by stimulating it (input) or observing it (output). The time-indexed inputs to systems are called *input trajectories*; likewise, their time-indexed outputs are called *output trajectories*. Ports are the only channels through which one can interact with the system. This means that system are *modular*. While Klir's framework can include dynamics, input/output ports and modularity, it is not dedicated to these concepts. However, understanding these concepts is critical to effectively solving the problems that arise in M&S.

Before discussing each level of the specification hierarchy in some detail, let's observe that we could have the very same real world object specified simultaneously at each of the levels. Thus there should be a way to associate the next lower level specification with any given one. This association concept is illustrated in Fig. 1.10. For example, if we have know the detailed structure at the Coupled Component level, then we ought to be able to construct the corresponding specification at the State Transition level. The hierarchy is set up to provide such an *association mapping* at each (other than the lowest) level. Indeed, this is the formal version of climbing down the levels just discussed. Since the association mapping is not necessarily one to one, many upper level specifications may map to the same lower level one. This is the underlying reason why climbing up the levels is much harder than climbing down the levels. Indeed, when we select one of the associated upper level specifications for a given lower level one, we are gaining knowledge we didn't have at the lower level.

1.4 THE SPECIFICATION LEVELS INFORMALLY PRESENTED

1.4.1 OBSERVATION FRAME

The Observation Frame specifies how to stimulate the system with inputs; what variables to measure and how to observe them over a time base.

As an example, Fig. 1.11 shows a forest subject to *lightning*, *rain* and *wind*, modeled as input ports and *smoke* produced from fire, represented as an output port. This is a level 0 or Observation Frame specification. Note the choice of variables we included as ports and their *orientation* (i.e., whether they

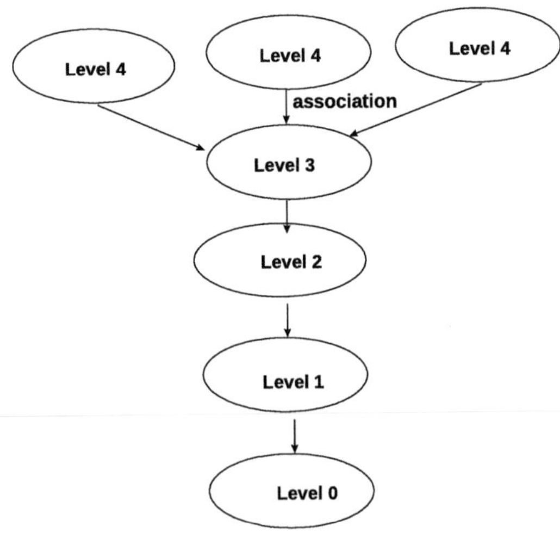

FIGURE 1.10

Association between levels of the System Specification Hierarchy.

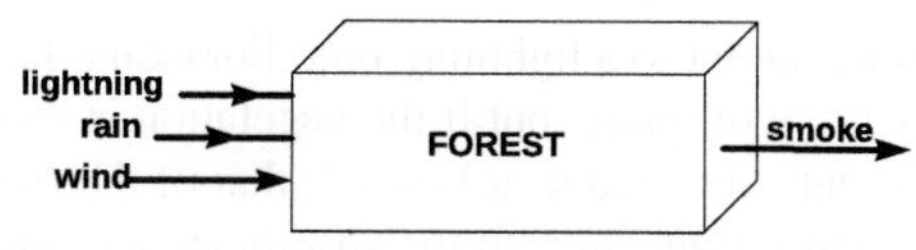

FIGURE 1.11

A forest specified as a system in the Observation Frame (Level 0).

are input or output ports). We could have chosen differently. For example, we could have included an output port to represent heat radiation. Moreover, rather than representing each variable by a single value, it can be distributed over space, i.e., represented by an array of values. Such choices depend on our modeling objectives and are specified through *experimental frames*, a concept which we discuss in Chapter 2.

Fig. 1.12 shows some examples of input and output trajectories. The input trajectory on the *lightning* port shows a bolt occurring at some particular time t0. Only one such bolt occurs in the time period shown. The *smoke* output trajectory, at the top, depicts a gradual build up of *smoke* starting at t0 (so presumably, caused by a fire started by the lightning bolt). The possible values taken on by *smoke*, called its range, would result from some appropriate measurement scheme, e.g., measuring density of particulate material in grams/cubic meter. The pair of input, and associated output, trajectories is called a *input/output* (or I/O) pair. Fig. 1.12 also displays a second I/O pair with the same input trajectory but different output trajectory. It represents the fact that there may be many responses to the same stimulus. In the second case, lightning did not cause a major fire, since the one that broke out quickly died. Such multiple output trajectories (for the same input trajectory) are characteristic of knowledge at Level 1. Knowing how to disambiguate these output trajectories is knowledge we will gain at the next level.

1.4.2 I/O BEHAVIOR AND I/O FUNCTION

The collection of all I/O pairs gathered by observation is called the *I/O Behavior* of a system. Returning to Table 1.3, this represents a system specification at Level 1. Now suppose that we are able to uniquely

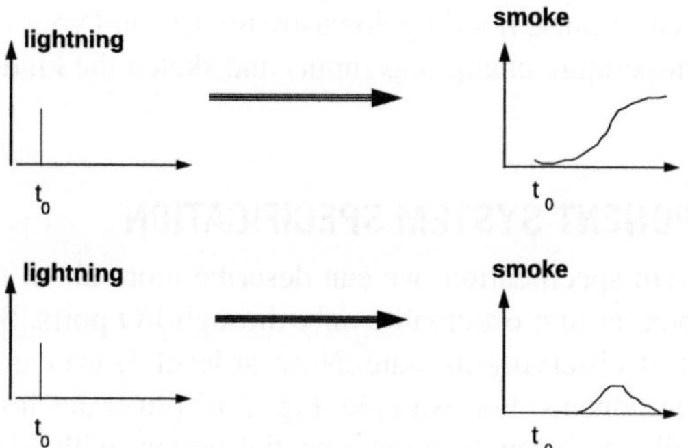

FIGURE 1.12

Some Input-Output Pairs for the Forest System Frame of Fig. 1.11.

predict the response of the smoke output to a lightning bolt. For example, suppose we know that if the vegetation is dry, then a major fire will ignite, but if the vegetation is moist then any fire will quickly die. Having such a factor represents knowledge at Level 2, that of the *I/O Function*. Here, in addition to lower level information, we add *initial states* to the specification – when the initial state is known, there is a functional relationship between input and output trajectories, i.e., the initial state determines the **unique** response to any input (Fig. 1.13A).

1.4.3 STATE TRANSITION SYSTEM SPECIFICATION

At the next level (3) of system specification, we can specify not only initial state information but also how the *state changes* as the system responds to its input trajectory. Fig. 1.13B and C illustrate this important concept. Fig. 1.13B presents the situation where the forest is in state (dry vegetation, unburned) when a lightning bolt occurs at time *t0*. The state that the forest is in at time *t1* when a second bolt occurs is (dry vegetation, burnt) reflecting the fact that a fire has ignited. Since the forest is in a different state, the effect of this second bolt is different from the first. Indeed, since there is little left to burn, there is no effect of the second bolt.

In contrast, Fig. 1.13C illustrates the situation where the forest is wet and unburned when the first bolt occurs. It does not cause a major fire, but it does dry out the vegetation so the resulting state is (dry, unburned). Now the second bolt produces a major fire, just as the first bolt did in Fig. 1.13B – since both the state and subsequent input trajectory are the same, the response of the system is the same.

Exercise. A watershed is a region like a valley or basin in which water collects and flows downward toward a river or sea. When it rains heavily the rain water starts to show up quite quickly at a measuring point in the river. For a lighter rain event very little of the rain water may be measured because it is absorbed in the ground. However, after several rain events the ground can get saturated and a light rain can send water quickly downstream. Sketch a set of input/output pairs similar to that of the forest fire to capture the dynamics of such rain events. What variable would you chose to represent that state of the system at the next level of specification.

Exercise. In climates where it can rain or snow, a watershed can have an even more interesting behavior. During winter snow from snow events might accumulate on the ground and there is no indication of any increase in water at the downstream measuring point. But as the temperature increases in spring, the melting snow can eventually cause flooding downstream. Expand your model of the last exercise to include snow events and temperature changes as inputs and sketch the kinds of input/output behaviors you would expect.

1.4.4 COUPLED COMPONENT SYSTEM SPECIFICATION

At the highest level of system specification, we can describe more about the internals of the system. Until now, it was a black box, at first observable only through I/O ports. Subsequently, we were able to peer inside to the extent of observing its state. Now, at level 4, we can specify how the system is composed of interacting components. For example, Fig. 1.14 illustrates how a forest system could be composed of interacting cells, each representing a spatial region, with adjacent cells interconnected. The cells are modeled at level 3, i.e., their state transition and output generation definitions are used to act upon inputs, and generate outputs, respectively to and from, other cells. The cells are coupled together using ports. The output ports of one cell are coupled to the input ports of its neighbors.

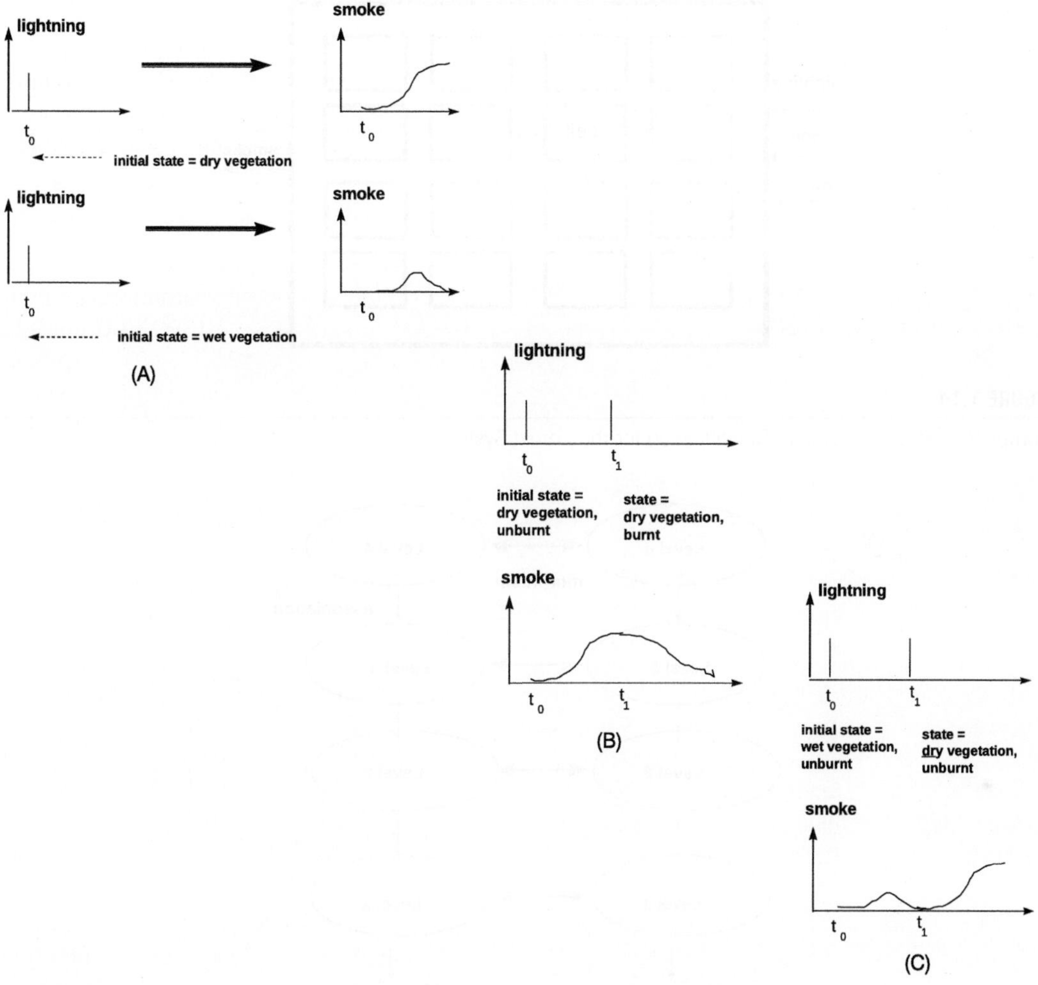

FIGURE 1.13

Initial State Concept: a specification at Level 3 (I/O Function) in which we have initial state knowledge about the forest.

1.5 SYSTEM SPECIFICATION MORPHISMS: BASIC CONCEPTS

The system specification hierarchy provides a stratification for constructing models. But, while constructing models is the basic activity in M&S, much of the real work involves establishing relationships between system descriptions. The system specification hierarchy also provides an orderly way of presenting and working with such relationships. Fig. 1.15 illustrates the idea that pairs of system can be related by morphism relations at each level of the hierarchy. A morphism is a relation that places elements of system descriptions into correspondence as outlined in Table 1.4.

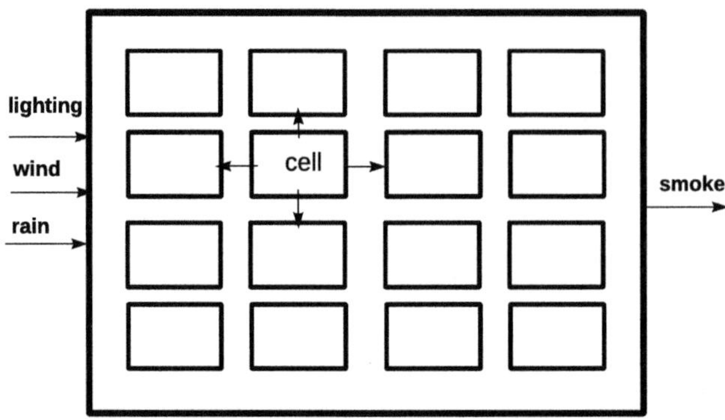

FIGURE 1.14

Component Structure System Specification for the Forest System.

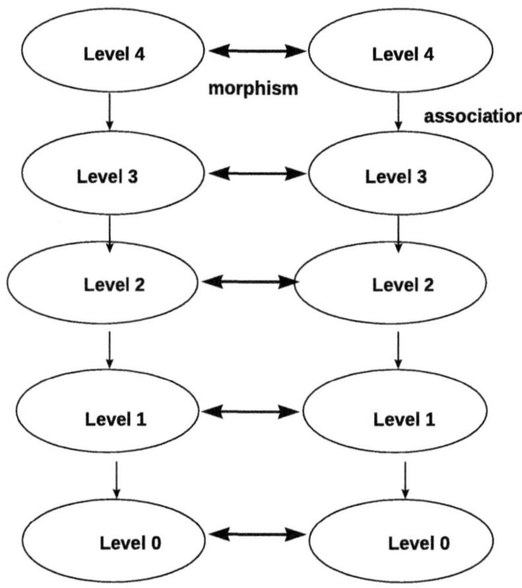

FIGURE 1.15

Morphism Concepts for System Specification Hierarchy.

For example, at the lowest level, two Observation Frames are *morphic*, if we can place their defining elements – inputs, outputs, and time bases into correspondence. Such Frames are *isomorphic* if their inputs, outputs, and time bases respectively, are identical. In general, the concept of morphism tries to capture similarity between pairs of systems at the same level of specification. Such similarity concepts have to be consistent between levels. When we associate lower level specifications with their respective

Table 1.4 Morphism relations between systems in System Specification Hierarchy and Klir's levels

Level	Specification Name	Two Systems are Morphic at this level if:
0	Observation Frame	their inputs, outputs and time bases can be put into correspondence
1	I/O Behavior	they are morphic at level 0 and the time-indexed input/output pairs constituting their I/O behaviors also match up in one-one fashion
2	I/O Function	they are morphic at level 0 and their initial states can be placed into correspondence so that the I/0 functions associated with corresponding states are the same
3	State Transition	the systems are homomorphic (explained below)
4	Coupled Component	components of the systems can be placed into correspondence so that corresponding components are morphic; in addition, the couplings among corresponding components are equal

upper level ones, a morphism holding at the upper level must imply the existence of one at the lower level. The morphisms defined in Table 1.4 are set up to satisfy these constraints.

The most important morphism, called **homomorphism**, resides at the State Transition level and is illustrated in Fig. 1.16. Consider two systems specified at level 3, S and S', where S may be bigger than S' in the sense of having more states. Later, we'll see that S could represent a complex model and S' a simplification of it. Or S could represent a simulator and S' a model it is executing. When S' goes

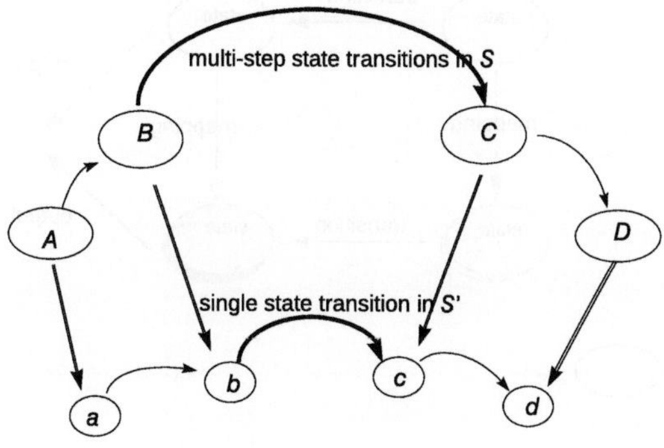

FIGURE 1.16

Homomorphism Concept. This figure illustrates the preservation of state transitions that a homomorphism requires.

through a state sequence such as a, b, c, d, then S should go through a corresponding state sequence say A, B, C, D. Typically, a simulator has a lot of apparatus, represented in its states, necessary to accommodate the whole class of models rather than a single one. Thus we don't assume that states of S and S' are identical – only that there is a predefined correspondence between them illustrated by the shaded connecting lines in the figure. Now to establish that this correspondence is a homomorphism requires that whenever S' specifies a transition, such as from state b to state c, then S actually makes the transition involving corresponding states B and C. Typically, the simulator is designed to take a number of *microstate* transitions to make the *macrostate* transition from B to C. These are computation steps necessary to achieve the desired end result. It is not hard to see that if such a homomorphism holds for all states of S', then any state trajectory in the S' will be properly reproduced in S.

Often, we require that the correspondence hold in a step-by-step fashion. In other words, that the transition from a to b is mirrored by a one-step transition from A to B. Also, as just indicated, we want the I/O Behavior's of homomorphic models specified at the I/O System level to be the same. Thus, as in Fig. 1.17, we require that the outputs produced from corresponding states be the same. In this type of homomorphism, the values and timing of the transitions and outputs of the base model are preserved in the lumped model. Thus, in this case, the state and output trajectories of the two models, when started in corresponding states, are the same.

1.6 EVOLUTION OF DEVS

Around 1960, the first use of a form of digital simulation appeared which we can roughly identify as event-oriented simulation. At its advent, event-oriented simulation was mainly thought to be a form of programming associated with the recent introduction of the digital computer and applied to operational research problems. In contrast, classical simulation was taken to be a form of numerical solution applicable to physics and related sciences whose speed could be greatly increased with mechanical, as opposed to, hand calculation. The concept of "system" was defined by Wymore (1967) as a ba-

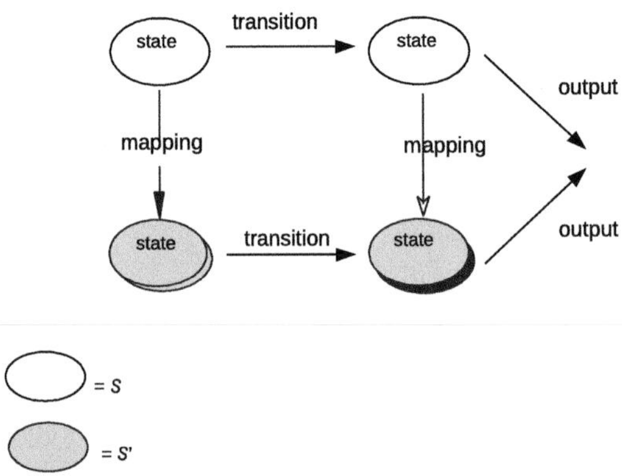

FIGURE 1.17

Homomorphism: a mapping preserving step-by-step state transition and output.

sis for unifying various forms of discrete and continuous model specification. About a decade after event-oriented simulation took hold, the Discrete Event System Specification (DEVS) formalism was defined as a specification for a subclass of Wymore systems that captured all the relevant features of the models underlying event-oriented simulations (Section 1.1.2). In contrast, Discrete Time Systems Specification (DTSS) and Differential Equation System Specification (DESS) were introduced to specify other common distinct subclasses of Wymore systems – the first, as a basis for discrete time models (including those specified by finite automata and cellular automata); the second to represent the continuous models underlying classical numerical solvers. K.D. Tocher appears to be the first to conceive discrete events as the right abstraction to characterize the models underlying the event-oriented simulation techniques that he and others were adopting in the mid-1950s. According to Hollocks (2008), Tocher's core idea conceived of a manufacturing system as consisting of individual components, or 'machines', progressing as time unfolds through 'states' that change only at discrete 'events'. Indeed, DEVS took this idea one step further in following Wymore's formalistic approach, both being based on the set theory of logicians and mathematicians (Whitehead and Russell, 1910, Bourbaki).

Some distinctive modeling strategies soon emerged for programming event-oriented simulation. They became encapsulated in the concept of world views: event scheduling, activity scanning, and process interaction. These world views were formally characterized in Zeigler (1984) showing that they could all be represented as subclasses of DEVS (Chapter 7), thus also suggesting its universality for discrete event model formalisms extending to other representations such as Timed Automata and Petri Nets (Fig. 1.18). Also at the same time the distinction between modular and non-modular DEVS was made showing that the world views all fit within the non-modular category. Moreover, while the modular class was shown to be behaviorally equivalent to that of the non-modular one, it better supported the concepts of modularity, object orientation, and distributed processing that were impending on the software engineering horizon.

An overview of some of the milestones in the development DEVS depicted in the figure below is given in Zeigler and Muzy (2017).

Classic DEVS is a formalism for modeling and analysis of discrete event systems can be seen as an extension of the Moore machine formalism, which is a finite state automaton where the outputs are determined by the current state alone (and do not depend directly on the input). Significantly, the extension associates a lifespan with each state and provides a hierarchical concept with an operation, called coupling, based on Wymore's system theory.

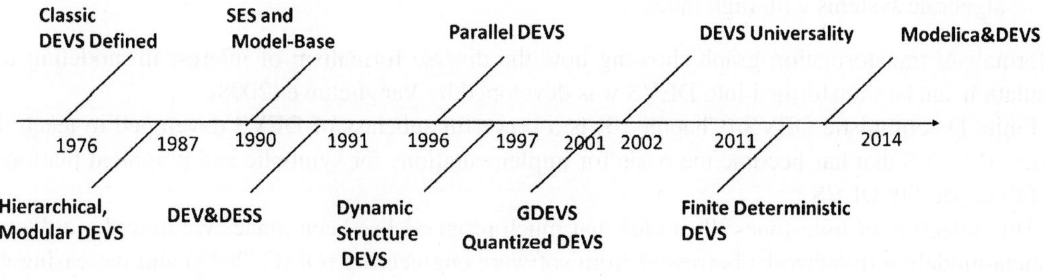

FIGURE 1.18

Timeline of some developments in DEVS.

Parallel DEVS (Chapter 4) revises the classic DEVS formalism to distinguish between transition collisions and ordinary external events in the external transition function of DEVS models, extends the modeling capability of the collisions. The revision also replaces tie-breaking of simultaneously scheduled events by a well-defined and consistent formal construct that allows all transitions to be simultaneously activated providing both conceptual and parallel execution benefits.

Hierarchical, Modular DEVS (Chapter 4) established the similarity and differences with, and implemented DEVS in, the Object-oriented programming (OOP) and modular programming paradigm, among the first in numerous implementations (Van Tendeloo and Vangheluwe, 2017).

System entity structure (SES) (Chapter 18 of TMS2000) is a structural knowledge representation scheme that contains knowledge of decomposition, taxonomy, and coupling of a system supporting model base management.

Dynamic Structure DEVS (Chapter 12), enables representing systems that are able to undergo structural change. Change in structure is defined in general terms, and includes the addition and deletion of systems and the modification of the relations among components.

DEVS considered as a universal computational formalism for systems (Chapter 18) found increasing implementation platforms that handled combined discrete and continuous models (also called co simulation, hybrid simulation, Chapter 12). Some of the milestones in this thread of development are:

- DEV&DESS (Discrete Event and Differential Equation System Specification) (Chapter 9) is a formalism for combined discrete-continuous modeling which based on system theoretical combines the three system specification formalisms-differential equation, discrete time, and the discrete event system specification formalism.
- Quantized State Systems (Chapter 19) are continuous time systems where the variable trajectories are piecewise constant and can be exactly represented and simulated by DEVS. The benefits of this approach in the simulation of continuous time systems are discussed in Chapter 19, including comparisons with conventional numerical integration algorithms in different domain applications.
- GDEVS (Giambiasi et al., 2001) (Generalized DEVS) organizes trajectories through piecewise polynomial segments utilizing arbitrary polynomial functions to achieve higher accuracies in modeling continuous processes as discrete event abstractions.
- Modelica&DEVS (Floros et al., 2011; Nutaro, 2014) transforms Modelica continuous models into DEVS thus supporting models with state and time events that comprise differential-algebraic systems with high index.

A formalism transformation graph showing how the diverse formalism of interest in modeling and simulation can be transformed into DEVS was developed by Vangheluwe (2008).

Finite Deterministic DEVS (Chapter 13) is a powerful subclass of DEVS developed to teach the basics of DEVS that has become the basis for implementations for symbolic and graphical platforms for full-capability DEVS.

This selection of milestones illustrates that much progress has been made. We note the influence of meta-modeling frameworks borrowed from software engineering (OMG, 2015) and increasing applied to development of higher level domain specific languages (Jafer et al., 2016). The confluence of such frameworks with the system-theory based unified DEVS development process (Mittal and Martín, 2013) may be increasingly important in the future simulation model development.

Over the years, DEVS has finding an increasing acceptance in the model-based simulation research community becoming one of the preferred paradigms to conduct modeling and simulation inquiries (Wainer and Mosterman, 2016). Following the approach proposed in the M&S framework, new variants, extensions and abstractions have been developed using the core of concepts defined by the original formalism. Several authors have improved the formalism capabilities in response to different situations, giving useful solutions to a wide range of simulation problems. Some of these solutions (not including listing those in milestones) are Cell-DEVS (Wainer, 2004), Fuzzy-DEVS (Kwon et al., 1996), Min-Max-DEVS (Hamri et al., 2006), and Vectorial DEVS (Bergero and Kofman, 2014). Moreover, the model/simulator separation of concerns inherent in the M&S framework of Chapter 2 allows researchers to develop alternative simulation algorithms in order to complement existent abstract DEVS simulators (Kim et al., 1998; Muzy and Nutaro, 2005; Shang and Wainer, 2006; Liu and Wainer, 2010). Franceschini et al. (2014) provide a survey of DEVS simulators with performance comparison. Kim et al. (2017) show how DEVS modeling for simulation greatly exceeds Big Data modeling techniques in predictive power when the usual kinds of deviations from the underlying state of the referent system come into play.

1.7 SUMMARY

We have outlined the basic concepts of systems theory: structure, behavior, levels of system specification and their associated morphisms. We have brought out the important distinctions that justify having different levels of specification. However, we have not considered all the possible distinctions and levels. For example, the important distinction between modular and non-modular systems has not been recognized with distinct levels. A more complete hierarchy will emerge as revisit the concepts introduced here in a more formal and rigorous manner in Chapter 5. We also have introduced the basic system specification formalisms and outlined the advances in the development of such formalisms that characterize the second edition, TMS2000 and reviewed some of the major milestones in the development of DEVS.

We now turn to a framework for modeling and simulation that identifies the key elements and their relationships. The systems specification hierarchy will provide the basis for presenting this framework. For example, we use specifications at different levels to characterize the different elements. The various system specification formalisms and their simulators provide the operational means to employ the framework in real world applications. We focus on real world application in the last part of the book.

1.8 SOURCES

The basic concepts of systems theory were developed by pioneers such as Arbib (1967), Zadeh and Desoer (1979) (later known more for his fuzzy sets theories), Klir (1985), Mesarovic and Takahara (1975) and Wymore (1977). Since the first edition of this book (Zeigler, 1976) there have been several trends toward deepening the theory (Mesarovic and Takahara, 1989; Wymore, 1993), extending its range of applicability with computerized tools (Pichler and Schwartzel, 1992) and going on to new more abstract formulations (Takahashi and Takahara, 1995). Also, somewhat independently, a new recognition of systems concepts within discrete event systems was fostered by Ho (1992). The

DEV&DESS formalism was introduced by Praehofer in his doctoral dissertation (Praehofer, 1991). The DEVS Bus originated in the research group of Tag Gon Kim (Kim and Kim, 1996). Quantized system theory was first presented in Zeigler and Lee (1998). A recent collection of systems concepts in computer science is given in Albrecht (1998).

DEFINITIONS, ACRONYMS, ABBREVIATIONS

- DEDS – Discrete Event Dynamic Systems.
- DESS – Differential Equation System Specification.
- DEVS – Discrete Event System Specification.
- DTSS – Discrete Time System Specification.
- DEV&DESS – Discrete Event and Differential Equation System Specification.
- M&S – Modeling and Simulation.
- TMS76 – 1976 Edition of Theory of Modeling and Simulation.
- TMS2000 – 2000 Edition of Theory of Modeling and Simulation.
- TMS2018 – 2018 Edition of Theory of Modeling and Simulation.

REFERENCES

Albrecht, R.F., 1998. On mathematical systems theory. Systems: Theory and Practice, 33–86.

Arbib, M., 1967. Theories of Abstract Automata. Prentice-Hall.

Bergero, F., Kofman, E., 2014. A vectorial DEVS extension for large scale system modeling and parallel simulation. Simulation: Transactions of the Society for Modeling and Simulation International 90 (5), 522–546.

Blas, S.J., Zeigler, B.P., 2018. A conceptual framework to classify the extensions of DEVS formalism as variants and subclasses. In: Winter Simulation Conference.

Floros, X., Bergero, F., Cellier, F.E., Kofman, E., 2011. Automated simulation of Modelica models with QSS methods: the discontinuous case. In: Proceedings of the 8th International Modelica Conference, number 063. March 20th–22nd, Technical University, Dresden, Germany. Linköping University Electronic Press, pp. 657–667.

Franceschini, R., Bisgambiglia, P.-A., Touraille, L., Bisgambiglia, P., Hill, D., 2014. A survey of modelling and simulation software frameworks using discrete event system specification. In: Neykova, R., Ng, N. (Eds.), 2014 Imperial College Computing Student Workshop. In: OpenAccess Series in Informatics (OASIcs). Schloss Dagstuhl—Leibniz-Zentrum fuer Informatik, Dagstuhl, Germany, pp. 40–49.

Giambiasi, N., Escude, B., Ghosh, S., 2001. GDEVS: a generalized discrete event specification for accurate modeling of dynamic systems. In: Proceedings of the 5th International Symposium on Autonomous Decentralized Systems. 2001. IEEE, pp. 464–469.

Hamri, M.E.-A., Giambiasi, N., Frydman, C., 2006. Min–Max-DEVS modeling and simulation. Simulation Modelling Practice and Theory 14 (7), 909–929.

Ho, Y.-C., 1992. Discrete Event Dynamic Systems: Analyzing Complexity and Performance in the Modern World. IEEE Press.

Hollocks, B.W., 2008. Intelligence, innovation and integrity—KD Tocher and the dawn of simulation. Journal of Simulation 2 (3), 128–137.

Jafer, S., Chhaya, B., Durak, U., Gerlach, T., 2016. Formal scenario definition language for aviation: aircraft landing case study. In: AIAA Modeling and Simulation Technologies Conference, p. 3521.

Kim, B.S., Kang, B.G., Choi, S.H., Kim, T.G., 2017. Data modeling versus simulation modeling in the big data era: case study of a greenhouse control system. Simulation 93 (7), 579–594. https://doi.org/10.1177/0037549717692866.

Kim, K.H., Kim, T.G., Park, K.H., 1998. Hierarchical partitioning algorithm for optimistic distributed simulation of DEVS models. Journal of Systems Architecture 44 (6–7), 433–455.

Kim, Y.J., Kim, T.G., 1996. A heterogeneous distributed simulation framework based on DEVS formalism. In: Proceedings of the Sixth Annual Conference on Artificial Intelligence, Simulation and Planning in High Autonomy Systems, pp. 116–121.

Klir, G.J., 1985. Architecture of Systems Complexity. Saunders, New York.

Kwon, Y., Park, H., Jung, S., Kim, T., 1996. Fuzzy-DEVS formalism: concepts, realization and applications. In: Proceedings AIS 1996, pp. 227–234.

Liu, Q., Wainer, G., 2010. Accelerating large-scale DEVS-based simulation on the cell processor. In: Proceedings of the 2010 Spring Simulation Multiconference. Society for Computer Simulation International, p. 124.

Mesarovic, M.D., Takahara, Y., 1975. General Systems Theory: Mathematical Foundations, vol. 113. Academic Press.

Mesarovic, M., Takahara, Y., 1989. Abstract Systems Theory, vol. 116. Springer-Verlag, NY.

Mittal, S., Martín, J.L.R., 2013. Netcentric System of Systems Engineering with DEVS Unified Process. CRC Press.

Muzy, A., Nutaro, J.J., 2005. Algorithms for efficient implementations of the DEVS & DSDEVS abstract simulators. In: 1st Open International Conference on Modeling & Simulation. OICMS, pp. 273–279.

Nutaro, J., 2014. An extension of the OpenModelica compiler for using Modelica models in a discrete event simulation. Simulation 90 (12), 1328–1345.

OMG, 2015. Documents associated with meta object facility version 2.5. Available via http://www.omg.org/spec/MOF/2.5/. (Accessed 2 November 2016).

Pichler, F., Schwartzel, H., 1992. CAST (Computer Aided System Theory) Methods in Modeling. Springer-Verlag, New York.

Praehofer, H., 1991. System theoretic formalisms for combined discrete-continuous system simulation. International Journal of General System 19 (3), 226–240.

Shang, H., Wainer, G., 2006. A simulation algorithm for dynamic structure DEVS modeling. In: Proceedings of the 38th Conference on Winter Simulation. Winter Simulation Conference, pp. 815–822.

Takahashi, S., Takahara, Y., 1995. Logical Approach to Systems Theory. Springer-Verlag, London.

Van Tendeloo, Y., Vangheluwe, H., 2017. An evaluation of DEVS simulation tools. Simulation 93 (2), 103–121.

Vangheluwe, H., 2008. Foundations of modelling and simulation of complex systems. Electronic Communications of the EASST 10.

Wainer, G.A., 2004. Modeling and simulation of complex systems with cell-DEVS. In: Proceedings of the 36th Conference on Winter Simulation. Winter Simulation Conference, pp. 49–60.

Wainer, G.A., Mosterman, P.J., 2016. Discrete-Event Modeling and Simulation: Theory and Applications. CRC Press.

Whitehead, A., Russell, B., 1910. Principia Mathematica 1, 1 ed. Cambridge University Press, Cambridge. JFM 41.0083.02.

Wymore, A.W., 1993. Model-Based Systems Engineering, vol. 3. CRC Press.

Wymore, W., 1967. A Mathematical Theory of Systems Engineering: The Elements. Wiley.

Wymore, W., 1977. A Mathematical Theory of Systems Engineering: The Elements. Krieger Pub Co.

Zadeh, L.A., Desoer, C.A., 1979. Linear System Theory. Krieger Publishing Co.

Zeigler, B.P., 1976. Theory of Modeling and Simulation. Wiley Interscience Co.

Zeigler, B., Muzy, A., 2017. From discrete event simulation to discrete event specified systems (DEVS). IFAC-PapersOnLine 50 (1), 3039–3044.

Zeigler, B.P., 1984. Multifacetted Modelling and Discrete Event Simulation. Academic Press Professional, Inc, London.

Zeigler, B.P., Lee, J.S., 1998. Theory of quantized systems: formal basis for DEVS/HLA distributed simulation environment. In: SPIE Proceedings, vol. 3369, pp. 49–58.

CHAPTER

FRAMEWORK FOR MODELING AND SIMULATION

2

CONTENTS

This chapter is devoted to establishing a *Framework for Modeling and Simulation* (M&S). The framework defines *entities* and their *relationships* that are central to the M&S enterprise. Understanding these concepts will help everyone involved in a simulation modeling project – analysts, programmers, managers, users – to better carry out their tasks and communicate with each other. Terms such as "model" and "simulator" are often loosely used in current practice but have a very sharp meanings in the framework we will discuss. Therefore, it is important to understand what is included and excluded by the definitions. (This is especially true if you have some experience in M&S and are likely to associate (or prefer) meanings that are different from those developed here.) As illustrated in Fig. 2.1, the basic entities of the framework are: *source system, model, simulator and experimental frame*. The basic inter-relationships among entities are the *modeling* and the *simulation relationships*, the entities are defined in Table 2.1. This table also characterizes the level of system specification that typically describes the entities. The level of specification is an important feature for distinguishing between the entities, which is often confounded in the literature. You can return to Fig. 2.1 and Table 2.1 to keep an overall view of the framework as we describe each of the components in the following presentation.

Theory of Modeling and Simulation. https://doi.org/10.1016/B978-0-12-813370-5.00010-9

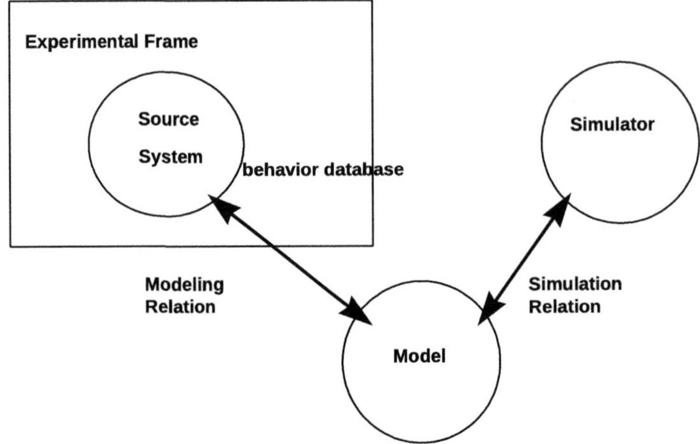

FIGURE 2.1

The Basic Entities in M&S and Their Relationships.

Table 2.1 Defining the Basic Entities in M&S and their usual Levels of Specification

Basic Entity	Definition	Related System Specification Levels
source system	real or artificial source of data	known at level 0
behavior database	collection of gathered data	observed at level 1
experimental frame	specifies the conditions under which system is observed or experimented with	constructed at levels 3 and 4
model	instructions for generating data	constructed at levels 3 and 4
simulator	computational device for generating behavior of the model	constructed at level 4

After several decades of development, there are still a wide variety of modeling and simulation terms, concepts with multiple interpretations. This variety derives from different streams of development which we describe in historical perspective at the end of this chapter. Based on the framework presented here, the basic issues and problems encountered in performing M&S activities and using their vocabulary can be better understood and coherent solutions developed.

2.1 THE ENTITIES OF THE FRAMEWORK

2.1.1 SOURCE SYSTEM

The *source system* (we will omit the 'source' qualifier, when the context is clear) is the real or virtual environment that we are interested in modeling. It is viewed as a *source of observable data*, in the form of time-indexed trajectories of variables. The data that has been gathered from observing or otherwise experimenting with a system is called the *system behavior database*. As indicated in Table 2.1, this concept of system is a specification at level 0 (or equivalently, Klir's source system) and its database is a specification at level 1 (or equivalently, Klir's data system). This data is viewed or acquired through experimental frames of interest to the modeler.

Applications of M&S differ with regard to how much data is available to populate the system database. In *data rich* environments, such data is abundant from prior experimentation or can easily be obtained from measurements. In contrast, *data poor* environments offer meager amounts of historical data or low quality data (whose representativeness of the system of interest is questionable). In some cases it is impossible to acquire better data (for example, of combat in real warfare); in others, it is expensive to do so (for example, topography and vegetation of a forest). In the latter case, the modeling process can direct the acquisition of data to those areas that have the highest impact on the final outcome.

2.1.2 EXPERIMENTAL FRAME

An *experimental frame* is a specification of the conditions under which the system is observed or experimented with. As such an experimental frame is the operational formulation of the objectives that motivate a modeling and simulation project. For example, out of the multitude of variables that relate to a forest, the set lightning, rain, wind, smoke represents one particular choice. Such an experimental frame is motivated by the interest in modeling the way lightning ignites a forest fire. A more refined experimental frame would add the moisture content of the vegetation and the amount of unburned material as variables. Thus, many experimental frames can be formulated for the same system (both source system and model) and the same experimental frame may apply to many systems. Why would we want to define many frames for the same system? Or apply the same frame to many systems? For the same reason that we might have different objectives in modeling the same system, or have the same objective in modeling different systems. More of this in a moment.

There are two equally valid views of an experimental frame. One, views a frame as a definition of the type of data elements that will go into the database. The second views a frame as a system that interacts with the system of interest to obtain the data of interest under specified conditions. In this view, the frame is characterized by its implementation as a measurement system or observer. In this implementation, a frame typically has three types of components (as shown in Fig. 2.2): *generator*, that generates input segments to the system; *acceptor* that monitors an experiment to see the desired experimental conditions are met; and *transducer* that observes and analyzes the system output segments.

OBJECTIVES AND EXPERIMENTAL FRAMES

Objectives for modeling relate to the role of the model in systems design, management or control. The statement of objectives serves to focus model construction on particular issues. It is crucial to formulate such a statement as early as possible in the development process. A firmly agreed upon statement of objectives enables project leaders to maintain control on the efforts of the team. Once the objectives are known, suitable experimental frames can be developed. Remember, that such frames translate the objectives into more precise experimentation conditions for the source system or its models. A model is expected to be valid for the system in each such frame. Having stated our objectives, there is pre-

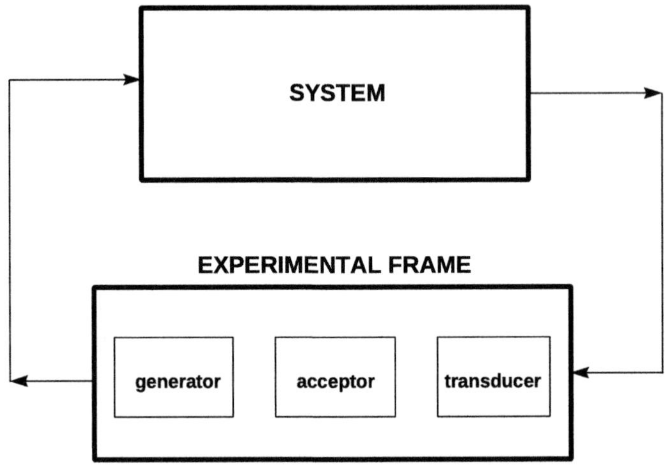

FIGURE 2.2

Experimental Frame and its Components.

sumably a best level of resolution to answer the questions raised. The more demanding the questions, the greater the resolution likely to be needed to answer it. Thus, the choice of appropriate levels of abstraction also hinges on the objectives and their experimental frame counterparts.

Fig. 2.3 depicts the process of transforming objectives into experimental frames. Typically modeling objectives concern system design. Here measures of the effectiveness of a system in accomplishing its goal are required to evaluate the design alternatives. We call such measures, outcome measures. In order to compute such measures, the model must include variables, we'll call output variables, whose values are computed during execution runs of the model. The mapping of the output variables into *outcome measures* is performed by the transducer component of the experimental frame. Often there may be more than one layer of variables intervening between *output variables* and outcome measures. For example, in military simulations, *measures of performance* are output variables that typically judge

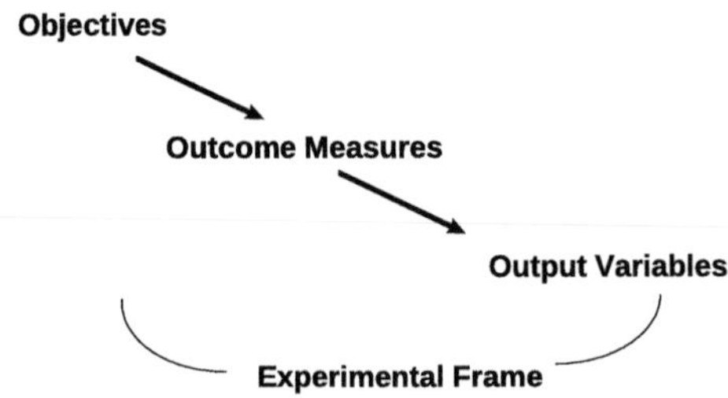

FIGURE 2.3

Transforming Objectives to Experimental Frames.

how well parts of a system are operating. For example, the success of a missile in hitting its target is a performance measure. Such measures enter as factors into outcome measures, often called *measures of effectiveness*, that measure how well the overall system goals are being achieved, e.g., how many battles are actually won by a particular combination of weapons, platforms, personnel, etc. The implication is that high performing components are necessary, but not sufficient, for highly effective systems, in which they must be coordinated together to achieve the overall goals.

Forest fire management is an interesting application domain. There are two quite different uses of M&S in this area: 1) that of fighting fires when they break out and 2) that of trying to prevent them from breaking out in the first place, or at least minimizing the damage when they do. Formulated as objectives for modeling these purposes lead to quite different experimental frames. Let's look at each of these frames.

In the first frame, real-time interdiction, which refers to on-the-spot firefighting, we require accurate prediction of where the fire will spread within a matter of hours. These predictions are used to allocate resources in the right places for maximum effectiveness. Because humans may be placed in great danger, highly reliable short-term predictions are required. A typical question asked here would be: is it safe to put a team of fire fighters on a particular ridge within reach of the current fire front for the next several hours? To improve the ability to make accurate predictions, the model state may be updated with satellite data to improve its correspondence with the real fire situation as it develops.

In fire prevention and mitigation the emphasis is less on short term prediction of spread than on answering "what-if" questions for planning purposes. For example land use planners might ask what should be the width of a fire break (area cleared of trees) around a residential area bordering a forest so that there is a less than 0.1% chance of houses catching fire. Here the model needs to be capable of working with a larger area of the landscape but the resolution it needs may be considerably less in order for useful comparisons of different planning alternatives to result. Indeed, a model might be capable of rank ordering alternatives without necessarily producing fire spread behavior with high accuracy.

As suggested the experimental frames that are developed for these contrasting objectives, interdiction and prevention, are quite different. The first (interdiction) calls for experimenting with a model in which all known prevailing fuel, wind and topographic conditions are entered to establish its initial state. The output desired is a detailed map of fire spread after say five hours within the region of interest.

The second experimental frame (prevention) calls for a wider scope, lower resolution representation of the landscape in which a range of expected lightning, wind, rain and temperature regimes may be injected as input trajectories. The model may then be placed into different states corresponding to different prevention alternatives, e.g., different fire break spatial regions might be investigated. The output for a particular run might be as simple as a binary variable indicating whether or not the residential area was engulfed by fire. The output summarized over all runs, might be presented as a rank ordering of alternatives according to their effectiveness in preventing fire spreading to the residential area (e.g., the percent of experiments in which the residential area was not engulfed by flame).

2.1.3 MODEL

In its most general guise, a *model* is a system specification at any of the levels discussed in Chapter 1. However, in the traditional context of M&S, the system specification is usually done at levels 3 and 4, corresponding to Klir's generative and structure levels. Thus the most common concept of a simulation model is that it is a *set of instructions, rules, equations, or constraints for generating I/O behavior*. In other words, we write a model with a state transition and output generation mechanisms (level 3) to accept input trajectories and generate output trajectories depending on its initial state setting. Such models form the basic components in more complex models that are constructed by coupling them together to form a level 4 specification.

There are many meanings that are ascribed to the word "model". For example, a model is conceived as any physical, mathematical, or logical representation of a system, entity, phenomenon, or process. The definition in terms of system specifications has the advantages that it has a sound mathematical foundation and is has a definite semantics that everyone can understand in unambiguous fashion. Like other formal definitions, it cannot capture all meanings in the dictionary. However, it is intended to capture the most useful concepts in the M&S context.

2.1.4 SIMULATOR

As a set of instructions, a model needs some agent capable of actually obeying the instructions and generating behavior. We call such an agent a simulator.[1] Thus, a simulator is any computation system (such as a single processor, a processor network, the human mind, or more abstractly an algorithm), capable of executing a model to generate its behavior. A simulator is typically specified at a high level since it is a system that we design intentionally to be synthesized from components that are off-the-shelf and well-understood. Separating the model and simulator concepts provides a number of benefits for the framework:

- The same model, expressed in a formalism, may be executed by different simulators thus opening the way for portability and interoperability at a high level of abstraction.
- Simulator algorithms for the various formalisms may be formulated and their correctness rigorously established.
- The resources required to correctly simulate a model afford a measure of its complexity.

2.2 PRIMARY RELATIONS AMONG ENTITIES

The entities – *system, experimental frame, model, simulator* – become truly significant only when properly related to each other. For example, we build a model of a particular system for some objective – only some models, and not others, are suitable. Thus, it is critical to the success of a simulation modeling effort that certain relationships hold. The two most fundamental are the modeling and the simulation relations (Table 2.2)

2.2.1 MODELING RELATION: VALIDITY

The basic modeling relation, *validity*, refers to the relation between a model, a system and an experimental *frame*. Validity is often thought of as the degree to which a model faithfully represents its system counterpart. However, it makes much more practical sense to require that the model faithfully

[1] TMS76 referred used the generic "computer" instead of the more specific "simulator".

Table 2.2 Primary Relationships among Entities

Basic Relationship	Definition	Related System Specification Levels
modeling relation replicative validity predictive validity structural validity	concerned with how well model-generated behavior agrees with observed system behavior	 comparison is at level 1 comparison is at level 2 comparison is at level 3, 4
simulation relation correctness	concerned with assuring that the simulator carries out correctly the model instructions	basic comparison is at level 2; involves homomorphism at levels 3 or 4

captures the system behavior only to the extent demanded by the objectives of the simulation study. In our formulation, the concept of validity answers the question of whether it is impossible to distinguish the model and system in the **experimental frame of interest**. The most basic concept, replicative *validity*, is affirmed if, for all the experiments possible within the experimental frame, the behavior of the model and system agree within acceptable tolerance. Thus replicative validity requires that the model and system agree at the I/O relation level 1 of the system specification hierarchy.

Stronger forms of validity are *predictive validity* and *structural validity*. In predictive validity we require not only replicative validity, but also the ability to predict as yet unseen system behavior. To do this the model needs to be set in a state corresponding to that of the system. Thus predictive validity requires agreement at the next level of the system hierarchy, that of the I/O function level 2. Finally, structural validity requires agreement at level 3 (state transition) or higher (coupled component). This means that the model not only is capable of replicating the data observed from the system but also mimics in step-by-step, component-by-component fashion, the way that the system does its transitions.

The term *accuracy* is often used in place of validity. Another term fidelity, is often used for a combination of both validity and detail. Thus, a high fidelity model may refer to a model that is both high in detail and in validity (in some understood experimental frame). However when used this way, beware that there may be a tacit assumption that high detail alone is needed for high fidelity, as if validity is a necessary consequence of high detail. In fact, it is possible to have a very detailed model that is nevertheless very much in error, simply because some of the highly resolved components function in a different manner than their real system counterparts.

2.2.2 SIMULATION RELATION: SIMULATOR CORRECTNESS

The basic *simulation* relation, simulator correctness, is a relation between a *simulator* and a *model*. A *simulator correctly simulates a model* if it is guaranteed to faithfully generate the model's output trajectory given its initial state and its input trajectory. Thus, simulator correctness requires agreement at the I/O function level (Zeigler, 1976). In practice, simulators are constructed to execute not just one model but a family of possible models. This flexibility is necessary if the simulator is to be applicable to a range of applications. In such cases, we must establish that a simulator will correctly execute a particular class of models. Since the structures of both the simulator and the model are at hand, it may be possible to prove correctness by showing that a *homomorphism* relation holds. Recall from Chapter 1 that a homomorphism is a correspondence between simulator and model states that is preserved under transitions and outputs.

2.3 OTHER IMPORTANT RELATIONSHIPS

Besides the two fundamental relationships, there are others that are important for understanding modeling and simulation work. These relations have to with the interplay and orderings of models and experimental frames.

MODELING AS VALID SIMPLIFICATION

The inescapable fact about modeling is that it is severely constrained by complexity limitations. Complexity, is at heart, an intuitive concept the feeling of frustration or awe that we all sense when things get too numerous, diverse, or intricately related to discern a pattern, to see all at once in a word, to comprehend. Generalizing from the boggled human mind to the overstressed simulator suggests that the complexity of model can be measured by the resources required by a particular simulator to correctly interpret it. As such, complexity is measured relative to a particular simulator, or class of simulators. However, as we will see in Chapter 16, properties intrinsic to the model are often strongly correlated with complexity independently of the underlying simulator. While computers continue to get faster and possess more memory, they will always lag behind our ambitions to capture reality in our models. Successful modeling can then be seen as *valid simplification*. We need to *simplify*, or reduce the complexity, to enable our models to be executed on our resource-limited simulators. But the simplified model must also be *valid*, at some level, and within some experimental frame of interest. As in Fig. 2.4, there is always a pair of models involved, call them the base and *lumped* models. Here, the base model is typically "more capable" and requires more resources for interpretation than the lumped model. By the term "more capable", we mean that the base model is valid within a larger set of experimental frames (with respect to a real system) than the lumped model. However, the important point is that within a **particular frame of interest** the lumped model might be just as valid as the base model. The concept of morphism introduced in Chapter 1 affords criteria for judging the equivalence of base and lumped models with respect to an experimental frame. In Chapter 16, we will discuss methods for constructing such morphisms.

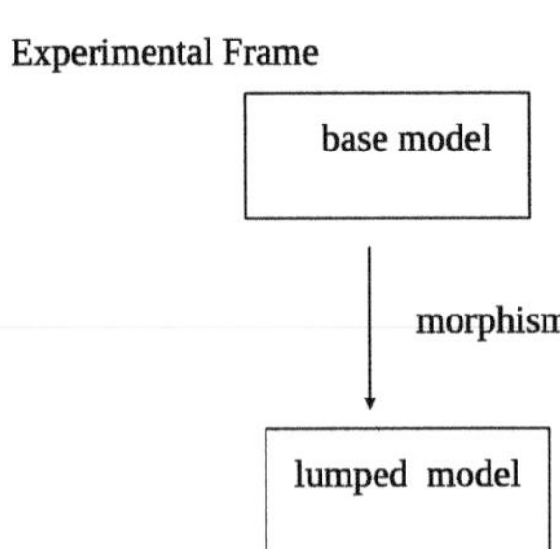

FIGURE 2.4

Base/Lumped Model Equivalence in Experimental Frame.

Table 2.3 Other M&S Relationships important when dealing with a model repository

Relationship	Definition
Experimental frame applies to a model (or 'is applicable to')	the conditions on experimentation required by the frame can be enforced in the model
Model accommodates experimental frame	frame is applicable to the model
Experimental Frame 1 is derivable from Experimental Frame 2	any model that accommodates Experimental Frame 2 also accommodates Experimental Frame 1

EXPERIMENTAL FRAME – MODEL RELATIONSHIPS

Assume that we have a whole repository of models and experimental frames that have been built up over years of experience. Then it is critical to have an ability to ask whether there are any experimental frames that meet our current objectives and whether there are models that can work within this frame. Only those models have a chance of providing valid answers to our current questions. The relation that determines if a frame can logically be applied to a model is called *applicability* and its converse, is called *accommodation* (Table 2.3). Notice that validity of a model in a particular experimental frame, requires, as a precondition, that the model accommodates the frame.

The degree to which one experimental frame is more restrictive in its conditions than another is formulated in the *derivability* relation. A more restrictive frame leaves less room for experimentation or observation than one from which it is derivable. So, as illustrated in Fig. 2.5, it is easier to find a model that is valid in a restrictive frame for a given system. It turns out that applicability may be reduced to derivability. To see this, define the *scope* frame of the model to represent the most relaxed conditions under which it can be experimented with (this is clearly a characteristic of the model). Then a frame is applicable to a model, if it is derivable from the scope frame of the model. This means that a repository need not support both applicability and derivability queries. Only the latter is sufficient if each model has an associated scope frame.

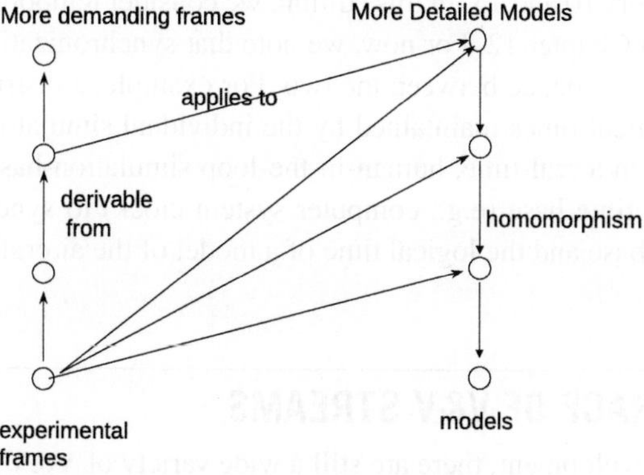

FIGURE 2.5

Illustrating Important M&S Relations Relevant to Model Repositories.

Table 2.4 A Time Taxonomy

		Logical/Physical Logical Time	Physical Time
Local/Global	Global Time	Global Logical: All components operate on the same abstract time base.	Global, Physical: All components operate on the same system clock.
	Local Time	Local, Logical: A component operates on its own abstract time base.	Local, Physical: A component operates on its own system clock.

2.4 TIME

Implicitly, until now, we have assumed that a *time base* is an abstract way of ordering observations made on a system. In Chapter 3, we will formally characterize a time base as an ordered set for indexing events that models the flow of actual time. If the interpretation of such a time base is left abstract in this manner, we refer to it as *logical* time. In contrast, when we consider events happening in the real world, in real time, we refer to a time variable as measured by an actual clock. Thus, *physical* time, also called metric time or wallclock time, is measured by ticks of physical clocks, while logical time is measured by ticks of a clock somehow embedded in a model Also, as relativity theory made clear, time, as perceived by observers at different locations may be different. Based on this distinction, time can be either *local* and *global*. The former is valid only within a component of a system; the latter is valid in the whole system. Thus, there are at least two dimensions for classifying time: one along the logical/physical axis and the other along the local/global axis. Consequently, a time base can be interpreted as falling in any one of the four combinations shown in Table 2.4.

Traditionally, modeling and simulation has considered mainly the first (global, logical) combination. That is, we assume all components of a modeled system have the same time frame of reference and we consider time as an abstract quantity. However, when a model is executing in a simulator, which may be distributed among computer nodes in a network and may also be interacting with the real world, it is hard to maintain this fiction. TMS2000 discussed such real-time and distributed simulation approaches in Chapters 10 and 13. In this edition, we consider temporal issues within the context of iterative specification in Chapter 12. For now, we note that synchronization between time bases requires maintaining a correspondence between the two. For example, a distributed simulation protocol synchronizes the local, logical times maintained by the individual simulator nodes. Another example of synchronization occurs in a real-time, human-in-the-loop simulation-based training. Here the simulator employs a physical time base (e.g., computer system clock) to synchronize between a pilot's physically perceived time base and the logical time of a model of the aircraft being simulated.

2.5 HISTORICAL TRACE OF V&V STREAMS

After several decades of development, there are still a wide variety of V&V terms, concepts, products, processes, tools and techniques (Roza et al., 2013) This variety derives from different streams of development and motivates the following effort to place the framework for more robust V&V methodology presented here in historical perspective.

Table 2.5 Conceptual definitions and MSF equivalents

Conceptual Definition of Object	MSF Formalization
A **simuland** is the real world system of interest. It is the object, process, or phenomena to be simulated.	Real world system is a source of data can be represented by a system specification at a behavioral level.
A **model** is a representation of a simuland, broadly grouped into conceptual and executable types. Model is a set of rules for generating behavior, can be represented by a system specification at a structural level.	A modeling formalism, e.g. DEVS, enables conceptual specification and is mapped to a simulation language for execution by a simulator.
Simulation is the process of executing a model over time.	A simulator is a system capable of generating the behavior of a model; simulators come in classes corresponding to formalisms, e.g., an abstract DEVS simulator describes implementable simulators of DEVS models.
The **results** of simulation are the output produced by a model during a simulation	The behavior of a model generated by a simulator constitutes a specification at the behavior level.

Table 2.6 Conceptual definitions and MSF equivalents

Conceptual Definition of Activity	MSF Formalization
Verification is the process of determining if an implemented model is consistent with its specification	There is a relation, called simulation correctness, between models and simulators. Verification is the process of proving this correctness in a simulator generating the behavior of the model. When this is done for a formalism, it certifies a simulator as correct for any model of the associated class.
Validation is the process of determining if a model behaves with satisfactory accuracy consistent with the study objectives within its domain of applicability to the simuland it represents[a]	There is a relation, called validity in a frame, between models and real systems within an experimental frame. Validation is the process of establishing that the behaviors of the model and real system agree in the frame in question. The frame can capture the intended objectives (extended to intended uses), domain of applicability, and accuracy requirements.
Abstraction is the omission or reduction of detail not considered necessary in a model.	Abstraction is the process of constructing a lumped model from a base model intended to be valid for the real system in a given experimental frame.

[a] *This definition is a synthesis of various definitions in the literature that separately relate the model to a real system, the purposes of model construction, the domain of applicability and the accuracy required. For example, Balci (1997) defines validation as the assessment of behavioral or representational accuracy and then later conditions accuracy on intended use. Our intent is to best represent the conceptual literature for the purposes of relating it to the MSF.*

2.5.1 INFORMAL V&V CONCEPTS AND PROCESSES

The work of Balci (1997, 2012), Balci and Ormsby (2000), Balci and Sargent (1981), and Kleijnen (1995) on V&V provided a starting point for many future efforts in the defense (Tolk, 2012; Pace, 2004, 2013) and other M&S communities (Obaidat and Boudriga, 2010). In this branch of the literature terminology is defined conceptually but not in the mathematically precise manner of the MSF presented here. A rough equivalence between objects in this literature and entities of the framework can be established in Tables 2.5 and 2.6.

Processes involved in M&S activities are likewise defined informally in this stream of literature. Since the MSF defines its entities as mathematical systems, it can define such processes as mathematical relations. A rough equivalence is given in Table 2.6.

2.5.2 THEORY-BASED AND MODEL-DRIVEN DEVELOPMENTS

Research and development on V&V based on the conceptual definitions has been focused on process rather than fundamentals of theory to supply solid foundations for such processes. On the more theoretical side, the use of simplified (reduced order, more abstract) meta-models in Verification, Validation and Accreditation (VV&A) was discussed in Caughlin (1995). Weisel (2011, 2012), Weisel et al. (2003) employed the idea of bi-simulation to ensure fundamental ordering relations and to evaluate them for existing systems. Multi-resolution modeling, in which a family of models at several levels of resolution is maintained (Davis and Bigelow, 1998) has been employed in distributed simulations of military systems (see e.g., Reynolds et al., 2002). Baohong (2007) formalized such multi-resolution model families employing a DEVS representation. He provided closure under coupling for such families which provides a foundation for research into consistency between levels of resolution, related to the problem of model simplification discussed here. Vangheluwe (2008) summarized the theory of M&S and emphasizes DEVS as a common basis for multi-formalism modeling using formalism transformation graphs. He emphasizes the notion that much of V&V can be done virtually by model-checking, simulation, and optimization prior to any actual "bending of metal", at least for systems design. This "doing it right the first time" leads to significant cost savings and improved quality. The mathematical foundations for model-based systems engineering originate with (Wymore, 1967) (see also Wymore, 1993). Model transformation as the basis for V&V as well as for simulation system development has been evaluated in great detail in the recent work at TU Delft (Cetinkaya, 2013; Cetinkaya et al., 2011). Cetinkaya et al. (2013) incorporated MDD4MS, a model drive development framework for modeling and simulation into the broader framework of the DEVS Unified Process (Mittal and Martín, 2013).

2.5.3 GENERIC METHODOLOGY PROCESSES AND BEST PRACTICE GUIDES

The wide variety of V&V terms, concepts, and processes, tools or techniques has a negative consequence in the same way that lack of standardization inhibits other areas of M&S. This was the motivation behind the recent development of the Generic Methodology for Verification and Validation (GM-VV) within the Simulation Interoperability Standards Organization (SISO). Paraphrasing Roza et al. (2013), as a generic methodology, the GM-VV comprises an abstract technical framework that consists of three parts: 1) the conceptual framework provides unifying terminology, concepts and principles, 2) the implementation framework translates these concepts into a building blocks for the development of concrete and consistent V&V solutions, and 3) the tailoring framework utilizes these building blocks to develop and cost-efficiently apply such V&V application instantiations. GM-VV can be tailored to more concrete guides and recommendations. The US Modeling and Simulation Coordination Office (US MSCO) compiled the VV&A Recommended Practice Guide (RPG) to facilitate the application of its directives and guidelines and to promote the effective application of VV&A (Petty, 2010). The High Level Architecture (HLA) standard entails the Federation Development Process (FEDEP) which provides definitions and descriptions of the steps, tasks, and activities that are standardized for the development of a federation. An extension of the standard was developed to provide an overlay to the FEDEP that adds the required phases and activities for verification, validation, and accreditation at the same level of detail (Tolk, 2012). This complies with Ören's (Ören et al., 1986) broader concept of Quality Assurance where any element or activity involve in the M&S enterprise can be subject to the kinds of consistency checking, evaluation and comparison activities, typically associated with verification and validation of models.

The GM-VV acceptance goal is to develop a recommendation that shows why an M&S asset is, or not, acceptable for the stakeholder. Intended use is a primary factor in such acceptability criteria. However, intended use is not defined in GM-VV nor in Roza's thesis (Roza, 2004) to which it refers for background. In relation to the MSF definition of experimental frame, GM-VV defines a V&V experimental frame as a set of experiments, tests and conditions used to observe and experiment with the M&S system to obtain V&V results which are the collection of data items produced by applying the frame. The implementation framework of the GM-VV specifies the use of an *argumentation structure* to capture the derivation of the V&V experimental frame specification from the criteria for acceptability of the M&S system which themselves derive from the goal of V&V.

2.6 SUMMARY

With this introductory exposition of the framework and the systems foundation in Chapter 1, the pattern of development underlying the book's table of contents should be readily apparent. The next two chapters continue, in an informal manner, to present the basic modeling formalisms and their simulation algorithms. We then present the levels of system specification and introduce the modeling formalisms and their simulators in a rigorous, but readable, manner. Part 2 introduces the material that is new to this edition relating to iterative system specification. Part 3 returns, in a spiral manner, to present a new perspective on distributed and parallel simulation. It then goes on to tackle the issues of modeling as valid simplification. A theme that this continues to be strongly emphasized in this edition, (one that could only be dimly perceived in TMS76) is that of representing systems, whatever their native formalism, in discrete event form. Part 4 breaks new ground with DEVS Markov modeling and simulation, emphasizing the importance of base-lumped model construction for simulation and analysis. Also iterative systems specification modeling is applied to neuronal systems. The book ends with discussion of Systems Dynamics and its relation to DEVS as a segue to open research. Indeed, we leave the reader at a superhighway exit ramp with many outward-bound roads to explore, guide book in hand, concepts and theory providing the compass.

2.7 SOURCES

This chapter established a Framework for Modeling and Simulation that defines entities and their relationships that are central to the M&S enterprise. The framework discussed here derives from the one in TMS76. It was subsequently elaborated and extended in Zeigler (1984) which contains much more in-depth development than can be provided here. The framework was presented in its current form in the book edited by Cloud and Rainey (Zeigler, 1997). Readers may consult the latter book for an integrated approach to the development and operation of models. An extension of the framework was presented in Zeigler and Nutaro (2016) and includes the more elaborated formulation of Intended Use, a characterization of modeler objectives that enhances the conceptual underpinning for experimental frame theory. Such enhancement significantly contributes to the clarity with which M&S methodology can be executed. Research is needed to extend the derivability and other relations defined for frames and models to support intended use constructs. The GM-VV argumentation structure described above could benefit from such support. Finally, it is important to consider V&V within the larger context of model engineering that is being addressed in the companion volume (Zhang et al., 2018). Model engineering is defined as the formulation of theories, methods, technologies, standards and tools relevant to a systematic, standardized, quantifiable engineering methodology that guarantees the credibility of the full lifecycle of a model with the minimum cost. Several chapters of the companion volume deal with evaluating the credibility of models and with the V&V process within such evaluation.

REFERENCES

Balci, O., 1997. Verification validation and accreditation of simulation models. In: Proceedings of the 29th Conference on Winter Simulation. IEEE Computer Society, pp. 135–141.

Balci, O., 2012. A life cycle for modeling and simulation. Simulation 88 (7), 870–883.

Balci, O., Ormsby, W.F., 2000. Well-defined intended uses: an explicit requirement for accreditation of modeling and simulation applications. In: Proceedings of the 32nd Conference on Winter Simulation. Society for Computer Simulation International, pp. 849–854.

Balci, O., Sargent, R.G., 1981. A methodology for cost-risk analysis in the statistical validation of simulation models. Communications of the ACM 24 (4), 190–197.

Baohong, L., 2007. A formal description specification for multi-resolution modeling based on DEVS formalism and its applications. The Journal of Defense Modeling and Simulation 4 (3), 229–251.

Caughlin, D., 1995. Verification, validation, and accreditation (VV&A) of models and simulations through reduced order metamodels. In: Simulation Conference Proceedings. 1995, Winter. IEEE, pp. 1405–1412.

Cetinkaya, D.K., 2013. Model Driven Development of Simulation Models: Defining and Transforming Conceptual Models Into Simulation Models by Using Metamodels and Model Transformations. PhD thesis. Technische Universiteit Delft.

Cetinkaya, D., Mittal, S., Verbraeck, A., Seck, M.D., 2013. Model driven engineering and its application in modeling and simulation. In: Netcentric System of Systems Engineering with DEVS Unified Process, pp. 221–248.

Cetinkaya, D., Verbraeck, A., Seck, M.D., 2011. MDD4MS: a model driven development framework for modeling and simulation. In: Proceedings of the 2011 Summer Computer Simulation Conference. Society for Modeling & Simulation International, pp. 113–121.

Davis, P.K., Bigelow, J.H., 1998. Experiments in Multiresolution Modeling (MRM). Technical report. RAND CORP, Santa Monica, CA.

Kleijnen, J.P., 1995. Verification and validation of simulation models. European Journal of Operational Research 82 (1), 145–162.

Mittal, S., Martín, J.L.R., 2013. Netcentric System of Systems Engineering with DEVS Unified Process. CRC Press.

Obaidat, M.S., Boudriga, N.A., 2010. Fundamentals of Performance Evaluation of Computer and Telecommunications Systems. Wiley-Interscience, New York, NY, USA.

Ören, T.I., Zeigler, B.P., Elzas, M.S., 1986. Modelling and Simulation Methodology in the Artificial Intelligence Era. North-Holland.

Pace, D.K., 2004. Modeling and simulation verification and validation challenges. Johns Hopkins APL Technical Digest 25 (2), 163–172.

Pace, D.K., 2013. Comprehensive consideration of uncertainty in simulation use. The Journal of Defense Modeling and Simulation 10 (4), 367–380.

Petty, M.D., 2010. Verification, validation, and accreditation. In: Modeling and Simulation Fundamentals: Theoretical Underpinnings and Practical Domains, pp. 325–372.

Reynolds, R., Iskenderian, H., Ouzts, S., 2002. Using multiple representations and resolutions to compose simulated METOC environments. In: Proceedings of 2002 Spring Simulation Interoperability Workshop.

Roza, M., Voogd, J., Sebalj, D., 2013. The generic methodology for verification and validation to support acceptance of models, simulations and data. The Journal of Defense Modeling and Simulation 10 (4), 347–365.

Roza, Z.C., 2004. Simulation Fidelity Theory and Practice: A Unified Approach to Defining, Specifying and Measuring the Realism of Simulations. PhD thesis. Technische Universiteit Delft.

Tolk, A., 2012. Engineering Principles of Combat Modeling and Distributed Simulation. John Wiley & Sons.

Vangheluwe, H., 2008. Foundations of modelling and simulation of complex systems. Electronic Communications of the EASST 10.

Weisel, E., 2011. Towards a foundational theory for validation of models and simulations. In: Proceedings of the Spring 2011 Simulation Interoperability Workshop, pp. 4–8.

Weisel, E.W., 2012. Decision theoretic approach to defining use for computer simulation. In: Proceedings of the 2012 Autumn Simulation Conference.

Weisel, E.W., Petty, M.D., Mielke, R.R., 2003. Validity of models and classes of models in semantic composability. In: Proceedings of the Fall 2003 Simulation Interoperability Workshop, vol. 9, p. 68.

Wymore, A.W., 1993. Model-Based Systems Engineering, vol. 3. CRC Press.

Wymore, W., 1967. A Mathematical Theory of Systems Engineering: The Elements. Wiley.

Zeigler, B.P., 1976. Theory of Modeling and Simulation. Wiley Interscience Co.

Zeigler, B., 1997. A framework for modeling & simulation. In: Cloud, D., Rainey, L. (Eds.), Applied Modeling & Simulation: An Integrated Approach to Development & Operation.

Zeigler, B.P., 1984. Multifacetted Modelling and Discrete Event Simulation. Academic Press Professional, Inc, London.

Zeigler, B.P., Nutaro, J.J., 2016. Towards a framework for more robust validation and verification of simulation models for systems of systems. The Journal of Defense Modeling and Simulation 13 (1), 3–16.

Zhang, L., Zeigler, B., Laili, Y., 2018. Model Engineering for Simulation. Elsevier.

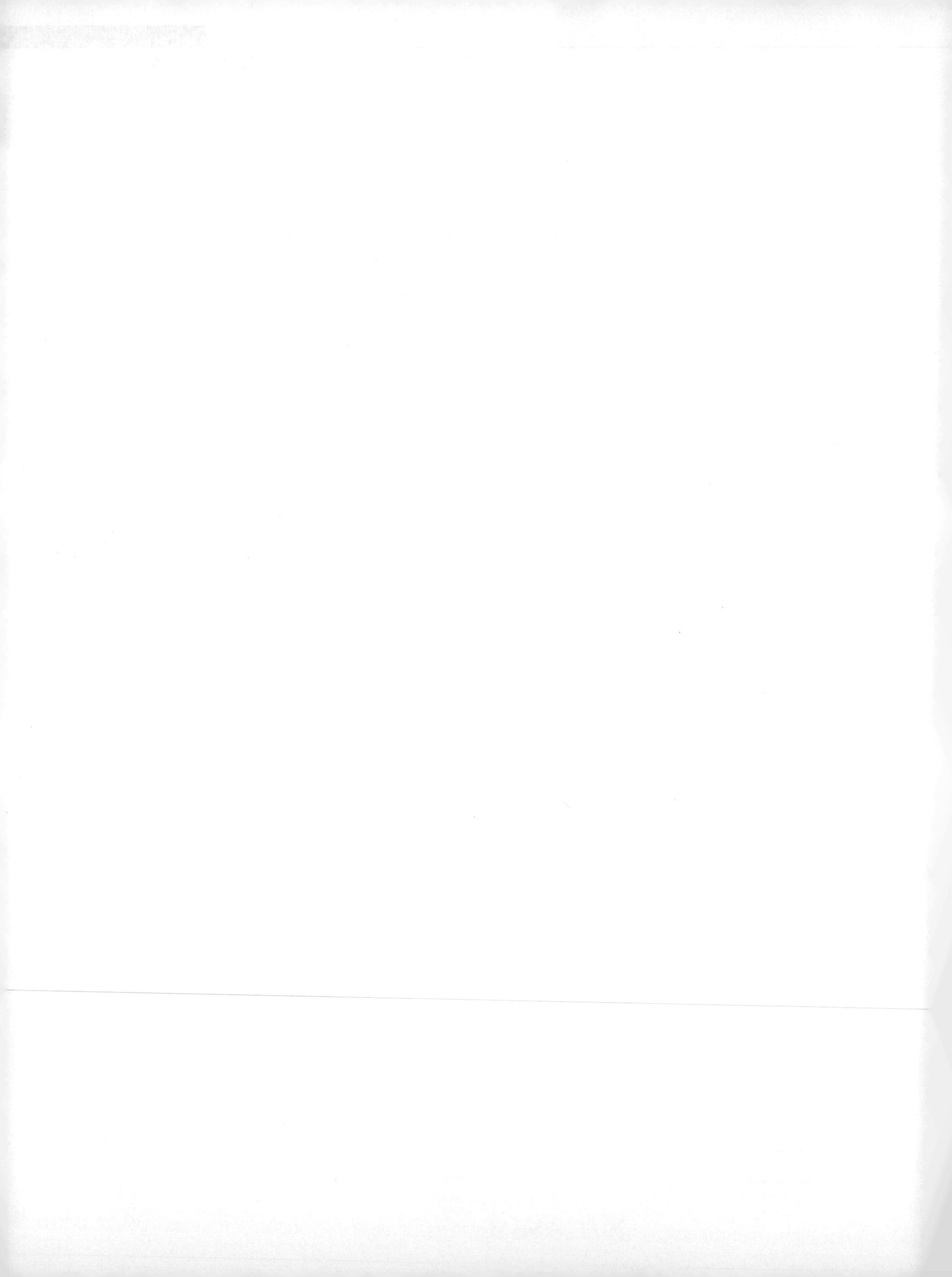

CHAPTER

MODELING FORMALISMS AND THEIR SIMULATORS

3

CONTENTS

Theory of Modeling and Simulation. https://doi.org/10.1016/B978-0-12-813370-5.00011-0

INTRODUCTION

This chapter presents, in an informal manner, the basic modeling formalisms for discrete time, continuous and discrete event systems. It is intended to provide a taste of each of the major types of models, how we express behavior in them, and what kinds of behavior we can expect to see. Each modeling approach is also accompanied by its prototypical simulation algorithms. The presentation employs commonly accepted ways of presenting the modeling formalisms. It does not presume any knowledge of formal systems theory and therefore serves as an independent basis for understanding the basic modeling formalisms that will be cast as basic system specifications (DESS, DTSS, and DEVS) after the system theory foundation has been laid. However, since the presentation is rather fast-paced you may want to consult some of the books listed at the end of the chapter for any missing background.

3.1 DISCRETE TIME MODELS AND THEIR SIMULATORS

Discrete time models are usually the most intuitive to grasp of all forms of dynamic models. As illustrated in Fig. 3.1, this formalism assumes a stepwise mode of execution. At a particular time instant the model is in a particular state and it defines how this state changes – what the state at the next time instant will be. The next state usually depends on the current state and also what the environment's influences currently are.

Discrete time systems have numerous applications. The most popular are in digital systems where the clock defines the discrete time steps. But discrete time systems are also frequently used as approximations of continuous systems. Here a time unit is chosen, e.g., one second, one minute or one year, to define an artificial clock and the system is represented as the state changes from one "observation" instant to the next. Therefore, to build a discrete time model, we have to define how the current state and the input from the environment determine the next state of the model.

The simplest way to define the way states change in a model is to provide a table such as Table 3.1. Here we assume there are a finite number of states and inputs. We write down all combinations of states and inputs and next states and outputs for each. For example, let the first column stand for the current state of the model and the second column for the input it is currently receiving. The table gives the next state of the model in column 3 and the output it produces in column 4. In Table 3.1, we have two states (0 and 1) and two inputs (also 0 and 1). There are 4 combinations and each one has an associated state and output. The first row says that if the current state is 0 and the input is 0, then the next state will be 0 and the output will be 0. The other three rows give similar information for the remaining state/input combinations.

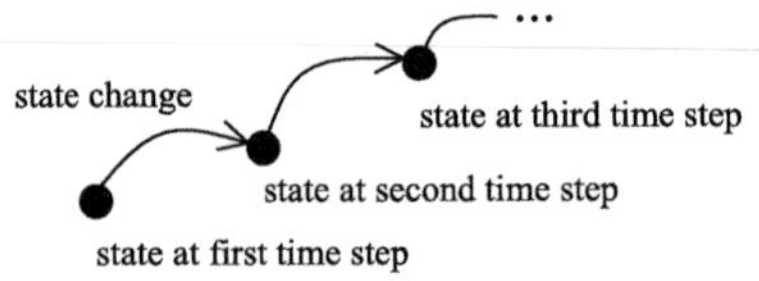

FIGURE 3.1

Stepwise execution of discrete time systems.

Table 3.1 Transition/Output table for a delay system

Current State	Current Input	Next State	Current Output
0	0	0	0
0	1	1	1
1	0	0	0
1	1	1	1

Table 3.2 State and output trajectories

time	0	1	2	3	4	5	6	7	8	9	
input trajectory	1	0	1	0	1	0	1	0	1	0	
state trajectory	0	1	0	1	0	1	0	1	0	1	0
output trajectory	1	0	1	0	1	0	1	0	1	0	

In discrete time models, time advances in discrete steps, which we assume are integers multiples of some basic period such as 1 second, 1 day or 1 year. The transition/output table just discussed would then be interpreted as specifying state changes over time, as in the following:

> if the state at time t is q and the input at time t is x then the state at time $t+1$ will be $\delta(q, x)$ and the output y at time t will be $\lambda(q, x)$.

Here δ is called the state transition function and is the more abstract concept for the first three columns of the table. λ is called the output function and corresponds to the first two and last columns. The more abstract forms, δ and λ constitute a more general way of giving the transition and output information. For example, Table 3.1 can be summarized more compactly as follows:

$$\delta(q, x) = x$$
$$\lambda(q, x) = x$$

which say the next state and current output are both given by the current input. The functions δ and λ are also much more general than the table. They can be thought about and described even when it is tedious to write a table for all combinations of states or inputs or indeed when they are not finite and writing a table is not possible at all. A sequence of states, $q(0), q(1), q(2), \ldots$ is called a state trajectory. Having an arbitrary initial state $q(0)$, subsequent states in the sequence are determined by:

$$q(t+1) = \delta(q(t), x(t)).$$

Similarly, a corresponding output trajectory is given by

$$y(t) = \lambda(q(t), x(t)).$$

Table 3.2 illustrates state and output trajectories (third and fourth rows, respectively) that are determined by the input trajectory in the second row.

Table 3.3 Computing state and output trajectories

time	0	1	2	3	4	5	6	7	8	9	
input trajectory	1	0	1	0	1	0	1	0	1	0	√
state trajectory	0										
output trajectory											√

Table 3.4 State and output trajectories

Current State	Current Input	Next State	Current Output
0	0	0	0
0	1	1	0
1	0	1	0
1	1	0	1

3.1.1 DISCRETE TIME SIMULATION

We can write a little algorithm to compute the state and output trajectories of a discrete time model given its input trajectory and its initial state. Such an algorithm is an example of a simulator (as defined in Chapter 2). Note that the input data for the algorithm corresponds to the entries in Table 3.3.

$T_i = 0, T_f = 9 - -$ the starting and ending times, here 0 and 9

$x(0) = 1, ..., x(9) = 0 - -$ the input trajectory

$q(0) = 0 - -$ the initial state

$t = T_i$

while ($t <= T_f$){

$y(t) = \lambda(q(t), x(t))$

$q(t+1) = \delta q(t), x(t))$

}

Executing the algorithm fills in the blanks in Table 3.3 (except for those checked).

Exercise 3.1. Execute the algorithm by hand to fill in Table 3.3. Why are the marked squares not filled in.

Exercise 3.2. Table 3.4 is for a model called a binary counter. Hand simulate the model in Table 3.4 for various input sequences and initial states. Explain why it is called a binary counter.

3.1.2 CELLULAR AUTOMATA

Although the algorithm just discussed seems very simple indeed, the abstract nature of the transition and output functions hide a wealth of potentially interesting complexities. For example, what if we connected together systems (as in Chapter 1) in a row with each system connected to its left and right neighbors, as shown in Fig. 3.2.

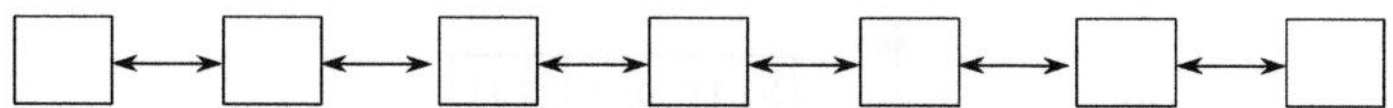

FIGURE 3.2

One dimensional cell space.

Table 3.5 A two input transition function

Current State	Current Left Input	Current Right Input	Next State
0	0	0	?
0	0	1	?
0	1	0	?
0	1	1	?
1	0	0	?
1	0	1	?
1	1	0	?
1	1	1	?

Imagine that each system has two states and gets the states of its neighbors as inputs. Then there are 8 combinations of states and inputs as listed in Table 3.5. We have left the next state column blank. Each complete assignment of a 0 or 1 to the eight rows results in a new transition function. There are $2^8 = 256$ such functions. Suppose we chose any one such function, started each of the components in Fig. 3.2 in one its states and applying an appropriate version of the above simulation algorithm. What would we observe?

The answer to such questions is the concern of the field of cellular automata. A *cellular automaton* is an idealization of a physical phenomenon in which space and time are discretized and the state sets are discrete and finite. Cellular automata have components, called *cells*, which are all identical with identical computational apparatus. They are geometrically located on a one-, two- or multi-dimensional grid and connected in a uniform way. The cells influencing a particular cell, called the *neighborhood* of the cell, are often chosen to be the cells located nearest in the geometrical sense. Cellular automata were originally introduced by von Neumann and Ulam (Burks, 1970) as idealization of biological self-production. Cellular automata show the interesting property that they yield quite diverse and interesting behavior. Actually, Wolfram (Wolfram et al., 1986) systematically investigated all possible transition functions of one-dimensional cellular automata. He found out that there exist four types of cellular automata which differ significantly in their behavior, (1) there are automata where soon any dynamic dies out, (2) automata which soon come to periodic behavior, (3) automata which show chaotic behavior, and (4), the most interesting ones, automata whose behaviors are unpredictable and non-periodic but which show interesting, regular patterns.

Conway's Game of Life in its original representation in Scientific American (Gardner, 1970) can serve as a fascinating introduction to the ideas involved. The game is framed within a two-dimensional cell space structure, possibly of infinite size. Each cell is coupled to its nearest physical neighbors both laterally and diagonally. This means for a cell located at point (0, 0) its lateral neighbors are at (0, 1), (1, 0), (0, −1), and (−1, 0) and its diagonal neighbors are at (1, 1), (1, −1), (−1, 1) and (−1, −1) as shown in Fig. 3.3. The neighbors of an arbitrary cell at (i, j) are the cells at

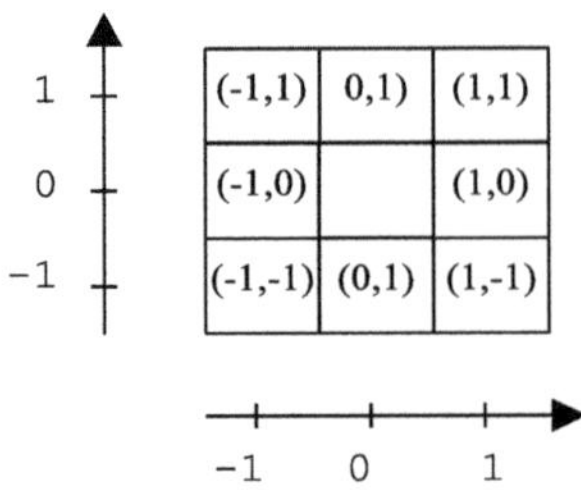

FIGURE 3.3

Cellular coupling structure for Game of Life.

$(i, j+1)$, $(i+1, j)$, $(i, j-1)$, $(i-1, j)$, $(i+1, j+1)$, $(i+1, j-1)$, $(i-1, j+1)$, and $(i-1, j-1)$ which can be computed from the neighbors at $(0, 0)$ by a simple translation. The *state* set of a cell consists of one variable which can take on only two values, 0 (*dead*) and 1 (*alive*). Individual cells survive (are alive and stay alive), are born (go from 0 to 1) or die (go from 1 to 0) as the game progresses. The rules as defined by Conway are:

1. A live cell remains alive if it has between 2 and 3 live cells in its neighborhood.
2. A live cell will die due to overcrowding if the has more than 3 live cells in its neighborhood.
3. A live cell will die due to isolation if it has less than 2 live neighbors.
4. A dead cell will become alive if it has exactly 3 alive neighbors.

When started from certain configurations of alive cells, the Game of Life show interesting behavior over time. As a state trajectory evolves, live cells form varied and dynamically changing clusters. The idea of the game is to find new patterns and study their behavior. Fig. 3.4 shows some interesting patterns.

The Game of Life exemplifies some of the concepts introduced earlier. It evolves on a *discrete time base* (time advances in steps 0, 1, 2, ...) and is a *multi-component system* (it is composed of *components* (cells) that are coupled together). In contrast to the *local* state (the state of a cell), the *global state* refers to the collection of states of **all** cells at any time. In Game of Life, this is a finite pattern, or configuration, of alive cells with all the rest being dead. Every such state starts a *state trajectory* (sequence of global states indexed by time) which either ends up in a cycle or continues to evolve forever.

Exercise 3.3. If an initial state is a finite configuration of alive cells, why are all subsequent states also finite configurations.

3.1.3 CELLULAR AUTOMATON SIMULATION ALGORITHMS

The basic procedure for simulating a cellular automaton follows the discrete time simulation algorithm introduced earlier. This is, at every time step, we scan all cells applying the state transition function to each, and saving the next state in a second copy of the global state data structure. When all next-states have been computed, they constitute the next global state and the clock advances to the next step. For example, let's start with the three live cells in a triangle shown in the upper left of Fig. 3.5. Analyzing the neighborhood of the lower corner cell, we see it is alive and has 2 alive neighbors, thus it survives

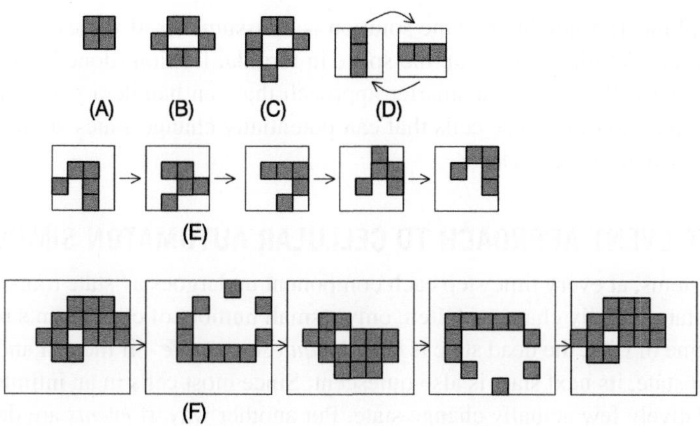

FIGURE 3.4

Patterns from Conway's Game of Life: Patterns (A) to (C) are stable, they don't change, (D) is an oscillating patterns, (E) and (F) are cycles of patterns which move.

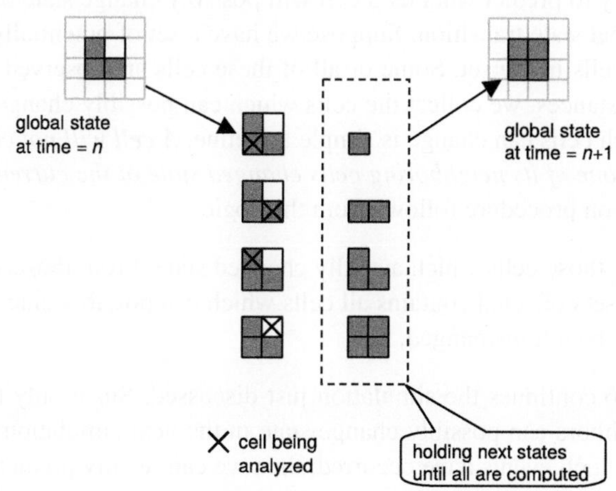

FIGURE 3.5

Cellular Automaton Simulation Cycle.

to the next generation. As shown, the other alive cells also survive. However, only one new cell is born. This is the upper right corner cell which has exactly three neighbors.

Of course as stated, the cell space is an infinite and we can't possibly scan all the cells in a finite amount of time. To overcome this problem, the examined part of the space is limited to a finite region. In many case, this region is fixed throughout the simulation. For example, a two-dimensional space might be represented by a square array of size $N(= 1, 2, ...)$. In this case, the basic algorithm scans all N^2 cells at every time step. Something must be done to take care of the fact that cells at the boundary lack their full compliment of neighbors. One solution is to assume fixed values for all boundary cells (for example, all dead). Another is to wrap the space in toroidal fashion (done by letting the index N also be interpreted as 0). But there is a smarter approach that can handle a potentially infinite space by limiting the scanning to only those cells that can potentially change states at any time. This is the discrete event approach discussed next.

3.1.4 DISCRETE EVENT APPROACH TO CELLULAR AUTOMATON SIMULATION

In discrete time systems, at every time step each component undergoes a "state transition"; this occurs whether or not its state actually changes. Often, only a small number of components really change. For example, in the Game of Life, the dead state is called a *quiescent state* – if the cell and all its neighbors are in the quiescent state, its next state is also quiescent. Since most cells in an infinite space are in the quiescent state, relatively few actually change state. Put another way, if *events* are defined as changes in state (e.g., births and deaths in the Game of Life), then often there are relatively few events in the system. Scanning all cells in an infinite space is impossible, but even in a finite space, scanning all cells for events at every time step is clearly inefficient. A discrete event simulation algorithm **concentrates on processing events** rather than cells, and is inherently more efficient.

The basic idea is to try to predict whether a cell will possibly change state or will definitely be left unchanged in a next global state transition. Suppose we have a set of potentially changing cells. Then we only examine those cells in the set. Some or all of these cells are observed to change their states. Then, in the new circumstances, we collect the cells which can possibly change state in the next step. The criterion under which cells can change is simple to define. *A cell will not change state at the next state transition time, if none of its neighboring cells changed state at the current state transition time.* The event-based simulation procedure follows from this logic:

In a state transition mark those cells which actually changed state. From those, collect the cells which are their neighbors. The set collected contains all cells which can possibly change at the next step. All other cells will definitely be left unchanged.

For example, Fig. 3.6 continues the simulation just discussed. Since only the black cell changed state, only its eight neighbors can possibly change state in the next simulation cycle. In other words, if we start with cells at which events have occurred, then we can readily predict where the next events can occur.

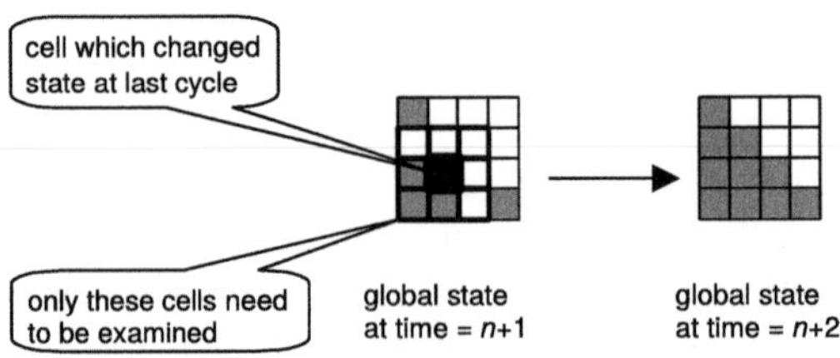

FIGURE 3.6

Collecting cells with possible events.

Table 3.6 State and output trajectories of a delay flip-flop

time	0	1	2	3	4	5	6	7	8	9	
input trajectory	1	0	1	0	1	0	1	0	1	0	
state trajectory	0	1	0	1	0	1	0	1	0	1	0
output trajectory	0	1	0	1	0	1	0	1	0	1	

Note that although we can limit the scanning to those cells that are candidates for events, we cannot *predict which cells will actually have events* without actually applying the transition function to their neighborhood states.

Exercise 3.4. Identify a cell which is in the can-possibly-change set but which does not actually change state.

Exercise 3.5. Write a simulator for one-dimensional cellular automata such as in Table 3.5 and Fig. 3.2. Compare the direct approach which scans all cells all the time with the discrete event approach in terms of execution time. Under what circumstances would it not pay to use the discrete event approach? (See later discussion in Chapter 17.)

Exercise 3.6. Until now we have assumed a neighborhood of 8 cells adjacent to the center cell. However, in general, a neighborhood for the cell at the origin is defined as any finite subset of cells. This neighborhood is translated to every cell. Such a neighborhood is called reflection symmetric if whenever cell (i, j) belongs to it then so does (j, i). For neighborhoods that are not reflection symmetric, define the appropriate set of influencees required for the discrete event simulation approach.

3.1.5 SWITCHING AUTOMATA/SEQUENTIAL MACHINES

Cellular automata are uniform both in their composition and interconnection patterns. If we drop these requirements but still consider connecting finite state components together we get another useful class of discrete time models.

Switching Automata (also called digital circuits) are constructed from flip-flop components and logical gates. The flip-flops are systems with binary states, the most straightforward example of which we have already met. This has the transition function shown in Table 3.1. However, instead of allowing the input to propagate straight through to the output, we will let the output be the current state (as in cellular automata). Table 3.6 shows how the output trajectory now lags the input trajectory by one time step. The difference, we'll see later, is between so-called Mealy and Moore models of sequential machines. In Mealy networks the effects of an input can propagate throughout space in zero time – even cycling back on themselves, causing "vicious circles" or ill-behaved systems.

To construct networks we can not only couple flip-flop outputs to inputs, but also connect them through so-called *gates*, which realize elementary Boolean functions. As we shall see later, these are instantaneous or memoryless systems. As shown in Fig. 3.7, their outputs are directly determined by their inputs with no help from an internal state. The clock usually used in synchronous machines determines a constant time advance. This enables transistor circuits to be adequately represented by discrete time models, at least for their functional behavior.

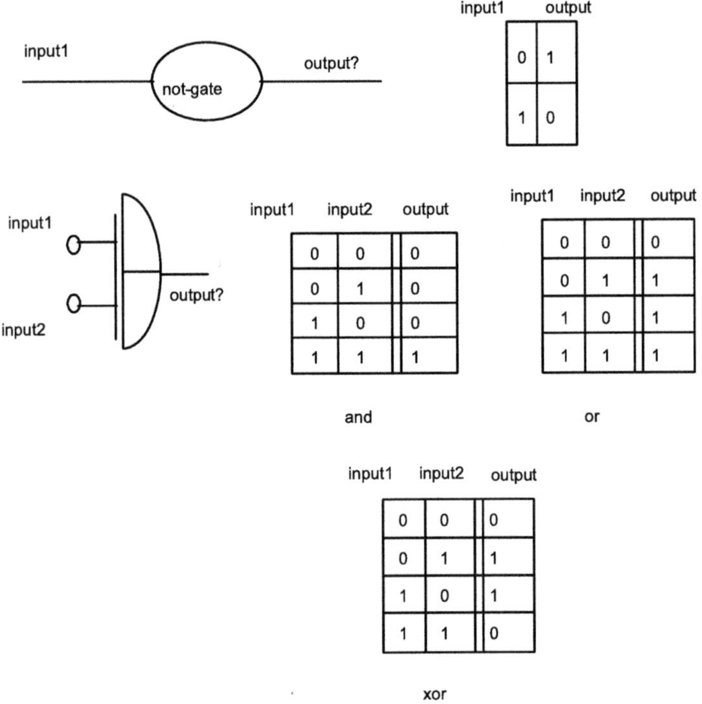

FIGURE 3.7

Gates Realizing Elementary Boolean Functions.

The switching automaton in Fig. 3.8 is made up of several flip-flop components coupled in series and a feedback function defined by a coupling of XOR-gates. Each flip-flop is an elementary memory component as above. Flip-flops q_1 to q_7 are coupled in series, that is, the state of flip-flop i defines the input and hence the next state of flip-flop i-1. This linear sequence of flip-flops is called a *shift-register* since it shifts the input at the first component to the right each time step. The latter input may be

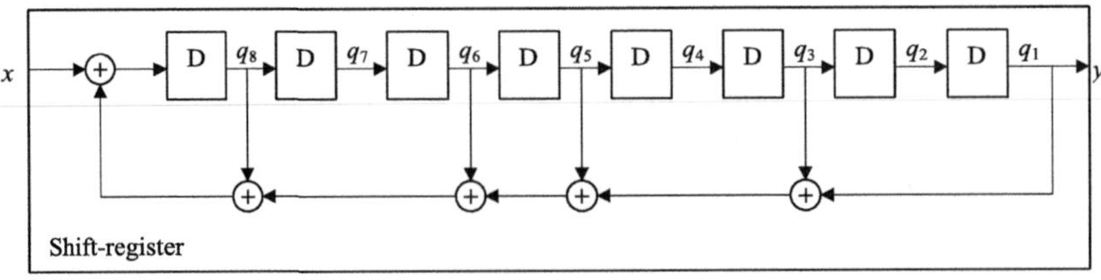

FIGURE 3.8

Shift register coupling flip-flop elements and logical gates.

computed using gates. For example, the input to flip-flop 8 is defined by an XOR-connection of memory values q_i and input value $x : x \oplus q_8 \oplus q_6 \oplus q_5 \oplus q_3 \oplus q_1$. (An XOR, $q_1 \oplus q_2$, outputs the exclusive-or operation on its inputs, i.e., the output is 0 if, and only if, both inputs are the same, otherwise 1.)

In Chapter 8 we will discuss simulators for such combinations of memoryless and memory-based systems.

3.1.6 LINEAR DISCRETE TIME NETWORKS AND THEIR STATE BEHAVIOR

Fig. 3.9 illustrates a Moore network with two delay elements and 2 memoryless elements. However, in distinction to the switching automaton above, this network now is defined over the reals. The structure of the network can be represented by the matrix:

$$\begin{bmatrix} 0 & g \\ -g & 0 \end{bmatrix}$$

where g and $-g$ are the gain factors of the memoryless elements. Started in a state represented by the vector $[1, 1]$, (each delay is in state 1), the next state $[q_1, q_2]$ of the network is $[-g, g]$ which can be computed by the matrix multiplication:

$$\begin{bmatrix} q_1 \\ q_2 \end{bmatrix} = \begin{bmatrix} 0 & g \\ -g & 0 \end{bmatrix} \begin{bmatrix} 1 \\ 1 \end{bmatrix} = \begin{bmatrix} g \\ -g \end{bmatrix}$$

The reason that a matrix representation and multiplication can be used is that the network has a *linear* structure. This means that all components produce outputs or states that are linear combinations of their inputs and states. A delay is a very simple linear component – its next state is identical to its input. A memoryless element with a gain factor is called a *coefficient* element and it multiplies its current input by the gain to produce its output. Both of these are simple examples of linearity. A summer, which a memoryless function that adds its inputs to obtain its output, is also a linear element.

A Moore discrete time system is in *linear matrix* form if its transition and output functions can be represented by matrices $\{A, B, C\}$. This means that its transition function can be expressed as:

$$\lambda(q, x) = Aq + Bx$$

Here q is a real n-dimensional state vector, and A is an n by n matrix. Similarly, x is a real m-dimensional input vector, and B has dimension m by n. Also the output function is

$$\lambda(q) = Cq$$

where, if the output y is a p-dimensional vector, then C is an n by p-dimensional matrix. A similar definition holds for Mealy discrete time system.

Exercise 3.7. Any discrete time network with linear structure can be represented by linear system in matrix form. Conversely, any linear discrete time system in matrix form can be realized by a discrete time network with linear structure. Develop procedures to convert one into the other. The state trajectory behavior of the linear system in Fig. 3.9 is obtained by interatively multiplying the matrix A into

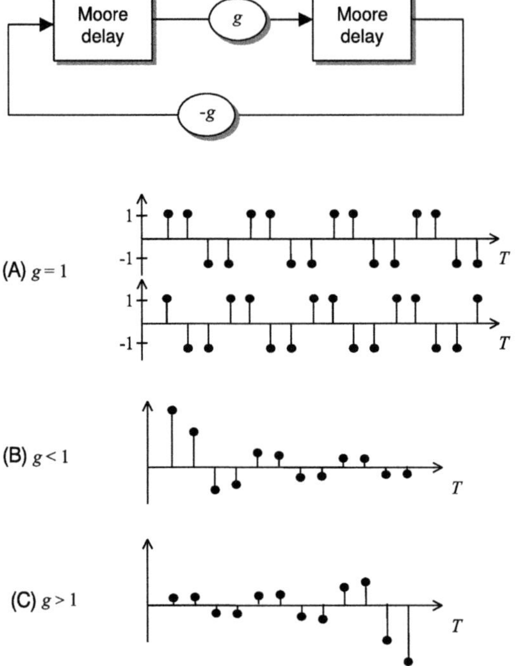

FIGURE 3.9

Simple Linear Moore Network and its Behavior.

successive states. Thus if $[g, -g]$ is the state following $[1, 1]$, then the next state is:

$$\begin{bmatrix} -g^2 \\ g^2 \end{bmatrix} = \begin{bmatrix} 0 & g \\ -g & 0 \end{bmatrix} \begin{bmatrix} g \\ -g \end{bmatrix}$$

Thus, the state trajectory starts out as:

$$\begin{bmatrix} 1 \\ 1 \end{bmatrix} \cdots \begin{bmatrix} -g^2 \\ g^2 \end{bmatrix} \cdots \begin{bmatrix} g \\ -g \end{bmatrix}$$

Exercise 3.8. Characterize the remainder of the state trajectory.

There are three distinguishable types of trajectories as illustrated in Fig. 3.9. When $g = 1$, we have a steady alternation between $+1$ and -1. When $g < 1$, the envelope of values decays exponentially; and finally, when $g > 1$, the envelope increases exponentially.

Actually the model above is a discrete form of an oscillating system. In the following sections we will learn to understand its better-known continuous system counterpart. In the discrete form above, oscillations (switching from positive to negative values) come into being due to the delays in the Moore components and the negative influence $-g$. In one step the value of the second delay is inverted giving a value with opposite sign to the first delay. In the next step, however, this inverted value in the first delay is fed into the second delay. The sign of the delay values is inverted every second step.

We will see that in the continuous domain to be discussed, oscillation also comes into being through feedback. However, the continuous models work with derivatives instead of input values. The delays in the discrete domain correspond to integrators in the continuous domain.

3.2 DIFFERENTIAL EQUATION MODELS AND THEIR SIMULATORS

In discrete time modeling we had a state transition function which gave us the information of the state at the next time instant given the current state and input. In the classical modeling approach of differential equations, the state transition relation is quite different. For differential equation models we do not specify a next state directly but use a derivative function to specify the rate of change of the state variables. At any particular time instant on the time axis, given a state and an input value, we only know the rate of change of the state. From this information, the state at any point in the future has to be computed.

To discuss this issue, let us consider the most elementary continuous system – the simple integrator (Fig. 3.10). The integrator has one input variable $u(t)$ and one output variable $y(t)$. One can imagine it as a reservoir with infinite capacity. Whatever is put into the reservoir is accumulated – but a negative input value means a withdrawal. The output of the reservoir is its current contents. When we want to express this in equation form we need a variable to represent the current contents. This is our state variable $z(t)$. The current input $u(t)$ represents the rate of current change of the contents which we express by equation

$$\frac{dz(t)}{dt} = u(t)$$

and the output $y(t)$ is equal to the current state $y(t) = z(t)$.

Fig. 3.11 shows input and output trajectories in a simple integrator.

Usually continuous time systems are expressed by using several state variables. Derivatives are then functions of some, or all, the state variables. Let $z_1, z_2, ..., z_n$ be the state variables and $u_1, u_2, ..., u_m$ be the input variables, then a continuous time model is formed by a set of first-order differential equations

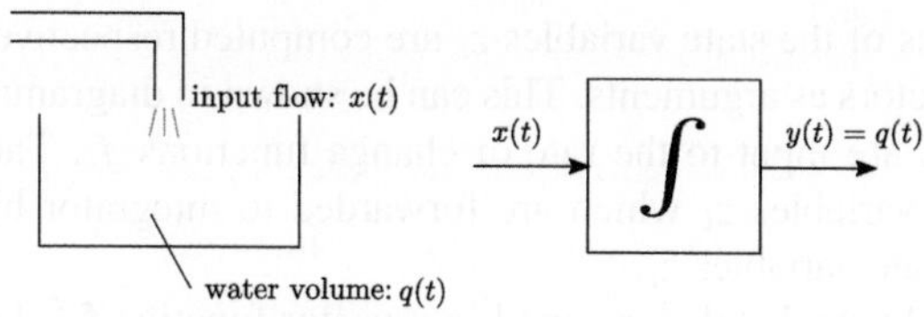

FIGURE 3.10

Integrator: A reservoir model and its representation.

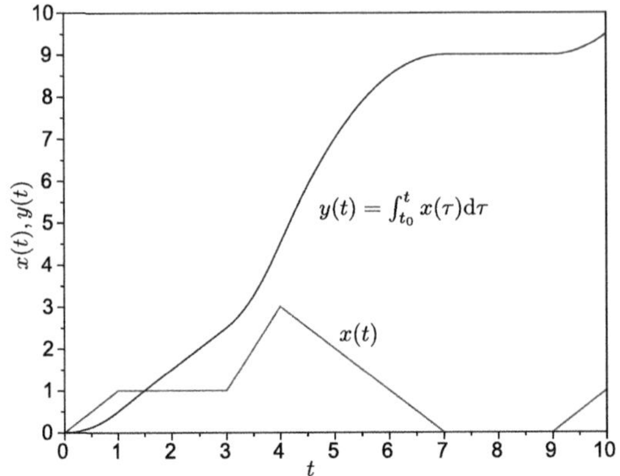

FIGURE 3.11

Integrator: Input and output trajectories.

(ODEs):

$$\begin{aligned}
\dot{z}_1(t) &= f_1(z_1(t), \cdots, z_n(t), u_1(t), \cdots, u_m(t)) \\
\dot{z}_2(t) &= f_2(z_1(t), \cdots, z_n(t), u_1(t), \cdots, u_m(t)) \\
&\vdots \\
\dot{z}_n(t) &= f_n(z_1(t), \cdots, z_n(t), u_1(t), \cdots, u_m(t))
\end{aligned}$$

where $\dot{z}_i$ stands for $\frac{dz_i}{dt}$. In order to have a more compact notation, the state variables are grouped in a state vector $\mathbf{z}(t) \triangleq [z_1(t), \cdots, z_n(t)]^T$. Similarly, the input variables for the input vector $\mathbf{u}(t) \triangleq [u_1(t), \cdots u_m(t)]$, so the equations are written as

$$\begin{aligned}
\dot{z}_1(t) &= f_1(\mathbf{z}(t), \mathbf{u}(t)) \\
\dot{z}_2(t) &= f_2(\mathbf{z}(t), \mathbf{u}(t)) \\
&\vdots \\
\dot{z}_n(t) &= f_n((\mathbf{z}(t), \mathbf{u}(t))
\end{aligned} \tag{3.1}$$

Note that the derivatives of the state variables z_i are computed respectively, by functions f_i which have the state and input vectors as arguments. This can be shown in diagrammatic form as in Fig. 3.12. The state and input vector are input to the rate of change functions f_i. Those provide as output the derivatives $\dot{z}_i$ of the state variables z_i which are forwarded to integrator blocks. The outputs of the integrator blocks are the state variables z_i.

Functions f_i in Eq. (3.1) can be also grouped in a vector function $\mathbf{f} \triangleq [f_1, \cdots, f_n]$, obtaining the classic *state equation* representation of ODEs:

$$\dot{\mathbf{z}}(t) = \mathbf{f}(\mathbf{z}(t), \mathbf{u}(t)) \tag{3.2}$$

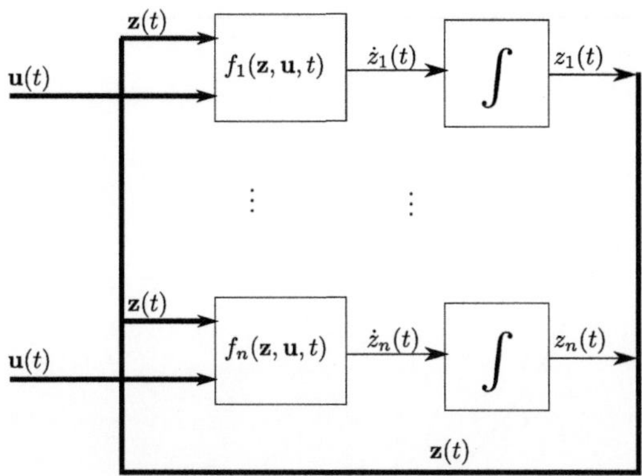

FIGURE 3.12

Block Diagram representation of Eq. (3.1).

Most continuous time models are actually written in (or converted to) the form of Eq. (3.2), a representation that does not provide an explicit value for the state $\mathbf{z}(t)$ after some amount of time. Thus, in order to obtain the state trajectories the ODE must be solved. The problem is that obtaining a solution for Eq. (3.2) can be not only very difficult but also impossible as only very few ODEs have analytical solution in terms of known functions and expressions. This is the reason why ODEs are normally solved using numerical integration algorithms that provide approximate solutions.

Before introducing the main concepts of numerical integration algorithms, we shall study the solution of Linear Time-Invariant (LTI) ODEs, as they can be analytically solved. For background, you may wish to consult some of the many texts that cover such systems and their underlying theory as listed at end of the chapter.

3.2.1 LINEAR ODE MODELS

Consider again the model of the tank represented by an integrator in Fig. 3.10, but suppose that now the tank has a valve in the bottom where the water flows away. Suppose also that the flow is proportional to the volume of water in the tank.

Under these new assumptions, the derivative of the water volume can be written as

$$\dot{z}(t) = u(t) - a \cdot z(t) \tag{3.3}$$

where $z(t)$ is the volume of water, $u(t)$ is the input flow, and $\lambda \cdot z(t)$ is the output flow.

Let us suppose that the input flow is null (i.e., $u(t) = 0$), and that the initial volume of water is $z(t_0) = z_0$. Then, it can be easily checked that the volume of water follows the trajectory:

$$z(t) = e^{-a \cdot t} \cdot z_0 \tag{3.4}$$

Notice that differentiating both terms of Eq. (3.4) with respect to t you obtain back the expression of Eq. (3.3). The trajectory given by Eq. (3.4) is depicted in Fig. 3.13.

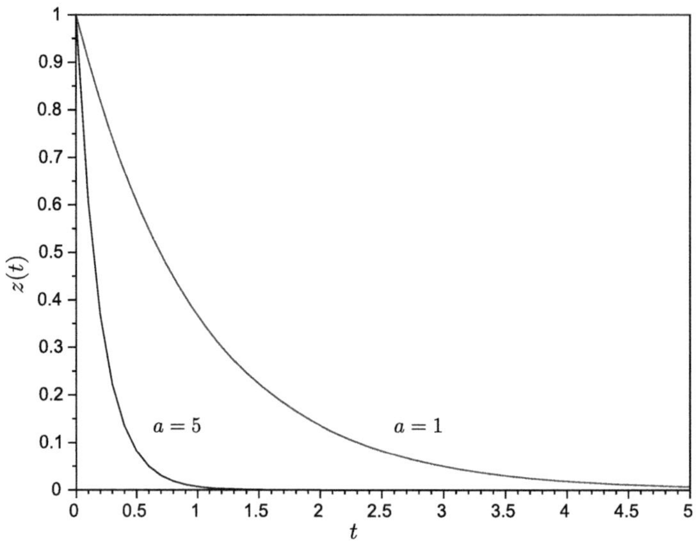

FIGURE 3.13

Solution of the ODE (3.3) for $z_0 = 1$, $a = 1$, and $a = 5$.

The constant $\lambda \triangleq -a$ is the ODE *eigenvalue*, and it provides the *speed* at which the solution converges to the final value. A large value of a implies that the solution converges fast, while a small value of a says that the trajectory slowly goes to its equilibrium value.

Consider now a different model that comes from population dynamics. Let us suppose that $z(t)$ is the number of individuals of certain specie. We may assume that the unit for $z(t)$ is given in billions of specimens so that non-integer numbers have some physical sense. We shall suppose that b is the birth rate per year, i.e., the number of individuals that are born each year is $b \cdot z(t)$. Similarly, d will be the death rate per year.

Thus, the population growth can expressed by the following ODE:

$$\dot{z}(t) = b \cdot z(t) - d \cdot z(t) = (b - d) \cdot z(t) \tag{3.5}$$

which is identical to Eq. (3.3) with $u(t) = 0$ and $a = d - b$. Thus, the solution is

$$z(t) = e^{(b-d)\cdot t} \cdot z_0 \tag{3.6}$$

Let us assume that the model corresponds to the world population of people. According to 2015 data, the birth rate is $b = 0.0191$ (i.e. there are 19.1 births every year per 1000 inhabitants), while the death rate is $r = 0.0076$. Taking into account that the world population of 2015 was about 7355 billion, if we consider 2015 as the initial time of simulation, we have $z_0 = 7355$. The trajectory given by Eq. (3.6) is depicted in Fig. 3.14.

Notice that this time, unlike Fig. 3.13, the solution diverges and after 100 years we may expect a population of 23,000 billion inhabitants (this is assuming that birth and death rates are kept constant).

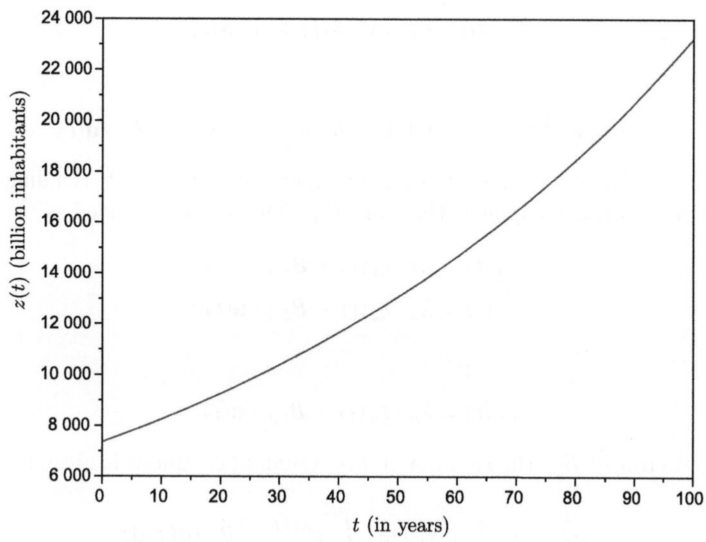

FIGURE 3.14

Solution of the ODE (3.5).

The fact that the solution diverges is mathematically explained by the fact that the eigenvalue $\lambda = b - d$ is now positive, so the expression $e^{\lambda \cdot t}$ goes to infinity as t advances. In conclusion, the sign of the eigenvalue λ dictates the *stability* of a linear time invariant system of first order. The solution is *stable* when $\lambda < 0$ and it is *unstable* when $\lambda > 0$.

Let us now consider a second order model that represents a spring–mass–damper mechanical system:

$$\begin{aligned} \dot{z}_1(t) &= z_2(t) \\ \dot{z}_2(t) &= -\frac{k}{m} z_1(t) - \frac{b}{m} z_2(t) + \frac{F(t)}{m} \end{aligned} \tag{3.7}$$

Here, the variables $z_1(t)$ and $z_2(t)$ are the position and speed, respectively. The parameter m is the mass, b is the friction coefficient, k is the spring coefficient and $F(t)$ is an input trajectory force applied to the mass.

Solving higher order LTI models is a bit more complicated than solving first order models. One way to do it is to introduce a change of variables that converts the n-th ODE in a set of n independent ODEs.

Given a LTI ODE of the form

$$\dot{\mathbf{z}}(t) = A \cdot \mathbf{z}(t) + B \cdot \mathbf{u}(t) \tag{3.8}$$

we introduce a new variable $\xi(t)$ so that $\mathbf{z}(t) = V \cdot \xi(t)$ where V is some $n \times n$ invertible matrix. Then, replacing this expression in Eq. (3.8), we obtain

$$V \cdot \dot{\xi}(t) = A \cdot V \cdot \xi(t) + B \cdot \mathbf{u}(t)$$

and then,

$$\dot{\xi}(t) = V^{-1}A \cdot V \cdot \xi(t) + V^{-1}B \cdot \mathbf{u}(t) \triangleq \Lambda\xi(t) + B_\xi \cdot \mathbf{u}(t) \tag{3.9}$$

If V is computed as an *eigenvector* matrix of A, then matrix $\Lambda \triangleq V^{-1}A \cdot V$ is a diagonal matrix with the eigenvalues of A in its main diagonal. That way, Eq. (3.9) can be expanded as

$$\begin{aligned}
\dot{\xi}_1(t) &= \lambda_1 \cdot \xi_1(t) + B_{\xi,1} \cdot \mathbf{u}(t) \\
\dot{\xi}_2(t) &= \lambda_2 \cdot \xi_2(t) + B_{\xi,2} \cdot \mathbf{u}(t) \\
&\vdots \\
\dot{\xi}_n(t) &= \lambda_n \cdot \xi_n(t) + B_{\xi,n} \cdot \mathbf{u}(t)
\end{aligned} \tag{3.10}$$

where $B_{\xi,n}$ is the i-th row of B_ξ. The solution of this system of equations is given by

$$\begin{aligned}
\xi_1(t) &= e^{\lambda_1 \cdot t} \cdot \xi_1(0) + \cdot \int_0^t e^{\lambda_1 \cdot (t-\tau)} B_{\xi,1}\mathbf{u}(\tau)d\tau \\
\xi_2(t) &= e^{\lambda_2 \cdot t} \cdot \xi_2(0) + \cdot \int_0^t e^{\lambda_2 \cdot (t-\tau)} B_{\xi,2}\mathbf{u}(\tau)d\tau \\
&\vdots \\
\xi_n(t) &= e^{\lambda_n \cdot t} \cdot \xi_n(0) + \cdot \int_0^t e^{\lambda_n \cdot (t-\tau)} B_{\xi,n}\mathbf{u}(\tau)d\tau
\end{aligned} \tag{3.11}$$

where $\xi(0) = V^{-1} \cdot \mathbf{z}(0)$. This solution can be brought back to the original state variables $\mathbf{z}(t)$ using the transformation $\mathbf{z}(t) = V \cdot \xi(t)$. Then, each component $z_i(t)$ follows a trajectory of the form

$$z_i(t) = \sum_{j=1}^{n} c_{i,j} \cdot e^{\lambda_j \cdot t} + \sum_{j=1}^{n} d_{i,j} \cdot \int_0^t e^{\lambda_j \cdot (t-\tau)} B_{\xi,j}\mathbf{u}(\tau)d\tau \tag{3.12}$$

for certain constants $c_{i,j}$ and $d_{i,j}$.

This way, the generic solution is governed by terms of the form $e^{\lambda_i \cdot t}$ depending exclusively on the eigenvalues of matrix A. The terms $e^{\lambda_i \cdot t}$ are called *system modes*.

Coming back the second order system of Eq. (3.7), we consider first the set of parameters $m = 1$, $b = 3$, $k = 2$, and, for the sake of simplicity, we assume that the input is null ($F(t) = 0$). Matrix A has the form

$$A = \begin{bmatrix} 0 & 1 \\ -k/m & -b/m \end{bmatrix} = \begin{bmatrix} 0 & 1 \\ -2 & -3 \end{bmatrix}$$

whose eigenvalues are the solutions of the determinant, $\det(\lambda) \cdot I - A = 0$, resulting $\lambda_1 = -1$, $\lambda_2 = -2$. Then, given an initial state $\mathbf{z}(0) = [z_1(0), z_2(0)]^T$, the solution of Eq. (3.12) results

$$\begin{aligned}
z_1(t) &= 2 \cdot (z_1(0) + z_2(0)) \cdot e^{-t} - (z_1(0) + z_2(0)) \cdot e^{-2 \cdot t} \\
z_2(t) &= (-2 \cdot z_1(0) - z_2(0)) \cdot e^{-t} + 2 \cdot (z_1(0) + z_2(0)) \cdot e^{-2 \cdot t}
\end{aligned} \tag{3.13}$$

Notice that this solution can be thought as the sum of the solutions of first order models like that of the tank in Eq. (3.3).

Let us consider now that the parameters are $b = m = k = 1$. This time, matrix A becomes

$$A = \begin{bmatrix} 0 & 1 \\ -k/m & -b/m \end{bmatrix} = \begin{bmatrix} 0 & 1 \\ -1 & -1 \end{bmatrix}$$

whose eigenvalues are now

$$\lambda_{1,2} = -\frac{1}{2} \pm i \cdot \frac{\sqrt{3}}{2} \tag{3.14}$$

where $i = \sqrt{-1}$ is the imaginary unit. In presence of complex eigenvalues like those of Eq. (3.14), the system modes have the form

$$e^{\lambda_i \cdot t} = e^{\mathbb{R}\mathrm{e}(\lambda_i) \cdot t} \cdot e^{i \cdot \mathbb{I}\mathrm{m}(\lambda_i) \cdot t} = e^{\mathbb{R}\mathrm{e}(\lambda_i) \cdot t} \cdot (i \sin(\mathbb{I}\mathrm{m}(\lambda_i) \cdot t) + \cos(\mathbb{I}\mathrm{m}(\lambda_i) \cdot t))$$

so the trajectories have oscillations with a frequency given by $\mathbb{I}\mathrm{m}(\lambda_i)$ and with an amplitude that goes to zero with the time when $\mathbb{R}\mathrm{e}(\lambda_i)$ is negative, or can become larger and diverge when $\mathbb{R}\mathrm{e}(\lambda_i) > 0$. A critical case is that of $\mathbb{R}\mathrm{e}(\lambda_i) = 0$, where sustained oscillations are obtained.

Coming back to the case of the second order system of Eq. (3.7) with $m = b = k = 1$, the solution for input $F(t) = 1$ and initial state $z_1(0) = z_2(0) = 0$ is given by

$$\begin{aligned} z_1(t) &= 1 - \frac{\sqrt{3}}{3} e^{-t/2} \sin(\frac{\sqrt{3}}{2} t) - e^{-t/2} \cos(\frac{\sqrt{3}}{2} t) \\ z_2(t) &= \frac{\sqrt{12}}{3} e^{-t/2} \sin(\frac{\sqrt{3}}{2} t) \end{aligned} \tag{3.15}$$

and depicted in Fig. 3.15.

If we set the friction coefficient $b = 0$, matrix A results

$$A = \begin{bmatrix} 0 & 1 \\ -k/m & -b/m \end{bmatrix} = \begin{bmatrix} 0 & 1 \\ -1 & 0 \end{bmatrix}$$

whose eigenvalues are now

$$\lambda_{1,2} = -i \tag{3.16}$$

and the trajectories for $F(t) = 1$ and null initial conditions are now the following undamped oscillations

$$\begin{aligned} z_1(t) &= 1 - \cos(t) \\ z_2(t) &= \sin(t) \end{aligned} \tag{3.17}$$

depicted in Fig. 3.16.

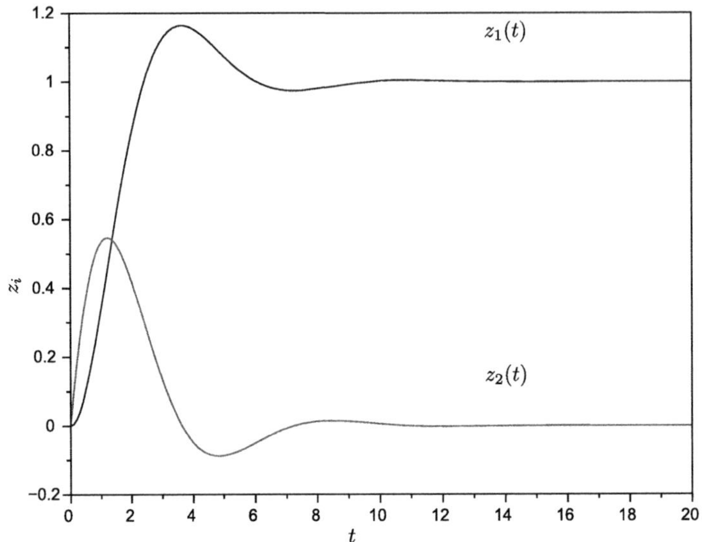

FIGURE 3.15

Solution of Eq. (3.7) for $b = m = k = 1$, and $F(t) = 1$.

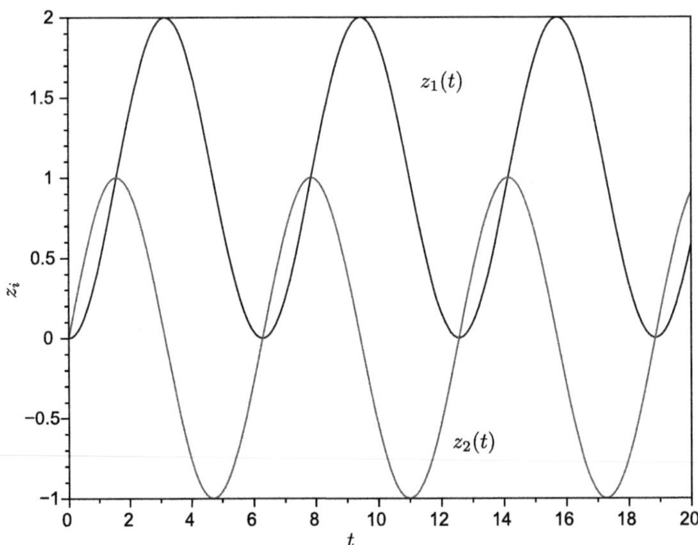

FIGURE 3.16

Solution of Eq. (3.7) for $m = k = 1$, $b = 0$, and $F(t) = 1$.

In conclusion, the solution of a LTI ODE is qualitatively dictated by the position of the eigenvalues of matrix A:

- A negative real valued eigenvalue produces a decaying term with a trajectory like those of Fig. 3.13.
- A positive real valued eigenvalue produces an exponentially growing term like that of Fig. 3.14.
- A pair of complex conjugated eigenvalues with negative real part produce damped oscillations like those of Fig. 3.15.
- A pair of pure imaginary conjugated eigenvalues produce undamped oscillations like those of Fig. 3.16.
- A pair of complex conjugated eigenvalues with positive real part produce oscillations whose amplitude exponentially grows with time.

The solution of a large system with several eigenvalues is ruled by the sum of the different modes. If all eigenvalues are negative (or have real negative part) then all the modes converge to zero and the solution is *asymptotically stable*. In there is one or more eigenvalues with positive real part then there are terms that diverge and the solution is *unstable*. Finally, the trajectories corresponding to eigenvalues with null real part are called *marginally stable*.

Non-Linear Models

While LTI system trajectories are limited to the sum of modes of the form $e^{\lambda_i \cdot t}$, in non-linear systems the solutions are far more complex. Moreover, in most cases the solutions cannot be expressed in terms of known functions.

However, in the vicinity of a given point of the solution trajectory, the behavior can be approximated by that of a LTI model. Thus, the solution of non-linear models can be thought as the concatenation of the solution of several LTI models. For this reason, many non-linear systems exhibit trajectories that resemble those of simple LTI models.

Nevertheless, there are several features of continuous systems that can only appear in non-linear models (chaotic trajectories, limit cycles, finite exit time, etc.).

While there exist some analysis tools to study qualitative properties of non-linear systems (Lyapunov theory is the preferred approach for these purposes), it is in most cases impossible to obtain quantitative information about the solutions. Thus, simulation is the only way to obtain the system trajectories.

3.2.2 CONTINUOUS SYSTEM SIMULATION

After describing the solutions of LTI systems, we are back to the problem of finding the solution of a general ODE:

$$\dot{\mathbf{z}}(t) = \mathbf{f}(\mathbf{z}(t), \mathbf{u}(t)) \tag{3.18}$$

with known initial state $\mathbf{z}(t_0) = \mathbf{z}_0$.

As we already mentioned, in most cases the exact solution cannot be obtained so we are forced to obtain numerical approximations for the values of the state at certain instants of time $t_0, t_1, \cdots, t_N$. These approximations are obtained using *numerical integration algorithms*, and traditionally the term *Continuous System Simulation* is tightly linked to those algorithms.

Nowadays, Continuous System Simulation is a wider topic. In fact, most continuous time models are not originally written like Eq. (3.18). That representation is the one used by numerical integration routines but, for a modeling practitioner, it is not comfortable (and it is sometimes impossible) to express a model in that way.

Modern continuous systems modeling languages like Modelica – the most widely accepted modeling standard language – allow representing the models in an object oriented fashion, reusing and connecting components from libraries. The resulting model is then an object oriented description that, after a flattening stage, results in a large set of differential and algebraic equations. Then, those *Differential Algebraic Equations* (DAEs) must be converted in an ODE like that of Eq. (3.18) in order to use numerical integration routines. This conversion requires different (and sometimes complex) algorithms and it constitutes itself an important branch of the Continuous System Simulation field (an extensive discussion is in Chapter 3 of TMS2000).

Many continuous time models describe the evolution of some magnitudes both in time and space. These models are expressed by *Partial Differential Equations* (PDEs) and their simulation constitute one of the most difficult problems in the discipline.

In this book we shall not cover the problems of DAE or PDE simulation. However, as both DAEs and PDEs can be converted or approximated by ODEs, the numerical algorithms for ordinary differential equations described next can still be applied for those problems.

3.2.3 EULER'S METHODS

The simplest method to solve Eq. (3.18) was proposed by Leonhard Euler en 1768. It is based on approximating the state derivative as follows

$$\begin{aligned}\dot{\mathbf{z}}(t) \approx \frac{\mathbf{z}(t+h)-\mathbf{z}(t)}{h} \frac{\mathbf{z}(t+h)-\mathbf{z}(t)}{h} \\ \Longrightarrow \mathbf{f}(\mathbf{z}(t),\mathbf{u}(t)) \approx \frac{\mathbf{z}(t+h)-\mathbf{z}(t)}{h}\end{aligned}$$

so we obtain

$$\mathbf{z}(t+h) \approx \mathbf{z}(t) + h \cdot \mathbf{f}(\mathbf{z}(t),\mathbf{u}(t))$$

where h is the integration *step size*.

Defining $t_k = t_0 + k \cdot h$ for $k = 1, \cdots, N$, Forward Euler method is defined by the following formula:

$$\mathbf{z}(t_{k+1}) = \mathbf{z}(t_k) + h \cdot \mathbf{f}(\mathbf{z}(t_k),\mathbf{u}(t_k)) \tag{3.19}$$

Thus, starting from the known initial state $\mathbf{z}(t_0)$, Forward Euler method allows us to compute a numerical approximation for $\mathbf{z}(t_1)$. Then, using this value, an approximation for $\mathbf{z}(t_2)$ can be obtained as well as the successive values until reaching the final simulation time.

Using this algorithm, we simulated the spring–mass–damper model of Eq. (3.7) for parameters $b = k = m = 1$, input $F(t) = 1$, and null initial state values. We used a step size $h = 0.1$ and simulated until $t = t_f = 15$. The numerical results are depicted in Fig. 3.17 together with the analytical solution.

In spite of using a relatively small step size of $h = 0.1$, Forward Euler method gives a result with a relatively large error. We shall explain this problem later in this chapter.

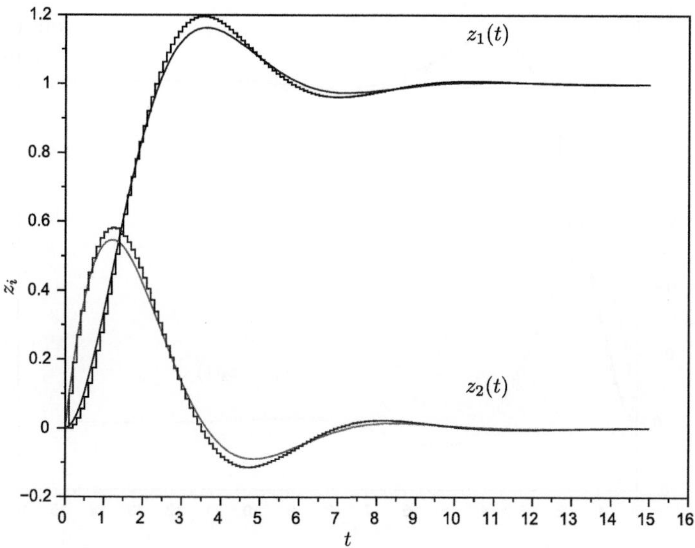

FIGURE 3.17

Forward Euler simulation of the spring–mass–damper model of Eq. (3.7). Numerical solution (stairs plot) vs. Analytical solution.

A variant of Forward Euler method is given by the following formula

$$\mathbf{z}(t_{k+1}) = \mathbf{z}(t_k) + h \cdot \mathbf{f}(\mathbf{z}(t_{k+1}), \mathbf{u}(t_{k+1})) \tag{3.20}$$

known as *Backward Euler* method. This is an *implicit formula*, where the unknown $\mathbf{z}(t_{k+1})$ is at both sides of the equation. Taking into account that function $\mathbf{f}()$ is usually non-linear, obtaining the next state value with this method requires solving a non-linear algebraic equation. This is usually accomplished making use of Newton's iterations.

Using this algorithm, we repeated the simulation of the spring–mass–damper system, obtaining the results depicted in Fig. 3.18.

The results are not significantly different from those obtained using Forward Euler in Fig. 3.17. The implicit nature of the algorithm and the iterations needed to find the next state value at each simulation step make this method far more expensive than its explicit counterpart. However, the comparison of the results suggest that we have gained nothing out of those additional computational costs.

The only difference between Figs. 3.17 and 3.18 is that the numerical solution shows larger oscillations using Forward Euler than using Backward Euler. We will see later that this small difference is connected to the main justifications for the use of implicit algorithms.

3.2.4 ACCURACY OF THE APPROXIMATIONS

Fig. 3.19 shows the results of using Forward Euler with different step sizes. Notice how the error grows with the value of h.

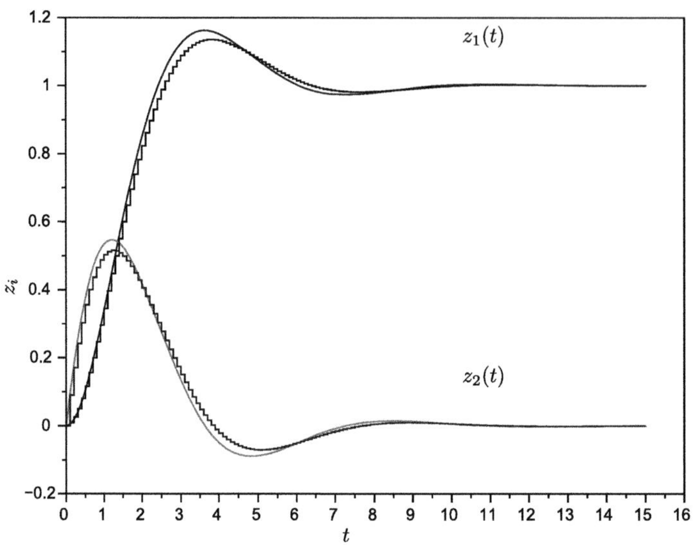

FIGURE 3.18

Backward Euler simulation of the spring–mass–damper model of Eq. (3.7). Numerical solution (stairs plot) vs. Analytical solution.

To confirm our intuition, we will analyze the approximation performed by Forward Euler method in order to find a formal proof that, as expected and observed, the error grows with the step size.

Let us assume that we know the state vector $\mathbf{z}$ at time t_k. Then, we can express the exact solution of the ODE $\dot{\mathbf{z}}(t) = \mathbf{f}(\mathbf{z}(t), \mathbf{u}(t))$ at time t_{k+1} using the following Taylor series

$$\mathbf{z}(t_{k+1}) = \mathbf{z}(t_{k+1}) + h \cdot \frac{\mathrm{d}\mathbf{z}}{\mathrm{d}t}(t_k) + \frac{h^2}{2!} \cdot \frac{\mathrm{d}^2\mathbf{z}}{\mathrm{d}t^2}(t_k) + \frac{h^3}{3!} \cdot \frac{\mathrm{d}^3\mathbf{z}}{\mathrm{d}t^3}(t_k) + \cdots$$

Replacing the state time derivative by function $\mathbf{f}$, we obtain

$$\mathbf{z}(t_{k+1}) = \mathbf{z}(t_{k+1}) + h \cdot \mathbf{f}(\mathbf{z}(t_k), t_k) + \frac{h^2}{2!} \cdot \frac{\mathrm{d}^2\mathbf{z}}{\mathrm{d}t^2}(t_k) + \frac{h^3}{3!} \cdot \frac{\mathrm{d}^3\mathbf{z}}{\mathrm{d}t^3}(t_k) + \cdots \tag{3.21}$$

that shows that Forward Euler method *truncates* the Taylor series of the analytical solution after the first term. For this reason, Forward Euler is a *first order* accurate algorithm.

The truncated terms form a series that start with the term of h^2. Thus, according to Taylor's Theorem, the remainder of the series can be approximated by some function of the state multiplied by h^2. This means that the error grows with h^2 and confirms our initial intuition that the larger the step size, the larger the error.

This error is called *Local Truncation Error*, as it is the difference observed between the analytical and the numerical solution after one step. In general, this error is proportional to h^{p+1} where p is the order of the method.

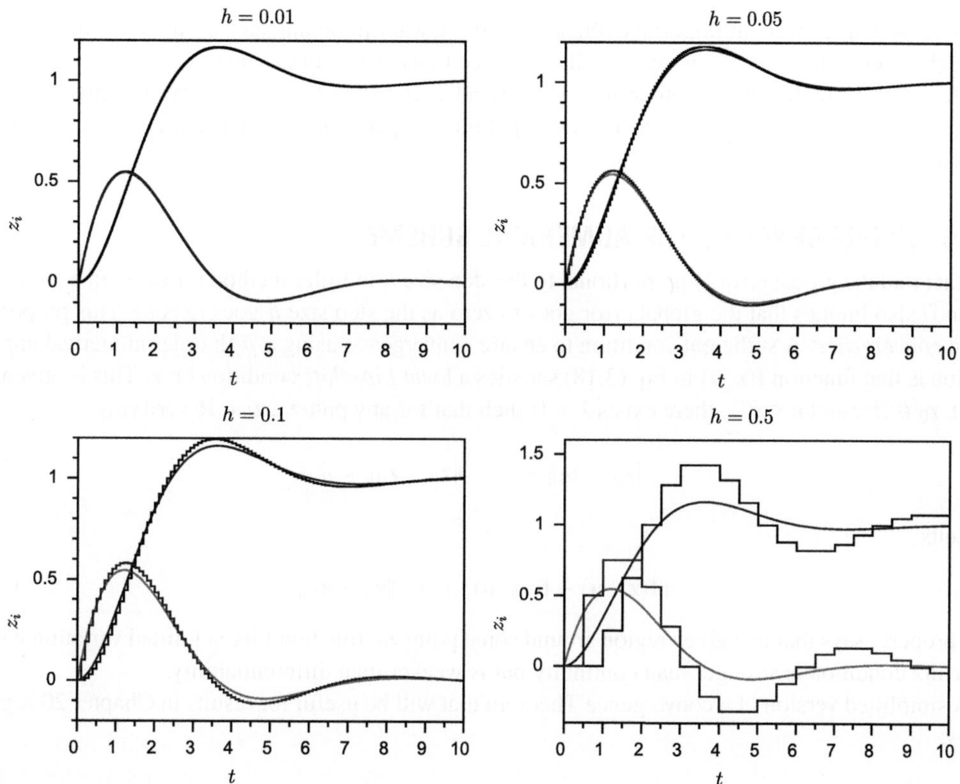

FIGURE 3.19

Accuracy of Forward Euler for different step sizes.

However, note that except for the first step, we do not know the analytical solution for $\mathbf{z}(t_k)$. For this reason, at time t_{k+1} we will have an additional error as we are starting from a non-exact value at time t_k. Thus the difference observed between the analytical and the numerical solution at any given instant of time is not the local truncation error. What is observed is called *Global Error* and it is the result of the accumulating local errors along the simulation run.

It can be proved that, under certain circumstances, the maximum global error has one order less than the local error. This is, a method of order p has a local error proportional to h^{p+1} but a global error proportional to h^p.

Forward and Backward Euler are both first order accurate methods, so their global errors are just proportional with h. Thus, if we reduce the step size h by a factor of 10,000, the error will also be reduced by the same factor. Posing the problem in the opposite direction, if we are not happy with the error obtained in a simulation and we want to obtain a result 10,000 times more accurate, we need to reduce the step size h by that factor, which implies performing 10,000 times more calculations.

If we use a fourth order method instead and we want to obtain a result 10,000 times more accurate, it is sufficient to reduce 10 times the step size h, which implies only 10 times more calculations. This last observation leads to an important conclusion: if we want to obtain very accurate results, we must use high order methods. Otherwise, we will be forced to use very small step sizes.

We will see later that higher order methods have a larger cost per step than Forward Euler. However, those higher costs are compensated by the capability of performing much larger steps with better accuracy.

3.2.5 CONVERGENCE OF THE NUMERICAL SCHEME

The fact that the global error is proportional to the step size h in Euler method (or to h^p in a p-th order method) also implies that the global error goes to zero as the step size h goes to zero. This property is called *convergence*. A sufficient condition to ensure convergence using a p-th order numerical approximation is that function $\mathbf{f}(\mathbf{z}, \mathbf{u})$ in Eq. (3.18) satisfies a local *Lipschitz* condition on $\mathbf{z}$. This is, given any $\delta > 0$, $\mathbf{z}_0 \in \mathbb{R}^n$, and $\mathbf{u} \in \mathbb{R}^m$, there exists $\lambda > 0$ such that for any pair $\mathbf{z}_a, \mathbf{z}_b \in \mathbb{R}$ verifying

$$\|\mathbf{z}_a - \mathbf{z}_0\| < \delta, \quad \|\mathbf{z}_b - \mathbf{z}_0\| < \delta \tag{3.22}$$

it results

$$\|\mathbf{f}(\mathbf{z}_a, \mathbf{u}) - \mathbf{f}(\mathbf{z}_b, \mathbf{u})\| \le \lambda \cdot \|\mathbf{z}_a - \mathbf{z}_b\| \tag{3.23}$$

This property says that in a given region around some point $\mathbf{z}_0$, function $\mathbf{f}$ has a limited variation with $\mathbf{z}$. Lipschitz condition is stronger than continuity but is weaker than differentiability.

A simplified version of a convergence Theorem that will be useful for results in Chapter 20 is given below.

Theorem 3.1. *Let $\mathbf{z_a}(t)$ be the analytical solution of the ODE*

$$\dot{\mathbf{z}}_{\mathbf{a}}(t) = \mathbf{f}(\mathbf{z_a}(t), \mathbf{u}(t)) \tag{3.24}$$

for $t \in [t_0, t_f]$, and consider the numerical solution

$$\mathbf{z}(t_{k+1}) = \mathbf{z}(t_k) + h \cdot \mathbf{f}(\mathbf{z}(t_k), \mathbf{u}(t_k)) + h^2 \cdot \gamma(\mathbf{z}(t_k), \mathbf{u}(t_k), h) \tag{3.25}$$

where $\gamma(\mathbf{z}(t_k), \mathbf{u}(t_k), h)$ represents the second order portion of the solution. Assume that $\mathbf{z_a}(t) \in D$ in that interval, where $D \subset \mathbb{R}^n$ is a compact set, and that $\mathbf{u}(t) \in U$ where $U \subset \mathbb{R}^n$ is also a compact set. Suppose that $\mathbf{f}(\mathbf{z_a}, \mathbf{u})$ is Lipschitz in D for all $\mathbf{u} \in U$.

Suppose also that

$$\|\gamma(\mathbf{z}(t_k), \mathbf{u}(t_k), h)\| \le \bar{\gamma} \tag{3.26}$$

(This bounds the effect of the state and input on the solution.) Then, the numerical solution $\mathbf{z}(t_k)$ obtained with $\mathbf{z}(t_0) = \mathbf{z_a}(t_0)$ converges to the analytical solution $\mathbf{z_a}(t_k)$ as h goes to 0 for $t \in [t_0, t_f]$.

Proof. From the Taylor series we know that

$$\mathbf{z_a}(t_{k+1}) = \mathbf{z_a}(t_k) + h \cdot \mathbf{f}(\mathbf{z_a}(t_k), \mathbf{u}(t_k)) + h^2 \cdot \mathbf{d}(t_k) \tag{3.27}$$

where the last term expresses the remainder of the series. The fact that $\mathbf{z_a}(t)$ is bounded in $t \in [t_0, t_f]$ implies that $\mathbf{d}(t_k)$ is also bounded, i.e.,

$$\|\mathbf{d}(t)\| \leq \bar{d} \tag{3.28}$$

for all $t \in [t_0, t_f]$.

The error at the first step is then, from Eqs. (3.25) and (3.27),

$$\|\mathbf{z}(t_1) - \mathbf{z_a}(t_1)\| = \|h^2 \cdot \gamma(\mathbf{z}(t_0), \mathbf{u}(t_0), h) - h^2 \cdot \mathbf{d}(t_0)\| \leq (\bar{\gamma} + \bar{d}) \cdot h^2 = c \cdot h^2 \tag{3.29}$$

where the last inequality comes from Eqs. (3.26) and (3.28) with $c \triangleq \bar{d} + \bar{\gamma}$.

In the $k+1$-th step, we have,

$$\begin{aligned}
&\|\mathbf{z}(t_{k+1}) - \mathbf{z_a}(t_{k+1})\| = \\
&= \|\mathbf{z}(t_k) + h \cdot \mathbf{f}(\mathbf{z}(t_k), \mathbf{u}(t_k)) + h^2 \cdot \gamma(\mathbf{z}(t_k), \mathbf{u}(t_k), h) - (\mathbf{z_a}(t_k) + h \cdot \mathbf{f}(\mathbf{z_a}(t_k), \mathbf{u}(t_k)) + h^2 \cdot \mathbf{d}(t_k))\| \\
&\leq \|\mathbf{z}(t_k) - \mathbf{z_a}(t_k)\| + h \cdot \|\mathbf{f}(\mathbf{z}(t_k), \mathbf{u}(t_k)) - \mathbf{f}(\mathbf{z_a}(t_k), \mathbf{u}(t_k))\| + c \cdot h^2 \\
&\leq \|\mathbf{z}(t_k) - \mathbf{z_a}(t_k)\| + h \cdot L \cdot \|\mathbf{z}(t_k) - \mathbf{z_a}(t_k)\| + c \cdot h^2
\end{aligned}$$

where we used the Lipschitz condition in the last step. Then,

$$\|\mathbf{z}(t_{k+1}) - \mathbf{z_a}(t_{k+1})\| \leq (1 + h \cdot L) \cdot \|\mathbf{z}(t_k) - \mathbf{z_a}(t_k)\| + c \cdot h^2 \tag{3.30}$$

Recalling that at the first step the error, according to Eq. (3.29), is bounded by $c \cdot h^2$, the recursive use of Eq. (3.30) leads to

$$\|\mathbf{z}(t_N) - \mathbf{z_a}(t_N)\| \leq c \cdot h^2 \cdot \sum_{k=1}^{N} (1 + h \cdot L)^k \tag{3.31}$$

Then, at a fixed time $t = t_N$, such that $h = t_N / N$, the last expression becomes

$$\|\mathbf{z}(t_N) - \mathbf{z_a}(t_N)\| \leq c \cdot \frac{t_N^2}{N^2} \cdot \sum_{k=1}^{N} \left(1 + \frac{(t_N \cdot L)}{N}\right)^k \tag{3.32}$$

Using the fact that

$$\lim_{N \to \infty} \left(1 + \frac{(t_N \cdot L)}{N}\right) = e^{t_N \cdot L} \tag{3.33}$$

then the sum in Eq. (3.32) is bounded by $N \cdot e^{t_N \cdot L}$, so the whole expression in Eq. (3.32) goes to 0 as $N \to \infty$, i.e., as $h \to 0$. □

This Theorem is valid for any numerical method that can be written in the form of Eq. (3.25). It then holds for Forward Euler (where $\gamma() = 0$) but also for Backward Euler and for higher order methods with a proper definition of function γ. Later we will see how discontinuities in the ODE violate the conditions required for Theorem 3.1 and require event-based control of solutions.

3.2.6 NUMERICAL STABILITY

A question that comes up after analyzing the difference between the local and global error is: does the latter keeps growing as the simulation advances?

Looking back to Fig. 3.19 we see that as the analytical solution approaches the equilibrium point, the numerical solution goes to the same value. Evidently, in this case, the global error tends to disappear after a certain time. In other words, what we see in Fig. 3.19 is that the numerical solution preserves the *stability* of the analytical solution. Here we say that a numerical solution preserves the stability when it does not diverge away from the analytical solution, provided that the latter is bounded.

Does this occur for any value of the step size h? Fig. 3.20 shows the trajectories (here we only draw the first component $z_1(t)$) obtained as the step size h becomes larger.

Evidently, the stability is not always preserved. For $h = 1$ the asymptotic stability is lost and we have a *marginally stable* numerical solution. For $h = 1.5$ the numerical solution is definitively unstable.

Preserving the numerical stability is a crucial problem in continuous system simulation, and, as we see, in the case of Forward Euler it strongly depends on the correct choice of the step size h.

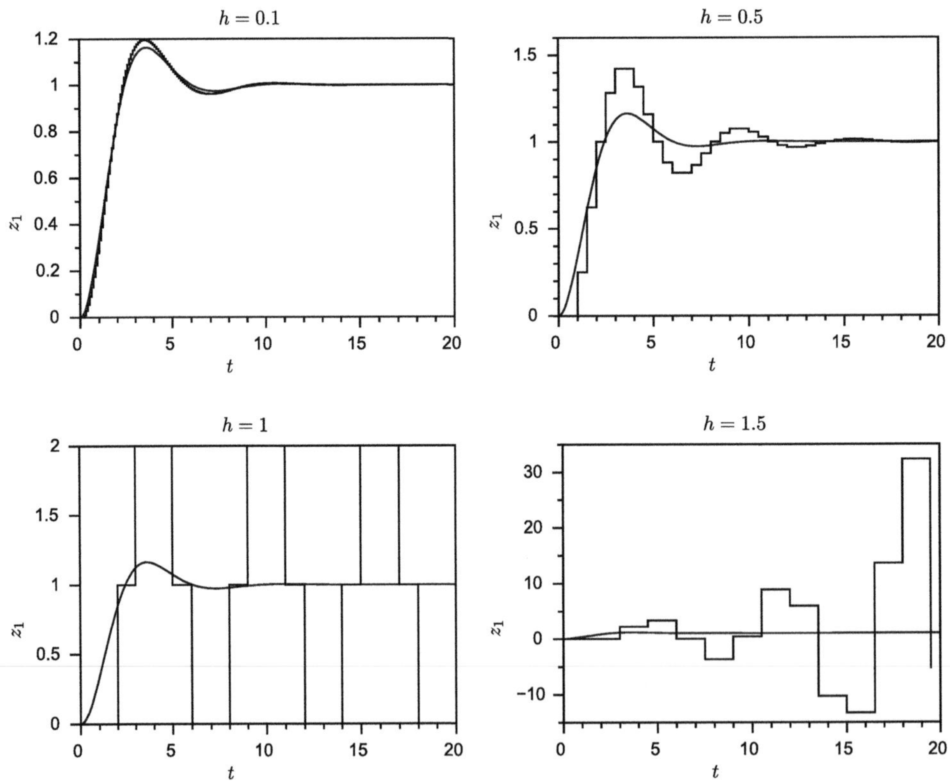

FIGURE 3.20

Stability of Forward Euler for different step sizes.

Let us see now what happens if we use Backward Euler instead of Forward Euler. Fig. 3.21 shows the results of repeating the experiments with the implicit method.

This time, the stability is always preserved. This feature is the main reason for using implicit algorithms. We will see later that for certain classes of systems, only methods that preserve stability like Backward Euler can be used.

In this empirical study on stability we observed two things. First, that Forward Euler becomes unstable when we increase h beyond certain value. Second, that Backward Euler does not seem to become unstable as h grows.

These observations can be easily justified analyzing the Forward and Backward Euler approximation of LTI systems. Given a LTI system

$$\dot{\mathbf{z}}(t) = A \cdot \mathbf{z}(t) + B \cdot \mathbf{u}(t)$$

Forward Euler approximates it by a LTI discrete time of the form

$$\mathbf{z}(t_{k+1}) = A_D \cdot \mathbf{z}(t_k) + B_F \cdot \mathbf{u}(t_k) \tag{3.34}$$

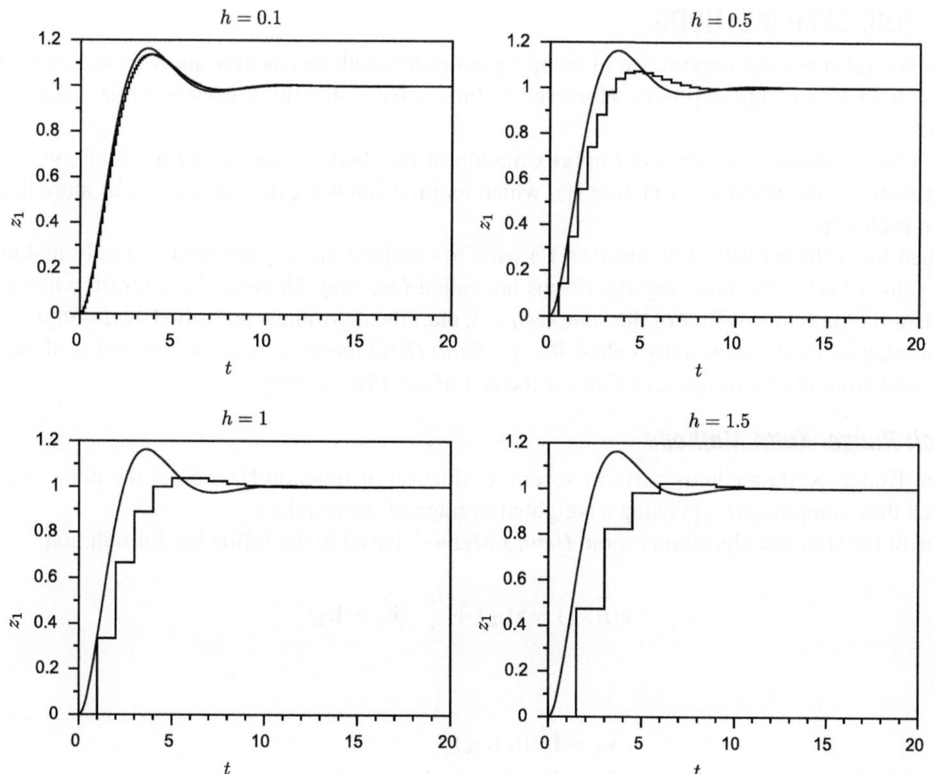

FIGURE 3.21

Stability of Backward Euler for different step sizes.

Let λ_i be the i-th eigenvalue of A. Then, the i-th eigenvalue of matrix A_D is $\lambda_i^D = \lambda_i \cdot h + 1$. The condition for the discrete time LTI system of Eq. (3.34) to be stable is that $|\lambda_i^D| < 1$ for all i. If the continuous time system is stable, then we know that $\mathbb{R}\mathrm{e}(\lambda_i) < 0$. Then, we see that

$$|\lambda_i^D| = |\lambda_i \cdot h + 1| < 1 \tag{3.35}$$

only when h is small. If the product $h \cdot \lambda_i$ becomes too large in absolute value, then the expression $|\lambda_i \cdot h + 1|$ becomes larger than 1 and the discrete time model becomes unstable.

In contrast, Backward Euler approximation produces a similar approximation to that of Eq. (3.34), but the eigenvalues are now

$$\lambda_i^D = \frac{1}{1 - \lambda_i \cdot h} \tag{3.36}$$

When $\mathbb{R}\mathrm{e}(\lambda_i) < 0$ (i.e., when the continuous system is stable) then it can be easily seen that $|\lambda_i^D| < 1$ for all $h > 0$. Thus, the numerical stability is always preserved.

3.2.7 ONE-STEP METHODS

We mentioned above the importance of using higher order methods, as they allow us to simulate with high accuracy using large step sizes. However, so far, we have only introduced two first order accurate algorithms.

In order to obtain a higher order approximation in the Taylor's series of Eq. (3.21), we need to approximate the derivatives of function $\mathbf{f}()$, which requires knowing the value of $\mathbf{f}()$ at more than one point at each step.

When the different values of function $\mathbf{f}()$ used to compute $\mathbf{z}(t_{k+1})$ are obtained only making use of the value of $\mathbf{z}(t_k)$, the resulting algorithms are called *One-Step Methods*. In contrast, when we use information of previous values of the state ($\mathbf{z}(t_{k-1})$, etc.) the algorithms are called *Multi-Step*.

One-Step methods are usually called *Runge–Kutta* (RK) methods because the first of these algorithms was proposed by Runge and Kutta at the end of the 19th century.

Explicit Runge–Kutta Methods

Explicit Runge–Kutta methods perform several evaluation of function $\mathbf{f}()$ around the point $(\mathbf{z}(t_k), t_k)$ and then they compute $\mathbf{z}(t_{k+1})$ using a weighted average of those values.

One of the simplest algorithms is the *Heun's Method*, based in the following formulation:

$$\mathbf{z}(t_{k+1}) = \mathbf{z}(t_k) + \frac{h}{2} \cdot (\mathbf{k}_1 + \mathbf{k}_2)$$

where

$$\mathbf{k}_1 = \mathbf{f}(\mathbf{z}(t_k), t_k)$$
$$\mathbf{k}_2 = \mathbf{f}(\mathbf{z}(t_k) + h \cdot \mathbf{k}_1, t_k + h)$$

Compared with Euler's, this method performs an extra evaluation of $\mathbf{f}()$ in order to compute $\mathbf{k}_2$. Notice that this is an estimate of $\dot{\mathbf{z}}(t_{k+1})$, but it is not used back in the next step. This price is compen-

Table 3.7 Maximum global error committed by Euler, Heun and RK4 methods

Step Size	Euler's Error	Heun's Error	RK4 Error
$h = 0.5$	0.298	0.0406	4.8×10^{-4}
$h = 0.1$	0.042	1.47×10^{-3}	6.72×10^{-7}
$h = 0.05$	0.0203	3.6×10^{-4}	4.14×10^{-8}
$h = 0.01$	3.94×10^{-3}	1.42×10^{-5}	6.54×10^{-11}

sated by the fact that Heun's performs a second order approximation obtaining more accurate results than Euler's.

One of the most used numerical integration algorithms is the fourth order Runge–Kutta (RK4):

$$\mathbf{z}(t_{k+1}) = \mathbf{z}(t_k) + \frac{h}{6} \cdot (\mathbf{k}_1 + 2 \cdot \mathbf{k}_2 + 2 \cdot \mathbf{k}_3 + \mathbf{k}_4)$$

where

$$\begin{aligned}
\mathbf{k}_1 &= \mathbf{f}(\mathbf{z}(t_k), t_k) \\
\mathbf{k}_2 &= \mathbf{f}(\mathbf{z}(t_k) + h \cdot \frac{\mathbf{k}_1}{2}, t_k + \frac{h}{2}) \\
\mathbf{k}_3 &= \mathbf{f}(\mathbf{z}(t_k) + h \cdot \frac{\mathbf{k}_2}{2}, t_k + \frac{h}{2}) \\
\mathbf{k}_4 &= \mathbf{f}(\mathbf{z}(t_k) + h \cdot \mathbf{k}_3, t_k + h)
\end{aligned}$$

This algorithm uses four evaluations of function $\mathbf{f}()$ at each step, obtaining a fourth order approximation.

Table 3.7 synthesizes the results of simulating the spring–mass–damper system of Eq. (3.7) using Euler, Heun and RK4 methods. There, the maximum global error is reported for each step size and for each method.

Notice that Euler's error is linearly reduced with the step size, while Heun's is quadratically reduced and RK4 is quartically reduced.

The literature on numerical integration contains hundreds of explicit Runge Kutta algorithms, including methods of more than 10th order. In practice, the most used RK methods are those of order between 1 and 5.

Regarding stability, explicit Runge–Kutta methods have similar features to those of Forward Euler. This is, they preserve stability provided that the step size h does not become too large. Thus, in practice, the use of high order RK methods allows us to increase the step size while still obtaining good accuracy but the stability of the algorithms establishes limits to the value of h.

Implicit One-Step Methods

We saw that Backward Euler was an implicit algorithm that preserves the numerical stability for any step size h. However, just like Forward Euler, it is only first order accurate.

There are several implicit One-Step methods of higher order that, like Backward Euler, preserve the stability. Among them, a widely used algorithm is the *Trapezoidal Rule* defined as follows:

$$\mathbf{z}(t_{k+1}) = \mathbf{z}(t_k) + \frac{h}{2} \cdot [\mathbf{f}(\mathbf{z}(t_k), t_k) + \mathbf{f}(\mathbf{z}(t_{k+1}), t_{k+1})] \tag{3.37}$$

This implicit method is second order accurate and it not only preserves stability but also marginal stability. This is, when a system with undamped oscillations is simulated using this algorithm, the numerical result also exhibit undamped oscillations.

Except for the trapezoid rule and two or three other algorithms, one-step implicit methods are not widely used in practice since multi-step implicit methods are normally more efficient.

3.2.8 MULTI-STEP METHODS

Multi-Step numerical algorithms use information from past steps to obtain higher order approximations of $\mathbf{z}(t_{k+1})$. Using *past information*, these methods do not need to recompute several times the function $\mathbf{f}()$ at each step as One-Step methods do. In consequence, Multi-Step algorithms usually have a lower cost per step.

Multi-Step Explicit Methods

The most used explicit Multi-Step methods are those of Adams–Bashforth (AB) and Adams–Bashforth–Moulton (ABM).

The second order accurate Adams–Bashforth method (AB2) is defined by the formula:

$$\mathbf{z}(t_{k+1}) = \mathbf{z}(t_k) + \frac{h}{2} \cdot (3\mathbf{f}_k - \mathbf{f}_{k-1}) \tag{3.38}$$

where we defined

$$\mathbf{f}_k \triangleq \mathbf{f}(\mathbf{z}(t_k), t_k)$$

The fourth order AB method (AB4), in turn, is defined as follows:

$$\mathbf{z}(t_{k+1}) = \mathbf{z}(t_k) + \frac{h}{24} (55\mathbf{f}_k - 59\mathbf{f}_{k-1} + 37\mathbf{f}_{k-2} - 9\mathbf{f}_{k-3}) \tag{3.39}$$

Notice that AB methods require a single evaluation of function $\mathbf{f}()$ at each step. The higher order approximation, as we already mentioned, is obtained using past values of function $\mathbf{f}()$.

One of the drawbacks of multi-step methods is related to the startup. The expression of Eq. (3.38) cannot be used to compute $\mathbf{z}(t_1)$ as it would require the value of $\mathbf{f}_{-1}$. Similarly, Eq. (3.39) can be only used to compute $\mathbf{z}(t_k)$ with $k \geq 4$. For this reason, the first steps of Multi-Step methods must be computed using some One-Step algorithm (e.g., Runge–Kutta).

Another problem of AB methods is that of stability. Like RK algorithms, AB only preserve numerical stability provided that h is small. However, the step size must be even smaller in AB than in RK. That feature is partially improved by Adams–Bashforth–Moulton methods, that have better stability features than AB at the price of performing two function evaluations per step.

Implicit Multi-Step Methods

As in One-Step methods, implicit approaches must be used in multi-step methods to obtain numerical solutions that preserve stability irrespective of the step size h.

The most used implicit multi-step methods are the family of *Backward Difference Formulae* (BDF). For instance, the third order BDF method (BDF3) is defined as follows:

$$\mathbf{z}(t_{k+1}) = \frac{18}{11}\mathbf{z}(t_k) - \frac{9}{11}\mathbf{z}(t_{k-1}) + \frac{2}{11}\mathbf{z}(t_{k-2}) + \frac{6}{11} \cdot h \cdot \mathbf{f}(t_{k+1}) \tag{3.40}$$

This algorithm has almost the same computational cost as Backward Euler, since both methods must solve a similar implicit equation. However, using three past state values, BDF3 is third order accurate.

There are BDF methods up to order 8, while the most used are those of order 1 (Backward Euler) to order 5.

3.2.9 STEP SIZE CONTROL

Up to here we always considered that the step size h was a fixed parameter that must be selected before starting the simulation. However, it is very simple to implement algorithms that automatically change the step size as the simulation advances. These *step size control* algorithms have the purpose of keeping a bounded simulation error. For that goal, they increase or decrease h according to the estimated error.

The use of step size control routines leads to *Variable Step Size Methods*, that are used in most modern continuous simulation tools.

One-Step Methods

Variable step size RK methods work as follows:

1. Compute $\mathbf{z}(t_{k+1})$ with certain step size h using the selected RK method.
2. Estimate the error in $\mathbf{z}(t_{k+1})$.
3. If the error is larger than the tolerance, reduce the step size h and recompute $\mathbf{z}(t_{k+1})$ going back to step 1.
4. Otherwise, accept the value $\mathbf{z}(t_{k+1})$, increase the step size h and go back to step 1 to compute $\mathbf{z}(t_{k+2})$.

While the idea is simple, the following questions come up from the procedure above:

- How is the error estimated?
- What should be the rule used to reduce or to increase the step size h?
- What value should be used as the initial value of h at the first step?

We answer those questions below:

Error estimation: The error is estimated using two methods of different order. Given $\mathbf{z}(t_k)$, the value of $\mathbf{z}(t_{k+1})$ can be computed using a fourth order method and then computed again with a fifth order method. Because the fifth order algorithm is far more accurate than the fourth order one, the difference between both results is approximately the error of the fourth order method.
In order to save calculations, RK algorithms that share some stages are commonly employed. One of the most used variable step size RK algorithms is that of Dormand–Prince (DOPRI), that

performs a total of six evaluations of function $\mathbf{f}()$ and uses them to compute a fourth and a fifth order approximation of $\mathbf{z}(t_{k+1})$.

Step Size Adjustment: Let us suppose that we used a p-th and a $p+1$-th order method to compute $\mathbf{z}(t_{k+1})$ with a step size h. Suppose that the difference between both approximations was

$$\|\mathbf{z}(t_{k+1} - \tilde{\mathbf{z}}(t_{k+1})\| = err(h)$$

Assume also that we wanted that this error was equal to the tolerance tol. The problem is then to compute the step size h_0 so that $err(h_0) = tol$.

For this goal, we recall that in a method of order p, the Taylor series expansion coincides with that of the analytical solution up to the term of h^p. For that reason, the difference between both approximations is proportional to h^{p+1}, i.e.,

$$err(h) = c \cdot h^{p+1} \tag{3.41}$$

where according to Taylor's Theorem, c is a constant depending on $\mathbf{z}(t_k)$. If we had used the right step size h_0 (still unknown), we would have obtained

$$err(h_0) = c \cdot h_0^{p+1} = tol \tag{3.42}$$

Then, dividing Eq. (3.41) and Eq. (3.42) we obtain

$$h_0 = h \cdot \sqrt[p+1]{\frac{tol}{err(h)}}$$

This expression allows us to adjust the step size according to the error obtained and the prescribed tolerance. In most variable step methods, the step size adjustment is made in a more conservative way, replacing the expression by

$$h_0 = C \cdot h \cdot \sqrt[p+1]{\frac{tol}{err(h)}} \tag{3.43}$$

for a (different constant) $C < 1$ (a typical value is $C = 0.8$). That way, the error is almost always kept below the tolerance.

Initial Step Size: Initially, most variable step size algorithms use a very small step size h. In this way, they obtain a valid value for $\mathbf{z}(t_1)$ and they can start increasing the step size with Eq. (3.43).

Fig. 3.22 shows the simulation result for the spring–mass–damper system of Eq. (3.7) using the variable step Runge–Kutta–Fehlberg algorithm, a method very similar to DOPRI that uses a fourth and a five order RK approximation with a total of six function evaluations per step.

The simulation was performed using a *relative error tolerance* $tol_{rel} = 10^{-3}$, i.e., the error tolerance was set to change with the signal values ($tol = 0.001 \cdot \|\vec{(t)}\|$). The algorithm completed the simulation after only 18 steps, and the difference between the analytical and the numerical solution at the time steps cannot be appreciated in Fig. 3.22.

The size of the steps is depicted in Fig. 3.23. There, we can see that the step size control algorithm slowly increases the step size as the trajectories approach the equilibrium.

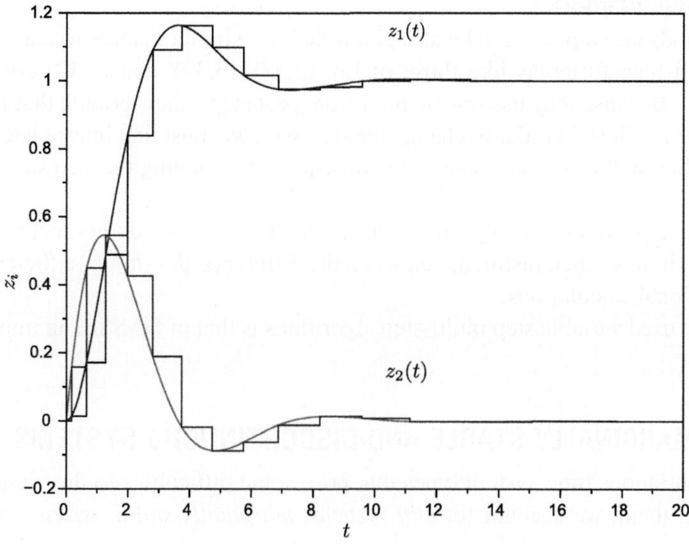

FIGURE 3.22

RKF45 simulation of the spring–mass–damper model of Eq. (3.7). Numerical solution (stairs plot) vs. Analytical solution.

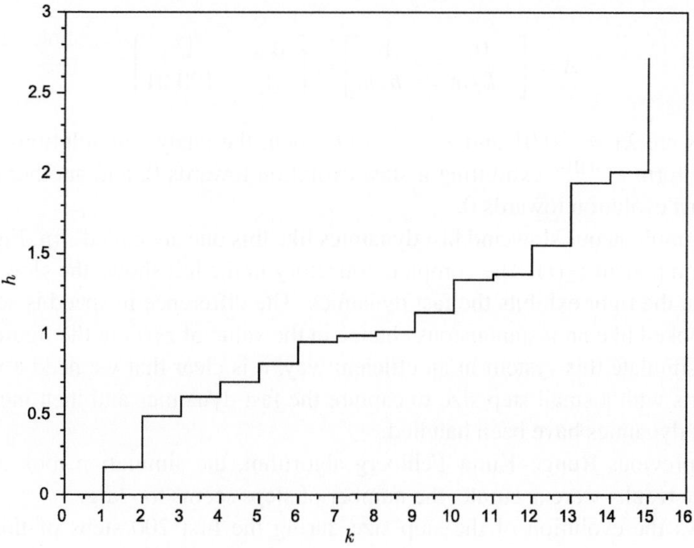

FIGURE 3.23

RKF45 step size evolution in the simulation of the spring–mass–damper model of Eq. (3.7).

3.2.9.1 Multi-Step Methods

In Multi-Step methods the step size can be also controlled in a similar manner to that of RK algorithms. However, the Multi-Step formulas like those of Eq. (3.38), (3.39), and (3.40) are only valid for a constant step size h. Because they use information from past steps, they assume that the information is equally spaced in time. Thus, in order to change the step size, we must first interpolate the required past information according to the new step size h. In consequence, changing the step size has an additional cost here.

For this reason, step size control algorithms for multi-step methods do not increase the step size at every step. They do it only when justified, i.e., when the difference $|h - h_0|$ is sufficiently large relative to the cost of additional calculations.

One of the most used variable step multi-step algorithms is that of DASSL, an implicit solver based on BDF formulas.

3.2.10 STIFF, MARGINALLY STABLE AND DISCONTINUOUS SYSTEMS

There are some continuous time system types that pose some difficulties to the numerical integration algorithms. Among them, we account for *stiff systems*, *marginally stable systems*, and systems with *discontinuities*.

Next, we start analyzing those problematic cases.

Stiff Systems

Let us consider once more the spring–mass–damper model of Eq. (3.7). Consider now the parameters $m = k = 1$, $b = 100.01$ and $F(t) = 1$. Using these parameter values, matrix A becomes

$$A = \begin{bmatrix} 0 & 1 \\ -k/m & -b/m \end{bmatrix} = \begin{bmatrix} 0 & 1 \\ -1 & -100.01 \end{bmatrix}$$

whose eigenvalues are $\lambda_1 = -0.01$ and $\lambda_2 = -100$. Then, the analytical solutions will contain two modes: one of the form $e^{-0.01 \cdot t}$ exhibiting a slow evolution towards 0, and another term of the form $e^{-100 \cdot t}$ having a fast evolution towards 0.

Systems with simultaneous slow and fast dynamics like this one are called *stiff*. Fig. 3.24 shows the trajectory of the component $z_2(t)$. The complete trajectory in the left shows the slow dynamics, while the startup detail in the right exhibits the fast dynamics. The difference in speed is so big that the rise of the trajectory looked like an instantaneous change in the value of $z_2(t)$ in the figure of the left.

If we want to simulate this system in an efficient way, it is clear that we need a variable step size algorithm that starts with a small step size to capture the fast dynamics and then increases h at some point once the fast dynamics have been handled.

Applying the previous Runge–Kutta–Fehlberg algorithm, the simulation took a total of 13,603 steps. Although the results were accurate, the number of steps seems too large.

Fig. 3.25 shows the evolution of the step size during the first 200 steps of this simulation. As expected, at the beginning of the simulation the algorithm used a small step size that captured the fast dynamics. However, then it failed to increase the step size beyond $h \approx 0.04$.

The problem is related to stability. For example, using Forward Euler, in order to preserve numerical stability, we must use a value of h so that Eq. (3.35) is true for all eigenvalues. In particular, for the fast

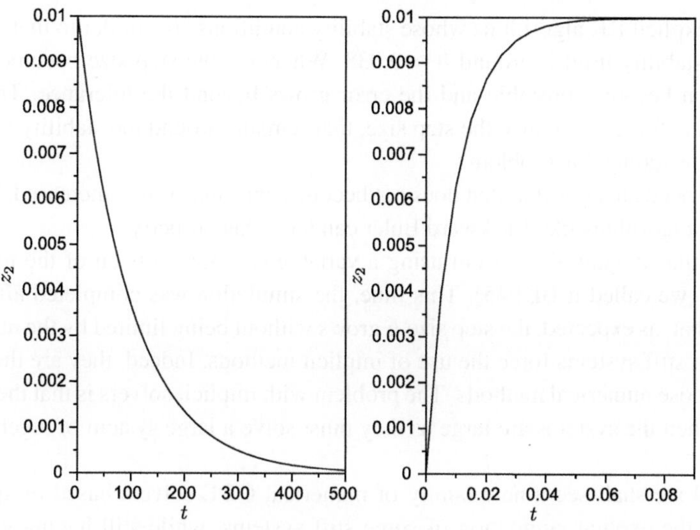

FIGURE 3.24

Trajectory of $z_2(t)$ for $b = 100$ (stiff case) in Eq. (3.7). Complete transient (left) and startup (right).

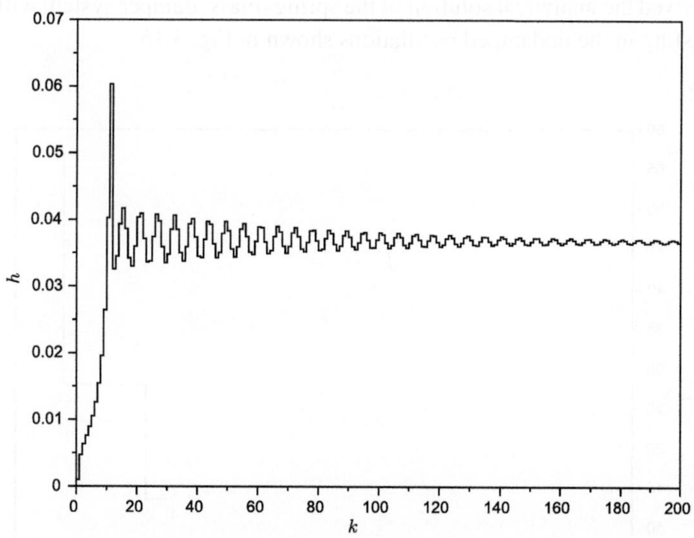

FIGURE 3.25

RKF45 step size evolution in the simulation of the stiff spring–mass–damper model of Eq. (3.7) with $b = 100.01$.

eigenvalue $\lambda_2 = -100$ we need that $|1 + h \cdot (-100)| < 1$, which is accomplished only if $h < 0.02$. For this reason, Forward Euler cannot employ a step size $h > 0.02$ without obtaining unstable results.

RKF45 uses explicit RK algorithms whose stability conditions are similar to that of Forward Euler. In this case, the stability limit is around $h \approx 0.038$. Whenever the step size becomes larger than this value, the solution becomes unstable and the error grows beyond the tolerance. Thus, the step size control algorithm is forced to reduce the step size, that remains around the stability limit.

How can we overcome this problem?

Evidently, we need an algorithm that does not becomes unstable as h is increased. We already know that only implicit algorithms like Backward Euler can have this property.

Thus, we simulated again the system using a variable step size version of the implicit multi-step BDF4 algorithm (we called it BDF45). This time, the simulation was completed after only 88 steps. Fig. 3.26 shows that, as expected, the step size h grows without being limited by the numerical stability.

So we see that stiff systems force the use of implicit methods. Indeed, they are the main reason for the existence of those numerical methods. The problem with implicit solvers is that they have expensive steps specially when the systems are large as they must solve a large system of algebraic equations in each step.

In Chapter 19 we shall see a new family of numerical ODE solvers based on quantization principles that allow the explicit simulation of some stiff systems, while still having good stability and computation properties.

Marginally Stable Systems

We had already derived the analytical solution of the spring–mass–damper system without friction (i.e., with $b = 0$), consisting in the undamped oscillations shown in Fig. 3.16.

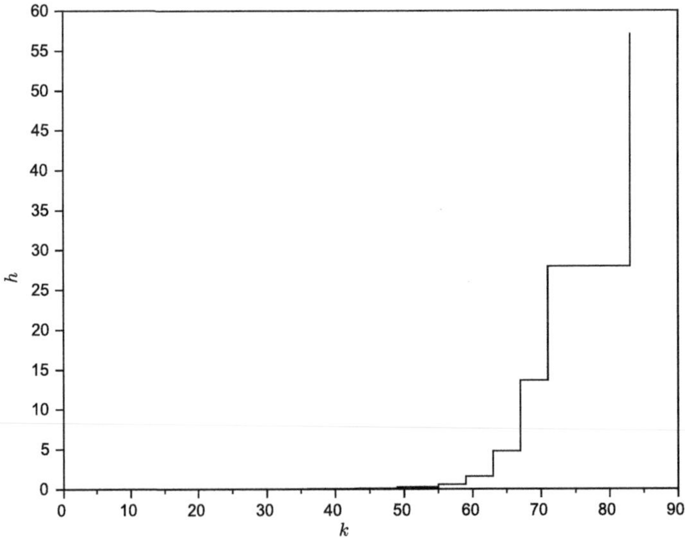

FIGURE 3.26

BDF45 step size evolution in the simulation of the stiff spring–mass–damper model of Eq. (3.7) with $b = 100.01$.

Systems containing undamped oscillations are called *Marginally Stable* and they can only be accurately simulated by some special methods that preserve the marginal stability in the numerical solution.

Fig. 3.27 shows the result of simulating this system using Forward and Backward Euler methods. The explicit Forward Euler method produces unstable oscillations that diverge in amplitude. In contrast, the implicit Backward Euler produces damped oscillations that tend to disappear. None of the algorithms preserve the marginal stability.

This fact can be easily explained from the stability conditions of Eq. (3.35) in Forward Euler and Eq. (3.36) in Backward Euler. In the first case, the moduli of the discrete eigenvalues are

$$|\lambda_i^D| = |\pm i \cdot h + 1|$$

which are always greater than 1 and is therefore unstable.

In the second case, the moduli of the discrete eigenvalues are

$$|\lambda_i^D| = \frac{1}{|1 \pm i \cdot h|}$$

that are always less than 1 so the numerical solution is always asymptotically stable.

We mentioned before that there is a method called the *Trapezoidal Rule*, defined in Eq. (3.37), that preserves the marginal stability. Fig. 3.28 shows the simulation results using this method. We see that, as expected, the marginal stability is preserved and the numerical solution is very close to the analytical solution.

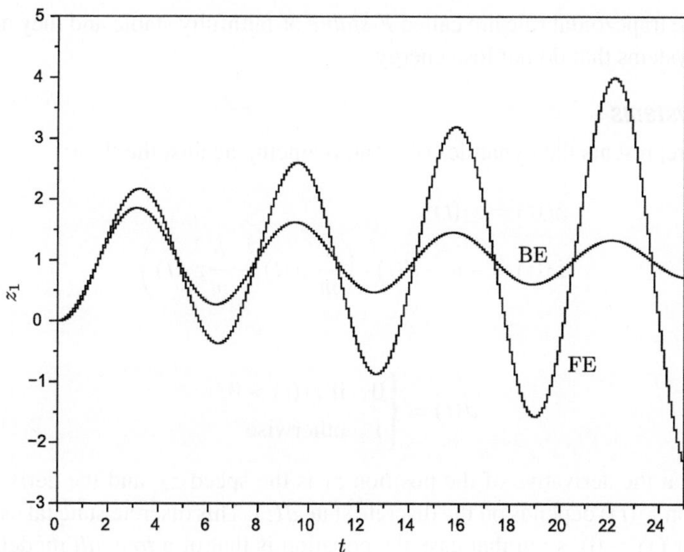

FIGURE 3.27

Forward (FE) and Backward Euler (BE) simulation of the spring–mass–damper model of Eq. (3.7) with $b = 0$ using a step size $h = 0.1$.

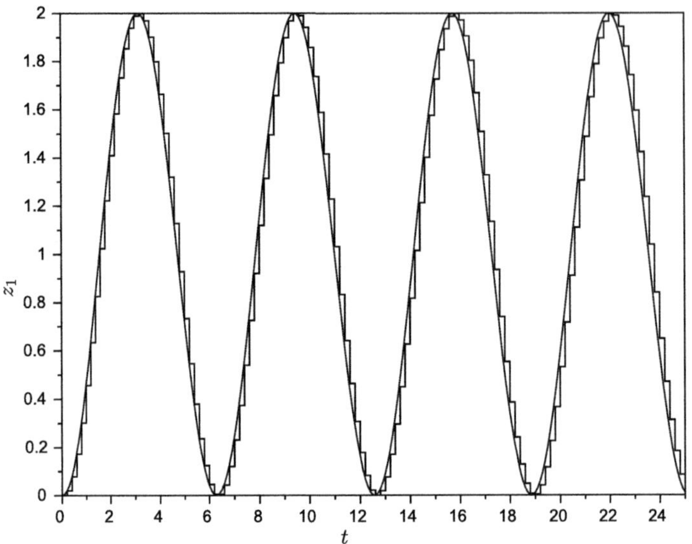

FIGURE 3.28

Trapezoidal rule simulation of the spring–mass–damper model of Eq. (3.7) with $b = 0$ using a step size $h = 0.2$. Numerical solution (stairs plot) vs. Analytical solution.

Methods like the trapezoidal rule are called *F-stable* or faithfully stable and they are very useful for the simulation of systems that do not lose energy.

Discontinuous Systems

The system below represents the dynamics of a ball bouncing against the floor:

$$\begin{aligned}\dot{z}_1(t) &= z_2(t) \\ \dot{z}_2(t) &= -g - d(t) \cdot \left(\frac{k}{m} z_1(t) + \frac{b}{m} z_2(t)\right)\end{aligned} \tag{3.44}$$

where

$$d(t) = \begin{cases} 0 & \text{if } z_1(t) > 0 \\ 1 & \text{otherwise} \end{cases} \tag{3.45}$$

This model says that the derivative of the position z_1 is the speed z_2, and the derivative of the speed (i.e., the acceleration $\dot{z}_2(t)$) depends on the discrete state $d(t)$. This discrete state takes the value 0 when the ball is in the air ($x_1 > 0$), so in that case the equation is that of a *free fall* model. Otherwise, $d(t)$ takes the value 1 when the ball is in contact with the floor ($x_1 \leq 0$), corresponding to a spring–damper model.

We adopted the following values for the parameters: $m = 1$, $b = 30$, $k = 10^6$, and $g = 9.81$. We also considered an initial state $z_1(0) = 2$, $z_2(0) = 0$.

We then simulated this system using the fourth order Runge–Kutta method (RK4) with three different small values for the step size: $h = 0.002$, $h = 0.001$, and $h = 0.0005$. The simulation results are depicted in Fig. 3.29.

In spite of the high order integration method used and the small step size h, the results are very different and there is not an obvious convergence to a reliable solution. Using variable step algorithms does not help much either.

The reason for the poor results is the presence of a discontinuity when $z_1(t) = 0$. Numerical integration methods are based on the continuity of the state variables and its derivatives. *This is in fact what allowed us to express the analytical solution as a Taylor series* in Eq. (3.21). In presence of discontinuities that series expansion is not valid and the numerical algorithms no longer approximate the analytical solution up their order of accuracy.

The solution to this problem requires the use of *event detection* routines, that detect the occurrence of discontinuities. Once a discontinuity is detected, the simulation is advanced until the location of the discontinuity, and it is restarted from that point under the new conditions after the event occurrence (this process is known as *event handling*). That way, the algorithms steps never integrate across a discontinuity. They just simulate a succession of purely continuous systems.

The use of event detection and event handling routines can add significant computational costs to numerical ODE solvers. For that reason, the simulation of systems with frequent discontinuities is still an important problem in the literature on numerical simulation.

In Chapter 19 we will introduce a new family of numerical ODE solvers based on state quantization principles that can considerably reduce the computational costs related to event detection and handling

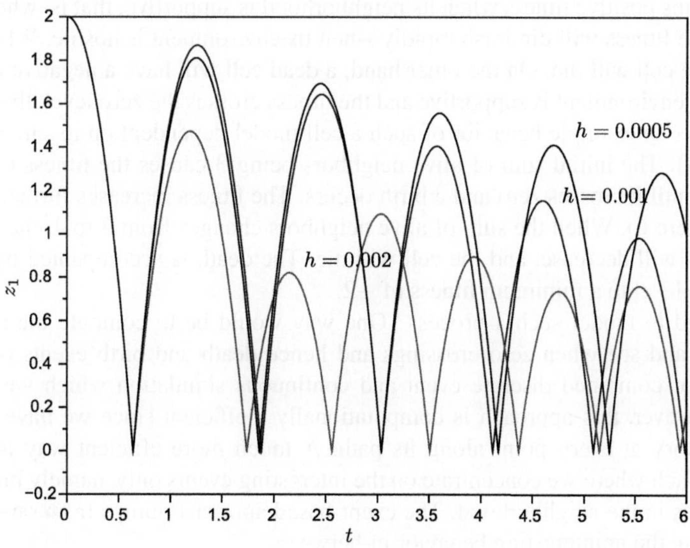

FIGURE 3.29

RK4 simulation of the bouncing ball model of Eq. (3.44) with $b = 0$ using different step sizes.

in discontinuous systems. This new family is an essential component supporting the iterative systems specification and DEVS-based hybrid simulation that are introduced in this third edition.

3.3 DISCRETE EVENT MODELS AND THEIR SIMULATORS

3.3.1 INTRODUCTION

We have already seen how a discrete event approach can be taken to obtain more efficient simulation of cellular automata. In this section, we will consider discrete event modeling as a paradigm in its own right. However, the basic motivation remains that discrete event modeling is an attractive formalism because it is intrinsically tuned to the capabilities and limitations of digital computers. Being the "new guy on the block", it still has not achieved the status that differential equations, with a three hundred year history, have. But it is likely to play more and more of a role in all kinds of modeling in the future. In keeping with this prediction, we start the introduction with a formulation of discrete event cellular automata which naturally follows from the earlier discussion. Only after having made this transition, will we come back to consider the more usual types of workflow models associated with discrete event simulation.

3.3.2 DISCRETE EVENT CELLULAR AUTOMATA

The original version of the Game of Life assumes that all births and deaths take the same time (equal to a time step). A more accurate representation of a cell's life cycle assumes that birth and death are dependent on a quantity that represents the ability of the cell to fit into its environment, called its fitness. A cell attains positive fitness when its neighborhood is supportive, that is, when it has exactly 3 neighbors. And the fitness will diminish rapidly when its environment is hostile. Whenever the fitness then reaches 0, the cell will die. On the other hand, a dead cell will have a negative initial fitness, lets say -2. When the environment is supportive and the fitness crosses the zero level, the cell will be born.

Fig. 3.30 shows an example behavior of such a cell model dependent on its environment (the sum of alive neighbors). The initial sum of alive neighbors being 3 causes the fitness of the dead cell to increase linearly until it crosses zero and a birth occurs. The fitness increases further until it reaches a saturation level (here 6). When the sum of alive neighbors changes from 3 to 4, the environment gets hostile, the fitness will decrease, and the cell will die. The death is accompanied by a discontinuous drop of the fitness level to a minimum fitness of -2.

What is needed to model such a process? One way would be to compute the trajectories of the fitness parameter and see when zero crossings and hence death and birth events occur. This would be the approach of combined discrete event and continuous simulation which we discuss in detail in Chapter 9. However, this approach is computationally inefficient since we have to compute each continuous trajectory at every point along its path. A much more efficient way is to adopt a pure event-based approach where we concentrate on the interesting events only, namely births and deaths as well as the changes in the neighborhood. The event-based approach jumps from one interesting event to the next omitting the uninteresting behavior in-between.

The prerequisite of discrete event modeling therefore is to have a means to determine when interesting things happen. Events can be caused by the environment, like the changes of the sum of alive neighbors. The occurrences of such *external* events are not under the control of the model component

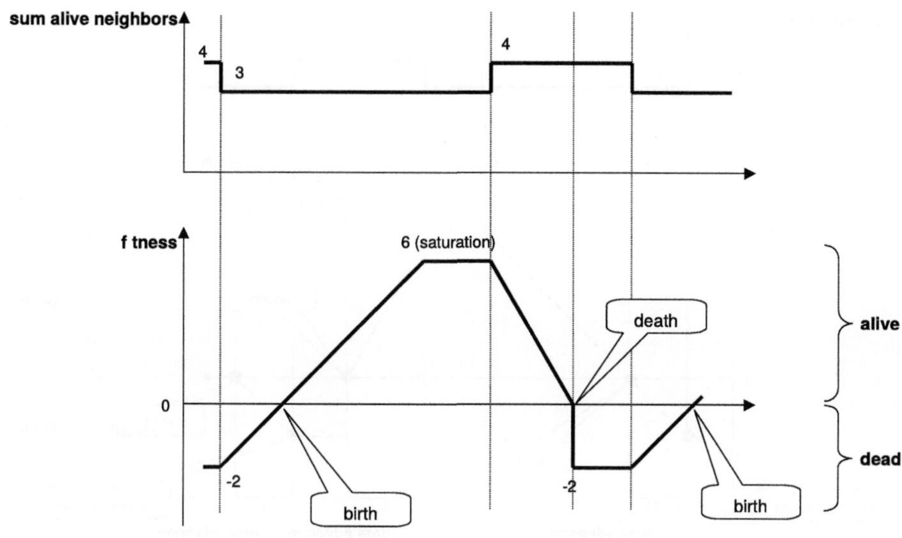

FIGURE 3.30

Behavior of GOL model with fitness.

itself. On the other side, the component may schedule events to occur. Those are called *internal* events and the component itself determines their time of occurrence. Given a particular state, e.g. a particular fitness of the cell, a time advance is specified as the time it takes until the next *internal event* occurs, supposing that no external event happens in the meantime. In our event-based GOL model the time advance is used to schedule the times when a birth or a death is supposed to happen (Fig. 3.31). As fitness changes linearly based on the neighborhood, the times of the events (times of the zero-crossings of fitness) can easily be predicted. Upon expiration of the time advance, the state transition function of a cell is applied, eventually resulting in a new time advance and a new scheduled event.

In discrete event simulation of the GOL model, one has to execute the scheduled internal events of the different cells at their event times. And since now cells are waiting, we must see to it that their time advances. Moreover, at any state change through an internal event we must take care to examine the cell's neighbors for possible state changes. A change in state may affect their waiting times as well as result in scheduling of new events and cancellation of events. Fig. 3.32 illustrates this process. Consider that all shaded cells in Fig. 3.32A are alive. Cell at $(0, 0)$ has enough neighbors to become alive and cell at $(-1, -1)$ only has 1 alive neighbor and therefore will die. Thus, a birth and a death are scheduled in these cells at times, say 1 and 2, respectively (shown by the numbers). At time 1, the origin cell undergoes its state transition and transits to the alive state shown by the thunderbolt in Fig. 3.32B. What effect does this have on the neighboring cells? Fig. 3.32B shows that as a result of the birth, cells at $(0, 1)$, $(0, -1)$ and $(1, 1)$ have three neighbors and must be scheduled for births at time 3. Cell $(-1, 1)$ has 4 alive neighbors now and a death will be scheduled for it at time 3. Cell $(-1, -1)$ has had a death scheduled before. However, due to the birth it has two neighbors now and the conditions for dying do not apply any more. Hence, the death at time 2 must be canceled. We see that the effect of

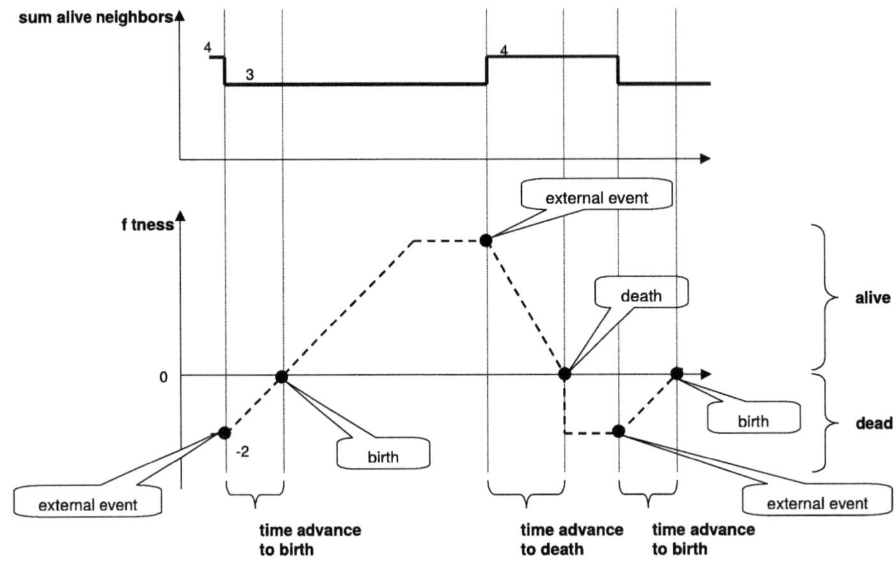

FIGURE 3.31

Behavior of the eventistic GOL model.

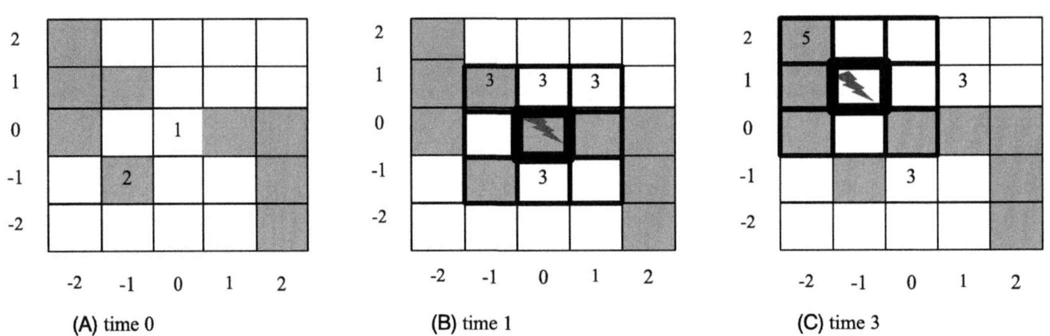

FIGURE 3.32

Event processing in Game of Life event model.

a state transition may not only be to schedule new events but also to cancel events that were scheduled in the past.

Now in Fig. 3.32B, the next earliest scheduled event is at time, 3. Therefore, the system can jump from the current time, 1, to the next event time, 3 without considering the times in-between. This illustrates an important *efficiency advantage in discrete event simulation* – during times when no events are scheduled, no components need be scanned. Contrast this with discrete time simulation in which scanning must be performed at each time step (and involves all cells in the worst case).

The situation at time 3 also illustrates a problem that arises in discrete event simulation – that of *simultaneous events*. From Fig. 3.32B we see that there are several cells scheduled for time 3, namely a birth at (0, 1), (1, 1) and (0, −1) and a death at (−1, 1). The question is who goes first and what is the result. Note, that if (−1, 1) goes first, the condition that (0, 1) has three live neighbors will not apply any more and the birth event just scheduled will be canceled again (Fig. 3.32C). However, if (0, 1) would go first, the birth in (0, 1) is actually carried out. Therefore, the results *will be different for different orderings of activation*. There are several approaches to the problem of simultaneous events. One solution is to let all simultaneous events undergo their state transitions together. This is the approach of parallel DEVS to be discussed in Chapter 6. The approach employed by most simulation packages and classic DEVS is to define a priority among the components. Components with high priority go first. To adopt this approach we employ a *tie-breaking* procedure, which selects one event to process out of a set of contending simultaneous events.

We'll return to show how to formulate the Game of Life as a well defined discrete event model in Chapter 7 after we have introduced the necessary concepts for DEVS representation. In the next section, we show how discrete event models can be simulated using the most common method called event scheduling. This method can also be used to simulate the Game of Life and we'll leave that as an exercise at the end of the section.

3.3.3 DISCRETE EVENT WORLD VIEWS

There are three common types of simulation strategies employed in discrete event simulation languages – *the event scheduling*, *the activity scanning* and *process interaction*, the latter being a combination of the first two. These are also called "world views" and each makes certain forms of model description more naturally expressible than others. In other words, each is best for a particular view of the way the world works. Nowadays, most simulation environments employ a combination of these strategies. In Chapter 7, we will characterize them as multi-component discrete event systems and analyze them in some detail. As an introduction to workflow modeling, here we will now discuss only the simplest strategy, event scheduling.

EVENT SCHEDULING WORLD VIEW

Event oriented models *preschedule* all events – there is no provision for conditional events that can be activated by tests on the global state. Scheduling an event is straightforward when all the conditions necessary for its occurrence can be known in advance. However, this is not always the case as we have seen in the Game of Life above. Nevertheless, understanding pure event scheduling in the following simple example of workflow modeling will help you to understand how to employ more complex world views later.

Fig. 3.33 depicts a discrete event model with three components. The generator *Gen* generates jobs for processing by a server component SS. Jobs are queued in a buffer called *Queue* when the server is busy. Component *Gen* continually reschedules a job output with a time advance equal to *inter-gen-time*. If the server is idle when a job comes in, then it is activated immediately to begin processing the job. Otherwise the number of waiting jobs in the queue is increased by one. When processing starts, the server is scheduled to complete work in a time that represents the job's *service-time*. When this time expires, the server starts to process the next job in the queue, if there is one. If not, it *passivates* in

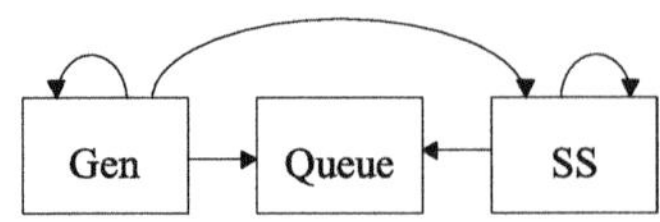

FIGURE 3.33

Generator, queue and server event scheduling model. The arrows indicate the effects of the event routines.

phase *idle*. To passivate means to set your time advance to infinity. The queue is a *passive* component (its time advance is always infinity) with a variable to store the number of jobs waiting currently.

Two event routines and a tie-breaking function express the interaction between these components:

```
Event: Generate_Job
  nr-waiting = nr-waiting + 1
  schedule a Generate_Job in inter-gen-time
  if nr-waiting = 1 then
          schedule a Process_Job in service-time


 Event: Process_Job
  nr-waiting = nr-waiting - 1
  if nr-waiting > 0 then
      schedule a Process_Job in service-time


// Note that the phrase "schedule an event in T"  means
// the same as "schedule an event  at time = clock time + T".

Break-Ties by: Process_Job then Generate_Job
```

As shown in Fig. 3.34, the event scheduling simulation algorithm employs a *list of events* that are ordered by increasing scheduling times. The event with the earliest scheduled time (e3 in Fig. 3.34) is removed from the list and the clock is advanced to the time of this imminent event (3). The routine associated with the imminent event is executed. A tie breaking procedure is employed if there are more than one such imminent events (recall the select function above). Execution of the event routine may cause new events to be added in the proper place on the list. For example *e*4 is scheduled at time 6. Also existing events may be rescheduled or even canceled (as we has seen above). The next cycle now begins with the clock advance to the earliest scheduled time (5), and so on. Each time an event routine is executed, one or more state variables may get new values.

Let's see how this works in our simple workflow example. Initially, the clock is set to 0 and a $Generate\_Job$ event is placed on the list with time 0. The state of the system is represented by $nr\text{-}waiting$ which is initially 0. Since this event is clearly imminent, it is executed. This causes two events to be scheduled: a $Generate\_Job$ at time $inter\text{-}gen\text{-}time$ and (since $nr\text{-}waiting$ becomes 1) a $Process\_Job$ at time $service - time$. What happens next depends on which of the scheduled times is earliest. Let's assume that $inter\text{-}gen\text{-}time$ is smaller than $service\text{-}time$. Then time is advanced to

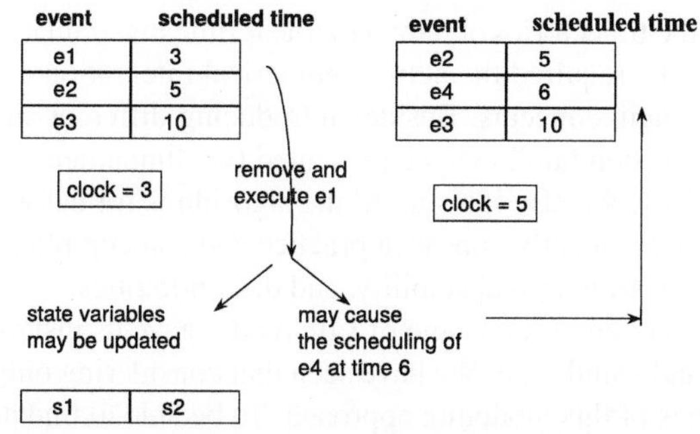

FIGURE 3.34

Event List Scheduling.

inter-gen-time and *Generate_Job* is executed. However, now only a new *Generate_Job* is scheduled – since the *nr-waiting* is 2, the *Process_Job* is not scheduled. Let's suppose that *service-time* $< 2*$*inter-gen-time* (the next event time for *Generate_Job*). Then the next event is a *Process_Job* and simulation continues in this manner.

Exercise 3.9. Hand execute the simulation algorithm for several cases. For example, 1) *inter-gen-time* $>$ *service-time*, 2) *inter-gen-time* $=$ *service-time*, 3) *inter-gen-time* $<$ *service-time* $< 2 *$ *inter-gen-time*, etc.

Exercise 3.10. Use the event scheduling approach to simulate the eventistic Game of Life introduced earlier in the section.

3.4 SUMMARY

In this chapter we have discussed the fundamental modeling formalisms which we will introduce in depth later in the book. Here we gained some insight into the intrinsic nature of the different modeling approaches. But the presentation can give us also some insight into the nature of dynamical systems in general. Let us summarize the modeling approaches in this sense.

The discrete time modeling approach which subsumes the popular finite automaton formalism as well as the difference equation formalism stands out through its simple simulation algorithm. It adopts a stepwise execution mode where all the components states are updated based on the state of the previous time step and the inputs.

The simplest form of this model type is the cellular automaton. In a cellular automaton a real system is reduced to basic coupling and communication between components. A cell on its own is quite boring. Only through couplings with other cells do quite astonishing complex and unpredictable behaviors result. So the cellular automaton is actually a model to study the complex behaviors that result from simple interactions among components. They are applied to study the emergence of group phenomena, like population dynamics, or spreading phenomena, like forest fires, avalanches or excitable media.

Continuous modeling and simulation is the classical approach of the natural sciences employing differential equations. Here, we focused only on ordinary differential equations and starting by studying the qualitative forms of the trajectories obtained in linear time invariant systems. We concluded that the simulation of general ODEs required the help of numerical integration methods and we developed a brief introduction to their main concepts. Besides introducing different simulation algorithms (like Euler, Runge Kutta and Multi-Step families), we presented two fundamental properties of the approximations: accuracy and stability. We also introduced the algorithms for automatic step size control and we studied some features that frequently appear in practice and can complicate the correct behavior of numerical ODE solvers: stiffness, marginal stability, and discontinuities.

Finally, the event-based version of the Game of Life model gave us insight into the intrinsic nature of discrete event modeling and simulation. We have seen that considering only the interesting points in time – the events – is the basis of this modeling approach. To be able to find and schedule the events in time is the prerequisite for applying discrete event simulation. A local event is scheduled by specifying a time advance value based on the current state of a component.

In the GOL model defining the event times was quite straightforward. Since the fitness parameter increases and decreases linearly, it is simple to compute the time when a zero crossing will occur – that is when an event occurs. However, in general this will not always be feasible and several different approaches to tackle this problem are known. A very common approach, which is heavily employed in work flow modeling, is to use stochastic methods and generate the time advance values from a random distribution. Another approach is to measure the time spans for different states and keep them in a table for later look-up. These approaches are discussed in Chapter 17. If none of these methods is appropriate, one has the possibility to step back to the more expensive combined simulation approach where the continuous trajectories are actually computed by continuous simulation and, additionally, events have to be detected and executed in-between. This combined continuous/discrete event modeling approach is the topic of Chapter 9.

The GOL event model also showed us two important advantages of discrete event modeling and simulation compared to the traditional approaches. First, only those times, and those components, have to be considered where events actually occur. Second, only those state changes must be reported to coupled components which are actually relevant for their behavior. For example in the GOL model, only changes in the alive state – but not changes in the fitness parameter – are relevant for a cell's neighbors. Actually, in discrete event modeling, the strategy often is to model only those state changes as events which are actually relevant to the environment of a component. As long as nothing interesting to the environment happens, no state changes are made. These two issues give the discrete event modeling and simulation approach a great lead in computational efficiency, making it most attractive for digital computer simulation. We will come back to this issue in Chapter 20 on DEVS representation of dynamical systems.

3.5 SOURCES

Research on cellular automata is summarized in Wolfram et al. (1986). Good expositions of continuous modeling are found in Cellier (1991), Bossel (1994), and Fritzson (2010), where the latter provides a complete reference about Modelica language. There is a vast literature on numerical integration methods. The main source of Section 3.2 is the book (Cellier and Kofman, 2006). A compact introduction to the most popular algorithms is given in Burden and Faires (1989) and algorithms in C can be found in Press et al. (1992). A more rigorous mathematical treatment of numerical algorithms for ODEs including the proofs of convergence can be found in the classic books (Hairer et al., 1993; Hairer and Wanner, 1991). Banks et al. (1995) and Law et al. (1991) give popular introductions to discrete event simulation. A control-oriented view of discrete event systems is given in Cassandras (1993).

REFERENCES

Bossel, H., 1994. Modeling and Simulations. Technical report. A K Peters, Ltd. ISBN 1-56881-033-4.

Burks, A.W., 1970. Essays on Cellular Automata. University of Illinois Press.

Cassandras, C., 1993. Discrete Event Systems: Modeling and Performance Analysis. Richard Irwin, New York, NY.

Cellier, F., 1991. Continuous System Modeling. Springer-Verlag, New York.

Cellier, F., Kofman, E., 2006. Continuous System Simulation. Springer, New York.

Fritzson, P., 2010. Principles of Object-Oriented Modeling and Simulation with Modelica 2.1. John Wiley & Sons.

Gardner, M., 1970. Mathematical games: the fantastic combinations of John Conway's new solitaire game "life". Scientific American 223 (4), 120–123.

Hairer, E., Nørsett, S., Wanner, G., 1993. Solving Ordinary Differential Equations. I, Nonstiff Problems. Springer-Verlag, Berlin.

Hairer, E., Wanner, G., 1991. Solving Ordinary Differential Equations. II, Stiff and Differential-Algebraic Problems. Springer-Verlag, Berlin.

Banks, J., Carson, C., Nelson, B.L., 1995. Discrete-Event System Simulation. Prentice Hall Press, NJ.

Law, A.M., Kelton, W.D., Kelton, W.D., 1991. Simulation Modeling and Analysis, vol. 2. McGraw-Hill, New York.

Press, W.H., Teukolsky, S.A., Vetterling, W.T., Flannery, B.P., 1992. Numerical Recipes in C: The Art of Scientific Computing Second.

Burden, Richard L., Faires, J.D., 1989. Numerical Analysis, fourth edition. PWS-KENT Publishing Company.

Wolfram, S., et al., 1986. Theory and Applications of Cellular Automata, vol. 1. World Scientific, Singapore.

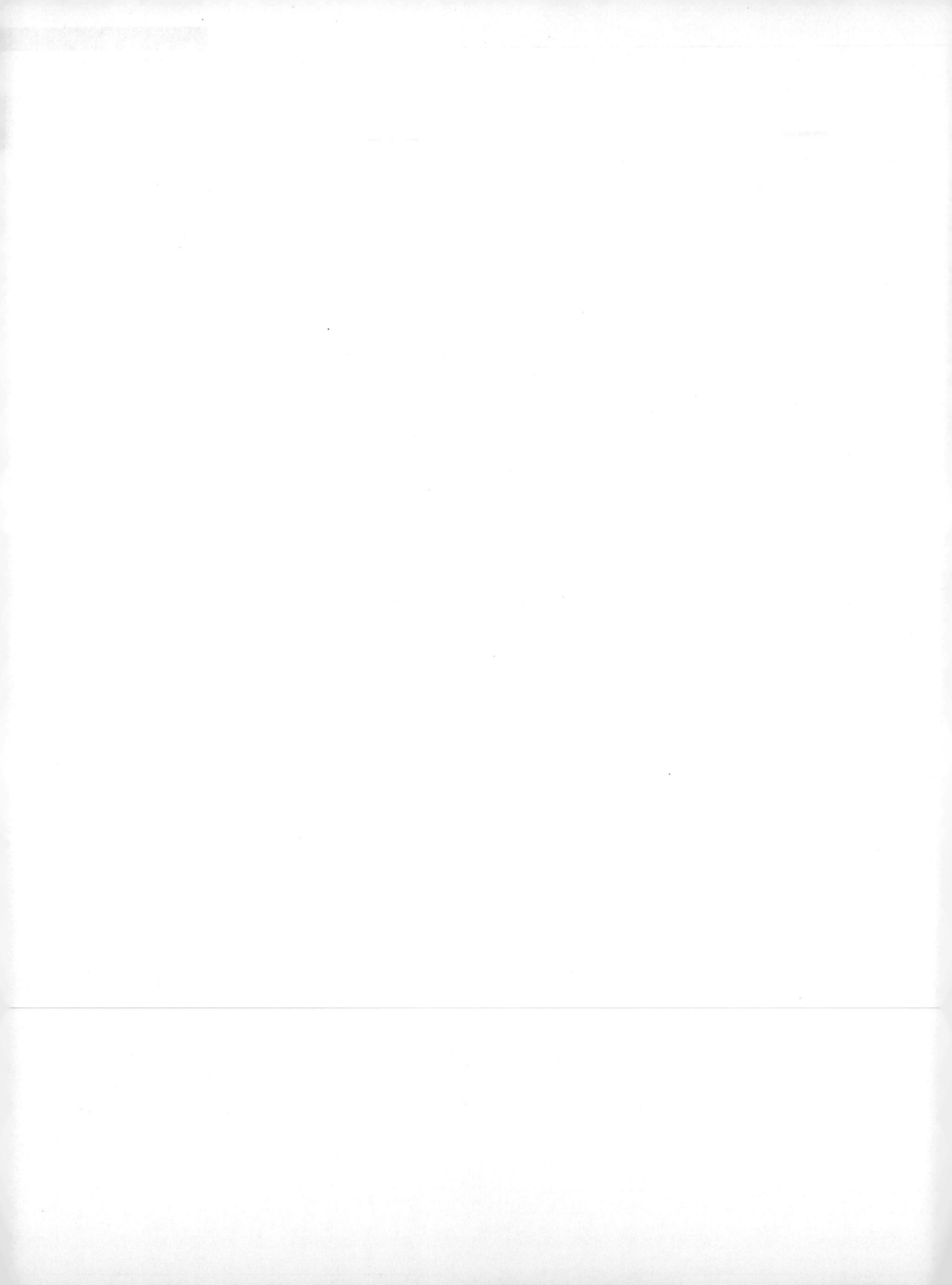

CHAPTER

INTRODUCTION TO DISCRETE EVENT SYSTEM SPECIFICATION (DEVS)

4

CONTENTS

Theory of Modeling and Simulation. https://doi.org/10.1016/B978-0-12-813370-5.00012-2

4.1 INTRODUCTION

This chapter introduces the DEVS formalism for discrete event systems. DEVS exhibits, in particular form, the concepts of systems theory and modeling that we have introduced already. DEVS is important not only for discrete event models, but also because it affords a computational basis for implementing behaviors that are expressed in the other basic systems formalisms – discrete time and differential equations. Particular features of DEVS are also present in more general form in the systems theory that we will see in the next chapter. So starting with DEVS also serves as a good introduction to this theory and the other formalisms that it supports.

In this chapter, we start with the basic DEVS formalism and discuss a number of examples using it. We then discuss the DEVS formalism for coupled models also giving some examples. The DEVS formalism that we start with is called Classic DEVS because after some fifteen years, a revision was introduced called Parallel DEVS. As we will explain later, Parallel DEVS removes constraints that originated with the sequential operation of early computers and hindered the exploitation of parallelism, a critical element in more modern computing. TMS2000 concluded this chapter with how DEVS is implemented in an object-oriented framework. In this edition, we add a section that discusses the fundamental computational model known as the Turing Machine in the context of DEVS and the hierarchy of system specifications. The approach leads to interesting insights into the statistical character of the Turing halting problem.

4.2 CLASSIC DEVS SYSTEM SPECIFICATION

A Discrete Event System Specification (DEVS) is a structure

$$M = \langle X, S, Y, \delta_{\text{int}}, \delta_{\text{ext}}, \lambda, \text{ta} \rangle$$

where

X is the set of input values

S is a set of states,

Y is the set of output values

$\delta_{\text{int}} : S \rightarrow S$ is the internal transition function

$\delta_{\text{ext}} : Q \times X \rightarrow S$ is the external transition function, where

$Q = \{(s, e) | s \in S, 0 \leq e \leq \text{ta}(s)\}$ is the total state set

e is the time elapsed since last transition

$\lambda : S \rightarrow Y$ is the output function

$\text{ta} : S \to \mathbb{R}^+_{0,\infty}$ is the time advance function

The interpretation of these elements is illustrated in Fig. 4.1. At any time the system is in some state, s. If no external event occurs the system will stay in state s for time $\text{ta}(s)$. Notice that $\text{ta}(s)$ could be a real number as one would expect. But it can also take on the values 0 and ∞. In the first case, the stay in state s is so short that no external events can intervene – we say that s is a *transitory* state. In the second case, the system will stay in s forever unless an external event interrupts its slumber. We say that s is a passive state in this case. When the resting time expires, i.e., when the elapsed time, $e = \text{ta}(s)$, the system outputs the value, $\lambda(s)$, and changes to state $\delta_{\text{int}}(s)$. Note output is only possible just before internal transitions.

If an external event $x \in X$ occurs before this expiration time, i.e., when the system is in total state (s, e) with $e \leq \text{ta}(s)$, the system changes to state $\delta_{\text{ext}}(s, e, x)$. Thus the internal transition function dictates the system's new state when no events have occurred since the last transition. While the external transition function dictates the system's new state when an external event occurs – this state is determined by the input, x, the current state, s, and how long the system has been in this state, e, when the external event occurred. In both cases, the system is then is some new state s' with some new resting time, $\text{ta}(s')$ and the same story continues.

The above explanation of the semantics (or meaning) of a DEVS model suggests, but does not fully describe, the operation of a simulator that would execute such models to generate their behavior. We will delay discussion of such simulators to later chapters. However, the behavior of a DEVS is well defined and can be depicted as in Fig. 4.2. Here the *input trajectory* is a series of events occurring at times such as $t0$ and $t2$. In between, such event times may be those such as $t1$ which are times of internal events. The latter are noticeable on the *state trajectory* which is a step-like series of states, which change at external and internal events (second from top). The elapsed *time trajectory* is a saw-tooth pattern depicting the flow of time in an elapsed time clock which gets reset to 0 at every event. Finally, at the bottom, the *output trajectory* depicts the output events that are produced by the output function just before applying the internal transition function at internal events. Such behaviors will be illustrated in the next examples.

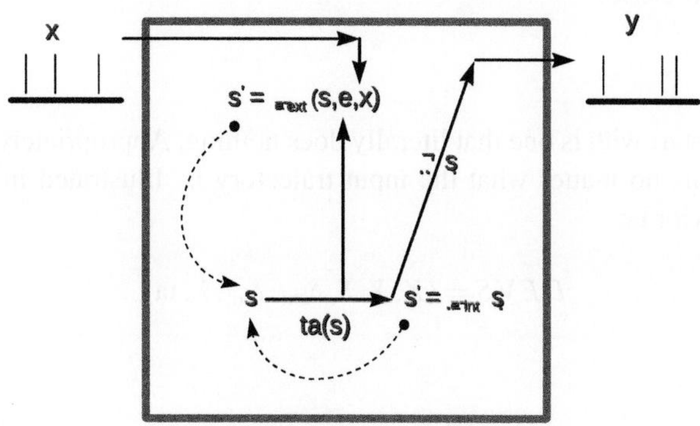

FIGURE 4.1

DEVS in action.

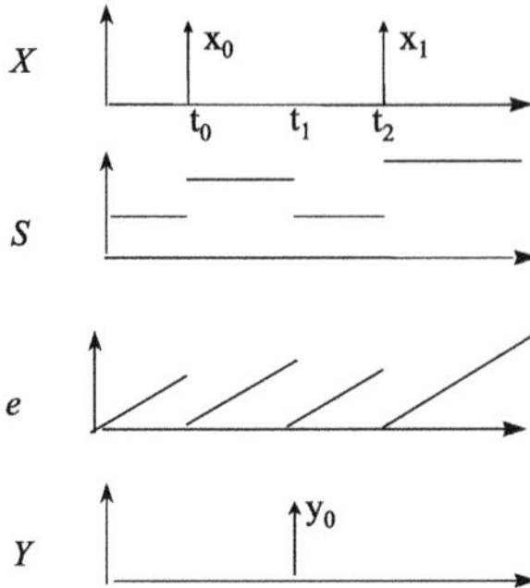

FIGURE 4.2

DEVS Trajectories.

4.2.1 DEVS EXAMPLES

We will present a number of models and discuss their behaviors to illustrate practical construction of DEVS models. Table 4.1 lists the models to be discussed. As will be detailed later, there are two classes of models in any object-oriented implementation of DEVS. *Atomic* models are expressed in the basic formalism. *Coupled* models are expressed using the coupled model specification – essentially providing component and coupling information.

We start with models that have only a single input and a single output, both expressed as real numbers.

Exercise. Write a simulator for any DEVS model that has scalar real values for its inputs, states and outputs. Pending discussion of DEVS simulators in Chapter 8, you can use the event scheduling approach of Chapter 3 in which the external events in an input trajectory are all prescheduled on the event list. The DEVS is started in a state s and an internal event notice is placed in the event list with a time advance given by $\text{ta}(s)$. This notice must be rescheduled appropriately every time an internal or external transition is processed.

PASSIVE

The simplest DEVS to start with is one that literally does nothing. Appropriately called passive, it does not respond with outputs no matter what the input trajectory is. Illustrated in Fig. 4.3, the simplest realization of this behavior is:

$$DEVS = (X, Y, S, \delta_{\text{ext}}, \delta_{\text{int}}, \lambda, \text{ta})$$

where

$$X = \mathbb{R}$$
$$Y = \mathbb{R}$$
$$S = \{\text{"passive"}\}$$

Table 4.1 Examples of DEVS models

Models	I/O Behavior Description
Atomic	
Classic DEVS	
passive	never generates output
storage	stores the input and responds with it when queried
generator	outputs a 1 in a periodic fashion
binaryCounter	outputs a 1 only when it has received an even number of 1's
nCounter	outputs a 1 only when it has received a number of 1's equal to n
Infinite Counter	outputs the number of input events received since initialization.
processor	outputs the number it received after a processing time
ramp	output acts like the position of a billiard ball that has been hit by the input
Parallel DEVS	
multiple input processor with queue	processor with queue with the capability to accept multiple simultaneous inputs
Coupled	
Classic DEVS	
pipeSimple	pipeline – processes successive inputs in assembly line fashion
netSwitch	outputs the input it receives after a processing time (internally input is alternated between two processors)
Parallel DEVS	
pipeSimple	same as before except for handling of event collisions
netSwitch	revised Switch and coupled model handling simultaneous inputs on different ports
experimental frame	generator and transducer encapsulated into a coupled model
Hierarchical DEVS	
frame-processor	experimental frame coupled to processor (could be an port compatible model e.g., processor with buffer, netSwitch, etc.)

$$\delta_{\text{ext}}(\text{“passive”}, e, x) = \text{“passive”}$$
$$\delta_{\text{int}}(\text{“passive”}) = \text{“passive”}$$
$$\lambda(\text{“passive”}) = \emptyset$$
$$\text{ta}(\text{“passive”}) = \infty$$

The input and output sets are numerical. There is only one state “passive”. In this state, the time advance given by ta is infinite. As already indicated, this means the system will stay in passive forever unless interrupted by an input. However, in this model, even such an interruption will not awaken it,

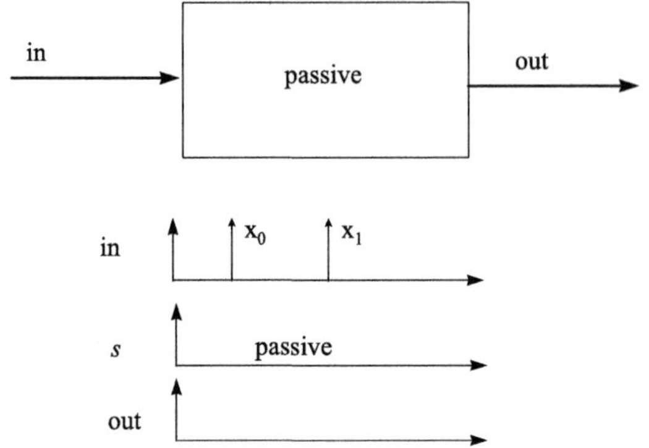

FIGURE 4.3

Passive DEVS.

since the external transition function does disturb the state. The specifications of the internal transition and output functions are redundant here since they will never get a chance to be applied.

STORAGE

In contrast to the Passive DEVS, the next system responds to its input and stores it forever, or until the next input comes along. This is not very useful unless we have a way of asking what is currently stored. So there is a second input to do such querying. Since there is only one input port in the current DEVS model, we let the input value of zero signals this query. As the first part of Fig. 4.4, shows, within a time, *response_time*, of the zero input arrival, the DEVS responds with the last stored non-zero input. To make this happen, the Storage is defined by:

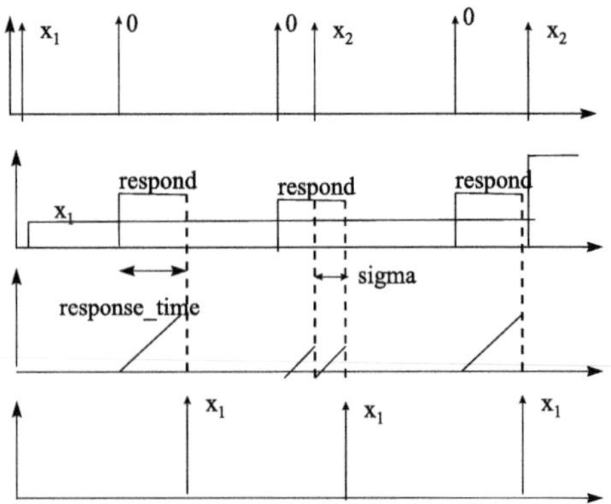

FIGURE 4.4

Trajectories for Storage.

$$DEVS_{\text{response\_time}} = (X, Y, S, \delta_{\text{ext}}, \delta_{\text{int}}, \lambda, \text{ta})$$

where

$$
\begin{aligned}
&X = \mathbb{R}\\
&Y = \mathbb{R}\\
&S = \{\text{"passive"},\text{"respond"}\} \times R_0^+ \times R - \{0\}\\
&\delta_{\text{ext}}(\text{"passive"}, \sigma, store, e, x) = \begin{cases} (\text{"passive"}, \sigma - e, x) & \text{if } x \neq 0\\ (\text{"respond"}, response\_time, store) & \text{if } x = 0 \end{cases}\\
&\delta_{\text{ext}}(\text{"respond"}, \sigma, store, e, x) = (\text{"respond"}, \sigma - e, store)\\
&\delta_{\text{int}}(\text{"respond"}, \sigma, store) = (\text{"passive"}, \infty, store)\\
&\lambda(\text{"respond"}, \sigma, store) = store\\
&\text{ta}(phase, \sigma, store) = \sigma
\end{aligned}
$$

There are three state variables: phase with values "passive", "respond", σ, having positive real values, and store having real values other than zero. We need the active "respond" phase to tell when a response is underway. As for σ, the definition of ta shows that it stores the time advance value. In other words, s is the time remaining in the current state. Note that when an external event arrives after elapsed time, e, s is reduced by e to reflect the smaller time remaining in the current state. When a zero input arrives, σ is set to response_time and the "respond" phase is entered. However, if the system is in phase "respond" and an input arrives, the input is ignored as illustrated in the second part of Fig. 4.4. When the *response_time* period has elapsed, the output function produces the stored value and the internal transition dictates a return to the passive state. What happens if, as in the third part of Fig. 4.4, an input arrives just as the response period has elapsed? Classic DEVS and Parallel DEVS differ in their ways of handling this collision as we shall show later.

Note that we subscript the DEVS with *response_time*, to indicate that the latter is a parameter of the model – it needs to be specified before execution but it doesn't change subsequently, at under the model's own actions.

GENERATOR

While the store reacts to inputs it does not do anything on its own. A simple example of a proactive system is a *generator*. As illustrated in Fig. 4.5, it has no inputs but when started in *phase* "active", it generates outputs with a specific period. The generator has the two basic state variables, phase and σ. Note that the generator remains in phase "active", for the period, after which the output function generates a one output and the internal transition function resets the phase to "active", and σ to *period*. Strictly speaking, restoring these values isn't necessary unless inputs are allowed to change them (which they aren't in this model).

$$DEVS_{period} = (X, Y, S, \delta_{\text{ext}}, \delta_{\text{int}}, \lambda, \text{ta})$$

where

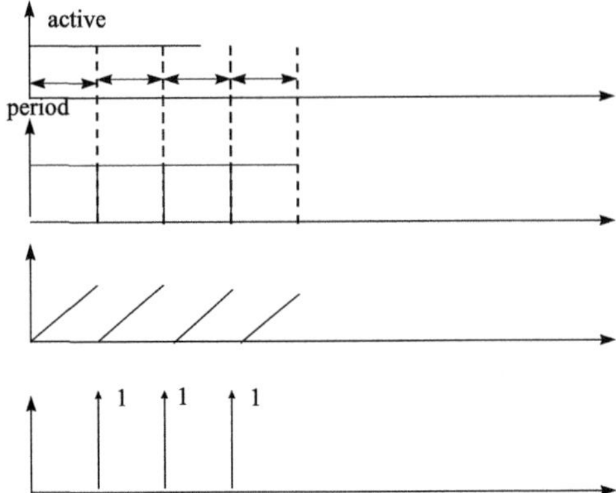

FIGURE 4.5

Generator Trajectories.

$$
\begin{aligned}
&X = \{\} \\
&Y = \{1\} \\
&S = \{\text{“passive”}, \text{“active”}\} \times R^{+} \\
&\delta_{\text{int}}(phase, \sigma) = (\text{“active”}, period) \\
&\lambda(\text{“active”}, \sigma) = 1 \\
&\text{ta}(phase, \sigma) = \sigma
\end{aligned}
$$

BINARY COUNTER

In this example, the DEVS outputs a “one” for every two “one”s that it receives. To do this it maintains a count (modulo 2) of the “one”s it has received to date. When it receives a “one” that makes its count even, it goes into a transitory phase, “active”, to generate the output. This is the same as putting $responce\_time := 0$ in the storage DEVS.

$$DEVS = (X, Y, S, \delta_{\text{ext}}, \delta_{\text{int}}, \lambda, \text{ta})$$

where

$$
\begin{aligned}
&X = \{0.1\} \\
&Y = \{1\} \\
&S = \{\text{“passive”}, \text{“active”}\} \times \mathbb{R}_0^{+} \times \{0.1\} \\
&\delta_{\text{ext}}(\text{“passive”}, \sigma, count, e, x) = \begin{cases} (\text{“passive”}, \sigma - e, count + x) & \text{if } count + x < 2 \\ (\text{“active”}, 0, 0) & \text{otherwise} \end{cases} \\
&\delta_{\text{int}}(phase, \sigma, count) = (\text{“passive”}, \infty, count)
\end{aligned}
$$

$$\lambda(\text{“active”}, \sigma, count) = 1$$
$$\text{ta}(phase, \sigma, count) = \sigma$$

Exercise. An nCounter generalizes the binary counter by generating a “one” for every n “one”s it receives. Specify an nCounter as an atomic model in DEVS.

Exercise. Define a DEVS counter that counts the number of non-zero input events received since initialization and outputs this number when queried by a zero valued input.

PROCESSOR

A model of a simple workflow situation is obtained by connecting a generator to a processor. The generator outputs are considered to be jobs to do and the processor takes some time to do them. In the simplest case, no real work is performed on the jobs, only the times taken to do them are represented.

$$DEVS_{processng\_time} = (X, Y, S, \delta_{\text{ext}}, \delta_{\text{int}}, \lambda, \text{ta})$$

where

$$X = \mathbb{R}$$
$$Y = \mathbb{R}$$
$$S = \{\text{“passive”},\text{“active”}\} \times R_0^+ \times \mathbb{R}$$
$$\delta_{\text{ext}}(phase, \sigma, job, e, x) = \begin{cases} (\text{“}busy\text{”}, processing\_time, job) & \text{if } phase = \text{“}passive\text{”} \\ (phase, \sigma - e, x) & \text{otherwise} \end{cases}$$
$$\delta_{\text{int}}(phase, \sigma, job) = (\text{“}passive\text{”}, processing\_time, job)$$
$$\lambda(\text{“}busy\text{”}, \sigma, job) = job$$
$$\text{ta}(phase, \sigma, job) = \sigma$$

Here, if the DEVS is passive (idle), when it receives a job, it stores the job and goes into phase “busy”. If it is already busy, it ignores incoming jobs. After a period, *rocessing_time*, it outputs the job and returns to “passive”.

Exercise. Draw typical input/state/elapsed time/output trajectories for the simple processor.

RAMP

As we shall see later, DEVS can model systems whose discrete event nature is not immediately apparent. Consider a billiard ball. As illustrated in Fig. 4.6, struck by a cue (external event), it heads off in a direction at constant speed determined by the impulsive force imparted to it by the strike. Hitting the side of the table is considered as another input that sets the ball off going in a well-defined direction. The model is described by:

$$DEVS_{step_time} = (X, Y, S, \delta_{\text{ext}}, \delta_{\text{int}}, \lambda, \text{ta})$$

where

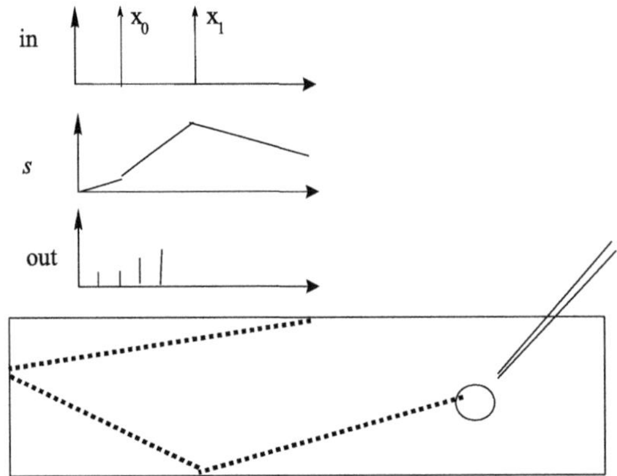

FIGURE 4.6

Ramp Trajectories.

$$X = \mathbb{R}$$
$$Y = \mathbb{R}$$
$$S = \{\text{“passive”}, \text{“active”}\} \times \mathbb{R}_0^+ \times \mathbb{R} \times \mathbb{R}$$
$$\delta_{\text{ext}}(phase, \sigma, position, input, e, x) = (\text{“}active\text{”}, \sigma - e, x, position + e * x)$$
$$\delta_{\text{int}}(phase, \sigma, input, position) = (\text{“}active\text{”}, step\_time, input, position + \sigma * input)$$
$$\lambda(phase, \sigma, input, position) = position + \sigma * input)$$
$$\text{ta}(phase, \sigma, input, position) = \sigma$$

The model stores its input and uses it as the value of the slope to compute the position of the ball (in a one-dimensional simplification). It outputs the current position every *step_time*. If it receives a new slope in the middle of a period, it updates the position to a value determined by the slope and elapsed time. Note that it outputs the position prevailing at the time of the output. This must be computed in the output function since it always called before the next internal event actually occurs. Note that we do not allow the output function to make this update permanent. The reason is that the output function is not allowed to change the state in the DEVS formalism. (A model in which this is allowed to happen may produce **unexpected and hard-to-locate** errors because DEVS simulators do not guarantee correctness for non-conforming models.)

4.2.2 CLASSIC DEVS WITH PORTS

Modeling is made easier with the introduction of input and output ports. For example, a DEVS model of a storage naturally has two input ports, one for storing and the other for retrieving. The more concrete DEVS formalism with port specifications is as follows:

$$DEVS = (X, Y, S, \delta_{\text{ext}}, \delta_{\text{int}}, \lambda, \text{ta})$$

where

$X = \{(p, v)|p \in InPorts, v \in X_p\}$ is the set of input ports and values;
$Y = \{(p, v)|p \in OutPorts, v \in Y_p\}$ is the set of output ports and values;
$S$is the set of sequential states;
$\delta_{\text{ext}} : Q \times X \rightarrow S$ is the *external state transition function*;
$\delta_{\text{int}} : S \rightarrow S$ is the *internal state transition function*;
$\lambda : S \rightarrow Y$ is the output function;
$\text{ta} : S \rightarrow R_0^+ \cup \infty$ is the time advance function;

with

$$Q := (s, e)|s \in S, 0 \leq e \leq \text{ta}(s) \text{ is the set of total states.}$$

Note that in classic DEVS, only one port receives a value in an external event. We shall see later that parallel DEVS allows multiple ports to receive values at the same time.

SWITCH

A switch is modeled as a DEVS with pairs of input and output ports, as shown in Fig. 4.7. When the switch is in the standard position, jobs arriving on port "in" are sent out on port "out", and similarly for ports "in1" and "out1". When the switch is in its other setting, the input-to-output links are reversed, so that what goes in on port "in" exits at port "out1", etc.

In the switch DEVS below, in addition to the standard *phase* and σ variables, there are state variables for storing the input port of the external event, the input value, and the polarity of the switch. In this simple model, the polarity is toggled between true and false at each input.

$$DEVS = (X, Y, S, \delta_{\text{ext}}, \delta_{\text{int}}, \lambda, \text{ta})$$

where

$InPorts = \{\text{"in"}, \text{"in1"}\}$, where $X_in = X_in1 = V$ (an arbitrary set),
$X = \{(p, v)|p \in InPorts, v \in X_p\}$ is the set of input ports and values;
$OutPorts = \{\text{"out"}, \text{"out1"}\}$, where $Y_out = Y_out1 = V$ (an arbitrary set),
$Y = \{(p, v)|p \in OutPorts, v \in Y_p\}$is the set of output ports and values;

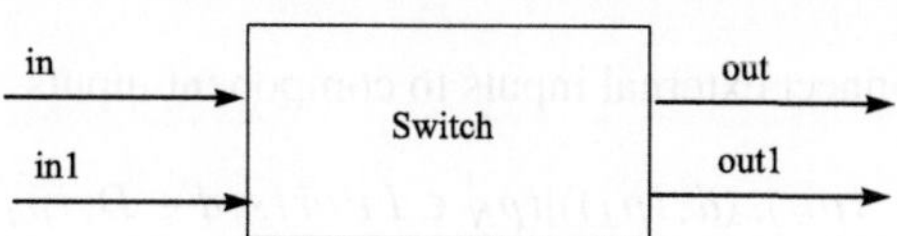

FIGURE 4.7

Switch with input and output ports.

$$S = \{\text{“passive”},\text{“active”}\} \times \mathbb{R}_0^+ \times \{in, in1\} \times V \times \{true, false\}$$

$$\delta_{\text{ext}}(phase, \sigma, inport, store, Sw, e, (p, v)) =$$

$$= \begin{cases} (\text{“busy”}, processing\_time, p, v, !Sw) & \text{if } phase = \text{ “passive” and } p \in in, in1 \\ (phase, \sigma - e, inport, store, Sw) & \text{otherwise} \end{cases}$$

$$\delta_{\text{int}}(phase, \sigma, inport, store, Sw, e, (p, v)) = (\text{“passive”}, \infty, inport, store, Sw)$$

$$\lambda(phase, \sigma, inport, store, Sw) =$$

$$= \begin{cases} (out, store) & \text{if } phase = \text{ “busy” and } Sw = true \text{ and } inport = in \\ (out1, store) & \text{if } phase = \text{ “busy” and } Sw = true \text{ and } inport = in1 \\ (out1, store) & \text{if } phase = \text{ “busy” and } Sw = false \text{ and } inport = in \\ (out, store) & \text{if } phase = \text{ “busy” and } Sw = false \text{ and } inport = in1 \end{cases}$$

$$\text{ta}(phase, \sigma, inport, store, Sw) = \sigma$$

4.2.3 CLASSIC DEVS COUPLED MODELS

The DEVS formalism includes the means to build models from components. The specification in the case of DEVS with ports includes the external interface (input and output ports and values), the components (which must be DEVS models), and the coupling relations.

$$N = (X, Y, D, M_d | d \in D, EIC, EOC, IC, \text{Select})$$

where

$X = \{(p, v) | p \in IPorts, v \in X_p\}$ is the set of input ports and values;
$Y = \{(p, v) | p \in OPorts, v \in Y_p\}$ is the set of output ports and values;
D is the set of the component names;

Component Requirements:
Components are DEVS models:
For each $d \in D$

$$M_d = (X_d, Y_d, S, \delta_{\text{ext}}, \delta_{\text{int}}, \lambda, \text{ta}) \text{ is a DEVS}$$
$$\text{with } X_d = (p, v) | p \in IPorts_d, v \in X_p$$
$$Y_d = (p, v) | p \in OPorts_d, v \in Y_p$$

Coupling Requirements:
external input coupling connect external inputs to component inputs:

$$EIC \subseteq \{((N, ip_N), (d, ip_d)) | ip_N \in IPorts, d \in D, ip_d \in IPorts_d\}$$

external output coupling connect component outputs to external outputs:

$$EOC \subseteq \{((d, op_d), (N, op_N)) | opN \in OPorts, d \in D, op_d \in OPorts_d\}$$

internal coupling connect component outputs to component inputs:

$$IC \subseteq \{((a, op_a), (b, ip_b)) | a, b \in D, op_a \in OPorts_a, ip_b \in IPorts_b\}$$

However, no direct feedback loops are allowed, i.e., no output port of a component may be connected to an input port of the same component i.e.,

$$((d, op_d), (e, ip_d)) \in IC \text{ implies } d \neq e.$$

(We omit constraints on the interface sets which are covered in Chapter 5.)

Select: $2^D \rightarrow D$, the tie-breaking function (used in Classic DEVS but eliminated in Parallel DEVS).

SIMPLE PIPELINE

We construct a simple coupled model by placing three processors in series to form a pipeline. As shown in Fig. 4.8, we couple the output port "out" of the first processor to the input port "in" of the second, and likewise for the second and third. This kind of coupling is called *internal coupling* (IC) and, as in the above specification, it is always of the form where an output port connects to an input port. Since coupled models are themselves usable as components in bigger models, we give them input and output ports. The pipeline has an input port "in" which we connect to the input port "in" of the first processor. This is an example of *external input coupling* (EIC). Likewise there is an external output port "out" which gets its values from the last processor's "out" output port. This illustrates *external output coupling* (EOC).

The coupled DEVS specification of the pipeline is as follows:

$$N = (X, Y, D, M_d | d \in D, EIC, EOC, IC)$$

where

$InPorts = \{\text{"in"}\}$,

$X_in = V$(an arbitrary set),

$X = \{(\text{"in"}, v) | v \in V\}$

$OutPorts = \{\text{"out"}\}$,

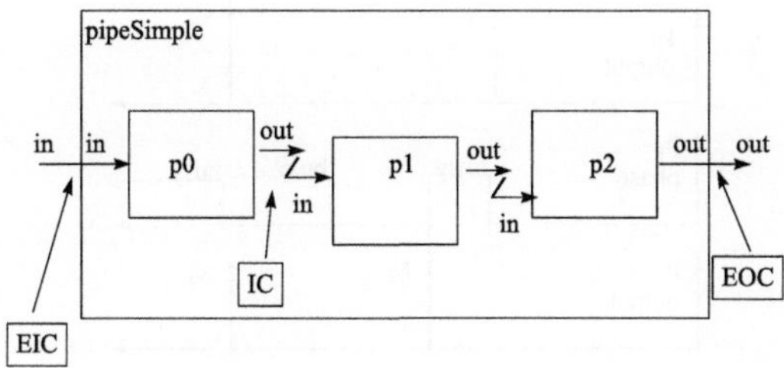

FIGURE 4.8

Pipeline Coupled Model.

$$Y_out = V$$
$$Y = \{(\text{"out"}, v) | v \in V\}$$
$$D = \{processor0, processor1, processor2\};$$
$$M_processor2 = M_{processor1} = M_{processor0} = Processor$$
$$EIC = \{((N, \text{"in"}), (processor0, \text{"in"})\}$$
$$EOC = \{((processor2, \text{"out"}), (N, \text{"out"})\}$$
$$IC = \{((processor0, \text{"out"}), (processor1, \text{"in"})), ((processor1, \text{"out"}), (processor2, \text{"in"}))\}$$
$$\text{Select}(D') = \text{ the processor in } D' \text{ with the highest index}$$

The interpretation of coupling specifications is illustrated in Fig. 4.9. An external event, x_1 arriving at the external input port of pipeSimple is transmitted to the input port "in" of the first processor. If the latter is passive at the time, it goes into phase "busy" for the processing time. Eventually, the job appears on the "out" port of $processor0$ and is transmitted to the "in" port of $processor1$ due to the internal coupling. This transmission from output port to input port continues until the job leaves the external output port of the last processor and appears at the output port of pipeSimple due to the external output coupling.

Now if the jobs arrive at the same rate that they are processed, the situation shown in the second part of Fig. 4.9 will arise. The outputs of $processor0$ and $processor1$ are generated at the same time with the output of $processor0$ appearing as the input to $processor1$. There are now two choices for the modeler to make. Should $processor1$ a) apply its *internal transition function* to return to phase "passive" in which it can accept the upstream input, or b) should it apply its *external transition function* first in which case it will lose the input since it is in phase "busy". The Select function specifies that if both $processor0$ and $processor1$ are imminent, then $processor1$ is the one that generates its output and carries out its internal transition. Thus it enforces alternative a). In parallel DEVS, we shall see that a similar solution can be achieved without serializing imminent component actions.

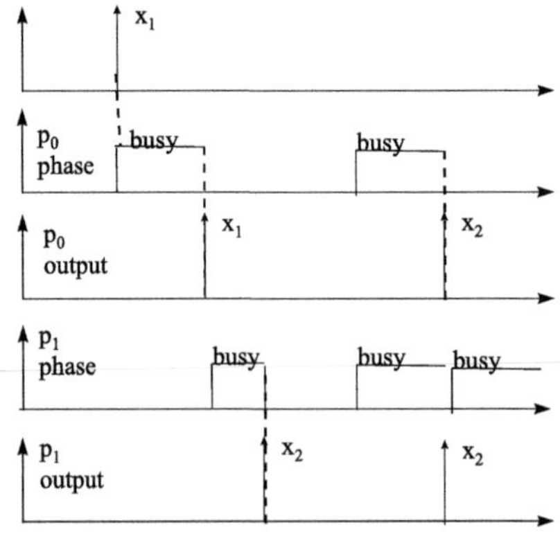

FIGURE 4.9

Message Transmission and Simultaneous Events.

SWITCH NETWORK

A second example of a coupled network employs the *switch* DEVS model defined before to send jobs to a pair of *processors*.

As shown in Fig. 4.10, the "out" and "out1" ports of the switch are coupled individually to the "in" input ports of the processors. In turn, the "out" ports of the processors are coupled to the "out" port of the network. This allows an output of either processor to appear at the output of the overall network. The coupled DEVS specification is:

$$N = (X, Y, D, \{M_d | d \in D\}, EIC, EOC, IC)$$

where

$InPorts = \{\text{"in"}\},$

$X_in = V(\text{an arbitrary set}),$

$X_M = \{(v) | v \in V\}$

$OutPorts = \{\text{"out"}\},$

$Y_out = V$

$Y_M = \{(\text{"out"}, v) | v \in V\}$

$D = \{Switch0, processor0, processor1\};$

$M_{Switch0} = Switch; M_{processor1} = M_{processor0} = Processor;$

$EIC = \{((N, \text{"in"}), (Switch0, \text{"in"})\}$

$EOC = \{((processor0, \text{"out"}), (N, \text{"out"})), ((processor1, \text{"out"}), (N, \text{"out"}))\}$

$IC = \{((Switch0, \text{"out"}), (processor0, \text{"in"})), ((Switch0, \text{"out1"}), (processor1, \text{"in"}))\}$

4.3 PARALLEL DEVS SYSTEM SPECIFICATION

Parallel DEVS differs from Classical DEVS in allowing all imminent components to be activated and to send their output to other components. The receiver is responsible for examining this input and

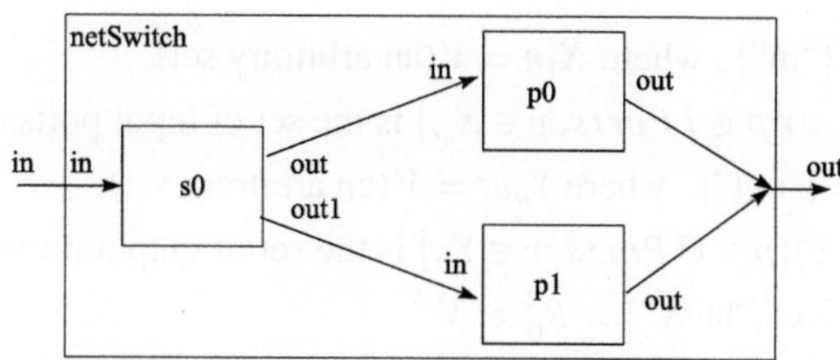

FIGURE 4.10

Switch Network.

properly interpreting it. Messages, basically lists of port – value pairs, are the basic exchange medium. This section discusses Parallel DEVS, and gives a variety of examples to contrast with Classical DEVS.

A basic parallel DEVS is a structure:

$$DEVS = (X_M^+, Y_M^+, S, \delta_{\text{ext}}, \delta_{\text{int}}, \delta_{\text{con}}, \lambda, \text{ta})$$

where

$X_M = \{(p, v) | p \in IPorts, v \in X_p\}$ is the set of input ports and values;
$Y_M = \{(p, v) | p \in OPorts, v \in Y_p\}$ is the set of output ports and values;
S is the set of *sequential* states;
$\delta_{\text{ext}} : Q \times X_M^+ \to S$ is the *external state transition function*;
$\delta_{\text{int}} : S \to S$ is the *internal state transition function*;
$\delta_{\text{con}} : Q \times X_M^+ \to S$ is the *confluent transition function*;
$\lambda : S \to Y^+$ is the output function;
$\text{ta} : S \to R_0^+ \cup \infty$ is the *time advance function*;

with

$$Q := (s, e) | s \in S, 0 \le e \le \text{ta}(s) \text{ is the set of total states.}$$

Note that instead of having a single input, basic parallel DEVS models have a *bag* of inputs. A bag is a set with possible multiple occurrences of its elements. A second difference is noted – this is the addition of a transition function, called *confluent*. It decides the next state in cases of collision between external and internal events. We have seen examples of such collisions earlier in examining classical DEVS. There is also a coupled model version of Parallel DEVS which matches the assumptions discussed above. We return to discuss it after discussing some examples.

PROCESSOR WITH BUFFER

A processor that has a buffer is defined in Parallel DEVS as follows:

$$DEVS_{processng\_time} = (X_M^+, Y_M^+, S, \delta_{\text{ext}}, \delta_{\text{int}}, \delta_{\text{con}}, \lambda, \text{ta})$$

where

$IPorts = \{\text{"in"}\}$, where $X_in = V$(an arbitrary set),
$X_M = \{(p, v) | p \in IPorts, v \in X_p\}$ is the set of input ports and values;
$OPorts = \{\text{"out"}\}$, where $Y_out = V$(an arbitrary set),
$Y_M = \{(p, v) | p \in OPorts, v \in Y_p\}$ is the set of output ports and values;
$S = \{\text{"passive"}, \text{"busy"}\} \times R_0^+ \times V^+$
$\delta_{\text{ext}}(phase, \sigma, q, e, ((\text{"in"}, x_1), (\text{"in"}, x_2), ..., (\text{"in"}, x_n))) =$

$$= \begin{cases} (\text{"busy"}, processing\_time, x_1, x_2, ..., x_n) & \text{if } phase = \text{"passive"} \\ (phase, \sigma - e, q, x_1, x_2, ..., x_n) & \text{otherwise} \end{cases}$$

$$\delta_{\text{int}}(phase, \sigma, q.v) = (\text{“passive”}, processing\_time, q)$$
$$\delta_{\text{con}}(s, \text{ta}(s), x) = \delta_{\text{ext}}(\delta_{\text{int}}(s), 0, x)$$
$$\lambda(\text{“busy”}, \sigma, q.v) = v$$
$$\text{ta}(phase, \sigma, q) = \sigma$$

Using its buffer the processor can store jobs that arrive while it is busy. The buffer, also called a queue, is represented by a sequence of jobs, $x_1 ... x_n$ in V^+ (the set of finite sequences of elements of V). In parallel DEVS, the processor can also handle jobs that arrive simultaneously on in its input port. These are placed in its queue and it starts working on the one it has selected to be first. Note that bags, like sets, are not ordered so there is no ordering of the jobs in the input bag. For convenience we have shown the job which is first in the written order in δ_{ext} as the one selected **as the one** to be processed. Fig. 4.11 illustrates the concurrent arrival of two jobs and the subsequent arrival of a third just as the first job is about to be finished. Note that the confluent function, δ_{con} specifies that the internal transition function is to be applied first. Thus, the job in process completes and exists. Then the external function adds the third job to the end of the queue and starts working on the second job. Classic DEVS has a hard time doing all this as easily!

Exercise. Write a parallel DEVS for the storage model with separate ports for store and query inputs. Also resolve the collision between external and internal transitions when a store input occurs.

Exercise. Write a parallel DEVS for a generator that has ports for starting and stopping its generation. Explore various possibilities for using the elapsed time when an stop input occurs. How does each of these choice impact the behavior of the start input.

4.3.1 PARALLEL DEVS COUPLED MODELS

Parallel DEVS coupled models are specified in the same way as in Classic DEVS except that the Select function is omitted. While this is an innocent looking change, its semantics are much different – they

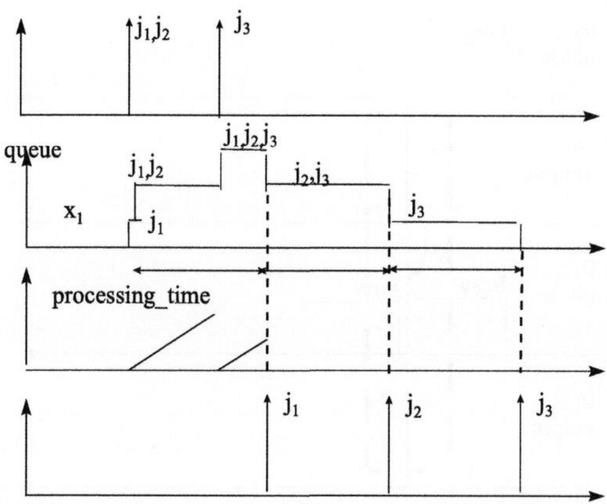

FIGURE 4.11

Parallel DEVS Process with Buffer.

differ significantly in how imminent components are handled. In Parallel DEVS there is no serialization of the imminent computations – all imminent components generate their outputs which are then distributed to their destinations using the coupling information. The detailed simulation algorithms for parallel DEVS will be discussed in later chapters. For now, let's revisit the simple pipeline in Fig. 4.8 to see how the interpretation underlying Parallel DEVS differs from that of Classic DEVS.

SIMPLE PIPELINE (PARALLEL DEVS)

Recall that the specification of the pipeline model is the same as in Classic DEVS except for omitting the Select function. In Fig. 4.12, $processor0$ and $processor1$ are imminent. In Classic DEVS, only one would be chosen to execute. In contrast, in Parallel DEVS, they both generate their outputs. $Processor0's$ output goes to $processor1's$ input. Since now $processor1$ has both internal and external events, the confluent transition function is applied. As in Fig. 4.11, the confluent function is defined to first apply the internal transition function and then the external one. As we have seen, this causes processor1 to complete the finished job before accepting the incoming one.

Now consider a pipeline in which a downstream processor's output is fed back to an upstream processor. For example, let processor1's output be fed back to precessor1's input. In Classic DEVS, the Select function must always make the same choice among imminent components. Thus either one or the other processor will lose its incoming job (assuming no buffers). However, in Parallel DEVS, both processors output their jobs and can handle the priorities between arriving jobs and just finished jobs in a manner specified by the confluent transition function.

Exercise. Define the coupled model for a cyclic pipeline and draw a state trajectory similar to that in Fig. 4.12.

Exercise. Write a Parallel DEVS basic model for a Switch that can accept simultaneous input on two ports as shown in Fig. 4.13. Write a Parallel DEVS coupled model as shown in the figure and draw its state trajectories.

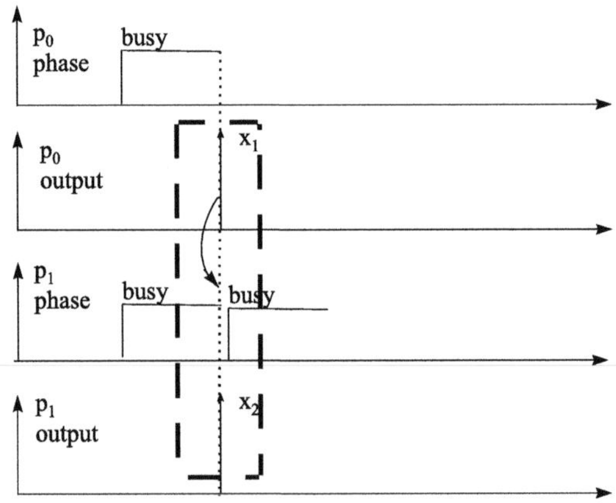

FIGURE 4.12

Illustrating handling of imminent components in Parallel DEVS.

4.4 HIERARCHICAL MODELS

Hierarchical models are coupled models with components that may be atomic or coupled models. As Fig. 4.14 illustrates, we encapsulate a generator and a transducer into a coupled model called an experimental frame, *ef*. The experimental frame can then be coupled to a processor to provide it with a stream of inputs (from the generator) and observe the statistics of the processor's operation (the transducer). The processor can be an atomic model (as is the processor with buffer, for example) or can itself be a coupled model (the coupled model in Fig. 4.13).

4.5 OBJECT-ORIENTED IMPLEMENTATIONS OF DEVS: AN INTRODUCTION

DEVS is most naturally implemented in computational form in an object-oriented framework. The basic class is *entity* from which class devs is derived (see Fig. 4.15). *Devs* in turn is specialized into classes *atomic* and *coupled*. *Atomic* enables models to be expressed directly in the basic DEVS formalism. *Coupled* supports construction of models by coupling together components. Class *content* is derived from *entity* to carry *ports* and their *values*, where the latter can be any instance of *entity* or its derived classes. Typically a modeler defines such derived classes to structure the information being

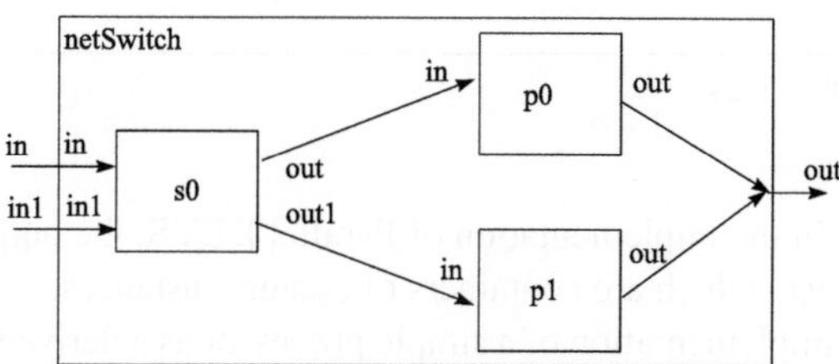

FIGURE 4.13

A switching net in Parallel DEVS.

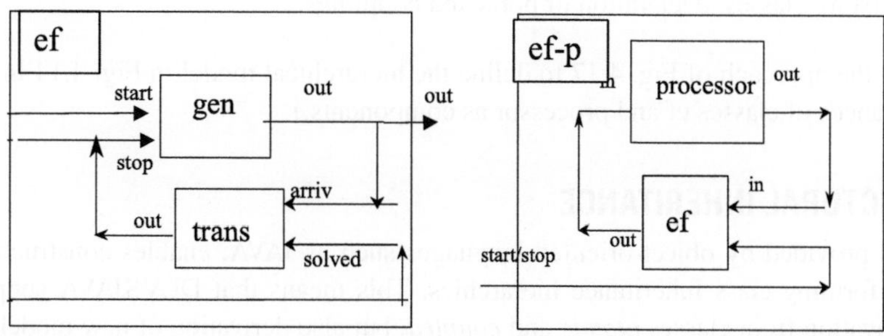

FIGURE 4.14

Hierarchical Model.

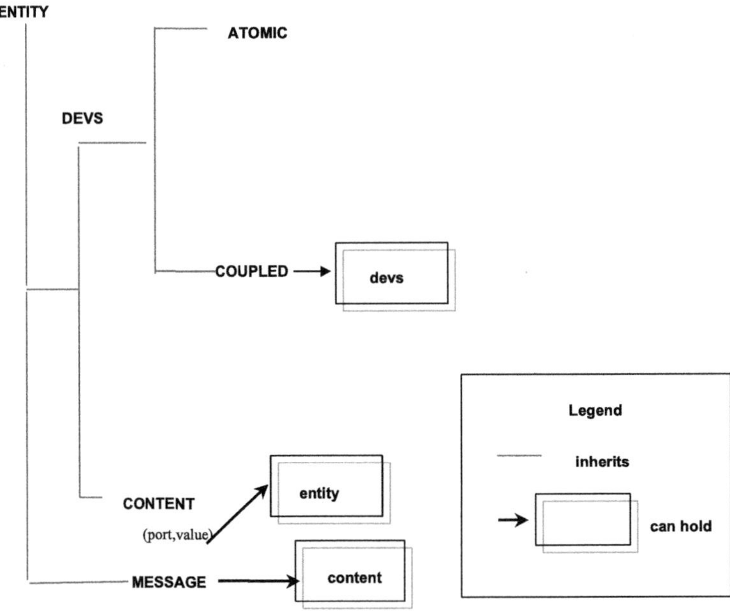

FIGURE 4.15

Basic Classes in Object Oriented DEVS.

exchanged among models. In the implementation of Parallel DEVS, the outputs of component models are instances of class *message*, which are containers of content instances.

Fig. 4.16 illustrates the implementation of a simple processor as a derived class of atomic in DEVS-JAVA, a realization of Parallel DEVS. Notice the one to one correspondence between the formal DEVS elements, e.g., external transition function, and their direct expression as methods in DEVSJAVA.

Fig. 4.17 suggests how an experimental frame class can be defined as a derived class of coupled. Notice the correspondence to the coupled model formalism of DEVS, viz., creation of components as instances of existing classes, declaration of ports and couplings.

Exercise. Use the approach of Fig. 4.17 to define the hierarchical model in Fig. 4.14 in DEVSJAVA. (Hint: add instances of classes ef and processor as components.)

4.5.1 STRUCTURAL INHERITANCE

Inheritance, as provided by object-oriented languages such as JAVA, enables construction of chains of derivations forming class inheritance hierarchies. This means that DEVSJAVA supports not just immediate derivation from classes *atomic* and *coupled*, but also derivation of new model classes from those already defined. This is called structural *inheritance* (Barros, 1996) because structure definitions, such as ports in atomic models and couplings in coupled models can be reused, and even modified, thus exploiting the polymorphism of object-orientation.

Class Processor *derived from* Atomic

external transition function

```
void  deltext(timetype e,message x)
{
  Continue();
  if (phase_is("passive"))
    for (int i=0; i< x.get_length();i++)
      if (message_on_port(x,"in",i)) {
        job = x.get_val_on_port("in",i);
        hold_in("busy",processing_time);
      }
}
```

output function

```
message out( )
{
  if (phase_is("busy"))
  message m = new message();
  entity val = job;
    m.add(make_content("out",val););
  return m;
}
```

internal transition function

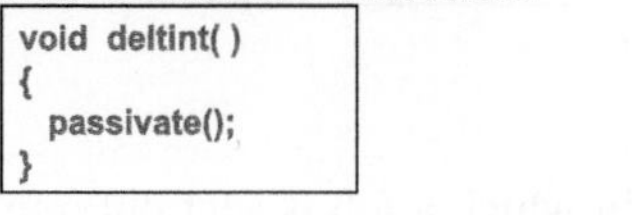

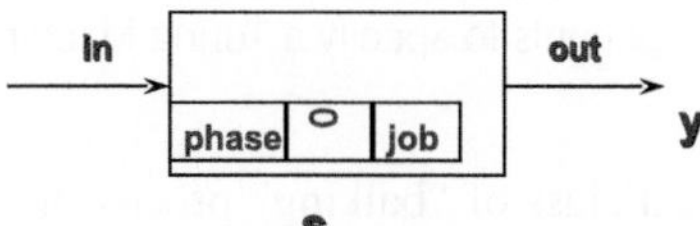

FIGURE 4.16

Processor as an atomic model in DEVSJAVA.

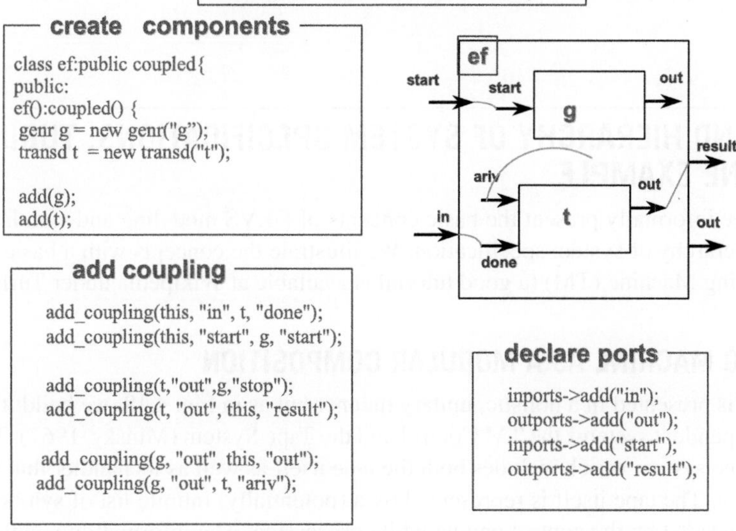

Class EF derived from Coupled
create components
class ef:public coupled{
public:
ef():coupled() {
genr g = new genr("g");
transd t = new transd("t");
add(g);
add(t);
add coupling
add_coupling(this, "in", t, "done");
add_coupling(this, "start", g, "start");
add_coupling(t,"out",g,"stop");
add_coupling(t, "out", this, "result");
add_coupling(g, "out", this, "out");
add_coupling(g, "out", t, "ariv");
ef
start
start
out
g
result
ariv
out
in
t
out
done
declare ports
inports->add("in");
outports->add("out");
inports->add("start");
outports->add("result");

FIGURE 4.17

Experimental Frame as a coupled model in DEVSJAVA.

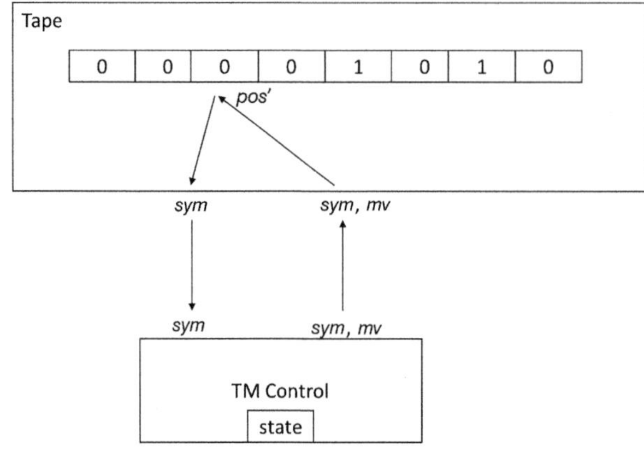

FIGURE 4.18

Composition of components to specify a Turing Machine.

Exercise. Define a class of "balking" processors in which a job is sent out immediately on a "balk" port if it cannot be processed upon arrival. Rewrite the this class as a subclass of class processor and assess the extent to which code is reusable.

Exercise. Define a class pipeSimple for a simple pipeline of processors (Fig. 4.8). Derive a class from pipeSimple in which the components are "balking" processors and whose "balk" ports are coupled to a "balk" external output port. Assess the reusability of code and the ease or difficulty in achieving such reuse.

4.6 DEVS AND HIERARCHY OF SYSTEM SPECIFICATIONS: TURING MACHINE EXAMPLE

In this section we informally present the basic concepts of DEVS modeling and simulation within the context of the hierarchy of system specification. We illustrate the concepts with a basic computational concept, the Turing Machine (TM) (a good tutorial is available at Wikipedia under Turing Machine).

4.6.1 TURING MACHINE AS A MODULAR COMPOSITION

Usually the TM is presented in a holistic, unitary manner but as in Fig. 4.18, we build it here from two stand-alone independent systems the TM Control and the Tape System (Minsky, 1967). The tape model represents the tape system which includes both the tape itself as well as its handler that interfaces with the control engine. The tape itself is represented by a (potentially) infinite list of symbols that records the information needed by the control engine while processing. The information is only accessed one symbol at a time. This is managed by the head at its current position which is part of the state. An external event inputs a pair consisting of a symbol to write and a move instruction. The logic writes the symbol in the current position and moves the head to the position specified in the input. It then reads and outputs the symbol at the new position. The control engine gets the new symbol through an external event and looks up its table that takes its current state and the new symbol to generate the output symbol that drives the tape system to its next state.

In the Turing machine example of Fig. 4.18, the tape system and control engine are each *atomic* Classic DEVS. For the tape system, a state is a triple $(tape, pos, mv)$ where the tape is an infinite sequence of zeros and ones (symbols), pos represents the position of the head, and mv is a specification for moving left or right. An external transition accepts a symbol, move pair, writes the symbol in the square of the current head position and stores the move for the subsequent internal transition that executes the specified move. For the control engine, a state is a pair (st, sym) where st is a control state and sym is a stored symbol. An external transition stores the received symbol for subsequent use. An internal transition applies the TM transition table (st, sym) to the pair and transitions to the specified control state.

4.6.2 TAPE SYSTEM

For the tape systems, a state is a triple $(tape, pos, mv)$ where $tape : I \to \{0.1\}$, $pos \in I$, $mv? \in \{-1, 1\}$ and I is the integers; in other words the tape an infinite sequence of bits (restricting the symbols to zero and one), represents the position of the head, and is a specification for moving left or right. An internal transition moves the head as specified; an external transition accepts a symbol, move pair, stores the symbol in the current position of the head and stores the move for subsequent execution. The slot for storing the move also can be null which indicates that a move has taken place.

$$M = \langle X, S, Y, \delta_{\text{int}}, \delta_{\text{ext}}, \lambda_{\lambda}, \text{ta} \rangle$$

where

$$\begin{aligned}
&X = \{0, 1\} \times \{-1, 0, 1\},\\
&S = \{(tape, pos, mv) : tape : I \to \{0.1\}\}, pos \in I, mv \in \{-1, 1\}\},\\
&Y = \{0, 1\}\\
&\delta_{\text{int}}(tape, pos, mv) = (tape, move(pos, mv), null)\\
&\delta_{\text{ext}}((tape, pos, null), e, (sym, mv)) = (store(tape, pos, sym), pos, mv)\\
&\text{ta}(tape, pos, mv) = 1\\
&\text{ta}(tape, pos, null) = \infty\\
&\lambda(tape, pos, mv) = getSymbol(tape, pos, mv)
\end{aligned}$$

where

$$\begin{aligned}
&move(pos, mv) = pos + mv\\
&store(tape, pos, sym) = tape\\
&\text{where } tape' \text{ is equal to tape everywhere except at } pos\\
&(tape'(pos) = sym, \text{ otherwise } tape'(i) = tape(i))\\
&getSymbol(tape, pos, mv) = tape'(pos)
\end{aligned}$$

4.6.3 TM CONTROL

For the control system, a state is a pair (st, sym) where st is a control state and sym is a stored symbol. An internal transition applies the TM transition table to the (st, sym) pair and transitions to the specified control state. An external transition stores the received symbol for subsequent use. In the following we assume the states are a subset of the alphabet and there are two input symbols 0 and 1.

$$M = \langle X, S, Y, \delta_{\text{int}}, \delta_{\text{ext}}, \lambda, \text{ta} \rangle$$

where

$$\begin{aligned}
&X = \{0, 1\} \\
&S = 2^{\{A,B,C,D,\ldots\}} \times \{0, 1\} \\
&Y = \{0, 1\} \times \{1, -1\} \\
&\delta_{\text{int}}(st, sym) = (TMState(st, sym), null) \\
&\delta_{\text{ext}}((st, null), e, sym) = (st, sym) \\
&\text{ta}(st, sym) = \begin{cases} 1 & \text{if } st \text{ is not a halt state} \\ \infty & \text{if } sym = null \text{ or } st \text{ is a halt state} \end{cases} \\
&\lambda(st, sym) = TMOutput(st, sym)
\end{aligned}$$

where

$$\begin{aligned}
&TMState(st, sym) = st' \\
&TMOutput(st, sym) = (sym', mv') \\
&\text{where } st', sym', mv' \text{ are the next state, symbol} \\
&\text{and move found in the lookup table under the current state and symbol}
\end{aligned}$$

TM EXAMPLE

A specific example of a Turing Machine atomic model here is a 3 state, 2 symbol TM.

$$\begin{aligned}
&S = \{A, B, C, H\} \times \{0, 1\}, \\
&TMState : \{A, B, C\} \times \{0, 1\} \rightarrow \{A, B, C\} \\
&TMOutput : \{A, B, C\} \times \{0, 1\} \rightarrow \{0, 1\} \times \{-1, 1\}
\end{aligned}$$

For the entry under state A and symbol 0, we might have:

$$TMState(A, 0) = A$$

Interpreted as: in the state A reading 0 on the tape then transition to A

$$TMOutput(A, 0) = (1, -1)$$

Interpreted as: In the state A reading 0 on the tape then output $sym = 1$ and $mv = -1$

$$TMState(A, 1) = H$$

Interpreted as: In the state A reading 1 on the tape then halt.

Recall that the components in a composition are specified in the DEVS formalism either at the atomic level or recursively, at the coupled model level which adds coupling information. Fig. 4.19 illustrates a composition of the components of control and tape to specify a Turing Machine.

$$N = \langle D, \{M_d\}, \{I_d\}, \{Z_i, d\}, \text{Selectfn}\rangle$$

with

$$D = \{TMControl, Tape\}$$

Referencing the atomic models noted above.

Each component influences the other

$$I_{Tape} = \{TMControl\}$$
$$I_{TMControl} = \{Tape\}$$

The outputs are directly mapped in to the inputs

$$Z_{TMControl,Tape} : Y_{TMControl} \rightarrow X_{Tape}$$

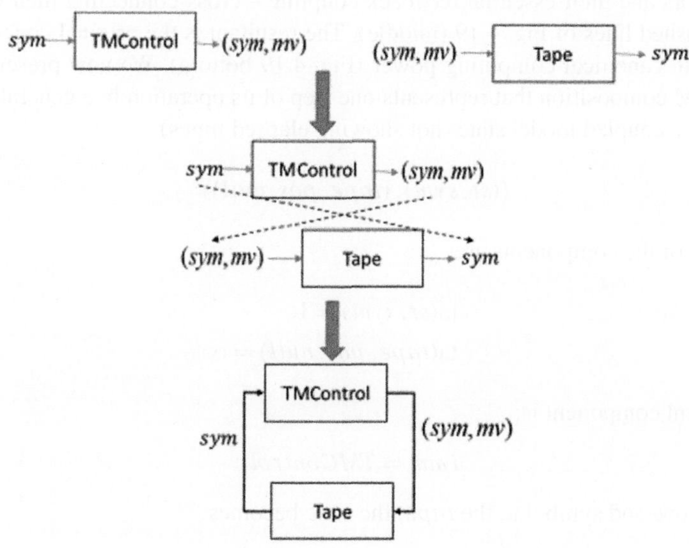

FIGURE 4.19

Composition of components to specify a Turing Machine.

$$\text{with: } Z_{TMControl,Tape}(sym, mv) = (sym, mv)$$
$$Z_{Tape,TMControl} : YTape \to XTMControl$$
$$\text{with: } Z_{Tape,TMControl}(sym) = sym$$

Choose Tape if both imminent

$$\text{Selectfn}(\{TMControl, Tape\}) \to Tape$$

4.6.4 SIMULATION OF THE TM COUPLED MODEL

Foo and Zeigler (1985) argued that the re-composition of the two parts was an easily understood example of emergence wherein each standalone system has very limited power but their composition has universal computation capabilities. Examining this in more depth, the Tape system shown in Fig. 4.19 (top right) is the dumber of the two, serving a memory with a slave mentality, it gets a symbol (sym) and a move (mv) instruction as input, writes the symbol to the tape square under the head, moves the head according to the instruction, and outputs the symbol found at the new head location. The power of the Tape system derives from its physicality – its ability to store and retrieve a potentially infinite amount of data – but this only can be exploited by a device that can properly interface with it. The TM Control (Fig. 4.19 top left) by contrast has only finite memory but its capacity to make decisions (i.e., use its transition table to jump to a new state and produce state-dependent output) makes it the smarter executive. The composition of the two exhibits "weak emergence" in that the resultant system behavior is of a higher order of complexity than those of the components (logically undecidable versus finitely decidable), the behavior that results can be shown explicitly to be a direct consequence of the component behaviors and their essential feedback coupling – cross-connecting their outputs to inputs as shown by the dashed lines of Fig. 4.19 (middle). The resultant is the original concept of Turing, an integral system with canonical computing power (Fig. 4.19 bottom). We now present a sample state trajectory of the TM composition that represents one step of its operation in a general way.

Starting from the coupled model state (not showing elapsed times)

$$((st, sym), (tape, pos, null))$$

The time advances of the components are:

$$\text{ta}(st, sym) = 1,$$
$$\text{ta}(tape, pos, null) = \infty$$

So that the imminent component is:

$$Imm = TMControl$$

After its sends a move and symbol to the *tape*, the state becomes:

$$((st', null)(tape', pos, mv))$$

where st' and $tape'$ are the new control and tape states.

Table 4.2 An Example Turing Control Table

State	Symbol	Next State A, B, C, H	Move {1, −1}	Print Symbol {0, 1}
A	0	B	1	1
A	1	H	x	x
B	0	C	1	1
B	1	x	x	x
C	0	x	x	x
C	1	A	1	0

The new time advances of the components are:

$$\text{ta}(st', null) = \infty,$$
$$\text{ta}(tape', pos, mv') = 1$$

So that

$$Imm = Tape$$

After *tape* sends a symbol to the control, the state becomes:

$$((st', sym'')(tape', pos', null))$$

Now we start again from new control and tape states.

4.6.5 EXAMPLE OF SIMULATION RUN

An example of a TM Control table is shown in Table 4.2. The "x" entries indicate choices not specified for this example.

The processing starts in Fig. 4.20 with the tape shown. It has all zeroes and a one in the 4th square on the right. The head is in state *A*. Successive coupled model states called tape configurations are shown which reflect the coupled model simulation discussed above. For example, in the first step the head reads a zero and the table dictates transitioning to state *B*, printing a one, and moving right. The processing continues until the state *A* reading a one is reached which transitions to the halt state and processing stops.

Exercises.

1. Based on the description given of the simulation of the TM write the resultant of the coupled model specification.
2. Our focus in this section has been on modularity and levels of specification. The usual TM is framed as an automaton in which both components operate together in lock step. The modular approach we have taken formulates a coupled model with modular standalone tape and control components. In this formulation each component can have its own timing expressed through their time advance functions. To understand this difference, redefine the time advances to have different values, then re-do the simulation take account of the new definition (refer to the DEVS simulation protocol in Chapter 8)

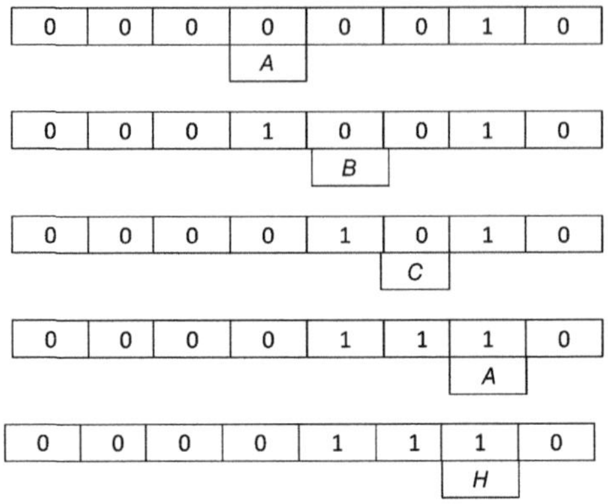

0	0	0	0	0	0	1	0
			A				

0	0	0	1	0	0	1	0
				B			

0	0	0	0	1	0	1	0
					C		

0	0	0	0	1	1	1	0
						A	

0	0	0	0	1	1	1	0
						H	

FIGURE 4.20

Successive Tape Configurations (coupled model states).

4.6.6 TURING MACHINE EXAMPLE OF HIERARCHY OF SYSTEM SPECIFICATIONS

Soon we will discuss some interesting results on stochastic simulation of Turing Machine computations. Before doing so, we discuss the levels of structure and behavior of such a system as a means to define its computations. Table 4.3 reviews the hierarchy of system specifications from Chapter 2 The column on the right briefly describes the TM system at each level of specification.

Generally, transitioning from higher to lower levels in the specification hierarchy, i.e., from structure to behavior, involves working out explicitly what is implicit in the given description. On the other hand going from lower to higher levels, i.e., from behavior to structure, presents an under-constrained problem, requires adding new structure to what is given, and can be characterized as a search rather than a computation. The systems specification hierarchy provides a basis for understanding the types of problems that are encountered in applying modeling and simulation to systems development, engineering, management, and testing. To fit the I/O Frame level, we extend the Turing machine coupled model so that it accepts an input that forms its tape and generates an output indicating whether the machine has halted:

I/O FRAME AT LEVEL 0

The Input/Output Frame:

$$IO = (TXY),$$

where

$$T = R_0^+, \text{ the time base}$$
$$X = Tapes = \{tape : I \rightarrow \{0.1\}\}, \text{ the input values set}$$
$$Y = \{true, false\} = \text{ the output values set}$$

Table 4.3 System specification hierarchy for Turing Machine

Level	Name	System Specification at this level	Turing Machine Example
0	I/O Frame	Input and output variables together with allowed values.	Input of tape configuration and output of halting result within 1000 time units.
1	I/O Behavior	Collection of input/output pairs constituting the allowed behavior of the system from an external Black Box view.	The union of the I/O functions in level 2.
2	I/O Function	Collection of input/output pairs constituting the allowed behavior partitioned according to initial state of the system.	The set consisting of functions computed as the I/O functions associated with each control state defined for the DEVS at level 3.
3	I/O System Structure	System with state and state transitions to generate the behavior.	DEVS basic model representing the resultant of the coupled model at level 5.
4	Coordinatized I/O System Structure	The system state is coordinatized by the states of the components.	The TM is usually presented in this holistic, unitary, non-modular manner. Computation steps are represented as change in the configuration of control state, head position, and tape symbol sequence.
5	Coupled Systems	System built from component systems with coupling recipe.	Construct the TM as a coupled model from components representing stand-alone modular systems the TM Control and the Tape System.

I/O RELATION OBSERVATION AT LEVEL 1

The Input/Output Relation Observation:

$$IORO = (TX, Y, \Omega, YR),$$

where

T, $X$and Y are the same as for IO frame and

$\Omega = DEVS(Tapes) -$ the set of allowable input segments

$R = \{\omega, \rho | \omega \in \Omega, \rho \in DEVS(\{true, false\}, (\exists st)[\rho = HaltFnst(tape\}]\}$

– the I/O relation, where

$HaltFn_{st}(tape)$ return true if the TM halts when started in control state st working on tape.

I/O FUNCTION OBSERVATION AT LEVEL 2

The Input/Output Function Observation:

$$IOFO = (TX, Y, \Omega, Y, F),$$

where

T, X, Y, Ω are the same as for IO frame and

$F = HaltFn_{st}|st \in A, B, C....$- the I/O function set

DEVS I/O SYSTEM SPECIFICATION AT LEVEL 4

In the non-modular representation the model state (not showing elapsed times)

$$M = \langle X, S, Y, \delta_{\text{int}}, \delta_{\text{ext}}, \lambda, \text{ta}\rangle$$

where

$$X = \{0, 1\}$$
$$S = \{A, B, C, D, ...\} \times \{0, 1\} \times Tapes \times I \times \{1, -1\}$$
$$Y = \{true, false\}$$
$$\delta_{\text{int}}(st, sym, tape, pos, null) = (st', null, tape', pos, mv)$$

where

$st' = TMState(st, sym$

and $tape'$ is the tape that records the new symbol TMOutput (st, sym)

$$\delta_{\text{int}}(st', null, tape', pos, mv) = (st', sym'', tape', pos', null)$$

$$\text{ta}(st, null, tape, pos, mv) = \begin{cases} \infty & \text{if } st = H \\ 1 \text{ otherwise} \end{cases}$$

$$\delta_{\text{ext}}((st, sym, tape, pos, mv), e, tape') = (st, tape'(pos), tape', pos, mv)$$

$$\lambda(st, sym, tape, pos, mv) = \begin{cases} true & \text{if } st = H \\ false \text{ otherwise} \end{cases}$$

The halting function is computed based on the I/O function defined for a DEVS as follows:

$$HaltFnst : Tapes \to \{true, false\}$$
$$HaltFnst(tape) = \beta_{st}(tape)$$

where

Table 4.4 2-symbol, 3-state Turing Machine Template

State	Symbol	Next State {A, B, C, D, ..., H}	Move {1, −1}	Print Symbol {0, 1}
A	0	x	x	x
A	1	x	x	x
B	0	x	x	x
B	1	x	x	x
C	0	x	x	x
C	1	x	x	x

Table 4.5 Empirical Frequency Distribution for halting step vs. Geometric Model

Halts at step N	Frequency of Occurrence	Geometric Model Probability Prediction
1	0.161	0.161
2	0.1	0.09
3	0.042	0.039
4	0.023	0.022
5	0.016	0.015
6	0.005	0.00488
7	0.004	0.00390
11	0.001	0.000990
16	0.001	0.000985
10,000	0.647	

$$\beta_{st}(tape) = \lambda(\delta((st, tape(pos), tape, 0, 1), \phi_{<1000>}))$$

where we allow 1000 transitions for the system to halt.

4.6.7 EMPIRICAL INVESTIGATION OF HALTING

We performed a statistical experiment sampling from 2-symbol, 3-state TMs. Each sampled TM fills in the $x's$ in Table 4.4 with values from the sets shown, where one of the rows is chosen at random and its next state changed to the halt state. The number of TMs in this class is $(2*2*3)^{2*3} = 12^6$. We generated 1000 samples randomly and simulated them until they halted or for a maximum of 1000 time steps.

The results are shown in the first two columns of Table 4.5. They seem to show that TMs break into two classes – those that halt within a small number of steps (e.g. 16) and those that do not halt. Indeed, the data in the table suggest that approximately one-third of the sampled TMs halt, while 2/3 do not. For those that halt, the probability of halting at the first step is greatest and then decreases exponentially. The third row shows the probabilities computed from a geometric success model using probability of success as 0.161. The agreement of the observed and predicted distributions are in close agreement (see Fig. 4.21). This suggests that for the set that will eventually halt, the probability of a TM halting at each step is approx. 1/6, given that it has not halted earlier.

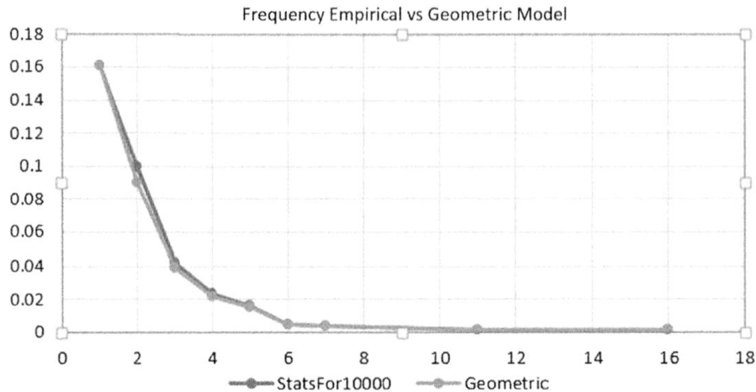

FIGURE 4.21

Chart of Frequency of Halting vs Geometric Model for Halting.

Summarizing, the existence of an iteratively specified system is not algorithmically decidable but there may be useful probabilistic formulations that can be applied to subclasses of such specifications. For example, it seems to be that with high probability the halting of a 2-symbol, 3-state TM can be decided within 16 simulation steps. Therefore, it might be tractable for an assemblage of components to find a way to solve its composition problems in a finite time for certain classes of components couplings.

Exercise.

1. Repeat the empirical simulation study just described for 2-symbol, 4-state Turing machines.

4.7 ARE DEVS STATE SETS ESSENTIALLY DISCRETE?

A possible misconception is that the sequential state set of a DEVS model, has to be discrete. However, the examples in this chapter (e.g., the Processor model discussed earlier), shows that the sequential state set may be defined as the set of reals or other non-discrete set. It is true that well-defined legitimate DEVS models are guaranteed to visit a finite set of sequential states in a finite time interval (Chapter 6). However, different subsets may be visited from different initial states which may cover the model's state set (Vicino et al., 2017). Moreover, coupling with different models may cause visits to different subsets of S. Moreover, typical models have states whose reachable subsets cannot be enumerated except by explicit simulation. In other words, there may be no practical way to define the states to be visited beforehand.

Exercise. Provide an example for which the states reached from a given initial state can't be enumerated except by explicit simulation. Hint: consider DEVS models of Turing Machines and the theory of recursive sets, related to the halting problem.

4.8 SUMMARY

The form of DEVS (discrete event system specification) discussed in this chapter provides a hierarchical, modular approach to constructing discrete event simulation models. In doing so, the DEVS formalism embodies the concepts of systems theory and modeling introduced in Chapter 1. We will see later, that DEVS is important not only for discrete event models, but also because it affords a computational basis for implementing behaviors that are expressed in DESS and DTSS, the other basic systems formalisms. This chapter introduced both Classic DEVS and the revision oriented to parallel and distributed simulation called Parallel DEVS. We also saw the essentials of how DEVS is implemented in an object-oriented framework. In a new section we discussed the Turing Machine in the context of DEVS and the hierarchy of system specifications. The approach led to a simple statistical model of the Turing halting problem.

Having a good understanding of how DEVS is used to model real systems should motivate us to ask deeper questions such as whether DEVS always specifies meaningful systems, how simulation of a DEVS model works, and how other formalisms can be embedded in DEVS. To answer such questions requires that we dig deeper in the underlying system theory and this is the focus of the next chapters of the book.

4.9 SOURCES

DEVS-based, object-oriented, modeling and simulation has been applied in many fields. Example applications are listed below (Barros, 1996; Sato and Praehofer, 1997; Ninios, 1994; Wainer and Giambiasi, 1998; Nidumolu et al., 1998; Vahie, 1996; Moon et al., 1996). Books on DEVS that have appeared since the last edition are listed below (Mittal and Martín, 2013; Nutaro, 2011; Sarjoughian and Cellier, 2013; Wainer, 2017; Wainer and Mosterman, 2016; Zeigler, 2018; Zeigler and Sarjoughian, 2017). Information on DEVSJAVA and DEVS-SUITE, a free version of the software are available on the web site: https://acims.asu.edu/. Numerous other implementations are available from sources like Source-Forge and Git-Hub. A commercial version is available from ms4systems.com.

REFERENCES

Barros, F.J., 1996. Structural inheritance in the delta environment. In: Proceedings of the Sixth Annual Conference on AI, Simulation and Planning in High Autonomy Systems. La Jolla (CA), pp. 141–147.

Foo, N., Zeigler, B.P., 1985. Complexity and emergence. International Journal of General Systems 10 (2–3), 163–168.

Minsky, M., 1967. Computation: Finite and Infinite Machines. Prentice Hall, NJ.

Mittal, S., Martín, J.L.R., 2013. Netcentric System of Systems Engineering with DEVS Unified Process. CRC Press.

Moon, Y., et al., 1996. DEVS approximation of infiltration using genetic algorithm optimization of a fuzzy system. Journal of Mathematical and Computer Modelling 23, 215–228.

Nidumolu, R., et al., 1998. Object-oriented business process modeling and simulation: a discrete-event system specification (DEVS) framework. Simulation Practice and Theory.

Ninios, P., 1994. An Object Oriented DEVS Framework for Strategic Modelling and Industry Simulation. London Business School, London.

Nutaro, J.J., 2011. Building Software for Simulation: Theory and Algorithms, with Applications in C++. John Wiley & Sons.

Sarjoughian, H.S., Cellier, F.E., 2013. Discrete Event Modeling and Simulation Technologies: A Tapestry of Systems and AI-Based Theories and Methodologies. Springer Science & Business Media.

Sato, R., Praehofer, H., 1997. A discrete event system model of business systems. IEEE Transactions on Systems, Man and Cybernetics 27 (1), 1–22.

Vahie, S., 1996. Dynamic neuronal ensembles: issues in representing structure change in object-oriented, biologically-based brain models. In: Intl. Conf. on Semiotics and Intelligent Systems. Washington, DC.

Vicino, D., Wainer, G., Dalle, O., 2017. An abstract discrete-event simulator considering input with uncertainty. In: Proceedings of the Symposium on Theory of Modeling & Simulation. Society for Computer Simulation International, p. 17.

Wainer, G.A., 2017. Discrete-Event Modeling and Simulation: A Practitioner's Approach. CRC Press.

Wainer, G., Giambiasi, N., 1998. Cell-DEVS with transport and inertial delay. In: SCS European MultiConference. Passau, Germany.

Wainer, G.A., Mosterman, P.J., 2016. Discrete-Event Modeling and Simulation: Theory and Applications. CRC Press.

Zeigler, B., 2018. Simulation-based evaluation of morphisms for model library organization. In: Zhang, L., et al. (Eds.), Model Engineering for Simulation. Elsevier.

Zeigler, B.P., Sarjoughian, H.S., 2017. Guide to Modeling and Simulation of Systems of Systems, 2nd edition. Springer.

CHAPTER

HIERARCHY OF SYSTEM SPECIFICATIONS

5

CONTENTS

In this chapter we provide basic system theoretical concepts which will serve as a rigorous foundation for the simulation modeling formalisms in subsequent chapters. In particular, we will formulate the hierarchy of system specifications, first introduced in Chapter 1, using mathematical set theory. Please refer to any book on discrete mathematics for the necessary set theory background. Also, don't hesitate to refer back to corresponding parts of Chapter 1 to augment the presentation here.

5.1 TIME BASE

Fundamental to the notion of a dynamic system is the passage of time. Time is conceived as flowing along independently, and all dynamic changes are ordered by this flow. A *time base* is defined to be a

Theory of Modeling and Simulation. https://doi.org/10.1016/B978-0-12-813370-5.00013-4

structure

$$time = (T, <)$$

where T is a set and $<$ is an *ordering relation on elements of* T with $<$ is *transitive*, *irreflexive* and *antisymmetric*.

An ordering is called *linear* or *total* if for every pair (t, t') either $t < t', t' < t$ or $t = t'$. Otherwise an ordering is called *partial*. We usually assume the total ordering, but sometimes a partial ordering offers a better representation of system knowledge, especially when uncertainty or multiplicity of trajectories is involved, as we shall see in Chapter 8 (Extended Formalisms).

The ordering relation enables us to use terms like *past*, *future* and *present*. If t is interpreted to be the present time, then the set $T_{t)} = \{\tau | \tau \in T, \tau < t\}$ denotes the past and the set $T_{(t} = \{\tau | \tau \in T, t < \tau\}$ *denotes* the future. The past and the future are disjoint as is required in a world where each moment counts and cannot be recaptured in the future once it is past. The set $Tt] = \{\tau | \tau \in T, \tau \leq t\}$ denotes the past including the present time and the set $T_{[t} = \{\tau | \tau \in T, t \leq \tau\}$ denotes the future including the present time. It is often not critical whether we deal with closed or open intervals. In this case, $T_{t>}$ either means $Tt)$ or $T_{t]}$. Similarly, $T_{<t}$ denotes $T_{(t}$ or $T_{[t}$. In like manner, the set $T_{[t_1,t_2)} = \{\tau | \tau \in T, t_1 \leq \tau < t_2\}$ denotes the *time interval* $[t_1, t_2)$ from starting time t_1 (included in the interval) and final time t_2 (excluded). Analogously, we define intervals $T_{(t_1,t_2)}$ or $(t_1, t_2) = \{\tau | \tau \in T, t_1 < \tau < t_2\}$, $T[t_1, t_2]$ or $[t_1, t_2] = \{\tau | \tau \in T, t_1 \leq \tau \leq t_2\}$, $T_{(t_1,t_2]}$ or $(t_1, t_2] = \{\tau | \tau \in T, t_1 < \tau \leq t_2\}$. $T \langle t_1, t_2 \rangle$ or $\langle t_1, t_2 \rangle$ refers to one of the above intervals.

For the interval $\langle t_1, t_2 \rangle$, t_1 is called the *initial* or *beginning* time and t_2 is called *final* or *ending* time. Note that the special interval $\langle t_1, t_1 \rangle$ refers either to the singleton t_1 or to the empty interval, depending on the type of interval we use.

Additional to the properties above, a time base may have additional structure. The time base may enjoy the *Abelian group property*, i.e., there is a binary operation + such that $(T, +)$ is an Abelian group with the identity element 0 and for every element t the inverse element $-t$. The addition operation $+$ is said to be *order preserving* if $t_1 < t_2 \Rightarrow t_1 + t_3 < t_2 + t_3$.

A time base may have a minimal element t_0, a lower bound, then all elements $t \in T$ are greater than t_0. If there is no lower bound, then $-\infty$ denotes the fictitious lower bound. If a time base has no upper bound, the fictitious upper bound is denoted by ∞. Note, that $-\infty$ and ∞ are not included in the time base. The infinity symbol satisfies $\infty + t = \infty$ for any $t \in T$.

Examples of useful time bases are the real number $\mathbb{R}$, the integers $\mathfrak{J}$, and all sets $c * \mathfrak{J}$, $c \in \mathbb{R}$ a constant, isomorphic to $\mathfrak{J}$ They have the Abelian group property and the $+$ operation is order preserving. Time base $\mathbb{R}$ represents the *continuous time base* and all time bases isomorphic to the integers $\mathfrak{J}$ are denoted to be discrete (Fig. 5.1).

5.2 SEGMENTS AND TRAJECTORIES

Having a time base we are in the position to describe how behavior occurs over time. Let A denote a set, for example an input, state, or output set of a model and let T be a time base as defined above. A function of the form

$$f : T \rightarrow A$$

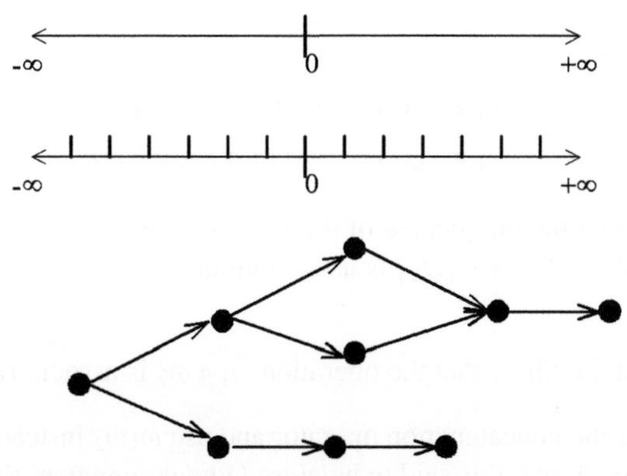

FIGURE 5.1

A continuous, a discrete and a partial ordered time base.

is called a *time function* (also called a *trajectory* or *signal*). The value of f at time t is given by $f(t)$. Restricting f to a subset T' of T is denoted by:

$$f|T' : T' \rightarrow A,$$

where

$$f|T'(t) := f(t) \text{ for all } t \in T'.$$

A restriction of a time function is also a time function.

Of special importance are restrictions to subsets $T_{t>}$, $T_{<t}$ and $T_{\langle t_1,t_2\rangle}$. With t denoting the present time, $f|Tt>$, also written $f_{t>}$, is called the past of f. Similarly, $f|T_{<t}$, also written $f_{<t}$, is called the future of f. A restriction of f to a time interval $\langle t_1, t_2\rangle$ is called a *segment* and we usually write

$$\omega : \langle t_1, t_2\rangle \rightarrow A \text{ or } \omega_{\langle t_1,t_2\rangle}.$$

The set of all segments over A and T is denoted by (A, T).

The trajectory $\omega : \langle t_1, t_2\rangle \rightarrow A$ describes a motion through the set A which begins at t_1 and ends at t_2, and at every $t \in \langle t_1, t_2\rangle$, $\omega(t)$ describes where the motion is at time t. For example, if a is a variable of a program with range A, a segment $\omega_a : \langle t_1, t_2\rangle \rightarrow A$ would describe a sequence of values $\omega_a(t_1)$, $\omega_a(t_1 + 1)$, ..., $\omega_a(t_2)$ assumed by the variable a during the run of the program beginning at time t_1 and ending at time t_2 (here $T = \mathfrak{J}$ and $\langle t_1, t_2\rangle = [t_1, t_2]$).

For segments we define a length operator $l : \Omega \rightarrow T_0^+$ by $l(\omega) = t_2 - t_1$, where $dom(\omega = \langle t_1, t_2\rangle$ with $dom(\omega)$ denoting the domain of ω. Thus $l(\omega)$ is the length of the observation interval on which ω is defined. The segment with empty domain $\langle t_1, t_1\rangle$ is an *empty segment* which usually is denoted by special symbol Φ. Obviously an empty segments has length 0.

A pair of segments $\omega_1 : \langle t_1, t_2\rangle \rightarrow A$ and $\omega_2 : \langle t_3, t_4\rangle \rightarrow A$ are said to be *contiguous* if their domains are contiguous, i.e. $t_2 = t_3$. For contiguous segments ω_1 and ω_2 we define the *concatenation operation* $\bullet$

$$\omega_1 \bullet \omega_2 : \langle t_1, t_4\rangle \rightarrow A$$

with

$$\omega_1 \bullet \omega_2(t) = \omega : (t) \text{ for } t \in \langle t_1, t_2 \rangle$$
$$\omega_1 \bullet \omega_2(t) = \omega_2(t) \text{ for } t \in \langle t_3, t_4 \rangle$$

(See Fig. 5.2.) We note that the uniqueness of the value of $\omega \bullet \omega_2$ at time t_2 depends on the type of interval used. If the interval $[t_1, t_2)$ or $(t_1, t_2]$ is used, a unique value is defined and the concatenation operation is associative.

Exercise 5.1. Let $<>$ mean $[)$. Show that the operation $\omega_1 \bullet \omega_2$ is associative.

Note, that we often omit the concatenation operator and use $\omega_1\omega_2$ instead.

A set Ω of segments over A and T is said to be *closed under concatenation* if for every contiguous pair $\omega_1, \omega_2 \in \Omega$ also $\omega_1 \bullet \omega_2 \in \Omega$.

Analogous to $f_{t>}$ and $f_{<t}$, for segments we define $\omega_{t>}$ or $\omega\langle t_1, t\rangle$ and $\omega_{t>}$ or $\omega\langle t, t_2\rangle$ with $t \in \langle t_1, t_2\rangle$ and call them *left segment* and *right segment* of ω (Fig. 5.3). We always choose the type of left segment and right segment operation in such a way that the resulting segment is of the same type as the original segment, e.g., for a segment ω defined over an interval $[t_1, t_2)$ we use left segments of the form $\omega_{t)})$ and right segments of the form $\omega_{[t}$. Similarly, for a segment defined over an interval $(t_1, t_2]$ we use $\omega_{t]}$ and $\omega_{(t}$.

It is only natural that having split a segment into parts, we should be able to put it back together again. Thus we require that

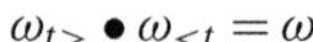

$$\omega_{t>} \bullet \omega_{<t} = \omega$$

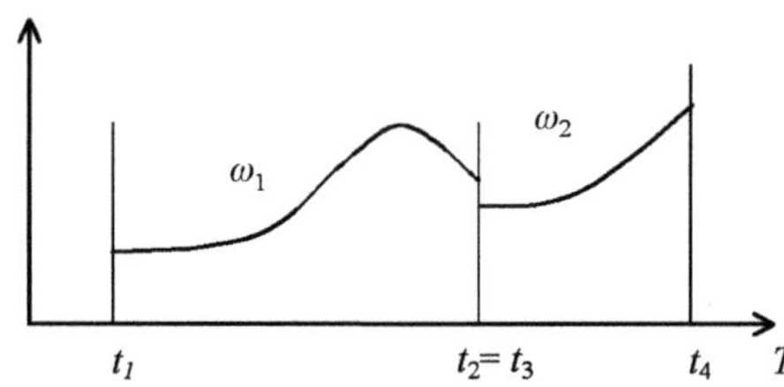

FIGURE 5.2

Concatenation of segments ω_1 and ω_2.

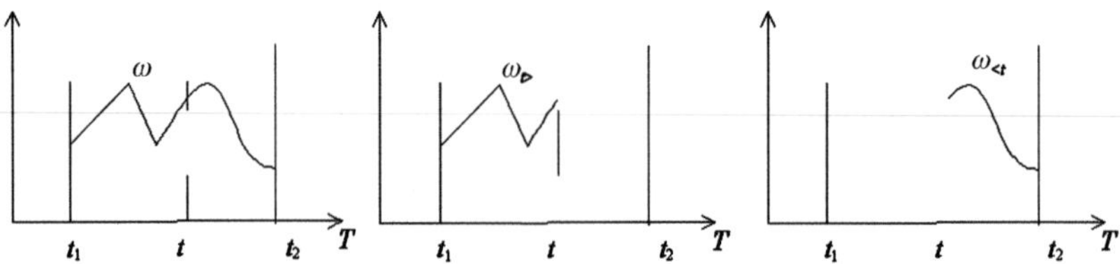

FIGURE 5.3

Segment ω, left segment $\omega_{t>}$, and right segment $\omega_{<t}$.

A set Ω of segments over A and T is said to be *closed under left segmentation* if every left segment is also in the set, i.e., if for every $\omega : \langle t_1, t_2\rangle \to A \in \Omega$ and $t \in \langle t_1, t_2\rangle$ we have $\omega_{t>} \in \Omega$. Analogously, a set Ω of segments over A and T is said to be *closed under right segmentation* if for every $\omega : \langle t_1, t_2\rangle \to A \in \Omega$ and $t \in \langle t_1, t_2\rangle$ we have $\omega_{<t} \in \Omega$.

Depending on the type of the time base and the type of the set A, we will be dealing with the different special types of segments, such as *continuous segments*, *piecewise continuous segments*, *piecewise constant segments*, *event segments* and *sequences*. We now discuss these classes and their applications.

5.2.1 PIECEWISE CONTINUOUS SEGMENTS

A *continuous* segment is a segment which moves smoothly through a real n-dimensional vector space, $\mathbb{R}^n$ for some $n = 1, 2, 3, ...$ In other words, a segment $\omega : \langle t_1, t_2\rangle \to \mathbb{R}^n$ over a continuous time base is a continuous segment if it is continuous at all points $t \in \langle t_1, t_2\rangle$.

A *piecewise continuous segment* is continuous at all points t except a finite number of points $t' \in \langle t_1, t_2\rangle$ as in Fig. 5.4.

Continuous and piecewise continuous segments often occur in differential equation modeling. Of special interest are here the subset of *bounded piecewise continuous* (bpc) segments, that is a piecewise continuous segment with a finite upper bound.

5.2.2 PIECEWISE CONSTANT SEGMENTS

The class of *piecewise constant segments* is a subclass of piecewise continuous segments in which each subsequence is a constant segment. A piecewise constant segment $\omega : \langle t_0, t_n\rangle \to A$ can be characterized as follows: there exist a finite set of time points $t_1, t_2, t_3, ..., t_{n-1} \in \langle t_0, t_n\rangle$ and values $c_1, c_2, ..., c_n \in A$, such that $\omega = \omega\langle t_0, t_1\rangle \bullet \omega\langle t_1, t_2\rangle \bullet ... \bullet \omega\langle t_{n-1}, t_n\rangle$ and $\omega\langle t_{i-1}, t_i\rangle(t) = c_i$, for all $t \in < t_{i-1}, t_i >$ (see Fig. 5.5).

Piecewise constant segments occur in discrete event models as state trajectories.

5.2.3 EVENT SEGMENTS

An event segment represents a sequence of events ordered in time. Formally, let $\omega : \langle t_0, t_n\rangle \to A \cup \{\emptyset\}$ be a segment over a continuous time base and an arbitrary set $A \cup \{\emptyset\}$. Here $\emptyset$ denotes the non-event and is an element not in A. Then ω is an event segment if there exists a finite set of times points

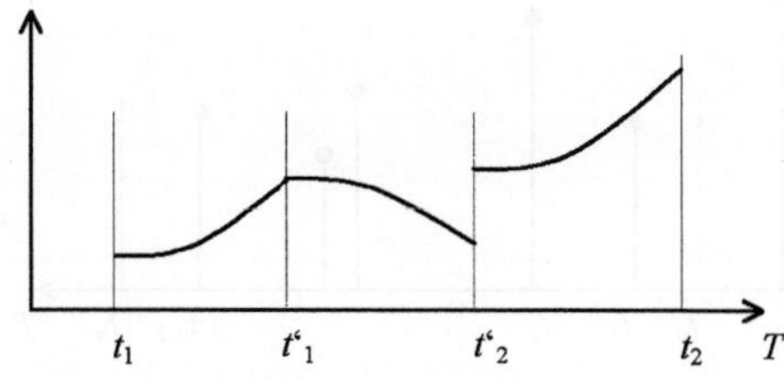

FIGURE 5.4

A piecewise continuous segment.

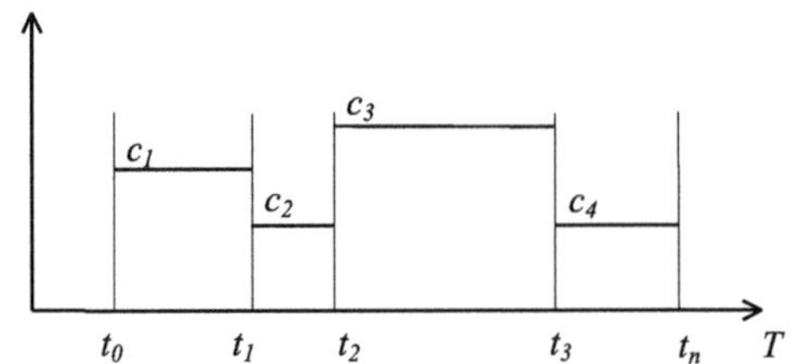

FIGURE 5.5

A piecewise constant segment.

$t_1, t_2, t_3, ..., t_{n-1} \in \langle t_0, t_n \rangle$ such that $\omega(t_i) = a_i \in A$ for $i = 1, ..., n-1$, and $\omega(t) = \emptyset$ for all other $t \in \langle t_0, t_n \rangle$. In an event segment, most of the domain is mapped to the non-event, with a finite number of events (possibly none) occurring otherwise (Fig. 5.6).

An event segment $\omega :< t_1, t_2 > \to A \cup \{\emptyset\}$ which does not contain any events is denoted by $\emptyset_{<t_1,t_2>}$. Event segments are employed in discrete event models as input and output trajectories.

CORRESPONDENCE BETWEEN PIECEWISE CONSTANT AND EVENT SEGMENTS

An event segment can be transformed into a piecewise constant segment and vice versa. Let $\omega : \langle t_0, t_n \rangle \to AA \cup \{\emptyset\}$ be an event segment with event times $t_1, t_2, t_3, ..., t_{n-1} \in \langle t_0, t_n \rangle$ as above. Then we define the corresponding piecewise constant segment $\omega' : \langle t_0, t_n \rangle \to A$ by $\omega'(t) = \omega(t_x)$ with t_x being the largest time in $\{t_1, t_2, t_3, ..., t_{n-1}\}$ with $t_x \leq t$. If such a number t_x does not exist, then $\omega'(t) = \emptyset$. Let $\omega' : \langle t_0, t_n \rangle \to A$ be a piecewise constant segment with event times $t_1, t_2, t_3, ..., t_{n-1} \in \langle t_0, t_n \rangle$ and constant values $c_1, c_2, ..., c_n \in A$ as defined above. Then we define the corresponding event trajectory $\omega : \langle t_0, t_n \rangle \to A \cup \{\emptyset\}$ by $\omega(t) = \omega'(t) = c_{i+1}$ if $t = t_i \in \{t_0, t_1, t_2, t_3, ..., t_{n-1}\}$ and $\omega(t) = \emptyset$ otherwise.

5.2.4 SEQUENCES

All segments over a discrete time base are called *finite sequences*, or for short just *sequences* (Fig. 5.7). Sequences are employed in discrete time models as input, state and output trajectories.

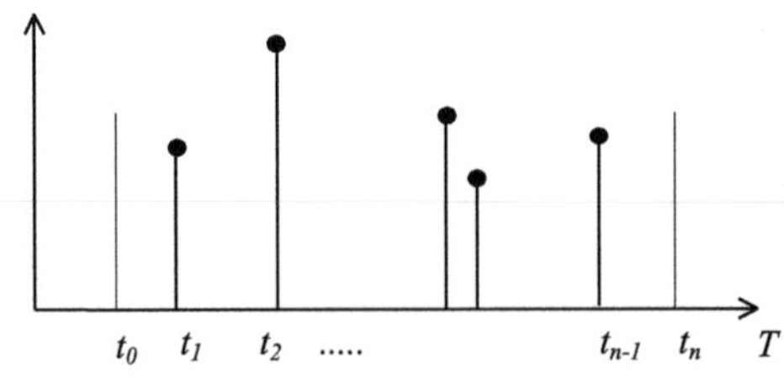

FIGURE 5.6

An event segment.

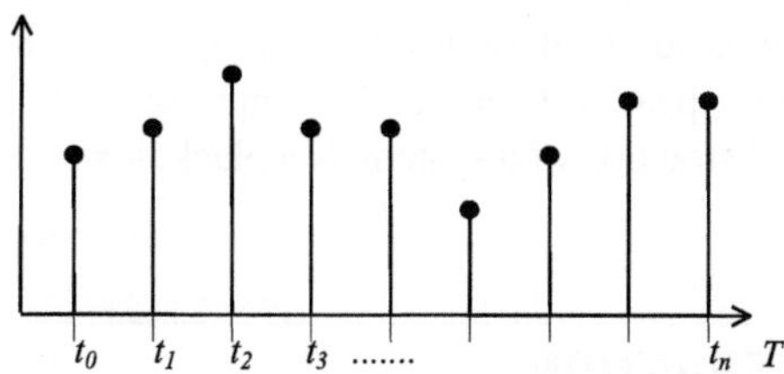

FIGURE 5.7

A sequence.

To test for the type of a segment, we introduce predicates *is-continuous-segment*(ω), *is-event-segment*(ω), and *is-sequence*(ω) which respond with true if ω is a continuous or piecewise continuous segment, an event segment, or a sequence, respectively.

5.3 I/O OBSERVATION FRAME

We start our elaboration of a hierarchy of system specifications at the most basic level with a means to describe the input/output interface of a system. This description of a system as a black box will be later elaborated into the concept of experimental frame which can characterize in more detail the observation and control constraints which can govern experimentation.

Suppose we have isolated a real system and have specified a set of input and output variables. An input variable is one we consider to influence the system but not to be directly influenced by it. An output variable is one which is controlled by the system and through which the system directly influences its environment. If $x_1, ..., x_n$ denote the input variables and $X_1, ..., X_n$ are their range sets, the cross product $X_1 \times X_2 \times ... \times X_n$ represents the set of all possible assignments to input variables. Let us use values coming from some subset X of this cross product to stimulate the system. We call X an *input* set. Even though we might know that X comes from some particular inputs, at the basic level of knowledge, where we are starting, we consider X to be an abstract set, independent of any particular representation as a subset of a crossproduct of input variables (in the words of Section 5.8 it is uncoordinatized or unstructured). The implication here is that in the abstract set theory we are not interested in any system properties that are tied to a particular structuring or representation of input set X. Likewise we consider an abstract set Y as *output* set.

To capture systems at that level, we employ the I/O observation frame (IO) as follows

$$IO = (T, X, Y)$$

where

T is a time base
X is a set – the input values set
Y is a set – the output values set.

Thus, very simply, we have as set X of values that can appear at the input of the system and an abstract set Y of values that can appear at its output. The input set and output set together with a time base constitute our most basic abstraction of a system as a black box.

5.4 I/O RELATION OBSERVATION

The I/O observation frame introduced above gives us a means for specifying the input/output interface of a system and segments introduced in Section 5.3 gives us a means for recording the behavior of variables over time. So equipped, we are able specify a system at the behavioral level. We can now conduct an experiment: we apply to the input a segment from (X, T) and we observe at the output a segment from (Y, T). Thus we have a picture as in Fig. 5.8. A segment $\omega \in (X, T)$ is called an *input segment*, a segment $\rho \in (Y, T)$ is called an *output segment*. We adopt the reasonable convention that the output segment observed in response to an input segment is recorded over the same observation interval (other conventions are possible). Thus if ω and ρ are to be associated, the domain of ω equals the domain of ρ. We call such a pair an *input-output* (I/O) pair.

If we continue experimenting, we will collect a set of such I/O pairs, which we call I/O *relation*. The set of all such pairs for all possible allowable input segments Ω defines all possible input/ output behaviors of a system. To render this information about a system we introduce the I/O *relation observation* as a structure

$$IORO = (T, X, \Omega, Y, R)$$

where

T, X, and Y are the same as for IO frame and
Ω is a set – the set of allowable input segments
R is a relation – the I/O relation

with the constraints that (a) $\Omega \subseteq (X, T)$ and (b) $R \subseteq \Omega \times (Y, T)$ where $(\omega, \rho) \in R \Rightarrow dom(\omega) = dom(\rho)$. Our definition does not require that R be finite because we want to deal with models that attempt to characterize the potential data observable from a real system. Such data are assumed to be infinite even though we can collect only a finite portion of the full amount.

Example. An IORO is often the underlying form used when the behavior of a system is described by means of a differential equation such as

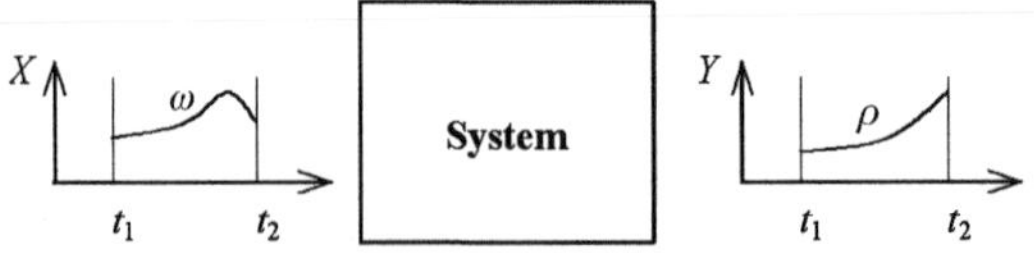

FIGURE 5.8

A system experiment.

$$\frac{d^3y}{dt^3}+\frac{2d^2y}{dt^2}+\frac{8dy}{dt}+7y=x$$

This equation implies that time is continuous and that the input and output sets are also one dimensional continuous variables. Thus in an explicit $IORO=(T,X,\Omega,Y,R)$, we have $T=X=Y=\mathbb{R}$. Since input segment set Ω is not specified by the equation, however, we are at liberty to choose it to meet the modeling situation at hand. One natural choice is the set of bounded piecewise continuous segments (bpc). Now it follows that

$$R=\{(\omega,\rho)|\omega\in bpc, dom(\omega)=dom(\rho)\} \text{ and}$$
$$\frac{d^3\rho(t)}{dt^3}+\frac{2d^2\rho(t)}{dt^2}+\frac{8d}{dt}\rho(t)+7\rho(t)=\omega(t) \text{ for all } t\in dom(\omega)$$

This means that we would pair an output segment ρ with an input segment ω over the same observation interval, if ω and ρ satisfy the differential equation at each point in time. Of course ρ would have to have derivatives up to third order at each point in the interval.

Exercise 5.2. Considering the differential equation above, argue why many output responses, ρ, will be paired with any one input segment, ω.

An IORO, as we have shown, summarizes what can be known about the system as a black box viewed externally. Two problems areas immediately suggest themselves. First we have the problem of *going from structure to behavior*: if we know what lies inside the box, we ought to be able to describe, in one or the other way, the behavior of the box as viewed externally. The second area relates to the reverse situation – *going from behavior to structure*: the problem of trying to infer the internal structure of a black box from external observations.

The first problem is a recurrent theme of this chapter. Each time we introduce a new, more structured level of system description, we show how to convert such a description into a description at a less structured level. Hence by a number of steps one may always convert a higher level structural description into one at the I/O relation observation level. The problem in the behavior-to-structure direction is more difficult and its consideration is deferred until Chapter 15.

5.5 I/O FUNCTION OBSERVATION

The I/O relation observation, just introduced, associates a set of output segments with allowable input segments. The output segments make up the possible system responses for the possible stimuli of the system. Depending on its inner (unknown) state, the system may respond with different output segments to a particular input segment. The system response is ambiguous when observed externally since it depends on the inner state information. In this section we develop concepts to characterize a system's initial state which will allow us to uniquely determine the transformation of input segments to output segments.

Starting with an I/O relation observation, suppose we have a means which allows us to record just as much of a system needed to uniquely associate one output segment with a particular input segment. We call this information the *initial state*. With this information we can partition the relation R into a set of functions $F = \{f_1, f_2, ...f_n...\}$ such that for each input segment ω each function f_i produces one unique output segment $\rho = f_i(\omega)$. In effect then, these functions represent the initial state information. Given an input segment and an initial state in the form of one of the functions, the system response is unambiguously defined.

In this sense we define an I/O function observation ($IOFO$) to be a structure

$$IOFO = (T, X, \Omega, Y, F)$$

where T, X, Y, Ω are the same as for $IORO$ and F is a set of I/O functions with the constraint that

$$f \in F \Rightarrow f \subseteq \Omega \times (Y, T) \text{ is a function,}$$
$$\text{and if } \rho = f(\omega), \text{ then } dom(\rho) = dom(\omega).$$

Note that f may be a partial function, that is, for some segments ω, $f(\omega)$ need not be defined, or equivalently, there is no pair of the form $(\omega, f(\omega))$.

Given an $IOFO(T, X, \Omega, Y, F)$, we associate with it an $IORO(T, X, \Omega, Y, R)$ where

$$R = \bigcup_{f \in F} f$$

In other words, by collecting together all the I/O segment pairs that were formerly partitioned into functional groupings, we obtain an I/O relation. In this process we lose the information about which individual functions were so united. This is what we would expect, since the $IOFO$ represents a higher level of knowledge than the $IORO$.

Example. In the case of behavior described by a differential equation such as

$$\frac{d^3y}{dt^3} + \frac{2d^2y}{dt^2} + \frac{8dy}{dt} + 7y = x$$

it is well known how to characterize an appropriate set of I/O functions. In our example, we have an $IOFO(T, X, \Omega, Y, F)$, where T, X, Ω, Y are as in the $IORO$, and

$$F = \{f_{a,b,c} | a, b, c \in \mathbb{R}\}$$

where $f_{a,b,c} : \Omega \to (Y, T)$ is defined by

$$f_{a,b,c}(\omega) = \rho$$

where

$$\frac{d^3\rho(t)}{dt^3} + \frac{2d^2\rho(t)}{dt^2} + \frac{8d}{dt}\rho(t) + 7\rho(t) = \omega(t) \text{ for } t \in dom(\omega) = dom(\rho) = \langle t_1, t_2 \rangle \text{ and}$$
$$\frac{2d^2\rho(t_1)}{dt^2} = a, \frac{2d\rho(t_1)}{dt} = b, \rho(t_1) = c$$

In general, given a differential equation operator L of order n, we require n parameters to specify an I/O function, corresponding to the specification of the initial values of the derivatives

$$\frac{d^{n-1}y}{dt^{n-1}}, \ldots, \frac{dy}{dt}, y.$$

Exercise 5.3. Given a differential operator L of order n, define the corresponding $IOFO$.

5.6 I/O SYSTEM

At the IOFO level of system description we have knowledge of the initial state of the system in the sense that if it is in a state represented by f then the input segment ω will yield a unique response $f(\omega)$. Note that at this level we are able to observe the initial state of the system (before the experiment), but not its immediate and final states. This section now extends these concepts and develops *the I/O* system as our most basic concept of a "system". This will employ a state set and state transition function for modeling the interior of a system. The state set is fundamental as it has to have the property to summarize the past of the system such that the future is uniquely determined by the current state and the future input. This property of the state set is called the *semigroup* or *composition property*.

The notion of concatenation of segments as defined in Section 5.2 allows us to formalize how we perform successive experiments on a system. We employ sets of segments closed under concatenation as our input segments. Thus if we consider the injection of contiguous segments ω and ω' separately, we can consider the concatenation $\omega \bullet \omega'$ also.

So equipped, we are ready to formulate the input/output system as

$$S = (T, X, \Omega, Y, Q, \Delta, \Lambda)$$

where $T, X, Y, and \Omega$ are the same as above and[1]

Q is a set, *the set of states*

$\Delta : Q \times \Omega \rightarrow Q$ is the global state transition function

$\Lambda : Q \times X \rightarrow Y (or\, \Lambda : Q \rightarrow Y)$ is the output function

subject to the following constraints

1. *Closure property*. Ω is closed under concatenation as well as under left segmentation
2. *Composition (or semigroup) property*. For every pair of contiguous input segments $\omega, \omega' \in \Omega$ $\Delta(q, \omega) = \Delta((\Delta(q, \omega), \omega')$.

The interpretation of the functions Δ and Λ is as illustrated in Fig. 5.9. Suppose that the system is in state $q \in Q$ at time $t_1 < t$ and we inject an input segment $\omega : \langle t_1, t_2 \rangle \rightarrow X$. The system responds to this segment and finishes in state $\Delta(q, \omega)$ at time t_2 Thus $\Delta(q, \omega)$ specifies the *final* state reached

[1]The output function Λ might be a function of state and input or only a function of state.

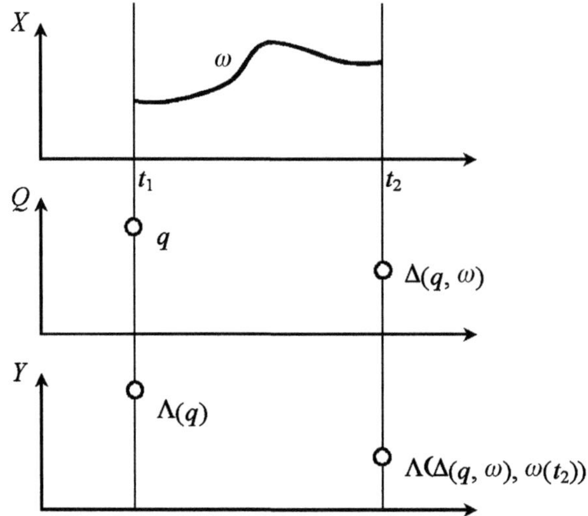

FIGURE 5.9

The basic system functions.

starting from q with input ω. It says nothing about the intermediate states traversed to arrive there. We observe the output of the system at time t_2 which is $\Lambda(\Delta(q,\omega),\omega(t_2)))$. Again $\Lambda(\Delta(q,\omega),\omega(t_2)))$ tells us the final output only and nothing about the interior outputs generated within the observation interval $\langle t_1, t_2\rangle$.

The important composition property of Δ guaranties that the state set Q is in the position to summarize all relevant history of the system. To interpret the composition property refer to Fig. 5.10. Suppose again that the system is in state $q \in Q$ at time t_1 and we inject an input segment $\omega : \langle t_1, t_2\rangle \to X$. The system finishes in state $\Delta(q,\omega)$ at time t_2. Now lets select any time $t \in \langle t_1, t_2\rangle$ to divide the seg-

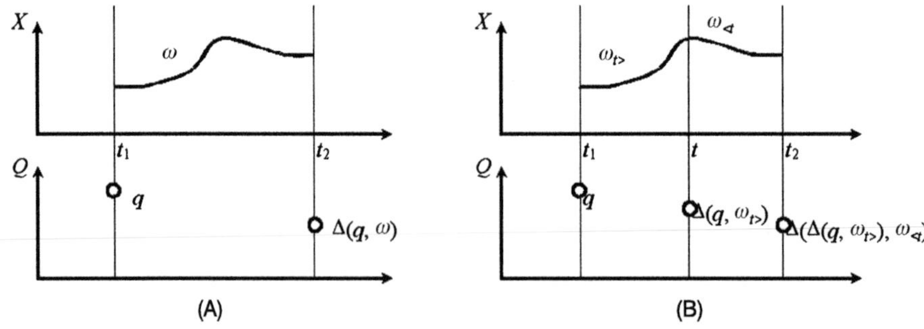

FIGURE 5.10

The composition property. (A) Results from a single composite experiment (B) Results from a two-part experiment.

ment ω into two contiguous segments $\omega_{t>}$ and $\omega_{<t}$. We inject $\omega_{t>} : \langle t_1, t\rangle \to X$ which transfers q into $\Delta(q, \omega_{t>})$ at time t. We immediately apply $\omega_{<t} : \langle t, t_2\rangle \to X$ and reach $\Delta(\Delta(q, \omega_{t>}), \omega_{<t})$ at time t_2. From the composition property it follows that $\Delta(\Delta(q, \omega_{t>}), \omega_{<t}) = \Delta(q, \omega)$. This means that the system can be interrupted at any point in time and its state recorded. Then restarting from that state with the continuation of the input segment will lead to the same final state as if no interruption had occurred.

5.6.1 GOING FROM SYSTEM STRUCTURE TO BEHAVIOR

Having an internal structure of a system, we ought to be able to generate its behavior. In terms of the knowledge levels, given the I/O system structure S, we ought to be able to associate with S an I/O function observation and an I/O relation observation which are complete in the sense of containing all possibly observable input/output pairs.

To do this we have to be able to describe the time course of state and output values in response to an input segment along the entire observation period. The global state transition function, however, only determines the final state after an input segment is applied and it says nothing about the intermediate states. However, since admissible input segments of I/O systems are closed under left segmentation, we are able to define a unique *state* and *output trajectory* given an input segment and an *initial state*.

For every $q \in Q$ and for every $\omega \in \Omega$, $\omega : \langle t_1, t_2\rangle \to X$, we define

$$STRAJ_{q,\omega} : \langle t_1, t_2\rangle \to Q$$

$$\text{with } STRAJ_{q,\omega}(t) = \Delta(q, \omega_{t>}) \text{ for all } t \in \langle t_1, t_2\rangle$$

$$\text{and } OTRAJ_{q,\omega} : \langle t_1, t_2\rangle \to Y$$

$$\text{with } OTRAJ_{q,\omega}(t) = \Lambda(STRAJ_q, \omega(t), \omega(t)) \text{ for all } t \in \langle t_1, t_2\rangle .$$

Fig. 5.11 demonstrates the relation between ω, $STRAJ_{q,\omega}$, and $OTRAJ_{q,\omega}$. Given an input trajectory ω which is left segmentable, we are in the position to compute the state value as it moves from initial state to final state. From the state value and input at every time point, we can directly compute the output value.

Now with each state $q \in Q$ we can associate a function $\beta_q : \Omega \to (Y, T)$, called the I/O *function of state* q where for each $\omega \in \Omega$

$$\beta_q(\omega) = OTRAJ_{q,\omega}.$$

The set of functions $B_S = \{\beta_q | q \in Q\}$ is called the I/O *behavior of* S. Thus with an I/O system $S = (T, X, \Omega, Y, Q, \Delta, \Lambda)$ we associate the I/O function observation $IOFO_s = (T, X, \Omega, Y, B_S)$.

By the union of all functions where $\beta_q \in B_S$ we derive the I/O relation observation $IORO_s = (T, X, \Omega, Y, R_s)$ associated with S

$$R_s = (\omega, \rho) | \omega \in \Omega, \rho = OTRAJ_{q,\omega}, q \in Q.$$

With each state $q \in Q$, we can also associate a function $\beta : \Omega \to Y$, called the *final output* function of state q, where for each $\omega \in \Omega$

$$\beta_q(\omega) = \Lambda(\Delta(q, \omega)).$$

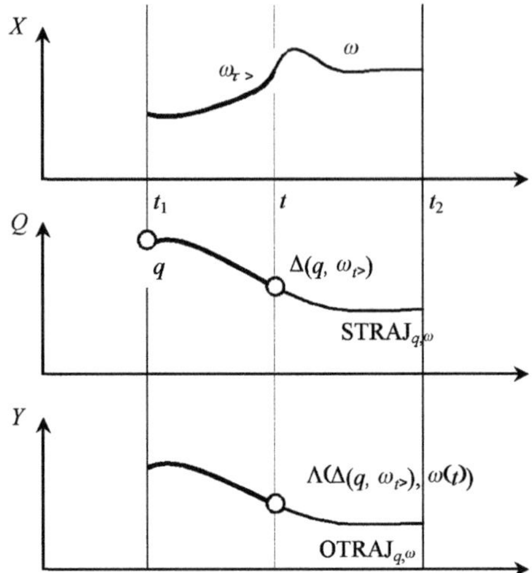

FIGURE 5.11

Generation of state and output trajectories.

The relationship between β_q and $\boldsymbol{\beta}_q$ is given by

$$\boldsymbol{\beta}_q(\omega_{t>}) = \beta_q(\omega)(t)$$

for $t \in dom(\omega)$.

Exercise 5.4. Prove this.

Knowing either one of the functions β_q or $\boldsymbol{\beta}_q$ is tantamount to knowing the other one. Thus the set of functions $B_S = \{\boldsymbol{\beta}_q | q \in Q\}$ carries the same information as the set B_S.

Exercise 5.5. Show how to define β_q given $\boldsymbol{\beta}_q$.

To summarize, given a system description $S = (T, X, \Omega, Y, Q, \Delta, \Lambda)$ we can always define the observable behavior it can produce. This behavior can be observed at two levels. If we have *no access* to the initial state of the system at the beginning of an experiment, we can collect the I/O pairs in R_s. If we *know* the initial state, we can separate the I/O pairs into functional relations, indexed by initial states, thus obtaining the set of functions B_S or the related set B_S.

5.6.2 TIME INVARIANT SYSTEMS

Until now we have not automatically allowed ourselves the freedom of sliding a segment around so that if $\omega : \langle t_1, t_2 \rangle \to X$ is applicable at time t_1, for example, we can also apply it at some other time t_3. To formalize this idea, we consider a class of unary operators on (X, T) called "translation" operators.

For each $\tau \in T$, we define a unary operator on segments

$$TRANS_\tau : (X, T) \to (X, T)$$

where if $\omega : \langle t_1, t_2\rangle \to X$ and $TRANS_\tau(\omega) = \omega$', then $\omega' : \langle t_1 + \tau, t_2 + \tau\rangle \to X$ and $\omega'(t + \tau) = \omega(t)$ for all $t \in \langle t_1, t_2\rangle$.

We say that $TRANS_\tau(\omega)$ is the τ-translate of ω and it has the same shape as ω expect that it has been translated by τ time units, as in Fig. 5.12.

We say that $\Omega \subseteq (X, T)$ is closed under translation if $\omega \in \Omega \Rightarrow$ for every $\tau \in T, TRANS_\tau(\omega) \in \Omega$.

Having the notion of translation of segments, we can now formally define time invariant systems. A System $S = (T, X, \Omega, Y, Q, \Delta, \Lambda)$ is *time invariant* if

A Ω is *closed under translation.*
B Δ is *time independent*: for every $\tau \in T, \omega \in \Omega$, and $q \in Q$
C $\Delta(q, \omega) = \Delta(q, TRANS_\tau(\omega))$

For a time invariant system, if a segment can be applied somewhere in time, it can be applied anywhere in time. Moreover, anywhere the same-shaped segment is applied to the same initial state, the same result will be achieved.

From here on, we restrict our treatment to time invariant systems. Most of our concepts carry through to time varying systems, perhaps with some modification, but the time invariance enables certain simplifications in notation which make exposition more straightforward.

Since for time invariant systems S, Ω is closed under translation, all segments can be represented by those beginning at some particular time – we pick $0 \in T$.

Formally, for $\omega \langle t_1, t_2\rangle \to X$ let the standard translation of ω, $STR(\omega) = TRANS_{-t_1}(\omega)$. Then $STR(\omega)$ is the translate of ω beginning at 0 ($STR(\omega) : \langle 0, t_2 - t_1\rangle \to X$). Let $\Omega_0 = STR(\Omega) = \{STR(\omega)|\omega \in \Omega\}$. Then Ω_0 is the set of all segments beginning at zero.

Exercise 5.6. Show that Ω is the translation closure of Ω_0 (i.e., the least set containing Ω_0 and closed under translation).

Recall the concatenation operator on segments from Section 5.2. With the concatenation operator it is possible to compose two contiguous segments. Here we define a natural extension of the concatenation operator for elements of Ω_0. When two segments beginning at zero are concatenated we translate the second so that it is contiguous to the first and then use the original concatenation operator.

Formally, we define the concatenation operation $\bullet_0 : \Omega_0 \times \Omega \to \Omega$ by

$$\omega \bullet_0 \omega' = \omega_1 \bullet_0 TRANS_{l(\omega)}(\omega')$$

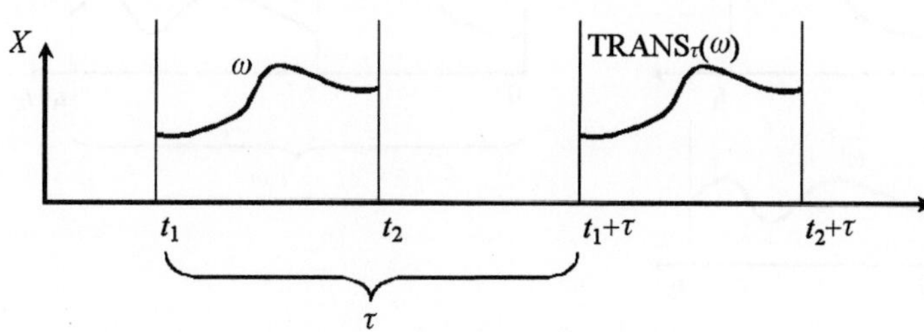

FIGURE 5.12

The translation operator.

for all ω, $\omega' \in \Omega_0$, and $l(\omega)$ is the length of ω (Section 5.2). The $\bullet_0$ is illustrated in Fig. 5.13. Usually, for $\omega \bullet_0 \omega'$ we write $\omega \bullet \omega'$ or just $\omega\omega'$, when the context is clear.

Since Δ is time independent, it can be fully defined simply be specifying its values for all segments in Ω_0 and we can introduce a short-hand form for specifying time invariant systems. A *time invariant system* in short form is a structure

$$S_0 = (T, X, \Omega_0, Y, Q, \Delta_0, \Lambda)$$

where T, X, Y, Q, and Λ are the same as before but Ω_0 is the set of segments beginning at time 0 and $\Delta_0 : Q \times \Omega_0 \to Q$ satisfies the composition property: for all $q \in Q$, ω, $\omega' \in \Omega_0$

$$\Delta_0(q, \omega \bullet \omega') = \Delta_0(\Delta_0(q, \omega), \omega')$$

The expanded form of S_0 is the structure

$$S = (T, X, \Omega, Y, Q, \Delta, \Lambda)$$

where Ω is the translation closure of Ω_0 and $\Delta : Q \times \Omega \to Q$ is defined by

$$\Delta(q, \omega) = \Delta_0(q, STR(\omega)) \text{ for all } q \in Q, \omega \in \Omega.$$

EXAMPLE: LINEAR SYSTEMS

Let $S^{cont}_{A,B,C} = (T, X, \Omega_0, Y, Q, \Delta_0, \Lambda)$, where X, Q, Y are finite-dimensional vector spaces over $\mathbb{R}$; $T = \mathbb{R}$; Ω_0 is the set of bpc segments beginning at 0; $\Delta_0 : Q \times \Omega_0 \to Q$ is given by

$$\Delta_0(q, \omega) = exp(Al(\omega))q + \int_0^{l(\omega)} exp(A(l(\omega) - \tau))B\omega(\tau)d\tau$$

and $\Lambda : Q \to Y$ is given by

$$\Lambda(q) = Cq$$

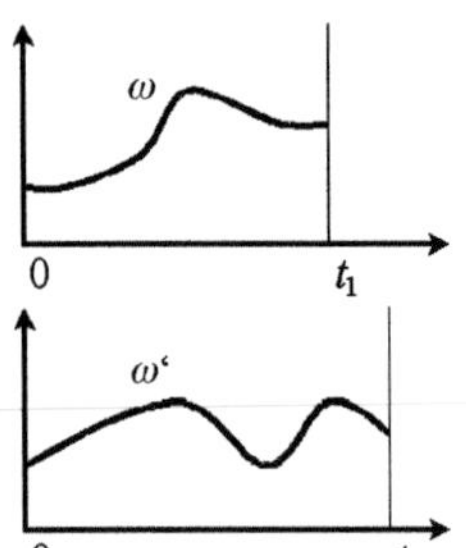

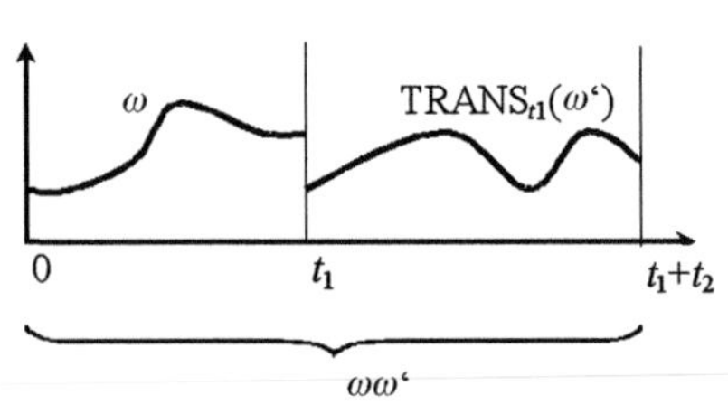

FIGURE 5.13

Composition of segments beginning at zero. The second segment is translated so that it is contiguous to the first.

where

$$A: Q \rightarrow Q$$
$$B: X \rightarrow Q$$
$$C: Q \rightarrow Y$$

are linear transformations. In addition

$$e^A = \sum_{i=0}^{\infty} \frac{A^i}{i!}$$

where

$$A^0 = I, A^{i+1} = A \cdot A^i$$

Let $S^{discr}_{A,B,C,F} = (T, X, \Omega_0, Y, Q, \Delta_0, \Lambda)$, where X, Q, Y are finite-dimensional vector spaces over a field F; $T = \mathfrak{J}$; $\Omega_0 = (X, T)_0$;$\Delta_0 : Q \times \Omega_0 \rightarrow Q$ is given by

$$\Delta_0(q, \omega) = A^{l(\omega)}q + \sum_{i=0} l(\omega) - 1A^{l(\omega)-1-i}B\omega(i)$$

and $\Lambda : Q \rightarrow Y$ is given by

$$\Lambda(q) = Cq$$

Exercise 5.7. Show that $S^{cont}_{A,B,C}$ and $S^{discr}_{A,B,C,F}$ are systems in short form and provide the corresponding expanded versions.

We say that $S^{A,B,C}_{cont}$ is the linear continuous time system specified by $\{A, B, C\}$. Similarly, $S^{discr}_{A,B,C,F}$ is the linear discrete time system specified by $\{A, B, C, F\}$.

Since we are restricting our attention to time invariant systems, we henceforth work only with systems in time-invariant form. Moreover, we drop the subscript zeros in the descriptions.

5.6.3 SPECIAL CASES: INPUT-FREE AND MEMORYLESS SYSTEMS

Defining additional constraints on the elements of a I/O system allows us to derive systems which have special properties. Introducing system formalism in the following chapters we will employ this procedure extensively. In the following sections, however, we discuss special types of systems which are of general interest independent from the particular formalism in hand.

INPUT-FREE SYSTEMS

By omitting the input set, we obtain a subclass of systems that cannot respond to input segments. These *input-free* systems are sometimes called *autonomous*. Such a system has the structure:

$$S = (T, Y, Q, \Delta, \Lambda)$$

with

$\Delta : Q \rightarrow Q$ is the *global state transition function*

$\Lambda : Q \rightarrow Y$ is the *output function*

The behavior, and other properties, that are defined for dynamic systems can also be defined for input-free systems. To do this, we extend the input-free structure to a full system structure by considering the input set to be a single-element set.

Exercise 5.8. Define the trajectories and the behavior on an input-free system by specializing these concepts for a corresponding system with a single-element input set. Show that the latter system satisfies the closure and semigroup properties required to be a system.

Input-free systems generate a unique output segment given a unique initial state. This means, given the initial state the response of the system is uniquely determined.

Exercise 5.9. Prove this.

Input-free systems are often used as generators for input signals of other systems, in particular for the generation of test segments in experimental frames.

MEMORYLESS SYSTEMS

Instead of omitting the input set, we can also omit the state set. Such systems are static rather than dynamic. Their state set degenerates to a single-element, which can be omitted entirely. Hence, we do not have a state transition function and the output function is a function of time and input only. The output is unambiguously determined by the current input. For this reason, they are often called *function specified systems*. We will use function systems together with differential equation specified and discrete time specified systems. Such specifications will be degenerated differential equation or discrete time systems. A *function specified system* is a structure

$$FNSS = (X, Y, \lambda)$$

where

X is the set of inputs

Y is the set of outputs

$\lambda : X \rightarrow Y$ is the output function

The function specified system is extended to a full I/O system specification dependent on the context employed in the following way. The time base T of a function specified system is arbitrary and is taken to be continuous when the function specified system is employed together with a differential equation specified system and discrete when employed together with a discrete time specified system. For the state set Q we take a single element 0. With that, the state transition function degenerates to a constant function always yielding 0, i.e., $\Delta(0, \omega[t_1, t)) = 0 forall \omega \in (X, T), all t \in dom(\omega)$.

5.7 MULTI-COMPONENT SYSTEM SPECIFICATION

Complex systems are often best modeled by collections of interacting components. It is usually much easier to identify simple components, model them independently, and then model the way in which they interact than to try to describe the overall behavior directly. To describe a multi-component system in this manner, one has to define the component systems and specify how these systems work together.

To provide formal means to specify a system as a set of interacting components, we introduce *multi-component systems* in this section and *networks of systems* (also called *coupled systems*) in the next. Whereas in networks of system specification individual component system are coupled by connecting their input and output interfaces in a modular way, components of multi-component systems influence each other directly through their state transition functions.

In multi-component system specification the system is structured into a set of interacting components, each with its own set of states and state transition function. Each component may be influenced by some other components – its *influencers* – and may influence other components – its *influencees* – by its state transition function. Furthermore, each component's state also might be influenced by the input of the system and it may contribute the output of the system. However, in contrast to coupled systems, components at this level are not stand-alone system specifications and do not employ their own input and output interface.

A *multi-component system* is a structure

$$MC = (T, X, \Omega, Y, D, \{M_d\})$$

where T, X, Y, Ω are the same as for I/O system and

D is a set, the *set of component references*

For all $d \in D$ are structures

$$M_d = (Q_d, E_d, I_d, \Delta_d, \Lambda_d)$$

is a component with

Q_d is the set of states of the component d
$I_d \subseteq D$ is the set components influencing d , its influencers
$E_d \subseteq D$ is the set of components influenced by$d$, its influencees
$\Delta d : \times_{i \in I_d} Q_i \times \Omega \to \times_{j \in E_d} Q_j$ is the state transition function of d
$\Lambda d : \times_{i \in I_d} Q_i \times \Omega \to Y$ is the output function of d.

A multi-component system is composed of a set of components D. Each component d has its own set of states Q_d and contributes to the overall state transition and output of the system by its individual state transition function, Δd, and output function, Λd. Each component additionally defines a set of influencers, I_d, the components influencing it, and a set influencees, E_d, which are the components influenced by it. The local state transition function takes the state set of the influencers and maps them into new states for the influencees. The local output functions takes the state sets of the influencers and defines its contribution to the overall output of the system.

A multi-component system embodies a higher level of system knowledge than a structured I/O system in the sense that not only state coordinates are known but also the state set and transition and output functions are decomposed and identified with individual components. Each component contributes to the overall system behavior with its individual transition and output functions. A system specification at the I/O system or structured I/O system level is easily derived by forming the crossproduct of the state sets and state transition and output functions of the individual components. However, to be meaningful, as we will see in the subsequent chapters, special constraints must be obeyed by the set of influencees of the components when considering a particular multi-component formalism. For example, in the discrete time and differential equation cases, a component is allowed to change its own state only so that the set of influencees, E_d, is restricted to contain d as its single element and the output set of the multi-component DTSS is a crossproduct of outputs sets Y_d belonging to components d. With this restrictions, the state set, state transition and output function of the I/O system can easily be defined as

$$Q = \times_{d \in D} q_d,$$

$$\Delta_d(q, \omega) = \Delta d(\times_{i \in I_d} q_i, \omega), \text{ and}$$

$$\Lambda_d(q, \omega) = \Lambda d(\times_{i \in I_d} q_i, \omega)$$

with d defining the part Y_d of the overall output Y which belongs to d

(see Chapter 7 for more details)

These restrictions do not apply in the case in the discrete event case, where a component may change the state of others. In Chapter 7 on coupled system formalisms we will learn to know the requirements for multi-component system specifications in detail and show that a multi-component system in a particular formalism defines a basic system in that formalism.

5.8 NETWORK OF SYSTEM SPECIFICATIONS (COUPLED SYSTEMS)

In the last section we introduced multi-component systems where the components of a system interact directly through their state and state transition functions. In this section we introduce a higher level of specification where component systems are coupled through their output and input interfaces. We call this type of coupling *modular* and the system specification so derived a *coupled system specification* or *network of system specifications.*

In this form of system specification, *the components are themselves systems with their own input and output interfaces*. They are coupled together by instantaneous functions that take the output values of the component's influencers, as well as the external inputs to the overall system, and compute the input values for the components.

Immediately the question arises what conditions on the individual component systems and on their coupling are needed to assure that the resulting assemblage is well defined. In terms of systems theory, we want a network of system specifications again to *specify a system*. However, in this general a form, conditions that would guarantee well-defined resultants are not known. Fortunately, for basic system formalisms, introduced in subsequent chapters, such sufficient conditions are well known. We say that a system formalism has the property that it is *closed under coupling* if the resultant of any network of systems specified in the formalism is itself a system specified in the formalism. Furthermore, we develop constructive methods for deriving a resultant system from a network of system specifications.

Through the closure under coupling property, hierarchical model construction becomes feasible. Since a network of system specifications of particular system type is again a system specification of that type, it can be used as a component in a (bigger) coupled system specification.

In the following we will first introduce a general structure for coupled system specification. Then, we will get familiar with a special form of coupling of structured systems which is most important in simulation practice.

5.8.1 COUPLED SYSTEM SPECIFICATION

A coupled system specification is a structure

$$N = (T, X_N, Y_N, D, \{M_d | d \in D\}, \{I_d | d \in D \cup \{N\}\}, \{Z_d | d \in D \cup \{N\}\})$$

where

X_N is the set of inputs of the network – the external inputs Y_N is the set of outputs of the network – the external outputs D is the set of component references

For all $d \in D$

$$M_d = (T, X_d, Y_d, \Omega, Q, \Delta, \Lambda) \text{ is an I/O system}$$

For all $d \in D \cup \{N\}$

$$I_d \subseteq D \cup \{N\} \text{ is the } \textit{set of influencers} \text{ of } d$$
$$Z_d : \times_{i \in I_d} YX_i \to XY_d \text{ is the interface map for } d$$

with

$$YX_i = \begin{cases} X_i & \text{if } i = N \\ Y_i & \text{if } i \neq N \end{cases}$$

$$XY_d = \begin{cases} Y_d & \text{if } d = N \\ X_d & \text{if } d \neq N \end{cases}$$

The most general multi-component system specification described above consists of the specification of the input/output interface of the network (X_N and Y_N), the specification of the components $d \in D$ which must be dynamic systems and the coupling scheme ($\{I_d\}$ and $\{Z_d\}$). The set of influencers I_d of a component d may also contain the network N. The interface map Z_d specifies how the input values of component d are derived from the outputs or external inputs of its influencers $i \in I_d$. Through the general form of the coupling specification scheme employing interface maps, arbitrary couplings can be realized.

Exercise 5.10. Instead of using the XY notation used above, write out the interface map domains for the four special cases that arise (where the {source, destination} may be a component or the network).

Fig. 5.14 shows an example of a two-component coupled system. The inputs of A are derived through interface map Z_A which has X_N and Y_A as its domain. The interface map Z_B takes the elements of Y_A. Finally the overall network output is defined by Z_N which takes the output sets Y_A cross Y_B as its domain.

5.8.2 COUPLED SYSTEM SPECIFICATION AT THE STRUCTURED SYSTEM LEVEL

Employing multi-variable sets for the input and output sets in a coupled system specification, allows us to derive a special coupled system formalism where coupling is done by directly connecting output and input ports. This way of coupling is the form, almost exclusively used in modeling practice. We will show how it maps to the general form above.

Input/output and components specification is the same as in the general multi-component specification formalism. However, the coupling now is specified by three parts:

- the *external input coupling* (EIC), couplings of network input ports to input ports of some components
- the *external output coupling* (EOC), couplings of component output ports to output ports of the network
- the *internal coupling* (IC), couplings of component output ports to input ports of components

and each coupling is a pair of pairs specifying the component and the port the coupling is beginning and a pair specifying the component and the port the coupling is ending. Fig. 5.15 gives an example of a coupled system in so-called *block diagram* representation.

A *coupled system specification at the structured system level* is a structure

$$N = (T, X_N, Y_N, D, \{M_d | d \in D\}, EIC, EOC, IC)$$

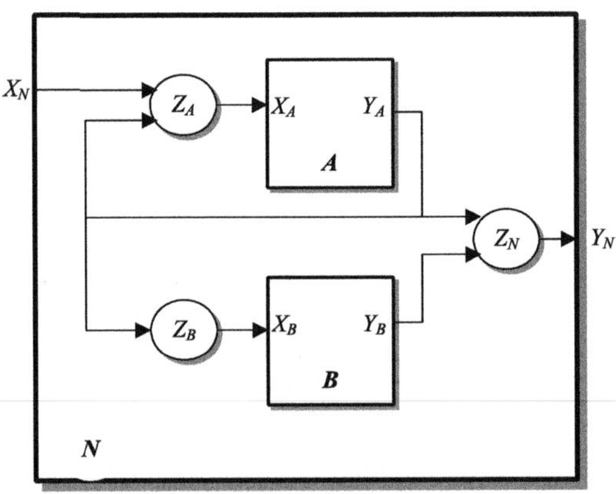

FIGURE 5.14

Example of a coupled system specification.

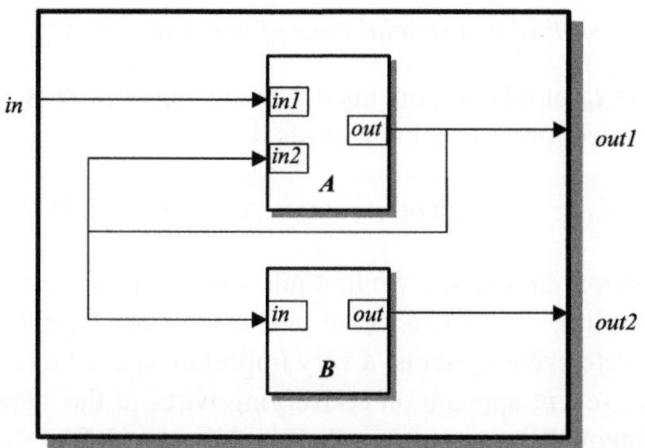

FIGURE 5.15

Block diagram representation of coupled system example.

where

$X_N = (IPorts_N, X_{N1} \times X_{N2} \times ..)$
is the multi-variable set of inputs of the network with variables $IportsN$
$YN = (OPorts_N, Y_{N1} \times Y_{N2} \times ..)$
is the multi-variable set of outputs of the network with variables $OportsN$
D is the set of the components, where $d \in D$ is a system, $M_d = (T, X_d, Y_d, \Omega, Q, \Delta, \Lambda)$
specified at the structured system level
with $X_d = (IPorts_d, X_{d1} \times X_{d2} \times ..)$ and $Y_d = (OPorts_d, Y_{d1} \times Y_{d2} \times ..)$
$EIC \subseteq \{((N, ip_N), (d, ip_d)) | ip_N \in IPorts_N, d \in D, ip_d \in IPorts_d\}$
is the external input coupling
$EOC \subseteq \{((d, op_d), (N, op_N)) | d \in D, op_d \in Oports_d, op_N \in OPorts_N\}$
is the external output coupling
$IC \subseteq \{((a, op_a), (b, ip_b)) | a, b \in D, op_a \in OPorts_a, ip_b \in IPorts_b\}$
is the internal coupling

with the constraints that the range of the to-port must be a subset of the range of the from-port, i.e.,

$$\forall((N, ip_N), (d, ip_d)) \in EIC : range_{ipN}(X_N) \subseteq range_{ipd}(X_d)$$
$$\forall((a, op_a), (N, op_N)) \in EOC : range_{op_a}(Y_a) \subseteq range_{op_N}(Y_N);$$
$$\forall((a, op_a), (b, ip_b)) \in IC : range_{op_a}(Y_a) \subseteq range_{ip_b}(X_b)$$

From this form of coupling specification we are able to derive the set of influencers $\{I_d\}$ and interface maps Z_d. We state this formally as:

Theorem 5.1. *Port-to-port coupling is a special case of general coupling.*

Proof. The set of influencers I_d of all components $d \in D$ and network N is easily derived by gathering all components from which a coupling to d or N exists, i.e.,

$$\forall d \in D \cup \{N\} : I_d = \{i | ((i, fromport), (d, toport)) \in EIC \cup EOC \cup IC\}$$

To define the interface maps Z_d and Z_N we first introduce a *resolution* operator $\oplus$. This is needed since we allow multiple couplings to a single input port or network output port. (Such couplings will be meaningful only for discrete event systems, a very important special case.) To simplify matters, we assume that at most one non-event appears on converging wires at the same time. Then the operator takes the value which is unequal the non-event $\emptyset$. Formally, we define $\oplus$ as an operator taking an arbitrary number of arguments by

$$\oplus x = x$$

$$\oplus x_1, x_2, ... x_n = \begin{cases} x_i & \text{if } x_i \neq \emptyset \wedge \forall j \neq i : x_j = \emptyset \\ \emptyset & otherwise \end{cases}$$

This means that we take the value of x if x is the single argument of $\oplus$, we take the value of argument x_i if x_i is unequal the non-event $\emptyset$ and all the other arguments are equal the non-event, and we take the non-event in all other cases. So equipped, we define the map Z_d as a function which always takes the value of the from-port of that coupling which is equal to the non-event, if such a one exists. For simplicity, we'll assume that only one destination port of a component or the network receives input at any time. Then

$$d \in D \cup \{N\} : Z_d : \times_{i \in I_d} YX_i \rightarrow XY_d$$

with

$$Z_d(..., yx_{i,fromport}, ...) = xy_{(d,toport)}$$

where

$$xy_{(d,toport)} = \oplus_{\{(i,fromport)((i,fromport),(d,toport)) \in EIC \cup EOC \cup IC\}} yx_{(i,fromport)} \qquad \square$$

In Chapter 7 we will adopt this type of system specification and introduce special coupled system formalisms which all have their unique restrictions. For these we will also show how a system at the I/O system level is derived from a coupled system specification.

5.9 SUMMARY

The hierarchy of levels at which a system may be known or specified is summarized in Table 5.1. As the level in the table increases, we move in the direction of increasing structural specificity (i.e., from behavior to structure). Let us briefly restate what new knowledge is introduced at each level in the hierarchy.

At level 0 we deal with the input and output interface of a system.

Table 5.1 Hierarchy of System Specifications

Level	Specification	Formal Object
0	I/O Frame	(T, X, Y)
1	I/O relation observation	(T, X, Ω, Y, R)
2	I/O function observation	(T, X, Ω, Y, F)
3	I/O system	$(T, X, \Omega, Y, Q, \Delta, \Lambda)$
4	Iterative specification	$(T, X, \Omega_G, Y, Q, \delta, \lambda)$
5	structured system specification	$(T, X, \Omega, Y, Q, \Delta, \Lambda)$ with X, Y, Q, Δ, and Λ are structured
6	non-modular coupled multi-component system	$(T, X, \Omega, Y, D, \{M_d = (Q_d, E_d, I_d, \Delta_d, \Lambda_d)\})$
7	modular coupled network of systems	$N = (T, X_N, Y_N, D, \{M_d \mid d \in D\}, \{I_d \mid d \in D \cup \{N\}\}, \{Z_d \mid d \in D \cup \{N\}\})$

At level 1 we deal with purely observational recordings of the behavior of a system. This is the set of I/O segment pairs called the I/O relation.

At level 2 we study the set of I/O functions that partitions a system's I/O relation. Each function is associated with a state of the system as it is described at level 3.

At level 3 the system is described by abstract sets and functions. The state space, transition, and output functions are introduced at this level. The transition function describes the state-to-state transitions caused by the input segments; the output function describes the state-to-observable-output mapping.

At level 4 we introduced a shorthand form of specifying a system at level 3 by presenting only its input generator segments and the state-to-state transitions caused by these segments. This level will serve us as a basis for introducing special basic modeling formalism in the subsequent chapters.

At level 5 the abstract sets and functions of lower levels are presented as arising as crossproducts of more primitive sets and functions. This corresponds to the form in which models are often informally described.

At level 6 a system is specified by a set of components each with its own state set and state transition function. The components are coupled non-modularly in that they influence each other directly through their state sets and state transition functions.

In contrast to the non-modular couplings of level 6, at level 7 a system is specified as a modular coupling of components. The components are system specifications on their own and coupling is done by connecting their output and input interfaces.

We showed how to convert a system specification at each level to one at the next lower level (expect level 7 which is postponed until we introduce special network formalisms). Thus by a multi-step process we can express the observable behavior associated with system specified at any of the levels.

The process of going from a system specification at the highest level to the behavioral level is the formal equivalent of simulation of a model by a computer. The machine computes, step-by-step, successive states and outputs; each step involves the component-by-component calculation of state variable values. Sequences of states and outputs constitute trajectories, and collections of such trajectories constitute the behavior of the model.

We have now a full set of levels at which models can be constructed. For simulation purposes the more structured levels – the network, multi-component, and structured system specifications – are most practical. However, the other levels play their role as well. This formal basis now will serve us for introducing well known modeling formalisms for simulation modeling in subsequent chapters. In particular, we will introduce basic differential equation, discrete time, and discrete event formalisms as specialization of iterative specifications. Likewise, non-modular and modular coupled differential equation, discrete time, and discrete event networks will be treated as specialization of multi-component and network specifications.

CHAPTER

BASIC FORMALISMS: DEVS, DESS, DTSS

6

CONTENTS

6.1 BASIC SYSTEM SPECIFICATION FORMALISMS

Simulation modeling is not done by writing out a dynamic system structure itself, but indirectly, by using a *system specification formalism*. A system specification formalism is a shorthand means of specifying a system. It implicitly implies constraints on the elements of the dynamic system and hence one only has to give the information necessary to distinguish this model from the others in the class. Such formalisms build subclasses of systems. We have already worked with these subclasses in Chapter 3 and we have seen the DEVS (*discrete event system specification*) formalism in Chapter 4. However, we have not seen how DEVS specifies a system nor when such a specification is valid. This is the purpose of this chapter. We also introduce the *differential equation specified system* (DESS) and the *discrete time specified system* (DTSS) as the other basic formalisms for model specification.

Basic system specification formalisms allow a *local description of the dynamic* behavior of the system, i.e., we have to specify how the state changes at any particular point in time. From this local description it is possible to derive the global dynamic behavior of the corresponding dynamic system, and we call it the *dynamic interpretation* of the system formalism. Another way to look at the same process is that each of the formalisms can be interpreted as an iterative system specification (Chapter 5). Recall that an iterative system specification provides a set of input generators such that if we specify the system dynamics for these generators, then the rest of the system specification can be easily generated.

Theory of Modeling and Simulation. https://doi.org/10.1016/B978-0-12-813370-5.00014-6

Thus if the DEVS, DTSS and DESS formalisms are cast as iterative system specifications, then the systems they specify are well characterized. In this chapter however, we follow the spirit, but not the letter, of this approach. Rather we will provide direct interpretations of the system behavior. The reason is that this provides a more understandable approach to the nature of the formalisms themselves. Nevertheless, in this edition we discuss the formulation of DEVS as an instance of the time scheduling iterative specification and the reader is encouraged to formulate the other formalisms as iterative system specifications through exercises.

At this point it will also help to examine the layout of the chapters in Part 2 as shown in Fig. 6.1. Here we see that this chapter introduces the DEVS, DTSS and DESS formalisms, which are called "basic" in the following two senses. In the first sense, they are *fundamental* because all the systems we shall deal with are derived from these formalisms. In the second sense, we distinguish "basic" from "coupled" system formalisms, which are introduced in Chapter 7. Coupled system formalisms enable us to create networks of systems as discussed in Chapter 5. (However, since the formalisms are closed under coupling, the resultant of the coupled system is a member of the same class of the systems and therefore has a basic specification for that class.)

Simulators for the formalisms are discussed in Chapter 8. When we implement a system specification by providing such a simulator, the basic formalism will be implemented as an *atomic* model class. To create an instance of such a class, the modeler uses the notation of the basic formalism. The coupled system formalism is then implemented by a *coupled* model class in which the modeler provides component and coupling information. Chapter 9 introduces a formalism to enable using all three basic formalisms together. Chapter 10 discusses a number of extensions to the DEVS formalism that provide more capabilities to model complex real systems.

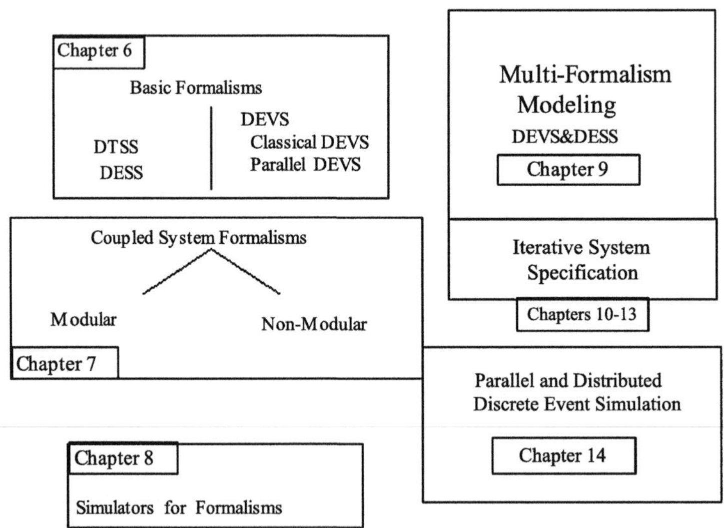

FIGURE 6.1

Layout of Chapters.

6.2 DISCRETE EVENT SYSTEM SPECIFICATION (DEVS)

Having already introduced DEVS in informal fashion in earlier chapters, we are now in a position to provide a more rigorous exposition in the context of system specification. We will review the basic formalism of classic DEVS and its extension to Parallel DEVS. In this edition we have already discussed iterative system specifications with the time scheduling extension. Now DEVS can be directly specified as a special instance of this time scheduling iterative specification. However, since DEVS is a more familiar and concrete concept you may want to first read the following development which relates DEVS directly to a system rather than through the intermediate iterative specification. After this discussion, we will provide the direct DEVS to the time scheduling iterative specification.

6.2.1 CLASSIC DEVS

A basic *discrete event system specification* is a structure

$$DEVS = (X, Y, S, \delta_{\text{ext}}, \delta_{\text{int}}, \lambda, \text{ta})$$

where

X is the set of inputs;
Y is the set of outputs;
S is the set of *sequential* states;
$\delta_{\text{ext}} : Q \times X \rightarrow S$ is the *external* state transition function;
$\delta_{\text{int}} : S \rightarrow S$ is the *internal state transition function*;
$\lambda : S \rightarrow Y$ is the output function;
$\text{ta} : S \rightarrow \mathbb{R}_0^+ \cup \infty$ is the *time advance function*;
$Q = \{(s, e)|s \in S, 0 \leq e \leq \text{ta}(s)\}$, the set of total states

6.2.2 STRUCTURE SPECIFIED BY DEVS

The true meaning of a DEVS is expressed by the system that it specifies.

So it is time to provide the mapping that will associate a system with a DEVS structure. Recall from Chapter 5 that a system is a structure $S = (T, X, \Omega, Y, Q, \Delta, \Lambda)$ where T is the time base, X is the input set, Y is the output set, Ω is the admissible set of segments over X and T, Q is a set, the set of states $\Delta : Q \times \Omega \rightarrow Q$ is the state transition function, and Λ: $Q \times X \rightarrow Y$ is the output function. The structure S specified by a DEVS is time-invariant and has the following characteristics:

- its time base T is the real numbers $\mathbb{R}$ (subsets such as the integers are also allowed).
- its input set is $X^{\infty} = X \cup \{\emptyset\}$, i.e., the input set of the dynamic system is the input set of the DEVS together with a symbol $\emptyset \notin X$ specifying the non-event.
- its output set is $Y^{\emptyset} = Y \cup \{\emptyset\}$, i.e., the output set of the dynamic system is the output set of the DEVS together with a symbol $\emptyset \notin Y_M$ specifying the non-event.
- its state set is the total state set Q of the DEVS;

- the set Ω of admissible input segments is the set of all DEVS segments over X and T;
- the state trajectories are piecewise constant segments over S and T;
- the output trajectories are DEVS segments over Y and T;
- The state trajectory is defined as follows: Let $\omega :< t_i, t_f] \rightarrow X^{\emptyset}$ be a DEVS segment and the state $q = (s, e)$ be the state at time t_i. Then we define

$$\Delta(q, \omega_{<t,t]}) = \begin{cases} (1) & (s, e + t_f - t_i))) \\ & \text{if } e + t_f - t_i < \text{ta}(s) \wedge \nexists t \in < t_i, t_f] : \omega(t) \neq \emptyset \\ & \text{(no internal and external events)} \\ (2) & \Delta((\delta_{\text{int}}(s), 0), \omega_{[t_i + ta(s) - e, t_f]}) \\ & \text{if } e + t_f - t_i = \text{ta}(s) \wedge \nexists t \in < t_i, t_i + \text{ta}(s) - e) : \omega(t) \neq \emptyset \\ & \text{(an internal event)} \\ (3) & \Delta((\delta_{\text{ext}}((s, e + t - t_i), \omega(t)), 0), \emptyset_{[t,t]} \bullet \omega_{[t,t_f]}) \\ & \text{if } t \in < t_i, min\{t_f, t_i + \text{ta}(s) - e\}] : \omega(t) \neq \emptyset \wedge \nexists t' \in < t_i, t) : \omega(t') \neq \emptyset \\ & \text{(an external event)} \end{cases}$$

- The output function Λ of the dynamic system is given by

$$(4) \quad \Lambda(q, x) = \begin{cases} \lambda(s) & \text{if } e = \text{ta}(s) \text{ and } \omega(t) = \emptyset \\ \lambda(s) \text{ or } \emptyset & \text{if } e = \text{ta}(s) \text{ and } \omega(t) \neq \emptyset \\ & \text{(only at internal events there is an output)} \\ \emptyset & \text{otherwise} \end{cases}$$

From the above definition we see that the dynamic behavior is not uniquely defined when both an internal events and an external event, $\omega(t)$ occur at the same time, t. For well-definition, we impose an order of processing where by the internal transition is executed first followed by the external transition. As mentioned earlier (Chapter 4), there are two approaches to deal with such collisions in DEVS-based simulation environments. In classical DEVS, the resolution is handled by the Select function which will allow only one component to be activated at any time, thus assuring that the collision can't happen. In Parallel DEVS there is an additional transition specification that deals with the collision directly.

The recursive definition of the dynamic behavior above has to be interpreted in the following way: Beginning at time t_i the system has been in state s for e time units. The variable e defines the *elapsed time* the systems has been in a particular state. The input values, the values of the input segment unequal to the non-event, determine *the external events* causing the execution of the external transitions. The time advance function determines the time of the execution of the internal transitions. Whenever the elapsed time becomes equal to the value of time advance function, an *internal event* is scheduled. After each event the elapsed time is set to zero.

If there is no input in the whole interval $< t_i, t_f]$ and the total elapsed time $e + t_f - t_i$ at time t_f is smaller than the value of the time advance function ta(s), then there is no event at all and only the elapsed time e has to be updated by the time elapsed during the interval (1). If there is no input until the execution of the first internal transition at time $t_i + \text{ta}(s) - e$, an internal event occurs and the internal transition function is executed (2). Note, that the condition, that no input occurs until the internal event, does not include the time of the internal event ($\nexists t \in= t_i, t_i + \text{ta}(s) - e) : \omega(t) \neq \emptyset$ and so an external event may also occur at the same time. Only at internal events does an output unequal the non-event also occurs (4). If there is an input in the interval before the first internal event, then an external transition has to be executed (3). After the internal as well as external events, the recursive application of Δ to the new state and the remaining segment defines what happens in the sequel.

Informally, we can describe the dynamic behavior as follows: *Internal events*: If at time t_i the DEVS has been in state s for $e \leq \text{ta}(s$ time units, then the DEVS is scheduled to undertake its internal transition at time $t_n = t_i + (\text{ta}(s) - e)$. When no external event input will arrive until time t_n, then the DEVS will stay in state s until that time so that it has been in s exactly ta(s) time units and then will transit from s to $s' := \delta_{\text{int}}(s)$ at time t_n and set the elapsed time e to 0. Right before undertaking the internal transition, the DEVS will put out the value $\lambda(s)$. After the internal transition, the DEVS is scheduled for the next internal transition at time $t_n + \text{ta}(s')$.

External events: Input values can arrive at arbitrary time instances at the input ports. If at time t_i the DEVS has been in state s for time $e \leq \text{ta}(s)$ and an external input arrives, then the DEVS has to execute its external transition function and the DEVS will transit from s to $s'' := \delta_{\text{ext}}(s, e, x)$ where it will stay either for ta(s'') time units or until the next external event (if that happens earlier). Again the elapsed time e is set to 0.

6.2.3 LEGITIMACY: WHEN IS THE STRUCTURE SPECIFIED BY A DEVS REALLY A SYSTEM?

We have associated a structure with a DEVS but how do we know that this structure is indeed a system? Since we employed recursion in the specification of the transition function of the system, we have to guarantee that this recursion always yields a unique result. For example, the DEVS could go into an infinite loop of internal events where time would not be advanced beyond a certain point. A property of a DEVS called legitimacy provides a necessary and sufficient condition for the system specified by a DEVS to be well defined.

To define legitimacy, we extend the internal transition function to its iterative form:

$$\delta^+_{\text{int}} : S \times I^+_0 \to S$$

defined recursively by

$$\delta^+_{\text{int}}(s, 0) = s$$
$$\delta^+_{\text{int}}(s, n + 1) = \delta_{\text{int}}(\delta^+_{\text{int}}(s, n))$$

$\delta^+_{\text{int}}(s, n)$ is the state reached after n iterations starting at state $s \in S$ if no external events intervene. The function $\sum(s, n)$ accumulates the time the system takes to make these n transitions.

$$\sum : S \times I^+_0 \to R^+_0$$

is defined recursively by

$$\sum(s, 0) = 0$$

$$\sum(s, n) = \sum_{i=0}^{n-1} \text{ta}(\delta_{\text{int}}^{+}(s, i))$$

Definition 6.1. A DEVS is *legitimate* if for each $s \in S$,

$$\lim_{n \to \infty} \sum(s, n) \to \infty$$

Theorem 6.1 (Legitimacy Conditions for Well Defined System Specification). *The structure specified by a DEVS is a system if, and only if, the DEVS is legitimate.*

Proof. Let us define $\#(q, \omega)$ to be the number of internal events, i.e., the events of execution of the internal transition function, in interval $< t_i, t_f]$ given the input segment $\omega :< t_i, t_f] \to X$ and the initial state q at time t_i. Then the system defined by the DEVS is well-defined if, and only if,

$$\forall \omega : [t_i, t_f) \to X \in \Omega, q \in Q : \#(q, \omega) < \infty,$$

i.e., the number of internal events in any state trajectory is finite.

The proof proceeds by decomposing the input segment into its generator segments. In each such segment the number of events is given by $\#(s, m)$, where s is the initial state at the beginning of the segment and m exists if $\lim_{n \to \infty}(s, n) \to \infty$.

Conversely, if for some s, there is a time t for which $\#(s, n) < t$ for $n = 0, 1, 2, \ldots$ then the state $\delta(s, 0, t])$ does not have an assigned value. □

A *transitory state* is a state, s for which $\text{ta}(s) = 0$.

Exercise 6.1. Show that a DEVS is not legitimate if there is a cycle in the state diagram of δ_{int} that contains only transitory states. (Hint: once entering this cycle, time will not advance.)

The deleterious effect of inadvertent zero-time-advance cycles is quite ubiquitous, especially in coupled, interactive models. They correspond to algebraic cycles in DTSS networks (see Chapter 16) and can arise in interactions of zero lookahead federates in distributed simulation (Chapter 11). Thus, legitimacy as the guarantor of a well-defined discrete event specification, is more than a theoretical necessity.

The sufficient conditions for illegitimacy can be turned into necessary conditions as well in the special case of finite DEVS.

Theorem 6.2 (Conditions for Legitimate DEVS). *A DEVS, M is legitimate under the following conditions:*

a) *M is finite (S is a finite set): Every cycle in the state diagram of δ_{int} contains a non-transitory state (necessary and sufficient condition).*

b) *M is infinite: There is a positive lower bound on the time advances, i.e., $\exists b, \forall s \in S, \text{ta}(s) > b$* (sufficient condition).

Proof. Left as an exercise. □

6.3 PARALLEL DEVS

The basic formalism of a *Parallel* DEVS model is:

$$DEVS = (X, Y, S, \delta_{\text{ext}}, \delta_{\text{int}}, \delta_{\text{con}}, \lambda, \text{ta})$$

with

X is the set of input events;
S is the set of sequential states;
Y is the set of output events;
$\delta_{\text{int}} : S \to S$ is the internal transition function;
$\delta_{\text{ext}} : Q \times X^b \to S$ is the external transition function,
where X^b is a set of bags over elements in X,
$\delta_{\text{con}} : S \times X^b \to S$ is the confluent transition function,
subject to $\delta_{\text{con}}(s, \phi) = \delta_{\text{int}}(s)$
$\lambda : S \to Y^b$ is the output function;
$\text{ta} : S \to \mathbb{R}_0^+ \cup \infty$ is the time advance function,
where $Q = \{(s, e) | s \in S, 0 < e < \text{ta}(s)\}$, and
e is the elapsed time since last state transition.

The basic formalism of Parallel DEVS extends classical DEVS by allowing bags of inputs to the external transition function. A bag of elements in X is similar to a set except that multiple occurrences of elements are allowed, e.g., $\{a, b, a, c\}$ is a bag. As with sets, bags are unordered. By using bags as the message structures for collecting inputs sent to a component we recognize that inputs can arrive in any order and that more than one input with the same identity may arrive from one or more sources.

Parallel DEVS also introduces the confluent transition function, δ_{con}. This gives the modeler complete control over the *collision* behavior when a component receives external events at the time of its internal transition, $e = \text{ta}(s)$ as in Fig. 6.2. Rather than serializing model behavior at collision times, through the Select function at the coupled model level, Parallel DEVS leaves this decision of what serialization to use, if any, to the individual component. Collision tie-breaking therefore becomes a more controllable local decision as opposed to the global decision that it was in classic DEVS. The default definition of the confluent function is:

$$\delta_{\text{con}}(s, x) = \delta_{\text{ext}}(\delta_{\text{int}}(s), 0, x)) \tag{6.1}$$

This is the same approach used in the defining the system specified by classic DEVS. Here the internal transition is allowed to occur and this is followed by the effect of the external transition function on the resulting state (so the elapsed time clock is reset to zero). For example, if a job is just about to finish processing when a second one arrives, we let the first one finish and then start on the new one. Another possibility is:

$$\delta_{\text{con}}(s, x) = \delta_{\text{int}}(\delta_{\text{ext}}(s, \text{ta}(s), x))$$

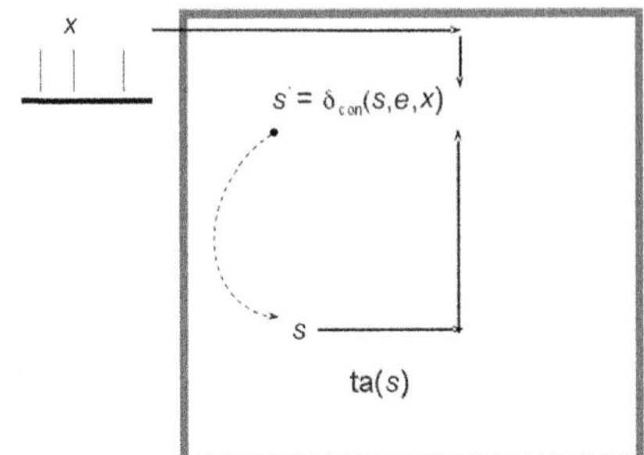

FIGURE 6.2

The confluent transition function in Parallel DEVS.

which expresses the opposite order of effects. Of course, in general, the modeler can define δ_{con} without reference to the other functions so as to express special circumstance that occur when a confluence of external and internal events arises.

Exercise 6.2. Since bags are input to δ_{ext} and δ_{con} we must consider the case where the input bags are empty. Show that for consistency we must have

$$\mathrm{ta}(\delta_{ext}(s, e, \phi)) = \mathrm{ta}(s) - e$$

and

$$\delta_{con}(s, \phi) = \delta_{int}(s).$$

6.3.1 SYSTEM SPECIFIED BY PARALLEL DEVS

Defining the system specified by a Parallel DEVS follows along the lines of the same process for classic DEVS. Indeed, we can proceed by treating the input set underlying the system-to-be-specified as X^b rather than X. Then the construction follows word-for-word except for the treatment of the case of confluent events, which of course, is now explicitly handled by δ_{con}.

Exercise 6.3. In the same manner, write the Time Scheduling Iterative System Specification for Parallel DEVS.

Exercise 6.4. Express a cell space model (Chapter 3) as a Parallel DEVS.

6.4 DISCRETE TIME SYSTEM SPECIFICATION (DTSS)

A *Discrete Time System Specification* is a structure

$$DTSS = (X, Y, Q, \delta, \lambda, c)$$

where

X is the set of inputs;

Y is the set of outputs;

Q is the set of states;

$\delta : Q \times X \rightarrow Q$ is *the state transition function*;

$\lambda : Q \rightarrow Y$ (Moore-type)

or

$\lambda : Q \times X \rightarrow Y$ (Mealy-type) is the output function;

c is a constant employed for the specification of the time base $c \bullet \mathfrak{J}$

In the case that the output function is only a function of time and state, we again speak of a *Moore*-type system and in the case it also depends on the input, we say likewise the system is of type *Mealy* (these names are taken from finite state machine theory context (Hartmanis and Stearns, 1966)).

The structure specified by the DTSS is a structure

$$S = (T, X, \Omega, Y, Q, \Delta, \Lambda)$$

where

- the time base T has to be the set $c \bullet \mathfrak{J}$ isomorphic to the integers;
- X, Y and Q can be arbitrary sets;
- the set Ω of admissible input segments is the set of all sequences over X and T;
- the state and output trajectories are sequences over Q and T and Y and T respectively.

A DTSS defines the following dynamic behavior:

- given an input segment $\omega : [t_1, t_2) \rightarrow X$ and an initial state q at time t_1, then we define the global transition function Δ of the dynamic system by:

$$\Delta(q, \omega) = \begin{cases} q & \text{if } t_1 = t_2 \\ \delta(q, \omega(t_1)) & \text{if } t_1 = t_2 - c \\ \delta(\Delta(q, \omega(t_2 - c)), \omega(t_2 - c)) & \text{otherwise,} \end{cases}$$

  where c is the constant in the definition of the time base $c \bullet \mathfrak{J}$.

- The output function Λ of the dynamic system is given by

$$\Lambda(q, x) = \lambda(q)$$

  in the case of Moore-type systems and

$$\Lambda(q, x) = \lambda(q, x)$$

  in the case of Mealy-type systems.

Exercise 6.5. Show that the structure is a system. (Hint: essentially this requires showing that Δ is well-defined.)

Exercise 6.6. Show that the DTSS is itself an iterative specification of the system it specifies.

Exercise 6.7. Express a finite state machine as a DTSS. Do the same for cellular automata.

6.5 DIFFERENTIAL EQUATION SYSTEM SPECIFICATION (DESS)

A *Differential Equation System* Specification is a structure

$$DESS = (X, Y, Q, f, \lambda)$$

where

X is the set of inputs;
Y is the set of outputs;
Q is the set of states;
$f : Q \times X \rightarrow Q$ is the rate of change function;
$\lambda : Q \rightarrow Y$ (Moore-type)
or
$\lambda : Q \times X \rightarrow Y$ (Mealy-type) is the output function.

Remark. The difference between Moore and Mealy-type systems is usually of interest only when the behaviors of coupled systems are under consideration due to the potential for cycles with zero delay (see e.g., Chapter 3). The structure associated with a $DESS$ is $S = (T, X, \Omega, Y, Q, \Delta, \Lambda)$ where

- the time base T has to be the real numbers $\mathbb{R}$;
- X, Y and Q are real valued vector spaces $\mathbb{R}^m$, $\mathbb{R}^p$, $\mathbb{R}^n$; respectively
- the set Ω of admissible input segments is the set of all bounded piecewise continuous input segments;
- the state and output trajectories are bounded piecewise continuous trajectories.

To define the state trajectory and state transition function we employ the iterative specification concept with the input generators taken to be the bounded continuous segments.

- given a bounded continuous input segment $\omega : \langle t_1, t_2 \rangle \rightarrow X$ and an initial state q at time t_1, the $DESS$ requires that the state trajectory $STRAJ_{q,\omega}$ of the dynamic system at any point in time $t \in \langle t_1, t_2 \rangle$ satisfies the given differential equation:

$$STRAJ_{q,\omega}(t_1) = q, \quad \frac{dSTRAJ_{q,\omega}(t)}{dt} = f(STRAJ_{q,\omega}(t), \omega(t));$$

Assuming the state trajectory is well-defined, the transition function can be defined for every state, q and bounded continuous input segment $\omega : \langle t_1, t_2 \rangle \to X$, by:

$$\Delta(q, \omega) = STRAJ_{q,\omega}(t_2)$$

- The output function Λ of the dynamic system is given by

$$\Lambda(q, x) = \lambda(q)$$

in the case of Moore-type systems and

$$\Lambda(q, x) = \lambda(q, x)$$

in the case of Mealy-type systems.

6.6 EXAMPLE OF DESS

Every ordinary differential equation is can be expressed as a DESS. As an example, consider the simple integrator representing the most elementary continuous system in Chapter 3, Fig. 3.10. Recall that the integrator has one input variable $u(t)$ and one output variable $y(t)$. The state variable is $z(t)$, the contents of the reservoir. The current input $u(t)$ represents the rate of current change of the contents which we express by equation

$$\frac{dz(t)}{dt} = u(t)$$

and the output $y(t)$ is equal to the current state $y(t) = z(t)$.

In this case we have:

$$DESS = (X, Y, Q, f, \lambda)$$

where

X is the set of reals, representing the input, u;

Y is the set of reals, representing the output, y;

Q is the set of reals, representing the states, z;

$f : Q \times X \to Q$ is the rate of change function, here;

$f(z, u) = u$,

$\lambda : Q \to Y$ (Moore-type, output function, here)

$\lambda(z) = z$.

In general, each of the input, output, and state sets are real valued vector spaces. Accordingly, the rate of change and output functions are defined on the corresponding products of these spaces.

Exercise 6.8. Consider the second order differential equation system in Chapter 3 that represents a damped spring with an external input:

$$\begin{aligned} \dot{z}_1(t) &= z_2(t) \\ \dot{z}_2(t) &= -\frac{k}{m} z_1(t) - \frac{b}{m} z_2(t) + \frac{F(t)}{m}. \end{aligned} \tag{6.2}$$

Write the DESS for this differential equation.

WHEN IS THE SYSTEM SPECIFIED BY A DESS WELL-DEFINED?

We have to go back and justify the assumption that there is a unique, well-defined state trajectory determined for every state and input generator segment. It is known that the following condition guaranties that such unique solutions exist. The *Lipschitz condition* is stated as:

$$||f(q, x) - f(q', x)|| < k||q - q'||$$

for all pairs $q, q' \in Q$, and input values $x \in X$, where k is a constant and $||\ ||$ is the Euclidean norm (Perko, 2013).

As you can see, we did not provide an explicit formula for the global transition function of the dynamic system. Instead, it was defined in terms of its state trajectories. However, when it is possible to find an analytical solution for the integral $F(t) = \int f(q, \omega)dt$, then we can explicitly specify the global transition function. In this case, Δ is defined by:

$$\Delta(q, \omega) = q + F(t2) - F(t1)$$

where ω has domain $\langle t_1, t_2 \rangle$. Practical models often do not have tractable analytical solutions. Indeed, in many cases the complexity of the input signal forbids analytical integration. As discussed in Chapter 3, using digital computers to simulate $DESS$, requires discrete approximation of the state trajectory and the state transition function.

Exercise 6.9. Express a linear time invariant system in the form of a set of matrices as a $DESS$. (Hint: show that the Lipschitz condition is satisfied and write the integral solution, F as above.)

Exercise 6.10. Show how to formulate the $DESS$ as an iterative system specification and characterize the system it specifies in this manner. (Hint: use the bounded continuous input segments as generators and the generator transition function as defined above. Show that the conditions required for the extension to piecewise continuous segments in Chapter 5 are satisfied.)

6.7 SUMMARY

This chapter is key to the development of the theory that follows. The three basic system specification formalisms, $DEVS$, $DTSS$ and $DESS$, were defined. The system classes that they specify were characterized based on the ideas underlying the iterative system specification. Indeed, iteration is the manner in which the system's global transition function is defined and also the way in which the system's state trajectories are generated. Except for the $DTSS$, specific conditions are required for the iteration to be well-defined – legitimacy in the case of $DEVS$, and the Lipschitz condition for $DESS$. In this edition, we will show how to formulate a $DEVS$ as an instance of the time scheduling iterative system specification in Chapter 12. For a view of the road ahead, please return to Fig. 6.1.

REFERENCES

Hartmanis, J., Stearns, R.E., 1966. Algebraic Structure Theory of Sequential Machines, vol. 147. Prentice-Hall, Englewood Cliffs, NJ.

Perko, L., 2013. Differential Equations and Dynamical Systems, vol. 7. Springer Science & Business Media.

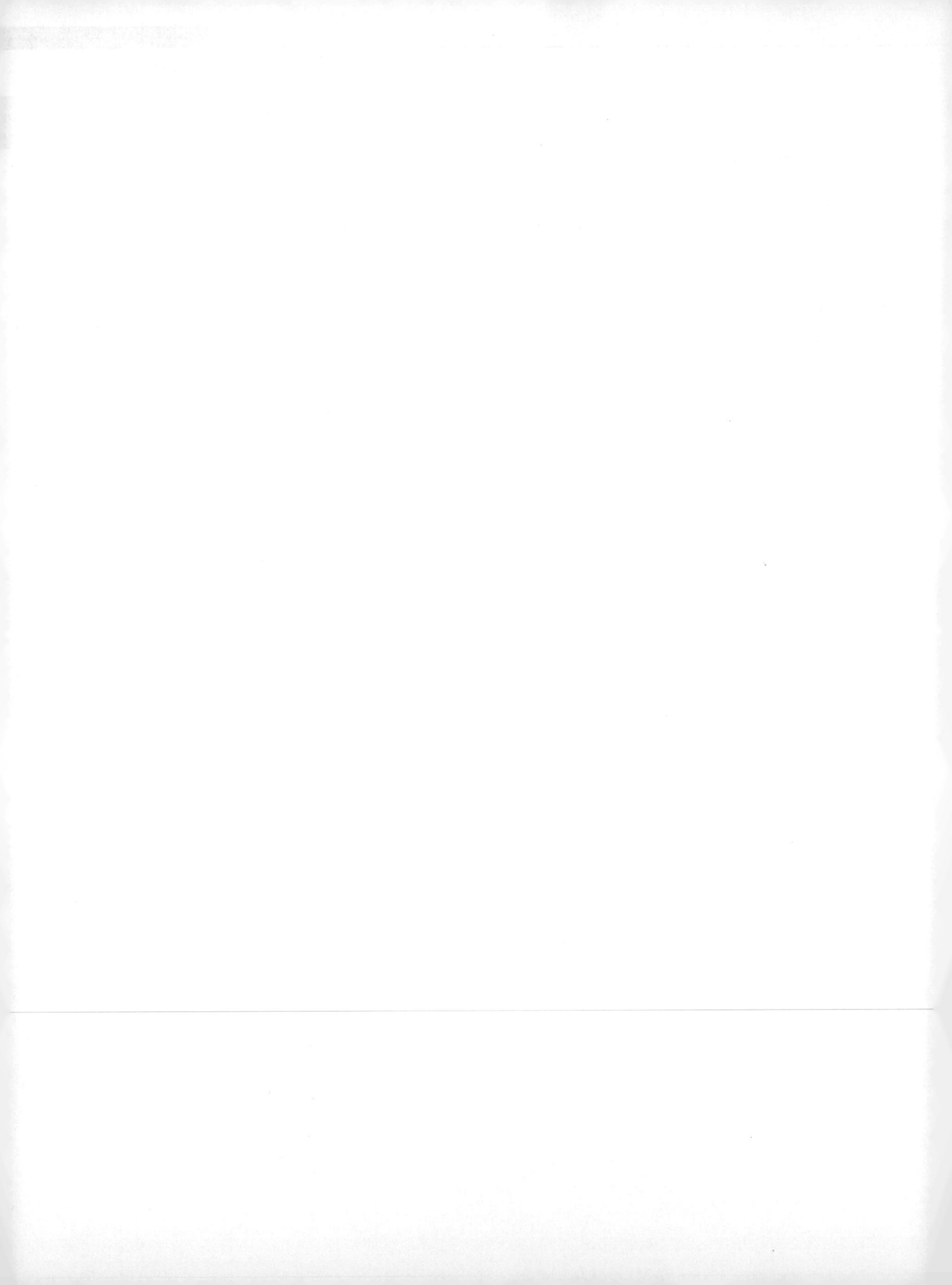

CHAPTER

BASIC FORMALISMS: COUPLED MULTI-COMPONENT SYSTEMS

7

CONTENTS

In Sections 5.7 and 5.8 we introduced the concept of system specification at the multi-component and network of systems (also called coupled systems) level. Both levels allow us to model systems

Theory of Modeling and Simulation. https://doi.org/10.1016/B978-0-12-813370-5.00015-8

by composing smaller systems together. In the multi-component system specification, composition is done non-modularly in that one component's state transitions can directly change another component's state. In contrast, the network of systems specification provides a means to couple standalone systems by connecting their output and input interfaces.

In Chapter 6 we introduced system specification formalisms as shorthand means to specify systems. A formalism defines a set of background conventions as features which are common to all systems of the class it specifies. We identified DESS, DTSS, and DEVS as the three fundamental system formalisms for simulation modeling.

The formalisms of Chapter 6 were at the I/O systems level or structured systems levels. In this chapter, we formulate the basic system formalisms at the *multi-component* and the *network of systems* (coupled system) level. Further, for each formalism, we will provide conditions that guarantee that a coupling of systems in this formalism defines a basic system in the same formalism – we say that the formalism is *closed under coupling*. Closure under coupling allows us to use networks of systems as components in a larger coupled systems, leading to hierarchical, modular construction.

Non-modular system specification is important because many traditional forms of modeling (e.g., the classical world views of discrete event simulation) are non-modular. However, the advantages of modular paradigms are decisive in the age of distributed simulation and model repositories. Thus, translation from non-modular to modular specifications is an important topic of this chapter.

7.1 DISCRETE EVENT SPECIFIED NETWORK FORMALISM

The Discrete Event Specified Network formalism (DEVN), also called the DEVS coupled model specification provides a basis for modular construction of discrete event models. Such modular construction, although, at the time of this writing, is not as common an approach as the non-modular alternative, is destined to be more and more dominant. The reason is that modularity is more natural for distributed simulation and supports reuse through model repositories. Interoperability through distributed simulation with repository reuse is the goal of the High Level Architecture (HLA) standard being promulgated of the U.S. Department of Defense.

In discrete event coupled networks, components are DEVS systems coupled exclusively through their input and output interfaces. Components do not have the possibility to access and influence the states or the timing of other components directly. All interactions have to be done by exchanging messages. More specifically, the events generated by one component at its output ports are transmitted along the couplings to input ports where they cause external events and state transitions at the influenced components.

Besides treating classical DEVS in the following discussion, we also will consider the revision called Parallel DEVS. We will discuss the closure under coupling of both classic and Parallel DEVS.

7.1.1 CLASSIC DEVS COUPLED MODELS

Recall the definition of coupled DEVS models in Chapter 4. Instead of repeating this definition here, we provide the more abstract version which does not explicitly work with input/output ports but corresponds to the abstract definition of network of systems of Chapter 5. The translation from structured to abstract version follows the lines discussed in Chapter 5. The definition for DEVS coupled models adopted here differs only slightly from the general definition in Chapter 5 as the interface map Z_d for a component d is broken up into several interface mappings $Z_{i,d}$, one for each influencing component i. In addition to the general definition of coupled models, classic DEVS coupled models employ a Select function which is used for tie-breaking in case of equal next event times in more than one component. The structure of the DEVN or *coupled* DEVS *model* is

$$N = \langle X, Y, D, \{M_d\}, I_d, Z_{i,d} Select \rangle$$

with

X a set of input events,
Y a set of output events, and
D a set of component references.

For each $d \in D$,

M_d is a classic DEVS model.

For each $d \in D \cup \{N\}$,

I_d is the influencer set of d: $I_d \subseteq D \cup \{N\}, d \notin I_d$

and for each $i \in I_d$

$Z_{i,d}$ is a function, the i-to-d output translation with
$Z_{i,d} : X \rightarrow X_d$, if $i = N$.
$Z_{i,d} : Yi \rightarrow Y$, if $d = N$.
$Z_{i,d} : Yi \rightarrow X_d$, if $d \neq N$ and $i \neq N$.

Note that we use the influencer sets to specify the influence propagation in order to be consistent with the approach of Chapter 5. Earlier definitions, in the first edition of TMS and elsewhere, employed the influencee sets, since as we see below, this is more natural for classic DEVS where only one imminent component is activated. However, either family of sets gives the same information and its more convenient to use the influencer sets for Parallel DEVS.

Exercise 7.1. Show that given the family of influencee sets we can derive the family of influencer sets, and conversely.

Finally, Select is a function

$$\text{Select} : 2^D \rightarrow D$$

with Select(E) $\in E$. Select is the tie-breaking function to arbitrate the occurrence of simultaneous events (see below).

Recall (Chapter 5) that the presence of N as a potential source or destination component is a device that makes it easy to represent external input and external output couplings.

CLOSURE UNDER COUPLING OF CLASSIC DEVS

Given a coupled model, N, with components that are classic DEVS, we associate with it a basic DEVS, called the *resultant*:

$$DEVSN = \langle X, Y, S, \delta_{\text{ext}}, \delta_{\text{int}}, \lambda, \text{ta}\rangle$$

where $S = \times_{d \in D} Q_d$, $\text{ta} : S \to \mathbb{R}_0^+ \cup \{\infty\}$ is defined by

$$\text{ta}(s) = minimum\{\sigma_d | d \in D\}$$

where for each component $d \sigma_d = \text{ta}_d(s_d) - -e_d$, i.e., σ_d is the time remaining to the next event in component d.

Let the set of $imminents$, $IMM(s) = \{d | d \in D \wedge \sigma_d = \text{ta}(s)\}$. $Imminents$ is the set of components that have minimum remaining time σ_d, i.e., they are candidates for the next internal transition to occur.

Let $d* = \text{Select}(IMM(s))$. This is the component whose output and internal transition functions will be executed because it was selected using the tie breaking function Select.

Now, we define $\delta_{\text{int}} : S \to S$. Let $s = (..., (s_d, e_d)...)$. Then $\delta_{\text{int}}(s) = s' = (..., (s'_d, e'_d)...)$, where

$$(s'_d, e'_d) = \begin{cases} (\delta_{\text{int}}, d(s_d), 0) & \text{if } d = d* \\ (\delta_{\text{ext}}, d((s_d, e_d + \text{ta}(s)), x_d), 0) & \text{if } d* \in I_d \wedge x_d \neq \emptyset \\ (s_d, e_d + \text{ta}(s)) & \text{otherwise} \end{cases}$$

where $x_d = Z_{d*,d}(\lambda_{d*}(s_{d*}))$. That means, the resultant's internal transition function changes the state of the selected imminent component $d*$ according to its $(d *' s)$ internal transition function and updates all influenced components according to the inputs produced by the output of $d*$. In all other components, d, only the elapsed times, e_d, are updated.

We define $\delta_{\text{ext}} : Q \times X \to S$, $\delta_{\text{ext}}((s, e), x) = s' = (..., (s'_d, e'_d)...)$ by

$$(sd', ed') = \begin{cases} (\delta_{\text{ext}}, d((s_d, e_d + e), x_d), 0) & \text{if } N \in I_d \wedge x_d \neq \emptyset \\ (sd, ed + e) & \text{otherwise} \end{cases}$$

where $x_d = Z_{N,d}(x)$. That is, all components influenced by the external input x change state according to the input x_d transmitted to them as converted by their respective interface mappings. All other components increment their elapsed times by e.

Finally, we define the resultant's output function, $\lambda : S \to Y$:

$$\lambda(s) = \begin{cases} Z_{d*,N}(\lambda_{d*}(s_{d*})) & \text{if } d* \in I_N \\ \emptyset & \text{otherwise.} \end{cases}$$

This says that we translate the output of the selected imminent component, $d*$, to an output of the coupled model through interface map $Z_{d*,N}$. This, of course, is only for the case that $d*$ actually sends external output.

7.1.2 PARALLEL DEVS COUPLED MODELS

The structure of a *coupled DEVS model* for Parallel DEVS is almost identical to that of the classical DEVS except for the absence of the Select function:

$$N = \langle X, Y, D, \{M_d\}, \{I_d\}, \{Z_{i,d}\} \rangle$$

with

X a set of input events,

Y a set of output events, and

D a set of component references.

For each $dinD$,

M_d is a *Parallel DEVS* model.

For each $d \in D \cup \{N\}$, I_d is the influencers set of d and $Z_{i,d}$ is the i-to-d output translation with the same definition as in classical DEVS (see above).

7.1.3 CLOSURE UNDER COUPLING OF PARALLEL DEVS

We consider a coupled model

$$N = \langle X, Y, D, \{M_d\}, \{I_d\}, \{Z_{i,d}\} \rangle$$

where each component, M_d is a Parallel DEVS.

We demonstrate closure under coupling by constructing the resultant of the coupled model and showing it to be a well-defined *Parallel DEVS*. The resultant is (potentially) a *Parallel DEVS*

$$DEVS_N = \langle X, S, Y, \delta_{\text{int}}, \delta_{\text{ext}}, \delta_{\text{con}}, \lambda, \text{ta} \rangle .$$

As with classic DEVS, the state set is a crossproduct, $S = \times_{d \in D} Q_d$. Likewise, the time advance function is defined by

$$\text{ta}(s) = minimum\{\sigma_d / d \in D\}, \text{ where } s \in S \text{ and } \sigma_d = \text{ta}(s_d) - e_d$$

The major difference with classic DEVS comes in the definitions of the remaining functions. To make these definitions, we partition the components into four sets at any transition. Let $s = (..., (s_d, e_d), ...)$. Then $IMM(s)$ are the imminent components whose outputs will be generated just before the next transition. Since there is no Select function, all of the imminents will be activated. $INT(s)$ is the subset of the imminents that have no input messages. $EXT(s)$ contains the components receiving input events but not scheduled for an internal transition. CONF(s) contains the components receiving input events and also scheduled for internal transitions at the same time. $UN(s)$ contains the remaining components. Formally,

$$IMM(s) = \{d / \sigma_d = \text{ta}(s)\} \text{ (the imminent components)},$$

$INF(s) = \{d | i \in I_d, i \in IMM(s) \wedge x_d^b \neq \Phi\}$ (components about to receive inputs),
where $x_d = \{Z_{i,d}(\lambda_i(s_i)) | i \in IMM(s) \cap I_d\}$
$CONF(s) = IMM(s) \cap INF(s)$ (confluent components),
$INT(s) = IMM(s) - INF(s)$ (imminent components receiving no input),
$EXT(s) = INF(s) - IMM(s)$ (components receiving input but not imminent),
$UN(s) = D - IMM(s) - INF(s)$.

OUTPUT FUNCTION

The output is obtained by collecting all the external outputs of the imminents in a bag:

$$\lambda(s) = \{Z_{d,N}(\lambda_d(s_d)) / d \in IMM(s) \wedge d \in I_N\}.$$

INTERNAL TRANSITION FUNCTION

The resultant internal transition comprises four kinds of component transitions: internal transitions of $INT(s)$ components, external transitions of $EXT(s)$ components, confluent transitions of $CONF(s)$ components, and the remainder, $UN(s)$, whose elapsed times are merely incremented by ta(s). (The participation of $UN(s)$ can be removed in simulation by using anabsolute time base rather than the relative elapsed time.) Note that, by assumption, this is an internal transition of the resultant model, and there is no external event being received by the coupled model at this time.

We define

$$\delta_{\text{int}}(s) = (..., (s'_d, e'_d), ...)$$

where

$$(s'_d, e'_d) = \begin{cases} (\delta_{\text{int}}, d(s_d), 0) & \text{for } d \in INT(s), \\ (s'_d, e'_d) = (\delta_{\text{ext}}, d(s_d, e_d + \text{ta}(s), x_d^b), 0) & \text{for } d \in EXT(s), \\ (s'_d, e'_d) = (\delta_{\text{con}}, d(s_d, x_d^b), 0) & \text{for } d \in CONF(s), \\ (s'_d, e'_d) = (s_d, e_d + \text{ta}(s)) & \text{otherwise} \end{cases}$$

where x_d^b is as defined above.

EXTERNAL TRANSITION FUNCTION

To construct δ_{ext} of the resultant, let

$$\delta_{\text{ext}}(s, e, x^b) = (..., (s'_d, e'_d), ...)$$

where $0 < e < \text{ta}(s)$ and

$$(s'_d, e'_d) = \begin{cases} (\delta_{\text{ext}}, d(s_d, e_d + e, x_d^b), 0) & \text{for } N \in I_d \wedge x_d^b \neq \Phi, \\ (s_d, e_d + e) & \text{otherwise,} \end{cases}$$

where

$$x_d^b = \{Z_{N,d}(x)/x \in x^b \wedge N \in I_d\}.$$

The incoming event bag, x^b is translated and routed to the event bag, x_d^b, of each influenced component, d. The resultant's external transition comprises all the external transitions of the influenced children. Note that by assumption, there are no internally scheduled events at this time ($e < \mathrm{ta}(s)$).

7.1.4 THE CONFLUENT TRANSITION FUNCTION

Finally, we construct the δ_{con} of the resultant. This is what happens in the coupled model when some of the components are scheduled to make internal transitions and are also about to receive 7 external events. Fortunately, it turns out that the difference between δ_{con} and the earlier defined, δ_{int}, is simply the extra confluent effect produced by the incoming event bag, x^b, at elapsed time $e = \mathrm{ta}(s)$. By redefining the influencee set $INF(s)$ as $INF'(s)$, which includes the additional influencees from the incoming couplings, I_N, we develop three similar groups for δ_{con}. Let

$$INF'(s) = \{d|(i \in I_d, i \in IMM(s) \vee N \in I_d) \wedge x_d^b \neq \Phi\}$$

where $x_d^b = \{Z_{i,d}(\lambda_i(s_i))|i \in IMM(s) \wedge i \in I_d\} \cup \{Z_{n,d}(x)|x \in x^b \wedge N \in I_d\}$ (this is a union of bags so common elements are retained. The first bag collects outputs from imminent components that are coupled to the receiving component, d. The second bag collects external inputs that are sent to the receiver through the external input coupling).

$$\begin{aligned} CONF'(s) &= \mathbf{IMM(s)} \cap INF'(s) \\ INT'(s) &= \mathbf{IMM(s)} - INF'(s), \\ EXT'(s) &= INF'(s) - \mathbf{IMM(s)}. \end{aligned}$$

We define

$$\delta_{\text{con}}(s, x^b) = (..., (s'_d, e'_d), ...),$$

where

$$(s'_d, e'_d) = \begin{cases} (\delta_{\text{int},d}(s_d), 0) & \text{for } d \in INT'(s), \\ (\delta_{\text{ext},d}(s_d, e_d + \mathrm{ta}(s), x_d^b), 0) & \text{for } d \in EXT'(s), \\ (\delta_{\text{con},d}(s_d, x_d), 0) & \text{for } d \in CONF'(s), \\ (s_d, e_d + \mathrm{ta}(s)) & \text{otherwise} \end{cases}$$

where x_d^b is defined above.

Hierarchical consistency is achieved here by the bag union operation that gathers all external events, whether internally or externally generated, at the same time into one single event group. By such consistency, we mean that a component would experience the same inputs independently of the manner in which we chose to decompose a model (see Chow, 1996 for more details).

From the definition of the $\delta_{\text{int}'}$, $\delta_{\text{con}'}$, and $\delta_{\text{ext}'}$, we see that they are special cases of one encompassing transition function, $\delta(s, e, x^b)$ applied to each component. As shown in Fig. 7.1, δ_{int} is applied

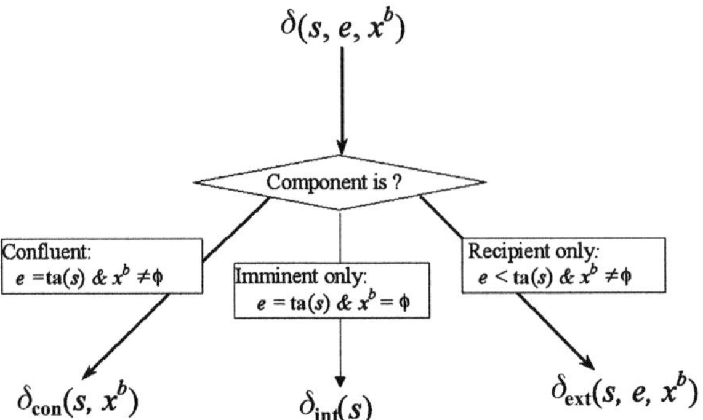

FIGURE 7.1

The overall transition function.

to imminent (non-confluent), i.e., when $(s, e, x^b) = (s, \mathrm{ta}(s), \Phi)$; δ_{con} is applied to confluent components, i.e., $(s, e, x^b) = (s, \mathrm{ta}(s), x^b)$ where $x_d^b \neq \Phi$; finally, δ_{ext} is applied to receiver (non-confluent) components, i.e., (s, e, x^b) where $0 < e < \mathrm{ta}(s)$ and $x_d^b \neq \Phi$.

The overall transition function is the basis for implementation of coupled Parallel DEVS. In terms of the special case functions, it is defined by:

$$\delta(s, e, x^b) = \delta_{\mathrm{ext}}(s, e, x^b) \text{ for } 0 < e < \mathrm{ta}(s), \text{ and } x^b \neq \Phi,$$
$$\delta(s, \mathrm{ta}(s), x^b) = \delta_{\mathrm{con}}(s, x^b) \text{ for } x^b \neq \Phi,$$
$$\delta(s, \mathrm{ta}(s), \Phi)) = \delta_{\mathrm{int}}(s).$$

Exercise 7.2. Write the generic transition function

$$\delta(s, e, x^b) = (..., (s'_d, e'_d), ...)$$

for the resultant of a coupled model in terms of the generic functions of its components.

7.2 MULTI-COMPONENT DISCRETE EVENT SYSTEM FORMALISM

Recall (Chapter 5) that in non-modular coupling, components directly influence each other through their state transitions. In DEVS multi-component non-modular coupled models, events 9 occurring in one component may result in state changes and rescheduling of events in other components. The simulation strategies realized in many commercial simulation languages and systems, known as "world views", fall into the category of multi-component DEVS. In this section, we will make the connection from the system theoretical DEVS approach to these discrete event simulation strategies. However, before we introduce the formalism for multi-component DEVS and formulate the GOL event model discussed in Chapter 3 as a multi-component DEVS.

A multi-component DEVS is a structure

$$multiDEVS = \langle X, Y, D, \{M_d\}, Select \rangle$$

where

X, Y are the input and output event sets, D is the set of component references, and Select : $2^D \rightarrow D$ with Select$(E) \in E$ is a tie-breaking function employed to arbitrate in case of simultaneous events.

For each $d \in D$

$$M_d = \left(S_d, I_d, E_d, \delta_{\text{ext},d}, \delta_{\text{int},d}, \lambda_d, \text{ta}_d\right)$$

where

S_d is the set of sequential states of d,

$Q_d = (s, e_d)|s \in S_d, e_d \in \mathbb{R}$ is the set of total states of d,

$I_d \subseteq D$ is the set of influencing components,

$E_d \subseteq D$ is the set of influenced components,

$\delta_{\text{ext},d} : \times_{i \in I_d} Q_i \times_{j \in E_d} X \rightarrow \times Q_j$ is the external state transition function,

$\delta_{\text{int},d} : \times_{i \in I_d} Q_e \rightarrow \times_{j \in E_d} Q_j$ is the internal state transition function,

$\lambda_d : \times_{i \in I_d} Q_i \rightarrow Y$ is the output event function, and

$\text{ta}_d : \times_{i \in I_d} Q_i \rightarrow \mathbb{R} + 0 \cup \infty$ is the time advance function.

A multiDEVS works in the following way: Internal events are scheduled individually by each component $d \in D$ by the individual time advance functions ta_d. When an event occurs in one of the components, it is executed and this results in state changes through the internal transition function $\delta_{\text{int},d}$ and eventually output events for the multiDEVS defined in the output function λ_d. Note, that the transition function depends on total states q_i of the influencing components I_d and change any total state q_j of the influencing components E_d, also including the elapsed time e_j. When events in different components are imminent, the Select function is used to arbitrate among them. External events at the multiDEVS's input interface can be handled by any of the external state transition functions $\delta_{\text{ext},d}$ of the components, d. However, a component may not react to external inputs if its external state transition function is not defined.

A multiDEVS defines an atomic DEVS as follows. Given a multiDEVS as defined above, we associate a basic

$$DEVS = \langle X, Y, S, \delta_{\text{ext}}, \delta_{\text{int}}, \lambda, \text{ta} \rangle$$

where $S = \times_{d \in D} Q_d$. We define

$$\text{ta}(s) = min\{\sigma_d | d \in D\}$$

with $\sigma_d = \text{ta}_d(..., q_i, ...) - e_d$ is the remaining time until the next scheduled event in d.

We define the next scheduled component $d* = \text{Select}(\{d|\sigma_d = \text{ta}(s)\}$ and

$$\delta_{\text{int}}((s_1, e_1), (s_2, e_2), ..., (s_n, e_n)) = ((s_1', e_1'), (s_2', e_2'), ..., (s_n', e_n'))$$

with

$$(s'_j, e'_j) = \begin{cases} (sj, ej + \mathrm{ta}(s)) & \text{if } j \notin E_{d*} \\ (s'_j, e'_j) = \delta_{\mathrm{int},d*}((..., (s_i, e_i + \mathrm{ta}(s)), ...)).j & \text{if } j \in E_{d*} \end{cases}$$

with $i \in I_d*$, i.e. the internal state transition of the DEVS is defined by the state transition of the next scheduled component $d*$.

The output of the system is defined by the output event of $d*$

$$\lambda(s) = \lambda_{d*}(..., q_i, ...), i \in I_d*$$

The external transition function δ_{ext} upon occurrence of an input event x is defined by the crossproduct of the external state transition functions

$$\delta_{\mathrm{ext},d}((..., q_i, ...), e, x)$$

of all components $d \in D$. We define the overall external transition function by

$$\delta_{\mathrm{ext},d}(((s_1, e_1), (s_2, e_2), ..., (s_n, e_n)), e, x) = ((s'_1, e'_1), (s'_2, e'_2), ..., (s'_n, e'_n))$$

with

$$(s'_j, e'_j) = \begin{cases} (s_j, e_j + e) & \text{if } j \notin E_d \\ \delta_{\mathrm{ext}}, d((..., (s_i, e_i + e), ...)).j & \text{if } j \in E_d. \end{cases}$$

7.2.1 CELLULAR AUTOMATA MULTI-COMPONENT DEVS OF GOL EVENT MODEL

We illustrate the general ideas of event coupling in multi-component systems by recalling the discrete event cell space model introduced in Chapter 3. Clearly such a model is a multi-component DEVS model with identical components, uniform arrangement and couplings. The time advance function of a cell takes care of scheduling its birth and death based on its current fitness and the sum of its neighbors' states. (Note, that the time advance is infinity in case that no event is scheduled to happen.) A state event in one cell can influence the time advance of its influencees, possibly causing them to be rescheduled. We will see in the next chapter how the simulation algorithm takes care to plan the events in time, in particular, how events in influenced components $E_{i,j}$ are rescheduled upon changes in a neighboring cell.

For the state set of each cell $M_{i,j}$ we consider the set of pairs,

$$S_{i,j} = \{(state, fitness) | state \in 0, 1, fitness \in ?\}$$

with variables $state$ to denote the actual dead or alive state of the cell and the $fitness$ parameter representing the current fitness of the cell. Additionally we employ a help variable sum representing the sum of the alive neighbors

$$sum = |\{(k,l)|(k,l) \in I_{i,j}, s_{k,l}.state = 1\}|,$$

and an auxiliary variable $factor$ representing the increase or decrease of the fitness parameter

$$factor = \begin{cases} 1 & \text{if } sum = 3 \\ -2 & \text{if } q_{i,j}.state = 1 \wedge (sum < 2 | sum > 3) \\ 0 & \text{otherwise} \end{cases}$$

CELLULAR AUTOMATA MULTI-COMPONENT DEVS OF GOL EVENT MODEL

We illustrate the general ideas of event coupling in multi-component systems by recalling the discrete event cell space model introduced in Chapter 3. Clearly such a model is a multi-component DEVS model with identical components, uniform arrangement and couplings. The time advance function of a cell takes care of scheduling its birth and death based on its current fitness and the sum of its neighbors' states. (Note, that the time advance is infinity in case that no event is scheduled to happen.) A state event in one cell can influence the time advance of its influencees, possibly causing them to be rescheduled. We will see in the next chapter how the simulation algorithm takes care to plan the events in time, in particular, how events in influenced components $E_{i,j}$ are rescheduled upon changes in a neighboring cell.

For the state set of each cell $M_{i,j}$ we consider the set of pairs,

$$S_{i,j} = \{(state, fitness)|state \in \{0,1\}, fitness \in\}$$

with variables $state$ to denote the actual dead or alive state of the cell and the $fitness$ parameter representing the current fitness of the cell. Additionally we employ a help variable sum representing the sum of the alive neighbors

$$sum = |\{(k,l)|(k,l) \in I_{i,j}, s_{k,l}.state = 1\}|,$$

and an auxiliary variable $factor$ representing the increase or decrease of the fitness parameter

$$factor = \begin{cases} 1 & \text{if } sum = 3 \\ -2 & \text{if } q_{i,j}.state = 1 \wedge (sum < 2 | sum > 3). \\ 0 & \text{otherwise} \end{cases}$$

Given the current $fitness$, and based on the value of $factor$, we easily can compute the time to the next zero crossing and hence the time advance. Note that only in the dead state, with a positive $factor$, and in the alive state with negative $factor$, are events scheduled. Formally,

$$ta_{i,j}(q_{i,j}, q_{i,j+1}, q_{i+1,j}, q_{i,j-1}, q_{i-1,j}, q_{i+1,j+1}, q_{i+1,j-1}, q_{i-1,j+1}, q_{i-1,j-1}) =$$

$$= \begin{cases} \dfrac{-fitness}{factor} & \text{if } ((state = 1 \wedge factor < 0) | (state = 0 \wedge factor > 0)) \\ \infty & \text{otherwise} \end{cases}$$

Finally, the state transition function has to consider time scheduled events (birth or death) as well as changes in the environment. These can be specified as given in the following

$$
\begin{aligned}
&\delta_{i,j}(q_{i,j}, q_{i,j+1}, q_{i+1,j}, q_{i,j-1}, q_{i-1,j}, q_{i+1,j+1}, q_{i+1,j-1}, q_{i-1,j+1}, q_{i-1,j-1}) = \\
&fitness = max\{6, fitness + factor \cdot e\} \\
&e_{i,j} = 0 \\
&\text{if } (state = 0 \land fitness = 0 \land factor \geq 0) \text{ then } state = 1 \\
&\text{else if } (state = 1 \land fitness = 0 \land factor \leq 0) \text{ then } state = 0,\ fitness = -2
\end{aligned}
$$

where $factor$ is defined as above. First the $fitness$ parameter for the current time is computed and the elapsed time of the component is set to 0. This applies to external changes as well as for internal events. Now, the increase in $fitness$ at any event is equal to $factor \cdot e$. (Since the last update of the $fitness$ value occurred at the last event and e is the time since the last event and the factor value did not change since the last event (why?).) Thus, we check if an internal event is to occur ($fitness = 0$) and execute a birth or death, respectively. We switch the state value from alive (1) to death (0) or vice versa and reinitialize the fitness with -2 in case of a death.

7.2.2 EVENT SCHEDULING MODELS

We come back now to discuss the world views of discrete event simulation, first considered in Chapter 3, in the context of multi-component DEVS. There are three common types of discrete event simulation strategies employed in discrete event simulation languages and packages – the *event scheduling*, the *activity scanning*, and the *process interaction* world view, the latter being a combination of the first two. A strategy makes certain forms of model description more naturally expressible than others. We will model the "pure" forms of these strategies, that is, the forms in which only the features inherent to the basic conception are employed.

Event oriented models work with prescheduling of all events and there is no provision for activating events by tests on the global state. Event scheduling strategy can be captured by the multi-component DEVS as introduced above.

IMPLEMENTING EVENT SCHEDULING SIMULATION SYSTEMS IN IMPERATIVE PROGRAMMING LANGUAGES

Due its simplicity, event scheduling simulation is the preferred strategy when implementing customized simulation systems in procedural programming languages. One takes the approach that the components' states are realized by arbitrary data records and the state transition functions by several procedures which operate on these data. Most important, each active component defines a sort of "control state" similar to the DEVS's phase variable to denote different events types it has to process. For each such event type, one procedure is coded which implements the state transition for the event type. We can say that the set of event types $S_d^{control} = \{ev_1, ev_2, ev_3, ...ev_n\}$ divides the state transition function δ_d of component d into n functions δ_d^{evi}, each describing the activity of the component d when started in one of its control states, ev_1. At an event the component branches to one of its event routines depending on the control state (Algorithm 1). On such a model implementation an event list simulation algorithm can be realized which sorts components into the event list depending on their event times and then takes them out in sequence to process.

```
Component d
$S_{control, d}$ = \{ ev_1 , ev_2 , ..., ev_n \}
$\delta_{ss}$ (($s_{control, d}$ , ... ))
  case $s_{control, d}$
    $ev_1$ : call event-routine_1
    $ev_2$ : call event-routine_2
    ....
    $ev_n$ : call event-routine_n
```

7.2.3 COMBINED EVENT SCHEDULING, ACTIVITY SCANNING SIMULATION STRATEGY

We now augment the event oriented scheme to allow for activating components by means of contingency tests. Referring to Fig. 7.2, we can understand the additional model description power attained by these means. Suppose that at time t component d is state s_d with a elapsed time e_d of 0 (meaning that d just entered state s_d with an event at time t). Then a time of next event is scheduled at time $t + ta_d((\ldots, q_i, \ldots))$. Unless d is rescheduled by the action of another imminent component, it will be activated at this time to undertake an internal transition. In the pure event scheduling strategy this will in either case result in a change of state in d and in a redefinition of its next activation time.

In contrast, in the activity scanning approach, events can be conditioned on a contingency test. When its time has expired and if the contingency test evaluates to true, the event will occur and the state will change. Otherwise, the event is said to be *due*. This means that the elapsed time component e_d is allowed to become greater than the time advance, so the total state set of a component d is $Q_d = \{(s_d, e_d) | s_d \in S_d, e_d \geq 0\}$. Once it becomes due, an event must be checked at any global change of state to see if its activation condition has become true. Whenever this condition becomes true, the event can be activated.

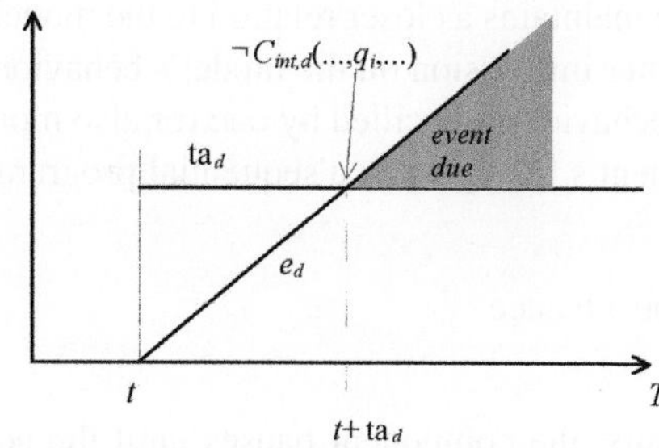

FIGURE 7.2

The meaning of due events.

To capture the activity scanning strategy, we introduce so-called *activity scanning multi-component* DEVS as follows:

$$ASDEVS = \langle X, Y, D, \text{Select} \rangle$$

with the same interpretation as for basic multi-component DEVS. However, the components $d \in D$ are

$$M_d = \left\langle S_d, E_d, I_d, \delta_{\text{ext},d}, C_{int,d}, \delta_{\text{int},d}, \lambda_d, \text{ta}_d \right\rangle$$

with Sd, Ed, I_d, $\delta_{\text{ext},d}$, $\delta_{\text{int},d}$, λ_d, ta_d the same as in the basic multi-component DEVS. Additionally, an activity DEVS employs an *event condition function* $Cd : \times_{i \in I_d} Q_i \rightarrow Bool$.

An activity scanning DEVS works in the following way: Each component has its own state transitions whose occurrence is conditioned by the event condition function. Whenever the time elapses and the event condition becomes true, the component is ready to be activated. The Select-function is used to arbitrate between components having true event conditions simultaneously. Formally, with an $ASDEVS = (X_N, Y_N, D, \text{Select})$ we associate the following DEVS

$$DEVS = \langle X, Y, S, \delta_{\text{ext}}, \delta_{\text{int}}, \lambda, \text{ta} \rangle$$

with $S = \times_{d \in D} Q_d$. In the definition of the time advance function, let $\sigma_d = \text{ta}_d(..., q_i, ...) - e_d$ be the time left to the next time in d. We distinguish two cases. For case (1) if $\exists d \in D : \sigma_d \leq 0 \wedge C_{int,d}(..., q_i, ...)$, i.e. at current simulation time t there is a component which is ready to be activated, then we define $\text{ta}(s) = 0$. Otherwise we have case (2) and we have to advance the time to the time of the next activatable component, i.e., we define $\text{ta}(s) = min\{\sigma_d | d \in D \wedge \sigma_d > 0 \wedge C_{int,d} * (..., q_i, ...)\})$. In either case, we define $d* = Select(\{d | \sigma_d \leq \text{ta}(s) \wedge C_{int,d}(..., q_i, ...)\})$ as the next activatable component. The internal transition function $\delta_{\text{int}}(s)$, the output function $\lambda(s)$ and the external transition function $\delta_{\text{ext}}(s, e, x)$ are defined in the same way as for event scheduling DEVS.

7.2.4 PROCESS INTERACTION MODELS

Process interaction simulation strategy is basically a combined event scheduling – activity scanning procedure. The distinguishing feature is that a model component description can be implemented as a unit rather than being separated into a number of unconnected events and activity routines. The advantage is that the program structure maintains a closer relation to the model structure and by scanning the source code, the reader gets a better impression on the model's behavior.

Each component's dynamic behavior is specified by one (or also more) routine which describes the state changes during the component's life cycle as a sequential program. This sequential program can be interrupted in two ways:

1. by the specification of time advance
2. by a contingency test

Whenever such an interrupt occurs, the component pauses until the activation condition is satisfied, i.e., either the specified time advance has elapsed or the contingency test has become true. Hence, an important state variable in a process is its *program counter* which defines its activation point and, therefore, the next state transition to be executed. The program counter represents a control state in each component. It usually shows the particular phase of processing that a component is in. Thus, it is comparable to the phase variable employed in atomic DEVS models.

The process interaction world view further breaks down into two views. Each view corresponds to a different assumption as to what are the active and passive components in building models of manufacturing or other processing systems. In the first view, the active components are taken to be the entities that do the processing, e.g., the machines, servers, and so on. In the second approach, the active components are the flowing elements, that is the customers, workpieces, packets, etc. Whereas the first is regarded to be the prototypical process interaction world view, the second is regarded as a variation of the first and often referred to as the *transaction world view*.

7.2.5 TRANSLATING NON-MODULAR MULTI-COMPONENT DEVS MODELS INTO MODULAR FORM

Multi-component systems with non-modular couplings, as discussed above, directly access and write on each other's state variables. In the sense of modular couplings, this means a violation of modularity. In the sequel, we give a procedure for translating non-modular coupled systems into modular form. Moreover, we will show that such a translation is always possible. The procedure works by identifying the dependencies between components and converting them into input and output interfaces and modular couplings. *Modularization* is the term used for such introduction of input/output interfaces.

We distinguish two forms of non-modular couplings which we have to treat differently. In case (1) component A has write access to a variable v of another component B (Fig. 7.3). In modular form, this has to be transformed into a communication from A to B through a modular coupling. The situation is modularized by giving A an output port, $vout$, and B an input port, vin, and by a coupling from $vout$ to vin. Whenever, in the original transition function, A wants to write to v, in the modular form, A has to initiate an output event at vout with the same output value. Through the coupling, this results in an input event at vin of B. The external transition function of B has to react by writing the received value into its variable v.

In case (2), component B has read access to a variable u of component A (Fig. 7.3). This situation is more complicated because B always has to have access to the actual value of the variable. The situation is solved in that B holds a copy of the state information of u in its own memory. Now, whenever A changes the value of u, B has to be informed of the change. This is accomplished by a coupling from

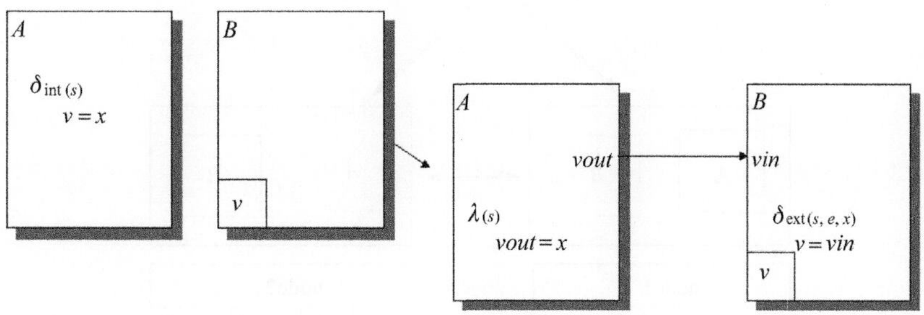

FIGURE 7.3

Component A has write access to variable v of component B.

A to B through ports *uout* and *uin* in a similar way as in case (1). In this way B always has an actual value of variable u in its own state variable *ucopy* (Fig. 7.4).

7.2.6 STATE UPDATING IN DISTRIBUTED SIMULATION

An important form of state updating occurs in distributed simulation, e.g., in the HLA run time infrastructure, as illustrated in Fig. 7.5. The components of a coupled model are mapped onto different computers, or nodes, in a network. To maintain knowledge of the states of other components, replicas of these components are retained on the node dedicated to a component. State updating, is employed to keep the local replicas, or proxies, up to date. In a more advanced concept, the proxies are not full copies of their counterparts but are smaller versions, that are valid simplifications (homomorphisms or endomorphisms) in the experimental frame of the simulation study.

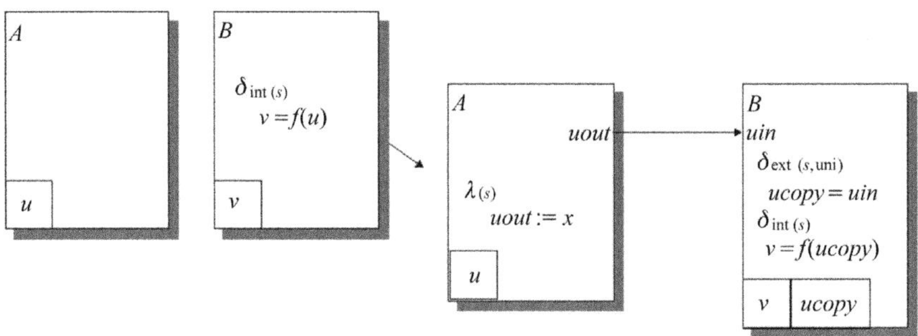

FIGURE 7.4

Component B has read access to variable u of component A.

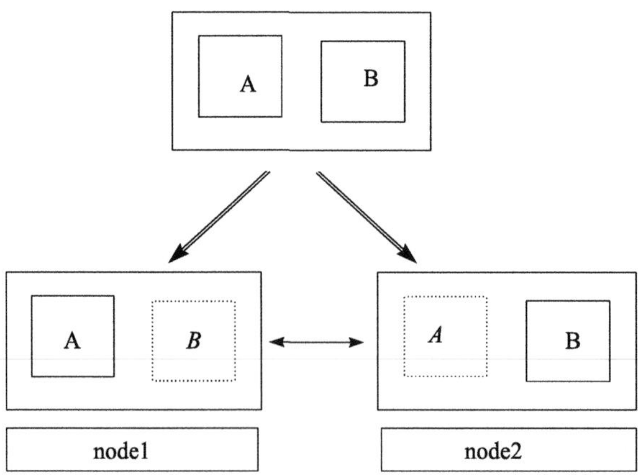

FIGURE 7.5

State Updating in Distributed Simulation.

7.3 DISCRETE TIME SPECIFIED NETWORK FORMALISM

In Chapter 3, we learned to know how to model digital sequential networks as built from elementary flip-flops and boolean gates. Digital networks are examples of networks of discrete time system components. In this section, we introduce the discrete time specified network formalism (DTSN) as a general discrete time formalism at the coupled system level. We specify discrete time networks as couplings of basic discrete time systems and of instantaneous functions, that is, memoryless models.

DELAY-FREE (ALGEBRAIC) CYCLES

Discrete time networks must obey certain constraints to fulfill closure under coupling. The most important constraint is that the network must be free of any *delay-less* (also called *algebraic*) loops, i.e., there is no cycle of output-to-input connections containing a component whose output feeds back to its input without going through some non-zero delay.

Fig. 7.6 will help to clarify. Suppose that we have a network model with three components A, B, and C, which are mutually interconnected. Additionally, A has an input coupling from the network input port. To compute the output of A we need the network input and the output from C. To compute the output of B we need the output from A. Finally to compute $C's$ output we need the output from B, thus forming a cycle of dependencies. Now suppose that there are no delays between input and output in any of the components. This means that all the components impose their input-to-output constraints on the ports at the same time. Then, given an input on the input port, we can try to solve for consistent values on the other ports. If there is an unique solution for every input then a consistent system can result. This approach is commonly adopted in differential algebraic equation models (Hairer and Wanner, 1991). However, we can obviate the problem by requiring that no algebraic loops are present. For example, we require that in the cycle of Fig. 7.6 there is at least one component, whose current output can be computed without knowledge of the current input.

Exercise 7.3. Give examples of choices for the components in Fig. 7.6 that create algebraic loops with consistent solutions. Do the same for inconsistent solutions.

Recall from Chapter 6, that in Mealy-type systems and in instantaneous functions, the output directly depends on the input as well as the state. Only Moore-type components compute their output

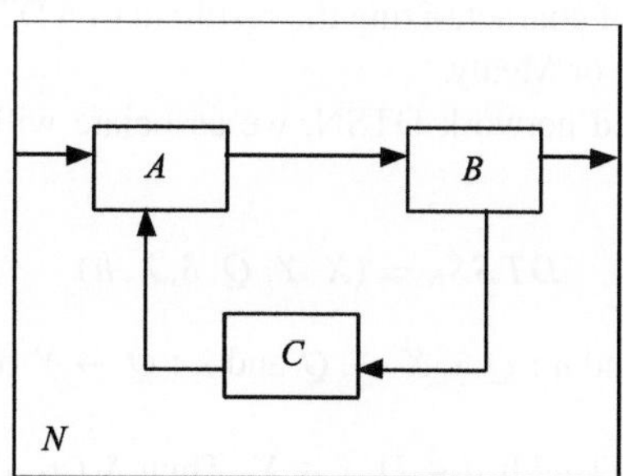

FIGURE 7.6

Dependencies in a network of DTSS.

solely on the state information and therefore impose a delay on the effect of the input on the output. *So to banish delay-less loops we require that in any feedback cycle of coupled components, there must be at least one component of type Moore.*

Exercise 7.4. Show that there are no delay-less loops if, and only if, the following is the case: when we remove all Moore type components then the remaining components form a set of directed acyclic subnetworks.

DEFINITION OF DISCRETE TIME COUPLED MODELS

A discrete time specified network (DTSN) is a coupled system

$$N = \langle X, Y, D, \{M_d\}, \{I_d\}, \{Z_d\}, h_N \rangle$$

where X, Y, D, $\{M_d\}$, $\{I_d\}$, $\{Z_d\}$ defined as in the general multi-component formalism in Section 5.7 and h_N is a constant time advance employed for the specification of the time base $hN \cdot \mathfrak{J}$. To be well defined, a discrete time network has to fulfill the following constraints:

- the components M_d are basic DTSS or FNSS,
- no delay-less cycles are allowed, i.e., in a feedback loop there has to be at least one component whose outputs can be computed without knowledge of its input or

$$\nexists d_1, d_2, ..., d_n \in D \cap \{d | d \in D \wedge d \text{ is a } FNSS \text{ or } d \text{ is of type } Mealy\} : d_1 \in I_{dn} \wedge d_i \in I_{di-1},$$
$$i = 2, ..., n$$

- the time base h_N of the network and all its components have to be identical.

CLOSURE UNDER COUPLING OF DTSS

We now will show that the DTSS formalism is closed under coupling under the conditions just given. Recall, that for the detection of algebraic loops it is crucial to decide if a component is of type Moore or Mealy. Therefore, in the process of characterizing the resultant of a DTSN we will develop a procedure to tell whether it is of type Moore or Mealy.

Given a discrete time specified network DTSN, we associate with it the following basic discrete time specified system

$$DTSS_N = (X, Y, Q, \delta, \lambda, h)$$

where $Q = \times_{d \in D} Q_d, h = h_N$, and $\delta : Q \times X \rightarrow Q$ and $\lambda : Q \rightarrow Y$ or $\lambda : Q \times X \rightarrow Y$ are defined as follows.

Let $q = (q_1, q_2, ..., q_d, ...) \in Q$ with $d \in D, x \in X$. Then $\delta((q_1, q_2, ..., q_d, ...), x) = q' = (q_1', q_2', ..., q_d', ...)$ is defined indirectly through

$$q_d' = \delta_d(q_d, x_d)$$

with $x_d = Z_d(..., y_j, ...)$, where for any $j \in I_d$

$$y_j = \begin{cases} \lambda_j(q_j)) & \text{if } j \text{ is of type Moore} \\ \lambda_j(q_j, x_j)) & \text{if } j \text{ is of type Mealy} \\ \lambda_j(x_j) & \text{if } j \text{ is of type function, respectively.} \end{cases}$$

Note that y_j has a well-defined value in the second and third cases because of the assumption that no algebraic loops are present.

Now the output function, $\lambda : Q \to Y$ or $\lambda : Q \times X \to Y$, where λ is defined as $Z_N(..., y_j, ...)$, $j \in I_N$ and y_j defined as above.

At this stage, we do not yet have a way of deciding whether we have a function $\lambda : Q \to Y$ depending just on state q or it is a function $\lambda : Q \times X \to Y$ also depending on the input x, i.e., if the resulting system is of type Moore or type Mealy. However, having such a decision procedure is essential in hierarchical networks due to the requirement that no delayless cycles exist. Recall that this requires at that at least one Moore-type component to be present in all feedback loops. Thus, in hierarchical construction, we need to know whether a network, that we are encapsulating as a component, is of type Mealy or Moore. The following procedure enables us to decide. We examine the subnetwork of all Mealy and memory-less components. If in this network, we find an input from the network that is connected to an output to the network through Mealy-type components or instantaneous functions, then the network is of type Mealy, otherwise, it is of type Moore. (Since there are no algebraic cycles, the connection from the network to itself really represents a coupling from the external input to the external output.) Formally, we present this procedure in the theorem in the Appendix.

7.4 MULTI-COMPONENT DISCRETE TIME SYSTEM FORMALISM

Recall from Section 5.7 that multi-component systems employ a set I_d of influencing components and a set E_d of components itself influences. We have seen an example of a multi-component discrete time system – the Game of Life cellular automaton in Chapter 3. In this case, an individual component has a set of influencing components, called its *neighbors*. Its state transition function defines new state values only for its own state variables and does not set other components' states directly. It only influences them indirectly because the state transitions of the other components depend on its own state. Therefore, the next state function of the overall multi-component system can be visualized as having all the components look at the state values of their individual influencers and simultaneously computing the next state values for their state variables. We have already seen how this inherently parallel computation can be done on a sequential computer. In Chapter 8 and 10 we will formulate appropriate simulators for each of the formalisms that are amenable to both sequential and parallel computation.

A *Multi-component Discrete Time System Specification* is a structure

$$multiDTSS = \langle X_N, D, \{M_d\}, h_N \rangle$$

with

X_N is an arbitrary set of input values, h_N is the time interval to define the discrete time base, D is the index set.

For each $d \in D$, the component M d is specified as

$$M_d = (Q_d, Y_d, I_d, \delta_d, \lambda_d)$$

where Q_d is an arbitrary set of states of d, Y_d is an arbitrary set of outputs of d, $I_d \subseteq D$ is the set of influencers of d, $\delta_d : \times_{i \in I_d} Q_i \times X \to Q_d$ is the state transition function of d, and $\lambda_d : \times_{i \in I_d} Q_i \times X \to Y_d$ is the local output function of d. (The set of influencees E_d of d usually employed in multi-component models – see Section 5.7 – is defined implicitly to be the set $E_d = \{d\}$, having d as the unique element.)

In a multiDTSS, the set of components jointly generate the dynamics of the system. Each of the components owns its local state set, a local output set and local state transition and output function. Based on the state of the influencing components $i \in I_d$ and on the current input value, the component determines its next state and its contribution to the overall output of the system. In this way, the resultant of a multiDTSS is a DTSS that is built from the crossproduct of the components.

Formally, a $multiDTSS = \langle X, D, \{M_d\}, h_N \rangle$ specifies a $DTSS = \langle X, Y, Q, \delta, \lambda, h \rangle$ at the I/O system level as follows: $Q = \times_{d \in D} Q_d$, $Y = \times_{d \in D} Y_d$, $\delta(q, x)$ is defined by

$$\delta(q, x).d = \delta_d((....., q_i, ...), x),$$

$\lambda(q)$ is defined by

$$\lambda(q).d = \lambda_d((...., q_i, ...))$$

with $i \in I_d$, and

$$h = h_N$$

SPATIAL DTSS: CELLULAR AUTOMATA

Recall the cellular automata discussed in Chapter 3. Those can be formulated as special cases of the multi-component DTSS formalism. For example, consider a two-dimensional automaton such as the Game of Life. The multi-component DTSS model can be specified as follows:

$$multiDTSS = \langle X, D, \{M_{i,j}\}, h_N \rangle$$

X is the empty set since there is no input to the model, $h_N = 1$, since the time base is the set of integers, $\{0, 1, 2, ...\}$, $D = \{(i, j) | i \in I, j \in I\}$ is the index set.

For each $(i, j) \in D$, the component, $M_{i,j}$ is specified as:

$$M_{i,j} = \langle Q_{i,j}, Y_{i,j}, I_{i,j}, \delta_{i,j}, \lambda_{i,j} \rangle$$

where $Q_{i,j} = \{0, 1\}$, $Y_{i,j}$ is empty set since there are no outputs. The set of influencers $I_{i,j}$ is defined by its neighbors $(i, j+1)$, $(i+1, j)$, $(i, j-1)$, $(i-1, j)$, $(i+1, j+1)$, $(i+1, j-1)$, $(i-1, j+1)$, and $(i-1, j-1)$. Let

$$sum = q_{i,j+1} + q_{i+1,j} + q_{i,j-1} + q_{i-1,j} + q_{i+1,j+1} + q_{i+1,j-1} + q_{i-1,j+1} + q_{i-1,j-1}$$

be the sum of all neighbors of a cell at i, j. Then we define $\delta_{i,j} : \times_{k,l \in I_{i,j}} Q_{k,l} \rightarrow Q_{i,j}$ by

$$\delta_{i,j}(q_{i,j}, q_{i,j+1}, q_{i+1,j}, q_{i,j-1}, q_{i-1,j}, q_{i+1,j+1}, q_{i+1,j-1}, q_{i-1,j+1}, q_{i-1,j-1}) =$$
$$= \begin{cases} 1 & \text{if } q_{i,j} = 1 \wedge (sum = 2 | sum = 3) \\ 0 & \text{if } q_{i,j} = 1 \wedge (sum < 2 | sum > 3) \\ 1 & \text{if } q_{i,j} = 0 \wedge sum = 3 \\ 0 & \text{if } q_{i,j} = 0 \wedge sum \neq 3 \end{cases}$$

$\lambda_{i,j}$ is not specified since there is no overall output.

In general, any cellular automaton has a multi-dimensional geometrical grid structure which is formalized as an Abelian group. A *neighborhood* template is specified as a subset of the group and defines the influencers of the cell at the origin, or zero element, of the group. By uniformity, the influencers of the cell at any other point can be obtained by translating the neighborhood template to that point, i.e., $I_{i,j} = N + (i, j)$, where N is the neighborhood template. The cells all have isomorphic component structures: basically, the same state sets and the same state transition functions. In other words, a cellular automaton can be specified by one neighborhood template and one component structure.

Exercise 7.5. Show that the Game of Life multi-component formalization above can be put into the space invariant form just described. (Hint: relate the influencers of cell (i, j) to those of cell $(0, 0)$, hence to a neighborhood template and show that the transition function $\delta_{i,j}$, can be expressed as a function of state values assigned to the appropriate elements in this template.)

Exercise 7.6. Formulate a space-invariant DEVS cell space formalism analogous to the DTSS formalism for cellular automata just discussed. Express the discrete event form of the Game of Life in this formalism.

7.5 DIFFERENTIAL EQUATION SPECIFIED NETWORK FORMALISM

Networks of differential equation specified systems (DEVN) are analogous to discrete time networks and the same ideas apply. Let us introduce networks of differential equation specified systems in the following by translating the considerations of discrete time networks to the continuous domain.

Recall from above that discrete time networks are specified as modular couplings of basic discrete time systems and memoryless instantaneous functions. In the continuous domain we have basic differential equation specified systems and again instantaneous functions. The whole network describes a basic differential equation as the crossproduct of the continuous components where the input rates are defined based on the influencers' outputs.

We introduce *differential equation specified network* (DESN), analogous to discrete time networks, as a coupled system

$$N = \langle X, Y, D, \{M_d\}, \{I_d\}, \{Z_d\} \rangle$$

with the following constraints:

- X and Y have to be real vector spaces,
- the components $d \in D$ have to be DESS or FNSS, and

- no algebraic cycles are allowed, i.e., in a feedback loop there has to be at least one component the output of which can be computed without knowledge of its input.

CLOSURE UNDER COUPLING OF DESS

Given a differential equation specified network DESN, we are able to associate with it the basic differential equation specified system, $DESS_N$

$$DESSN = \langle X, Y, Q, f, \lambda \rangle$$

where $Q = \times_{d \in D} Q_d$, and $f : Q \times X \to Q$ and $\lambda : Q \to Y$ or $\lambda : Q \times X \to Y$ defined similar to δ and λ in DTSN as follows.

Let $q = (q_1, q_2, ..., q_d, ...) \in Q$ with $d \in D$, $x \in X$. Then $f((q_1, q_2, ..., q_d, ...), x) = q' = (q'_1, q'_2, ...q'_d, ...)$ is defined indirectly through $q'_d = f_d(q_d, x_d)$ with $x_d = Z_d(..., y_j, ...)$, $j \in I_d$ and $y_j = \lambda_j(q_j)$ or $y_j = \lambda_j(q_j, x_j)$ or $y_j = \lambda_j(x_j)$ if the component is of type Moore, of type Mealy, or an instantaneous function, respectively. The rest of the proof that f is well-defined follows along the lines of the closure proof for DTSS.

However, not only must f be well defined, it must also satisfy the Lipschitz condition.

$$||f(q, x) - f(q', x)|| \le k||q - q'||$$

The proof that this is true will follow from the observation that f is composed of a sequence of coordinate functions, f_d, taking the form of the non-modular DESS specification to be discussed below. We will show that if each of the coordinate functions in the latter specification satisfies a Lipschitz condition, then so does the composite function. Since by assumption, in the present case, the f_d are the rate-of-change functions of well-defined DESS specifications, they do indeed satisfy the Lipschitz condition. Thus, on the basis of a proof yet to come, f satisfies the Lipschitz condition.

7.6 MULTI-COMPONENT DIFFERENTIAL EQUATIONS SPECIFIED SYSTEM FORMALISM

In a similar way to the DTSS case, we formulate a multi-component differential equation system specification, *multiDESS*, with non-modular coupling. Recall that the basic DESS formalism does not define a next state function directly but only through rate-of-change functions for the individual continuous state variables. In the multi-component case, the individual components define the rate of change of their own state variables based on the state values of their influencers. Let us first define the general formalism and then discuss the modeling approach by considering partial differential equations – a special model type showing much resemblance to the cellular automata in the discrete time domain.

A *multi-component differential equation system specification* is a structure

$$multiDESS = \langle X, D, \{M_d\} \rangle$$

where X is the set of inputs, a real valued vector space $\mathbb{R}^m$ and D is the index set. For each $d \in D$, the component M_d is specified as

$$M_d = \langle Q_d, Y_d, I_d, f_d, \lambda_d \rangle$$

where Q_d is the set of states of d, a real valued vector space $\mathbb{R}^n$, Y_d is the set of outputs of d, a real valued vector space $\mathbb{R}^p$, $I_d \subseteq D$ is the set of influencers of d, $f_d : \times_{i \in I_d} Q_i \times X \to Q_d$ is the rate of change function for state variables of d, $\lambda_d : \times_{i \in I_d} Q_e \times X \to Y_d$ is the local output function of d. The set of influencees E_d of d is again defined to be the set $\{d\}$. We require that each f_d satisfies a Lipschitz condition:

$$||f_d(q, x) - f_d(q', x)|| \leq k_d ||q - q'||$$

In a multiDESS, the derivative function of each component defines the rate of change of its local state variables. Formally a $multiDESS = \langle X_N, D, \{M_d\} \rangle$ specifies a $DESS = \langle X, Y, Q, f, \lambda \rangle$ at the I/O system level in the following way: $Q = \times_{d \in D} Q_d$, $Y = \times_{d \in D} Y_d$, $f(q, x)$ is defined by

$$f(q, x).d = f_d((..., q_i, ...), x),$$

and $\lambda(q)$ is defined by

$$\lambda(q).d = \lambda_d((..., q_i, ...)),$$

with $i \in I_d$.

Now, we must show that the resultant derivative function satisfies the Lipschitz condition:

$$||f(q, x) - f(q', x)|| \leq k ||q - q'||$$

This will follow from the fact that each of its coordinate functions satisfies such a condition by the constraint placed on these functions given before. We demonstrate how this works using two coordinates only:

$$\begin{aligned}
&||f(q_1, q_2, x) - f(q_1', q_2', x)|| \\
&= ||(f_1(q_1, q_2, x) - f_1(q_1', q_2', x), f_2(q_1, q_2, x) - f_2(q_1', q_2', x))|| \\
&\leq ||f_1(q_1, q_2, x) - f_1(q_1', q_2', x)|| + ||f_2(q_1, q_2, x) - f_2(q_1', q_2', x)|| \\
&\leq k_1 ||(q - q'|| + k_2 ||(q - q'|| \\
&\leq (k_1 + k_2) ||q - q'||
\end{aligned}$$

7.6.1 SPATIAL DESS: PARTIAL DIFFERENTIAL EQUATION MODELS

Partial differential equation models emerge from an extension of differential equation where space coordinates, besides time, are introduced as independent variables. A partial differential equation specified system therefore shows variations in time as well as in space.

Partial differential equation systems require a science of their own and a whole discipline deals with the solution of such differential equation systems. We will only give an short glimpse of them here to place them into our framework of simulation modeling formalisms.

For our exposition, let us consider a simple example of a general flux-conservative equation in one variable u. The equation

$$\frac{\partial u}{\partial t} = -v\frac{\partial u}{\partial x}$$

expresses that the change of the variable u in time is equal to the negative velocity, $-v$, multiplied by the change of the variable u in the spatial dimension x. The result of this equation is a wave which propagates with velocity v along the x dimension.

The approach to solve such problems, which is a representative of the so-called *hyperbolic* partial differential equation, leads to the discretization of the space and time dimensions. Let us first introduce a discretization of space. The entire observation interval $[x_0, x_l]$ with length l is divided into k equal pieces each $\Delta x = l/k$ wide. Then we get k *mesh points* for which we set up equations to express the changes over time. In the so-called *Forward Time Centered Space* (FCTS) approach, this is done for each mesh point j by replacing the spatial derivative $\frac{\partial u_j}{\partial x}$ of u at point j by the difference of the neighboring states divided by the length of the spatial interval

$$\frac{u_{j-1} - u_{j+1}}{2\Delta x}$$

(note the similarity to the Euler integration method) giving an equation for the time derivative of variable u at point j

$$\frac{\partial u_j}{\partial t} = -v\frac{u_{j-1} - u_{j+1}}{2\Delta x}$$

for each mesh point j. Obviously, we have a multiDESS with k components and influencers set $I_j = \{j-1, j+1\}$ for each component j, as well as derivative functions as above.

Usually, when solving partial differential equations the model is set up by discretizing also the time dimension. When we apply the same difference method for discretizing the time dimension, namely dividing the time interval into intervals of equal length Δt, we can replace the time derivative of u at spatial point j and time point $n+1$ by the difference of the value at time $n+1$ minus the value at time n divided by Δ_t (Euler integration)

$$\frac{u_j^{n+1} - u_j^n}{\Delta t}$$

With that we finally obtain an equation for state at mesh point j for time $n+1$:

$$u_j^{n+1} = u_{j-1}^n - v\frac{u_j^{n-1} - u_j^{n+1}}{2\Delta x}\Delta t.$$

What have we accomplished finally? Starting with a partial differential equation with derivatives in time and space dimension, we have discretized space and time. With the discretization of space we obtained a continuous multi-component model in cellular form with equal derivative functions for the cells. With the discretization of space we finally have obtained a cellular automaton with neighborhood $\{j-1, j+1\}$, time step Δ_t, and equal next state function for cell j as above.

7.7 MULTI-COMPONENT PARALLEL DISCRETE EVENT SYSTEM FORMALISM

In the second edition parallel DEVS was introduced and applied to PDEVS networks, i.e., coupled models with modular components. As Vicino et al. (2015) point out, there was no discussion of how the equivalent parallel multi-component networks whose components are non-modular. Essentially, this means that we must omit the Select function from the specification and handle the resulting collisions when more than one component wants to change the state of another component. Recall that each component in a non-modular network can have two subsets of components: the influencers and influencees. When only one component is imminent it looks at the states of its influencers and uses them to compute new states for its influencees. So the question arises in the parallel case were there is no selection of one actor from the imminents: how should the desires of multiple imminents be resolved into unique next states for their collective influencees? Foures et al. (2018) define a multi-component Parallel DEVS formalism, called multiPDEVS, in which the imminents are considered to "propose" new states for their influences which have their own "reaction" function that implements an autonomous approach to uniquely resolving the multiple proposals. MultiPDEVS employs PDEVS constructs to collect state collisions in a bag and manage them explicitly without increasing message exchanges. Important points of the multiPDEVS formalism:

- A multiPDEVS network is shown to be equivalent to a well-defined atomic PDEVS model supporting hierarchical construction
- An abstract simulator is defined and provides implementation perspective.
- The CellSpace as studied by Wainer (Wainer and Giambiasi, 2001) can be considered as a restriction of multiPDEVS.
- implementation showed significant speedup for highly communicative models with tight coupling but also for modeling paradigms falling under bottom-up approaches.
- MultiPDEVS works at the model level as opposed to flattening of hierarchical structure (Bae et al., 2016) which produces direct coupling at the lowest level to reduce message routing. The non-modular approach eliminates I/O ports so that components are able to influence each other directly.

Extension of the multiPDEVS approach can produce a non-modular equivalent of the modular DTSS network (Section 7.5) which could support Multi-agent System modeling with multiPDEVS as agents and multiPDTSS as environment.

7.8 SUMMARY

After defining modular and non-modular coupled models and showing how a non-modular system can be transformed into modular form for DEVS, let us conclude this chapter with a comparison of the two approaches.

From the procedure for translating non-modular couplings into modular form, it is clear that non-modular systems are easier to build in the conception phase, because the whole system can be built as one gestalt without breaking up the interrelations of components. Components do not have to maintain information about their neighbors since they are allowed to access others' state information directly.

However, this deceptive advantage soon becomes undermined when one considers such issues as testing components or components' reusability. Components in modular coupled models are systems by themselves. From their interface description one has a clear understanding what the inputs to this systems can be and what one can expect as outputs. The input and output interfaces unambiguously show what each component needs to get from, and what it provides to other components. Components' dynamic behavior only depends on its inner state and its inputs. Such a module is much easier to comprehend and test for correctness. And modeling coupled systems in a bottom-up approach with well tested components is much less error-prone than developing complex systems as one big unit.

Also, modular system components lend themselves for reusability. Again the interface description gives a clear understanding where a component can be used. An interface-based classification of components (Thoma, 1990) is useful for organizing families of compatible components. We will come back to this question of reusability of simulation models when we discuss the organization of model bases and the architecture of modeling and simulation environments in the third part of the book.

Finally, modular coupled models best fit the needs of distributed simulation since their components can be assigned to network nodes in a straightforward manner.

The third edition added new Section 7.7 on multiPDEVS which fills the hole opened with the introduction of Parallel DEVS and its application in a modular but not non-modular way. The text summarizes (Foures et al., 2018) article which provides a comprehensive treatment of the introduced formalism addressing points such as well-definition and abstract simulators that are of interest whenever a new formalism is introduced.

Event routing and message sending overhead in DEVSRuby is reduced using techniques similar to Himmelspach and Uhrmacher (2006), and Vicino et al. (2015).

7.9 SOURCES

Kiviat (1971) was the first to characterize the different world views of discrete event simulation while Tocher (1969) was the first to provide a formal representation of the activity scanning world view. Nance (1971) characterized the time flow mechanisms for discrete event simulation. Renewed interest in activity scanning strategy is emerging in distributed simulation contexts (see Chapter 11). Process interaction world view discussions can be found in Cota and Sargent (1992) and Franta (1977). The hierarchical, modular approach to discrete event simulation was introduced by Zeigler (1984) and its relation to testing, reusability, and distributed simulation is discussed in Zeigler (1990).

APPENDIX 7.A

The following theorem gives the condition under which a network is of type Moore.

Theorem 7.1 (Characterization of Moore-type Networks). *Let* $IMealy_d = \{e | e \in I_d \text{ and } e \text{ is of type } Mealy \text{ or } Function \text{ or } e = N\}$ *be those influencers in set* I_d *which are of type Mealy, Functions or the network itself and let*

$$IMealy_d^+ = IMealy_d \cup \bigcup_{e \in IMealy_d} IMealy_e^+$$

be the transitive closure of $IMealy_d$*. Then the network* N *is of type Moore if and only if* $N \notin IMealy_N^+$.

Proof. Let $IMealy_N^0 = \{N\}$ and for $i = 1, 2, ...$ let

$$IMealy_d^i = (\cup_{\{e \in IMealy_N^{i-1}\}} I_e) \cap \{d | d \text{ is of type Mealy or FNSS or } d = N\}$$

be the indirect influencers of type Mealy of network output y N at level i. As there are no algebraic cycles allowed and the set of components is finite, $\exists i \in \mathfrak{J}^+$ such that $IMealy_N^i = \{\}$ and let n be the maximum i with $IMealy_N^i \neq \{\}$. Obviously,

$$IMealy_N^+ = \cup_{i=1,...,n} IMealy_N^i$$

For $i = 1, ..., n+1$, let

$$IMoore_N^i = \cup_{e \in IMealy_N^{i-1}} I_e \cap \{d | d is of type Moore\}$$

be the influencers of type Moore of the Mealy influencers at level $i-1$. Then for $i = 1, ..., n$ and $\forall e \in IMealy_N^{i-1}$ it follows that $I_e \subseteq IMealy_N^i \cup IMoore_N^i$. Now, since $IMealy_N^{n+1} = \{\} \forall e \in IMealy_N^n$ we have $I_e \subseteq IMoore_N^{n+1}$.

/A) We first show if N is of type Moore, it is required that $N \notin IMealy_N^+$ and then from above follows $N \notin IMealy_N^i$ for $i = 1, ..., n$.

The output of network N is defined as a function Z_N dependent on the output of the Mealy and Moore influencers at level 1, $y_N = Z_N(..., \lambda_{d1}(q_{d1}, x_{d1}), ..., \lambda_{e_1}(q_{e1}), ...)$, for all $d_1, e_1 \in I_N, d1 \in IMealy_N^1$, $e_1 \in IMoore_N^1$. And for all $d_1 \in Imealy_N^1$ the output $\lambda_{d1}(q_{d1}, x_{d1})$ is computed using their input x_{d1}. However inputs x_{d1} are computed by functions Z_{d1} which are dependent on influencers of d_1 which are in the level 2 Mealy or Moore influencers of N, $x_{d1} = Z_{d1}(..., \lambda_{d2}(q_{d2}, x_{d2}), ..., \lambda_{e2}(q_{e2}), ...)$ for all $d_2, e_2 \in I_{d1}, d_2 \in IMealy_N^2$, $e_2 \in IMoore_N^2$. The inputs x_{d2} of the Mealy influencers d_2 are again computed in the same way using the influencers of d_2 which are in the level 3 Mealy and Moore influencers of N. We continue this process until we come to level n components' inputs. For these, however, the set of influencers of type Mealy is empty and the inputs are definitely only dependent on the Moore type components, $x_{dn} = Z_{dn}(..., \lambda_{en+1}(q_{en+1}), ...)$ for all $e_{n+1} \in I_{dn}$, $e_{n+1} \in IMoore_N^{n+1}$.

This shows that if $N \notin IMealy_N^+$, following the lines of computations of outputs of Mealy type influencers of the network output above, the network output y_N is finally only a function of the state vector $(..., q_d, ...)$, $d \in D$.

(b) The proof that, if $N \in IMealy_N^+$, the output of N is also dependent on the input x_N of N and the network therefore is of type Mealy, follows the lines of proof (a). □

REFERENCES

Bae, J.W., Bae, S.W., Moon, I.-C., Kim, T.G., 2016. Efficient flattening algorithm for hierarchical and dynamic structure discrete event models. ACM Transactions on Modeling and Computer Simulation (TOMACS) 26 (4), 25.

Chow, A.C.H., 1996. Parallel DEVS: a parallel, hierarchical, modular modeling formalism and its distributed simulator. Transactions of the Society for Computer Simulation 13 (2), 55–68.

Cota, B.A., Sargent, R.G., 1992. A modification of the process interaction world view. ACM Transactions on Modeling and Computer Simulation (TOMACS) 2 (2).

Foures, D., Franceschini, R., Bisgambiglia, P.-A., Zeigler, B.P., 2018. MultiPDEVS: a parallel multicomponent system specification formalism. Complexity 2018.

Franta, W., 1977. A Process View of Simulation. North-Holland, New York.

Hairer, E., Wanner, G., 1991. Solving Ordinary Differential Equations. II, Stiff and Differential-Algebraic Problems. Springer-Verlag, Berlin.

Himmelspach, J., Uhrmacher, A.M., 2006. Sequential processing of PDEVS models. In: Proceedings of the 3rd EMSS.

Kiviat, P.J., 1971. Simulation languages. In: Computer Simulation Experiments with Models of Economic Systems, pp. 397–489.

Nance, R.K., 1971. On time flow mechanisms for discrete system simulation. Management Science 18 (1), 59–73.

Thoma, J.U., 1990. Bondgraphs as networks for power and signal exchange. In: Simulation by Bondgraphs. Springer, pp. 9–40.

Tocher, K., 1969. Simulation languages. In: Aronsky, J. (Ed.), Progress in Operations. Wiley, NY.

Vicino, D., Niyonkuru, D., Wainer, G., Dalle, O., 2015. Sequential PDEVS architecture. In: Proceedings of the Symposium on Theory of Modeling & Simulation: DEVS Integrative M&S Symposium. Society for Computer Simulation International, pp. 165–172.

Wainer, G.A., Giambiasi, N., 2001. Application of the cell-DEVS paradigm for cell spaces modelling and simulation. Simulation 76 (1).

Zeigler, B.P., 1990. Object-Oriented Simulation with Hierarchical Modular Models. Academic Press, Boston.

Zeigler, B.P., 1984. Multifacetted Modelling and Discrete Event Simulation. Academic Press Professional, Inc, London.

CHAPTER

SIMULATORS FOR BASIC FORMALISMS

8

CONTENTS

In the view of the hierarchy of system specifications (Chapter 5), simulation is actually a transformation from higher-level system structure knowledge to the I/O relation observation level (Fig. 8.1). Typically, we are given a system specification (e.g., a coupled model) together with the initial state values for all the state variables and time segments for all input ports at the model's interface. The task of a simulator is to generate the corresponding state and output trajectories. To do this, it has to realize the background conventions associated with the particular formalism in which the model is expressed. For example, a simulator, which is given a differential equation, must be written to accept piecewise continuous input trajectories and produce such trajectories for the state and output variables. In this chapter we construct corresponding simulation objects and algorithms for the DEVS, DTSS and DESS modeling formalisms.

We erect a framework that exploits the hierarchical, modular construction underlying the high level of system specifications. While each formalism requires its specific interpretation, the framework is

Theory of Modeling and Simulation. https://doi.org/10.1016/B978-0-12-813370-5.00016-X

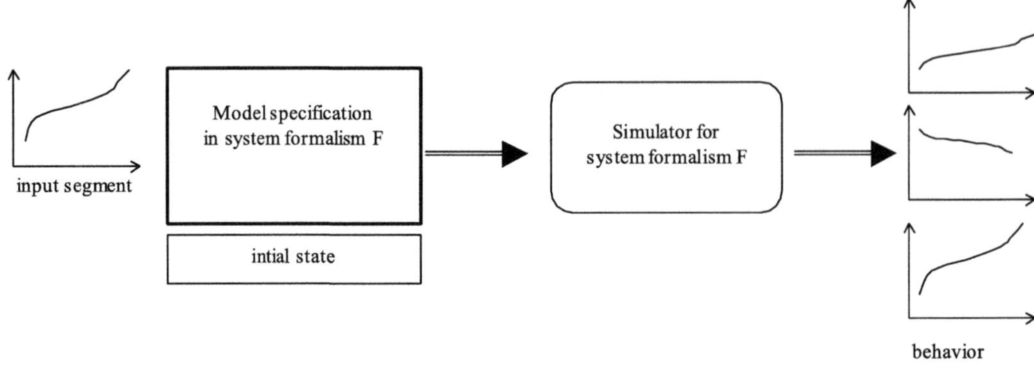

FIGURE 8.1

A Simulator realizes the background conventions of its modeling formalism.

generic and is based on the same concepts we have already seen in the context of system formalisms (Chapter 5 and 7):

- The basic specification structure of each formalism is supported by a class of atomic models. For example, we can write the elements of a basic DEVS in the *atomic* class.
- Each such atomic class has its own *simulator* class.
- The coupled model structure of each formalism is supported by a class of coupled models. For example, we can write the elements of a coupled DEVS in the *coupled* class.
- Each such coupled model class has its own simulator class, called *coordinator*.
- Simulator objects adhere to a generic message protocol that allows others to coordinate with them to execute the simulation.
- Assignment of coordinators and simulators follows the lines of the hierarchical model structure, with simulators handling the atomic level components, and coordinators handling successive levels until the root of the tree is reached.

Fig. 8.2 shows the mapping of a hierarchical model to an abstract simulator associated with it. Atomic simulators are associated with the leaves of the hierarchical model. Coordinators are assigned to the coupled models at the inner nodes of the hierarchy. At the top of the hierarchy a *root-coordinator* is in charge to initiate the simulation cycles.

Although the framework employs message passing, it can be realized in different ways, and it can be optimized depending on the performance requirements. Indeed, we may refer such simulators as *abstract simulators* in the sense that they characterize what has to be done to execute atomic or coupled models with hierarchical structure, not necessarily how it is to be done.

The ability to break down the simulation of hierarchical DEVS and other formalisms in terms of hierarchical simulators ultimately depends on the closure under coupling of such formalisms. Later (Chapter 11) we exploit closure under coupling to develop another approach to simulation of DEVS models, particularly Parallel DEVS models.

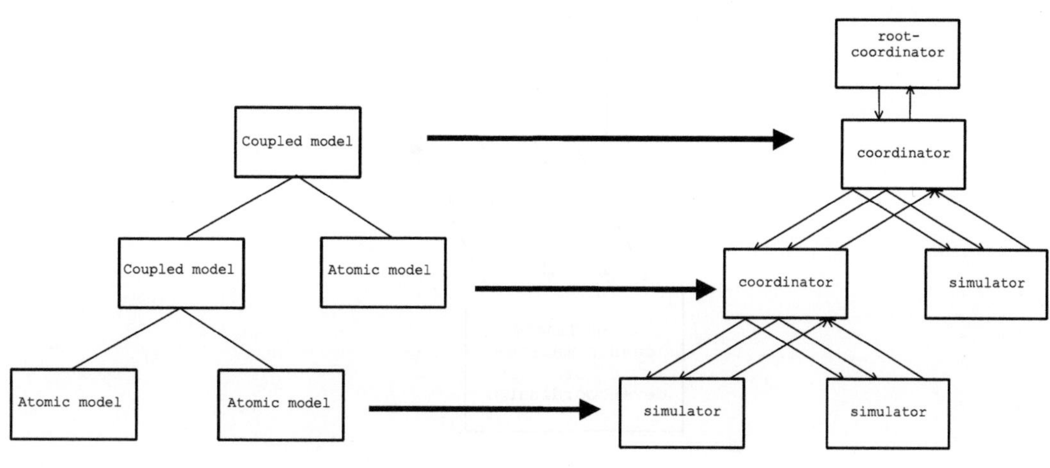

FIGURE 8.2

Mapping a Hierarchical Model onto a Hierarchical Simulator.

8.1 SIMULATORS FOR DEVS

In Chapter 3 we have seen that discrete event simulation works by maintaining a list of events sorted by their scheduled event times. Events are executed by taking them from the event list in sequential order and doing the state transitions. Executing events results in new events being scheduled and inserted into the event list as well as events being canceled and removed from the event list.

Based on these general considerations, we formulate simulation schemes for modular hierarchical DEVS. We formulate simulators for atomic DEVS and coordinators for coupled DEVS. Then we show that a hierarchical structure of simulators and coordinators correctly simulates a DEVS coupled model. The event list in such architectures takes on a hierarchical structure which has an intrinsically efficient time management scheme. Moreover, this approach has the advantage that hierarchical DEVS can be readily transformed into executable simulation code. Finally, it has the advantage that it *supports formalism interoperability*. To explain: we will formulate simulators for multi-component DEVS in non-modular form which capture major world views (Chapter 7). Due to the polymorphic protocols, these simulators appear to coordinators as if they were basic DEVS. Generalizing this DEVS-Bus concept, we will show that it allows the integration of modular components formulated in different modeling approaches, not just those restricted to the ones considered so far.

A hierarchical simulator for hierarchical DEVS coupled model models consists of *devs-simulators* and *devs-coordinators* and uses three types of messages (Fig. 8.3). An initialization method

$$(i, t)$$

is sent at the initialization time from the parent simulator object to all its subordinates. The scheduling of events is done by the internal state transition message $(*, t)$ and are sent from the coordinator to its imminent child. An output message (y, t) is sent from the subordinates to their parents to notify them of output events. The input message (x, t) is sent from the coordinator to its subordinates to cause external events.

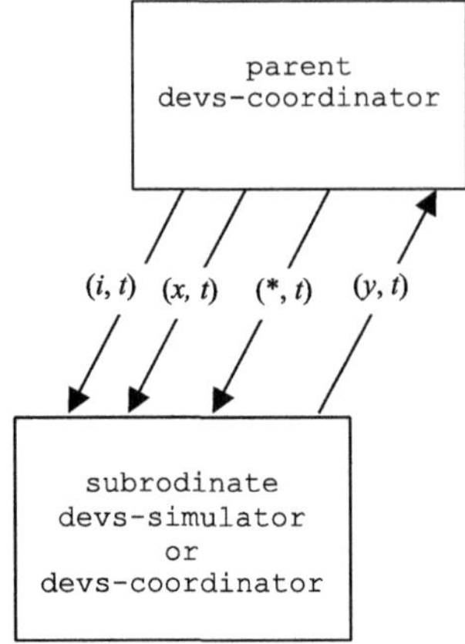

FIGURE 8.3

DEVS simulator protocol.

8.1.1 SIMULATOR FOR BASIC DEVS

An interpretation of the dynamics of a DEVS is given by considering the *devs-simulator* for DEVS. The simulator employs two time variables tl and tn. The first holds the simulation time when the last event occurred and the second holds the scheduled time for the next event. From the definition of the time advance function of DEVS it follows that

$$tn = tl + \mathrm{ta}(s).$$

Additionally, if it is given the global current simulation time t, the simulator can compute from these variables the elapsed time since the last event

$$e = t - tl$$

and the time left to the next event

$$\sigma = tn - t = \mathrm{ta}(s) - e.$$

The time of next event, tn is sent to its simulator's parent to allow for correct synchronization of events.

As shown in Algorithm 1, for correct initialization of the simulator a initialization message

$$(i, t)$$

has to be sent at the beginning of each simulation run. When a devs-simulator receives such an initialization message, it initializes its time of last event tl by subtracting the elapsed time e from the time t provided by the message. The time of next event tn is computed by adding the value of the time advance $\mathrm{ta}(s)$ to the time of last event tl. This time tn is sent to the parent coordinator to tell it when the first internal event should be executed by this component simulator.

Algorithm 1 Simulator for basic DEVS.

```
Devs-simulator
variables :
    parent                                  ▷ parent coordinator
    tl                                      ▷ time of last event
    tn                                      ▷ time of next event
    DEVS                                    ▷ associated model with total state (s, e)
    y                                       ▷ current output value of the associated model
when receive i-message (i, t) at time t
    tl = t − e
    tn = tl + ta(s)
when receive ∗-message (∗, t) at time t
    if t ≠ tn then
        error: bad synchronization
    end if
    y = λ(s)
    send y-message (y, t) to parent coordinator
    s = δint(s)
    tl = t
    tn = tl + ta(s)
when receive x-message (x, t) at time t with input value x
    if not (tl ≤ t ≤ tn) then
        error: bad synchronization
    end if
    e = t − tl
    s = δext(s, e, x)
    tl = t
    tn = tl + ta(s)
end Devs-Simulator
```

An internal state transition message $(*, t)$ causes the execution of an internal event. When a $*$-message is received by a devs-simulator, it computes the output and carries out the internal transition function of the associated DEVS. The output is sent back to the parent coordinator in an output message (y, t). Finally the time of the last event tl is set to the current time and the time of the next event is set to the current time plus the value of the time advance function $\text{ta}(s)$.

An input message (x, t) informs the simulator of an arrival of an input event, x, at simulation time, t. This causes the simulator to execute the external transition function of the atomic DEVS given x and the computed value of elapsed time, e. As with internal events, the time of the last event tl is set to the current time and the time of the next event is set to the current time plus the value of the time advance function $\text{ta}(s)$.

In Chapter 6 we defined the dynamic interpretation of DEVS by defining the global state transition Δ in points (1), (2), (3) and output function Λ in point (4) of the I/O system associated with the DEVS.

In the following we state and prove the theorem that the devs-simulator interprets the DEVS as defined by the I/O system.

Theorem 8.1. *Under the assumption that at initial time t_i the system has been in state s for e time units and the parent coordinator provides an i-message at initial time t_i before any other message, x-messages for every event-input of the current input segment $\omega :< t_i, t_f] \rightarrow X$, and $*$-messages when the simulation time t is equal the value tn of the simulator in a timely ordered sequence, then the state $(s, t - tl)$ at any event time t corresponds to the state trajectory of the I/O system associated with the DEVS. Likewise, the outputs sent out in y-messages corresponds to the output trajectory.*

Proof. The proof follows the dynamic definition of the global state transition and output function of the DEVS (compare Chapter 6). In particular, we show that, if at time $t = tl$, $(s, t - tl) = (s, 0)$ is correct then the state after the execution of the next x-or $*$-message is correct. We also show that tn always represents the time of the next scheduled event.

Let (s, e) be the correct state at time $t = t_i$. After the i-message at time $t = t_i$, variable tl is set to $t_i - e$. Obviously, state $(s, tl - tl) = (s, 0)$ was the correct state at time $t = tl$. Also variable tn is set to $tl + \mathrm{ta}(s)$ as required.

Let $(s, 0)$ be the correct state at time tl and let $tn = tl + \mathrm{ta}(s)$ be the time of the next scheduled event. Then, by the assumption above, either an x-message or a $*$-message has to be received next, depending on whether an input event or internal event will occur next.

Let the next message be an x-message with input value $x = \omega(t)$ at time $t \leq \mathrm{ta}(s)$. Then following the commands of the x-message (Algorithm 1), the state is set to $s' = \delta_{\mathrm{ext}}(s, t - tl, x)$, tl is set to t and finally, tn is set to $tl + \mathrm{ta}(s)$. Obviously $(s', t - tl) = (s', 0)$ finally corresponds the global state transition function $\Delta(s, \omega_{(tl,t]})$ according to point (3) $(\Delta((\delta_{\mathrm{ext}}(s, e + t - t_i, x), 0), \emptyset_{(t,t)}) = (\delta_{\mathrm{ext}}(s, t - tl, x), 0)$, with $t_i = tl$ and $e = 0$.) Also tn is set to $tl + \mathrm{ta}(s)$ which by definition is the time of the next scheduled event.

In the case that there is no input event until time tn, then by assumption above the next message is a $*$-message at time tn. As $tn = tl + \mathrm{ta}(s)$, $t - tl = \mathrm{ta}(s)$ as required by point (2) in Chapter 6. Following the commands of the $*$-message one sees that the state is set to $s' = \delta_{\mathrm{int}}(s)$ and $tl = t$. This corresponds to the definition of $\Delta(s, \omega_{(tl,t]})$ according to points (2) of Chapter 6. Also tn is set to $tl + \mathrm{ta}(s)$.

The devs-simulator guarantees that it sends out an output $\lambda(s)$ in a y-message exactly when there is an internal transition at time tn. Hence the trajectory constructed from the y-messages (y, t) sent out from the devs-simulator at internal event times tn and constant value $\emptyset$ between these events corresponds to the output function definition Λ in point (4) of the I/O system specification of DEVS in Chapter 6. □

8.1.2 SIMULATORS FOR MODULAR DEVS NETWORKS

In a simulator for a DEVS coupled model, each model component is handled by its own simulator which is assumed to correctly simulate it. A coordinator, assigned to the network, is responsible for the correct synchronization of the component simulators and handling of external events arriving at the network inputs. To perform its task, the coordinator makes use of the protocols of its subordinates and provides the same protocol to its parent.

The coordinator implements an event list algorithm to take care for the correct synchronization of its components. It maintains an event list which is a list of pairs of components simulators with their times of next event. This list is sorted by the event times tn and, when the tn of components are equal, by the Select function of the coupled model. The first component in the event list defines the next internal event within the scope of the coordinator. This minimum next event time

$$tn = min\{tn_d | d \in D\}$$

is provided to the parent of the coordinator as its next event time. In a similar way, the time of last event of the coordinator is computed by

$$tl = max\{tl_d | d \in D\}$$

as the last time when an event occurred in one of the subordinate simulators. Additionally, a coordinator has to handle input events arriving at the network's input interface.

As mentioned above, the coordinator responds to, and sends, the same types of messages as do devs-simulators. As shown in Algorithm 2, it may receive a message

$$(i, t)$$

for initialization purposes, internal state transition messages $(*, t)$ to execute internal events, and input messages (x, t) for external events. It sends the events at the network output interface back to its parent in output messages (y, t). Additionally, the coordinator has to handle the y-messages received from its subordinates.

When a devs-coordinator receives an i-message it forwards it to each of its children. When all the children have handled their i-messages, it sets its time of last event tl to the maximum of the times of last event of the children and its time of next event tn to the minimum of the times of next event of its children.

When a devs-coordinator receives a $*$-message, the message is transmitted to its imminent child which is the first in the event list. The imminent child executes its state transition and may send output messages (y, t) back. After the completion of the internal event and all the external events caused by the output of the imminent component, the time of the next event is computed as the minimum of the next event times of its children.

When a coordinator receives an output message (y_d*, t) which carries the output information of its selected imminent child, it consults the external output coupling of the DEVS network to see if it should be transmitted to its parent coordinator and the internal couplings to obtain the children, and their respective input ports, to which the message should be sent. In the first case, an output message (y_N, t) carrying the external network output y_N is sent to the parent coordinator. In the second case, the output message (y_{d*}, t) is converted into input messages $(x_r(y_d*), t)$ with $x_r = Z_{d*,r}(y_{d*})$, indicating the arrival of external events at the influenced components, r.

When a coordinator itself receives an input message (x, t) from its parent coordinator, it consults the external input couplings of the DEVS network to generate the appropriate input message for the subordinates influenced by the external event. Input messages $(Z_{N,r}(x), t)$ are sent to all children, r, influenced by the external input x. After that, the event times are updated.

In Section 7.1 we defined how a discrete event network DEVN defines an atomic DEVS system. Now we will show that the devs-coordinator implements a devs-simulator for the $DEVS_N$ associated with DEVN. This allows us to treat a coordinator of a DEVS network in the same way as a simulator for an atomic DEVS. Also, we have to guarantee that the coordinator sends messages to the subordinates as required by the assumptions in the correctness theorem for DEVS simulator.

Algorithm 2 Coordinator for DEVS coupled model.

```
Devs-coordinator
variables :
    DEVN = (X, Y, D, {M_d}, {I_d}, {Z_i,d}, Select)        ▷ the associated network
    parent                                                 ▷ parent coordinator
    tl                                                     ▷ time of last event
    tn                                                     ▷ time of next event
    event-list          ▷ list of elements (d, tn_d) sorted by tn_d and Select
    d*                                                     ▷ selected imminent child
when receive i-message (i, t) at time t
    for dinD do
        send i-message (i, t) to child d
    end for
    sort event-list according to tn_d and Select
    tl = max{tl_d | d ∈ D}
    tn = min{tn_d | d ∈ D}
when receive *-message (*, t) at time t
    if t ≠ tn then
        error: bad synchronization
    end if
    d* = first(event − list)
    send *-message (*, t) to d*
    sort event − list according to tn_d and Select
    tl = t
    tn = min{tn_d | d ∈ D}
when receive x-message (x, t) at time t with external input x     ▷ consult external input coupling to get
children influenced by the input
    if not (tl ≤ t ≤ tn) then
        error: bad synchronization
    end if
    receivers = {r | r ∈ D, N ∈ Ir, Z_N,r(x) ≠ Φ}
    for r in receivers do
        send x-messages (x_r, t) with input value x_r = Z_N,r(x) to r
    end forsort event − list according to tn_d and Select
    tl = t
    tn = min{tn_d | d ∈ D}
when receive y-message (y_d*, t) with output y_d* from d*     ▷ check external coupling to see if there is an
external output event
    if d* ∈ IN ∧ Z_d*,N(y_d*) ≠ Φ then
        send y-message (y_N, t) with value y_N = Z_d*,N(y_d*) to parent     ▷ check internal coupling to
get children influenced by output y_d* of d*
    end if
    receivers = {r | r ∈ D, d* ∈ I_r, Z_d*,r(y_d*) ≠ Φ}
    for r in receivers do
        send x-messages (x_r, t) with input value x_r = Z_d*,r(y_d*) to r
    end for
end Devs-coordinator
```

Theorem 8.2. *Under the assumption that at initial time t_i the components states are (s_d, e_d), meaning that component d has been in state s_d for e_d time units and the superior coordinator provides an i-message at initial time t_i before any other message, x-messages for every input event of the current input segment $\omega :< t_i, t_f] \rightarrow X$ arriving at the network inputs, and ∗-messages when the simulation time t is equal the value t_n of the coordinator in a timely ordered sequence, then the devs-coordinator*

(i) *implements the correct devs-simulator operations for the $DEVS_N$ with total state $(..., (s_d, t - tl_d), ...), t - tl)$ associated with the coupled system N, and*
(ii) *provides the correct synchronization for the subordinate simulators associated with components d.*

Proof. **(i)** *Correct implementation*: We show that (i) holds by showing that the commands in the i-, x-, and ∗-messages of the coordinator are equivalent to the messages of a simulator for the $DEVS_N$ associated with a coupled DEVS.

(i.1) To implement i-messages, a devs-coordinator has to implement commands $tl = t - e$, $tn = tl + \text{ta}(s)$ given the state $s = (..., (s_d, e_d), ...)$ at initial time t_i. Since the last state change in component d was e_d time units prior to time t_i, the last state change within the scope of the network N was $e = min\{e_d | d \in D\}$ prior to t_i. After sending i-messages to the children d, tl_d is equal to $t - e_d$. The variable tl is set to $max\{tl_d | d \in D\}$. Now, since $tl = max\{t - e_d | d \in D\}$, we have $tl = t - min\{e_d | d \in D\}$ or $tl = t - e$, as required by a devs simulator.
In the coordinator, tn is set to $min\{tn_d | d \in D\}$. Let $e_d = tl_d - tl$ be the elapsed time at time tl, then $\sigma_d = \text{ta}_d(s_d) - e_d$ is the time remaining for component d at time tl. $DEVS_N$ requires that $\text{ta}(s) = min\{\sigma_d | d \in D\} = min\{\text{ta}_d(s_d) - e_d | d \in D\}$ and according to the devs-simulator, $tn = tl + \text{ta}(s)$. Hence, we have to show that $min\{tn_d | d \in D\} = tl + min\{\text{ta}_d(s_d) - e_d | d \in D\}$. Now, $min\{tn_d | d \in D\} = min\{tl_d + \text{ta}_d(s_d) | d \in D\} = min\{tl - e_d + \text{ta}_d(s_d) | d \in D\} = min\{tl + \text{ta}_d(s_d) - e_d | d \in D\} = tl + min\{\sigma_d | d \in D\}$ as required.

(i.2) To implement ∗-messages, a devs-coordinator has to implement commands to compute the output $\lambda(s)$ and send it back to the parent, compute the next state using $\delta_{\text{int}}(s)$ and set tl to t and tn to $tl + \text{ta}(s)$ of the $DEVS_N$ (compare Section 7.1.1). Following the operations of the devs-coordinator, we see that $d*$ is computed by computing the imminents as $\{d | d \in D \wedge tn_d = min\{tn_c | c \in D\}\}$ and selecting one of them using function Select. This corresponds to the selection of the child $d*$ in $DEVS_N$ by Select($IMM(s)$) with $IMM(s) = \{d | d \in D \wedge \sigma_d = \text{ta}(s)\}$. When a ∗-message is sent to the coordinator, this results in y-message being sent back carrying the output $y_{d*} = \lambda_{d*}(s_{d*})$. This output is converted to $Z_{d*,N}(y_{d*})$, providing the required output $\lambda(s)$ of the $DEVS_N$. If it is unequal to the non-event, it is sent as an output of the DEVN to the parent coordinator which is in correspondence to the operation of the devs-simulator. After that, x-messages with inputs $x_r = Z_{d*,r}(y_{d*})$ are sent to the influencees r of $d*$; this results in computation of $s'_r = \delta_{\text{ext},r}(s_r, t - tl_r, x_r)$ and setting of tl_r to t resulting in total state $(s'_r, t - tl_r) = (s'_r, 0)$ for component r, $r \in D$ with $d* \in I_r$. This corresponds to the second point of δ_{int} of $DEVS_N$. In the ∗-message of $d*$, the internal transition function computes $s'_{d*} = \delta_{\text{int},d*}(s_{d*})$ of $d*$ and tl_{d*} is set to t resulting in state

$(\delta_{\text{int},d*}(s_{d*}), t - tl_{d*}) = (\delta_{\text{int},d*}(s_{d*}), 0)$ as required in the first point of the definition of δ_{int}. All other components have not received messages and no change occurred. For those, by definition e_d should be increased by $\text{ta}(s)$ according to the third point of δ_{int}. Now, e_d was equal to $tl - tl_d$ before the $*$-message arrived and $t = tn$, $tn - tl = \text{ta}(s)$ at arrival of the $*$-message. As tl is set to $t(= tn)$, tl is increased by $\text{ta}(s)$ and hence $e_d = tl - tl_d$ is increased by $\text{ta}(s)$ as required. Also, tn is set to $tn = min\{tn_d | d \in D\}$ which is equivalent to $tn = t + \text{ta}(s)$ as shown in (i.1).

(i.3) To implement receiving of x-messages at time t with input x, a devs-coordinator has to implement commands to compute the next state by $\delta_{\text{ext}'}$ set tl to t and tn to $tl + \text{ta}(s)$ (compare Section 7.1.1). By following the operations of the devs-coordinator, we see that the x-messages with inputs $x_r = Z_{N,r}(x)$ are sent to the influencees of N to compute $s'_r = \delta_{\text{ext},r}(s_r, t - tl_r, x_r)$. After computation of the new state s'_r and setting of tl_r to t the state of r is equal to $(s'_r, t - tl_r) = (s'_r, 0)$ as required in the first point of the definition of δ_{ext} for $DEVS_N$. All other components have not received messages, so no changes occurred (point 2 of δ_{ext}). Again, since $e = t - tl$ and setting tl to t, $e_d = tl - tl_d$ is increased by e as required. Again tn is set to $tn = min\{tn_d | d \in D\}$ which is equivalent to $tn = t + \text{ta}(s)$ as shown in (i.1).

(ii) *Correct synchronization*: The devs-coordinator, by assumption, receives an i-messages at time t_i prior to every other message and sends i-messages to its children; hence the subordinate simulators receive correct i-messages at time t_i. Now, $*$-messages are sent at times $t = tn$ to component $d*$ which is one of imminents, $\{d | d \in D \wedge tn_d = min\{tn_e | e \in D\}\}$. As tn is defined by $min\{tn_d | d \in D\}$ it follows that $t = tn = tn_d*$ as required. When receiving x-messages at times t, $tl \leq t \leq tn$, and when receiving y-messages at times $t = tn$; since tn is defined by $min\{tn_d | d \in D\}, t \leq tn$ implies $t \leq tn_d$. At the first i-message tl is defined by $max\{tl_d | d \in D\}$ hence $tl_d \leq tl$ is true after the i-message. At every message at time t, tl is set to t and only those tl_d's of children d which received message are changed to t. The others are left unchanged, hence, never $tl_d > tl$. Hence, $tl_d \leq tl \leq t \leq tn \leq tn_d$ as required to guarantee correct timely order of messages. □

8.1.3 THE ROOT-COORDINATOR

The root-coordinator implements the overall simulation loop (Algorithm 3). It accomplishes its task by sending messages to its direct subordinate, corresponding to the all inclusive coupled model, to initiate their simulation cycles. In our terminology, we call the message to initiate a new simulation cycle for an event a $*$-message. The subordinate of the root coordinator responds by reporting back its time of next state transition (tn). Thus the root-coordinator first sends a message to initialize the simulator (i-message) and then in a loop sends $*$-messages with time tn to its subordinate to initiate the simulation cycles until some termination condition is met. From the root-coordinator we require that it provides the correct synchronization for its subordinate.

Theorem 8.3. *The devs-root-coordinator provides the correct synchronization for the subordinate devs-simulator.*

Proof. Directly follows by the operations of the root-coordinator. □

Algorithm 3 Root-coordinator for DEVS abstract simulator.

```
devs-root-coordinator
variables :
    t                       ▷ current simulation time
    child                   ▷ direct subordinate devs-simulator or -coordinator
t = t0
send initialization message (i, t) to subordinate
    t = tn of its subordinate
loop
    send (*, t) message to child
    t = tn of its child
end loop
until end of simulation
end devs-root-coordinator
```

8.2 DEVS BUS

After introducing abstract simulator concepts for classic DEVS, we will show how other discrete event modeling approaches can be integrated into the framework. The DEVS bus concept for interoperating with diverse discrete event modeling formalisms is illustrated in Fig. 8.4. Any formalism to be integrated must be wrapped in a DEVS form through a suitably defined simulator. Its models can then be coupled to other such models and executed through a standard coordinator. Moreover, as we shall show in Chapter 9, DEVS is capable of expressing models formulated in the other major formalisms (DTSS and DESS), so these can also be integrated into an all-DEVS simulation environment supporting multi-formalism modeling.

We begin by developing simulators for event scheduling and activity scanning/process interaction formalisms. These formalisms have already been characterized as non-modular multi-component DEVS (Chapter 7). However, we made them compatible with basic DEVS by adding the requisite in-

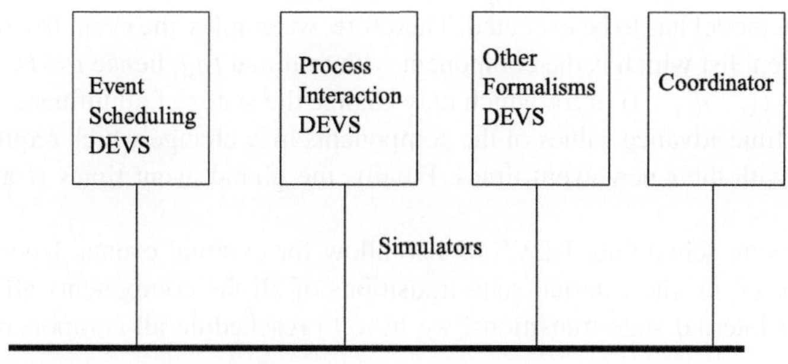

FIGURE 8.4

DEVS Bus.

put/output interface features (basically, external transition and output functions). The simulators exploit these extensions to adhere to the DEVS bus protocol.

8.2.1 SIMULATOR FOR EVENT SCHEDULING MULTI-COMPONENT DEVS

Earlier we defined *event scheduling multi-component DEVS* as a formalism for models formulated in the event scheduling strategy. Recall that event scheduling models are built from several components which are coupled in a non-modular way. In distinction to the other formalisms, in event scheduling the only means to bring events to execution is by prescheduling them in the future.

In the following we present a simulator for multi-component DEVS with event scheduling strategy which realizes the general simulation scheme for this modeling strategy. Moreover, the formulation as a simulator object and the provision of input/output interfaces allows using such models as components in hierarchical simulators together with components formulated in other strategies. We present the simulator object first and then show that it realizes a simulator for basic DEVS associated with multi-component DEVS.

The simulator for event scheduling DEVS is similar to simulators of basic DEVS, but realizes its own event handling for the model components. For this purpose, it uses variables to hold its time of last event tl and time of next event tn, as well as variables tl_d and tn_d for all its components. Furthermore an event list is used to sort the components based on their event times and Select order.

As shown in Algorithm 4, upon the receipt of an initialization message (i, t), first of all the event times tl_d and tn_d are set to

$$tl_d = t - e_d$$

and

$$tn_d = tl_d + \mathrm{ta}_d((..., q_i, ...)),$$

respectively, where $(..., q_i, ...)$ denote the state vector formed by the states of the influencing components $i \in I_d$. Based on the next event times tn_d and the Select order, the components d are inserted into the event list. Finally the global last event time tl of the multi-component model is the maximum of the last event times of the components and the global next event time tn is the minimum of the next event times of the components.

Upon receipt of a state transition message $(*, t)$ the next state transition within the components of the multi-component model has to be executed. Therefore, we employ the event list and retrieve the first element $d*$ in the event list which is the component with minimal tn_d, hence $t = tn_{d}*$. We execute the state transition $\delta_{\mathrm{int},d*}((..., q_i, ...))$ of $d*$ which may change the states of all influenced components $j \in E_{d*}$. As a result the time advance values of the components may change which requires the reinsertion of the components with their new event times. Finally, the global event times tl and tn are updated appropriately.

We formulated event scheduling DEVS to also allow for external events. Upon receipt of an external input message (x, t), the external state transitions of all the components affected by the input are executed. As for internal state transitions, we have to reschedule all components changed by the external event.

Theorem 8.4. *Under the assumption that at initial time ti the components of the system have been in state s_d for e_d time units and the superior coordinator provides an i-message at initial time t_i before*

Algorithm 4 Simulator for event scheduling multi-component DEVS.

```
Esdevs-simulator
variables :
    parent                                              ▷ parent coordinator
    tl                                                  ▷ time of last event
    tn                                                  ▷ time of next event
    ESDEVS = (X, Y, D, {M_d}, Select)
    with components M_d = (S_d, I_d, E_d, δ_ext,d, δ_int,d, λ_d, ta_d)
    with states q_d = (s_d, e_d)
    time of last event tl_d and time of next event tn_d
    and local outputs y_d
    event-list = list of pairs (d, tn_d) sorted by tn_d and Select
when receive i-message (i, t) at time t
    for components d ∈ D do
        tl_d = t − e_d
        tn_d = tl_d + ta_d((..., q_i, ...))
    end for
    for all components d ∈ D do
        sort (d, tn_d) into event-list
    end for
    tl = max{tl_d | d ∈ D}
    tn = min{tn_d | d ∈ D}
when receive ∗-message (∗, t) at time t
    if t ≠ tn then
        error: bad synchronization
    end if
    (d∗, tn_d∗) = first(event − list)
    (..., q_j, ...) = δ_int,d∗((..., q_i, ...))
    for all components j ∈ E_d∗ do
        tl_j = t − e_j
        tn_j = tl_j + ta_j((..., q_i, ...))
        remove (j, tn_j) and
        reinsert (j, tn_j) into event-list
    end for
    tl = t
    tn = min{tn_d | d ∈ D}
when receive x-message (x, t) at time t with input value x
    if not (tl ≤ t ≤ tn) then
        error: bad synchronization
    end if
    for all components d ∈ D if defined δ_ext,d do
        (..., q_j, ...) = δ_ext,d((..., q_i, ...), t − tl_d, x)
    end for
    for all components j ∈ E_d do
        tl_j = t − e_j
        tn_j = tl_j + ta_j((..., q_i, ...))
        remove (j, tn_j) and
        reinsert (j, t + tn_j) into event-list
    end for
    tl = t
    tn = min{tn_d | d ∈ D}
end Esdevs-Simulator
```

any other message, x-messages for every event input of the current input segment $\omega :< t_i, t_f] \rightarrow X$, *and ∗-messages when the simulation time t is equal the value tn of the simulator in a timely ordered sequence, then the simulator for an event scheduling multi-component DEVS realizes a simulator for the basic DEVS associated with the event scheduling DEVS as defined in Chapter 7.*

Proof. The proof follows the lines of the definition of the basic DEVS associated with the event scheduling DEVS in Section 7.2.1 and is similar to the proof for the devs-coordinator. □

8.2.2 SIMULATOR FOR ACTIVITY SCANNING AND PROCESS INTERACTION MULTI-COMPONENT DEVS

In addition to prescheduling of events, activity scanning and process interaction models also allow for conditional events, that is, events whose execution is dependent on a test of the global state. Recall that events can become due when the event time has come but the event condition does not evaluate to true. The event has to wait until some influencing states change so that the event condition allows the event to be executed. A simulator for activity scanning DEVS, therefore, has to check the conditions of due events at any state change to see if they are ready to be executed.

A simulator object for activity scanning DEVS employs two lists – 1) the *future event list* (FEL), which serves to store the events scheduled for the future (essentially, similar to the events-list in the event scheduling simulator) and 2) the *current-events-list* (CEL), which is used to store the components whose event times have come. Events are sorted in one of those lists depending on their event times – if the time is greater than the current time t, the event is inserted into the FEL, if it is equal or smaller the current simulation time the event goes into the CEL. Hence, the CEL also contains all the due events. At any state change all events in the CEL have to be checked to see if they are ready to be executed. While the FEL is sorted by the event time, the CEL is sorted by the Select order. This guarantees that the events are always executed according to the priority specified by Select.

In its remaining features, the simulator is similar to that of the event scheduling DEVS. Let us look at the main differences. In the initialization message, the components are inserted into the FEL or the CEL depending on their event times. Upon receiving a state transition message $(*, t)$ with a new event time t, all events with that event time are removed from the FEL and inserted into the CEL. Then the CEL is scanned from the beginning, that is, in Select order, to find the first component with true event condition. This event is executed. The influenced components are rescheduled, either in the FEL or the CEL. Then, the CEL is again scanned from the beginning for the next executable event. Finally, when no executable events are left, the next event time tn is computed as the smallest scheduled time in the FEL (see Algorithm 5). In the same way as for event scheduling multi-component DEVS, we can formulate a theorem which proves the simulator correctness, i.e., it interprets the model in conformity with the DEVS associated with the activity scanning DEVS formalism. Finally, having constructed DEVS protocol compliant simulators for event scheduling, activity scanning, and process interaction strategies we can state:

Corollary 8.1. *Simulators exist for non-modular multi-component DEVS that enable models in such formalisms to interoperate in hierarchical, modular DEVS models as if they were basic atomic DEVS models.*

Algorithm 5 Simulator for activity scanning multi-component DEVS.

```
Asdevs-simulator
variables :
    parent                                                      ▷ parent coordinator
    tl                                                          ▷ time of last event
    tn                                                          ▷ time of next event
    ASDEVS = (X, Y, D, {M_d}, Select)
    with components M_d = (S_d, I_d, E_d, δ_ext,d, C_int,d, δ_int,d, λ_d, ta_d)
    with states q_d = (s_d, e_d)
    time of last event tl_d and time of next event tn_d
    and local outputs y_d
    FEL = List of pairs (d, tn_d) sorted by tn_d and Select
    CEL = List of pairs (d, tn_d) sorted by Select
when receive i-message (i, t) at time t
    for all components d ∈ D do
        tl_d = t − e_d
        tn_d = tl_d + ta_d((..., q_i, ...))
    end for
    for all components d ∈ D do
        if tn_d ≥ t then
            insert (d, tn_d) into FEL
        else
            insert (d, tn_d) into CEL
        end if
    end for
    tl = max{tl_d | d ∈ D}
    tn = min{tn_d | d ∈ FEL}
when receive *-message (*, t) at time t
    if t ≠ tn then
        error: bad synchronization
    end if
    take out all elements (d, tn_d) with tn_d = t from FEL and
    reinsert (d, tn_d) into CEL
    SCAN:                                                       ▷ try to execute the due events
    for (d*, tn_d*) in CEL do
        if C_int,d*((..., q_i, ...)) then
            (..., q_j, ...) = δ_int,d*((..., q_i, ...))
            for j ∈ E_d* do
                remove (j, tn_j)
                tl_j = t − e_j
                tn_j = tl_j + ta_j((..., q_i, ...))
                if tn_j > t then
                    insert (j, tn_j) into FEL
                else
                    else insert (j, tn_j) into CEL
                end if
            end for
            exit to SCAN
        end if
    end for
    tl = t
    tn = min{tn_d | d ∈ D}
when receive x-message (x, t) at time t with input value x   ▷ similar as for event scheduling DEVS but with current event
    scanning
end Asdevs-Simulator
```

8.3 SIMULATORS FOR DTSS

The simulator for hierarchical discrete time system has to implement the fixed step algorithm to compute the state and output values given an input value at each instant of the discrete time base. Recall (Chapter 6) that for a DTSS one has to specify a constant h which gives a constant time advance at each simulation step. Also, recall that a DTSS employs a next state function which computes the state value at the next time step given the state value at the current time and the input at the current time. An output function computes the output at any time instant based on the current state.

Hence, a DTSS simulator is a simple algorithm which jumps from one simulation step to the next and computes the next state and output values. However, through the dependencies between the different components in a coupled system, namely, that the output of one component define the input for others, the right sequence for executing the state transitions and output functions in the individual components is crucial. Recall that components of DTSS networks can be of three different types: DTSS models of Moore-type, DTSS models of Mealy-type and memoryless FNSS, the main difference being that for Moore-type systems the output is computed solely based on the current state, whereas for the output of Mealy-type and FNSS models, we need the current input to compute the current output. The Moore-type components are those which break algebraic loops. A scheduling sequence for a network of DTSS, therefore, has to compute the outputs of the Moore-type components solely based on the current state and then, subsequently, go to the others to compute the current output when the current input is available. At the end of the sequence we come back to the Moore-type components to compute the next states.

In the following we present a hierarchical simulator for discrete time systems which assumes networks consisting only of Moore-type components and memoryless functions. This leads to a simpler scheme compared to networks also containing Mealy-type components. (In Praehofer, 1991 a more general scheduling algorithm for hierarchical models is presented which also works with Mealy-type systems and Mealy-type networks. It is also optimized in requiring only a single traversal of components in each simulation step.)

A hierarchical simulator for hierarchical, coupled DTSS models consists of *dtss-simulators* and *dtss-coordinators* and uses three types of messages (Fig. 8.5). An initialization method (i, t) is sent at the initialization time from the parent simulator object to all its subordinates. The parent sends a $*$-message to tell its subordinates to compute their outputs. A state transition message (x, t) is sent to the subordinates to trigger the execution of the state transition. The subordinates send an output message (y, t) to their parent.

8.3.1 SIMULATOR FOR ATOMIC DTSS

The *dtss-simulator* implements the simulation algorithm for an atomic DTSS (of the Moore-type). To take care of the feedback loops, a dtss-simulator provides two messages at each time step. The $*$-message is used to generate the output and the x-message is used to execute the state transition. Note that viewed as a DEVS, the $*$-message causes an output but no internal transition, while the x-message causes an external transition. So, as we shall see later, a DTSS can be represented as a externally stimulated DEVS with no internal autonomy.

A dtss-simulator (Algorithm 6) has to have access to the model description and employs variables to store the sequential state q of the DTSS, the current output y, and the simulation time tn of its next state transition. It implements the protocol messages as follows. The initialization message (i, t)

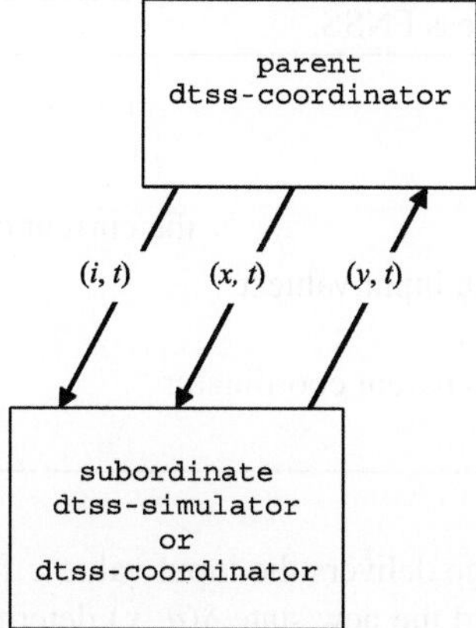

FIGURE 8.5

DTSS simulator protocol.

Algorithm 6 Simulator for Moore-type DTSS.

```
Dtss-Moore-simulator
variables :
    DTSS = (X, Y, Q, δ, λ, h)          ▷ associated model with state q
    y                                  ▷ the current output value of the associated model
    tn                                 ▷ time of next state transition
when receive message (i, t) at time t
    tn = t
when receive message (*, t) at time t
    if t ≠ tn then
        synchronization error
    end if
    y = λ(q)
    send y in message (y, tn) to parent
when receive message (x, t) with input value x
    if t ≠ tn then
        synchronization error
    end if
    q = δ(q, x)
    tn = tn + h
end Dtss-Moore-simulator
```

is intended to be called at the beginning of the simulation experiment which sets the time of the first state transition. In every simulation cycle at a state transition time t, we suppose that first a $*$-message is received to compute the output using $\lambda(q)$ based on the current state q. This value is sent back in a

Algorithm 7 Simulator for memoryless FNSS.

```
Fnss-simulator                                   ▷ for memoryless models
variables :
    Fnss = (X, Y, λ)                             ▷ associated memoryless model
    y                 ▷ the current output value of the associated model
when receive message (x, t) with input value x
    y = λ(x)
    send y in y-message (y, t) to parent coordinator
end Fnss-simulator
```

y-message. Afterwards, an x-message delivers the input value x for time t. Based on the input value, the state transition can be applied and the new state $\delta(q, x)$ determined. The time is set to the time of the next state transition, $t + h$.

Theorem 8.5. *For input segments $\omega : [t_1, tm) \rightarrow X$, for initial state q_0 at time t_1 and under the assumption that the superior coordinator provides an i-message at initial time t_1 prior to any other message, exactly one $*$- and one x-message for times $t_i \in [t_1, t_m)$, with $t_i = t_i - 1 + h$, with input $\omega(t_i)$ in a timely ordered sequence, then the state q at any time t_i corresponds to the state, and the outputs y sent out in y-messages, corresponds to the output trajectory of the I/O system associated with the DTSS.*

Proof. The proof follows the line of the definition of the I/O system associated with the DTSS in Chapter 6.

We prove by induction that if state q before the execution of an x-message at time $t_n = t_i$ is the correct state, $\Delta(q_0, \omega_{[t1,t_n)})$, then the state after the execution of an x-message is the correct state, $\Delta(q_0, \omega_{[t_1,t_n)})$ where now $t_n = t_i + h$. After the i-message at time t_1, the state q_0 by definition represents the correct state and hence $q = \Delta(q_0, \omega_{[t_1,t_1)}) = \Delta(q_0, \omega_{[t_1,t_n)})$. Following the commands of the x-message at time $t_i = t_n$ with input $\omega(t_i)$, q is set to $\delta(q, \omega(t_i)) = \delta(\Delta(q_0, \omega_{[t_1,t_i)}), \omega(t_i)) = \delta(\Delta(q_0, \omega_{[t_1,t_i)}), \omega(t_i)) = \Delta(q_0, \omega_{[t_1,t_i+h)})$ by the definition of Δ. After setting of $t_n = t_i + h$, we have $q = \Delta(q_0, \omega_{[t_1,t_n)})$. As state $q = \Delta(q_0, \omega_{[t_1,t_{i+1})})$ is correct state at time t_i and by assumption a $*$-message is received before an x-message at time t_i, the output $\lambda(q) = \Lambda(q, x)$, as required by the I/O system for DTSS. □

8.3.2 SIMULATOR FOR INSTANTANEOUS FUNCTIONS

Recall from Chapter 6 that function specified systems (FNSS) are elementary systems which do not own a state but compute the output values based on the current input values. Also recall from Chapter 7 that FNSS can occur as components in networks of DTSS. We therefore define a simulator for FNSS which occur as components in DTSS networks. The *fnss-simulator* (Algorithm 7) simply has to react to x-messages at any time step t by the computation of the output $\lambda(x)$.

Theorem 8.6. *For input segments $\omega : [t_1, t_m) \rightarrow X$ and under the assumption that the superior coordinator sends an x-message for time $t_i \in [t_1, t_m)$ with input $\omega(t_i)$ in a timely ordered sequence, then the outputs $\lambda(x)$ corresponds to the I/O system associated with the FNSS.*

Proof. Directly follows from the definition of the I/O system for FNSS in Chapter 6. □

Algorithm 8 Simulator for multi-component DTSS.

```
Mcdtss-simulator
variables :
    MCDTSS = (X, D, {M_d}, h)
    with d ∈ D, M_d = (Q_d, Y_d, I_d, δ_d, λ_d)          ▷ associated model
    {q_d | d ∈ D}                                        ▷ states of components d
    {q'_d | d ∈ D}      ▷ temporary variables to hold new states of components d
    {y_d | d ∈ D}                                ▷ the outputs of the components d
    tn                                               ▷ time of next state transition
when receive message (i, t) at time t
    tn = t
when receive message (*, t) at time t
    if t ≠ tn then
        synchronization error
    end if
    for all components d ∈ D do
        y_d = λ_d(..., q_i, ...), i ∈ I_d
    end for
    send message (y, tn) with y = (×_{d∈D} y_d) to parent
when receive message (x, t) with input value x
    if t ≠ tn then
        synchronization error
    end if
    for all components d ∈ D do
        q'_d = δ_d((..., q_i, ...), x), i ∈ I_d
    end for
    tn = tn + h
    for all components d ∈ D do
        q_d = q'_d
    end for
end Mcdtss-simulator
```

8.3.3 SIMULATOR FOR NON-MODULAR MULTI-COMPONENT DTSS

Multi-component DTSS are DTSS with several components, each with its own set of states and state transition function. A simulation algorithm for a multi-component DTSS has to traverse the components to update their states by executing their state transition functions. Also it has to apply the components' *individual* output functions which together define the output of the multi-component system.

A simulator for multi-component DTSS (Algorithm 8) looks similar to a simulator for Moore-type DTSS. In the $*$-message, the output values of all components are computed and the overall output is sent back to the parent in a y-message. In the x-message, the new states are computed and stored in temporary variables, q'_d. Then, the time is advanced and the state variables are set to the values stored in the temporary variables.

Exercise 8.1. Show as an exercise that a simulator for multi-component DTSS as described above realizes a dtss-simulator for the basic DTSS associated with the multi-component DTSS.

Exercise 8.2. Investigate conditions under which saving of states in temporary variables is not needed. In general, develop an algorithm for finding a sequence of component traversal that minimizes the number of temporary variables needed. (Hint: see TMS76, Section 16.6, Min Cycle Break Set.)

8.3.4 SIMULATORS FOR COUPLED DTSS

After having defined simulators for basic and multi-component DTSS, we now define a *coordinator* for networks of DTSS. A coordinator for a DTSS network uses the protocol of its subordinates, that is, it sends i, $*$, and x-messages, and receives y-messages from them, and it provides the same protocol to its parent. Therefore, a coordinator can serve as a subordinate simulator for a coordinator of a larger encompassing network. Seen through the identical protocol, coordinators and the simulators are indistinguishable to superior coordinators.

The main task of the dtss-coordinator (Algorithm 9) is to schedule the components in the right sequence. The scheme is data-driven – only when all component's inputs are available, the component is ready to go.

A $*$-message of a dtss-coordinator has to compute the output of the network (to provide the same protocol as a dtss-simulator). It does this by forwarding the $*$-message to all the Moore-type components to cause them to compute their outputs. The outputs are received in the y-messages and are stored (in variables xy_d) to be used for the computations of the inputs for components and the network outputs. After all Moore-type components have sent their outputs, a recursion takes place in which all instantaneous functions which have their inputs ready are sent x-messages to compute their outputs. We require in this scheme that the network is of type Moore and this guarantees that the output of the network can be determined and sent back (see Theorem in Chapter 7, Appendix).

The ensuing x-message, which delivers the network input x_N for the time step, causes all other functions to be scheduled. Finally, all the Moore-type components can be sent x-messages with their respective input values to undertake their state transitions.

We now state and show that a dtss-coordinator realizes a dtss-simulator for the $DTSS_N$ associated with the coupled DTSS. Again the proof directly follows from the definition of the DTSS associated with the coupled system and from the comparison of the message code of the dtss-coordinator and dtss-simulator. Also the coordinator propagates the correct synchronization to its subordinate simulators.

Theorem 8.7. *For input segments $\omega : [t_1, t_m) \rightarrow X$, for initial state $q_0 = (..., q_{d0}, ...)$, $d \in D$, at time t_1 and under the assumption that the superior coordinator provides an i-message at initial time t_1 prior to any other message, exactly one $*$- and one x-message for times $t_i \in [t_1, t_m)$, with $t_i = t_{i-1} + h$, with input $x = \omega(t_i)$ in a timely ordered sequence, then (i) the dtss-coordinator implements the correct dtss-simulator operations for the atomic DTSS with total state $(..., q_d, ...)$ associated with the DTSS network N, (ii) provides the correct synchronization for the subordinate simulators associated with components d.*

Proof. **(i)** To prove (i) we show the correspondence of the messages of the coordinator and the simulator for the $DTSS_N$ associated with network N.

The i-message of the coordinator clearly corresponds because both set t_n to t.

Algorithm 9 Coordinator for coupled DTSS.

```
Dtss-coordinator
variables :
    N = (X_N, Y_N, D, {M_d}, {I_d}, {Z_d}, h)          ▷ the network associated with the coordinator
    MOORE = {d ∈ D | d is of type Moore}
    FNSS = {d ∈ D | d is of type FNSS}
    {xy_d | d ∈ D ∪ {N}}    ▷ variables to store the output values for components d and the network input
    t_n                                                 ▷ time of next state transition
when receive message (i, t) at time t
    for d ∈ D do
        send i-message (i, t) with time t to d
    end for
    t_n = t
when receive message (*, t) with time t
    if t ≠ t_n then
        synchronization error
    end if
    for d ∈ MOORE do
        send message (*, t) with time t to d
    end for
    while (∃d ∈ FNSS : ∀i ∈ I_d : xy_i at time t is available ∧ d not scheduled for time t) do
        send x-message (x_d, t) with x_d = (Z_d(..., xy_i, ...), t), i ∈ I_d, to d
        mark d scheduled for time t
    end while
when receive x-message (x_N, t) with input value x_N
    if t ≠ t_n then
        synchronization error
    end if
    set xy_N to x_N
    while ( ∃d ∈ D : ∀i ∈ I_d : xy_i at time t is available ∧ d not scheduled for time t) do
        send x-message (x_d, t) with x_d = (Z_d(..., xy_i, ...), t), i ∈ I_d, to d
        mark d scheduled for time t
    end while
when receive y-message (y_d, t) with output value y_d from component d
    set xy_d to y_d
    if ( ∀i ∈ I_N : xy_i at time t is available ∧ network output not sent for time t) then
        send y-message (y_N, t) with y_N = (Z_N(..., xy_i, ...), t), i ∈ I_N, to parent
        mark network output sent for time t
    end if
end Dtss-coordinator
```

The $*$-message of a dtss-simulator computes the output $\lambda(q)$ and sends it back. The output λ of the $DTSS_N$ is defined as $Z_N(..., y_i, ...)$, $i \in I_N$. Hence the task of the $*$-message of the coordinator is to compute $Z_N(..., y_i, ...)$ and send it back in a y-message. The $*$-message causes $*$-messages to be forwarded to all Moore-type components which compute their outputs and send them back in y-messages. After that, all $FNSS$ for which all inputs are available are sent x-messages with input $Z_d(..., y_i, ...)$, $i \in I_d$ to compute their outputs. As the network is of type Moore, it is guaranteed that finally all the influencers' outputs of the network are available and the output $Z_N(..., y_i, ...)$ can be sent back.
Upon receipt of an x-message with a network input x, the simulator computes the next state $\delta(q, x)$. The $DTSS_N$ requires that the state transitions $\delta_d(q_d, x_d)$ of all the individual components are done. This is accomplished by sending x-messages with the respective inputs $Z_d(..., y_i, ...)$ to all the simulators in a sequence as outputs of influencers become available. As algebraic loops are not allowed, it is guaranteed that such a sequence always exists.

(ii) Since by assumption, a coordinator receives a *-message followed by an x-message at every time step and forwards these messages to its subordinates, it is guaranteed that the subordinates also receive the messages in correct order. □

Exercise 8.3. The main problem in simulating hierarchical DTSS networks which also contain Mealy-type components arises from the fact that the output of the network might be defined by Mealy-type components but it might still be able to compute it without the knowledge of the network input. The input for a Mealy type component might come from a Moore-type component (or from another Mealy-type component whose input however comes from a Moore-type) With computing the outputs of the Moore-types and scheduling the Mealy-type components with available inputs one time step ahead, it is possible to compute the output of a network without the knowledge of the input. This, however, requires that the network coordinator handles components operating at different time steps. Extend the coordinator for DTSS networks to allow also handling of Mealy-type components.

8.3.5 THE ROOT-COORDINATOR

The root-coordinator for a DTSS abstract simulator has to initiate the stepwise execution of the overall DTSS simulation cycles. It accomplishes its task by sending $*$-messages followed by x-messages (with possibly empty input value x) to its subordinate every state transition time, t_n (Algorithm 10).

From the root-coordinator we require that it provides the correct synchronization for its subordinate.

Theorem 8.8. *The dtss-root-coordinator provides the correct synchronization for the subordinate simulator.*

Proof. Directly follows by the operations of the dtss-root-coordinator. □

8.4 SIMULATORS FOR DESS

After defining simulation schemes for discrete time systems, let us now make a step towards simulation of continuous systems. Simulation of discrete time systems on digital computers is relatively simple as they employ a discretized time base. This allows a next state function to determine how the state at a

Algorithm 10 Root-coordinator for DTSS abstract simulator.

```
dtss-root-coordinator
variables :
    t                                          ▷ current simulation time
    child            ▷ direct subordinate simulator or coordinator
t = t_0
send initialization message (i, t) to child
t = t_n of its child
loop
    send (*, t) message to child
    send (x, t) message to child
    t = t_n of its child
end loop
until end of simulation
end dtss-root-coordinator
```

particular time instant is computed given the state and the input at the previous time instant. In Chapter 3, we have stated the fundamental problem of simulating continuous models on digital computers – how to compute continuous time behavior in discrete steps. We have identified numerical integration methods as the principal means to accomplish this task. We also have identified two main categories of integration methods – *causal* and *non-causal* methods. Recall that causal methods use state values at past and present time instants to compute the state value at the next time instant. Non-causal methods can also use estimates of state values at time instants later the current time.

In this section now we formulate simulation concepts for DESS systems to correspond with the abstract simulators already introduced for DEVS and DTSS. In fact, an abstract simulator for DESS – for causal as well as for non-causal methods – shows much resemblance to the abstract simulator for DTSS. In particular, coordinators for DESS networks can employ the same scheduling scheme. The main difference is that in coordinators for DTSS, the $*$- and x-messages are used for the computation of outputs and state transitions in one simulation step, while in the coordinators of DESS each phase in an integration step requires its own $*$- and x-messages. Therefore, we will present the simulators for basic DESS – for the causal and non-causal integration methods – in the following and the reader is referred to the DTSS coordinator as well as the simulator for FNSS to formulate the simulation architecture for coupled systems.

8.4.1 CAUSAL SIMULATOR FOR DESS

Recall that a causal method uses m past state and derivative values to estimate the state at the next time instant. With these considerations in mind, we formulate a general scheme for a causal simulator for atomic DESS (Algorithm 11). To implement a causal method of order m, a simulator must save the m past values of the model state values, q, derivatives, r, and inputs, x. With them, and the simulation step size h, the causal integration method determines an estimate for the state at the time instant $t_{i+1} = t_i + h$ by

$$q(t_{i+1}) = CausalMethod([q(t_{i-m-1}), ..., q(t_i)], [r(t_{i-m-1}), ..., r(t_i)], [x(t_{i-m-1}), ..., x(t_i)], h)$$

Algorithm 11 Causal simulator for basic DESS.

Dess-causal-simulator
variables :
  $DESS = (X, Y, Q, f, \lambda)$ ▷ associated model
  $[q(t_{i-m-1}), ..., q(t_i)]$ ▷ is a vector of state values
  $[r(t_{i-m-1}), ..., r(t_i)]$ ▷ is a vector of derivatives values
  $[x(t_{i-m-1}), ..., x(t_i)]$ ▷ is a vector of inputs
  h ▷ integration stepsize
when receive message (i, t_i) at time t_i
  initialize $[q(t_{i-m-1}), ..., q(t_i)]$,
  $[r(t_{i-m-1}), ..., r(t_i)]$,
  $[x(t_{i-m-1}), ..., x(t_i)]$
  using initialization specified by CausalMethod.
when receive message $(*, t_i)$ at time t_i
  $y = \lambda(q(t_i))$
  send y in message (y, t_i) to parent
when receive message (x, t_i) with input value x
  $x(t_i) = x$
  $q(t_{i+1}) = q(t_i + h) = CausalMethod([q(t_{i-m-1}), ..., q(t_i)],$
  $[r(t_{i-m-1}), ..., r(t_i)],$
  $[x(t_{i-m-1}), ..., x(t_i)], h)$
end Dess-causal-simulator

with $CausalMethod$ being a function (usually linear) dependent on the particular method in hand. The state value in turn directly gives an estimate for the output value by applying the output function

$$y(t_i) = \lambda(q(t_i)).$$

As already indicated in Chapter 3, and clearly from the simulator definition, at start of the simulation, past state values may not be available. This requires special provision to be taken for startup. We do not specify any particular scheme here, but assume that this is covered by the integration method.

Besides these issues, the simulator appears similar to the simulator of atomic DTSS.

Exercise 8.4. Formulate a causal simulator for multi-component DESS.

8.4.2 NON-CAUSAL SIMULATOR FOR DESS

Recall from Chapter 3 that non-causal integration methods work in phases. In predictor phases, estimates of future values are computed. In the corrector phase, the state values for the time instant are finally determined. A DESS simulator with a non-causal method (Algorithm 12), therefore, has to provide for these different phases.

As for a causal method, a simulator using a non-causal method may employ vectors to store the m past state, derivative, and input values. In addition, it also needs vectors to store n predicted values. In

Algorithm 12 Non-causal simulator for basic DESS.

```
Dess-non-causal-simulator
variables :
    DESS = (X, Y, Q, f, λ)                          ▷ associated model
    [q(t_{i-m-1}), ..., q(t_i)]                      ▷ is a vector of past state values
    [r(t_{i-m-1}), ..., r(t_i)]                      ▷ is a vector of past derivative values
    [x(t_{i-m}), ..., x(t_i)]                        ▷ is a vector of past input values
    [q'(1), ..., q'(n)]                              ▷ is a vector of predicted state values
    [r'(1), ..., r'(n)]                              ▷ is a vector of predicted derivative values
    [x'(1), ..., x'(n)]                              ▷ is a vector of predicted input values
    h                                                ▷ integration stepsize
    k                                                ▷ indicator for the phase of the integration
when receive message (i, t_i) at time t_i
    initialize [q(t_{i-m-1}), ..., q(t_i)],
    [r(t_{i-m-1}), ..., r(t_i)],
    [y(t_{i-m-1}), ..., y(t_i)]
    using initialization specified by integration method.
    k = 0
when receive message (*, t_i)
    if k = 0 then
        y = λ(q(t_i))
    else
        y = λ(q'(k))
    end if
    send y in message (y, t_i) to parent
when receive message (x, t_i)
    if k = 0 then
        x(t_i) = x
    else
        x'(k) = x
    end if
    k = (k + 1)mod(n + 1)
    if k > 0 then
        q'(k) = kth − Predictor([q(t_{i-m-1}), ..., q(t_i), q'(1), ..., q'(k − 1)],
        [r(t_{i-m-1}), ..., r(t_i), r'(1), ..., r'(k − 1)],
        [x(t_{i-m-1}), ..., x(t_i), x'(1), ..., x'(k − 1)],
        h)
    else
        q(ti + 1) = Corrector([q(t_{i-m-1}), ..., q(t_i), q'(1), ..., q'(n)],
        [r(t_{i-m-1}), ..., r(t_i), r'(1), ..., r'(n)],
        [x(t_{i-m-1}), ..., x(t_i), x'(1), ..., x'(n)],
        h)
        t_i = t_i + h
    end if
end Dess-non-causal-simulator
```

one simulation cycle, in n predictor phases the vectors $q'(i)$ for $i = 1, ..., n$ are filled by

$$q'(k+1) = kth - Predictor([q(t_{i-m-1}), ..., q(t_i), q'(1), ..., q'(k)], \\ [r(t_{i-m-1}), ..., r(t_i), r'(1), ..., r'(k)], [x(t_{i-m-1}), ..., x(t_i), x'(1), ..., x'(k)], h)$$

and then in the final corrector phase these values are used to compute the corrected state and output value for time t_{i+1}

$$q(t_{i+1}) = Corrector([q(t_{i-m-1}), ..., q(t_i), q'(1), ..., q'(n)], \\ [r(t_{i-m-1}), ..., r(t_i), r'(1), ..., r'(n)], [x(t_{i-m-1}), ..., x(t_i), x'(1), ..., x'(n)], h).$$

In the simulator of Algorithm 12, we use a counter k to distinguish between the different predictor and corrector phases. Based on the counter, we decide if an output has to be computed based on the current state or a state estimate, if an input received represents the $k\text{-}th$ estimate or corrected value, and if we should make the next prediction or the final correction.

A simulator is supposed to start an integration loop by first receiving an $*$-message to compute the output. At the first $*$-message, k is 0 and the output is computed based on the state $q(t_i)$. In a later message with a k value unequal 0, we have to compute the output for the $k\text{-}th$ state estimate $q'(k)$.

If an input is received in an x-message with $k = 0$, it represents an input stemming from some corrected (and not estimated) output y at time t_i and, therefore, the value is stored as normal input $x(t_i)$. Otherwise the input represents an estimate and the value is stored as an input estimate $x'(k)$. Then k is increased by 1 and taken modulo $n + 1$. This allows us to undertake exactly n predictions before getting 0 again and making the correction. With a k which is between 1 and n, the k predicator values are computed. With a value of $k = 0$, we have already computed the n predictor values and can compute the corrected value for the next time step t_{i+1}.

Exercise 8.5. Formulate a non-causal simulator for multi-component DESS.

As we have indicated already, for DESS coupled models, the same scheduling scheme used in DTSS coordinators can be employed to schedule the components in the correct computational sequence. In a simulator with a non-causal integration method, the main difference is, that the coordinator is supposed to receive $n + 1$ $*$- and x-messages and forward those to its subordinates before the integration cycle is completed and the time can be advanced.

Exercise 8.6. Formulate a causal and non-causal coordinators for coupled DESS models.

8.5 SUMMARY

In this chapter we have erected a framework of abstract simulators which parallels the framework of system formalisms introduced in the previous chapters. We have shown that the simulators for atomic models correctly generate their behavior. And we have shown the correctness of the coordinators by showing that each coordinator realizes a simulator for the basic model associated with the coupled model (the resultant under closure of coupling).

The coordinators and simulators obey to the same polymorphic protocol which allows them to work together in an transparent way. A coordinator need not know of the type of its subordinate, whether coordinator or simulator. The polymorphic simulator protocols, however, will also serve a further purpose. They will allow us to define simulation schemes for a multi-formalism modeling and simulation methodology. Multi-formalism models are built by coupling together models in different formalisms. They may also be simulated by coordinating the different abstract simulators in the DEVS bus concept to be described. This will be the topic of the next chapter.

Simulation is a computationally hard task and has to be implemented efficiently. The simulation protocols defined in this chapter have to be seen as abstract schemes without considering implementation details and performance requirements. In particular this is true for the distribution of outputs and translation to inputs according to the coupling schemes in coupled models. Especially in continuous simulation, where values have to be exchanged multiple times within one simulation cycle, communication must be implemented most efficiently. We will return to communication issues when discussing distributed simulation in later chapters.

8.6 SOURCES

The abstract simulator concept for DEVS first appeared in Zeigler (1984) and was extended to the other basic formalisms by Praehofer (1991). The DEVS bus was defined and prototyped by Kim and Kim (1996) in relation to distributed simulation and the High Level Architecture (HLA).

REFERENCES

Kim, Y.J., Kim, T.G., 1996. A heterogeneous distributed simulation framework based on devs formalism. In: Proceedings of the Sixth Annual Conference on Artificial Intelligence, Simulation and Planning in High Autonomy Systems, pp. 116–121.

Praehofer, H., 1991. System theoretic formalisms for combined discrete-continuous system simulation. International Journal of General System 19 (3), 226–240.

Zeigler, B.P., 1984. Multifacetted Modelling and Discrete Event Simulation. Academic Press Professional, Inc, London.

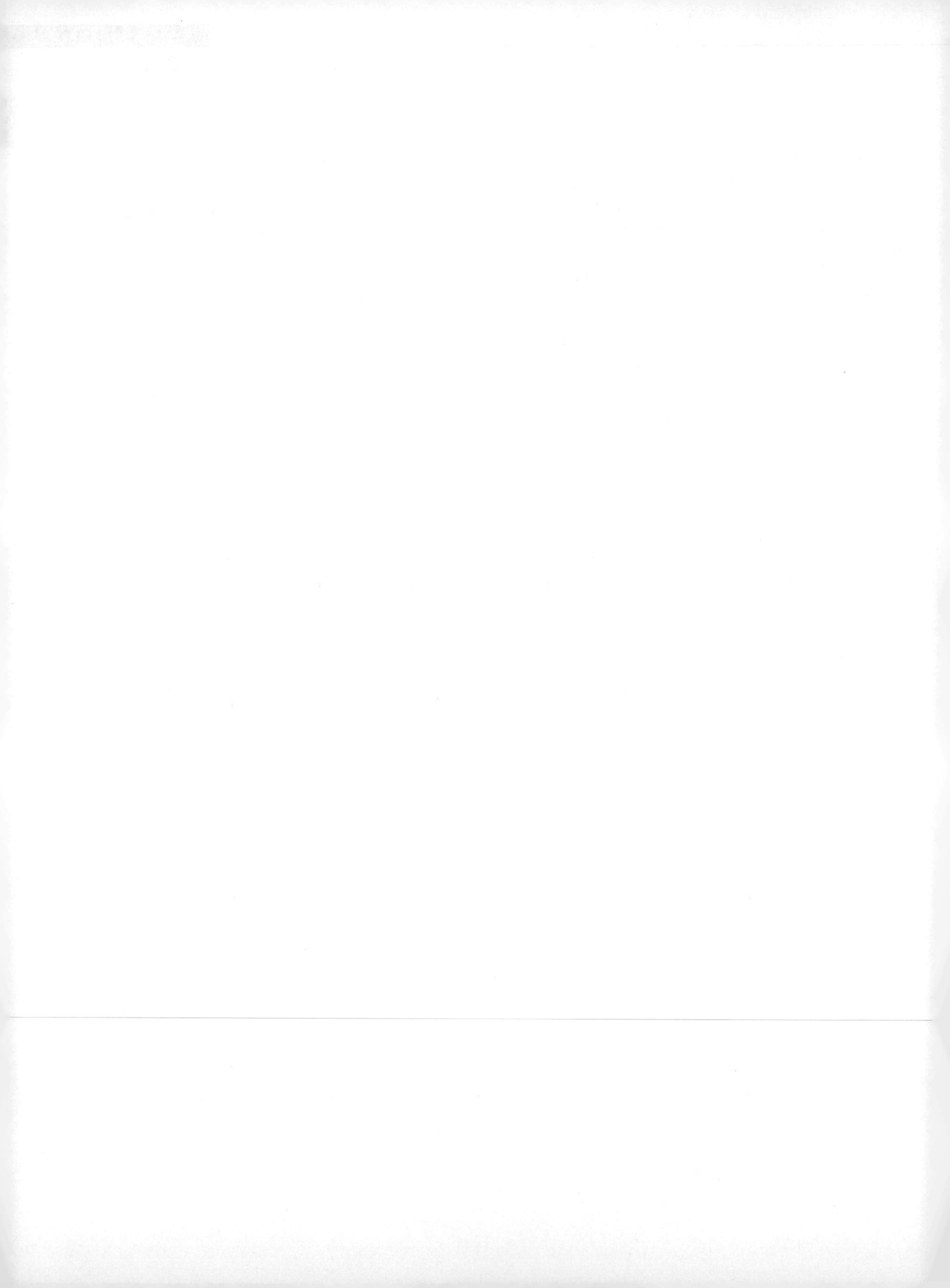

CHAPTER

MULTI-FORMALISM MODELING AND SIMULATION

CONTENTS

We have introduced three basic formalisms (DEVS, DTSS and DESS) all based on the unifying framework afforded by the general dynamic systems formalism. Although these formalisms describe subclasses of systems, they are, nevertheless, generic formalisms in the sense that encompass a wide variety of systems. In contrast, a multitude of more *specialized* formalisms have been developed. Such formalisms are proposed, and sometimes receive acceptance, because they provide convenient means to express models for particular classes of systems and problems. Formalisms such as Petri Nets and Statecharts, on the discrete event side, and Systems Dynamics and Bond Graphs, on the continuous

Theory of Modeling and Simulation. https://doi.org/10.1016/B978-0-12-813370-5.00017-1

side are widely employed. However, many real world phenomenon cannot be fit into the Procustrean bed of one formalism at a time. More generally, the ambitious systems now being designed, such as an automated highway traffic control system, cannot be modeled by a pure discrete or continuous paradigm. Instead, they require a combined discrete/continuous modeling and simulation methodology that supports a *multi-formalism* modeling approach and the simulation environments to support it. Before continuing to discuss multi-formalism modeling – the use of many formalisms in the same model – we provide brief introductions to some of the more influential ones. We will focus on the strengths, and also the limitations, of each formalism, for it is these attributes that motivate the integrative approach that systems formalisms provide.

9.1 BRIEF INTRODUCTION TO SPECIALIZED FORMALISMS

We will briefly consider some popular continuous formalisms and discrete event formalisms.

9.1.1 DESS SUBFORMALISMS: SYSTEMS DYNAMICS AND BOND GRAPHS

Systems Dynamics, developed by Forrester (1972) is widely used in natural systems and business and social systems modeling. It conceptualizes differential equations in stylized terms such as auxiliary, rate and level variables patterned couplings with an associated graphical representation. In terms of a DESS network, level variables are the outputs of integrators, while auxiliary and rate variables are the outputs of memoryless functions (see Chapters 3 and 7).

Bond graphs, originated by Paynter (1961), provide another graphically based approach to continuous system modeling, yet with an entirely different targeted class of systems and user group. Bond graphs generalize the current/voltage relationships of electrical circuits to other physical phenomena such as fluid flow and mechanical interactions. Bond graph models actually do not describe differential equations directly but represent the dependencies between the system variables in a non-directed manner. For example, current can be considered to cause voltage across a resistor or the other way around, depending on whether the energy is coming from a current or voltage source. By fixing which variables are given (input) and which variables of the system should be computed (state and output), one particular DESS can be derived from a set of possibilities. Bond graphs are thus capable of representing dynamic systems at a higher level of abstraction than the DESS.

The current/voltage metaphor is very useful in analyzing and designing system of components built from different energetic media since it affords a common basis for interfacing such components. However, the metaphor may need to be strained to fit the accepted ways of understanding phenomena in a field, thereby making it hard to adopt. Table 9.1 summarizes the main features of these formalisms, emphasizing the objectives, advantages and limitations of each.

While traditionally continuous systems have been the targets of modeling and simulation, in recent years there has emerged a multiplicity of formalisms for discrete event dynamic systems (Ho, 1992).

9.1.2 DEVS SUBFORMALISMS: PETRI NETS AND STATECHARTS

The classical *Petri net model* is a 5-tuple (P, T, I, O, M). P is a finite set of places (often drawn as circles) representing conditions. T is a finite set of transitions (often drawn as bars) representing events.

Table 9.1 Basic Features of Systems Dynamics and Bond Graphs

Target Systems	Social/Industrial/Ecological Systems	Engineering Systems with a variety of electrical, mechanical, thermal elements
Structure	Network of auxiliary, rate, and level elements	Vertices and arcs labeled in various ways
Main Feature	Methodology for continuous system model construction	Generalized current/voltage variables; non-deterministic
Analysis method	Simulation	Solution based on causal direction assignments
Analysis objective	Growth of populations, use of resources, feedback effects	Equilibrium solutions, derivation of dynamic state equations
Advantages	Simplifies differential equation model building with graphical elements	Supports multidisciplinary modeling
Limitations of the Basic Formalisms	Graphical elements can capture limited subset of all dynamic behaviors	Non-intuitive representations, electrical metaphor may be strained in many cases

I and O are sets of input and output functions mapping transitions to bags of places. M is the set of initial markings. Places may contain zero or more tokens at a moment in time. Tokens in Petri nets model dynamic behavior of systems. The execution of Petri nets is controlled by the number and distribution of the tokens. A transition is enabled if each of its input places contains at least as many tokens as there exists arcs from that place to the transition. When a transition is enabled it may fire. When a transition fires, all enabling tokens are removed from its input places, and a token is deposited in each of its output places. Petri nets have been employed in a wide variety of modeling applications. Peterson (1981) is a good introduction. Fishwick et al. (1995) discusses Petri nets in the context of modeling and simulation. Due to their limited expressive power, there have been myriad extensions to overcome particular limitations. Berthomieu and Diaz (1991) provides an example of the extension to timed nets. Sanders and Malhis (1992) developed an stochastic extension for simulation-based performance evaluation.

Statecharts describe a system's behavior over time, control, and timing behavior, and conditions and events that cause changes in a system's operation. Statecharts evolved from state-transition diagrams originally developed to depict the behavior of relatively simple digital systems. Such diagrams use a combination of arrows and boxes to indicate transitions among a system's states. Statecharts contain features that permit depiction of hierarchical states, concurrent or parallel processes, and timing. These features are essential to depicting the behavior of complex systems, allowing a designer to express even subtle semantics of system logic concisely and unambiguously. See Harel et al. (1990) for an introduction to Statecharts.

Summaries of the Petri net and Statechart formalisms are provided in Table 9.2.

9.2 MULTI-FORMALISM MODELING

Two approaches to multi-formalism modeling are depicted in Fig. 9.1. To *combine* formalisms we have to come up with a new formalism that subsumes the original ones. This is the case with DEV&DESS, which combines DEVS and DESS, the *discrete event* and *differential equation* system specification formalisms. Another approach to multi-formalism modeling is to *embed* one formalism into another.

Table 9.2 Basic Features of Petri nets and Statecharts

Formalism	Petri Nets	Statecharts/STATEMATE
Target Systems	Systems with interacting concurrent components	Real Time Software Systems
Structure	Place, Transition, Input function, Output function, Initial markings	State Diagrams, Module charts, Block diagrams
Main Feature	Event-conditional transitions	State generalization, Orthogonality, Broadcasting
Analysis method	Reachability analysis	Simulation, Reachability analysis
Analysis objective	Logical analysis, Performance evaluation (with stochastic, timed extensions)	Performance evaluation, Logical analysis
Advantages	Large research community. Simple and common graphical notations	Expressive power greater than Petri nets. Widely-used in industry
Limitations of the Basic Formalisms	Not modular Time is not considered. Not suitable for implementation	Partially modular, Complex

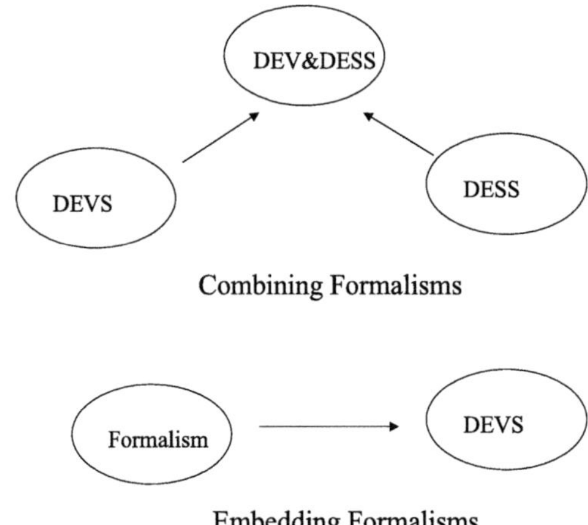

FIGURE 9.1

Multi-formalism Approaches.

For example, it is not hard to see that the systems dynamics and bond graph formalisms can be embedded in the DESS formalism – indeed, each is a way of writing a set of ordinary differential equations. Similarly, it can be shown that the Petri net and Statecharts formalisms can be embedded into the DEVS formalism. Indeed, any discrete event behavior can be expressed as a DEVS model – we leave the demonstration of this fact for later (Chapter 16).

Now let's consider embeddings of one major formalisms in another other. Traditionally, differential equations have been solved with numerical integration in discrete time. In formal terms, this means that DESS has been embedded into DTSS. However, we will show later (Chapter 17) that, especially for parallel and distributed execution, event-based simulation of continuous systems has greater potential for efficient processing at comparable accuracy. For example, recall the event-based game of life model from Chapter 3. In this example the continuous behavior of the fitness parameter was abstracted to an event model by only considering the relevant events. What justified this error-free abstraction was the fact that each cell had piecewise constant input and output trajectories (the birth/death state). As we will see later (Chapter 17), piecewise constancy of input and output is a sufficient condition for any system to have an exact DEVS representation. Further, since the continuous dynamics of the fitness function were quite simple (piecewise linear) we were able to easily predict the times to the next life cycle events.

Embedding the other formalisms (DESS and DTSS) into DEVS is attractive since then both discrete and continuous components are naturally included in the same environment. But, with continuous and discrete components interacting, we still need the DEV&DESS formalism to provide the basics of how such interaction should occur and to provide a formalism in which embedding DESS into DEVS can be rigorously analyzed. Moreover, for combined models that cannot be feasibly embedded into DEVS, we must employ the DEV&DESS formalism.

9.3 DEV&DESS: COMBINED DISCRETE EVENT AND DIFFERENTIAL EQUATION SPECIFIED SYSTEMS

In this section we extend our modeling framework by introducing a system formalism for combined discrete/continuous modeling. The formalism comes into being by a combination of the DEVS and the DESS formalism and, therefore, is called *discrete event and differential equation specified system* formalism (DEV&DESS).

Fig. 9.2 illustrates the modeling concept which has both DEVS and DESS elements working together. Input ports of X^{discr} accept event segments and input ports of X^{cont} accepts piecewise continuous or piecewise constant segments. The latter influences both model parts, while the event input

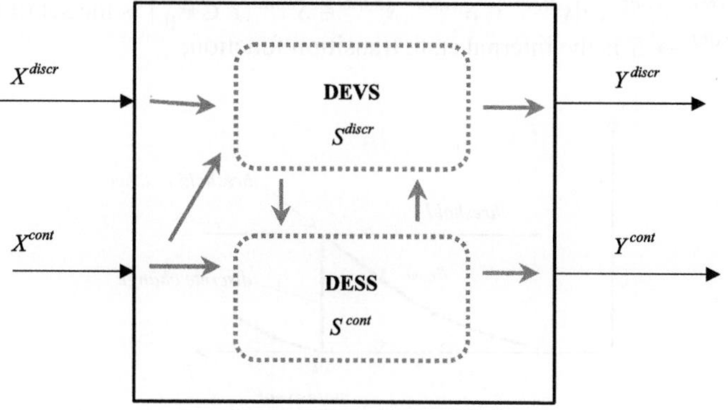

FIGURE 9.2

DEVS and DESS combined model.

only influences the DEVS part. Each part produces its own corresponding outputs, Y^{discr} and Y^{cont}, respectively. The parts can also influence each other's states.

Fundamental for the understanding of combined modeling is how the discrete part is affected by the continuous part, that is, how the DESS part causes events to occur. Fig. 9.3 illustrates this. In the intervals between events, the DESS input, state, and output values change continuously. In a combined model, events occur whenever a condition specified on the continuous elements becomes true. Typically, the condition can be viewed as a continuous variable reaching and crossing a certain threshold or whenever two continuous variables meet (in which case, their difference crosses zero). In such situations, an event is triggered and the state is changed discontinuously. An event which is caused by the changes of continuous variables is called a *state event* (in distinction to internal events of pure DEVS which are scheduled in time and are called *time events*).

We will now define the *combined discrete event and differential equation system specification* (DEV&DESS). For simplicity and clarity we define the formalism *without allowing for time events.* However, such an extension can easily be achieved by adding time scheduling of events from the basic DEVS formalism. Moreover, we will show that a fully functional DEVS behavior can be embedded into the DEV&DESS formalism as it stands, so nothing is lost from the behavior point of view (although simulation efficiency will suffer).

$$DEV\&DESS = (X^{discr}, X^{cont}, Y^{discr}, Y^{cont}, S, \delta_{\text{ext}}, C_{int}, \delta_{\text{int}}, \lambda^{discr}, f, \lambda^{cont})$$

where

- X^{discr}, Y^{discr} are the set of event input and output from DEVS,
- $X^{cont} = \{(x_1^{cont}, x_2^{cont}, ...) | x_1^{cont} \in X_1^{cont}, x_2^{cont} \in X_2^{cont}, ...\}$ is the structured set of continuous value inputs with input variables x_i^{cont},
- $Y^{cont} = \{(y_1^{cont}, y_2^{cont}, ...) | y_1^{cont} \in Y_1^{cont}, y_2^{cont} \in Y_2^{cont}, ...\}$ is the structured set of continuous value outputs with output variables y_i^{cont},
- $S = S^{discr} \times S^{cont}$ is the set of states as a Cartesian product of discrete states and continuous states,
- $\delta_{\text{ext}} : Q \times X^{cont} \times X^{discr} \rightarrow S$ is the external state transition function, where $Q = \{(s^{discr}, s^{cont}, e) | s^{discr} \in S^{discr}, s^{cont} \in S^{cont}, e \in \mathbb{R}_0^+\}$ is the set of total states,
- $\delta_{\text{int}} : Q \times X^{cont} \rightarrow S$ is the internal state transition function,

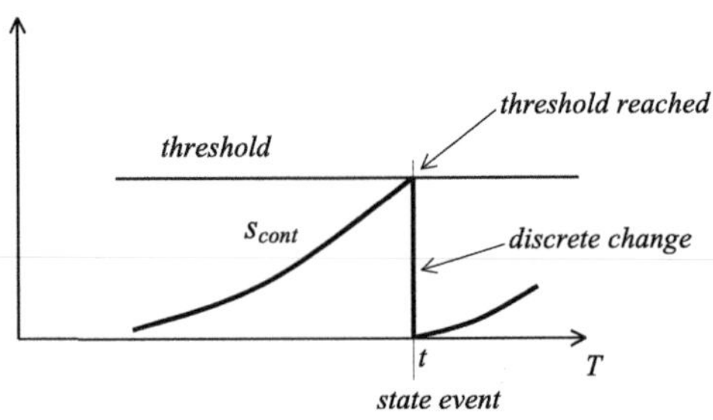

FIGURE 9.3

State event.

- $\lambda^{discr} : Q \times X^{cont} \rightarrow Y^{discr}$ is the event output function,
- $\lambda^{cont} : Q \times X^{cont} \rightarrow Y^{cont}$ is the continuous output function to define the continuous output variable values,
- $f : Q \times X^{cont} \rightarrow S^{cont}$ is the rate of change function and
- $C_{int} : Q \times X^{cont} \rightarrow Bool$ is the *state event condition function* for conditioning the execution of state events.

The semantics of a DEV&DESS are informally described as follows:

1. **Intervals $\langle \mathbf{t_1}, \mathbf{t_2} \rangle$ with no events:** Only the continuous states and the elapsed time e change. The derivative function together with the value output function specify the continuous behavior. The continuous states at the end of the interval are computed from the state at the beginning plus the integral of the derivative function f along the interval (see behavior of DESS in Chapter 6). The elapsed time e is increased by the total elapsed time $t_2 - t_1$. The output function of the continuous part is operating throughout the interval.
2. **A state event occurs first at time t** in interval $\langle \mathbf{t_1}, \mathbf{t_2} \rangle$: At time t the internal event condition function $C_{int}(s(t), x^{cont}(t))$ evaluates to true – a state event occurs. Here, the internal transition function is executed to define a new state. The continuous states at the time of the transition are computed from the state at the beginning plus the integral of the derivative function f along the interval until t. The elapsed time is set to 0 and the output function of the discrete part is called to generate an output event at t.
3. **A discrete event at the external input port occurs first at time t in interval $\langle \mathbf{t_1}, \mathbf{t_2} \rangle$** : Here, the external transition function is executed to define a new state at instant, t. The continuous states at the time of the transition are computed from the state at the beginning plus the integral of the derivative function f along the interval until t. Likewise, the continuous output is generated until t. The elapsed time is set to 0 at t.

9.3.1 A SIMPLE EXAMPLE: DEV&DESS MODEL OF A BARREL FILLER

Manufacturing typically involves processes that combine both continuous and discrete event phenomena. Fig. 9.4 and 0 illustrates DEV&DESS modeling using a filling process. Filling a barrel is a continuous process, however, control input and output is discrete. Control comprises loading the barrel in place, turning on the liquid flow and stopping the flow when the barrel fills up. The model has

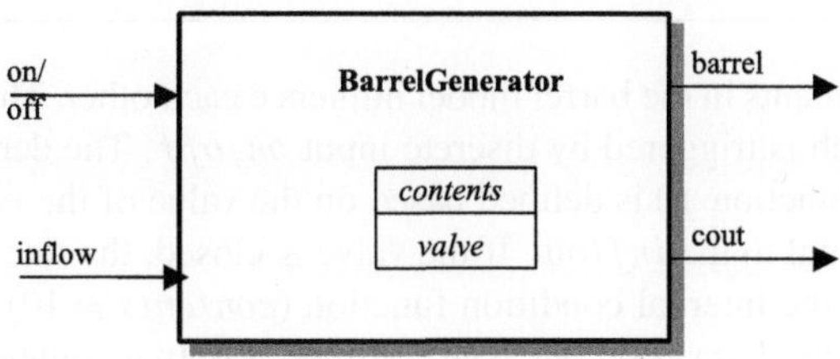

FIGURE 9.4

Barrel generator model.

one continuous input port $inflow$ and one discrete input port on/off. It has one discrete output port, $barrel$ and one continuous output port, $cout$, for observing the contents. There is also a continuous state variable $contents$ and a discrete state variable valve. The derivative of the continuous state variable $contents$ is given by the input value of $inflow$ when the system $valve$ is $open$, and it is zero when the $valve$ is *closed*. The system switches between these two phases according to the discrete input at the discrete input port *on/off*. Barrels are sent out at the discrete output port $barrel$ whenever the variable $contents$ reaches the value 10, a state event. At that point, contents is re-initialized to 0.

BarrelFiller = $(X^{cont}, X^{discr}, Y^{cont}, Y^{discr}, S^{cont}, S^{discr}, \delta_{\text{ext}}, C_{int}, \delta_{\text{int}}, \lambda^{discr}, f, \lambda^{cont})$
$X^{cont} = \{inflow | inflow \in \mathbb{R}\}$
$X^{discr} = \{on/off | on/off \in \{on, off\}\}$
$Y^{cont} = \{cout | cout \in \mathbb{R}\}$
$Y^{discr} = \{barrel | barrel \in \{10\text{-}liter\text{-}barrel\}\}$
$S^{discr} = \{valve | valve \in \{open, closed\}\}$
$S^{cont} = \{contents | contents \in \mathbb{R}\}\}$
$\delta_{\text{ext}}((contents, valve), e, inflow, on/off)$
if $on/off = on$ **then**
    $valve := open$
else $on/off = off$
    $valve := closed$
end if
$C_{int}((contents, valve), e, inflow)$
$contents == 10$
$\delta_{\text{int}}((contents, valve), e, inflow)$
$contents = 0$
$\lambda^{discr}((contents, valve), e, inflow)$
$barrel = 10\text{-}liter\text{-}barrel$ – send out $10\text{-}liter\text{-}barrel$ at port $barrel$
$f((contents, valve), e, inflow)$
case $valve$
    open: $dcontents/dt = 0$
    closed: $dcontents/dt = inflow$
$\lambda^{cont}((contents, valve), e, inflow)$
$cout = contents$
Example of a DEV&DESS model: Barrel Filler

Fig. 9.5 depicts how elements in the barrel model influence each other. The phase is set by the external transition function which is triggered by discrete input on/off. The derivative for the continuous state $contents$ (derivative function f) is defined based on the value of the $valve$ variable. If the valve is open, the derivative is equal to the $inflow$. If the valve is closed, the rate of change is 0. The internal events are triggered by the internal condition function ($contents = 10$) and change the $contents$ variable as well as produce the $barrel$ output. The continuous output contents is solely determined by the state variable $contents$.

Fig. 9.6 shows the trajectories of a sample experiment of the barrel filler system. Initially the $valve$ is $closed$ and the value of $contents$ is 0. After some time filling is turned on by an input signal on.

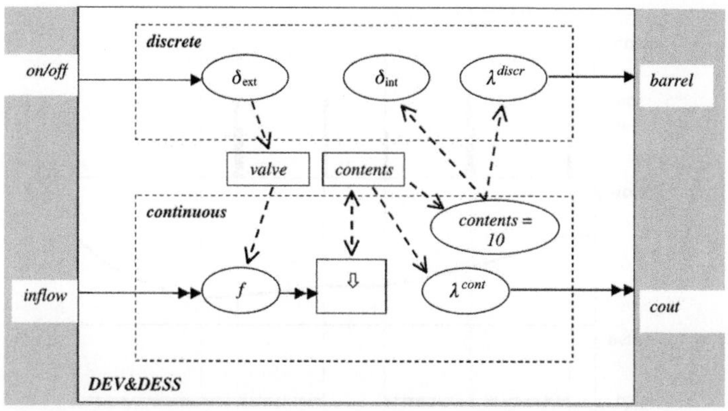

FIGURE 9.5

Functions dependencies of barrel model.

The continuous input $inflow$ is considered to be dependent on the level of an inflow tank and will decrease as the tank empties. The contents increases dependent on the inflow until it reaches 10 and a state event occurs. Filling is stopped (input off), the first barrel is released, $contents$ is set back to 0, and filling is resumed immediately by an input signal on. The filling of the second barrel is interrupted by an external event at the discrete input port on/off and filling is stopped. After some time there is a second external event which activates the model again and the filling resumes. The barrel is released at time 28. During the filling of the third barrel, there is a discontinuity of the inflow resulting from the tank being filled up again.

Exercise 9.1. Consider that the inflow in the example above is constant. Then the system can be modeled by a classic DEVS (Hint: see the GOL event model in Section 3.3 and Section 7.2). Build the DEVS model for the barrel filler problem.

Exercise 9.2. Extend the barrel filler model so that it detects when the inflow is too low and at that points stops the filling process and also issues an output notice that warns of this condition.

For a complete example of DEV&DESS modeling and simulation see Praehofer et al. (1993).

9.3.2 SYSTEM SPECIFIED BY A DEV&DESS

As always, the true meaning of a specification is the I/O system it specifies. Determining this meaning requires three steps:

- Constructing the putative I/O System, $S = (T, X, \Omega, Y, Q, \Delta, \Lambda)$ from the specification
- Working out the conditions that guarantee that the system is well defined
- Proving that the conditions are indeed sufficient for the system to exist.

Appendix 9.A contains details of steps 1 and 3. Let's consider step 2. Clearly, for the combination to exist, the DESS portion must specify a well-defined system on its own. Recall that to determine unique continuous state trajectories, the derivative function has to satisfy the Lipschitz condition (Chap-

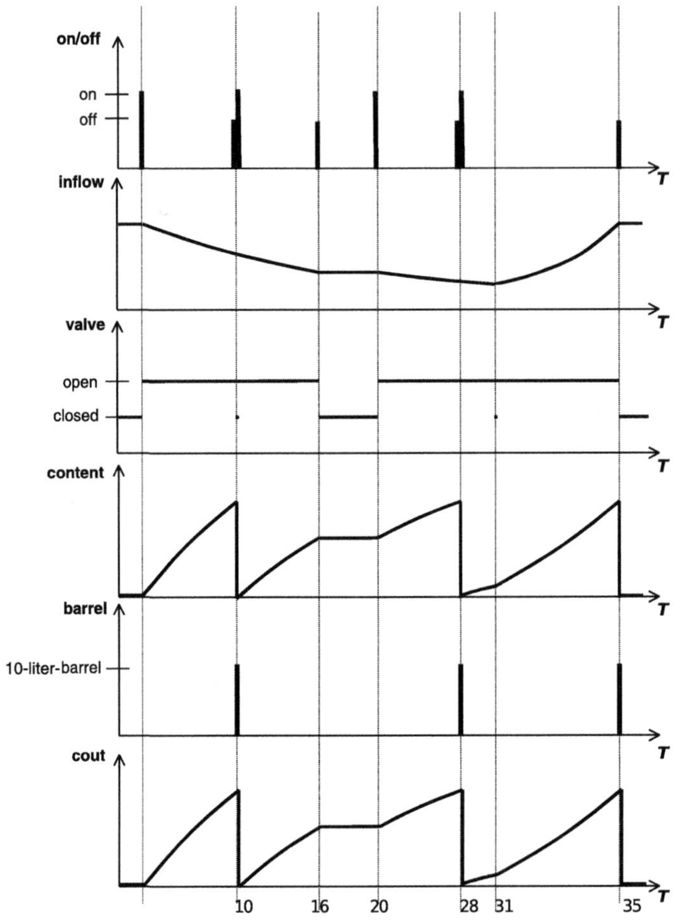

FIGURE 9.6

Barrel generator trajectories.

ter 6). However, the property of legitimacy defined for time events for DEVS must be adapted for DEV&DESS. We define the property of *state-event-legitimacy* in the same way legitimacy is defined for pure DEVS.

Definition: State-event-legitimacy: Let us define $\sum'(q,\omega)$ to be the number of state events, i.e., the events of execution of the internal transition function, in interval $< t_1, t_2]$ given the input segment $\omega :< t_1, t_2] \rightarrow X$ and the initial state q at time t_1. Then we say that a DEV&DESS is *state-event-legitimate*, if and only if

$$\omega :< t_1, t_2] \rightarrow X \in \Omega, q \in Q : \sum{}'(q,\omega) < \infty,$$

i.e., the number of state events is finite for all admissible input segments.

Steps 1 and 2 enable us to state and prove the:

Theorem 9.1. *A DEV&DESS specifies an I/O System if and only if the derivative function f has unique solutions and if the system is state-event-legitimate.*

The proof is in the Appendix.

9.4 MULTI-MODELING WITH DEV&DESS

Many real world phenomenon cannot be modeled by one single model but require a set of complementary models which together are able to describe the whole process. A model which subsumes several different models is termed a *multi-model*. The DEV&DESS formalism is an appropriate means to implement multi-models.

In a multi-model, depending on a current state the process is in, one out of the set of models will be applicable and will describe the process appropriately. The transitions from a state to a next state with a different applicable model signifies a qualitative change in the dynamic behavior. We call such qualitative states *phases* and qualitative changes in behavior *phase transitions*.

Phase transitions either can be caused from outside or arise inside the system when particular situations occur. Let us clarify the ideas by considering two rather simplified versions of transmission systems, viz. a vehicle with a stick operated transmission system and a vehicle with a rudimentary automatic transmission system which changes gears at particular speeds only. The phases of the system obviously are given by the different gears which determine different system behaviors, that is, different reactions of the vehicle. In the model of the hand-operated system, the gears are determined from outside. Thus the phase transitions are determined by the external transition function. In the model of the automatic system, however, the gear changes occur when the speed reaches certain thresholds. The phase transitions can be modeled by state events. The different gears are directly associated to certain subsets of the continuous speed variable as depicted in Fig. 9.7. The different phases and phase transitions define a partition of the continuous state space into mutual exclusive blocks which show qualitatively different behaviors.

Exercise 9.3. Elaborate a model of a real transmission system where the gear changes are not only dependent on the current speed but also on current acceleration. Also consider that changes to higher gears and lower gears occur at different thresholds.

In a DEV&DESS representation of a multi-model, the phase transitions are modeled either by external events, if determined from outside, or internal state events, if caused internally. Let us concentrate

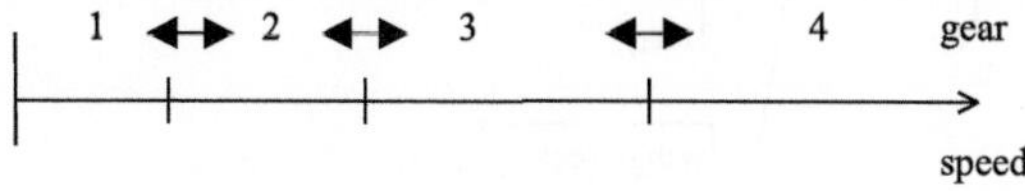

FIGURE 9.7

State Partitioning of Automated Transmission System.

on modeling phase transitions by internal events in the following. In such a multi-model, the underlying n-dimensional continuous state space is partitioned into different blocks where the state events for the phase transitions define the block boundaries. Each block in a partition corresponds to a particular phase and each phase may have associated with it a unique continuous dynamic behavior. The current phase value implies the currently applicable continuous model and the phase transitions define the qualitative changes in system behavior (Fig. 9.8).

The continuous state space of the DEV&DESS multi-model then is an n-dimensional real vector space

$$S^{cont} = \{(s_1, s_2, ..., s_i, ..., s_n) | s_i \in \mathbb{R}\}.$$

The discrete state space has phase variables $phase_i$ for each dimension i of the continuous state space. Each phase variables has a discrete range (e.g. the integers $\mathfrak{J}$) used to denote the blocks of the partition

$$S^{discr} = \{(phase_1, phase_2, ..., phase_i, ..., phase_n) | phase_i \in \mathfrak{J}\}.$$

Also, for each dimension i, a set of threshold values

$$TH_i = \{th_{i,1}, th_{i,2}, ..., th_{i,p}, ..., th_{i,ni} | th_{i,j} \in \mathbb{R}\}$$

define continuous state values where phase transitions have to occur. Let's agree that if $s_i \in [th_{i,p}, th_{i,p+1})$, then $phase_i = p$. The internal event condition evaluates to true whenever $phase_i$ is equal to p and the continuous variable s_i cross a threshold and leaves block $[th_{i,p}, th_{i,p+1})$, i.e.,

$$C_{int}(s, x^{cont}) = \begin{cases} true & \text{if } \exists i : phase_i = p \wedge s_i \notin [th_{i,p}, th_{i,p+1}) \\ false & \text{otherwise.} \end{cases}$$

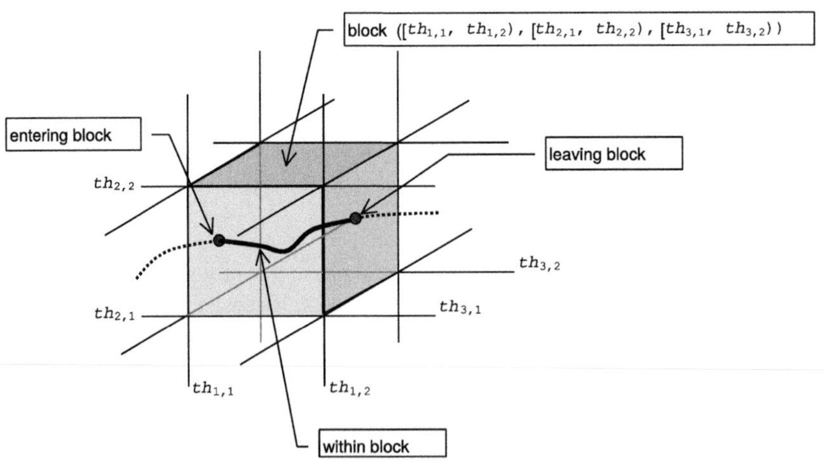

FIGURE 9.8

3-Dimensional state space with thresholds.

The internal state transition function defines the new phase in case of a threshold crossing, i.e., $\delta_{\text{int}}(s, x^{cont})phase_i = phase'_i$ with

$$phase'_i = \begin{cases} phase_i + 1 & \text{if } phase_i = p \wedge s_i \geq th_{i,p+1} \\ phase_i - 1 & \text{if } phase_i = p \wedge s_i < th_{i,p} \\ phase_i & \text{otherwise} \end{cases}$$

The discrete state vector $(phase_1, phase_2, ..., phase_i, ..., phase_n)$ of phase variables is employed to define individual continuous behaviors for the blocks in the state partition. In each block denoted by phase variables values $(p_1, p_2, ..., p_i, ...)$, a different derivative may be defined for dimension i by a unique function $f_{i,(p_1,p_2,...,p_i,...)}$, i.e.,

$$\frac{ds_i}{dt} = f_{i,(p_1,p_2,...,p_i,...)}(s, x^{cont}) \text{ if } (phase_1, phase_2, ...phase_i, ...) = (p_1, p_2, ..., p_i, ...).$$

9.4.1 EXAMPLE: POT SYSTEM WITH COMMAND INPUTS AND THRESHOLD VALUE OUTPUTS

Let us now look at an example from hybrid systems domain. Actually, the model below shows elements of output-partitioning as well as multi-modeling. Fig. 9.9 shows a two-dimensional system of a pot which can be heated and cooled, filled and emptied. The two dimensions are the *temp* and the *level* dimension representing the liquid level and the liquid temperature, respectively. The system has two discrete command inputs – the *heat-com* and the *fill-com* input – with three different commands for each, *heat-it, cool-it, stop-heating* and *fill-it, empty-it, stop-filling*, respectively. The system's discrete outputs are given by four simple threshold sensors, viz. *is-cold, is-hot, is-empty*, and *is-full*. The values for output sensors are *on* and *off* and they react at particular threshold values of the two continuous state variables *level* and *temp*. Each state dimension is partitioned into three mutual exclusive blocks according to the threshold sensor output values. Fig. 9.10 shows the state partitioning together with a sample state trajectory. There are 3 times 3 mutual exclusive blocks which have different sensor output

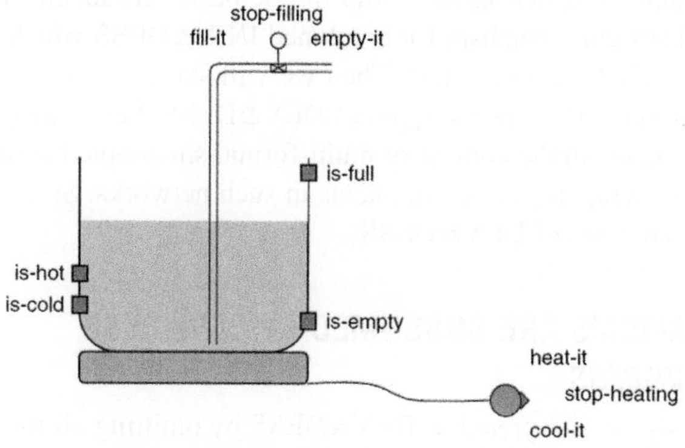

FIGURE 9.9

Pot model.

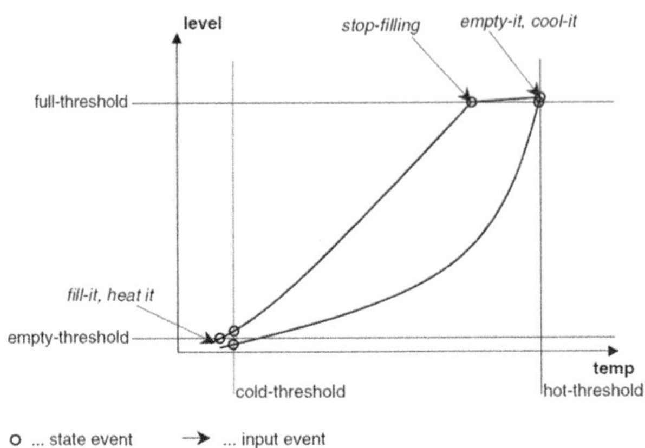

FIGURE 9.10

Pot state space partitioning.

values and which are denoted by the phases *cold*, *t-betw*, *hot* and *emtpy*, *l-betw*, *full*, respectively. With different blocks we have different derivatives. For example, in the whole block *hot* the *temp* and *level* derivatives change because the liquid begins to boil. The temperature does not increase any more but the level decreases through evaporation.

We come back to thresholds and phase partitioning of continuous processes in Chapters 18 and 19 when discussing DEVS representation of continuous systems and event based control, respectively.

9.5 COUPLED DEV&DESS: NETWORK OF MULTI-FORMALISM MODELS

Having the basis of the DEV&DESS formalism puts us in a position to use DEVS&DESS as basic components in suitably defined networks as we did for the basic formalisms in Chapter 7. So now we will introduce a coupled system formalism for combined DEV&DESS which supports modular, hierarchical construction of DEV&DESS models. Then we will show how the basic formalisms, DESS, DTSS, and DEVS are interpreted as special types of DEV&DESS. Such "wrappings" give us interpretations of the basic formalisms in the context of multi-formalism coupled systems. Moreover, we also are justified in using these wrappings as components in such networks. So we turn to interpreting the basic formalisms as special cases of DEV&DESS.

9.5.1 BASIC FORMALISMS ARE SUBSUMED BY DEV&DESS

DESS Are Special DEV&DESS

Basic DESS can obviously be interpreted as DEV&DESS by omitting all the event parts of the combined formalism. Therefore, a basic differential equation specified system

$$DESS = (X_{dess}, Y_{dess}, Q_{dess}, f_{dess}, \lambda_{dess})$$

defines a special type of combined DEV&DESS

$$DEV\&DESS = (X^{discr}, X^{cont}, Y^{discr}, Y^{cont}, S, \delta_{\text{ext}}, C_{int}, \delta_{\text{int}}, \lambda_{discr}, f, \lambda_{cont})$$

with $X^{cont} = X_{dess}$, $Y^{cont} = Y_{dess}$, $S^{cont} = Q_{dess}$, $f = f_{dess}$, $\lambda_{cont} = \lambda_{dess}$, and all the event parts can be omitted, i.e., $X^{discr} = \{\}$, $Y^{discr} = \{\}$, $S^{discr} = \{\}$, $\delta_{\text{int}}(s, e), x^{cont}) = s$, $\delta_{\text{ext}}((s, e), x^{cont}) = s$, $\lambda_{discr}((s, e), x^{cont}) = \emptyset$ and $C_{int}((s, e), xc) = false$.

DTSS Can Be Represented by Equivalent DEV&DESS

We construct a DEV&DESS which is equivalent to a DTSS by employing variable e, which represents the elapsed time since the last state transition, to schedule the execution of the state transitions at constant time intervals h. The state transitions are scheduled employing the state event condition function. Whenever, the elapsed time e reaches the step size h, the condition function evaluates to true and a state transition occurs. In the internal state transition function, the state is set according to the next state function of the DTSS (recall that the elapsed time e is set to 0 at each state transition). Input and outputs of the DTSS are interpreted to be piecewise constant.

Formally, we construct a combined

$$DEV\&DESS = (X^{discr}, X^{cont}, Y^{discr}, Y^{cont}, S, \delta_{\text{ext}}, C_{int}, \delta_{\text{int}}, \lambda_{discr}; f, \lambda_{cont})$$

from a

$$DTSS = (X_{dtss}, Y_{dtss}, Q_{dtss}, \delta_{dtss}, \lambda_{dtss}, h_{dtss})$$

in the following way: $X^{cont} = X_{dtss}$, $X^{discr} = \{\}$, $S^{cont} = \{\}$, $S^{discr} = Q_{dtss} \times \{h\}$ with $h = h_{dtss}$, $Y^{cont} = \{\}$, $Y^{discr} = Y_{dtss}$, the state event condition function $C_{int} : Q \times X_c \to Bool$ is defined by

$$C_{int}((q_{dtss}, h), e, x^{cont}) = \begin{cases} true & \text{if } e \geq h \\ false & \text{otherwise,} \end{cases}$$

The internal transition function $\delta_{\text{int}} : Q \times X^{cont} \to S$ is built by

$$\delta_{\text{int}}((q_{dtss}, h), e, x_{dtss}) = (\delta_{dtss}(q_{dtss}, x_{dtss}), h)$$

The discrete output function $\lambda_{discr} : Q \times X^{cont} \to Y^{cont}$ is given by

$$\lambda_{discr}((q_{dtss}, h)e, x_{dtss}) = \lambda_{dtss}(q_{dtss}, x_{dtss})$$

for Mealy-type systems and

$$\lambda_{discr}((q_{dtss}, h), e, x_{dtss}) = \lambda_{dtss}(q_{dtss})$$

for Moore-type systems.

The derivative function $f : Q \times X^{cont} \to \{\}$ and the continuous output function $\lambda_{cont} : Q \times X^{cont} \to \{\}$ are omitted.

Theorem 9.2. *A DEV&DESS so constructed is behaviorally equivalent to the original DTSS.*

The theorem can easily be proven by showing that there exists a system isomorphism from the state component q_{dtss} of the DEV&DESS and the state of the original DTSS. We come back to system isomorphism in the third part of the book and will prove this theorem there.

DEVS Can Be Represented by Equivalent DEV&DESS

In similar way as for DTSS we describe a procedure how to construct a DEV&DESS from a basic DEVS. In particular, we again employ the elapsed time variable e and the state event condition function to realize the scheduling of internal events. The internal condition function evaluates to true whenever the elapsed time reaches the value of the time advance given by the DEVS.

We construct a combined

$$DEV\&DESS = (X^{discr}, X^{cont}, Y^{discr}, Y^{cont}, S, \delta_{\text{ext}}, C_{int}, \delta_{\text{int}}, \lambda_{discr}, f, \lambda_{cont})$$

from a

$$DEVS = (X_{devs}, Y_{devs}, S_{devs}, \delta_{\text{ext,devs}}, \delta_{\text{int,devs}}, \lambda_{devs}, \text{ta}_{devs})$$

in the following way: $X^{cont} = \{\}$, $X^{discr} = X_{devs}$, $S^{cont} = \{\}$, $S^{discr} = S_{devs}$, $Y^{cont} = \{\}$, $Y^{discr} = Y_{devs}$.

The state event condition function $C_{int} : Q \times X^{cont} \rightarrow Bool$ is defined by

$$C_{int}((s_{devs}, e)) = \begin{cases} true & \text{if } e \geq \text{ta}_{devs}(s_{devs}) \\ false & \text{otherwise.} \end{cases}$$

The internal transition function $\delta_{\text{int}} : Q \times X^{cont} \rightarrow S$ is built by

$$\delta_{\text{int}}((s_{devs}, e)) = \delta_{\text{int,devs}}(s_{devs}).$$

The external transition function $\delta_{\text{ext}} : Q \times X^{discr} \rightarrow S$ is equivalent to that of the DEVS, $\delta_{\text{ext}}((s_{devs}, e), x_{devs}) = \delta_{\text{ext}}, devs((s_{devs}, e), x_{devs})$, and the discrete output function $\lambda_{discr} : Q \rightarrow Y^{discr}$ is equivalent to that of the DEVS, $\lambda_{discr}((s_{devs}, e)) = \lambda_{devs}(s_{devs})$. The derivative function $f : Q \rightarrow \{\}$ and the continuous output function $\lambda_{cont} : Q \rightarrow \{\}$ are omitted.

Theorem 9.3. *A DEV&DESS so constructed is behaviorally equivalent to the original DEVS.*

Once again the prove requires constructing a system isomorphism between the DEV&DESS and the DEVS. In particular, it has to be shown that the internal state transitions are scheduled at the same times.

Exercise 9.4. Show that a clock working in continuous time with a reset port can be realized as DEV&DESS without using the elapsed time component, *e*. (Hint: use a constant derivative.) This shows that the role of the elapsed time component can be played by such a clock.

9.5.2 COUPLED DEV&DESS FORMALISM

We have shown that basic DESS are special DEV&DESS and also how to embed DEVS or DTSS into DEV&DESS. So now we can use a basic formalism wherever we can use a DEV&DESS. With that capability we introduce networks of DEV&DESS in order to get a method supporting the construction of networks of components specified in any one of the formalisms – this is a *multi-formalism modeling methodology*.

Let us first examine how we would intuitively like to interpret the couplings of multi-formalism components. Then we will show how this is naturally achieved in the coupled DEV&DESS formalism. Fundamental to this enterprise is the interpretation of couplings of discrete outputs to continuous inputs and vice versa. Recall (Chapter 5) that event segments can be translated to piecewise constant segments and vice versa. Hence, we have a means for interpretation of couplings of event outputs to continuous inputs and piecewise constant outputs to event inputs. However, *arbitrary couplings of continuous outputs* to discrete inputs cannot be allowed. The reason is that a continuously changing segment would mean an infinite number of events – and this is not allowed for event segments. Therefore, we allow only couplings of *piecewise constant continuous* outputs to discrete inputs.

Fig. 9.11 shows the couplings of the different basic formalisms. Each box shown should be interpreted as its wrapped DEV&DESS version. The interpretation is as follows:

- *Coupling of DTSS to DEVS*: In a coupling of an output of a DTSS to a DEVS, each output at regular intervals from the DTSS means an external event at the DEVS. As the outputs of the DTSS are interpreted as discrete in context of multi-formalism systems, the output segment of a DTSS is an event segment with events at regular intervals which result in external events at the state transitions times of the DTSS (Fig. 9.11A).
- *Coupling of DEVS to DTSS*: In a coupling of an output of a DEVS to an input of a DTSS, the value of the last output event that occurred prior to the state transition of the DTSS should define the input value at its next state transition time. In context of multi-formalism models, the input of DTSS is continuous. Hence, the event output of the DEVS is translated into a piecewise constant input segment for the DTSS which results in the required behavior (Fig. 9.11B).
- *Coupling of DEVS to DESS*: In a coupling from an event output of a DEVS to a continuous input of a DESS, we would like that the events define discrete changes of the continuous input values. The translation of event segments to piecewise constant segments accomplishes this. The input of a DESS coming from a DEVS is piecewise constant, a subclass of the class of piecewise continuous segments allowed (Fig. 9.11C).
- *Coupling of DESS to DEVS*: The output of a DESS is a continuous segment, i.e., for every point in time a value is defined. Such a segment cannot be interpreted as an event segment because it would mean an infinite number of events. Thus, arbitrary couplings of continuous outputs of DESS to inputs of DEVS are not allowed. However, we allow couplings when the output is guaranteed to be piecewise constant. Piecewise constant segments are translated into event segments (Fig. 9.11D).
- *Coupling of DTSS to DESS*: In couplings of DTSS to DESS, the discrete outputs of the DTSS at regular intervals should mean a new constant output value until the next output occurs. The translation of the event segment of the DTSS into piecewise constant segments at the continuous interface of the DESS accomplishes just this (Fig. 9.11E).
- *Coupling of DESS to DTSS*: In a coupling of a DESS to a DTSS, the current continuous output value of the DESS is supposed to define the input value for the state transition of the DTSS. As the input of the DTSS is interpreted to be continuous in multi-formalism networks, the DTSS just takes the current continuous value at any particular state transition time (Fig. 9.11F).

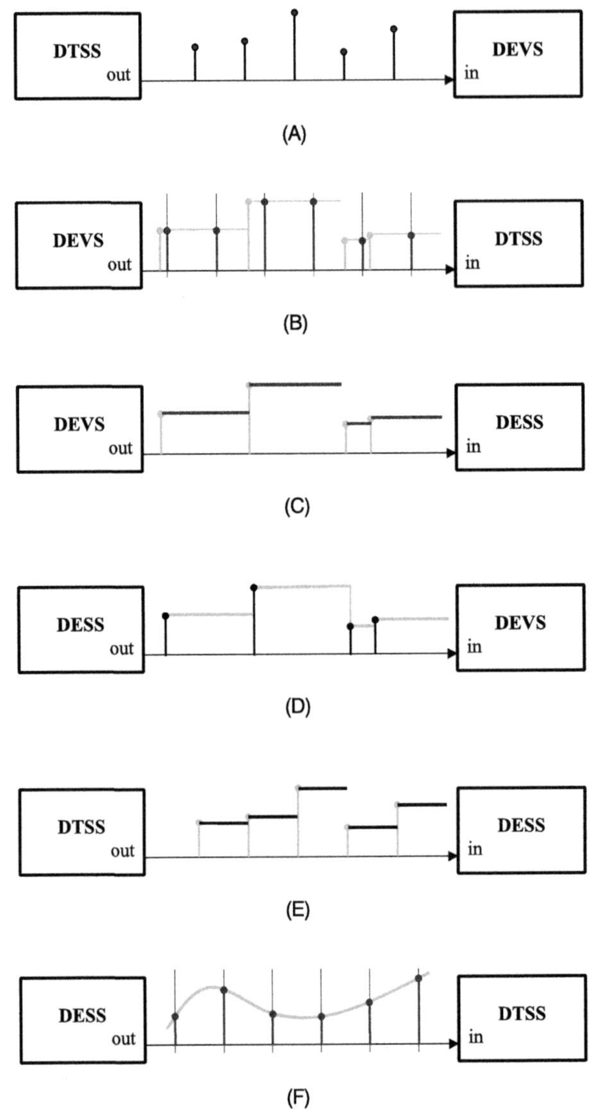

FIGURE 9.11

Multi-formalism system couplings interpretations. (A) DTSS – DEVS coupling. (B) DEVS – DTSS coupling. (C) DEVS – DESS coupling. (D) DESS – DEVS coupling (only piecewise constant allowed!!). (E) DTSS – DESS coupling. (F) DESS – DTSS coupling.

Exercise 9.5. Consider the coupling interface between two DTSSs with different time step sizes. In particular consider using an averaging function in the mapping of fast output to slow input. (Hint: see Laffitte, 1994.)

Formally we define a multi-formalism network as a coupled system

$$MFN = \left\langle X^{discr} \times X^{cont}, Y^{discr} \times Y^{cont}, D, \{M_d\}, \{I_d\}, \{Z_d\}, \text{Select} \right\rangle$$

where

- $X^{discr} \times X^{cont}$ and $Y^{discr} \times Y^{cont}$ are the input and output set, which consist of discrete input and outputs sets, X^{discr} and Y^{discr}, and continuous input and output sets, X^{cont} and Y^{cont}, respectively.
- The components' references D and components systems $\{M_d\}$ as well as influencers set $\{I_d\}$ and coupling definitions $\{Z_d\}$ are defined as in the general coupled system formalism in Section 5.8 with Z_d divided into functions $Z_d^{discr} : \times_{i \in I_d} XY_i \rightarrow YX_d^{discr}$ and $Z_d^{cont} : \times_{i \in I_d} Y_i \rightarrow X^{cont} d$ according to YX_d^{discr} and YX_d^{cont}.
- The tie-breaking function Select is as for DEVS coupled models.

To be well defined, a multi-formalism network has to fulfill the following constraints:

- the components M_d must be of the DEVS, DEVN, DTSS, DTSN, DESS, DTSN, DEV&DESS, or MFN type
- no algebraic cycles are allowed, i.e., in a feedback loop there has to be at least one component the output of which can be computed without knowledge of its input (see DTSS networks in Chapter 7)
- coupling from continuous outputs to inputs of DEVS components to restricted to piecewise constant segments.

A network of DEV&DESS obeying these constraints itself specifies a basic DEV&DESS, therefore, DEV&DESS are closed under coupling. We provide a procedure for deriving a basic DEV&DESS from a network of DEV&DESS in Appendix 9.B.

9.6 SIMULATOR FOR DEVS&DESS

The abstract simulator concepts for basic models formulated in Chapter 8 give us a basis for the definition of simulators for multi-formalism models. The abstract simulator for atomic and coupled DEV&DESS can be defined using a combination of the simulators for DEVS and DESS with additional handling of state events. This means, we have to detect state events, report state events, and execute them. This will require several additions in the continuous as well as the discrete parts of the abstract simulators.

The structure of an abstract simulator contains as discrete part an extended DEVS simulation protocol and as continuous part an extended DESS simulation protocol. These two parts are synchronized by a special root coordinator. In simulation, the two parts alternate in model execution. While the discrete part executes the state transition at the event times, the continuous part computes the state trajectories in between.

The abstract simulator for DEV&DESS has to allow the simulation of multi-formalism models consisting of DEVS, DESS and DTSS components. A simulator for multi-formalism models is built up by simulators for basic formalisms as introduced in Chapter 8 and with a combined devs&dess-coordinator

as their overall integrator. Additionally, the conversion of input and output segments between the different type of models is accomplished by special *interface objects*. We will present the dev&dess-simulators and dev&dess-coordinators first and then introduce interfaces for basic model types in the sequel.

9.6.1 THE DEV&DESS-SIMULATOR AND -COORDINATOR

Handling of state events is the main extension necessary for the simulation of combined models. This includes:

- the detection of state events,
- the determination of their exact times, and
- scheduling such events for execution.

Such state event handling has to occur during integration and therefore has to be undertaken by the continuous part. The execution of the state events, once determined, is identical to the exection of internal scheduled events in pure DEVS.

Recall that state events occur when continuous variables reach certain values, e.g. a threshold, or when continuous variables meet, e.g. when two objects hit. The state event then is specified by an equality or inequality expression like

$$expr_1 = expr_2, expr_1 \langle expr_2, expr_1 \rangle \, expr_2.$$

Inequality expressions are usually transformed by subtracting one side from the other so that there is zero on one side, i.e.,

$$expr_1 - expr_2 > / < / = 0.$$

State event detection then can be accomplished by checking for zero-crossing, i.e., whenever the $expr_1 - expr_2$ changes sign in an integration step, a state event must have occurred in-between. An approximation method is then used to determine to the exact time of the event.

Several different methods are used for the approximation of zero-crossings. The best known method is the bisection method. Basically, it employs a binary search mechanism at the integration interval the state event occurred (Fig. 9.12). When a state event occurred during an integration step, the interval is halved and a new integration step is undertaken with the half stepsize. If the state event still occurs, one knows that it is in the first half and this half is considered further. Otherwise, one knows that the state event is in the second half and the second half is investigated further. This process is continued until the time of the state event is approximated to a predefined accuracy.

Let us discuss now the extensions which are necessary for a continuous simulator to handle state events. Recall from Chapter 8 that a dess-simulator (we discuss the causal version here, extension to non-causal integration methods is straightforward) upon the receipt of an x-message computes the state for next time step $t + h$. In the combined simulator, we have to check for the occurrence of a state event in each integration step. Whenever a zero-crossing for the new state at $t + h$ has occurred (which is recognized by the state event condition function C_{int}) a state event has been detected and is reported in a *state event message (se-message)* to the discrete part. The discrete part then schedules the execution of a state event and reports this state event upward to the parent coordinator. Finally the message arrives at the root-coordinator. In the combined simulation scheme the root-coordinator has important duties.

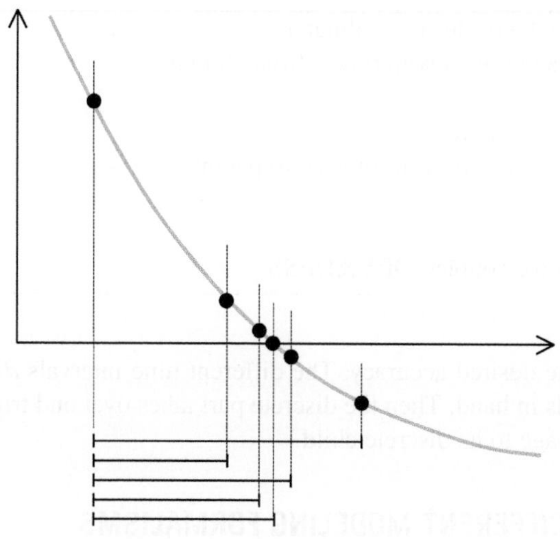

FIGURE 9.12

Bi-section method to search for state event times.

```
dev&dess-simulator extends devs-simulator, dess-causal-simulator
when receive message (x, t_i) with input value x
    q(t_{i+1}) = q(t_i + h) = CausalMethod([q(t_{i-d-1}), ..., q(t_i)], [r(t_{i-d-1}), ..., r(t_i)], [x(t_{i-d-1}),
    ..., x(t_i)], h)
    if (C_int(q(t + h), x(t + h))) then
        tn := t + h
    else
        tn = ∞
    end if
    if tn has changed then then
        send se-message (se, tn) with time tn to discrete parent
    end if
end dev-dess-simulator
Causal simulator for basic DEV&DESS
```

Not only is it responsible for the alternating execution of the continuous and discrete part, but it also has to take care that the state events are hit with desired accuracy. It implements the overall event loop where it first integrates to the time of the next event tn and executes the internal event at time tn. The integration part is done in a loop where it sends integration messages to the continuous child with integration stepsize dh which computes as the normal integration stepsize h or the time interval to the next internal event $(tn - t)$, whatever is smallest.

After each integration step, the root coordinator checks if an se-message has been received from the discrete part and a state event has occurred. Then, in a further loop the root-coordinator iterates to the state event by successive integration steps with different time intervals dh until the time of the state

```
dev-dess-coordinator extends devs-coordinator
when receive se-message (se, t) with time t from child c
    tn = t
    if tn has changed then then
        send se-message (se, tn) with time tn to parent
    end if
end dev-dess-coordinator
Coordinator extension for coupled DEV&DESS
```

event has been hit with the desired accuracy. The different time intervals dh are computed based on the approximation methods in hand. Then the discrete part takes over and triggers the execution of the internal event by a $*$-message to its discrete child.

9.6.2 INTEGRATING DIFFERENT MODELING FORMALISMS

The dev&dess-coordinator can serve as an integrator for multi-formalism model simulation. However, interfaces are needed to transform the messages of the simulators and coordinators for the different model types (Fig. 9.13). In particular, they have the task to transform the input messages as required by the interpretation of the modeling formalisms in a multi-formalism network as defined in Section 9.5 In the following we define the interfaces for the messages for the different model formalisms. All other messages not specified in the interfaces should be regarded to be forwarded directly to the simulator or coordinator object.

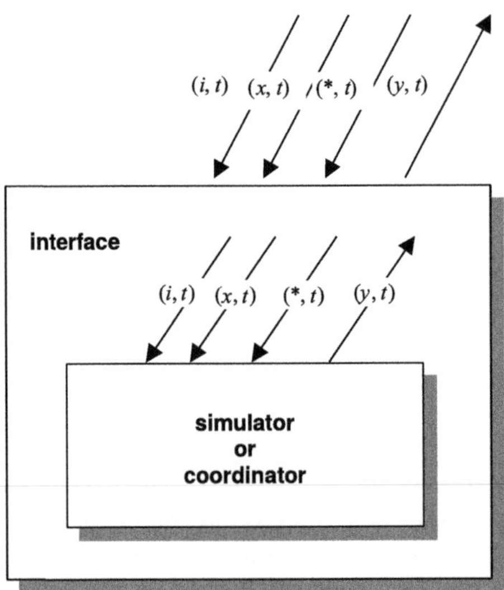

FIGURE 9.13

Interface object in multi-formalism abstract simulator.

```
root-coordinator
variables :
    child – direct subordinate simulator or coordinator
    tn – time of next event of overall child
    h – desired time step
    t – current simulation time
t = t0
send initialization message (i, t) to discrete and continuous child
tn = tn of its child
loop                                                                    ▷ integration loop
    while t < tn do
        dh = if (t + h < tn) then dh = h else dh = tn − t              ▷ make-integration step
        state-event-detected = false
        send (∗, t, dh) to continuous child
        send (x, t, dh) to continuous child
        if state-event-detected then                                    ▷ iterate to state event
            loop
                dh = compute new dh based on approximation method       ▷ make integration step
with new time step dh
                state-event-detected = false
                send (∗, t, dh) to continuous child
                send (x, t, dh) to continuous child
            end loop
        end if
        until state-event-detected ∧ accuracy reached
        t = t + dh
    end while                                                           ▷ execute event
end loop
send (∗, tn) message to continuous child
t = tn
until end of simulation
when receive se-message (se, tnc) with time tn_c from discrete child
state-event-detected = true
if tn_c < tn then
    tn = tn_c
end if
end root-coordinator
Root-coordinator in DEV&DESS abstract simulator
```

DEVS-INTERFACE

The interface for DEVS receives the continuous input values. It checks if they are different from the last received. If so it sends an x-message to its simulator or coordinator reporting the change event in the input value (recall that the inputs have to be piecewise constant so that only a finite number of changes occur).

```
devs-interface
xs^cont                                                  ▷ variable to save last continuous inputs
when receive input message (x^cont, t) with continuous input x^cont
    when x^cont ≠ xs^cont then
        send (x^cont, t) to devs-simulator or devs-coordinator
        xs^cont = x^cont
end devs-interface
devs-interface
```

DTSS-INTERFACE

The interface for DTSS may receive inputs from DEVS, from DESS, and also from different DTSS components, possibly operating with different step sizes h. All these inputs are stored for the next state transitions. When a dtss-interface receives a $*$-message at the time of its next state transition, tn, it sends a $*$-message to its dtss-simulator or -coordinator to compute the output and then an x-message with stored input values to undertake the state transition.

```
dtss-interface
xs                                                       ▷ variable to save last input
when receive input message (x, t)
    xs = x
when receive (*, t)                                      ▷ execute state transitions
    send (*, t) to dtss-simulator or dtss-coordinator    ▷ compute output
    send (xs, t) to dtss-simulator or dtss-coordinator   ▷ state transition
    tn = tn + h
end dtss-interface
dtss-interface
```

DESS-INTERFACE

When a dess-interface receives an discrete input event, it stores the values to interpret its inputs as piecewise constant trajectories.

```
dess-interface
xs^discr                                                 ▷ variable to save last discrete inputs
when receive input message (x^discr, t) with discrete input x^discr
    when x^discr ≠ xs^discr then
        xs^discr = x^discr
end dess-interface
dess-interface
```

Exercise 9.6. Show that the interface objects together with their simulators or coordinators realize the simulation protocol as required for the basic formalisms in multi-formalism networks (Section 9.5.1).

9.7 UPDATE ON DEV&DESS AND CYBER-PHYSICAL SYSTEMS

Interest in modeling and simulation of combined discrete event and continuous systems has burgeoned with the rise of interest in cyber-physical systems (CPS). The definition of CPS has become widely accepted as a tight combination of, and coordination between, computational and physical elements. As such it transcends traditional embedded electronic systems, the architecture of Internet of Things, and ultimately invokes the unification of computer science and control theory. A prime example of current focus is energy systems integration which combines energy carriers, including electricity, with energy users, to maximize efficiency and minimize waste, such systems are studied as system-of-systems whose component systems are brought together in a virtual testbed also called co-simulation (Mittal and Martin, 2017). With its formal combination of DEVS and DESS, the DEV&DESS formalism provides a rigorous framework for CPS modeling and simulation. However, DEV&DESS does not provide a computational framework for simulating such models. Therefore, Zeigler (2006) proposed a formal approach to embedding DEV&DESS into DEVS. This would provide the framework to leverage DEVS simulation environments for CPS applications. Note that the DEVS quantization of DESS in Chapter 16 does not provide the extension to DEV&DESS since the interfaces between continuous and discrete event parts are not directly considered.

The just mentioned embedding deals specifically with three behaviors mediated by continuous-discrete interface. These are that discrete Events, both state and external, can 1) control the derivative function, thus causing an instantaneous change in the derivative of the continuous state trajectory, 2) control the event detection condition, thus instantaneously changing the thresholds governing event detection, and 3) instantaneously change the continuous system state (causing discontinuities from the point of view of differential equation systems). Complete implementations of DEV&DESS must support such interfaces.

Preyser et al. (2015) recognized the need for an efficient simulator able to rigorously implement and simulate a DEV&DESS. Guided by the embedding they implemented a generic DEV&DESS block in PowerDEVS (Chapter 19) similar to Parallel DEVS. An alternative implementation of DEV&DESS, called HDEVS, was provided by Hee Lee (2017) in the HDEVSim++ environment. As opposed to energy systems, HDEVS is focused on the challenges of CPS underlying advanced air combat systems. Nutaro et al. (Chapter 1) defined a DEVS-based hybrid formalism implemented in a combination of ADEVS (C++) and Open Modelica.

Such implementations demonstrate the utility of the DEV&DESS formalism for CPS co-simulation. However, deeper theoretical investigations and formal proofs of the correctness are needed. To this end, we will provide a more fundamental characterization of DEV&DESS and hybrid modeling and simulation using iterative systems specification in Chapter 20.

9.8 SOURCES

State events are an integral part in many continuous simulation languages. A thorough investigation of the problems associated with state events was done by Cellier (1979, 1986). The system framework for combined modeling was first introduced in Praehofer (1991), Praehofer et al. (1993). The term multi-modeling stems from Ören (1991) and has been elaborated in Fishwick and Zeigler (1992). A multi-modeling approach also has been developed to study automated highways control architectures

(Eskafi, 1996; Deshpande et al., 1996). Hybrid systems modeling and analysis is part of an effort to establish a methodology for event based intelligent control of continuous systems (Antsaklis et al., 1989; Meystel, 1989; Grossmann et al., 1993). Some references to widely adopted formalisms are Forrester (1972), Harel et al. (1990), Paynter (1961) and Peterson (1981). Cellier (1986) provides an in-depth discussion of systems dynamics and bond graphs. An example of employing Statecharts within a DEVS simulation environment for co-design of hardware and software is given in Schulz et al. (1998).

APPENDIX 9.A THE SYSTEM SPECIFIED BY A DEV&DESS

The DEV&DESS system specification $= (X^{discr}, X^{cont}, Y^{discr}, Y^{cont}, S, \delta_{\text{ext}}, C_{int}, \delta_{\text{int}}, \lambda^{discr}, f, \lambda^{cont})$ relies on the following background conventions and these become the corresponding elements in the I/O system $S = (T, X, \Omega, Y, Q, \Delta, \Lambda)$ that it specifies:

- The time base T is the real numbers $\mathbb{R}$
- $X = X^{cont} \times (X^{discr} \cup \emptyset$ is the set of inputs of the system being specified, with the symbol $\emptyset \notin X^{discr}$ specifying the non-event
- X^{discr} can be an arbitrary set
- X^{cont} is a structured set $X^{cont} = \{(x_1^{cont}, x_2^{cont}, ..., x_i^{cont}, ..., x_n^{cont}) | x_i^{cont} \in X_i^{cont}\}$ with X_i^{cont} are the reals
- $Y = Y^{cont} \times (Y^{discr} \cup \emptyset$ is the set of outputs of the dynamic system
- Y^{discr} can be an arbitrary set
- Y^{cont} is structured set $Y^{cont} = \{(y_1^{cont}, y_2^{cont}, ..., y_i^{cont}, ..., y_n^{cont}) | y_i^{cont} \in Y_i^{cont}\}$ with Y_i^{cont} are the reals
- $Q = Scont \times Q_discr$, is the set of states of the specified system. This system state set has a typical element $q = (s, e) = (s^{cont}, s^{discr}, e)$, i.e., a tuple containing the DESS state, s^{cont}, the DEVS state, s^{discr}, the elapsed time, e
- S^{discr} can be an arbitrary set
- S^{cont} is structured set $S^{cont} = \{(s_1^{cont}, s_2^{cont}, ..., s_i^{cont}, ..., s_n^{cont}) | s_i^{cont} \in S_i^{cont}\}$ with S_i^{cont} are the reals
- The set Ω^{discr} of admissible input segments of the discrete inputs is the set of all event segments over X^{discr} and T
- The allowable input segment Ω_i^{cont} for input variable x_i^{cont} are piecewise continuous segments over X_i^{cont} and T
- The state trajectories of the discrete states S^{discr} are piecewise constant segments over S^{discr} and T
- The state trajectories of the continuous states S^{cont} are piecewise continuous segments over S^{cont} and T
- The output trajectories of the discrete outputs Y^{discr} are event segments over Y^{discr} and T
- The allowable output segment for output variable y_i are piecewise continuous segments over Y_i^{cont} and T.

The global transition function, Δ, of the I/O system is defined as follows: Let $\omega :< t_1, t_2] \rightarrow X$ be the input segment and the state $q = (s, e) = (s^{cont}, s^{discr}, e)$ be the state at time t_1. Then we define Δ by

1. Intervals $\langle t_1, t_2\rangle$ with no events:

$$\begin{aligned}\Delta(q,\omega) &= (s^{cont}(t_2), s^{discr}(t_2), e(t_2))\\ &= ((s^{cont}(t_1) + \int_{t_1}^{t_2} f(s(\tau),\omega(\tau))d\tau, s^{discr}(t_1), e(t_1) + t_2 - t_1)\\ &\text{if } \nexists t \in < t_1, t_2] : C_{int}(s(t), x^{cont}(t)) \wedge \nexists t \in < t_1, t_2] : \omega^{discr}(t) \neq \Phi\end{aligned}$$

2. An state event occurs first at time t in interval $\langle t_1, t_2\rangle$:

$$\begin{aligned}\Delta(q,\omega_{t]}) &= (s^{cont}(t_2), s^{discr}(t_2), e(t))\\ &= (\delta_{\text{int}}((s^{cont}(t_1) + \int_{t_1}^{t} f(s(\tau),\omega(\tau))d\tau, s^{discr}(t_1)), x^{cont}(t)), 0)\\ &\text{if } \exists t \in < t_1, t_2] : C_{int}(s(t), \omega^{cont}(t)) \wedge \nexists t' \in < t_1, t] : \omega^{discr}(t') \neq \Phi\end{aligned}$$

3. An event input at a discrete input occurs first at time t in interval $\langle t_1, t_2\rangle$:

$$\begin{aligned}\Delta(q,\omega_{t]}) &= (s(t), e(t))\\ &= (\delta_{\text{ext}}((s^{cont}(t_i) + \int_{t_1}^{t} f(s(\tau),\omega(\tau))d\tau, s^{discr}(t_1)), x^{cont}(t), e(t_1) + t - t_1, \omega(t)), 0)\\ &\text{if } \exists t \in < t_1, t_2] : \omega^{discr}(t) \neq \Phi \wedge \exists t' \in < t_1, t] : C_{int}(s(t'), \omega^{cont}(t'))\end{aligned}$$

The output function Λ is defined by:

$$\Lambda^{cont}(q,x) = \lambda(q, x^{cont})$$

$$\Lambda^{discr}(q,x) = \begin{cases}\lambda^{discr}(q, x^{cont}) & \text{if } e = 0\\ \Lambda^{discr}(q,x) = \Phi & \text{otherwise}\end{cases}$$

After executing the external or state events, the same considerations apply to the new state and the remaining segment.

APPENDIX 9.B THE SYSTEM SPECIFIED BY A MULTI-FORMALISM SYSTEM – CLOSURE UNDER COUPLING OF NETWORKS OF DEV&DESS

DEV&DESS systems are closed under coupling, i.e., networks of DEV&DESS are themselves DEV&DESS systems. The resulting system can be regarded as a cross-product of the component systems. The sequential state, inputs, and outputs are defined by the cross-product of the states, inputs and outputs of the system description of the components. The state event condition function is the disjunction (logical *or*) of the state event condition functions of all components. This means that a state event occurs whenever one of the components' state event condition evaluates to true. The internal and external state transition functions are defined analogously to network of DEVS (Chapter 7).

Theorem 9.4 (Closure under coupling of DEV&DESS). *The resultant of coupled DEV&DESS is well-defined and is itself a DEV&DESS.*

Proof. Given a DEV&DESS coupled model,

$$MFN = \left\langle X^{discr} \times X^{cont}, Y^{discr} \times Y^{cont}, D, \{M_d\}, \{I_d\}, \{Z_d\}, \text{Select} \right\rangle$$

we associate with it the following basic DEV&DESS:

$$DEV\&DESS = (X^{discr}, X^{cont}, Y^{discr}, Y^{cont}, S, \delta_{\text{ext}}, C_{int}, \delta_{\text{int}}, \lambda^{discr}, f, \lambda^{cont})$$

where

$$X^{discr} = X_N^{discr}, X^{cont} = X_N^{cont}, Y^{discr} = Y_N^{discr}, Y^{cont} = Y_N^{cont}, S = \times_{d \in D} Q_d.$$

A continuous input of a component d in the network is defined by the couplings of the continuous outputs of its influencers, i.e., $x_d^{cont} = Z_d^{cont}(..., \lambda_i^{cont}(q_i^{cont}, x_i^{cont}), ...)$ with $i \in I_d$. Then the derivative function $f : Q \times X^{cont} \to S$ is the crossproduct over the derivative functions f_d of the components $d \in D$, i.e. $f(q, x_N^{cont}).d = f_d(q_d, x_d^{cont})$, the continuous output function $\lambda^{cont} : Q \times X^{cont} \to Y^{cont}$ is derived from the external output couplings of the continuous output variables, i.e., $Z_N^{cont}(..., \lambda_i^{cont}(q_i, x_i^{cont}), ...)$ with $i \in I_N$.

The state event condition function $C_{int} : Q \times X_c \to Bool$ is defined by the disjunction of the condition functions of the components, i.e. $C_{int}(q, x^{cont}) = \vee_{d \in D} C_{int,d}(q_d, x_d^{cont})$, and the state transition and output functions are as follows:

Let $IMM(s) = \{d | d \in D \wedge C_{int,d}(q_d, x_d^{cont})\}$ the set of components that have a condition evaluating to true and $d* =$ Select($Imminent(s)$) select one of them using the tie breaking function Select, then we define $\delta_{\text{int}} : Q \times X_N^{cont} \to S$, $\delta_{\text{int}}((s, e), x_N^{cont}) = s' = (..., (s_d', e_d')...)$ by

$$(sd', ed') = \begin{cases} (\delta_{\text{int}}, d(q_d, x_d^{cont}), 0) & \text{if } d = d* \\ (\delta_{\text{ext}}, d(s_d, e_d + e, x_d^{cont}, x_d^{discr}), 0) & \text{if } d* \in I_d \wedge x_d^{discr} \neq \emptyset \\ (s_d, e_d + e) & \text{otherwise} \end{cases}$$

where

$$x_d^{discr} = Z_d^{discr}(\emptyset, \emptyset, ...\lambda_{d*}^{discr}(s_{d*})..., \emptyset, \emptyset).$$

We define $\delta_{\text{ext}} : Q \times X^{cont} \times X^{discr} \to S$,

$$\delta_{\text{ext}}((s, e), x^{cont}, x^{discr}) = s' = (..., (s_d', e_d')...) by$$

$$(s_d', e_d') = \begin{cases} ((\delta_{\text{ext},d}(s_d, e_d + e, x_d^{cont}, x_d^{discr}), 0) & \text{if } N \in I_d \wedge x_d^{discr} \neq \emptyset \\ (s_d, e_d + e) & \text{otherwise} \end{cases}$$

where $x_d^{discr} = Z_d^{discr}(\emptyset, \emptyset, ...x^{discr}..., \emptyset, \emptyset)$. And we define $\lambda^{discr} : S \to Y$ by

$$\lambda^{discr}(s) = \begin{cases} Z_N^{discr}(\emptyset, \emptyset, ..., \lambda_{d*}^{discr}(s_{d*})..., \emptyset, \emptyset) & \text{if } d* \in I_N \\ \emptyset & \text{otherwise} \end{cases}$$

The rest of the proof shows that the resultant is well defined. □

REFERENCES

Antsaklis, P.J., Passino, K.M., Wang, S., 1989. Towards intelligent autonomous control systems: architecture and fundamental issues. Journal of Intelligent & Robotic Systems 1 (4).

Berthomieu, B., Diaz, M., 1991. Modeling and verification of time dependent systems using time Petri nets. IEEE Transactions on Software Engineering 17 (3).

Cellier, F., 1979. Combined Discrete/Continuous System Simulation by Use of Digital Computers. Techniques and Tools.

Cellier, F., 1986. Combined discrete/continuous system simulation-application, techniques and tools. In: Proceedings of the 1986 Winter Simulation Conference, SCS.

Deshpande, A., Göllü, A., Varaiya, P., 1996. SHIFT: a formalism and a programming language for dynamic networks of hybrid automata. In: Hybrid Systems IV.

Eskafi, F.H., 1996. Modeling and Simulation of the Automated Highway System. California Partners for Advanced Transit and Highways (PATH).

Fishwick, P.A., Fishwick, P.A., Fishwick, P.A., Fishwick, P.A., 1995. Simulation Model Design and Execution: Building Digital Worlds, vol. 432. Prentice Hall, Englewood Cliffs, NJ.

Fishwick, P.A., Zeigler, B.P., 1992. A multimodel methodology for qualitative model engineering. ACM Transactions on Modeling and Computer Simulation (TOMACS) 2 (1).

Forrester, J., 1972. World Dynamics.

Grossmann, R.L., Nerode, A., Raun, A., Ritschel, H., 1993. Hybrid Systems. Lecture Notes in Computer Science, vol. 736. Springer-Verlag, Berlin.

Harel, D., Lachover, H., Naamad, A., Pnueli, A., Politi, M., Sherman, R., Shtull-Trauring, A., Trakhtenbrot, M., 1990. Statemate: a working environment for the development of complex reactive systems. IEEE Transactions on Software Engineering 16 (4), 403–414.

Hee Lee, J., 2017. Development of air combat HDEVS model implemented in HDEVSim++ environment. In: July 2017 Conference: SummerSim'17 Proceedings of the Summer Simulation Multi-Conference. San Diego, CA, USA.

Praehofer, Herbert, Auernig, F., Reisinger, G., 1993. An environment for DEVS-based multi-formalism modeling and simulation. Discrete Event Dynamic Systems 3, 119–149.

Ho, Y.-C., 1992. Discrete Event Dynamic Systems: Analyzing Complexity and Performance in the Modern World. IEEE Press.

Laffitte, J.A., 1994. Interfacing Fast and Slow Subsystems in the Real-Time Simulation of Dynamic Systems. PhD thesis.

Meystel, A., 1989. Intelligent control: a sketch of the theory. Journal of Intelligent & Robotic Systems 2 (2).

Mittal, S., Martin, J.L.R., 2017. Simulation-based complex adaptive systems. In: Mittal, S., Durak, U., Oren, T. (Eds.), Guide to Simulation-Based Disciplines: Advancing Our Computational Future. Springer AG.

Ören, T.I., 1991. Dynamic templates and semantic rules for simulation advisors and certifiers. Knowledge Based Simulation: Methodology and Application, 53–76.

Paynter, H.M., 1961. Analysis and Design of Engineering Systems. MIT Press.

Peterson, J.L., 1981. Petri Net Theory and the Modeling of Systems.

Praehofer, H., 1991. System theoretic formalisms for combined discrete-continuous system simulation. International Journal of General System 19 (3), 226–240.

Preyser, F.J., Hafner, I., Rössler, M., 2015. Implementation of hybrid systems described by DEV&DESS in the QSS based simulator PowerDEVS. Simulation Notes Europe 25 (2).

Sanders, W.H., Malhis, L.M., 1992. Dependability Evaluation Using Composed SAN-Based Reward Models, vol. 15. Elsevier.

Schulz, S., Rozenblit, J.W., Mrva, M., Buchenriede, K., 1998. Model-based codesign. Computer 31 (8), 60–67.

Zeigler, B.P., 2006. Embedding DEV&DESS in DEVS. In: DEVS Symposium. Huntsville, Alabama.

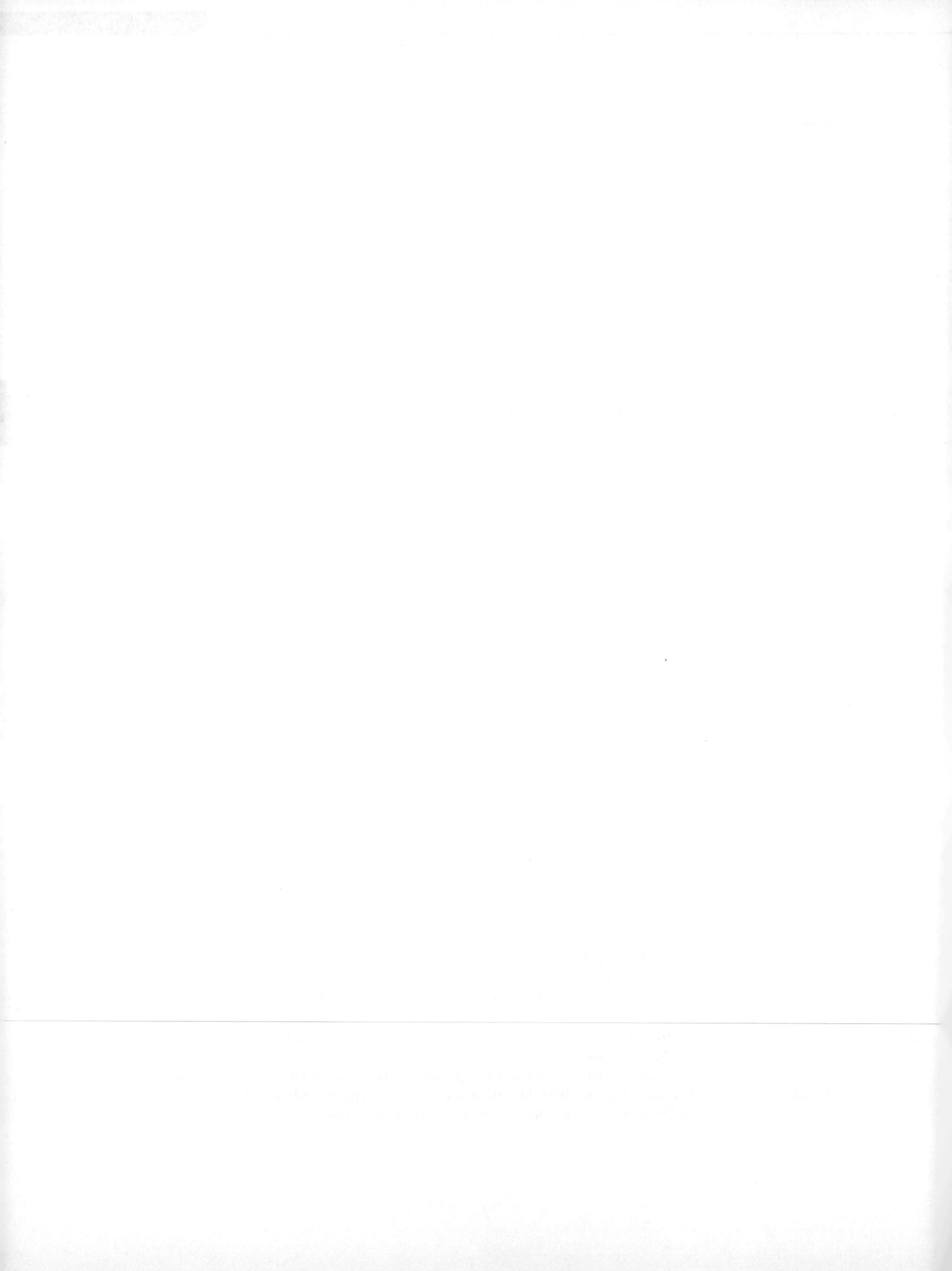

PART 2

ITERATIVE SYSTEM SPECIFICATION

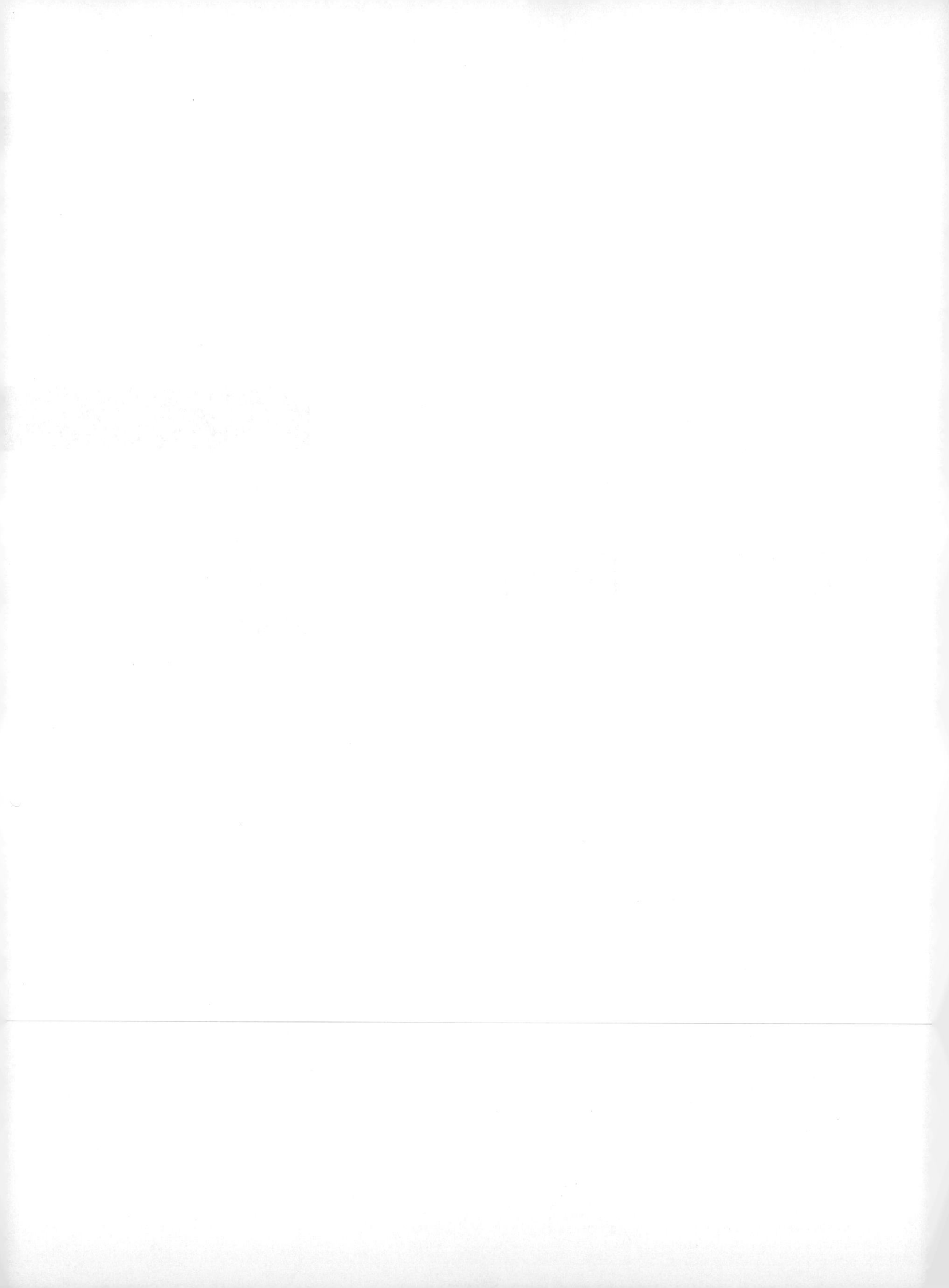

CHAPTER

INTRODUCTION TO ITERATIVE SYSTEM SPECIFICATION 10

CONTENTS

10.1 OVERVIEW OF ITERATIVE SYSTEM SPECIFICATION

In this edition, we focus on iterative system specification as a constructive formalism for general systems specification that is computationally implementable as extension to DEVS-based modeling and simulation environments. Iterative system specification enables deriving fundamental conditions for "well-defined system".[1]

[1]Our analysis is fundamental and helps explain common approaches to fixed point and zenoness issues.

Theory of Modeling and Simulation. https://doi.org/10.1016/B978-0-12-813370-5.00019-5

The main question we pose is: given components that are "well-defined systems", when can we be sure that their composition is also "well-defined"? This kind of question will show up frequently and often in the form of the problem of closure under coupling of a class of systems. In this exposition, one simple, yet non-trivial form of interaction, namely active/passive compositions will anchor an easily understood example of the problem and its solution. We mention application to emergence of natural language – showing a novel well-defined active/passive system. Other examples and applications are discussed.

10.2 ABSTRACTION, FORMALIZATION, AND IMPLEMENTATION

The rationale for focusing on iterative system specification in this edition stems from the interplay of abstraction and concreteness in advancing the theory and practice of modeling and simulation. Indeed, this interplay and the associated interplay of the generic and the specific also impel more broadly, the development of knowledge. Abstraction is defined as the drawing out of a general quality or characteristic, apart from concrete realities, specific objects, or actual instances. To formalize an idea or abstraction is to develop and state rules in symbolic form that characterize the abstraction. Important advances in human knowledge have been made through the progression from an abstraction to its formalization and subsequently to implementation in reality as illustrated in Fig. 10.1. This progression will help put our focus on iterative system specification in this book in context of the broader progress of knowledge in modeling and simulation.

An abstraction focuses on an aspect of reality and, almost by definition, greatly reduces the complexity of the reality being considered. Subsequent formalization makes it easier to work out implications of the abstraction and implement them in reality. Implementation can be considered as providing a concrete realization of the abstraction (often called "reduction to concrete form"). Such implementation reintroduces "messy" real world complexity into the situation and may stimulate another round of abstraction to address the emerging problems.

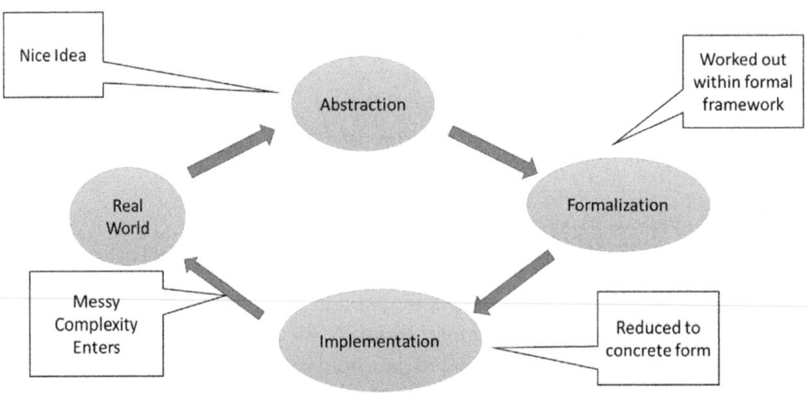

FIGURE 10.1

Progression of abstraction to implementation.

A few instances of the abstraction progression of Fig. 10.1 may help to understand this idea. One can see the eventual breakthrough innovation in computer fabrication technology as stemming from the abstraction of propositional logic derived from common sense informal reasoning (Fig. 10.2). "An Investigation of the Laws of Thought on Which are Founded the Mathematical Theories of Logic and Probabilities" was authored by George Boole in 1854. Subsequently the very simple rules of Boolean algebra and the representation of 0's and 1's through thresholding of continuous valued voltages enabled the design of digital computers and eventual mass volume production of Very Large Scale Integrated circuits. None of this could have happened had the benefit of computation based on 0's and 1's not been realized by engineers a century after Boole's abstraction.

A second example of the progression in Fig. 10.1 is that from the invention of the fundamental model of computation by Turing to today's powerful programming languages and software systems (Fig. 10.3). Turing Machines gave a formal basis for understanding the heretofore informal concepts of algorithmic computation. This abstraction led to recursive function theory and the Lambda Calculus (among other frameworks for computation) that led to the explosive growth of software. The time-limited ability of an abstraction to survive real-word complexities is illustrated by the current problematic state of software sharability. This raises the question of whether return to the simplicity of the Turing Machine could ameliorate the situation (Hinsen, 2017).

As discussed in Chapter 1, K.D. Tocher appears to be the first to conceive discrete events as the right abstraction to characterize the models underlying the event-oriented simulation techniques that he and others were adopting in the mid 1950s. Although there was clearly a similarity in purpose across the United Steels' steel plants that Tocher was attempting capture with the limited computational facilities of the era for any accurate modeling to be possible, the various technologies, equipment and layouts would have to be taken into account. Nevertheless, to have any hope of meeting the challenging complexity, Tocher had to conceive a framework that would address the steel plant problem more generically and exploit the commonality of purpose. According to Hollocks (2008), Tocher's core idea conceived of a manufacturing system as consisting of individual components, or 'machines', progressing as time unfolds through 'states' that change only at discrete 'events'. As in Fig. 10.4, Tocher's abstraction of discrete events, states, and components was formalized in DEVS and led to implementations in simulation environments based on DEVS (see Zeigler, 2017 and Zeigler et al., 2013 for

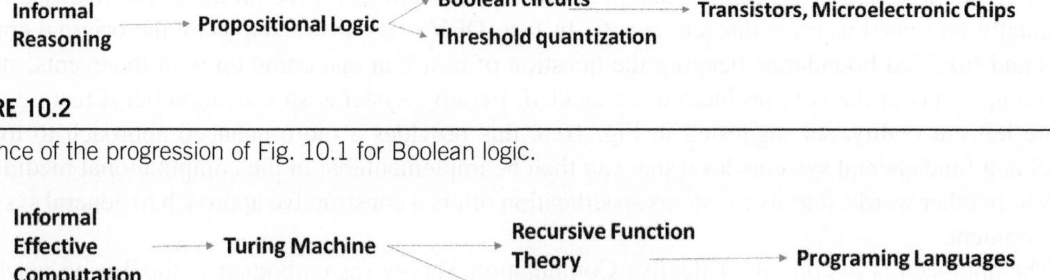

FIGURE 10.2

Instance of the progression of Fig. 10.1 for Boolean logic.

FIGURE 10.3

Instance of the progression of Fig. 10.1 for effective computation.

Early Computerized Simulation 40s-60s → Event Based Simulation → Discrete Events States → DEVS Family of Models → DEVS Simulation Systems

FIGURE 10.4

Instance of the progression of Fig. 10.1 for evolution of DEVS.

discussion of DEVS-based environments and the companion volume on model engineering (Zhang et al., 2018)).

As in spiral software system development, the progression in Fig. 10.1 can spawn a second (or more) round of abstraction, formalization, and implementation. Fig. 10.2 illustrates how this idea applies to iterative system specification. The maturing of DEVS-based environments led to demands for application of DEVS to an expanded range of problems beyond the operations research and early applications that discrete event simulation were aimed at.

The expanded range of problems is characterized by problems that require multiple disciplines to solve. Tolk (2017) distinguishes between interdisciplinary, multidisciplinary and transdisciplinary approaches to such problems. Although only transdisciplinary solutions can result in completely consistent representations of truth, the concurrent use of different paradigms within multi- and interdisciplinary environments must also be supported in practice. Powell and Mustafee (2016) distinguish between hybrid simulation and hybrid M&S. Hybrid simulation is the use of multiple M&S techniques in the model implementation, while hybrid M&S refers to the application of methods and techniques from different disciplines to one or more stages of a simulation development. DEVS is one of the key formalisms (along with UML) that they include within their recommended framework for hybrid studies (which typically include hybrid simulation) due to its general systems basis.

10.3 DERIVING ITERATIVE SYSTEM SPECIFICATION

In the light of Fig. 10.5, Iterative System specification is intended to be an abstraction that provides an approach to hybrid M&S at the transdisciplinary level. This abstraction is more fundamental than DEVS in that it serves as a precursor that provides representations for the preimages of reality that are eventually presented as crisp discrete events. In fact, DEVS has abstracted away the original continuous and fuzzified boundaries begging the question of how can one come up with the events, states, and components in the next problem to be tackled. Iterative systems specification helps restore more of the left out reality. As suggested in Fig. 10.6, this provides a more nuanced approach to hybrid M&S at a fundamental systems level that can then be implementable in the computational medium of DEVS. In other words, iterative systems specification offers a constructive approach to general systems development.

The approach is to combine Effective Computation Theory (as embodied in the Turing machine, for example) with Wymore Systems Theory (1967). The first focuses on finite states, state transitions, iteration and recursion as the fundamental mechanisms of computation. Turing introduced the notion of halting, i.e., indication of when a result is ready to be reported which in recursive function theory is formally the equivalent of full versus partial definition. These essential concepts of effective

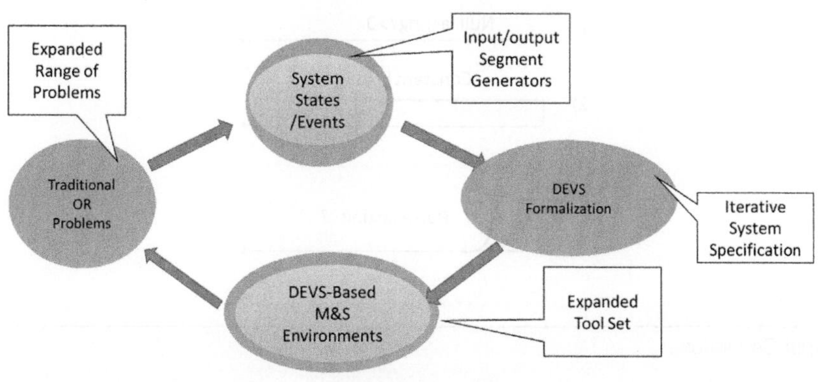

FIGURE 10.5

Second spiral of progression of abstraction to.

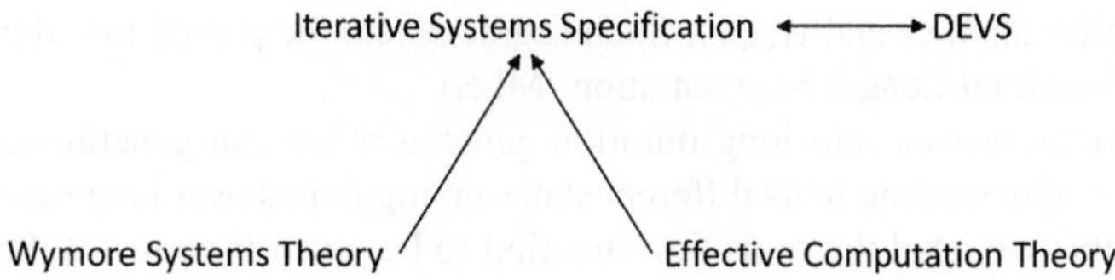

FIGURE 10.6

Iterative specification based on systems and computation theory.

computation bring along both the strength of computing as well as the associated halting problem, the insurmountable limitation of real world computation. On the other hand, Wymore Systems Theory brings to the table the use of real-value temporal ordering, as opposed to abstract sequences, using e.g., continuous time bases along with continuous states, global transition mapping, the structure/behavior distinction, and well-defined systems and their compositions. Combining the two offers a way to specify systems in an event-like computational manner that is ultimately implementable in DEVS-based M&S environments.

In the Turing Machine example of DEVS in Chapter 4, the machine was treated as an example of unitary design (hence at the level of non-modular system specification) that could however, be decomposed into simple components and reconstructed in a composition with simple coupling. The kind of passive/active interaction between the components exemplified by the system's operation will help to illustrate some of the basic concepts of iterative systems specification.

10.4 INPUT GENERATORS

Input generators are relatively short segments (time functions) that can be concatenated to form longer segments. For example, Fig. 10.7 shows three types of generators: 1) null segments that can last any

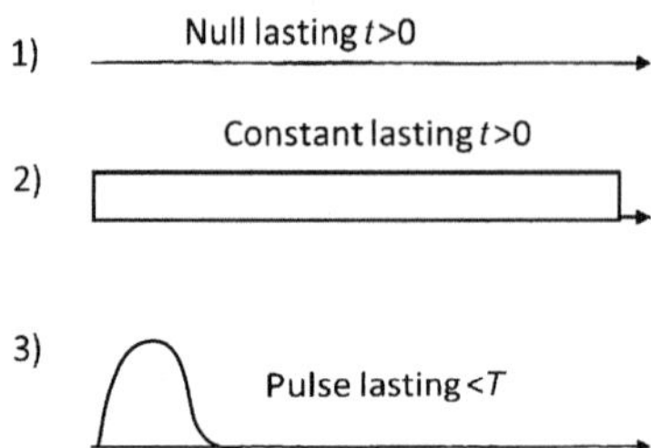

FIGURE 10.7

Illustrating Input Generators.

finite duration, 2) constant segments that can last any finite duration and 3) pulse segments lasting at most a duration T (e.g., 10 seconds).

As illustrated in Fig. 10.8, we consider composite inputs segments composed of concatenated generators placed one after another end-to-end. Such segments can be parsed into their component parts by a process called Maximal Length Segmentation (MLS).

With input segments broken into long duration generators we can generate the state trajectories associated with the system starting in its different states taking jumps over long time intervals as shown in Fig. 10.9. To do this we need the transition function to be given in terms of the generators so that given a starting state and an input generator we compute a state at the end of the generator. Doing things this way can save a lot of computation that would otherwise be done with small steps as suggested in the figure. Of course to compile the state transition function for the generators it might be necessary to do the small step computation. However, this needs to be done only once and there is still much utility in so doing, since the compiled form can be used every time the generator appears in an input segment. Also if there is an analytic solution for the generator input's effect on the state then we can

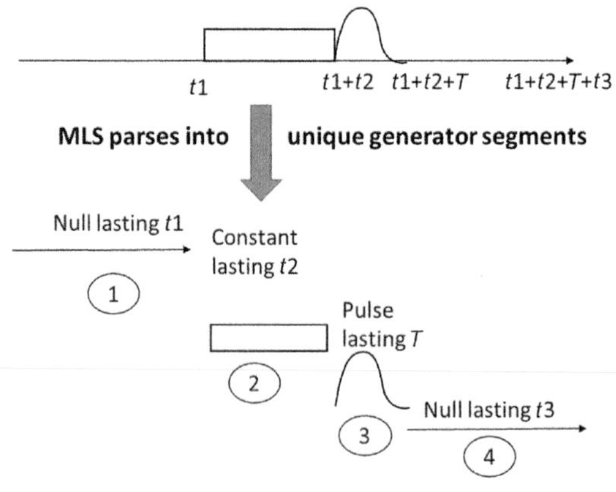

FIGURE 10.8

Maximal length segmentation.

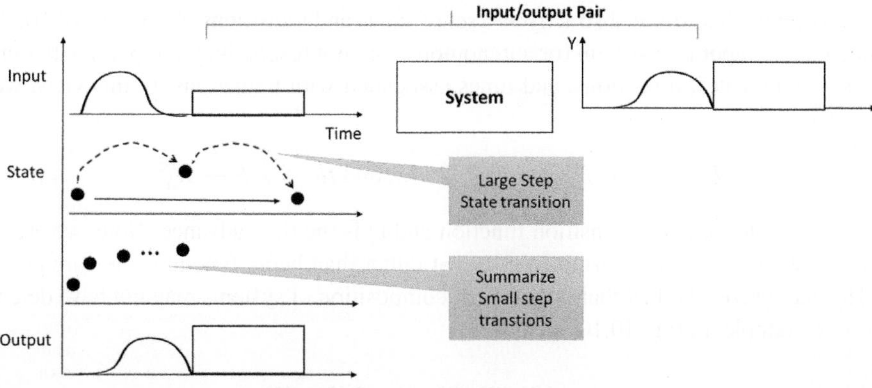

FIGURE 10.9

Computing with Generators.

use that within the large step-based computation. Since the output trajectory is computed from that state trajectory, we can compute the output trajectories corresponding to the input generators as well. The computation for atomic and coupled models can roughly be described as follows:

Atomic component system – compute I/O function

1. Apply MLS to get finite sequence of input generators
2. Given starting state, apply the transition function and output function to get next state and output for successive generators
3. Collect input/output pairs into I/O function

Composition of component systems

1. Compute I/O function of each component
2. Given starting states, guess input generators
3. Apply I/O functions to get output generators
4. Apply the coupling relations, to get input generators
5. Repeat step 3 until consistent input/output relation is obtained
6. Advance time using the minimum MLS break point (where shortest generator ends) at each stage

Note that these computations must terminate to produce effective results – recall our earlier discussion of effective computation, algorithms, and Turing machines. We shall now see the particular conditions that guarantee such effective computation are called *progressivity* and *well-definition.*

10.5 PROGRESSIVITY AND WELL-DEFINITION OF SYSTEMS

Recall that the resultant of a composition of systems is itself, potentially, a system. At the coupled level of system specification, the state of the system is a vector-like listing of its component states. So now let's look at this vector-like space abstractly as a set of states and transitions. As in Fig. 10.10, this leads us to a concept of an input free system (or autonomous, i.e., not responding to inputs) with outputs not considered and with states, transitions, and times associated with transitions. In this view, we have a transition system

$$M = \langle S, \delta, ta \rangle, \text{ where } \delta \subseteq S \times S, \text{ and } ta : S \times S \rightarrow R_0^\infty$$

where, S is set of states, δ is the transition function and ta is the time advance. Note, we are allowing the transition system to be non-deterministic so that rather than being functions they are presented as relations. This recognizes the fact that our original composition of systems may not have deterministic transitions. For example, in Fig. 10.10, we have

$$S = \{S1, S2, S3, S4, S5, S6, S7\},$$
$$\delta = \{(S1, S3), (S3, S4), \ldots\}$$
$$ta(S1, S3) = 1, ta(S3, S4) = 1, \ldots$$

There are transitions from state $S1$ to state $S3$ and from $S3$ to $S4$, each of which take 1 time unit and there is a cycle of transitions involving $S4 \ldots S7$ each of which take zero time. There is a self-transition involving $S2$ which consumes an infinite amount of time (signifying that it is passive, remaining in that state forever). This is distinguished from the absence of any transitions out of $S8$.

A *state trajectory* is a sequence of states following along existing transitions, e.g., $S1$, $S3$, $S4$ is such a trajectory.

This example gives us a quick understanding of the conditions for system existence at the fundamental level:

We say that the system is

- *Not defined* at a state, e.g., $S8$, because there is no trajectory emerging from it. More precisely, there is no transition pair with the state as its left member, there is no pair $(S8, x)$ for x in S.

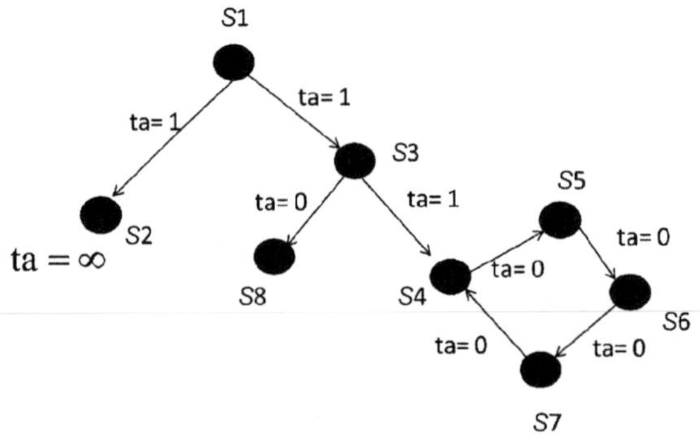

FIGURE 10.10

Example non-deterministic timed transition system.

- *Non-deterministic* at a state, e.g., $S1$, because there are two distinct outbound transitions defined for it. More precisely, the state is a left member of two transition pairs, e.g., for $S1$ we have $(S1, S2)$ and $(S1, S3)$.
- *Deterministic* at a state when there is exactly one outbound transition. For example, $S2$ and $S4$ are deterministic because there is only one outbound transition for each.

We say that the system is *well-defined* if it is defined at all its states.

These conditions relate to static properties, i.e., they relate to states per-se not how the states follow one another over time. In contrast, state trajectories relate to dynamic and temporal properties. When moving along a trajectory we keep adding the time advances to get the total traversal time, e.g., the time taken to go from $S1$ to $S4$ is 2. Here a trajectory is said to be *progressive* in time if time always advances as we extend the trajectory. For example, the cycle of states $S4 \ldots S7$ is not progressive because as we keep adding the time advances the sum never increases. Conceptually, let's start a clock at 0 and, starting from a given state, we let the system evolve following existing transitions and advancing the clock according to the time advances on the transitions. If we then ask what the state of the system will be at some time later, we will always be able to answer if the system is well-defined and progressive. A well-defined system that is not progressive signifies that the system gets stuck in time and after some time instant, it becomes impossible to ask what the state of the system is in after that instant. Zeno's paradox offers a well-known metaphor where the time advances diminish rapidly so that time accumulates to a point rather than continues to progress. It also offers an example showing that the pathology does not necessarily involve a finite cycle. This concept of progressiveness generalizes the concept of legitimacy for DEVS (Chapter 6) and deals with the "zenoness" property which has been much studied in the literature (Lee, 1999).

Thus we have laid the conceptual groundwork in which a system has to be *well-defined* (static condition) and *progressive* (temporal dynamic condition) if it is to be a bonafide result of a composition of components.

10.6 ACTIVE/PASSIVE COMPOSITIONS

An interesting form of the interaction among components may take on a pattern found in numerous information technology and process control systems. In this interaction, each component system alternates between two phases, active and passive. When one system is active the other is passive only one can be active at any time. The active system does two actions: 1) it sends an input to the passive system that activates it (puts it into the active phase), and 2) it transits to the passive phase to await subsequent reactivation. For example, in a Turing Machine (Chapter 4), the TM control starts a cycle of interaction by sending a symbol and move instruction to the tape system then waiting passively for a new scanned symbol to arrive. The tape system waits passively for the symbol/move pair. When it arrives it executes the instruction and sends the symbol now under the head to the waiting control.

Active/passive compositions provide a class of systems from which we can draw intuition and examples for generalizations about system emergence at the fundamental level. We will employ the pattern of active/passive compositions to illuminate the conditions that result in ill defined deterministic, non-deterministic and probabilistic systems. They provided sufficient conditions, meaningful especially for feedback coupled assemblages, under which iterative system specifications can be composed to create a well-defined resultant system.

10.7 HOW CAN FEEDBACK COUPLED COMPONENTS DEFINE A SYSTEM?

We examine the problem of emergence of a well-defined system from a coupling of a pair of component systems (the smallest example of the multi-component case). As in Fig. 10.11, the time base is a critical parameter on which input and output streams (functions of time) are happening. The cross-coupling imposes a pair of constraints as shown that the output time function of one component must equal the input time function of the other. The problem is given that the system components have fixed input/output behavior how can the constraints be solved so that their composition forms a well-defined system? We must recognize that inputs and outputs are occurring simultaneously in continuous time so that the constraints must be simultaneously satisfied at every instant. One step toward a solution is to break the time functions into parts or segments that allow restricting the problem to segments rather than complete streams.

We can apply the insight from the discussion of well-definition for timed transition systems (Section 10.5) to the coupled system case by recognizing that the abstract state of the latter represented the vector-like state of the former. This leads us to say that for any pair of states $(s1, s2)$ of components $S1$ and $S2$ in Fig. 10.11, determinism requires that there be a unique trajectory emerging from it. This in turn, requires that for all generator segments, 1) the I/O relations are satisfied and 2) the coupling relations are satisfied. In other words,

If (ω_1, ρ_1) is an input/output pair starting from s1
and (ω_2, ρ_2) is an input/output pair starting from s2
then $\omega_1 = \rho_2$ and $\omega_2 = \rho_1$

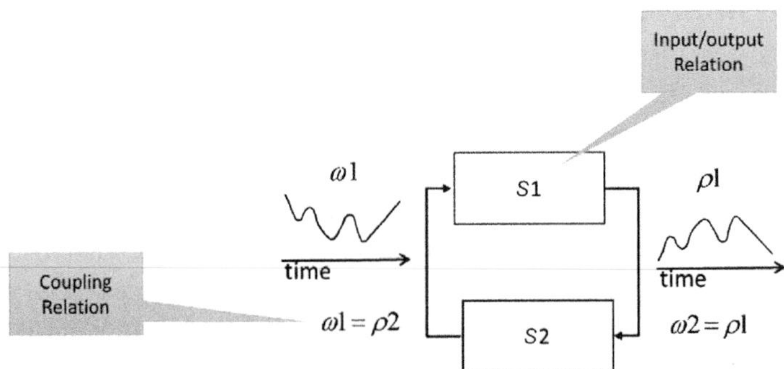

FIGURE 10.11

Interaction of two components attempting to form a composite system.

Furthermore, the time advances obtained by the maximal length segmentations as we go from one generator to the next must satisfy the progressivity requirement.

Now, the way that active/passive compositions help can be explained as in Fig. 10.12. In this interaction, each component system alternates between two phases, active and passive. When one system is active the other is passive because only one can be active at any time. The active system does two actions:

1) It sends an input to the passive system that activates it (puts it into the active phase), and
2) It transits to the passive phase to await subsequent reactivation.

Moreover, during any one such segment, only the active component sends a non-null output segment to the passive component. Notice that in Fig. 10.12, $S1$ starts out passive and eventually receives an activation from $S2$ at which point it becomes active. Meanwhile, $S2$ is active and unaffected by $S1$'s null input. At some point it decides to go passive and activates $S1$. The active/passive roles reverse and the cycle continues.

So we see that the input/output constraints are satisfied as long as the active component's output segment is accepted by the passive component and that component outputs only null segments while passive. The Turing Machine system of Chapter 4 satisfies this requirement. Also the fact that its time step is a constant means that the generators can be fixed at that size and the progressivity condition is satisfied.

Exercise 10.1. Show that the components and coupling of the Turing Machine are active/passive systems satisfying the input/output constraints of Fig. 10.11.

Exercise 10.2. Show that the components of the Turing Machine system of Chapter 4 always take the same time in the active state before going passive.

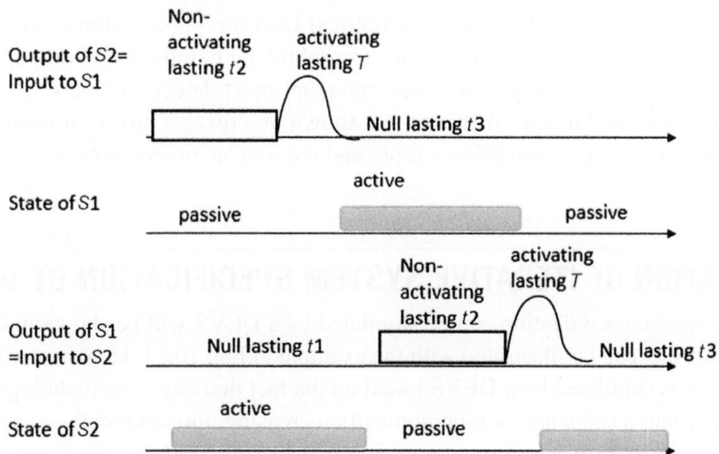

FIGURE 10.12

Active/passive systems as S_1 and S_2 in Fig. 10.11.

Exercise 10.3. Explore the limitations of DEVS in the following circumstances for the Turing Machine System. What if tape elements are not discrete symbols but have continuous durations and profiles? What if such a "symbol" comes in while control is still processing a previous "symbol"? Start with an approach based on that of the iterative systems specification.

Now let's look at another example where active/passive composition may solve a part of the enigma of how humans came to develop conversational language.

10.8 EXAMPLE: EMERGENCE OF HUMAN CONVERSATIONAL LANGUAGE INTERACTION

Among the many steps that might have occurred in the emergence of human language must have been the emergence of a duplex channel allowing bi-directional communication between agents (Zeigler, 2017). However, the establishment of a discipline or protocol for when to speak and when to listen may be more novel and problematic. The problem might be suggested by talking with a friend on the phone (perhaps a successful instance of agent-agent interaction) or pundits from opposite sides of the political spectrum attempting to talk over each other (a negative instance). From the system modeling perspective, we return to the problem, illustrated in Fig. 10.11, that components with cyclic (looped) – as opposed to acyclic – coupling face in forming a well-defined system. We have seen that such solutions are difficult to realize due to the concurrency of interaction and must overcome the potential to get stuck in Zeno-like singularities. On the other hand, we see that the passive/active alternation interaction for a cross-coupled pair can more easily support the conditions for a well-defined system. Let's identify listening with being passive (although mental components are active, they do not produce output) and speaking with being active (output production). Then see that alternation between speaking and listening would satisfy the conditions for establishing a well-defined coupled system from a pair of agents. Fig. 10.13 illustrates a simplified DEVS representation of such a speak/listen active/passive system. The active agent transitions from thinking to talking and outputting a command that starts the passive agent listening. The receiving agent must detect the end of the command and commence with an associated action. Although not shown, the reverse direction would have the roles reversed with the second agent transmitting a reply and the first agent receiving it.

10.9 SIMULATION OF ITERATIVE SYSTEM SPECIFICATION BY DEVS

That any iterative system specification can be simulated by a DEVS will be shown in Chapter 20. The concepts behind this fact can be illustrated with the example in Fig. 10.14. Here a system with continuous input and output is simulated by a DEVS based on the fact that any (non-pathological) continuous curve can be broken into a sequence of monotonically increasing, monotonically decreasing, and constant segments. If follows that the three classes just mentioned form the generators for an iterative specification.

An exact simulation can be had by encoding an input segment as a DEVS event segment with the event capturing the segment itself and its length equal to that of the segment. The DEVS uses the

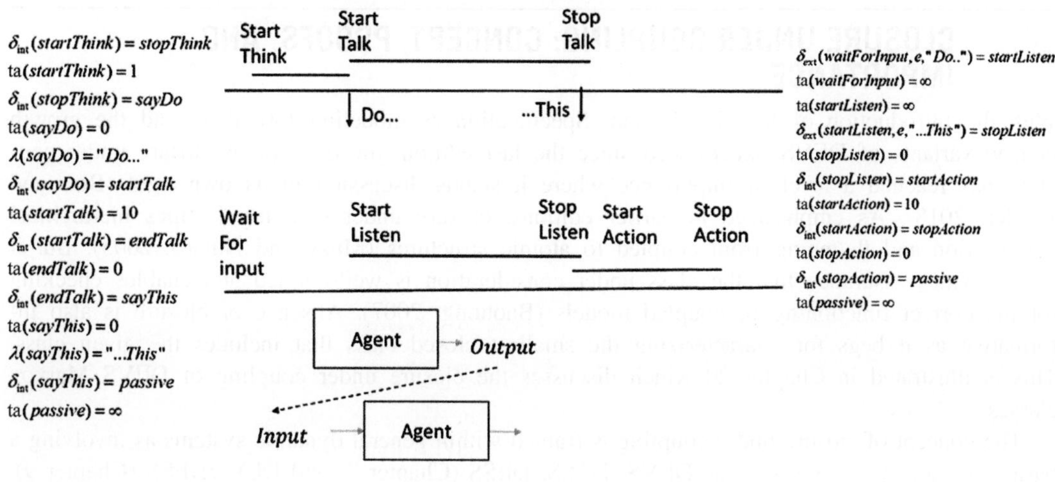

FIGURE 10.13

DEVS modeling of one way agent-agent communication.

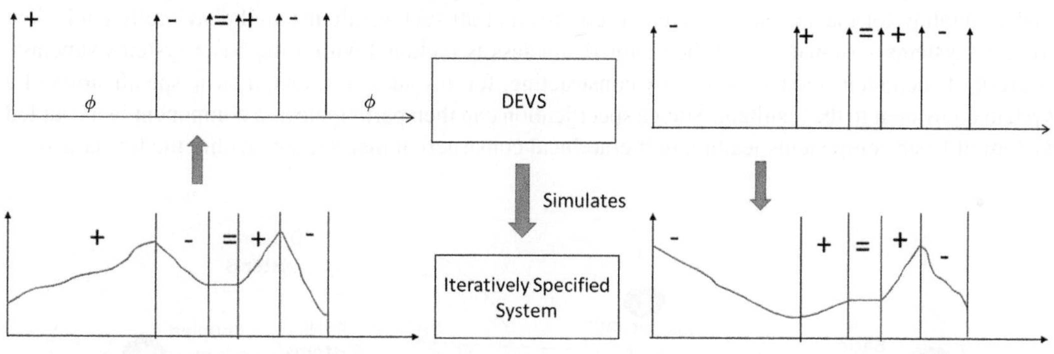

FIGURE 10.14

Iteratively Specified System can be simulated by a DEVS.

generator transition function of the iterative specification to compute the next state and generate the corresponding output segment as an event. After holding for a time equal to the segment length, the DEVS receives the next encoded segment and the cycle continues. More parsimonious encoding are possible. For example, the initial and final values of an input segment can be encoded as the event and with its length. This is enough information to compute the activity of an overall segment which is the sum of the absolute differences in these pairs (see Appendix 10.A). Depending on the iterative specification, the DEVS simulation might prove to provide an acceptable approximation to the activity of the overall output segment and of the complete I/O relation.

10.10 CLOSURE UNDER COUPLING: CONCEPT, PROOFS, AND IMPORTANCE

With the introduction of Iterative System Specification as modeling formalism and the growth in new variants of DEVS experienced since the last edition, the concept of closure under coupling has reached a level of importance where it stands discussion in its own right (Blas and Zeigler, 2018). As emphasized in earlier editions, closure under coupling justifies hierarchical construction and flattening from coupled to atomic structures (Muzy and Nutaro, 2005). But it also provides assurance that the class under consideration is well-defined and enables checking for the correct functioning of coupled models (Baohong, 2007). Absence of closure is also informative as it begs for characterizing the smallest closed class that includes the given class. This is illustrated in Chapter 20 which discusses the closure under coupling of DEVS Markov classes.

The concept of closure under coupling is framed within general dynamic systems as involving a basic subclass of systems such as DEVS, DTSS, DESS (Chapter 7) and DEV&DESS (Chapter 9). Referring to Fig. 10.15 as with the just mentioned system specifications, a basic formalism (1) is assumed that specifies (2) the subclass of basic systems. Also a coupled subclass (3) is assumed that has a specification (4) that is based on coupling (5) of basic specifications. The coupled specification is assumed to produce a resultant (6) which is a coupled system of basic component systems. Closure under coupling for the subclass of interest requires that all such resultants are behaviorally equivalent to basic systems – shown as that the coupled subclass is included within the basic systems subclass. A proof of such inclusion can be had by constructing, for any such resultant, a basic specification of a system equivalent to the resultant. Such a specification can then participate as a component in a coupled system of basic components leading to hierarchical construction that remains within the formalism.

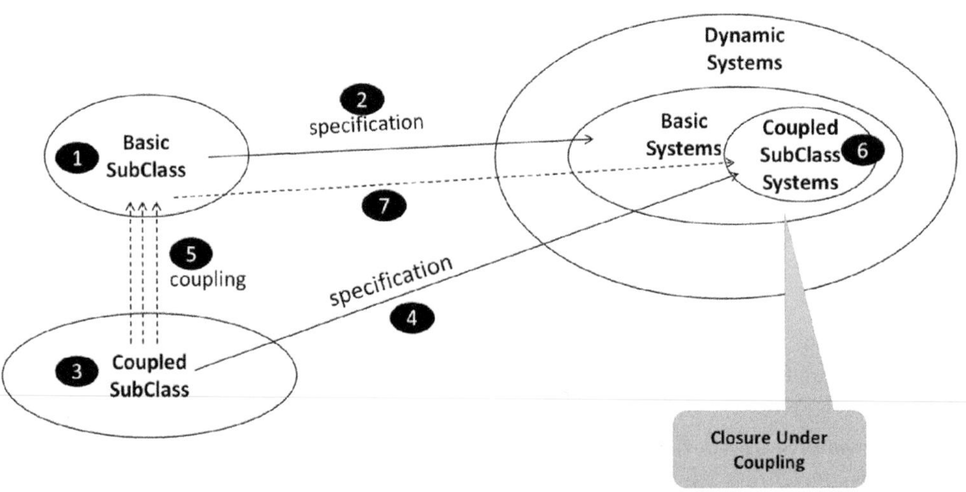

FIGURE 10.15

Closure under coupling.

For DEVS, we have a structure M (1) that specifies (2) a System, SM and a Network structure, N (3) that, based on a coupling recipe (5), specifies (4) a resultant System SN (6). A DEVS MN is constructed from the structure of N which specifies a System that is behaviorally equivalent to SN. Thus any valid coupling of a basic DEVS components results in a system equivalent to a basic DEVS – showing closure under coupling.

Besides closure under coupling, two types of questions arise for such formalisms: 1) are they subsets of DEVS, behaviorally equivalent to DEVS but more expressive or convenient, or bring new functionality to DEVS, and 2) have simulators been provided for them to enable verification and implementation? We address such problems with relation to iterative system specifications in Chapter 12 (see Table 12.1 for a summary). Here we discuss 3 examples of recent DEVS-based formalisms where these questions arise. Overall, we will see that such proofs can be exercises in reducing the additional functionality to that available in DEVS itself by expanding the state set sufficiently to support their explicit operation in DEVS. Besides supporting hierarchical construction and flattening from coupled to atomic structures (Muzy and Nutaro, 2005), such exercises can push the designer toward well-definition of the formalism itself in an iterative development process. In the sequel, the first example, Multi-Level DEVS references the original paper by Steniger and Uhrmacher (2016) to point to a complete proof of closure under coupling. The second example points out that although a network of Routed DEVS (Blas et al., 2017) components is a well-defined DEVS, closure under coupling requires proof that it is equivalent to a basic Routed DEVS. The third example Min-Max DEVS (Hamri et al., 2006) illustrates how consideration of closure under coupling can bring up new questions concerning the definition of the formalism.

10.10.1 EXAMPLE: MULTI-LEVEL DEVS

The Multi-Level DEVS (ML-DEVS), based on Parallel DEVS, follows the reactive systems metaphor with particular application to computational biology (Uhrmacher et al., 2007). It supports modeling combinations of different levels of organization and the explicit description of interdependencies between those levels. ML-DEVS supports MICRO-DEVS (atomic) and MACRO-DEVS (coupled) models, respectively. Different from DEVS, MACRO-DEVS coupled models have a state and behavior of their own thereby obviating an additional atomic model that otherwise is needed to represent the macro-level behavior. A recent extension (Steniger and Uhrmacher, 2016) adds an expressive intensional (i.e., implicit) coupling mechanism which employs dynamic structure to selectively enforce explicit couplings depending on the state of the model (see also Park and Hunt, 2006). A formal proof that ML-DEVS is closed under such coupling shows that a MACRO-DEVS model can be specified as a MICRO-DEVS model (i.e., a basic model in the formalism as discussed above). So to construct a MICRO-DEVS model that is behaviorally equivalent to any given MACRO-DEVS, Steniger and Uhrmacher (2016) must take account of the additional structure at the Macro-DEVS level. In turn, this requires considering "vertical" couplings expressing upward and downward causation between micro and macro levels as well as the usual "horizontal" couplings in DEVS coupled models. It also requires accounting for the intensional coupling mechanism, as well as other features such as interfaces that can belong to components, active and inactive ports, and accessible and private states. The reader is referred to the Appendix of Steniger and Uhrmacher (2016) for details of this proof which shows explicitly how the resultant's (qua basic model) state and characteristic functions are defined to account for all the behavior introduced in its coupled model incarnation.

10.10.2 EXAMPLE: ROUTED DEVS

Routed DEVS (RDEVS) is a subclass of DEVS intended to hide the handling of routing information among DEVS component models in a DEVS network (Blas et al., 2017). The RDEVS formalism is based on DEVS and defines a set of DEVS atomic models that use routing information to authenticate senders, receivers, and transactions. Interpreting this incoming information, routing models only accept specified input messages before passing on the message payload to an associated model for processing. Also, routing models are capable of directing processing results to a specific set of receivers. Blas et al. (2017) define the associated model (called essential model) as embedded within the routing model. A RDEVS Network is defined with all-to-all coupling of routing models. Note that although the network is a well-defined DEVS coupled model based on DEVS closure under coupling, it is not necessarily equivalent to a routing model. Hence, closure under coupling of RDEVS shows how to define an equivalent routing model for any such network. This requires that the structure of a routing model is extracted from the network structure and shown to be behaviorally equivalent to it.

To illustrate the closure under coupling proof we adopt the representation of a routing model in Fig. 10.16A.

Here the Input and Output Handlers take care of the routing information in the incoming message and pass on the operative content to the Essential Model for processing. The coupling from the Input to Output Handler is for passing on the authentication and routing information which depends on the

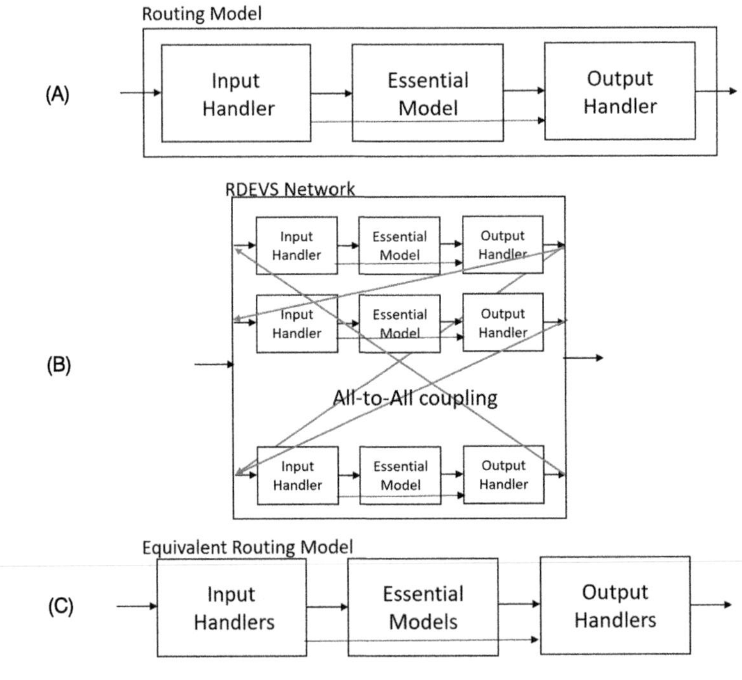

FIGURE 10.16

RDEVS formulation to illustrate closure under coupling.

processing state. The Routing Network is then a coupled model of Routing Models with all-to-all output to input coupling in Fig. 10.16B. The equivalent Routing Model of the Routing Network is then constructed first by defining equivalent models for the groups of Input, Output Handlers and Essential Models. Then these are coupled as in Fig. 10.16C to define the equivalent Routing Model of the Routing Network. One advantage of this approach is that all models are explicit and not embedded in others. Of course the original may be faster in processing. The construction of the equivalent routing model for any routing network demonstrates closure under coupling of the RDEVS formalism and therefore hierarchical construction that stays within the formalism. This allows the use of the RDEVS formalism as a "layer" above the DEVS formalism that provides routing functionality without requiring the user to "dip down" to DEVS itself for any functions.

10.10.3 EXAMPLE: MIN-MAX DEVS

Min-Max DEVS introduces the concept of ambiguous delay into the DEVS formalism by extending the time advance function to yield an interval (tmin, tmax) that represents all that is knowable about the timing of the associated state transition (Hamri et al., 2006). Other approaches used in digital circuits to represent the imprecise knowledge of real gate delays are based on probabilistic models and fuzzy-logic models and involve complex simulations. However, they do not necessarily give more information than in the min-max approach because the probability or possibility distributions are generally unknown. The extended DEVS atomic state keeps duplicates of the state of the modeled system so that it can track both slow and fast evolution of the system simultaneously. In addition it has a state variable that takes on values: fast, slow, autonomous, and passive that helps make transition decisions. The basic defining rule is: fast events are applied to the faster model (minimum delay) and slow events are applied to the slower model (maximum delay). In order to avoid a combinatorial explosion, the totally unknown state is introduced, in which the model remains forever once entered. The Min-Max-DEVS simulator carries out a single simulation which is equivalent to multiple simulation passes in which the lifetime of transitory states corresponds to all the possible values of the real system. Min-Max-DEVS is still in a state of development so that coupled models and consequently, closure of the formalism under coupling have not been addressed. However, the basic model is a full-fledged DEVS (with input and output sets and functions) and therefore can support coupling into network models. As discussed above the proof of closure under coupling would present interesting challenges: Does the state of a coupling of Min-Max-DEVS components define a well-defined time advance of the form (tmin, tmax)? Can such a global state, whose components each contain the transition decision support variable, be mapped to a single variable of the same kind?

Exercise 10.4. Define the network (coupled) model formalism of the Min-Max-DEVS formalism and confirm (by proof) or disconfirm (by example) its closure under coupling. Does your thought process suggest any iterative improvements on the design that could better resolve the issue?

10.11 ACTIVITY FORMALIZATION AND MEASUREMENT

The use of active/passive compositions in this chapter and the earlier references to activity scanning concepts (Chapter 8) raises the need for a more formal definition of activity. The Appendix provides a brief introduction to Activity theory for modeling and simulation. It aims at providing canonical (because very simple) definitions, in the context of both general systems and discrete event systems.

Usually, in simulation, *(qualitative) activities* of systems consist of phases, which "start from an event and end with another" (Balci, 1988). Information about the dynamics of the system is embedded into phases $p \in P$ corresponding to strings ("burning", "waiting", etc.). Mathematically, an event ev_i is denoted by a couple (t_i, v_i), where $t_i \in \mathbb{R}^{+,*}$ is the timestamp of the event, and $v_i \in V$ is the value of the event. Therefore, usual qualitative activities have values in P. Each activity consists of a triple $a = (p, ev_i, ev_{i+1})$, with $v_i = p$ for $t_i \leq t < t_{i+1}$, with $i \in \mathbb{N}$. An example of qualitative activity sequence is depicted in Fig. 10.17.

The definitions presented in the appendix provide different (*quantitative*) definitions of activity. These include metrics, for example, of continuous changes, number of transitions, number of state changes.

10.12 SUMMARY

Iterative system specification was here introduced as a constructive formalism for general systems specification that is computationally implementable as extension to DEVS-based modeling and simulation environments. Iterative system specification enables deriving fundamental conditions for "well-defined systems" as involving transition determinism and progressive time advance. Such conditions come into play when coupling systems that are well-defined so to form a resultant system that is itself well-defined. In this exposition active/passive compositions provided easily understood examples of the problem and its solution. We discussed how the necessity of these conditions for active/passive systems can offer insight into the emergence of natural language. Formalization of activity and its measurement were briefly introduced. The companion volume (Zhang et al., 2018) discusses performance profiling of DEVS models using activity metrics among others.

APPENDIX 10.A ACTIVITY DEFINITIONS

10.A.1 ACTIVITY OF CONTINUOUS SEGMENTS

Considering a continuous function $\Phi(t)$ (*cf.* in Fig. 10.18) and related extrema m_n, model *continuous activity* $A_c(T)$ (Muzy et al., 2011) of this trajectory, over a period of time T, consists of kind of

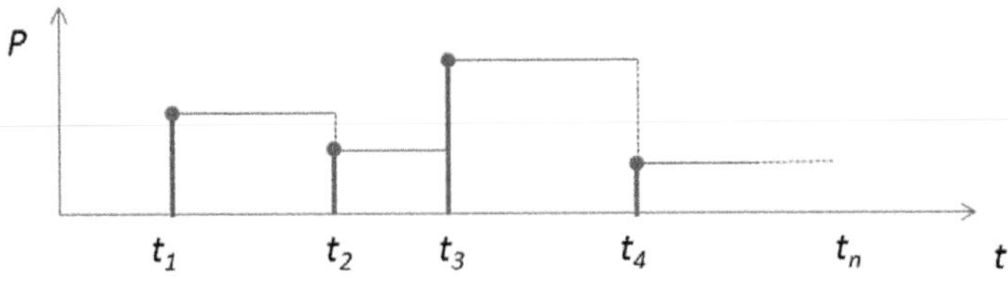

FIGURE 10.17

An example of usual qualitative activity definition.

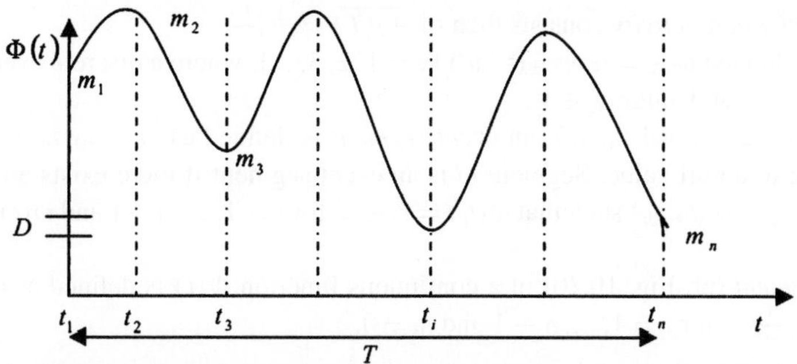

FIGURE 10.18

Continuous trajectory with extrema.

"distance":

$$A_c(T) = \int_0^T \left| \frac{\partial \Phi(t)}{\partial t} \right| dt \simeq \Sigma_{i=1}^n |m_i - m_{i+1}|$$

Average continuous activity consists then of $\overline{A_c(T)} = \frac{A_c(T)}{T}$.

Now considering a significant change of value of size $D = \left|\Phi^{n+1} - \Phi^n\right|$, called a *quantum*, the *discretization activity* $A_d(T)$ (Muzy et al., 2008), corresponding to the *minimum number of transitions* necessary for discretizing/approaching the trajectory of $\Phi(t)$ (*cf.* Fig. 10.19) is:

$$A_d(T) = \frac{A_c(T)}{D}$$

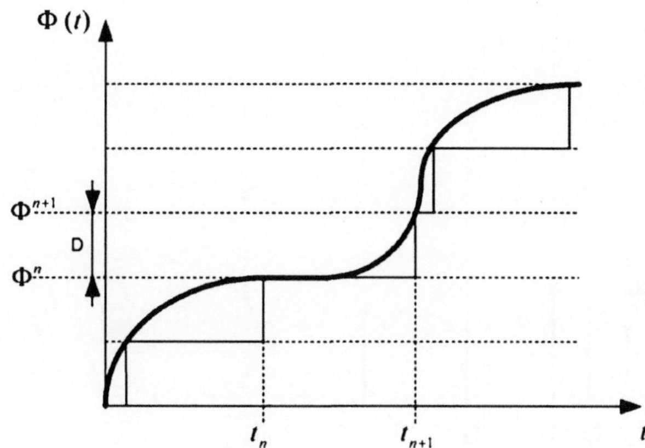

FIGURE 10.19

Discretization activity.

Average discretization activity consists then of $\overline{A_d(T)} = \frac{A_d(T)}{T}$.

An *event set* is defined as $\xi = \{ev_i = (t_i, v_i) \mid i = 1, 2, 3, ...\}$, where a discrete event ev_i is a couple of timestamp $t_i \in \mathbb{R}^{+,*}$ and value $v_i \in V$.

Considering a time interval $\langle t_0, t_n \rangle$, an *event segment* is defined as $\omega :< t_0, t_n > \rightarrow V \cup \{\phi\}$, with "$\phi$" corresponding to a null value. Segment ω is an event segment if there exists a finite set of times points $t_1, t_2, t_3, ..., t_{n-1} \in \langle t_0, t_n \rangle$ such that $\omega(t_i) = v_i \in V$ for $i = 1, ..., n-1$ and $\omega(t) = \phi$ for all other $t \in < t_0, t_n >$.

An *activity segment* (cf. Fig. 10.20) of a continuous function $\Phi(t)$ is defined as an event segment such that $\omega(t_i) = \frac{m_i}{t_i - t_{i-1}}$ for $i = 1, ..., n-1$ and $t_0 = 0$.

10.A.2 EVENT-BASED ACTIVITY

We consider here the activity as a measure of the number of events in an event set $\xi = \{ev_i = (t_i, v_i) \mid i = 1, 2, 3, ..., n-1\}$, for $0 \leq t_i < T$ and $v_i \in V$.

Activity in a Discrete Event Set

Event-based activity $A_\xi(T)$ (Muzy et al., 2010) consists of:

$$A_\xi(T) = |\{ev_i = (t_i, v_i) \in \xi \mid 0 \leq t_i < T\}|$$

Average event-based activity consists then of $\overline{A_\xi(T)} = \frac{A_\xi(T)}{T}$.

For example, assuming the event trajectory depicted in Fig. 10.21, the average event-based activity of the system corresponds to the following values for different time periods: $\overline{A_\xi(10)} = 0.3$, $\overline{A_\xi(20)} = 0.15$, $\overline{A_\xi(30)} \simeq 0.133$, $\overline{A_\xi(40)} = 0.175$.

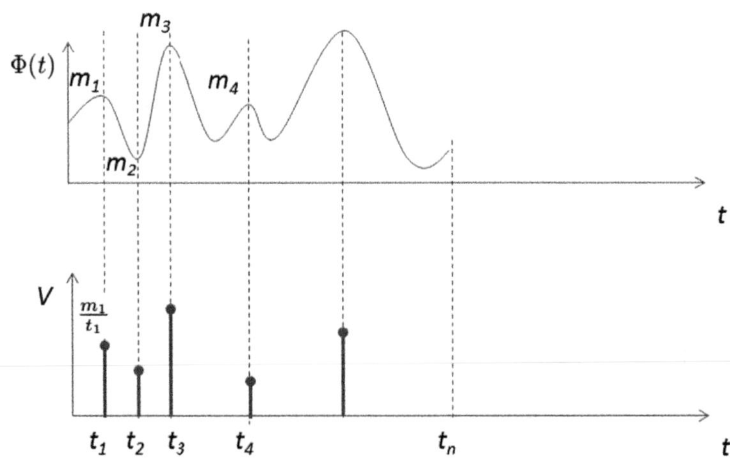

FIGURE 10.20

Activity segment of a continuous trajectory with extrema.

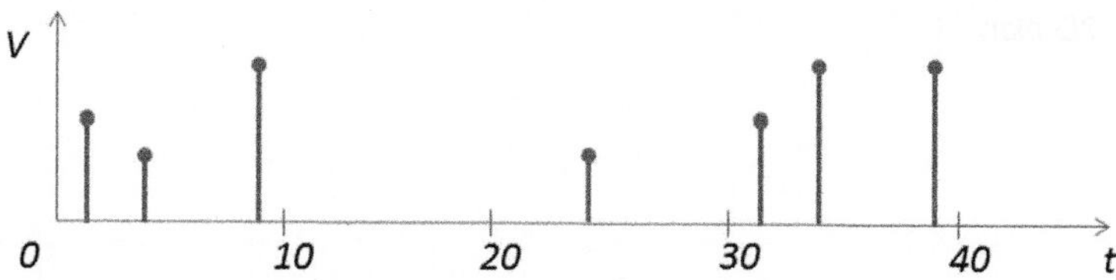

FIGURE 10.21

An example of event trajectory.

Event-Based Activity in a Cartesian Space

Activation and *non-activation* can be used to partition the set of positions $p \in \mathcal{P}$ in a Cartesian space. *Activation* is simply defined as an *event-based activity* $A_\xi(T) > 0$ while *non-activation* is defined as an *event-based activity* $A_\xi(T) = 0$. Related partitions are called *activity* and *inactivity regions* (Muzy et al., 2010):

- Activity region in space:

$$\mathcal{AR}^{\mathcal{P}}(T) = \{p \in \mathcal{P} \mid A_{\xi,p}(T) > 0\}$$

  where $A_{\xi,p}(T)$ corresponds to the event-based activity at position $p \in \mathcal{P}$.
- Inactivity region in space:

$$\overline{\mathcal{AR}^{\mathcal{P}}(T)} = \{p \in \mathcal{P} \mid A_{\xi,p}(T) = 0\}$$

A function of reachable states can be considered in time and space as $r : \mathcal{P} \times \mathcal{T} \to Q$, where Q is the set of states of the system and $\mathcal{T}$ is the time base. The set of all reachable states in the state set Q, through time and space, can be defined *universe* $\mathcal{U} = \{r(p,t) \subseteq Q \mid p \in \mathcal{P}, t \in \mathcal{T}\}$. Considering that all reachable states in time and space can be active or inactive, an activity-based partitioning of space $\mathcal{P}$ can be achieved: $\forall t \in \mathcal{T}, \mathcal{P} = \mathcal{AR}^{\mathcal{P}}(T) \cup \overline{\mathcal{AR}^{\mathcal{P}}(T)}$.

Fig. 10.22 depicts activity values for two-dimensional Cartesian coordinates $X \times Y$. This is a neutral example, which can represent whatever activity measures in a Cartesian space (fire spread, brain activity, etc.).

In spatialized models (cellular automata, L-systems, ...), components are localized at Cartesian coordinates in space $\mathcal{P}$. Each component c is assigned to a position $c_p \in \mathcal{P}$.

Applying the definition of activity regions in space to components, we obtain:

$$\mathcal{AR}^{\mathcal{C}}(T) = \left\{c \in \mathcal{C} \mid c_p \in \mathcal{AR}^{\mathcal{P}}(T)\right\}$$

$\mathcal{AR}^{\mathcal{P}}(T)$ specifies the coordinates where event-based activity occurs. Consequently, active components, over time period T, correspond to the components localized at positions p, with $A_{\xi,p}(T) > 0$, while inactive components have a null event-based activity $A_{\xi,p}(T) = 0$.

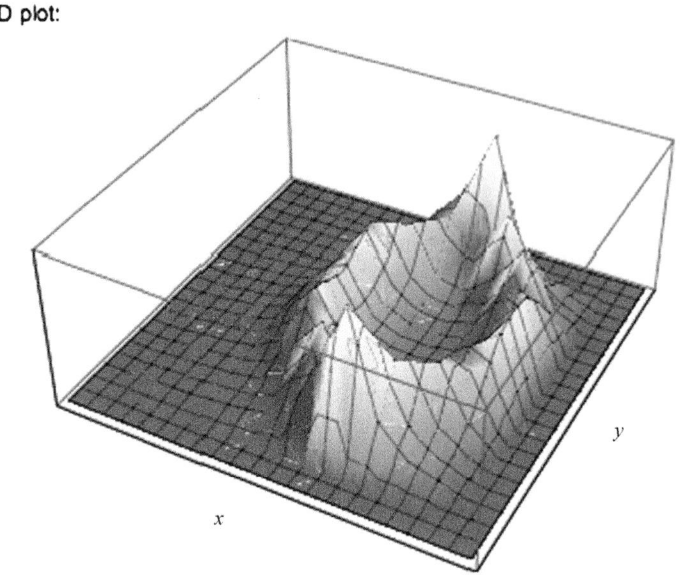

FIGURE 10.22

2D and 3D visualization of event-based activity in a 2D space. *x* and *y* represent Cartesian coordinates. The event-based activity of each coordinate is represented in the third dimension.

10.A.3 ACTIVITY IN DISCRETE EVENT SYSTEM SPECIFICATION (DEVS)

DEVS allows separating model and simulator (called the *abstract simulator*). The latter is in charge of *activating* the transitions of the model. This allows counting the number of transition executions (activations). This measure is the *simulation activity* (Muzy and Zeigler, 2012). Each transition can be also *weighted* (Hu and Zeigler, 2013).

Model

The dynamics of a component can be further described using a Discrete Event System Specification (DEVS). The latter is a tuple, denoted by $DEVS = < X, Y, S, \delta, \lambda, \tau >$, where X is the *set of input values*, Y is the *set of output values*, S is the *set of partial sequential states*, $\delta : Q \times (X \cup \{\phi\}) \to S$ is the *transition function*, where $Q = \{(s, e) \,|\, s \in S, 0 \leq e \leq \tau(s)\}$ is the *total state set*, e is the *time elapsed* since the last transition, ϕ is the *null input value*, $\lambda : S \to Y$ is the *output function*, $\tau : S \to \mathbb{R}^{+}_{0,\infty}$ is the *time advance function*.

If no event occurs in the system, the latter remains in partial sequential state s for time $\tau(s)$. When $e = \tau(s)$, the system produces an output $\lambda(s)$, then it changes to state $(\delta(s, e, x), e) = (\delta(s, \tau(s), \emptyset), 0)$, which is defined as an *internal transition* $\delta_{int}(s)$. If an external event, $x \in X$, arrives when the system is in state (s, e), it will change to state $(\delta(s, \tau(s), x), 0)$, which is defined as an *external transition* $\delta_{ext}(s, e, x)$.

Activity-Based Abstract Simulator

Modifications of usual abstract simulator for atomic models (Muzy and Zeigler, 2012) is presented here:

Algorithm 13 Modified abstract simulator for activity.

```
variables
    parent – parent coordinator
    tl – time of last event
    tn – time of next event
    DEVS – associated model with total state (s, e)
    y – output message bag
    n_int – number of internal transitions
    n_ext – number of external transitions

when receive i-message (i, t) at time t
    tl = t − e
    tn = tl + τ(s)
when receive *-message (*, t) at time t
    if (t = tn) then
        y = λ(s)
        send y-message (y, t) to parent coordinator
        s = δ_int(s)
        n'_int = n_int + 1
when receive x-message (x, t)
    if (x ≠ ∅ and tl ≤ t ≤ tn) then
        s = δ_ext(s, x, e)
        n'_ext = n_ext + 1
tl = t
tn = tl + τ(s)
```

Activity measures consist of:

- *External activity* A_{ext}, related to the counting n_{ext} of *external transitions* $\delta_{ext}(s,x) = (\delta(s,\tau(s),x),0)$, over a time period $[t,t']$:

$$\begin{cases} s' = \delta_{ext}(s,e,x) \Rightarrow n'_{ext} = n_{ext} + 1 \\ A_{ext}(t'-t) = \frac{n_{ext}}{t'-t} \end{cases}$$

- *Internal activity* A_{int}, related to the counting n_{int} of *internal transitions* $\delta_{int}(s) = \big(\delta(s,\tau(s),\emptyset),0\big)$, over a time period $[t,t']$:

$$\begin{cases} s' = \delta_{int}(s,e) \Rightarrow n'_{int} = n_{int} + 1 \\ A_{int}(t'-t) = \frac{n_{int}}{t'-t} \end{cases}$$

- *Simulation (total) activity* $A_s(t'-t)$ is equal to:

$$A_s(t'-t) = A_{ext}(t'-t) + A_{int}(t'-t)$$

- *Average simulation activity* $\overline{A_s(t'-t)}$ is equal to:

$$\overline{A_s(t'-t)} = \frac{A_{ext}(t'-t) + A_{int}(t'-t)}{t'-t}$$

Here simulation activity is simply a count of the number of internal and external transition events. However, when events have different impacts, weighted activity is introduced.

Exercise 10.5. Compute the total activity of a binary counter with N stages as it goes through a full cycle of transitions. Note that the least significant bit changes with every global transition while the most significant bit changes only once halfway through the cycle. This illustrates how there can be a range of activities in components of a model.

Abstract Simulator for Weighted Activity

Weighted simulation activity $A_w(T)$ has been defined in Hu and Zeigler (2013). It is related to a modified abstract simulator:

Algorithm 14 Modified abstract simulator for weighted activity.

variables
 $parent$ – parent coordinator
 tl – time of last event
 tn – time of next event
 $DEVS$ – associated model with total state (s, e)
 y – output message bag
 n_{int} – number of internal transitions
 n_{ext} – number of external transitions

when receive i-message (i, t) at time t
 $tl = t - e$
 $tn = tl + \tau(s)$
when receive *-message $(*, t)$ at time t
 if $(t = tn)$ **then**
  $y = \lambda(s)$
  send y-message (y, t) to parent coordinator
  $s = \delta_{int}(s)$
  $n'_{int,w} = n_{int,w} + wt_{int}(s)$
when receive x-message (x, t)
 if $(x = \emptyset$ and $tl \leq t \leq tn)$ **then**
  $s = \delta_{ext}(s, x, e)$
  $n'_{ext,w} = n_{ext,w} + wt_{ext}(s, e, x)$
$tl = t$
$tn = tl + \tau(s)$

Activity measures consist of:

- *External weighted activity* $A_{ext,w}$, related to the counting $n_{ext,w}$ of *external transitions* $\delta_{ext}(s, x) = (\delta(s, \tau(s), x), 0)$, over a time period $[t, t']$:

$$\begin{cases} wt_{ext} : X \times Q \rightarrow \mathbb{N}^0 \\ s' = \delta_{ext}(s, e, x) \Rightarrow n'_{ext,w} = n_{ext,w} + wt_{ext}(s, e, x) \\ A_{ext,w}(t' - t) = \frac{n_{ext,w}}{t'-t} \end{cases}$$

- *Internal weighted activity* $A_{int,w}$, related to the counting $n_{int,w}$ of *internal transitions* $\delta_{int}(s) = (\delta(s, \tau(s), \emptyset), 0)$, over a time period $[t, t']$:

$$\begin{cases} wt_{int} : S \rightarrow \mathbb{N}^0 \\ s' = \delta_{int}(s, e) \Rightarrow n'_{int,w} = n_{int,w} + wt_{int}(s) \\ A_{int,w}(t' - t) = \frac{n_{int,w}}{t'-t} \end{cases}$$

- *Simulation (total) weighted simulation activity* $A_w(t' - t)$ is equal to:

$$A_w(t' - t) = A_{ext,w}(t' - t) + A_{int,w}(t' - t)$$

- *Average weighted simulation activity* $\overline{A_w(t' - t)}$ is equal to:

$$\overline{A_w(t' - t)} = \frac{A_{ext,w}(t' - t) + A_{int,w}(t' - t)}{t' - t}$$

REFERENCES

Balci, O., 1988. The implementation of four conceptual frameworks for simulation modeling in high-level languages. In: WSC'88: Proceedings of the 20th Conference on Winter Simulation. ACM, New York, NY, USA, pp. 287–295.

Baohong, L., 2007. A formal description specification for multi-resolution modeling based on DEVS formalism and its applications. The Journal of Defense Modeling and Simulation: Applications, Methodology, Technology 4 (3), 229–251.

Blas, M.J., Gonnet, S., Leon, H., 2017. Routing structure over discrete event system specification: a DEVS adaptation to develop smart routing in simulation models. In: Chan, W.K.V., D'Ambrogio, A., Zacharewicz, G., Mustafee, N., Wainer, G., Page, E. (Eds.), Proceedings of the 2017 Winter Simulation Conference, pp. 774–785.

Blas, S.J., Zeigler, B.P., 2018. A conceptual framework to classify the extensions of DEVS formalism as variants and subclasses. In: Winter Simulation Conference.

Hamri, E.A., Giambiasi, N., Frydman, C., 2006. Min–Max-DEVS modeling and simulation. Simulation Modelling Practice and Theory 14, 909–929.

Hinsen, K., 2017. A dream of simplicity: scientific computing on Turing machines. Computing in Science & Engineering 19 (3), 78–85.

Hollocks, B.W., 2008. Intelligence, innovation and integrity—KD Tocher and the dawn of simulation. Journal of Simulation 2 (3), 128–137.

Hu, X., Zeigler, B.P., 2013. Linking information and energy—activity-based energy-aware information processing. Simulation 89 (4), 435–450.

Lee, E.A., 1999. Modeling concurrent real-time processes using discrete events. Annals of Software Engineering 7, 25–45.

Muzy, A., Jammalamadaka, R., Zeigler, B.P., Nutaro, J.J., 2011. The activity-tracking paradigm in discrete-event modeling and simulation: the case of spatially continuous distributed systems. Simulation 87 (5), 449–464.

Muzy, A., Nutaro, J.J., 2005. Algorithms for efficient implementations of the DEVS&DSDEVS abstract simulators. In: 1st Open International Conference on Modeling & Simulation. OICMS, pp. 273–279.

Muzy, A., Nutaro, J.J., Zeigler, B.P., Coquillard, P., 2008. Modeling and simulation of fire spreading through the activity tracking paradigm. Ecological Modelling 219 (1), 212–225.

Muzy, A., Touraille, L., Vangheluwe, H., Michel, O., Kaba Traoré, M., Hill, D., 2010. Activity regions for the specification of discrete event systems. In: Spring Simulation Multi-Conference Symposium on Theory of Modeling and Simulation (DEVS), pp. 176–182.

Muzy, A., Zeigler, B.P., 2012. Activity-based credit assignment (ACA) in hierarchical simulation. In: Proceedings of the 2012 Symposium on Theory of Modeling and Simulation-DEVS Integrative M&S Symposium. Society for Computer Simulation International, p. 5.

Park, S., Hunt, C.A., 2006. Coupling permutation and model migration based on dynamic and adaptive coupling mechanisms. In: The Proceedings of the 2006 DEVS Symposium.

Powell, J.H., Mustafee, N., 2016. Widening requirements capture with soft methods: an investigation of hybrid M&S studies in healthcare. Journal of the Operational Research Society.

Steniger, A., Uhrmacher, A., 2016. Intensional coupling in variable structure models: an exploration based on multi-level DEVS. TOMACS 26 (2).

Tolk, A., 2017. Multidisciplinary, interdisciplinary and transdisciplinary federations in support of new medical simulation concepts: harmonics for the music of life. In: Computational Frameworks. Elsevier, pp. 47–59.

Uhrmacher, A.M., Ewald, R., John, M., Maus, C., 2007. Combining micro and macro-modeling in DEVS for computational biology. In: Henderson, S.G., Biller, B., Hsieh, M.-H. (Eds.), Proceedings of the 2007 Winter Simulation Conference.

Zeigler, B.P., 2017. Emergence of human language: a DEVS-based systems approach. In: Proceedings of the Symposium on Modeling and Simulation of Complexity in Intelligent, Adaptive and Autonomous Systems. Society for Computer Simulation International, p. 5.

Zeigler, B.P., Sarjoughian, H.S., Duboz, R., Souli, J.-C., 2013. Guide to Modeling and Simulation of Systems of Systems. Springer.

Zhang, L., Zeigler, B., Laili, Y., 2018. Model Engineering for Simulation. Elsevier.

CHAPTER

BASIC ITERATIVE SYSTEM SPECIFICATION (ITERSPEC) 11

CONTENTS

Having introduced iterative specification informally in the last chapter, we introduce here the fundamental mathematical definitions of the basic iterative specification and how to implement such a specification with the help of generators such as discussed in the previous chapter (and introduced in Muzy et al., 2017). In earlier editions, generators were introduced in a general manner. We show here how to consider classes of generators exhibiting particular properties (e.g., as having their length less or equal to a fixed length) obtained by maximal length segmentation under *specific conditions* and how these classes are combined through the intersection, union and disjoint union of generator classes. This combination allows generating complex trajectories as compositions of generators. Finally, we show how appropriate properties of these generator classes lead to sufficient conditions for the iterative specification of a general I/O System.

A more compact notation will be used to simplify the mathematical descriptions at basic and composition levels. Indeed, with more and more focus on system-of-systems (SoS) in the current evolution of modeling, we expect that this new notation will simplify further developments in SoS. More precisely, in a mathematical structure, instead of noting both sets and functions (using the sets), only functions are noted, e.g., instead of a structure (X, Y, D, f, g) with functions $f : X \rightarrow Y$ and $g : D \rightarrow Y$, the structure is simply noted (f, g). Notice that this is consistent with the fact that a function is identified uniquely by its name embedding *implicitly* its set definitions. This notation is employed in this chapter and some new chapters while retaining the earlier notation in most others.

Theory of Modeling and Simulation. https://doi.org/10.1016/B978-0-12-813370-5.00020-1

11.1 BASIC ITERATIVE SYSTEM SPECIFICATION: ITERSPEC DEFINITION

Simulating a model requires generating and keeping track of the different values of inputs, states and outputs using a computer. Recall that a state is abstractly the minimal information needed to compute the response of a system to its input. Often the state is represented by the past and current values of descriptive variables to enable us to compute their future values. However, keeping its minimal information properties in mind, a state does not have to describe all the variable values in a continuous fashion. A state can be considered as a snapshot of the dynamics of the system, valid for a certain period of time. During this interval it is known that the variable values will not change drastically the dynamics of the model (e.g., a system in phases *active* or *passive*). Conceived in this way, a state is consistent with the discrete nature of digital computers as well as a the cognitive way people perceive the different phases of the dynamics of a system.

Consistent with the notion of states is then the notion of segments of a trajectory of a dynamic system. As shown in Fig. 11.1, segments of a trajectory can be concatenated. Reversely, a trajectory can be cut into pieces *(segments)*. This last operation is called a *segmentation.*

As basic formal element of our framework, we define a segment as a function depending on time. More precisely,

Definition 11.1. *A segment is a map* $\omega : <0, t_f> \rightarrow Z \cup \{\phi\}$, with $t_f \in \mathbb{R}^+$ the *duration of* ω, where Z *is an arbitrary set of values,* ϕ a *null value*, and $<>$ referring indifferently to open or closed intervals.

Remark 11.1. System's segments can share the same duration but not necessarily the same domain, e.g., each input segment $\omega_X \in \Omega_X$ can be defined over a continuous time base ($T_c = dom(\omega_X) = <0, t_f> \cap \mathbb{R}$) while each output segment $\omega_Y \in \Omega_Y$ can be defined over a discrete time base ($T_d = dom(\omega_Y) = <0, t_f> \cap \mathbb{N}$). The resulting time base then consists of $T = T_c \times T_d$. The domain of each segment is a projection of the time base.

As depicted in Fig. 10.9, in a segment-based approach, a computational model of a system consists of: (i) segmenting its input trajectory, (ii) injecting the resulting first input segment into the model, being in a current state, (iii) computing the new state, and (iv) generating its output segment. The whole process is achieved iteratively for each subsequent segment.

Definition 11.2. A *basic Iterative System Specification (IterSpec)* is a structure

$$\text{IterSpec} = (\delta, \lambda)$$

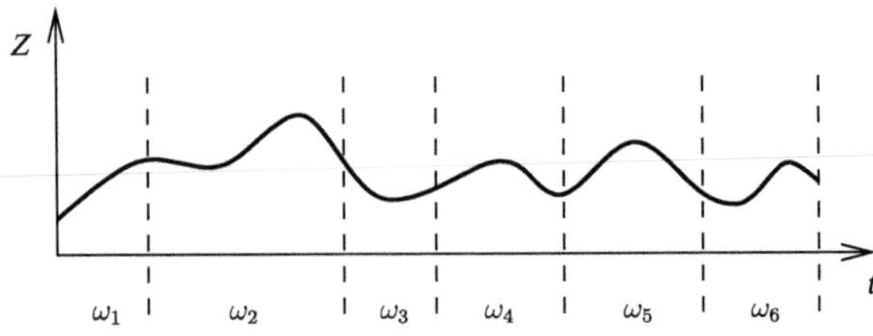

FIGURE 11.1

Basic segments in a trajectory.

where

- $\delta : Q \times \Omega_G^X \rightarrow Q$ is the *generator transition* function with Q the *set of states* and $\Omega_G^X \subseteq X^T$ the set of *admissible input segment generators* (a subset of all functions from *time set* T to *input set* X).
- $\lambda : Q \times \Omega_G^X \rightarrow \Omega_G^Y$ is the *output* function with $\Omega_G^Y \subseteq Y^T$ the set of *admissible output segment generators* (a subset of all functions from *time set* T to *input set* Y).

11.2 COMPOSITION PROCESS

At the iterative specification level, the concatenation (or composition) operation of segments needs to ensure the composition of the resulting continuously defined segments at the general system level. Based on an iterative specification, the problem is to find a consistent segmentation that is unique and able to produce iteratively the whole system dynamics by concatenating the subsegments into complete segments at the general system level. Making an analogy with automata theory, $(\delta, \lambda) \xrightarrow{+} (\Delta, \Lambda)$ is a compact notation showing that an iterative specification (δ, λ) can be extended to the concatenation closure process where an alphabet is concatenated until closure is obtained. Such closure puts together letters to make strings (as languages) in a system (Δ, Λ) (as defined in Wymore, 1967).

The following material relies on background from TMS2000 that is reviewed in Appendix 11.A. You may wish to visit this appendix before proceeding.

The IterSpec transition function is extended to account for concatenation/composition process:

Definition 11.3 (TMS2000). *A concatenation/composition process is a structure*

$$COPRO = (\delta^+, \lambda^+)$$

where $\delta^+ : Q \times \Omega_G^{X+} \rightarrow Q$ is the *the extension of* δ *to* Ω_G^{X+} such that:

$$\begin{aligned}\delta^+(q, \omega) &= \delta(q, \omega) && \text{if } \omega \in \Omega_G^X \\ &= \delta^+(\delta(q, \omega_1), \omega_2 \bullet \ldots \bullet \omega_n) && \text{if } \omega \in \Omega_G^{X+},\end{aligned}$$

with $\omega_1 \bullet \omega_2 \bullet \ldots \bullet \omega_n$ the MLS decomposition for ω, Ω_G^{X+} the *generated input set* (or concatenated input segments), and $\lambda^+ : Q \times \Omega_G^{X+} \rightarrow \Omega_G^{Y+}$ the *extension of* λ *to* Ω_G^{X+} defined the same way:

$$\begin{aligned}\lambda^+(q, \omega) &= \lambda(q, \omega) && \text{if } \omega \in \Omega_G^X \\ &= \lambda(q, \omega_1)\lambda^+(\delta(q, \omega_1), \omega_2 \bullet \ldots \bullet \omega_n) && \text{if } \omega \in \Omega_G^{X+}.\end{aligned}$$

Finally the general system structure is obtained:

Definition 11.4 (Based on Chapter 5). *A Deterministic Input-Output General System is a structure*

$$SYS = (\Delta, \Lambda)$$

where $\Delta : Q \times \Omega_X \rightarrow Q$ is the *state transition function*, and $\Omega_X \subseteq X^T$ is the *set of admissible input segments*, and $\Lambda : Q \times \Omega_X \rightarrow \Omega_Y$ is the *output* function.

To sum up, following an MLS, an iterative specification allows computing (eventually by simulation) the whole state trajectory along the entire input segment following the requirements:

Theorem 11.1 (Modified from TMS2000, p. 121). *Sufficient conditions for iterative specification: An iterative specification $G = (\delta, \lambda)$ can be associated to a system $SYS_G = (\Delta, \Lambda)$ if the following conditions hold:*

1. Existence of longest prefix input segments: $\omega \in \Omega_G^{X+} \Longrightarrow max\{t \mid \omega_{t)} \in \Omega_G^X\}$ *exists*
2. Closure under right input segmentation: $\omega \in \Omega_G^X \Longrightarrow \omega_{[t} \in \Omega_G^X$ *for all* $t \in dom(\omega)$
3. Closed under left input segmentation: $\omega \in \Omega_G^X \Longrightarrow \omega_{t)} \in \Omega_G^X$ *for all* $t \in dom(\omega)$
4. Consistency of composition: $\omega_1, \omega_2, \ldots, \omega_n \in \Omega_G^X$ *and* $\omega_1 \bullet \omega_2 \bullet \ldots \bullet \omega_n \in \Omega_G^{X+}$ $\Longrightarrow \delta^+(q, \omega_1 \bullet \omega_2 \bullet \ldots \bullet \omega_n) = \delta(\delta(\ldots \delta(\delta(q, \omega_1), \omega_2), \ldots), \omega_n)$.

Requirements 2 and 3 are quite strong. They can be relaxed:

Remark 11.2. According to requirements 2 and 3, any input generator can be segmented at any time point leading to left and right subsegments. From (TMS2000, p. 123), left and right segmentability can be *weakened* while still enabling Theorem 11.1 to hold (the stated conditions are *sufficient* but not *necessary*). To do so, both conditions are replaced by

$$\text{For } \omega \in \Omega_G^{X+} \text{ if } t^* = max\{t \mid \omega_{t)} \in \Omega_G^X\} \text{ and } t_k < t^* < t_{k+1} \text{ then } \omega_{[t_k, t^*)} \in \Omega_G^X \text{ and } \omega_{[t^*, t_{k+1})} \in \Omega_G^{X+}.$$

This weakened condition does away with the need for right segmentation of the generator and requires only that any remaining right segment (if it exists) continues to be segmentable at the end of each cycle of the segmentation process.

A particular example is where all the generators are positive ramps. The corresponding set of input generators then consists of $\Omega_G^R = \{\omega \mid \exists a > 0 : \omega(t) = at,\ \omega(0) = 0,\ t \in dom(\omega)\}$. Note that a ramp is closed under left segmentation *but not* under right segmentation (as a right segment of a ramp does not start from *0*). Nevertheless, the generated set is uniquely segmentable since MLS consumes the first generator completely. Thus, there is no right segment remaining and the segmentation can continue from the next breakpoint.

11.3 SPECIFIC MAXIMAL LENGTH SEGMENTATIONS

In this section, previous segmentation requirements from Theorem 11.1 are extracted to enable us to define proper generated sets obtained by segmentation (cf. Definition 11.7). MLS, as presented in TMS2000, consists of a general principle to allow a sound and formal building of a system specification hierarchy. Here we will apply this concept to the vast class of computational models. To achieve this goal, specific MLS processes will be proposed (cf. Definition 11.9) which can be based on easily understood factors, such as the length of segments and the number of events they contain, and combined through intersection, union and disjoint union operations. Examples of specific MLS applications will illustrate different combinations.

In the following sections we say that Ω is a *proper* set of generators if Ω is an admissible set of generators for Ω^+. In other words, Ω is a *proper* set of generators if the set it generates is such that every element has a unique MLS decomposition by Ω. Likewise, Ω is *improper* if there is at least one segment that it generates that does not have a unique MLS decomposition by Ω.

11.3.1 DEFINITION OF SPECIFIC MAXIMAL LENGTH SEGMENTATIONS

The set of generated segments can be *improper*, i.e., every generated segment $\omega \in \Omega^+$ may not be segmentable to elements of the generator set, $\omega \in \Omega_G$.[1] This happens when the segmentation is not in coherence with the generator set Ω_G, i.e., the latter is missing generators of the generated segments and conversely, the generated set (and corresponding system) will not be proper. Therefore, requirements need to be set to ensure that the generated set, Ω_G^+, is proper by segmentation. To achieve this goal, three definitions of propriety of generator set, Ω_G, are proposed, requiring increasing segmentation constraints. The usual Theorem 11.1 concerns the closure of the generator set Ω_G without detailing explicitly implications on the generated set Ω_G^+. Here, the following definitions are based on the principle that disposing of a proper generator set Ω_G ensures a proper corresponding generated set Ω_G^+.

First, a general definition requires *only* the right segment left over to be recursively *proper.*

Definition 11.5. *A generator set* Ω_G *is proper* when, satisfying a maximal length segmentation, it follows the two requirements:

1. *Existence of longest prefix segments*: $\omega \in \Omega_G^+ \implies max\{t \mid \omega_{t)} \in \Omega_G\}$ exists, with $\omega_{t)} \in \Omega_G$, the left generator,
2. *Propriety of the right segment:* The right segment left over $\omega_{[t} \in \Omega_G^+$, is *proper*.

As in Theorem 11.1, segmentation closure requirements can be *weak*.

Definition 11.6. *A generator set* Ω_G is *weakly proper* when, satisfying a maximal length segmentation, it follows the two requirements:

1. *Existence of longest prefix segments*: $\omega \in \Omega_G^+ \implies max\{t \mid \omega_{t)} \in \Omega_G\}$ exists,
2. *Weakened segmentability:* For $\omega \in \Omega_G^+$ *if* $t^* = max\{t \mid \omega_{t)} \in \Omega_G\}$ and $t_k < t^* < t_{k+1}$ *then* $\omega_{[t_k,t^*)} \in \Omega_G$ and $\omega_{[t^*,t_{k+1})} \in \Omega_G^+$.

As in Theorem 11.1, segmentation closure can be required over both left and right segments.

Definition 11.7. *A generator set* Ω_G is *strongly proper* when, satisfying a maximal length segmentation, it follows the three requirements:

1. *Existence of longest prefix segments*: $\omega \in \Omega_G^+ \implies max\{t \mid \omega_{t)} \in \Omega_G\}$ exists
2. *Closure of the generated set under left segmentation*: $\omega \in \Omega_G \implies \omega_{t)} \in \Omega_G$ for all $t \in dom(\omega)$
3. *Closure of the generated set under right segmentation*: $\omega \in \Omega_G \implies \omega_{[t} \in \Omega_G$ for all $t \in dom(\omega)$

Remark 11.3. If Definition 11.7 holds for a set of generators, then Definition 11.6 and likewise Definition 11.5 hold as well. Closure under left (resp. right segmentation) of a set of generators implies

[1] Notice that we use for simplification the notation Ω_G (instead of Ω_G^X or Ω_G^Y) when there is no need to distinguish input and output sets.

the same closure of a set of generated segments, but not conversely. Therefore, Definition 11.7 is our preferred means of working with a set of generators to achieve the desired condition in Definition 11.5. However, Definition 11.5 and Definition 11.6 allow us to proceed where Definition 11.7 doesn't hold in certain cases such as the example of the ramps (Section 11.2).

Definition 11.8. *A generated set* Ω_G^+ is *(simply, weakly or strongly) proper* if corresponding generator set Ω_G is respectively (simply, weakly or strongly) proper.

Previous definitions are fundamental for the system theory of computational systems. They bring a new light and open new theoretical fields. The usual Definition 11.7 inherited from original TMS2000 requires both closure under left and right segmentation closure. Both are strong but not necessary requirements. Left segmentation closure is a fundamental assumption of the theory stipulating that at any time, the state of a system can be determined. This allows concatenating together generated segments to *compose* trajectories at system level. Relaxing these requirements, as we will see in the next sections, allows composing low level generators (e.g., neuronal spikes) into higher level generators (e.g., neuronal bursts) while still leading to unique segmentations.

Different specific MLS processes are then possible, each one having its own *segmentation condition*. Compared with general MLS as defined in Subsection 10.4, the following derivation is set:

Definition 11.9. *A specific maximal length segmentation* is based on both generator set Ω_G and generated set Ω_G^+, fulfilling *segmentation condition* $c : \Omega_G \to \mathbb{B}$, i.e., $\Omega_G = \{\omega \,|\, c(\omega) = true\}$ and $\Omega_G^+ = \{\omega \,|\, c(\omega_{t\rangle}) = true\}$. General function MLS$(\omega \in \Omega)$ is derived as a specific MLS adding: (i) the segmentation condition such that MLS$(\omega \in \Omega,\, c(\omega_{t\rangle}))$, and (ii) the maximum time breakpoint detection such that $t \leftarrow max\{t \,|\, \omega_{t\rangle} \in \Omega_G \wedge c(\omega_{t\rangle}) = true\}$.

Now disposing of a proper generator set Ω_G allows concatenating generators into higher level generators. To ensure this property the closure under composition of an admissible set of generators Ω_G can be checked:

Definition 11.10. *A generator set* Ω_G, admissibly generating a *proper* set of generated segments Ω_G^+, is *closed under composition* if for every pair of contiguous generators $\omega_{t\rangle} \in \Omega_G$ and $\omega_{\langle t} \in \Omega_G$: $\omega = \omega_{t\rangle} \bullet \omega_{\langle t}$, where $\omega \in \Omega_G$ is also a generator, $\bullet : \Omega_G \times \Omega_G \to \Omega_G$ is the *generator concatenation/composition operation*, and concatenation respects segmentation condition, i.e., $t \leftarrow max\{t \,|\, \omega_{t\rangle} \in \Omega_G \wedge c(\omega_{t\rangle}) = true\}$.

The set of segments Ω is now restricted to a simple (general) case of *discrete event segments.* Discrete events consist of discrete values at precise time points and no values at other time points. Therefore, discrete events are used to focus on value changes in a system (at input, state or output levels). The set of discrete event segments can be formally defined as:

Definition 11.11 (TMS2000). *The set of all discrete event segments* consists of $\Omega_E = \{\omega \,|\, \omega : T_{[t_0,t_n)} \to Z \cup \{\phi\}\}$, with ω a *map*, Z an *arbitrary set*, ϕ the *null value*. There exists a finite set of time points $t_0, t_1, \ldots, t_{n-1} \in [t_0, t_n)$ such that $\omega(t_i) = z_i$ for $i = 0, \ldots, n-1$ and $\omega(t) = \phi$ for all other $t \in [t_0, t_n)$.

Concatenating discrete event segments requires relying on a set of discrete event generators. Furthermore, defining basic discrete event generators to be used in various segmentations simplifies the search of segmentation solutions and their combination:

Definition 11.12. *The set of basic discrete event generators* Ω_G^E, corresponding to the set of discrete event segments Ω_E consists of $\Omega_G^E = \Omega_G^Z \cup \Omega_G^\phi$ with *the set of event generators* such that $\Omega_G^Z = \{z_i \mid z_i(0) = z\}$, where Z is an *arbitrary set*, and *the set of non-event generators is* $\Omega_G^\phi = \{\phi_i \mid \phi_i : [0, t_i) \to \{\phi\}, t_i \in \mathbb{R}^+\}$, where t_i is *the end time of corresponding non-event generator* ϕ_i.

Two specific MLS examples will now be discussed. The first one is based on event counting segmentation while the second one is based on time length segmentation. This is a fundamental aspect of many computational systems that can be modeled either based on state change (e.g., discrete event models) or on time change (e.g., discrete time models).

Example 11.1 (Segmentation by counting events to a fixed number). The set of all event segments having a number of events less or equal to N (for $N \geq 0$), is defined as a subset of all discrete event segments as $\Omega_N \subseteq \Omega_E$ (cf. Fig. 11.2).

The generated set Ω_G^{N+} is *strongly proper*, i.e., following Definition 11.7:

1. Existence of longest initial segments: Considering any discrete event segment in Ω_N, an initial segmentation of the latter (being a concatenation of the basic discrete event generators of Definition 11.12) can be set easily based on the generator set $\Omega_G^N = \{\omega \mid c(\omega) = true\}$, with segmentation condition $c(\omega_{t\rangle}) = \begin{cases} true & if\, n_E \leq N \\ false & otherwise \end{cases}$, where $n_E = \Sigma_{\hat{t}\in[0,t)} \chi_{\omega_{t>}(\hat{t})\neq\phi}$ is the *number of events* in segment $\omega_{t\rangle}$, with χ the *indicator function*.
2. Closure under left and right segmentations: Subsegments of a segment have no more events than the segment itself.
3. Based on both basic discrete event generators of Definition 11.12, using MLS, and a fixed number segmentation condition c, every possible discrete event segment $\omega \in \Omega_E$ can be segmented.

Example 11.2 (Segmentation by measuring segment lengths to a fixed length). The set of all event segments having a length less or equal to T (for $T \geq 0$), is defined as a subset of all discrete event segments as $\Omega_T \subseteq \Omega_E$ (cf. Fig. 11.3). It is obvious that the set of generated discrete event segments Ω_G^{T+} is also *strongly proper*.

Example 11.3 (Segmentation by fixed segments). The set of all event segments having a length equal to $\overline{T}$ (for $\overline{T} \geq 0$), is defined as a subset of all discrete event segments as $\Omega_{\overline{T}} \subseteq \Omega_E$ (cf. Fig. 11.3). The set of generated discrete event segments $\Omega_G^{\overline{T}+}$ is *proper*, depending on the size of $\overline{T}$ compared

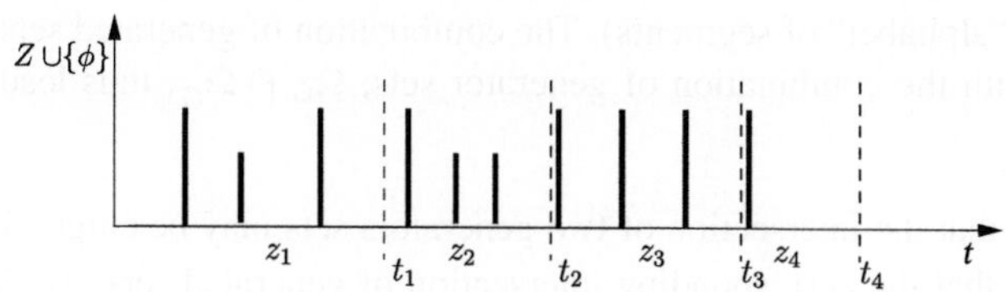

FIGURE 11.2

Segmentation by counting the number of events such that $N = 3$.

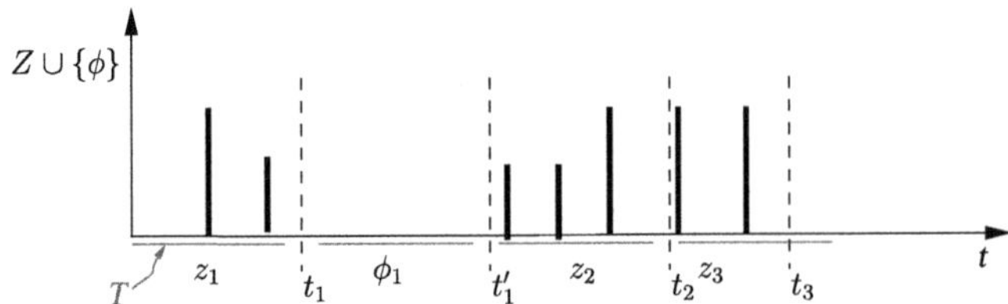

FIGURE 11.3

Segmentation by measuring time intervals such that $l(\omega) \leq T$. Notice that only the last generator is not exactly of size T.

with the total size of segment $l(\omega)$. Notice that it cannot be strongly proper as generators cannot be split into subgenerators. Let us see here the closure property of the set of admissible generators $\Omega_G^{\overline{T}}$ corresponding to the proper generated set $\Omega_G^{\overline{T}+}$. We prove it by induction. At a first segmentation step 1, it is obtained $\omega = \omega_{t)}^1 \bullet \omega_{[t}^1$ with $t = \overline{T}$, $\omega_{t)}^1 \in \Omega_G^{\overline{T}}$ and $\omega_{[t}^1 \in \Omega_G^{\overline{T}+}$. The same process is achieved until obtaining decomposition $\omega_1 \bullet ... \bullet \omega_n$, with ω_n the last right segment. Then, $\Omega_G^{\overline{T}+}$ is *proper*.

The same proof arguments can be extended to the variable segment length segmentation of Example 11.3. By induction, at a first segmentation step 1, it is obtained $\omega = \omega_{t)}^1 \bullet \omega_{[t}^1$ with $t = \alpha T$ and $\alpha \in (0, 1)$ and $\omega_{[t}^1 \in \Omega_G^{\overline{T}+}$. The same process is achieved until obtaining decomposition $\omega_1 \bullet ... \bullet \omega_n$, with ω_n the last right segment. Considering the total length of *n-1* concatenated segments as $L = \Sigma_{i \in \{1,...,n-1\}} l(\omega_i) \leq T - \alpha T$, this means that the last right segment ω_n can be segmented into $\omega_{t)}^n \bullet \omega_{[t}^n$ (both of length less than T) and previous *n-1* segments concatenated with $\omega_{t)}^n$ while still respecting segmentation condition $\Sigma_{i \in \{1,...,n-1\}} l(\omega_i) + l(\omega_{t)}^n) \leq T$. This means that $\Omega_G^{\overline{T}+}$ is *proper*.

11.3.2 COMBINATION OF SPECIFIC MAXIMAL LENGTH SEGMENTATIONS

Definitions 11.5 to 11.7 exposed the requirements for a generated set Ω_G^+ (resp. for a generator set Ω_G) to be proper. Here, we explore formally the different combinations/operations (by intersection, disjoint union and union) of proper generated sets (resp. generator sets) and verify if the combination is also proper. First specific combination examples are shown for each operation to discuss their propriety. After these results are sum up and generalized.

Segmentations use two different sets: the generated set Ω_G^+ (a concatenation of segments) and a generator set Ω_G (an "alphabet" of segments). The combination of generated sets, $\Omega_G^+ \cap \Omega_G'^+$, should not be confounded with the combination of generator sets, $\Omega_G \cap \Omega_G'$, thus leading to the following remark.

Remark 11.4. Notice that the intersection of two generators sets may be empty, i.e., $\Omega_G \cap \Omega_G' = \{\phi\}$, which does not mean that the corresponding intersection of generated sets $\Omega_G^+ \cap \Omega_G'^+$, does not exist and/or is not proper.

As shown previously different specific segmentations lead to different generated segments. Let us see a simple example of combination (intersection) of two specific segmentations and how proper it is.

Example 11.4 (Intersection of two proper generated sets that is proper, cf. Fig. 11.4)**.** The *intersection* of two proper generated sets, $\Omega_G^+ \cap \Omega_G'^+$, obtained by specific MLS decomposition of Example 11.1, with condition

$$c(\omega_{t>}) = \begin{cases} true & if\, n_E \leq N \\ false & otherwise \end{cases},$$

and by specific MLS decompositions of Example 11.3, with condition

$c'(\omega_{t>}) = \begin{cases} true & if\, t \leq T \\ false & otherwise \end{cases}$, is also proper. The resulting MLS consists of segmenting when the number of events in the left segment $\omega_{t)} \in \Omega_G \cap \Omega'_G$ reaches N or segmenting when the length of the left segment $\omega_{t)} \in \Omega_G \cap \Omega'_G$ reaches T, whichever occurs first.

However, the intersection of two proper generated sets is not always proper as shown in following example:

Example 11.5 (Intersection of two proper generated sets that is not proper)**.** The *intersection* of two proper generated sets, $\Omega_G^+ \cap \Omega_G'^+$, obtained by specific MLS decomposition of segments having a number of events greater than or equal to one, with condition

$$c(\omega_{t>}) = \begin{cases} true & if\, n_E \geq 1 \\ false & otherwise \end{cases},$$

and by specific MLS decompositions of Example 11.3, with condition

$c'(\omega_{t>}) = \begin{cases} true & if\, t \leq T \\ false & otherwise \end{cases}$, is not proper.

Following the arguments of Example 11.3 for segments of length less than *T*, it can be seen that the last right segment ω_n could not be segmented into $\omega^n_{t)} \bullet \omega^n_{[t}$ (both of length less than T) because right segment $\omega^n_{[t}$ could have no events leading second condition *c'* to be false. Notice, that replacing first condition c by fixed length condition (and corresponding generator set $\Omega_G^{\overline{T}}$), the intersection is then proper (each segment being of fixed length they are not concatenated together whereas satisfying the intersection condition).

Hence, not all combinations of generated sets are necessarily proper, as it is also the case for the union of generated sets.

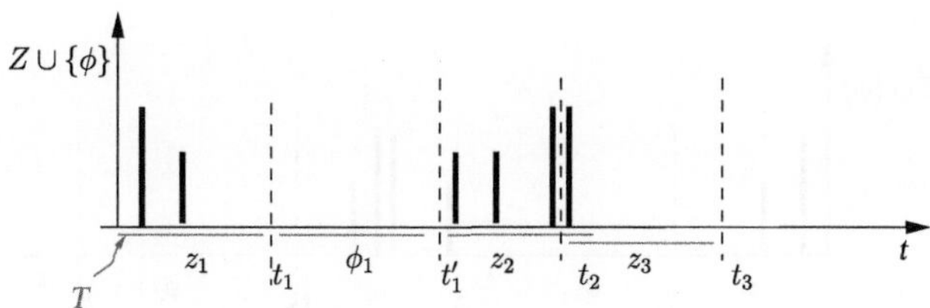

FIGURE 11.4

Intersection of segmentation by counting events such that $N \leq 3$ and segmentation by measuring time intervals such that $l(\omega) \leq T$.

Remark 11.5. *The union of two proper subsets,* Ω_G^+ and $\Omega_G'^+$, *is not necessarily proper* even if the two subsets are proper. Applying MLS to the union of the two generated sets Ω_G^+ and $\Omega_G'^+$, MLS will stop at which ever condition occurs last, i.e., $t^* = max\{t\}$ such that $c_1(\omega_{t\rangle}) \vee c_2(\omega_{t\rangle}) = max\{t_1, t_2\}$. Let us consider a set of generators as $\Omega_G = \{\omega, \omega'\}$ with $\omega(t) = \begin{cases} 1 & for\ t = 0 \\ \phi & elsewhere \end{cases}$ defined over $[0, t_f]$ and $\omega'(t) = \begin{cases} 1 & for\ t = t'_f \\ \phi & elsewhere \end{cases}$ defined over $[0, t'_f]$. Concatenating the second generator at the end of the first extends the first except that it leaves the end point without a generator (e.g., the remainder of a concatenation $\omega \bullet \omega'$ is just a single event point).

Let us see finally a case of disjoint union of two proper sets:

Example 11.6 (Disjoint union of two proper generated sets is weakly proper, cf. Fig. 11.5)**.** Let us consider segmentation condition

$$c(\omega_{t\rangle}) = \begin{cases} true & if\, n_E \geq N \wedge t \leq T \\ false & otherwise \end{cases}, \text{ with } N \geq 1,$$

and segmentation condition

$$c'(\omega_{t\rangle}) = \begin{cases} true & if\, n_E = 0 \wedge t > T \\ false & otherwise \end{cases}.$$

The intersection of both corresponding generators Ω_G and Ω'_G is empty. Then, since each generated set Ω_G^+ and $\Omega_G'^+$ is proper, it follows that their *disjoint union* $\Omega_G^+ \amalg \Omega_G'^+$, is *weakly proper.*

Indeed, condition *c* guarantees closure under left segmentation whereas condition *c'* does not guarantee closure under left *and* right segmentations (for some $t \in dom(\omega)$ of segments satisfying the second condition, left and right segments do not necessarily satisfy this condition). Finally, the generated set consists of segments with a number of events greater or equal to N within interval T separated by longer non-event intervals. At the operational level, for any segmentation, only one condition can be true.

Table 11.1 sums up previous examples and adds information. Concerning the operations on *generator sets*, they can be summarized as follows:

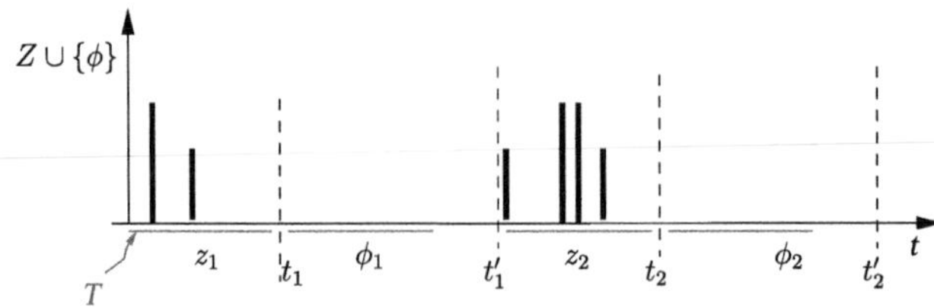

FIGURE 11.5

Disjoint union of MLS segmentations with $N = 2$.

Table 11.1 propriety of operations on generated and generator sets

	proper generator sets	proper generated sets
intersection	proper	not nec. proper
union	not nec. proper	not nec. proper
disj. union	not nec. proper	at least weakly proper

- $\Omega_G \cap \Omega'_G$ (as defined for generated sets) is proper, the latter intersection of the two generators being a strong requirement determining the minimum size segment being also proper.
- $\Omega_G \cup \Omega'_G$ (as defined for generated sets) can be improper, following Remark 11.5, it can be seen that the union of the two segments is improper.
- $\Omega_G \amalg \Omega'_G$ (as defined for generated sets) can be improper, following also Remark 11.5, it can be seen that the disjoint union of the two generators is improper.

Concerning the operations on *generated sets*, they can be summarized as follows:

- Intersection: $\Omega_G^+ \cap \Omega_G'^+ = \{\omega \,|\, (c(\omega_{t)}) \wedge c'(\omega_{t')})) = true,\ l(\omega) = min\{t, t'\}, \omega_{t)} \in \Omega_G, \omega_{t')} \in \Omega'_G\}$, with $(c(\omega_{t)}) \wedge c'(\omega_{t')})) = true$ meaning that both conditions *c* and *c'* are true. Example 11.5 shows that $\Omega_G^+ \cap \Omega_G'^+$ is not necessarily proper: $\Omega_G^{T+} \cap \Omega_G^{N_{sup}+}$, with $\Omega_G^{N_{sup}} = \{\omega \,|\, n_E > 0\}$.
- Union: $\Omega_G^+ \cup \Omega_G'^+ = \{\omega \,|\, (c(\omega_{t)}) \vee c'(\omega_{t')})) = true,\ l(\omega) = max\{t, t'\}, \omega_{t)} \in \Omega_G, \omega_{t')} \in \Omega'_G\}$, with $(c(\omega_{t)}) \vee c'(\omega_{t')})) = true$ meaning that one or both conditions *c* and *c'* are true. Remark 11.5 shows that $\Omega_G^+ \cup \Omega_G'^+$ is not necessarily proper.
- Disjoint union: $\Omega_G^+ \amalg \Omega_G'^+ = \{\omega \,|\, (c(\omega_{t)}) \veebar c'(\omega_{t')})) = true, \omega_{t)} \in \Omega_G, \omega_{t')} \in \Omega'_G\}$, with $l(\omega) = \begin{cases} t & if \quad c(\omega_{t)}) = true \\ t' & otherwise \end{cases}$ and $(c(\omega_{t)}) \veebar c'(\omega_{t')})) = true$ meaning that only one condition *c* or *c'* is true. Example 11.6 shows that $\Omega_G^+ \amalg \Omega_G'^+$ can be weakly proper. Theorem 11.2 stipulates that it is *always* at least weakly proper.

Let us see that the union of two generated sets Ω_G^+ and $\Omega_G'^+$ is always at least weakly proper by following theorem:

Theorem 11.2. The disjoint union of two proper generated sets, $\Omega_G^+ \amalg \Omega_G'^+$, is *at least weakly proper.*

Proof. Let us consider the set of generators $\Omega_G = \{\omega \,|\, c(\omega) = true\}$, corresponding to the first specific MLS and $\Omega'_G = \{\omega' \,|\, c'(\omega') = true\}$ corresponding to the second specific MLS. Remember now that an MLS should find the longest generator (of length $l(\omega) = t^*$) in a generated set Ω_G^+. Applying MLS to the disjoint union of both generated segments, only one condition can be true, i.e., $t^* = \begin{cases} t & if \quad c(\omega_{t)}) = true \\ t' & otherwise \end{cases}$. Each generated set being proper, accounting for Example 11.6, their disjoint union is then at least weakly proper. □

11.3.3 ITERATIVE SYSTEM SPECIFICATION

Based on previous propriety definitions and examples, Theorem 11.1 can be generalized as follows:

Theorem 11.3. *Sufficient conditions for iterative specification of a general system,* $(\delta, \lambda) \xrightarrow{+} (\Delta, \Lambda)$*:*

1. *The input generator set* Ω_G^X *is proper (cf. Definitions 11.5 to 11.7),*
2. Consistency of composition*:* $\omega_1, \omega_2, \ldots, \omega_n \in \Omega_G^X$ *and* $\omega_1, \omega_2, \ldots, \omega_n \in \Omega_G^{X+}$
$\Longrightarrow \delta^+(q, \omega_1 \bullet \omega_2 \bullet \ldots \bullet \omega_n) = \delta(\delta(\ldots\delta(\delta(q, \omega_1), \omega_2), \ldots), \omega_n)$.

Proof. Replacing condition 1) by strong propriety Definition 11.7 is equivalent to Theorem 11.1. Strong propriety can be weakened as shown at the end of Subsection 11.3.1. Weak and strong propriety can be generalized as simple propriety as shown in Example 11.3 by induction. □

11.4 COMPOSITION OF SEGMENTS

Even if the length of a segment $\omega \in \Omega$ is bounded, the number of possible values is possibly infinite, considering that $dom(\omega) \subset \mathbb{R}_0^+$. The question then is how to split a segment in a finite number of subsegments. To achieve this goal, each segment $\omega \in \Omega$ can be segmented using a specific segmentation condition such that the set of segmented (or conversely generated) segments is $\Omega^+ = \{\omega_{t)} \,|\, t = max\{t \,|\, \omega_{t)} \in \Omega_G \wedge c(\omega_{t)}) = true\}\}$, with t the time break obtained by MLS and $c(\omega_{t)})$ the specific segmentation condition. Also, corresponding generator set should be proper.

Once disposing of a finite number of proper segments, composite segments can be obtained using a new specific segmentation condition $c'(\omega)$ again showing the propriety of the new generated segments. Fig. 11.6 shows how to compose segments in a composite segment.

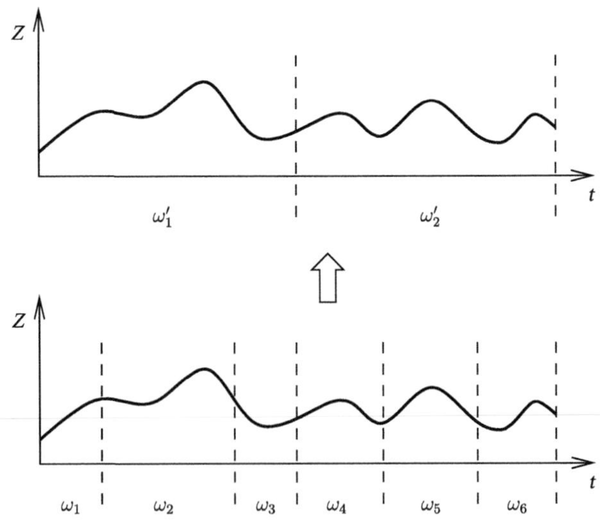

FIGURE 11.6

Composite segment.

11.5 DILATABLE GENERATOR CLASSES

Fig. 11.7 illustrates the time dilation operation. Formally, every point in the domain is multiplied by the same factor, $\tau > 0$, while not changing the range value associated with it. Fig. 11.7A shows the original segment and its MLS decomposition into generators, B shows stretching (multiplying by $\tau > 1$) and C shows contraction (multiplying by $\tau < 1$). Note that $\tau = 1$ is allowed and corresponds to the identity operation. From order preservation we immediately have that dilation preserves segmentability properties. In particular, we have that if a set of generators is proper and the same dilation is applied to all of its members then the resultant set is also proper.

Formally, a segment ω is dilated by τ to a segment v if for all $t \in dom(\omega)$, $v(\tau t) = \omega(t)$. We also say that v is a dilation of ω by τ.

Exercise 11.1. Consider the conditions for which $dom(v) = \tau dom(\omega)$.

Remark 11.6. Dilation can be applied to any segment.

Exercise 11.2. Show that dilation establishes an equivalence relation on all segment, i.e., two segments are equivalent if one can be transformed into the other with some dilation.

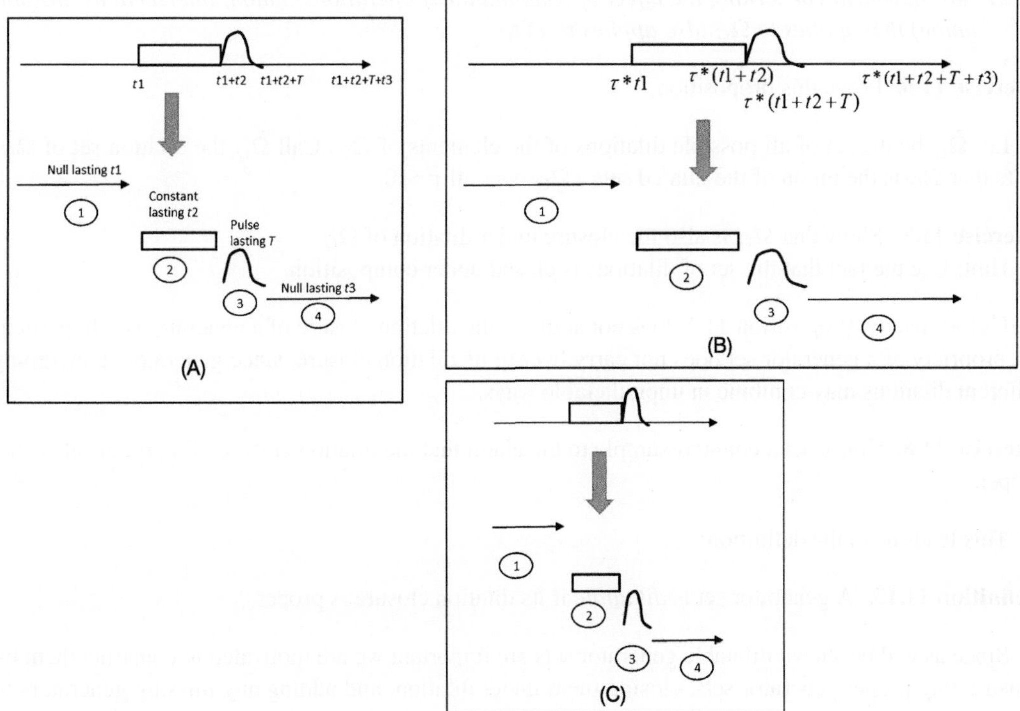

FIGURE 11.7

Illustrating dilation of segments and generators: (A) originals (B) stretching by τ (C) contraction.

Remark 11.7. Dilation preserves ordering of events on the time base.

Exercise 11.3. Prove this remark. Hint: multiplication preserves the $<$ relation.

From order preservation we immediately have that dilation preserves segmentability properties.

Exercise 11.4. Show that if $\omega = \omega_1\omega_2$ then $\upsilon = \upsilon_1\upsilon_2$ where υ, υ_1, and υ_2 are dilations of $\omega, \omega_1, \omega_2$ by the same factor, resp.

Exercise 11.5. Show that if t^* is an MLS breakpoint of ω then τt^* is an MLS breakpoint of υ, for a dilation of ω by τ.

In particular, we have that if a set of generators is proper and the same dilation is applied to all of its members then the resultant set is also proper. Formally,

Proposition 11.1. *Let Ω_G be a set of generators and $\tau\Omega_G$ be dilations of the elements of Ω_G by the factor $\tau > 0$. Then,*

1. *Ω_G is (simply, weakly, strongly) proper $\Rightarrow \tau\Omega_G$ is (simply, weakly, strongly) proper.*
2. *Any statement concerning the effect of combinational operations: union, intersections, disjoint union) that applies to Ω_G also applies to $\tau\Omega_G$.*

Exercise 11.6. Prove this proposition.

Let $\tilde{\Omega}_G$ be the set of all possible dilations of the elements of Ω_G. Call $\tilde{\Omega}_G$ the dilation set of Ω_G. Note that $\tilde{\Omega}_G$ is the union of the dilated sets $\tau\Omega_G$ over all $\tau > 0$.

Exercise 11.7. Show that $\tilde{\Omega}_G$ is also the closure under dilation of Ω_G.
Hint: Use the fact that the set of dilations is closed under composition.

Unfortunately, Proposition 11.1 does not apply to the dilation closure of a generator set. In particular, propriety of a generator set does not carry over to its dilation closure, since generators undergoing different dilations may combine in unpredictable ways.

Exercise 11.8. Construct a counterexample to the claim that the dilation closure of a proper set is also proper.

This leads us to the definition:

Definition 11.13. A generator set is *dilatable* if its dilation closure is proper.

Since as will be shown dilatable generator sets are important we are motivated to construct them by constructing proper generator sets, closing them under dilation, and adding any missing generators to produce proper results.

Exercise 11.9. Prove that Ω_G^τ is proper. Hint: For all $\omega \in \Omega_G^\tau$, if ω is proper so are all corresponding dilations sharing the same rate τ.

Examples of applications consist of:

- Timed automata where dilatable generators are considered in time,
- DEVS Turing machine where the time to read symbols is considered,
- Input/output segments (behaviors) of systems where patterns of values (a, c, d, ...) are considered with different interdelays.

The set of generated segments can then be studied by a morphism from an iterative specification G to an iterative specification G′. For example, the subclass of dilatable generators consists of all generators contracted with rate $\tau < 1$ (i.e., with the same sequence of values but occurring at closer time points) (see Fig. 11.8).

Using a dilation is a simplification of an iterative specification morphism as there is an *identity map* $Id_{\overline{Q}}$ between state states $\overline{Q}$ and $\overline{Q'}$ and corresponding input/output sets (of generators or generated) are proper (as shown in last exercise).

A *dilation morphism* $\mu_\tau = (\overline{d_\tau^X}, id_{\overline{Q}}, \overline{d_\tau^Y})$ from an iterative specification G to an iterative specification G′ is such that:

1. $d_\tau^X : \Omega_G^{X_\tau} \to \Omega_G^{X_\tau'}$,
2. $id_{\overline{Q}} : \overline{Q} \to \overline{Q}$, with $\overline{Q} \subseteq Q$ and $\overline{Q} = Q'$,
3. $d_\tau^Y : \Omega_G^{Y_\tau'} \to \Omega_G^{Y_\tau}$, and preservation of output/transition functions, $\forall q \in \overline{Q}, \forall \omega \in \Omega_G^{X_\tau}$:
4. $\delta(q, \omega) = \delta'(q, d_\tau^X(\omega))$,
5. $d_\tau^Y(\lambda(q, \omega)) = \lambda'(q, d_\tau^X(\omega))$.

The same dilation rate τ (stretch or contraction) is carried from any detailed input/output generator to any abstract input/output generator.

The behavior at system level can then be investigated with respect to the impact of the input/output dilations such that: $\forall q \in Q, \forall t \in dom(\Omega_X)$,

$$\beta_q = \{(\omega_{t\rangle}^X, \omega_{t\rangle}^Y) \mid \omega_{t\rangle}^X \in \omega_G^X, \omega_{t\rangle}^Y \in \omega_G^Y, \omega_{t\rangle}^Y = \lambda(\delta(q, \omega_{t\rangle}^X), \omega_{t\rangle}^X)\}$$

Exercise 11.10. Prove that the class of dilatable input/output generators leads to equivalent behaviors.

Hint: Use congruence relation of Section 12.4.2 for behavior and state equivalences as well as a dilation morphism.

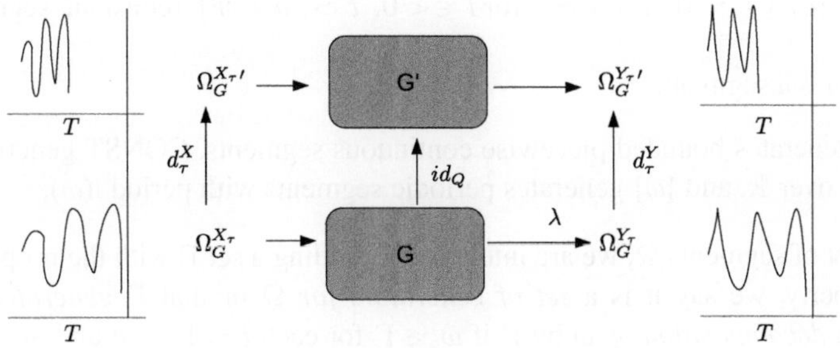

FIGURE 11.8

Contraction of generators between two iterative specifications.

11.6 SUMMARY

Several specific generator classes (based on their length and number of events) have been presented here as simple examples. We have shown how to combine these classes into higher level classes while still being able to iteratively specify general systems. To achieve this goal, we have defined a *proper generator class* $\Omega_G = \{\omega \,|\, c(\omega) = true\}$ such that corresponding *generated set* $\Omega_G^+ = \{\omega \,|\, c(\omega_{t\rangle}) = true \wedge t = max\{t \,|\, \omega_{t\rangle} \in \Omega_G\}$ *is proper* and *time breakpoint* $t \in \mathbb{R}_0^{+\infty}$ obtained by maximal length segmentation. In the next chapter we will see how generator classes can be used to model classes of iterative specifications (timed, finite, real time, etc.) by simply defining particular generator classes.

APPENDIX 11.A

The following is taken from Section 4.11.1 of TMS2000.

11.A.1 GENERATOR SEGMENTS

Consider (Z, T) the set of all segments $\{\omega : <0, t_1> \rightarrow Z \mid t_1 \in T\}$, which is a semigroup under concatenation. For a subset Γ of (Z, T), we designate by Γ^+ the *concatenation closure* of Γ (also called the *semigroup generated* by Γ). Then Γ^+ can be constructed as follows. Let $\Gamma^1 = \Gamma$ and let $\Gamma^{i+1} = \Gamma^i \cdot \Gamma = \{\omega\omega' \mid \omega \in \Gamma^i \text{ and } \omega' \in \Gamma\}$. Then $\Gamma^+ = \bigcup_{i \in \mathbb{N}} \Gamma^i$. The proof is given in the following exercise.

Exercise 11.11. Show that $\Gamma^i = \{\omega_1, \omega_2, ..., \omega_i \mid \omega_j \in \Gamma$ for j = 1, ..., i$\}$. Then show that as the least semigroup containing Γ, Γ^+ is included within $\bigcup_{i \in \mathbb{N}} \Gamma^i$. On the other hand, since Γ^+ contains Γ, it also contains Γ^2; hence by induction Γ^2, Γ^3, ..., Γ^i for each i $\in \mathbb{N}$, and $\bigcup_{i \in \mathbb{N}} \Gamma^i \subseteq \Gamma^+$. Conclude that $\Gamma^+ = \bigcup_{i \in \mathbb{N}} \Gamma^i$.

Exercise 11.12. Consider the following set of segments:

1. bc = bounded continuous segments.
2. CONST = $\{a_\tau \mid \tau > 0, a_\tau(t) = a \text{ for } t \in <0, \tau>, a \in \Re\}$ (constant segments of variable length).
3. $\{\omega\}$ where ω is a segment.

Show that bc generates bounded piecewise continuous segments, CONST generates the piecewise constant segments over $\mathbb{R}$, and $\{\omega\}$ generates periodic segments with period $l(\omega)$.

Now given a set of segments Ω, we are interested in finding a set Γ with the property that $\Gamma^+ = \Omega$. If Γ has this property, we say it is a *set of generators for* Ω or that Γ *generates* Ω. We say that $\omega_1, \omega_2, ..., \omega_n$ is a *decomposition of* ω by Γ if $\omega_i \in \Gamma$ for each $i = 1, ..., n$ and $\omega = \omega_1 \cdot \omega_2 \cdot ... \cdot \omega_n$. It happens in general that if Γ generates Ω, we *cannot* expect each $\omega \in \Omega$ to have a unique decomposition by Γ; that is, there may be distinct decompositions $\omega_1, \omega_2, ..., \omega_n$ and $\omega_1', \omega_2', ..., \omega_n'$ such that $\omega_1 \cdot \omega_2 \cdot ... \cdot \omega_n = \omega$ and $\omega_1' \cdot \omega_2' \cdot ... \cdot \omega_n' = \omega$.

Exercise 11.13. Show that unique decompositions exist in case of $\{\omega\}$ but not in cases bc and CONST. Consider what properties Γ might have to ensure that each $\omega \in \Gamma^+$ has a unique decomposition.

Of the many possible decompositions, we are interested in selecting a single representative, or *canonical* decomposition. We go about selecting such a decomposition by a process called *maximal length segmentation*. First we find ω_1, the longest generator in Γ which is also a left segment of ω. This process is repeated with what remains of ω after ω_1 is removed, generating ω_2, and so on. If the process stops after n repetitions, then $\omega = \omega_1\omega_2 \ldots \omega_n$ and $\omega_1, \omega_2, \ldots \omega_n$ is the decomposition sought. We call a decomposition obtained by this process a maximal length Segment (MLS) decomposition.

More formally, a decomposition $\omega_1, \omega_2, \ldots \omega_n$ is an MLS decomposition of ω by Γ if for each i = 1,..,n, ω' e Γ is a left segment of $\omega_i\omega_{i+1\ldots}\omega_n$ implies ω' is a left segment of ω_i. In other words, for each i, ω_i the longest generator in Γ that is a left segment of $\omega_i\omega_{i+1\ldots}\omega_n$.

It is not necessarily the case that MLS decompositions exist, (i.e., that the just-mentioned processes will stop after a finite number of repetitions). Thus we are interested in checkable conditions on a set of generators which will guarantee that every segment generated has an MLS decomposition. Fortunately, a segment can have at most one MLS decomposition.

Exercise 11.14. Show that ω is a left segment of ω' and ω' is a left segment of ω if and only if $\omega = \omega'$. Hence show that if ω has a MLS decomposition by Γ then it is unique.

Let us say that Γ is an *admissible* set of generators for Ω if Γ generates Ω and for each ω e Ω, a unique MLS decomposition of by Γ exists. (We also say Γ admissibly generates Ω.) We now provide a set of conditions guaranteeing that Γ is an admissible set of generators. It should be noted that these conditions are too strong in some cases – for example, when all segments have unique decompositions.

Exercise 11.15. Check whether conditions a) and b) of the following theorem hold for bc.

Theorem 11.4 (Sufficient conditions for admissibility). *If Γ satisfies the following conditions, it admissibly generates Γ^+:*

(a) *Existence of longest segments,*
(b) *Closure under right segmentation.*

The proof is given in Section 4.11.1 of TMS2000.

REFERENCES

Muzy, A., Zeigler, B.P., Grammont, F., 2017. Iterative specification as a modeling and simulation formalism for I/O general systems. IEEE Systems Journal, 1–12. https://doi.org/10.1109/JSYST.2017.2728861.

Wymore, W., 1967. A Mathematical Theory of Systems Engineering: The Elements. Wiley.

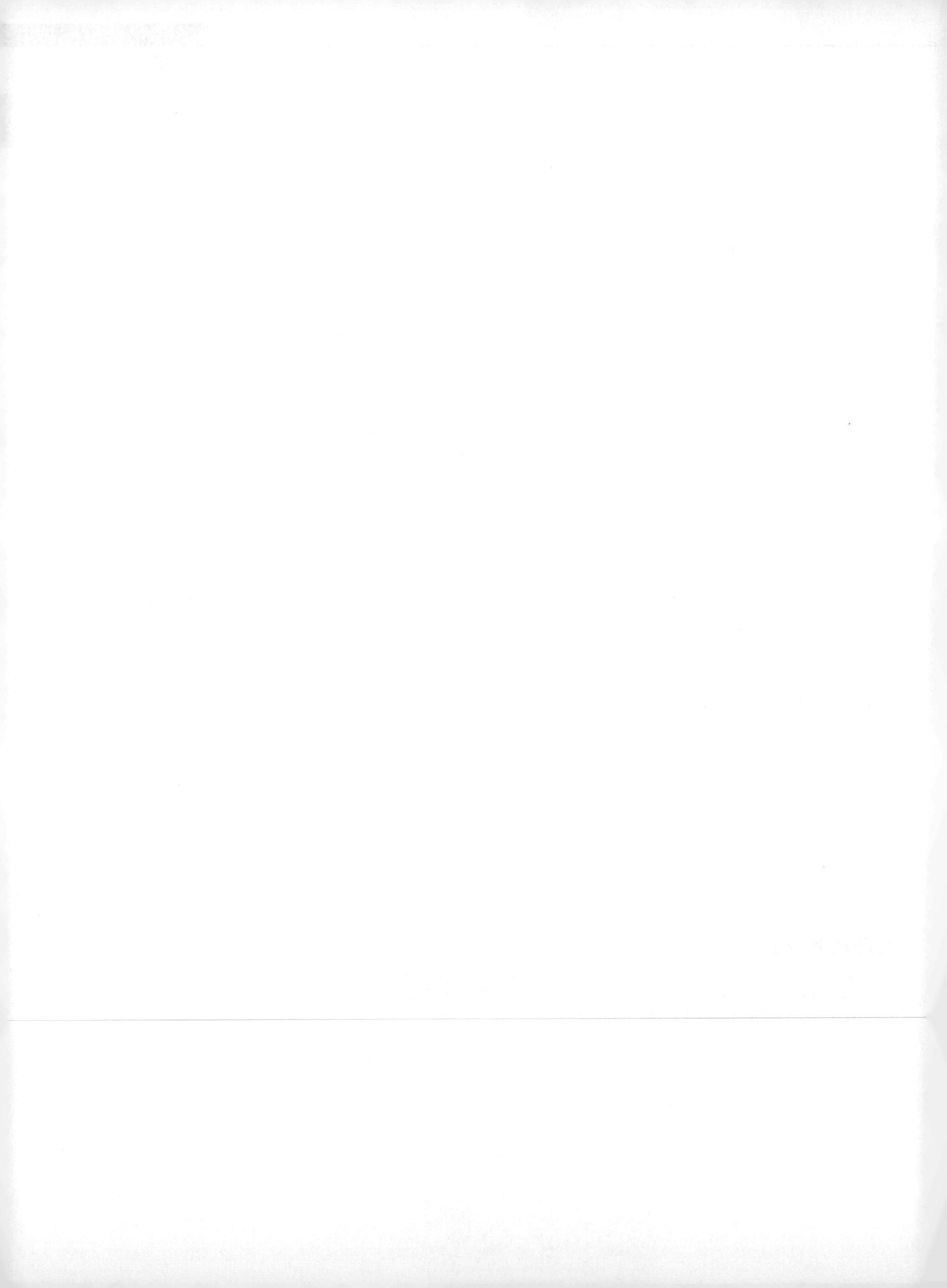

CHAPTER

ITERATIVE SPECIFICATION SUBFORMALISMS 12

CONTENTS

We saw in previous chapters how to define iterative specifications for modeling general systems. In this chapter we present an ecosystem of iterative subformalisms that can fit the right abstraction level of specific systems and how DEVS can be used to simulate them. Also we will show how convenient it is to achieve modeling at such abstract levels. Fig. 12.1 shows a toy example of a traveling drone from a point A to a point B, reacting to a transmitter call in A and B.

Fig. 12.2 introduces a first representation of the drone example with a phase transition diagram on the left in A. At the initial phase, the drone is in phase "waiting" either in A or B (step 0). When it receives an input generator segment "signal" from the emitter at opposite point A or B (step 1), it goes to phase "moving" (step 2) and emits an output generator segment "sound" until it reaches the opposite

Theory of Modeling and Simulation. https://doi.org/10.1016/B978-0-12-813370-5.00021-3

FIGURE 12.1

A toy example of a traveling drone from a point A to a point B, reacting to a transmitter call in B and vice versa. Free transmitter and drone images from https://pixabay.com.

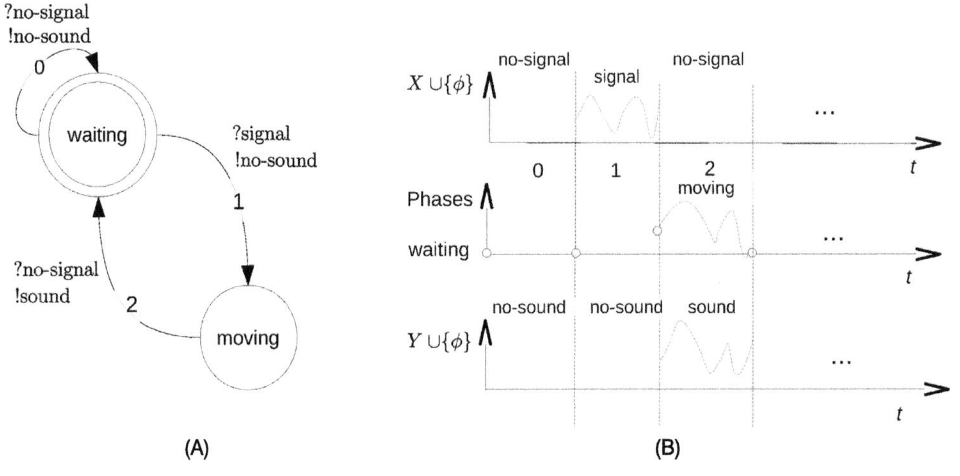

FIGURE 12.2

Phase transition diagram (A) and trajectories (B) corresponding to the drone example of Fig. 12.1. First steps 0, 1, 2 are indicated in both parts (A) and (B). Initial phase is indicated with a double circle. Inputs are indicated by a symbol ? and outputs by a symbol !. States are indicated with white circles.

point thus returning to the "waiting" phase. The process can be then repeated when the emitter in the opposite point calls the drone.

This example is convenient for representing iterative specification modeling. On the other hand many elements remain unspecified at this level. For example, notice the dashed input/output generators. Many generators can be used in this way, so the question is: How to characterize the generator sets and the properties of the generators? Besides, this representation requires a finite set of transitions and phases and, as in Fig. 12.2, therefore a *finite* iterative specification being a finite abstraction of the usual generated segments from an iterative specification.

12.1 CLASS MAPPING OF ITERATIVE SYSTEM SPECIFICATIONS

After formally defining and considering the concept of iterative system specification itself in more depth, we will go on to consider several specializations and elaborations that establish its utility as a constructive formalism for general systems specification. Moreover, this formalism should be computationally implementable in DEVS-based modeling and simulation environments.

Fig. 12.3 maps out some of the subformalisms and places them in relation to each other and to the general system. Two relations are shown. One in which a formalism *simulates* another and the other in which one *specifies* another. We have seen that DEVS *simulates* iterative systems specifications in the sense that given an iterative systems specification, there is a DEVS that can simulate it, as informally suggested in Chapter 10. This simulation capability represents a computational means to manipulate any systems specifications that are specialized versions of the Iterative Specification.

We will consider a *formal specification to specify another if there is a way to construct an instance of the second given an instance of the first*. In other words, there is a constructive mapping from the objects and relations of the first to those of the second. Of course, the basic concept of iterative systems specification specifies a system in this sense as shown in Fig. 12.3.

Exercise 12.1. Verify that the specifies relation is transitive. Hint: argue that if A specifies B, then the class of systems specified by A is included in those specified by B, hence transitivity follows from set theory. Argue that the converse is not necessarily true.

Exercise 12.2. Show that transitivity implies that DEVS can simulate any specialized iterative systems specification.

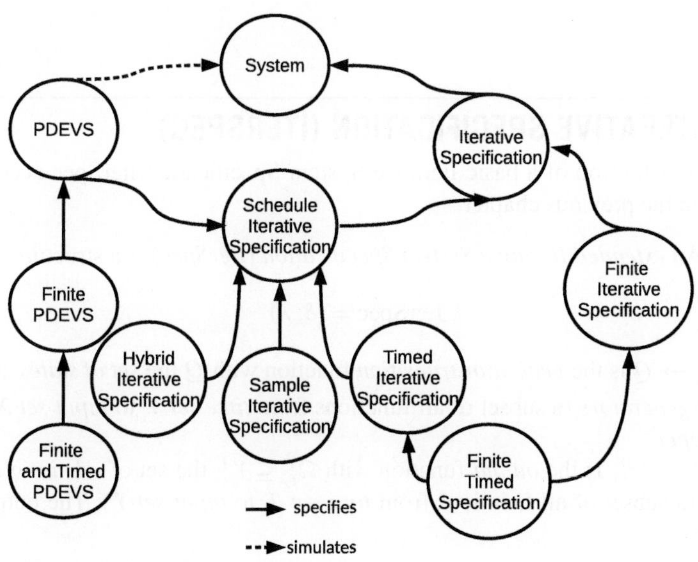

FIGURE 12.3

Class Mapping of Iterative System Specifications.

Starting on the left of Fig. 12.3, the Iterative Specification and System classes it specifies are the same as earlier introduced. In addition to PDEVS (Parallel Discrete Event Systems Specification), three other iterative specifications are shown and will be considered in more detail. The scheduled iterative systems specification, Scheduled Iterative Specification, introduces a time scheduling concept explicitly into the basic iterative systems specification. This allows the modeler to employ a time advance function that generalizes the time advance concept of PDEVS. It does not extend the class of systems that are specified since the modeler can already define such a function on a case-by-case basis to determine when an internal event should happen. In contrast, the time advance function in Scheduled Iterative Specification provides this capability explicitly giving it a well-defined semantics. PDEVS then turns out to be a specific kind of Scheduled Iterative Specification and of the general iterative specification (by transitivity).

Two formalisms are then shown as further specializations of Scheduled Iterative Specification. Hybrid Iterative Specification is a formulation of the Hybrid system specification first defined by Nutaro et al. (2012) to support hybrid co-simulation (Section 9.7). Hybrid Iterative Specification will then be shown to be a specialization of Scheduled Iterative Specification. Similarly, Sample Based Iterative Specification will be shown to be a specialization of Scheduled Iterative Specification. Sample Based Iterative Specification is a formulation of the HyFlow formalism introduced by Barros (2002) which allows a system to sample its input in a single atomic operation, a capability that can be conceptually and operationally appealing. This facility can be captured within the Scheduled Iterative Specification, as will be explained later, but not directly within PDEVS. Indeed, PDEVS lacks the ability to directly "pull" a sample value from a source; it must instead request and wait in a two step sequence. However, in keeping with PDEVS' ability to simulate any iterative specification, it can simulate the Sample Based Iterative Specification, and we will provide a direct simulation based on the request/wait equivalence.

12.2 BASIC ITERATIVE SPECIFICATION (ITERSPEC)

We extend here the definition of a basic Iterative System Specification IterSpec according to the elements introduced in the previous chapter

Definition 12.1. An *extended Iterative System Specification (IterSpec)* is a structure

$$\text{IterSpec} = (\delta, \lambda)$$

- $\delta : Q \times \Omega_G^X \to Q$ is the *generator transition* function with Q the *set of states* and $\Omega_G^X \subseteq X^T$ the set of *input generators* (a subset of all functions from *time set* T to *input set* X) where Ω_G^{X+} is *strongly proper*.
- $\lambda : Q \times \Omega_G^X \to \Omega_G^Y$ is the *output* function with $\Omega_G^Y \subseteq Y^T$ the set of *admissible output segment generators* (a subset of all functions from *time set* T to *input set* Y). The output functions has constraints:

  1. $l(\lambda(q, \omega)) = l(\omega)$
  2. $\lambda(q, \omega_{t>}) = \lambda(q, \omega)_{t>}$
  3. $\lambda(q, \omega\omega') = \lambda(q, \omega)\lambda(\delta(q, \omega), \omega')$.

Theorem 12.1. *An IterSpec $G = (\delta, \lambda)$ specifies a system $S = (\Delta, \Lambda)$ where $\Omega_X = \Omega_G^{X+}$ and $\Omega_Y = \Omega_G^{Y+}$.*

Proof. The system transition function $\delta : Q \times \Omega_G^{X+} \to Q$ is defined recursively:

Given $\omega \in \Omega_G^{X+}$, let the *time advance* from state q *be* $t_{MLS}(\omega)$ where $t_{MLS}(\omega)$ *is the* MLS *break-point.* Then

$$\Delta(q, \omega) = \begin{cases} \delta(q, \omega)) & \text{if } l(\omega) \le t_{MLS}(\omega) \\ \Delta(\delta(q, \omega_{t_{MLS}(\omega)>}), \omega_{<t_{MLS}(\omega)}) & \text{otherwise} \end{cases}$$

Since Ω_G^{X+} is *strongly proper*, the MLS of $\omega_{<t_{MLS}(\omega)}$ *continues and sets up* the recursive specification for the transition function.

This process converges provided that the time advances are *progressive* as discussed in Section 10.5. *Similarly,* $\Lambda : Q \times \Omega_X \to \Omega_Y$ is defined recursively:

$$\Lambda(q, \omega) = \begin{cases} \lambda(q, \omega)) & \text{if } l(\omega) \le t_{MLS}(\omega) \\ \lambda(q, \omega_{t_{MLS}(\omega)>})\Lambda(\delta(q, \omega_{t_{MLS}(\omega)>}), \omega_{<t_{MLS}(\omega)}) & \text{otherwise} \end{cases} \qquad \square$$

Exercise 12.3. Show that the constraints satisfied by the iterative specification continue to hold for the system it specifies, i.e.,

1. $l(\Lambda(q, \omega)) = l(\omega)$
2. $\Lambda(q, \omega_{t>}) = \Lambda(q, \omega)_{t>}$
3. $\Lambda(q, \omega\omega') = \Lambda(q, \omega)\Lambda(\delta(q, \omega), \omega')$

12.3 SCHEDULED ITERATIVE SYSTEM SPECIFICATION

The Scheduled Iterative Specification introduces another element, the time schedule function, with its non-negative real value, that plays the same role, in more general form, as the time advance function does in DEVS. Roughly, this introduces time scheduling for internal events into the Iterative Specification and thus supports a more direct specification of discrete event systems. We will also introduce a Hybrid Iterative Specification that represents a standard approach to co-simulation of continuous and discrete models that is a specialized form of Scheduled Iterative Specification. The hybrid specification is also shown to be a subclass of DEVS. Finally, we will introduce a Sample-Based Iterative Specification that is another specialized form of Scheduled Iterative Specification which captures *pulling* of samples from other component systems in contrast with the usual *pushing* of inputs that DEVS components employ to move data from one to another. Such pulling of inputs is not representable directly in DEVS but is specifiable as a Scheduled Iterative Specification, hence as an Iterative Specification and a System.

Definition 12.2. *A Scheduled Iterative System Specification (SchedIterSpec) is defined by*

$$G = (\delta, \lambda, \text{ts})$$

where

- $\Omega_G^X \subset X^T$ is the *input generator set* where Ω_G^{X+} is *proper*,
- $\delta : Q \times \Omega_G^X \to Q$ is the *generator transition function*, with composition property,
- $\lambda : Q \to Y$ is the output function,
- $ts : Q \to \mathbb{R}_0^{+\infty}$ is the schedule function, with $\text{ts}(q)$ = time to next internal event.

Note that SchedItSpec adds the schedule function to the Iterative Specification.

Theorem 12.2. *A SchedIterSpec, $G_{\text{ts}} = (\delta_{\text{ts}}, \lambda_{\text{ts}}, \text{ts})$ specifies an IterSpec $G = (\delta, \lambda)$.*

Proof. The main task is to define the IterSpec transition function in terms of the given SchedIterSpec transition function. □

Given $\omega \in \Omega_G^X$, $\delta(q, \omega)$ is defined recursively as follows:
Let the time advance from state q with input ω be the minimum of $ts(q)$ and $l(\omega)$. Then

$$\delta(q, \omega) = \delta(q_{t^*}, \omega_{<t^*})$$

where

$$q_{t^*} = \begin{cases} \delta_{\text{ts}}(q, \omega_{t^*>}) & \text{if } t^* < l(\omega) \ 1) \\ \delta_{\text{ts}}(q, \omega) & \text{otherwise} \quad 2) \end{cases}$$

Line 1) is the case where t^* *occurs within the input segment. In this case, the left segment up to* t^* is applied as input to get the state, q_{t^*} that pertains at t^*. The left segment exists due the propriety of Ω_G^X. Then the remaining right segment is applied *recursively* to get the state at the end of the overall segment. There may be more than one internal transitions developed in this manner within the span of the input generator.

Line 2) concerns the case where the next event is scheduled to occur at or beyond the end of the overall segment. In this case, the overall segment is applied. When $t^* = l(\omega)$ the internal event occurs at the end of the segment. Note that the state q_{t^*} at the end of the segment provides a new scheduled time $\text{ts}(q_{t^*})$ which will apply to the next generator segment in an MLS state computation process.

The recursive definition for the transition function converges provided that the time advances are *progressive* as discussed in Section 10.5. We need to show that it has the composition property. But this follows from the assumed composition property of δ_{ts}.

Exercise 12.4. Prove the assertion about composition property.

The output function is defined on input segments by extension:

$$\lambda : Q \times \Omega_X \to Y$$
$$\lambda(q, \omega) = \lambda_{\text{ts}}(\delta(q, \omega))$$

12.3.1 DEVS IS A SCHEDULED ITERATIVE SYSTEM SPECIFICATION

Since DEVS is a special case of Iterative System Specification we can show that DEVS also is a Scheduled Iterative System Specification if we can interpret the time schedule function as determining when the internal events occur and rewrite the transition function accordingly. Now an internal event starting from state (s, e) at time $= 0$ occurs at time $= \mathrm{ta}(s) - e$. So we can define the time schedule function as $\mathrm{ts}(s, e) = \mathrm{ta}(s) - e$. In Appendix 12.G we prove this yields a definition that is consistent with the definition of the time schedule function in the Scheduled Iterative System Specification.

12.4 SAMPLE-BASED ITERATIVE SYSTEM SPECIFICATION

Sampling is an efficient way to obtain time-dependent data. For example, highly detailed trajectories of aircraft simultaneously engaged in an air battle may be gathered over an extended time period. Stored in a file, these data can be replayed as input in simulations that test different strategies, tactics, and equipment for combat under realistic conditions. The natural representation of such a situation is to consider this data production as the generator component within an experimental frame. One approach is to "push" every event out to the model components at the simulation time it occurs. Another approach is to let components "pull" data elements from the generator as they need them. In this section, we show how such an approach that is not directly represented as a DEVS but can be formulated as an iterative system specification.

Before proceeding we briefly review the Hybrid Flow System Specification (HyFlow) that was defined by Barros (2002) as a formalism for computational modeling and simulation of hybrid systems with time-variant topologies. While both HyFlow and DEVS have their origin in general system theory, HyFlow differs in that it achieves the representation of continuous systems with novel concepts of generalized sampling (both component and time varying sampling) and the description of arbitrary dense output functions (Barros, 2002). HyFlow provides a unifying representation for a large variety models, including digital controllers, digital filters, fluid stochastic Petri Nets, sliding mode controllers, zero-cross detectors, stochastic solvers for molecular reactions, geometric numerical integrators, and exponential integrators (Barros, 2017b). Since its original definition, Barros has gone on to develop HyFlow in numerous directions including abstract simulators, determinism in the presence of simultaneous events on a hyperreal time base, and guarantees of seamless interoperability of all models represented in the formalism (Barros, 2016, 2017b).

In this section we show how the "pull" semantics that underlies HyFlow's sampling concepts can be characterized as an iterative system specification. Although HyFlow pull semantics (sampling), can be mapped into DEVS push semantics, we note that pull semantics provides more compact descriptions while enforcing modular models (Barros, 2017a). As with DEVS, closure under composition and hierarchical modular construction of HyFlow support a formal basis for co-simulation. The formulation of pull semantics within an iterative system specification shown here opens the way for much productive research into co-simulation that applies the developments in HyFlow and DEVS in various directions such as just mentioned.

The sample-based iterative system specification will lead to formalization of push and pull approaches and a better understanding of the difference between them. We will also show how the sample-based iterative system specification can be realized by extending the DEVS abstract simulator, thus suggesting a practical implementation of the concept.

Fig. 12.4A and B illustrate how outputs are "pushed" from sender (A) to receiver (B) in DEVS coupled models. The output occurs first, as an output event, and is transmitted via coupling as input via an external event. Although output occurs before input the time at which they occur are the same.

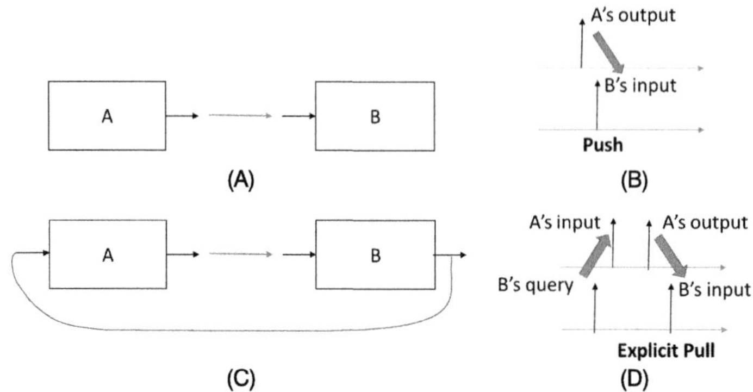

FIGURE 12.4

Push Semantics – the output of sender A is coupled to receiver B (A) and the sequence of events (B). Explicit Pull Semantics component B sends a query to A (C) and waits for the output to be pushed to it (D).

Fig. 12.4C illustrates how outputs are "pulled" by the receiver (B) from the sender (A) in DEVS coupled models. Here B has to first send an input to A to request that sometime later A should push an output as in Fig. 12.4D.

Fig. 12.5 illustrates how a sample of a component's output (A) can be pulled by a component (B) without explicitly sending a request for it. Here B undergoes a sample event that immediately calls on A's output to appear as B's input. This is called "implicit" pull to distinguish from the explicit pull described in Fig. 12.4C. We note that this form of interaction cannot be directly represented in DEVS with the closest realization being that in Fig. 12.4C. However, we now show how such behavior (semantics) can be expressed in the larger class of systems as an iterative system specification in a manner similar to that of the HyFlow formalism.

Exercise 12.5. Write a DEVS coupled model with atomic models such as A and B in Fig. 12.4A exhibiting the push semantics.

Exercise 12.6. Write a DEVS coupled model with atomic models such as A and B in Fig. 12.4A exhibiting the explicit pull semantics.

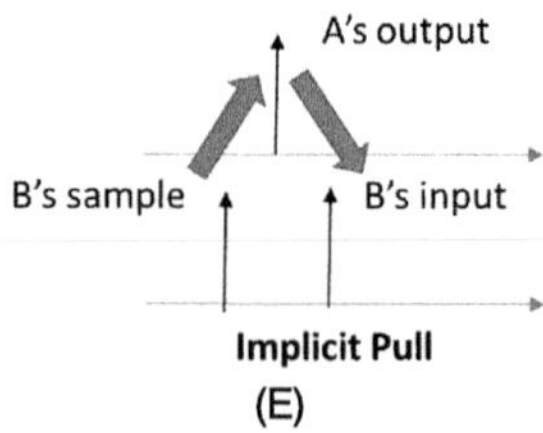

FIGURE 12.5

(E): Implicit Pull Semantics component B samples A's output without explicitly requesting it as in Fig. 12.4C.

12.4.1 FORMAL DEFINITION OF SAMPLE-BASED ITERATIVE SYSTEM SPECIFICATION

We present the formal definition as

Definition 12.3. Sample-based system specification (SampleIterSpec)

$$G = (\delta_0, \delta_\Phi, \lambda, \text{ts})$$

where

- $\delta_0 : Q \times X \to Q$ is the *external transition function* providing the immediate effect of an input on the state
- $\delta_\Phi : Q \times \Omega_\Phi \to Q$ is the *extended transition function*, where Ω_Φ is the *set of all null event segments*, to be explained in a moment. δ_Φ *satisfies the composition property*: $\delta_\Phi(\delta_\Phi(q, \phi_{t>}), \phi_{t'>}) = \delta_\Phi(q, \phi_{t+t'>})$.
- $\lambda : Q \to Y$ is the *output function*, $\text{ts} : Q \to \mathbb{R}_0^{+\infty}$ is *time to next input sampling function*;
- $\text{ts}(q)$ is the time in state $q \in Q$ until the next sampling of the input.

We require that *ts* satisfies the preservation of schedule under update condition.

12.4.2 PRESERVATION OF SCHEDULED TIME UNDER UPDATE

Fig. 12.6 illustrates the situation where a sample is requested from a system. This circumstance should not affect the time that this system was scheduled to issue a request for its own sample.

From the figure we can see the requirement:

$$\text{ts}(q) = t + \text{ts}(\delta(q, \omega_{t>}))$$

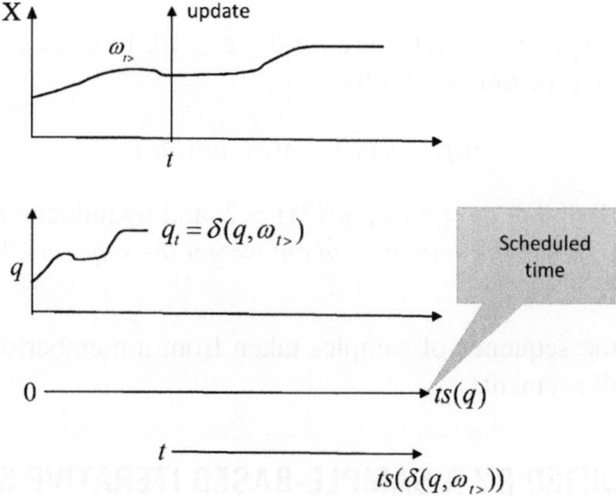

FIGURE 12.6

Preservation of scheduled time under update.

which is equivalent to the final requirement:

$$\mathrm{ts}(\delta(q, \omega_{t>})) = \mathrm{ts}(q) - t$$

Exercise 12.7. A subclass of sample based systems is obtained where the state remains constant between samples.

$$\delta_\Phi : Q \times \Omega_\Phi \to Q$$
$$\delta_\Phi(q, \phi_{t>}) = q$$

Show that δ_Φ satisfies the composition property.

12.4.3 EXAMPLE OF SAMPLE-BASED ITERATIVE SYSTEM SPECIFICATION

$$SampleIterSpec = (\delta_0, \delta_\Phi, \lambda, \mathrm{ts})$$

where

$X = Q = Y = \mathbb{R}$

$\Omega =$ *piecewise continuous segments,with* $\omega : \langle 0, t\rangle \to \mathbb{R}$

$\delta_0 : Q \times X \to Q$ is defined by $\delta_0(q, x) = x$,

i.e., *a sampled value x is stored as the next state*

$\delta_\Phi : Q \times \Omega_\Phi \to Q$ is defined by $\delta_\Phi(q, \phi_{t>}) = q$

i.e., *the state remains constant at the last value between samples*

$\lambda : Q \to Y$ is defined by $\lambda(q) = q$,

i. e, *the last sample is the output*

$\mathrm{ts} : Q \to \mathbb{R}_0^{+\infty}$ *is defined by* $\mathrm{ts}(q) = 1$, i.e., *samples are taken every time unit.*

As an example of an input segment, let $\omega(t) = t$ for $0 \le t \le 10$. In this example, a sample is taken of the input, here a ramp, every time unit. Formally,

$$\delta(q, \omega_{\langle 0,10\rangle}) = \delta(q_1, \omega_{\langle 1,10\rangle})$$

where $q_1 = \delta_0(q, \omega(1)) = 1$ so that $q_2 = \delta_0(q_1, \omega(2)) = 2$, and by induction we can show that, $q_i = i$. So that $\delta(q, \omega\langle 0, 10\rangle) = 10$. Since *the state is read out we get the sequence* 1, 2, ···, 10 *output at times* 1, 2, ···, 10.

Exercise 12.8. Compute the sequence of samples taken from a member of the set generated by the ramps together with the null segments.

12.4.4 SYSTEM SPECIFIED BY A SAMPLE-BASED ITERATIVE SPECIFICATION

The transition function of the system specified by a sample-based iterative specification must realize the behavior implied by sampling its input at the time scheduled by the time schedule function. Fig. 12.7 illustrates how this is done. We see that at the scheduled time, the state is updated to store the value

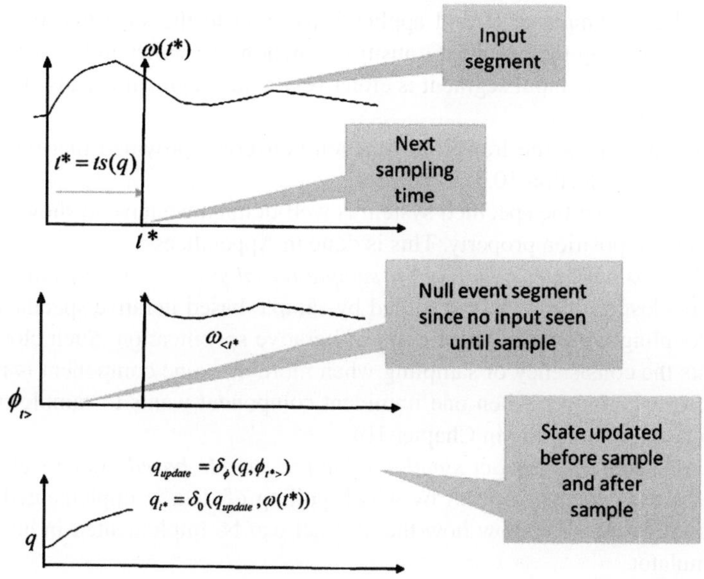

FIGURE 12.7

Definition of sampled-based iterative specification transition function.

of the input segment at that time. Moreover, since there is no effect of the input until that time the effective input segment until that time is set to the null event segment.

The system specified by a sample-based iterative specification is developed as follows:

Theorem 12.3. *A* $SampleIterSpec = (\delta_0, \delta_\Phi, \lambda, \mathrm{ts})$ *specifies a system,*

$$S_M = (\delta, \lambda)$$

where the system transition function

$$\delta : Q \times \Omega \to Q$$

is defined recursively: Let $t^* = \mathrm{ts}(q)$ *be the time to the next sample operation.*

$$\delta(q, \omega) = \delta(q_{t^*}, \omega_{<t^*})$$

where

$$q_{t^*} = \begin{cases} \delta_0(q, \omega(0)) & \text{if } t^* = 0 \quad 1) \\ \delta_\Phi(q, \Phi_{l(\omega)>}) & \text{if } t^* \notin dom(\omega) \quad 2) \\ \delta_0(\delta_\Phi(q, \phi_{t^*>}), \omega(t^*)) & \text{otherwise} \quad 3) \end{cases}$$

Line 1) concerns an immediate sample $\omega(0)$ taken and applied as input to the current state. Line 2) concerns the case where the next sample is scheduled to occur beyond the end of the current input segment. Line 3) takes a sample at t^* and applies it as input to the state that pertains at this time which is obtained by applying the extended transition function. As shown in Fig. 12.11 at t^* the input sample is taken. Until then the input segment is effectively a null segment because the model does not experience the actual input.

The recursive definition for the transition function converges provided that the time advance is *progressive* as discussed in Section 10.5.

To complete the proof that the specified system is well-defined we have to show that the transition function satisfies the composition property. This is done in Appendices.

We also consider *closure under coupling for sample-based systems* in Appendices. Closure under coupling here refers closure of the class specified by sample-based iterative specification as opposed to closure under coupling within the larger class of iterative specification. Such closure is important here because it tests the consistency of sampling when more than one component is involved in feedback coupling and/or when more when one imminent component wants to sample from another (cf. discussion of closure under coupling in Chapter 10).

Appendices also sketch an *abstract simulator for the sample-based systems* class. The abstract simulator formalizes the two-step process by which pulling of data is implemented by request/wait protocol as discussed earlier. We show how the abstract can be implemented in an extension of the DEVS abstract simulator.

12.5 HYBRID ITERATIVE SYSTEM SPECIFICATION

A Hybrid Iterative System Specification, HybridIterSpec, is a Scheduled Iterative System Specification, with additional condition/action specification for state events.

HybridIterSpec is defined by

$$G = (\delta, \lambda, ts, C, A)$$

where

$\Omega_G^X = DEVS(X)$, *the DEVS generator segments on X*
$\delta : Q \times \Omega_G^X \to Q$ *is the generator transition function, with the composition property*
$\lambda : Q \to Y$ *is the output function*
$ts : Q \to \mathbb{R}_0^{+\infty}$ *is the schedule function,* $\mathrm{ts}(q) =$ *time to next internal event.*
$C : Q \to Bool$, *the state event predicate*
$A : Q \to Q$, *the action function*

Theorem 12.4. *A HybridIterSpec,* $G_{ts} = (\delta_{\mathrm{ts}}, \lambda, \mathrm{ts}, C, A)$ *specifies an IterSpec* $G = (\delta, \lambda)$.

Proof. The main task is to define the IterSpec transition function in terms of the given HybridIterSpec transition function.

Given $\omega \in \Omega_X, \delta(q, \omega)$ is defined recursively as follows:

First let $\tau(q, \omega) = min\{t | C(\delta(q, \omega_{t>})) = true\}$, time to the next *state event*. In hybrid terms, a state event occurs when conditions enabling an action become true. Such an event is distinguished from a *scheduled event* due to the schedule function. The scheduled time associated with state q is $ts(q)$ Thus the time of the actual event to occur, from state q with input ω, is

$$t^* = min\{\text{ts}(q), \tau(q, \omega), l(\omega)\}$$

Then

$$\delta_G(q, \omega) = \delta_G(q_{t^*}, \omega_{<t^*})$$

where

$$q_{t^*} = \begin{cases} \delta_{\text{ts}}(q, \omega_{t^*>}) & \text{if } t^* = \text{ts}(q) \quad 1) \\ A(\delta_{\text{ts}}(q, \omega_{t^*>})) & \text{if } t^* = \tau(q, \omega) \; 2) \\ \delta_{\text{ts}}(q, \omega) & \text{otherwise} \quad\quad 3) \end{cases}$$

Lines 1) and 2) are the cases where t^* *occurs within the input segment.* Line 1) is *the case where the scheduled time occurs first, and the left segment up to* t^* is applied as input to get the state, q_{t^*} that pertains at this time. *The left segment exists due the propriety of* Ω_X. Then the remaining right segment is applied *recursively* to get the state at the end of the overall segment. Line 2) is *the case where the state event occurs first. This case is similar to line 1) except that after the state is updated to* q_{t^*} *the action function is applied to it to get the new state at that time.* Line 3) concerns *the case where the next event is scheduled to occur at or beyond the end of the overall segment.* In this case, the overall segment is applied.

The recursive definition for the transition function converges provided that the time advances are *progressive* as discussed in Section 10.5.

We also need to show that it has the composition property. This can be done by induction on the number of state events. Consider $P(n) \equiv$ the composition property holds for all input segments containing n state events. Now $P(0)$ follows from the composition property of δ_{ts}. Assume $P(n)$ and consider $P(n+1)$. Then for any segment with $n+1$ state events, the state at the nth state event is uniquely defined independently of any composition of subsegments until the nth state event. But now that state can be uniquely "projected" to the last state event because there are no intervening state events and the composition property of δ_{ts} applies in the interval between the two state events. □

Exercise 12.9. Give a formal inductive proof of the composition property.

The output function is treated in the same manner as with the schedule iterative specification.

12.5.1 EXAMPLE OF HYBRID BARREL FILLING ITERATIVE SPECIFICATION

To illustrate hybrid iterative system specification, we discuss an example of barrel filling from Chapter 9 and analyze its typical behavior.

A hybrid specification of barrel filling HybridIterSpec is defined by

$$G = (\delta, \lambda, \text{ts}, C, A)$$

where

- $X = \{fillRate\} = \mathbb{R}_0^+$, set the rate with external events,
- $Q = \{level\} \times \{fillRate\}$ where $\{level\} = \mathbb{R}_0^+$, *the state is a pair of dynamic level and a stored fill rate.*

- Y={"full"}, *output when full,*
- $\delta : Q \times \Omega_G^X \rightarrow Q$ *is the generator transition function,*
- $\delta(level, r', r_{t>}) = \delta(level, r, \phi_{t>})$, *external event sets rate* with r the *fillRate*,
- $\delta(level, r, \phi_{t>}) = (level + r_{t>}, r)$, with $r \geq 0$ the rate of filling,
- $ts(level, r) = \begin{cases} 2 & \text{if } level < 10 \\ 0 & \text{otherwise} \end{cases}$

Note that the time advances by constant time steps until the barrel is full.

$$C(level, r) = [level = 10], \text{ recognizes a state event as occurring when full}$$
$$A : Q \rightarrow Q, \text{ action function}$$
$$A(level, r, \phi) = \begin{cases} (level, r) & \text{if } level < 10 \\ (0, 0) & \text{otherwise} \end{cases}$$

This expresses the rule that there is no state action when filling; Once full, the action is to empty the container and stop the filling.

$$\lambda : Q \rightarrow Y$$
$$\lambda(level, r) = \begin{cases} \phi & \text{if } level < 10 \\ full & \text{otherwise} \end{cases}$$

The output function outputs a "full" symbol when the state event occurs.

Exercise 12.10. Verify that δ satisfies the composition property.

We generate the state trajectory as illustrated in Fig. 12.8. The external event is applied immediately storing a new rate,

$$\delta(level, r', r_{t>}) = \delta_G(level, r, \phi_{t>})$$

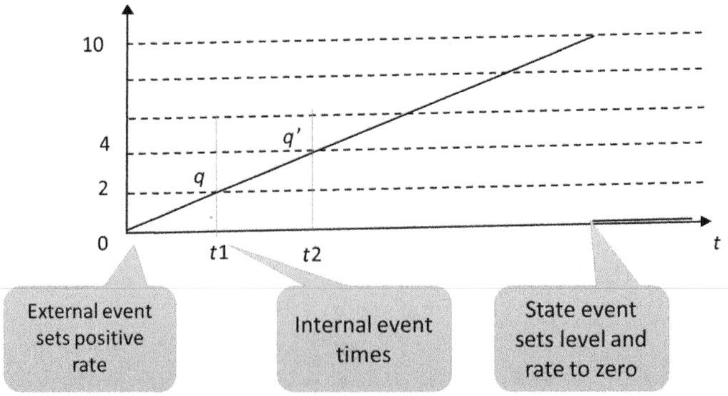

FIGURE 12.8

Typical state trajectory of barrel filling hybrid model.

then $\delta_G(level, r, \phi_{t>}) = (level + r_{t>}, r)$. Let $\tau_{level,r} = (10 - level)/r$, filling time from level $level$, so $\tau_{0,r} = 10/r$, the time to fill from empty ($level = 0$). Also while $\tau_{x,r} > 2$, we advance with time step 2,

$$level_i + 1 = level_t + 2r$$

i.e., $level_i = 2ri$ *The state event occurs when* $C[level, r] \equiv level = 10$, *so*

$$\delta((0, r), \Phi_{10/r>}) = A(10, r) = (0, 0)$$

and filling stops and we remain at empty until a next external event The output event when full is

$$\lambda(10, r) = \text{"full"}$$

12.6 COUPLED ITERATIVE SPECIFICATIONS

Although closure of coupling holds for DEVS, the generalization to iterative specification more generally does not immediately follow. The problem is illustrated in Fig. 12.9 where two iterative system specifications, ISP1 and ISP2, are cross-coupled such as exemplified by Fig. 10.11 of Chapter 10. The cross-coupling introduces two constraints shown by the equalities in the figure, namely, the input of ISP2 must equal the output of ISP1, and the input of ISP1, must equal the output of ISP2. Here we are referring to input and output trajectories over time as suggested graphically in Fig. 12.10. Since each system imposes its own constraints on its input/output relation, the conjunction of the four constraints (2 coupling-imposed, 2 system-imposed) may have zero, one, or multiple solutions.

Definition 12.4. A *Coupled Iterative Specification* is a network specification of components and coupling where the components are iterative specifications at the I/O System level.

As indicated, in the sequel we deal with the simplified case of two coupled components. However, the results can be readily generalized with use of more complex notation. We begin with a definition of the relation of input and output segments that a system imposes on its interface. We need this definition to describe the interaction of systems brought on through the coupling of outputs to inputs.

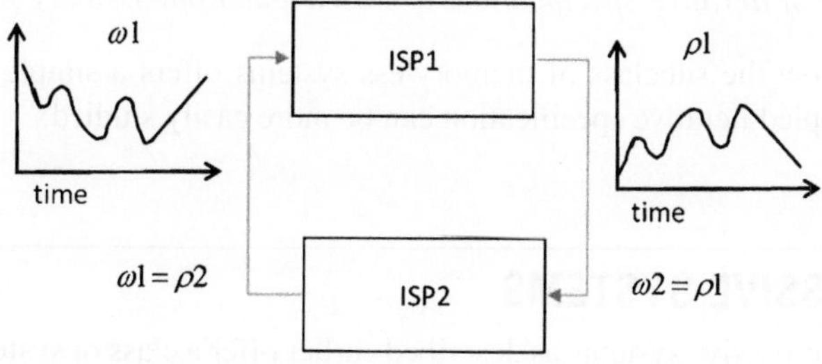

FIGURE 12.9

Input and output trajectories of coupled systems.

Definition 12.5. The I/O Relation of an iterative specification is inherited from the I/O Relation of the system that it specified. Likewise, the set of I/O Functions of an iterative specification is inherited from the system it specifies.

Formally, let $S = (\Delta, \Lambda)$ be the system specified by an IterSpec $G = (\delta, \lambda)$.
Then the I/O Functions associated with G are given by

$$\beta : Q \times \Omega_G^{X+} \rightarrow \Omega_G^{Y+}$$

where for $q \in Q, \omega \in \Omega_G^X$

$$\beta(q, \omega) = \Lambda(\Delta(q, \omega))$$

(See Definition of I/O Behavior in Chapter 5.)

Let β_1 and β_2 represent the I/O functions of the iterative specifications, $G1$ and $G2$, resp. Applying the coupling constraints expressed in the equalities above, we make the definition:

Definition 12.6. A pair of output trajectories (ρ_1, ρ_2) is a consistent output trajectory for the state pair (q_1, q_2) if $\rho_1 = \beta_1(q_1, \rho_2)$ and $\rho_2 = \beta_2(q_2, \rho_1)$.

Definition 12.7. A Coupled Iterative Specification has *unique solutions* if there is a function,

$$F : Q_1 \times Q_2 \rightarrow \Omega \text{ where}$$
$$F(q_1 . q_2) = (\rho_1, \rho_2)$$

where there is exactly one consistent pair (ρ_1, ρ_2) with infinite domain for every initial state (q_1, q_2). The infinite domain is needed for convenience in applying segmentation.

Definition 12.8. A Coupled Iterative Specifications is *admissible* if it has unique solutions.

The following theorems are proved in Appendices of this chapter.

Theorem 12.5. *An admissible Coupled Iterative Specification specifies a well-defined Iterative Specification at the I/O System level.*

Theorem 12.6. *The set of Iterative Specifications is closed under admissible coupling.*

Appendices show how the subclass of memoryless systems offers a simple context in which the well-definition of a coupled iterative specification can be more easily studied.

12.7 ACTIVE-PASSIVE SYSTEMS

We now show that active-passive systems as described earlier offer a class of systems for which the iterative specifications of components satisfy the admissibility conditions specified in the above Theorem. As in Fig. 12.10 consider a pair of cross-coupled systems each having input generators that represent null and non-null segments.

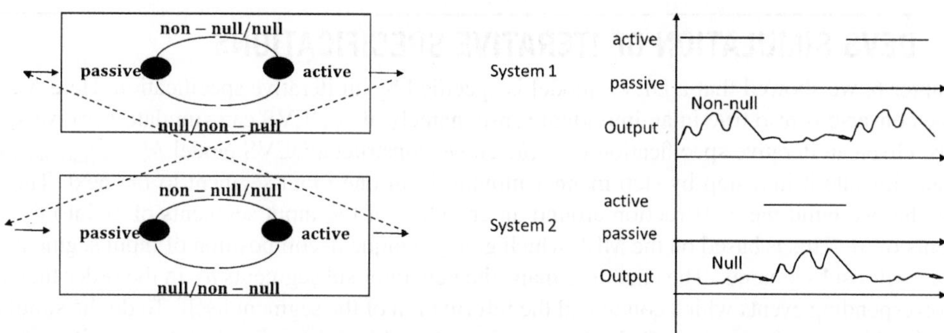

FIGURE 12.10

Active-Passive Example of Admissible Progressive Coupled Iterative Specifications.

A null generator represents the output of a passive system whereas a non-null generator represents the output of an active system. For example in the Turing Machine (TM) case, a non-null generator is a segment that represents transmission of a symbol by the tape unit and of a symbol, move pair for the TM control. As in Fig. 12.10B let $S1$ and $S2$ start as active and passive, respectively. Then they output non-null and null generators respectively. Since the null generator has infinite extent, the end-time of the non-null generator determines the time of next event, t^* (as defined earlier) and we apply the transition functions in Fig. 12.6A to find that at t^* the systems have reversed phases with S1 in passive and S2 in active.

Define a triple $\langle Q, \delta, \lambda \rangle$ by

$$Q = \{\text{active, passive}\} \times \{\text{active, passive}\}$$

with

$$t^*(active, passive) = \text{endtime of the non-null generator segment}$$

and

$$F(\text{active, passive}) = (\text{non-null, null}).$$

$$\delta(\text{active, passive}) = (\Delta(\text{active, null}_{t^*>}), \Delta(\text{passive, non-null}_{t^*>}) = (\text{passive, active}).$$

$$\Lambda(\text{active, passive}) = (\text{non-null, null})$$

Exercise 12.11. Let the non-null and null generators stand for sets of concrete segments that are equivalent with respect to state transitions and outputs. Then show that such a scheme can define a deterministic or non-deterministic resultant depending on the number of consistent pairs of (non-null, null) generator segments possible from the (active, passive) state, and *mutatis-mutandis* for the state (passive, active).

12.8 DEVS SIMULATION OF ITERATIVE SPECIFICATIONS

In Chapter 6, we showed that a DEVS model is specified by an iterative specification. Here we show that the converse is also true in an important sense, namely, that DEVS can simulate iterative specifications. Given an iterative specification $G = (\delta, \lambda)$, we construct a DEVS model $M = (\delta_{\text{int}}, \delta_{\text{ext}}, \lambda, \text{ta})$ that can simulate it in a step-by-step manner moving from one input segment to the next. The basic idea is that we build the construction around an encoding of the input segments of G into the event segments of M. This is based on the MLS which gives a unique decomposition of input segments to G into generator subsegments. The encoding maps the generator subsegments ω_i in the order they occur into corresponding events which contain all the information of the segment itself. To do the simulation, the model M stores the state of G. It also uses its external transition function to store its version of $G's$ input generator subsegment when it receives it. M then simulates G by using its internal transition function to apply its version of $G's$ transition function to update its state maintaining correspondence with the state of G.

We note that the key to the proof is that iterative specifications can be expressed within the explicit event-like constraints of the DEVS formalism, itself defined through an iterative specification. Thus DEVS can be viewed as the computational basis for system classes that can be specified in iterative form satisfying all the requirements for admissibility.

The statement that a DEVS atomic model can simulate an Iterative Specification is formalized and proved in Appendix 12.G.

Theorem 12.7. *A DEVS coupled model can component-wise simulate a coupled Iterative Specification.*

Proof. The coupled model has components which are DEVS representations of the individual Iterative Specs (according to the previous theorem) and also a coordinator as shown in Fig. 12.12. The coordinator receives the current states of the components and applies the F function to compute unique consistent output segments. After segmentation using the MLS as in the previous proof, it packages each as a single event in a DEVS segment as shown. Each DEVS component computes the state of its Iterative Specification at the end of the segment as in the theorem. Then it sends this state to the coordinator and the cycle repeats. This completes an informal version of the proof which would formally proceed by induction. □

The solution function F represents an idealization of the fixed point solutions required for DESS and the local solution approaches of DTSS (Chapter 6) and Quantized DEVS (Chapter 19). GDEVS polynomial representation of trajectories (Giambiasi et al., 2001) is the closest example of such representation but the approach opens the door to realization by other trajectory prediction methods.

12.9 SUMMARY

Table 12.1 summarizes the iterative system specification and its specializations that were discussed in this chapter. Also shown are aspects of such specifications that are relevant to their well definition. The composition property is a required condition for the well definition of any system transition function. Closure under coupling within a subclass of systems is not a required condition but tests the correctness of the interactions and can be informative in establishing the power of representations (cf. DEVS

Table 12.1 Iterative System Specifications and their properties

	Composition Property	Closure under Coupling	Abstract Simulator
Iterative System Specification			Shows how DEVS can simulate iterative specifications
Scheduled Iterative Specification	Checks the well definition of the transition function under scheduling		
Sample Based Iterative Specification		Checks the correct functioning of feedback coupled sample-based systems	Shows how to implement sampling by extending the DEVS abstract simulator
Hybrid Iterative Specification	Checks the well definition of the transition function under state events.		

Markov models in Chapter 21). The abstract simulator shows how to implement specific behaviors of the class.

APPENDIX 12.A PROOF THAT DEVS IS A SCHEDULED ITERATIVE SYSTEM SPECIFICATION

The system specified by a $DEVS_M = (\delta_{\text{ext}}, \delta_{\text{int}}, \lambda_M, \text{ta})$, $G(M) = (\delta, \lambda)$, is more particularly specified by a

$$\text{SchedIterSpec } G_{sched}(M) = (\delta_{sched}, \lambda, \text{ts}).$$

Proof. We define $G(M)$ to have the same sets as G_{sched}. Thus the generators of $G(M)$ are defined

$$\Omega_X = \langle DEVS_G(X)\rangle = \Omega_X \cup \Omega_\Phi$$
$$\Omega_X = \{x_{t>} | x \in X, t \in \mathbb{R}^+_{0,\infty}\}$$
$$\Omega_\Phi = \{\Phi_{t>} | t \in \mathbb{R}^+_{0,\infty}\}$$
$$\Omega_Y = \langle DEVS_G(Y)\rangle$$

We define the same state set:

$$Q = \{(s, e) | s \in S, 0 \le e \le \text{ta}(s))\}$$

and the response to external events:

$$\delta_{sched}(s, e, x_{t>}) = \delta_{sched}(\delta_{\text{ext}}(s, e, x), 0, \phi_{t>})$$

which agrees with δ.

Then the schedule function for G_{sched}, ts : $Q \to \mathbb{R}_0^{+\infty}$ is defined by:

$$\text{ts}(s, e) = \text{ta}(s) - e$$

Now the main task is to define the internal transition function based on ts. Given $\phi_{t>} \in \Omega_\Phi$, let $t^* = \text{ts}(s, e) = \text{ta}(s) - e$ Define δ_{sched} to match δ_G

$$\delta_{Sched}(s, e, \phi_{t>}) = \begin{cases} (s, e+t) & \text{if } e+t \geq \text{ta}(s) \\ \delta_{sched}(\delta_{\text{int}}(s), \phi_{e+t-\text{ta}(s)>}) & \text{otherwise} \end{cases}$$

We note that:

$$t^* \leq t \Leftrightarrow \text{ta}(s) - e \leq t \Leftrightarrow \text{ta}(s) \leq t + e$$

so that $\text{ts}(s, e) \leq t \Leftrightarrow \text{ta}(s) \leq t + e$ and we can redefine

$$\delta_{sched}(s, e, \Phi_{t>}) = \begin{cases} (s, e+t) & \text{if ts}(s, e) \leq t \\ \delta(\delta_{\text{int}}(s), \phi_{t-\text{ts}(s,e))>}) & \text{otherwise} \end{cases}$$

which agrees with the internal event occurring when $\text{ts}(s, e) = t$ at the end of the segment. □

APPENDIX 12.B COUPLED ITERATIVE SPECIFICATION AT THE I/O SYSTEM LEVEL

Theorem 12.8. *An admissible Coupled Iterative Specification specifies a well-defined Iterative Specification at the I/O System level.*

Proof. Given two iterative specifications, G_i and their state sets, Q_i, $i = 1, 2$, let $Q = Q_1 \times Q_2$, eventually the state set of the iterative specification to be constructed. For any pair (q_1, q_2) in Q, let (ρ_1, ρ_2) be a consistent output trajectory for (q_1, q_2), i.e. $\rho_1 = \beta(q_1, \rho_2)$ and $\rho_2 = \beta(q_2, \rho_1)$ and $F(q_1, q_2) = (\rho_1, \rho_2)$.

Define an autonomous Iterative Specification $\langle Q, \delta_G, \lambda \rangle$ by

$$\delta(q_1, q_2) = (\Delta(q_1, \rho_{2_{t^*(q_1,q_2)>}}), \Delta(q_2, \rho_{1_{t^*(q_1,q_2)>}}))$$

where

$$t^*(q_1, q_2) = min\{t_1^*, t_2^*\}$$

where

t_1^*, t_2^* are the times of the MLS for ρ_1 and ρ_2.

In other words, since each of the component iterative specifications have maximum length segmentations we take the time of next update of the constructed specification to be determined by earliest of the times of these segmentations for the consistent pair of output trajectories. This allows us to define a step-wise transition for the constructed transition function. Closure under composition for this transition function can be established using induction on the number of generators in the segments under consideration. Similarly, the output function is defined by:

$$\lambda(q_1, q_2) = (\lambda(q_1), \lambda(q_2)). \qquad \square$$

Theorem 12.9. *The set of Iterative Specifications is closed under admissible coupling.*

Proof. This theorem is a corollary of the previous theorem. □

APPENDIX 12.C COMPOSITION PROPERTY FOR SAMPLE-BASED SYSTEMS

We *must* show *that* composition property holds for the transition function defined for the sample-based iterative specification, i.e.,

$$\delta(q, \omega\overline{\omega}) = \delta(\delta(q, \omega), \overline{\omega})$$

The essence of the proof is that, as illustrated in Fig. 12.11, the breakpoint between ω and $\overline{\omega}$ occurs at the sampling points or between them. If the latter then using the composition property of δ_Φ we continue the state from the sample point to the breakpoint and to the next sample point.

Exercise 12.12. Prove the case for the case of the breakpoint at a sample point.

APPENDIX 12.D CLOSURE UNDER COUPLING OF SAMPLE-BASED ITERATIVE SPECIFICATION

Closure under coupling here refers closure of the class specified by sample-based iterative specification as opposed to closure under coupling within the larger class of iterative specification. Such closure is important here because it tests the consistency of sampling when more than one component is involved in feedback coupling and/or when more when one imminent component wants to sample from another.

For considering such issues in the context of closure under coupling, we consider the simplified case of two Sample-Based Iterative Specifications with crosscoupling (as in Section 10.6). Consider two SampleIterSpecs:

$$G_i = (\delta_{\Phi,i}, \lambda_i, \mathrm{ts}_i), i = 1, 2.$$

Define a structure to be shown to be an input free iterative specification

$$M_S = < Q_S, \delta_S, \mathrm{ta} >$$

where

$$Q_S = Q_1 \times Q_2$$
$$\mathrm{ta}(q_1, q_2) = min\{\mathrm{ts}_1(q_1), \mathrm{ts}_2(q_2)\} = t^*$$

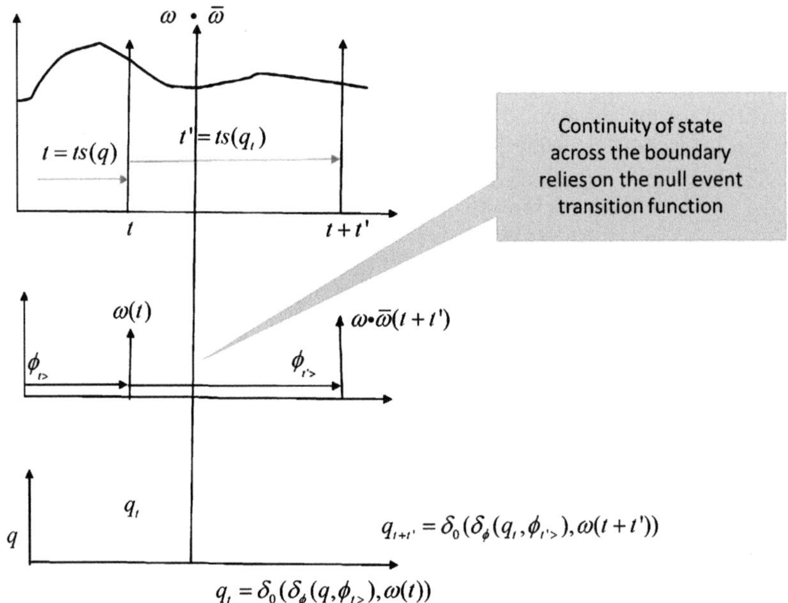

FIGURE 12.11

Composition property for Sample-based Systems.

Consider a single imminent

$$Imm = \{i | \mathrm{ts}_i = t^*\}$$

Let

$$\begin{aligned}
&Imm = \{1\},\ \text{the component that samples the other's output}\\
&q_{t^*>,1} = \delta_{\Phi,1}(q_1, \phi_{t^*>}),\ \textit{the state of } G_1 \textit{ just before the sample}\\
&q_{t^*>,2} = \delta_{\Phi,2}(q_2, \phi_{t^*>}),\ \textit{the state of } G_2 \textit{ just before the sample}\\
&q_{t^*,1} = \delta_{0,1}(q_{t^*>,1}, \omega_1(t^*)),\ \textit{the state of } G_1 \textit{ just after the sample}
\end{aligned}$$

where

$$\omega_1(t^*) = \lambda_2(q_{t^*>,2})$$

and

$$\delta_S(q_1, q_2) = (q_{t^*,1}, q_{t^*>,2})$$

Note the state of the sampled component, $i = 2$, has been updated to its value but that due to preservation of schedule under update

$$\mathrm{ts}_2(q_{t^*>,2}) = \mathrm{ts}_2(\delta_{\Phi,2}(q_2, \phi_{t^*>})) = \mathrm{ts}_2(q) - t^*$$

which is the same schedule time that would obtain had it not been updated.

This case easily extends to $Imm = \{1, 2\}$ since both outputs depend on state just before sampling time.

Exercise 12.13. Extend the proof to the simultaneous sampling case.

So far we have created an input free system that cannot be a sample-based iterative specification because of the obvious absence of input. To extend the proof to result in a system with input we consider a coupled model with external input to either or both of the components. This situation raises the situation where more than one sample request can arrive from different sources and more than one port can be available for sampling. Both of these situations can be handled with the same approach as given in the proof.

Exercise 12.14. Consider the general case of a coupled model of two sample-based iteratively specified systems. Prove that the resultant is can be specified by a sample-based iterative system specification.

APPENDIX 12.E ABSTRACT SIMULATOR FOR SAMPLE-BASED ITERATIVE SPECIFICATION

While the semantics of pull sampling can be expressed within DEVS, the implicit version in Fig. 12.4D does not require extra operations. We show how the sampling performed by model B in Fig. 12.4D can be implemented in an extension of the DEVS abstract simulator. In Fig. 12.12, let

$$C = \text{coordinator for coupled modeling}$$

The steps that must be taken "behind the scenes" to get the implicit pull implemented are outlined as:

1. C gets t_N's from S_A and S_B, takes the minimum, t^*, and finds that B is imminent
2. C consults the coupling to find that A is the influencee of $B's$ input
3. C tells simulator S_A to get $A's$ output with time t^*,
4. S_A applies $A's$ transition function to get the state at t^*,

$$q_{t^*>,2} = \delta_{\Phi,2}(q_2, \phi_{t^*>}) \text{ and}$$

output function to get the output,

$$\omega_1(t^*) = \lambda_2(q_{t^*>,2})$$

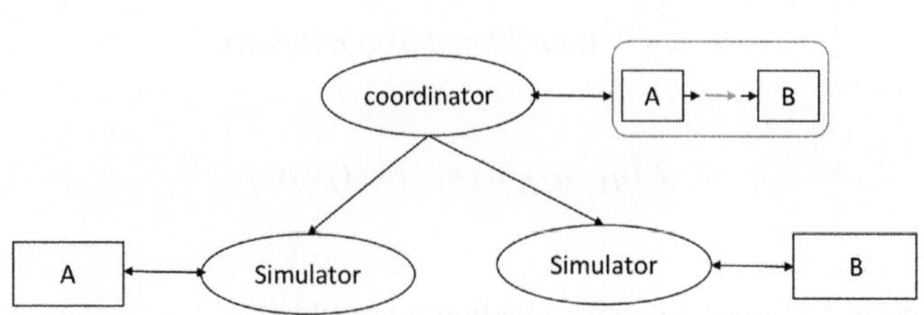

FIGURE 12.12

Abstract Simulator for Sample-based Iterative Specification.

5. S_A sends the sampled value, $\omega_1(t^*)$ to C
6. C tells S_B to execute $B's$ external transition function with input $\omega_1(t^*)$ at time t^*

Exercise 12.15. Design an abstract simulator for a general coupled model of sample-based iteratively specified systems by extending the given approach and explicitly employing the coupling information available to the coupled model. Prove that your design is correct in that all properties of the sample-based iterative specification are satisfied.

APPENDIX 12.F EXAMPLE OF CLOSURE UNDER COUPLING: MEMORYLESS SYSTEMS

Consider the case where each component's output does not depend on its state but only on its input. Let gr represent the ground state in which the device is always found (all states are represented by this state since output does not depend on state). In the following R_1 and R_2 are the I/O relations of systems 1 and 2, resp. In this case they take special forms:

$$\rho_1 = \beta_1(gr, \omega_1) \Leftrightarrow (\omega_1, \rho_1) \in R_1$$
$$\rho_2 = \beta_2(gr, \omega_2) \Leftrightarrow (\omega_2, \rho_2) \in R_2$$
$$\omega_2 = \rho_1$$
$$\omega_1 = \rho_2$$
$$\rho_2 = \beta_2(gr, \omega_2) \Leftrightarrow (\rho_1, \omega_1) \in R_2 \Leftrightarrow (\omega_1, \rho_1) \in R_2^{-1}$$
$$\text{i.e., } (\rho_2, \rho_1) \textit{ is consistent } \Leftrightarrow (\rho_2, \rho_1) \in R_1 \cap R_2^{-1}$$

Let f and g be defined in the following way

$$f(\rho_2) = \beta 1(gr, \rho_2)$$
$$g(\rho_1) = \beta_2(gr, \rho_1)$$

so for any ρ_1,

$$\rho_1 = f(\rho_2) = f(g(\rho_1))$$

so

$$g = f^{-1}(\textit{considered as a relation})$$

and

$$F(q_1 \cdot q_2) = (\rho_1, f^{-1}(\rho_1))$$

Therefore, $F(q_1 \cdot q_2)$

- has no solutions if f^{-1} does not exist, yielding no resultant
- has a unique solution for every input if f^{-1} exists, yielding a deterministic resultant and
- has multiple solutions for a given segment ρ_1 if $f^{-1}(\rho_1)$ is multivalued, yielding a non-deterministic resultant

Exercise 12.16. Consider an adder that always adds 1 to its input, i.e.,

$$\beta(gr, \rho) = f(\rho) = \rho + 1$$
$$\text{i.e., } \forall t \in T, \beta(gr, \rho)(t) = \rho(t) + 1$$

Show that cross-coupling a pair of adders does not yield a well-defined resultant. However, coupling an adder to a subtracter, its inverse yields a well-defined deterministic system.

Exercise 12.17. Consider combinatorial elements whose output is a logic function of the input and consider a pair of gates of the same and different types (AND, OR, NOT) connected in a feedback loop. Characterize the resultants as underdetermined, deterministic or non-deterministic.

APPENDIX 12.G PROOF THAT A DEVS ATOMIC MODEL CAN SIMULATE AN ITERATIVE SPECIFICATION

Theorem 12.10. *Given an iterative specification there is a DEVS model that can simulate it according to the definition of system simulation in Chapter 5.*

Proof. Given an iterative specification $G = (\delta, \lambda)$, we construct a DEVS model

$$M = (\delta_{\text{int}}, \delta_{\text{ext}}, \lambda, \text{ta})$$

$g : \Omega_G^X \to \Omega_{DEVS}^X$ maps the generators of G into the DEVS generator segments as defined by:

$$g(\omega) = [\omega]_{l(\omega)>} = [\omega]\phi_{l(\omega)>}$$

where as in Fig. 12.13 in, $[\omega]$ denotes the actual segment encased in braces for easy reading. Thus the length of $g(\omega)$ is the length of ω and its initial value is ω itself treated as a unitary object.

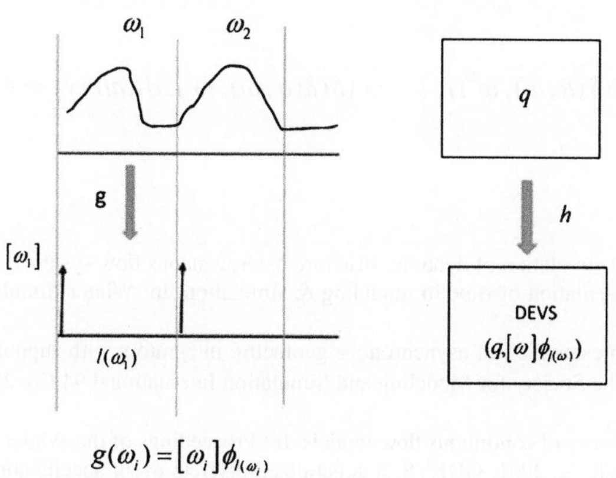

FIGURE 12.13

DEVS Simulation of an Iterative Specification.

Definition of the simulating DEVS:

$$S = Q \times \Omega_G$$
$$\delta_{\text{int}}(q, \omega) = (\delta(q, \omega), dummy)$$
$$\text{ta}(q, \omega) = l(\omega)$$
$$\delta_{\text{ext}}((q, \omega), e, \omega') = (\delta(q, \omega_{e>}), \omega')$$
$$\lambda(q, dummy) = \lambda(q)$$

The simulation map is defined by:

$$h : Q \to Q \times \Omega_G h(q) = (q, dummy)$$

Now, in the Iterative Specification, the transition proceeds from one generator segment to the next:

$$\delta^+(q, \omega\omega') = \delta(\delta(q, \omega), \omega')$$

We want the simulating DEVS to mimic this process. Let δ_{DEVS} be the transition function of the system specified by the DEVS. We will show that the homomorphism holds:

$$h(\delta^+(q, \omega\omega')) = \delta_{DEVS}(h(q), g(\omega\omega'))$$

$$\begin{aligned}
\delta_{DEVS}(h(q), g(\omega\omega')) &= \delta_{DEVS}((q, dummy), g(\omega)g(\omega')) \\
&= \delta_{DEVS}((q, dummy), [\omega]_{l(\omega)>}, [\omega']_{l(\omega')>}) \\
&= \delta_{DEVS}(\delta_{\text{ext}}((q, dummy), [\omega]), \phi_{l(\omega)>}), \omega'_{l(\omega')>}) \\
&= \delta_{DEVS}((q, [\omega]), \phi_{l(\omega)>}), \omega'_{l(\omega')>}) \\
&= \delta_{DEVS}(\delta_{\text{int}}(q, [\omega]), \phi_{l(\omega)>}), \omega'_{l(\omega')>}) \\
&= \delta_{DEVS}((\delta(q, \omega), dummy), \omega'_{l(\omega')>}) \\
&= (\delta(\delta(q, \omega), \omega'), dummy)
\end{aligned}$$

So

$$h(\delta^+(q, \omega\omega')) = h(\delta(\delta(q, \omega), \omega')) \qquad = (\delta(\delta(q, \omega), \omega'), dummy) = \delta_{DEVS}(h(q), g(\omega\omega')) \quad \square$$

REFERENCES

Barros, F.J., 2002. Modeling and simulation of dynamic structure heterogeneous flow systems. Simulation 77 (1), 552–561.

Barros, F.J., 2016. On the representation of time in modeling & simulation. In: Winter Simulation Conference (WSC), 2016. IEEE, pp. 1571–1582.

Barros, F.J., 2017a. Modular representation of asynchronous geometric integrators with support for dynamic topology. SIMULATION: Transactions of The Society for Modeling and Simulation International 94 (3), 259–274. https://doi.org/10.1177/0037549717714613.

Barros, F.J., 2017b. Towards a theory of continuous flow models. In: Proceedings of the Winter Simulation Conference.

Giambiasi, N., Escude, B., Ghosh, S., 2001. GDEVS: a generalized discrete event specification for accurate modeling of dynamic systems. In: Proceedings of the 5th International Symposium on Autonomous Decentralized Systems. 2001. IEEE, pp. 464–469.

Nutaro, J., Kuruganti, P.T., Protopopescu, V., Shankar, M., 2012. The split system approach to managing time in simulations of hybrid systems having continuous and discrete event components. Simulation 88 (3), 281–298.

CHAPTER

FINITE AND TIMED ITERATIVE SPECIFICATIONS AND SIMULATIONS

13

CONTENTS

Iterative specification can be used to formally study digital systems. On the other hand, digital systems are characterized by a finite set of values allowing real time systems to operate safely based on the automatic verification of their expected behaviors. Restrictions of the iterative specification (to a finite set of values and to timing constraints of event occurrences) are presented here. Their implementation in PDEVS is also described. The whole framework consists of many levels (from more general and formal to less general and formal) that can be used to implement and study digital systems as well as to model classes of system behaviors. The behaviors are caught by system phases leading to a simple graphical representation of timed state transitions.

13.1 TIME MANAGEMENT

Section 2.4 presented a first classification of the different time notions. Here we introduce the time managements used for finite-time and continuous-time specifications. We will see that both notions can be related using a simple time step (or a granule).

Fig. 13.1 describes a combination of some different time management concepts used in simulation and their corresponding time bases:

1. A *real time base* $T_R = \mathbb{N} * \Delta t$, based on a clock of precision $\Delta t \in \mathbb{Q}$, the *time granule* (where each clock tick is an event that can occur or not) (1),
2. A *logical time base* $T_L = \mathbb{N}$ where each element corresponds to the execution of a process (3). Each process is executed sequentially and its duration is measured on a real time base $T'_R =$

Theory of Modeling and Simulation. https://doi.org/10.1016/B978-0-12-813370-5.00022-5

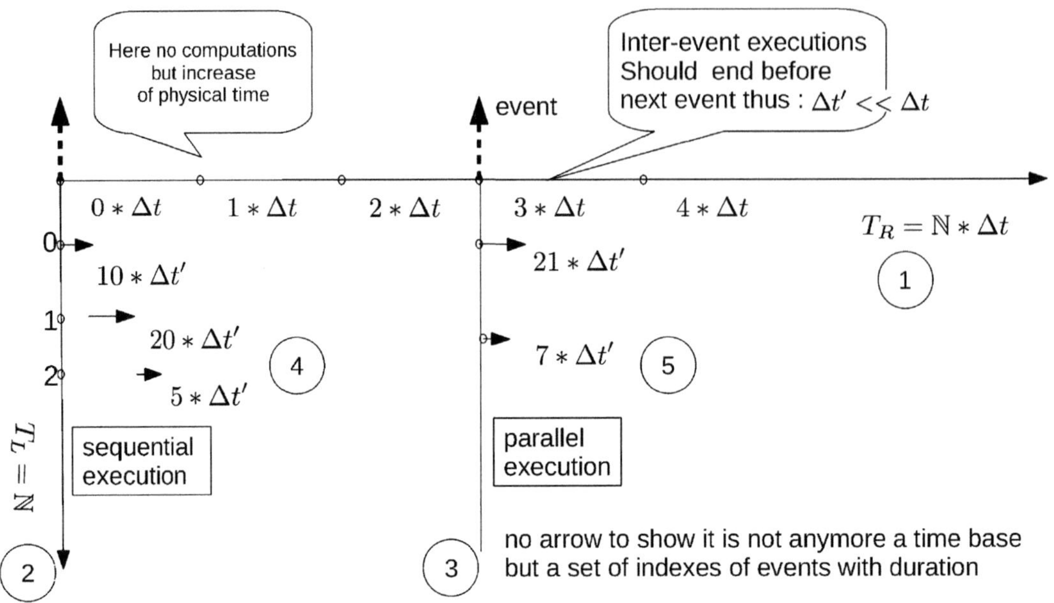

FIGURE 13.1

Representation of the usual time bases in a simulation, with: a real time base T_R (1), a logical time base T_L (2), a set of processes executed in parallel from time step $3 * \Delta t$ (3). The sequential (2) and the parallel execution (3) of the processes based on a real time base $T'_R = \mathbb{N} * \Delta t'$.

$\mathbb{N} * \Delta t'$. To guarantee the temporal consistency of the whole system, the sum of all process durations is required to be less than the time granule Δt of the system.

3. A set of processes executed in parallel from time step $3 * \Delta t$ (3).

The last case consists of a *simulation time base* piloted by the events: $T_S = \mathbb{R}_0^+$. The time base of the events used here is the real time base. However, notice that this real time is not related to the time elapsed in the real world (that requires the use of a clock). The use of nested time bases is related to a *super dense time* (Liu et al., 2006).

Table 13.1 sums up the different iterative specifications and the corresponding time bases used at the implementation level. The last kind of iterative specification is the Finite Timed IterSpec (FinTimedSpec). As we will see this specification adds timing to FinIterSpec thus requiring a clock with *time granule* Δt. Notice that both logical and physical time bases are isomorphic (the former being obtained by a simple product of the latter by Δt).

13.2 BASIC FINITE ITERATIVE SPECIFICATION (FINITERSPEC)

We introduce a finite simplification of iterative specification. This structure supports abstract reasoning (as we saw in the introduction) as well as electronic and computer implementations. At the electronic

Table 13.1 Relationships between time and iterative specification formalisms from an implementation point of view

Specification	Time base	Implementation
IterSpec	$\mathbb{R}_0^{+\infty}$	simulation time
FinIterSpec	$\mathbb{N}$	logical time
FiniTimedSpec	$\mathbb{N} * \Delta t$	physical time

level, the mathematical finiteness meets hardware finiteness. Relying on this level, computers can run automatic model verification, checking exhaustively all the possible states and transitions.

Definition 13.1. A *Finite Iterative Specification (FinIterSpec)* is a structure

$$\text{FinIterSpec} = (\delta_F, \lambda_F)$$

where $\delta_F : P \times \mathbb{N} \times X \rightarrow P \times \mathbb{N}$ is the *finite generator transition function* and $\lambda_F : P \times \mathbb{N} \times X \rightarrow Y$ is the *finite output function*, with P a *finite set of phases* and (p, e) the *set of total states* representing the *elapsed time* $e \in \mathbb{N}$ in phase $p \in P$, X the *finite set of input values* and Y the *finite set of output values.*

A Finite Iterative Specification, $FinIterSpec = (\delta_F, \lambda_F)$ is an Iterative Specification, $IterSpec = (\delta, \lambda)$, where $\delta :\ Q \times \Omega_G \rightarrow Q$ and Q is the set of total states just defined, and

$$\delta((p,e), \omega(0)\omega(1)...\omega(n)) = \begin{cases} \delta((p, e+1), \omega(1)...\omega(n)) & if\ \omega(0) = \phi \quad 1) \\ \delta(\delta_F((p,e), \omega(0)), 0), \omega(1)...\omega(n)) & otherwise \quad 2) \end{cases}$$

Note that since a generator $\omega \in \Omega_G$ is defined on the natural number time base $\mathbb{N}$, we can express $\omega :< 0, n > \rightarrow X$ as a sequence $\omega(0)\omega(1)...\omega(n)$. Line 1 applies when the first input is null; it keeps the phase fixed and increments the elapsed time up by 1. Line 2) applies when the first input is not null; it applies the transition function δ_F to the total state (p, e) and the first input, then reseting the elapsed time clock. Both lines use recursion to define the effect of the remaining sequence. Since the input segment is eventually fully consumed the recursion always terminates.

Exercise 13.1. Explain what is different about non-finite iterative specifications that makes termination problematic.

Exercise 13.2. Define the output function of the iterative specification and show that it satisfies the 3 constraints of Definition 12.1.

Choosing the time base T as the set of natural numbers $\mathbb{N}$ guarantees a finite number of transition over a finite duration. Fig. 13.2 presents a phase transition diagram of a finite iterative specification. The figure describes a generic two-phase finite iterative specification alternating between phases "active" and "inactive". (Recall active/passive compositions in Chapter 10.) A phase is attached to a generator class defined by constraints on the values (greater or lower and equal to a threshold θ) and/or on the durations (lower or equal to a duration T). As we saw in Chapter 11, constraints can be combined to generate complex trajectories. For example, in transition 2, "non-null" generators have values greater than θ and durations lower or equal to T). Notice that this approach is very general allowing constraints to be applied and combined on segments providing that the generated set is proper.

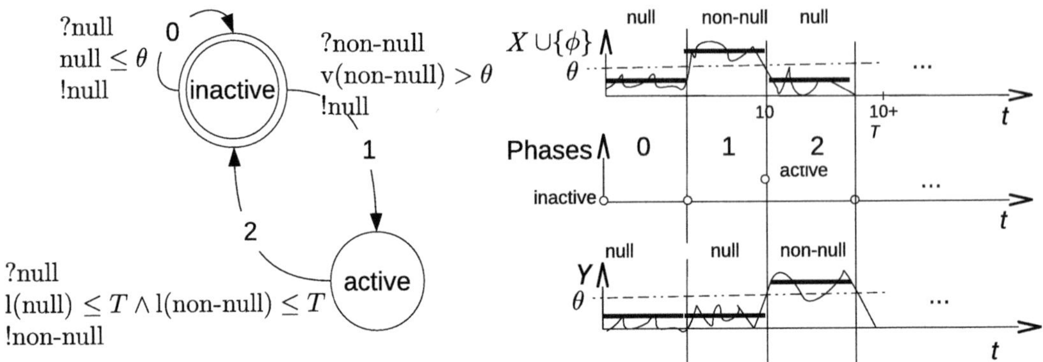

FIGURE 13.2

State transition diagram of a generic two-phase finite iterative specification (on the left) with corresponding generated trajectories (on the right). Transitions are numbered 0, 1 and 2. Constraints on input/output generators are indicated on transition labels, appearing between input labels and output labels. Input labels begin with a symbol '?'. Output labels begin with a symbol '!'.

A transition concerns a class of generators (simply indicated by a name, e.g., "null") and possibly a constraint on the generators of the class. For example, "l(null) $\leq T$" represents the set of segments $\Omega_G^{\text{null},\leq T} = \{\omega \mid c(\omega) = true\}$ with $c(\omega) = \begin{cases} true & \text{if } l(\omega) \leq T \\ false & otherwise \end{cases}$.

13.3 FINITE PDEVS

A finite iterative specification can be simulated by a Finite PDEVS:

Definition 13.2. A *Finite PDEVS* (*FiniPDEVS*) is a structure

$$\text{FiniPDEVS}=(\delta_{ext}, \delta_{int}, \delta_{con}, \lambda, ta)$$

where $\delta_{ext} : Q \times X^b \to S$ is the *external transition function*, with $Q = \{(s, e) \mid s \in S, 0 \leq e \leq ta(s)\}$, the *finite set of total states*, $e \in \mathbb{N}$ the *elapsed time since the last transition*, S the *finite set of states*, X^b the *finite set of bags over input values*, $ta : S \to \mathbb{N}$ the *time advance function*, $\delta_{int} : S \to S$ the *internal transition function*, $\delta_{con} : X^b \times S \to S$ the *confluent transition function*, $\lambda : S \to Y^b$ the *output function*, and Y^b the *finite set of bags over the output values*.

Now that since we employ a finite state set at the PDEVS level, it is possible to represent it as a state transition diagram (cf. Fig. 13.3). The objective of this type of transition diagram is to *propose a graphical representation of states and corresponding timed transitions with inputs/outputs representation that is as simple as possible*. Here, starting from initial state A, the two first kinds of *timed transition* are defined by:

1. *Internal transition*: Outputting the event value $!z$ at time $ta(A)$,
2. *External transition*: Receiving the event value $?v$ at *elapsed time* $e < ta(A)$.

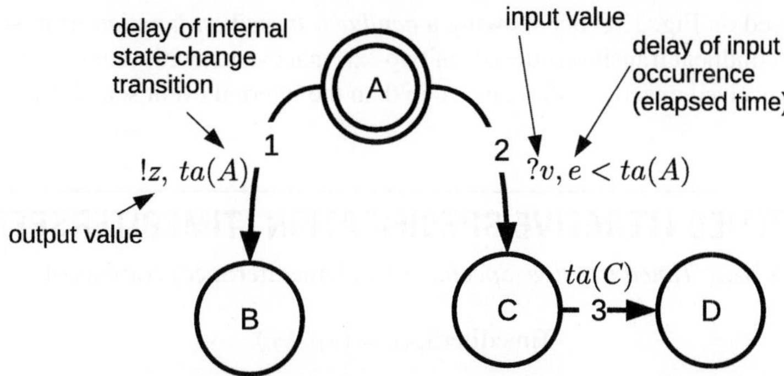

FIGURE 13.3

State transition diagram of a FiniPDEVS: A fundamental example where the input event changes the schedule (only the state).

Conceptually, the first transition choice (between transitions 1 and 2) represents the fundamental mechanism of a timed state transition in PDEVS: Phase A will last $ta(A)$ (transition 1) *unless* an event occurs before $ta(A)$ is *consumed* (transition 2). Transition 2 brings the FiniPDEVS immediately to a new state C at $e < ta(A)$. Finally, in phase C, if the event is expected to *change the schedule*, the FiniPDEVS will go to a new state D after $t + ta(C)$.

Fig. 13.4 is based on Fig. 13.3 with state C being now in transition 3, an intermediate state lasting the duration remaining from the initial scheduled time $ta(A)$: $ta(C) = ta(A) - e$, where e is the time elapsed in state A. In Fig. 13.3 the initial scheduled time $ta(A)$ was changed to a new value $ta(C)$.

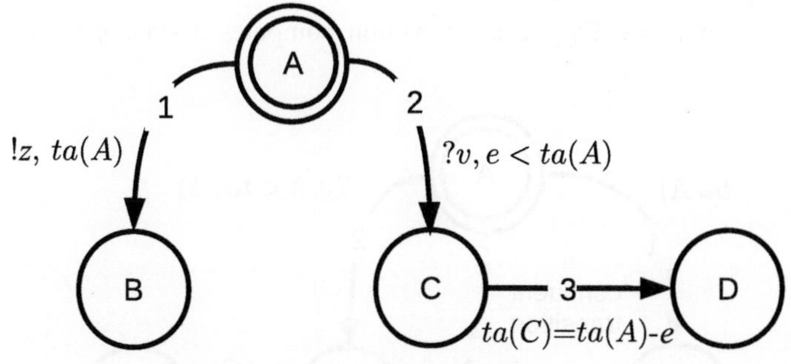

FIGURE 13.4

State transition diagram of a FiniPDEVS: A fundamental example where the input event does not change the schedule.

Fig. 13.5 is based on Fig. 13.3 but showing a *confluent transition function* from state A to state C (transition 4). The confluent transition consists of two external events, with values $?u$ and $?v$ (embedded in the input bag), received exactly at the same time than the internal event scheduled at time $ta(A)$.

13.4 BASIC TIMED ITERATIVE SPECIFICATION (TIMEDITERSPEC)

Definition 13.3. A basic *Timed Iterative Specification* (TimedIterSpec) consists of:

$$\text{TimedIterSpec} = (\delta_T, \lambda_T)$$

with $\delta_T : Q \times \Omega_G \to Q$ the *timed generator transition function*, Q the *set of possibly infinite states*, $\Omega_G = \cup_{i \in \{0,...,n\}} \Omega_{G_i}$ the *set of timed generators* with $\Omega_{G_i} = \{\omega \mid c_i(\omega)\}$, with $c_i(\omega) = \begin{cases} true & \text{if } l(\omega) \text{ less, greater or equal to } T_i \\ false & otherwise \end{cases}$ the *timing constraint condition*, with $T_i \in \mathbb{R}_0^{+\infty}$ the *timing constraint*, the *time base* $T = \mathbb{R}_0^{+\infty}$, and $\lambda_T : Q \times \Omega_G \to \Omega_G$ the *output function*, with X and Y the *possibly infinite sets of input/output values*.

Fig. 13.6 describes a set of constrained generators $\omega_k \in \Omega_{G_i}$ sharing the same timing constraint $T_i \in \mathbb{R}_0^{+\infty}$ and therefore following the same sequence of values but at different time points. Section 11.5.

A TimedIterSpec can be studied using the dilatable generators presented in

Theorem 13.1. *A TimedIterSpec* (δ_T, λ_T) *specifies a SchedIterSpec* (δ_S, λ, ts).

Proof. For each generator set Ω_{G_i}, the schedule function ts is set to the timing constraint value as $t^*_{int} = ts(q) = T_i$, which means that during the interval $[0, T_i]$, the *time break* t^*_{MLS} obtained by MLS should occur before the timing constraint to do not be *illegal*:

$q_{t^*_{MLS}} = \begin{cases} \delta_S(q, \omega_{t^*_{MLS}>}) & if \quad t^*_{MLS} \leq ts(q) \\ \text{illegal} & otherwise \end{cases}$. Transitions need not to be schedule preserving. For each generator $\omega \in \Omega_{G_i}$, each transition computes a state $q = (s, e)$ to a new state

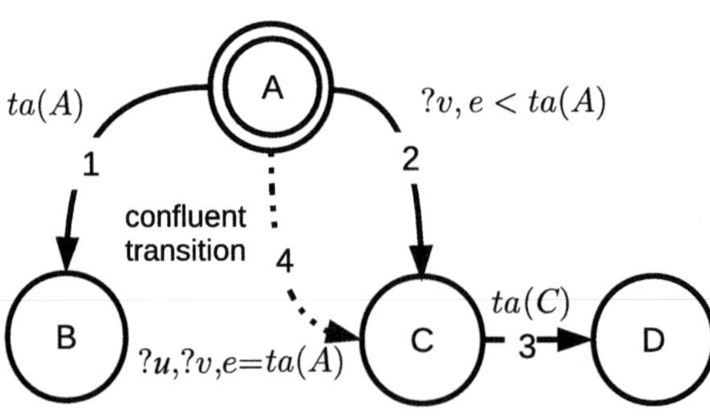

FIGURE 13.5

State transition diagram of a FiniPDEVS: A fundamental example of confluent function.

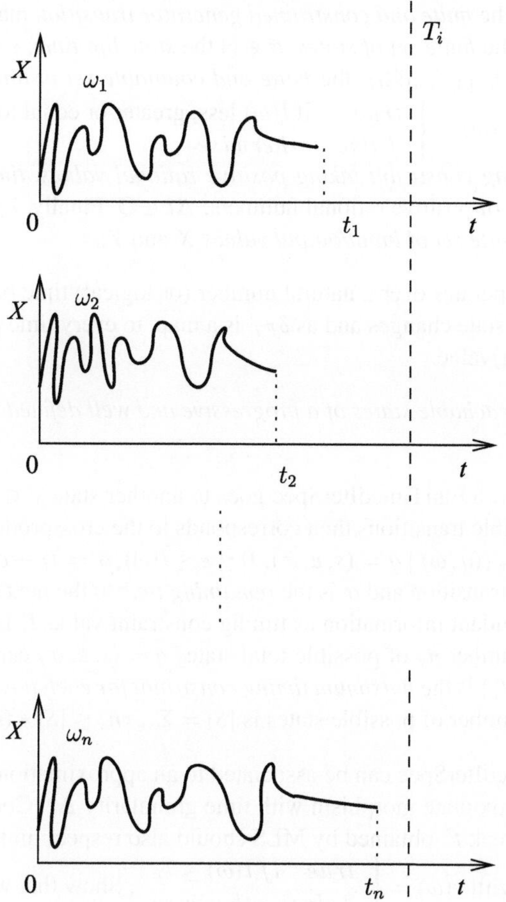

FIGURE 13.6

Example of constrained input generators. Notice that the sequence of values is always the same but values occur at different time points.

$q' = (s', e')$ where elapsed times e, e' take values in the interval $[0, T_i]$. Therefore, a TimedItSpec (δ_T, λ_T) specifies a SchedIterSpec (δ_S, λ, ts), with $t^*_{int} = t^*_{sched} = T_i$ and $t^*_{ext} = t^*_{MLS}$. □

13.5 BASIC FINITE TIMED ITERATIVE SPECIFICATION (FINITIMEDITERSPEC)

Definition 13.4. A basic *Finite Timed Iterative Specification* consists of:

$$FiniTimedIterSpec = (\delta_{FT}, \lambda_{FT})$$

with $\delta_{FT} : Q \times \Omega_G \to Q$ the *finite and constrained generator transition* map, with $Q = \{(s, \sigma, e)\}$ the *finite set of states*, with S the *finite set of states*, $\sigma \in \mathbb{N}$ the *state life-time*, $e \in \mathbb{N}$ the *time elapsed since the last transition*, $\Omega_G = \cup_{i \in \{0,\ldots,n\}} \Omega_{G_i}$ the *finite and countable set of timed generator classes* with $\Omega_{G_i} = \{\omega \mid c_i(\omega)\}$, with $c_i(\omega) = \begin{cases} true & \text{if } l(\omega) \text{ less, greater or equal to } T_i \\ false & otherwise \end{cases}$ the *timing constraint condition*, $T_i \in \mathbb{Q}$ the *timing constraint taking positive rational values*, *time base* $T = \mathbb{N} * \Delta t$, with Δt taking values in the set of positive rational numbers: $\Delta t \in \mathbb{Q}$. Finally, $\lambda_{FT} : Q \times \Omega_G \to \Omega_G$ is the *output function*, with the *finite set of input/output values* X and Y.

A FiniTimedIterSpec operates over a natural number (or logical) time base. This logical time base corresponds to an order of state changes and as δ_{FT} is a map, to every time point corresponds an input (non)value and a state (non)value.

Lemma 13.1. *The set of reachable states of a progressive and well defined FiniTimedIterSpec is finite and bounded.*

Proof. For each state $s \in S$, a FiniTimedIterSpec goes to another state $s' \in S$ for each input generator in Ω_{G_i}. The set of all possible transitions then corresponds to the crossproduct of total states and input generators, i.e., $Q \times \Omega_G = \{(q, \omega) \mid q = (s, e, \sigma), 0 \le e \le l(\omega), \sigma = T_i - e, \omega \in \Omega_{G_i}\}$, where e is the *elapsed time* since the last transition and σ is the *remaining time* to the next transition. Notice that $\sigma = T_i - e$ is indicated as redundant information as timing constraint value T_i is known. The specification being deterministic, the number n_s of possible total states $q = (s, e, \sigma)$ can then be bounded as $n_s \le max\{T_i\} + 1$, where $max\{T_i\}$ is the *maximum timing constraint for each possible transition* (q, ω) with $\omega \in \Omega_{G_i}$. Then the total number of possible states is $|S| = \Sigma_{s \in S} n_s \le |S| * (max\{T_i\} + 1)$. □

Exercise 13.3. A FiniTimedIterSpec can be associated to an approximation of a general system.

Hint: Consider an approximate morphism with time granularity Δt. Considering a system equivalence it means that time break t^* obtained by MLS should also respect timing constraint, i.e., $t^* \le T_i$, $\Omega_{G_i} = \{\omega \mid c(\omega) = true\}$ with $c(\omega) = \begin{cases} true & if\, l(\omega) \le T_i \\ false & otherwise \end{cases}$. Show that when an input occurs it follows the concatenation of $\phi_i \bullet x_{i+1}$ with $l(\phi_i \bullet x_{i+1}) \le T_i$ where $l(\phi_i) = \Delta t$.

13.6 EVENT BASED CONTROL AND FINITE TIMED PDEVS

Using our graphical representation, an event based control (Zeigler, 1989; Luh and Zeigler, 1993) can be represented as in Fig. 13.7. The system starts in the phase "TOO-EARLY". At step 1, if it receives the input value $?v$ before the time advance "ta(TOO-EARLY)", the system will output the information "TOO-EARLY" and after returns to phase "OFF" for an infinite time advance. Still in step 1, if the system does not receive the input value $?v$ before the time advance "ta(TOO-EARLY)", it goes to phase "ON-TIME". At step 2, if the system receives the input value $?v$ before the time advance "ta(TOO-EARLY)", it will output the information "ON-TIME". If the system does not receive the input value $?v$ before the time advance "ta(TOO-EARLY)", it will output the information "TOO-LATE".

Event based control allows implementing a timing constraint $\varphi \in \mathbb{Q}$ with $\mathbb{Q}$ the *set of positive rational numbers*. In timed automata (Alur and Dill, 1994), clocks allow checking input value occurrences

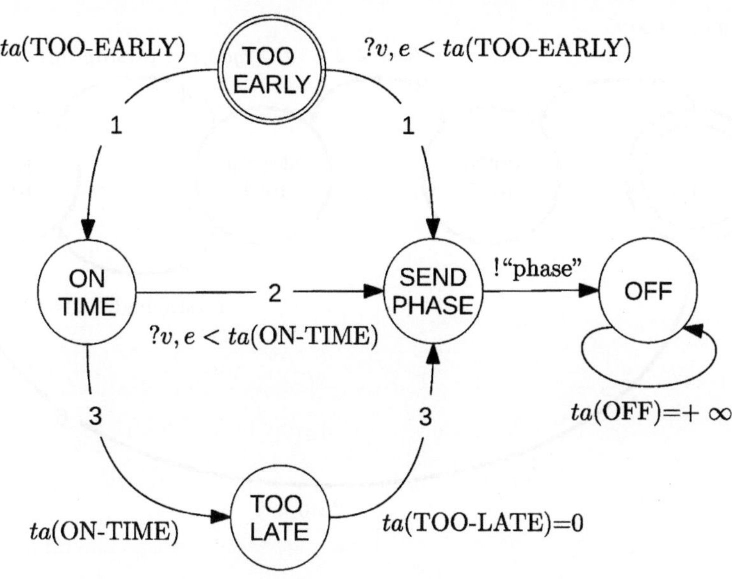

FIGURE 13.7

Event-based control implemented in FiniPDEVS. Confluent transitions occur between ON-TIME and SEND-PHASE states.

with respect to real value timing constraint. A formal comparison between timed automata and Finite and Deterministic DEVS (FDDEVS) can be found in (Hwang, 2011). Considering an input value $?v$, $?v < \varphi$ means that the value should occur before reaching time boundary φ. Such approach can be implemented by an event based control mechanism with special time advance values:

- A timing constraint $?v < \varphi$ consists of both time advances ta(TOO-EARLY) $= 0$ and ta(ON-TIME) $= \varphi$.
- A timing constraint $?v > \varphi$ consists of both time advances ta(TOO-EARLY) $= \varphi$ and ta(ON-TIME) $= 0$.
- A timing constraint $?v = \varphi$ consists of both time advances ta(TOO-EARLY) $= 0$ and $e = ta$(ON-TIME) $= \varphi$.

Event based control is used in the following example.

Example 13.1. State transition diagram of a simple elevator.

In Fig. 13.8, a very simple elevator is described as being able to carry people from level *L0* to level *L1* and *vice versa*. The behavior is the following the transitions (assuming for example a time base in minutes):

1. The elevator remains in state "stayingInL0" for infinity, unless
2. The elevator receive an input of value $?go1$ and goes to state "goingInL1",
3. The elevator should arrive at level 1 in less than 2 minutes then going to state "stayingInL1". This detection is achieved by a sensor providing the value $?in1$ if the elevator arrives at level 1.

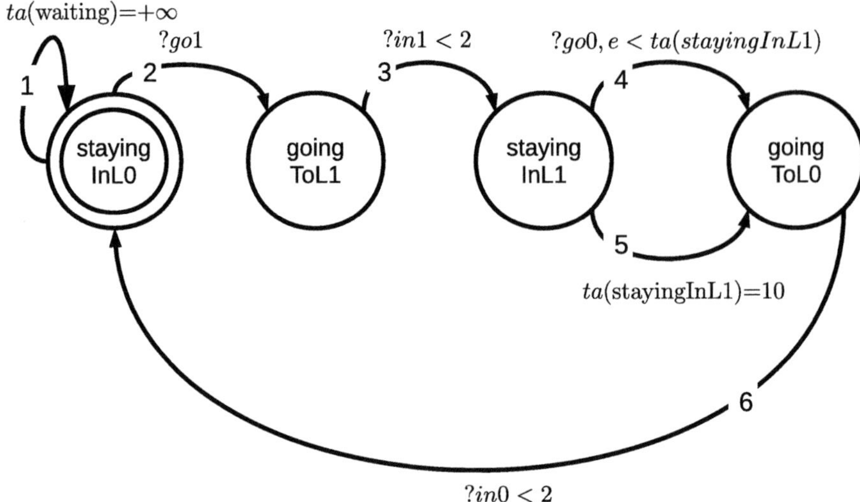

FIGURE 13.8

State transition diagram of a simple elevator, with double circle as initial state.

Otherwise it is facing a breakdown. This uncontrolled event is detected by an event control mechanism[1] as shown before, with no state "TOO-EARLY" and with $ta(ON-TIME)=2$.

4. The elevator receives an input event with value $?go0$ and then goes to state "goingToL0", otherwise
5. The elevator remains in state "stayingInL1" for a time $ta(stayingInL1) = 10$ minutes and goes automatically to state "goingToL0",
6. The elevator should arrive at level 1 in less than 2 minutes then going to state "stayingInL1". As in transition 3, this is an event-based control mechanism.

FiniTimedIterSpec allows studying formally real time systems. A finite and timed iterative specification (FiniTimedIterSpec) can be simulated by a Finite and Timed PDEVS (FiniTimedPDEVS). An event based control mechanism can be implemented by the FiniTimedPDEVS.

Definition 13.5. A *Finite and Timed PDEVS* (*FiniTimedPDEVS*) is a FiniPDEVS implementing the event based control presented in Fig. 13.7.

13.7 SUMMARY

Abstractions of iterative specification and corresponding PDEVS implementation have been presented focusing on both finite set of values and timing restrictions on event time occurrences. Both restrictions

[1]Notice that the FiniTimedPDEVS controlled mechanism is indicated implicitly as $?in1 < 2$ (instead of the usual FiniPDEVS mechanism shown in transition 4).

allow to graphically represent timed state changes and event occurrences. Besides, these restrictions open new perspectives for studying formally and for automatically verifying system behaviors.

REFERENCES

Alur, R., Dill, D.L., 1994. A theory of timed automata. Theoretical Computer Science 126 (2), 183–235.

Hwang, M.H., 2011. Taxonomy of DEVS subclasses for standardization. In: Proceedings of the 2011 Symposium on Theory of Modeling & Simulation: DEVS Integrative M&S Symposium. Society for Computer Simulation International, pp. 152–159.

Liu, X., Matsikoudis, E., Lee, E.A., 2006. Modeling timed concurrent systems. In: CONCUR 2006 – Concurrency Theory. Springer, pp. 1–15.

Luh, C.-J., Zeigler, B.P., 1993. Abstracting event-based control models for high autonomy systems. IEEE Transactions on Systems, Man, and Cybernetics 23 (1), 42–54.

Zeigler, B.P., 1989. DEVS representation of dynamical systems: event-based intelligent control. Proceedings of the IEEE 77 (1), 72–80.

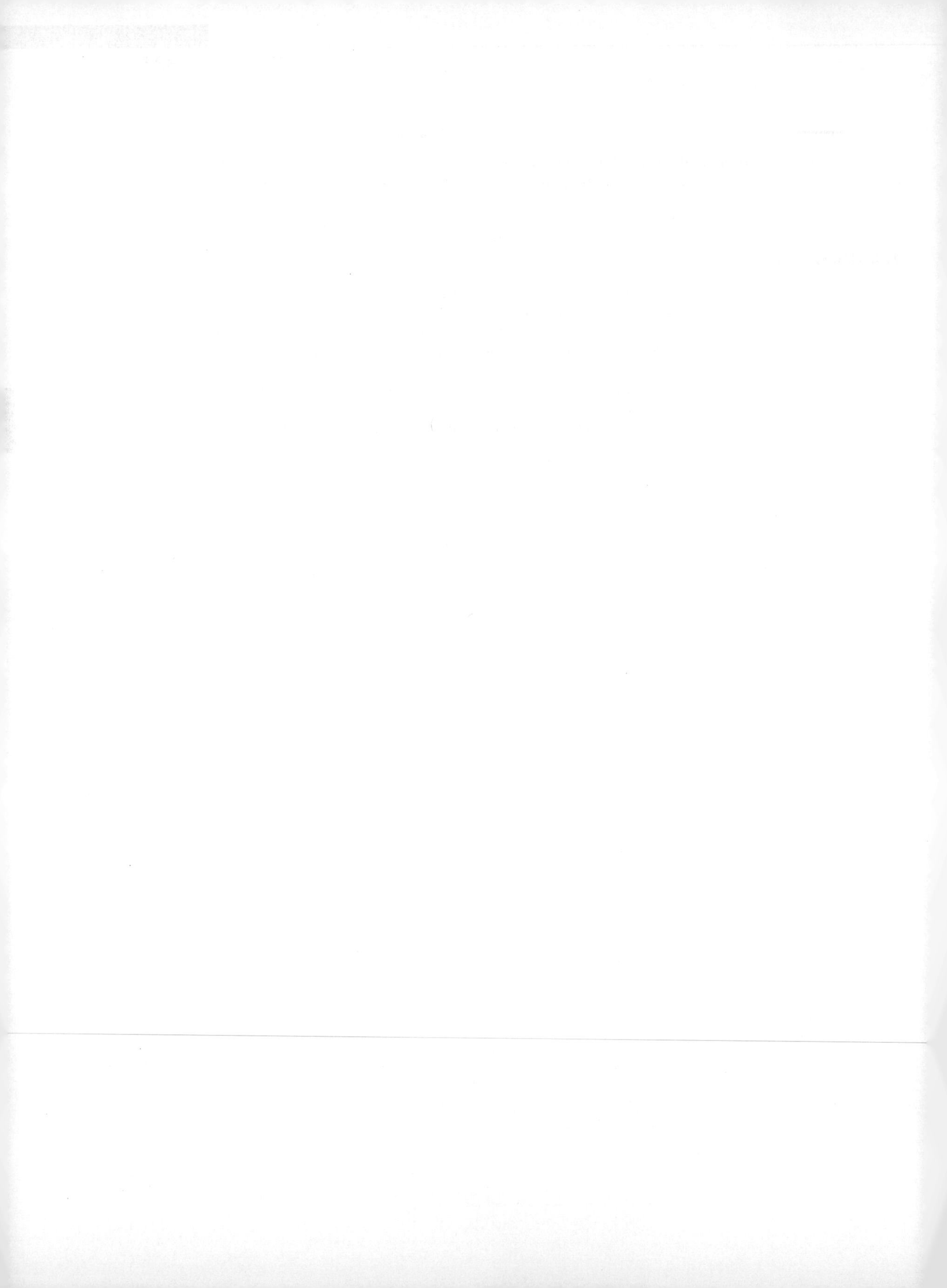

PART

3

SYSTEM MORPHISMS: ABSTRACTION, REPRESENTATION, APPROXIMATION

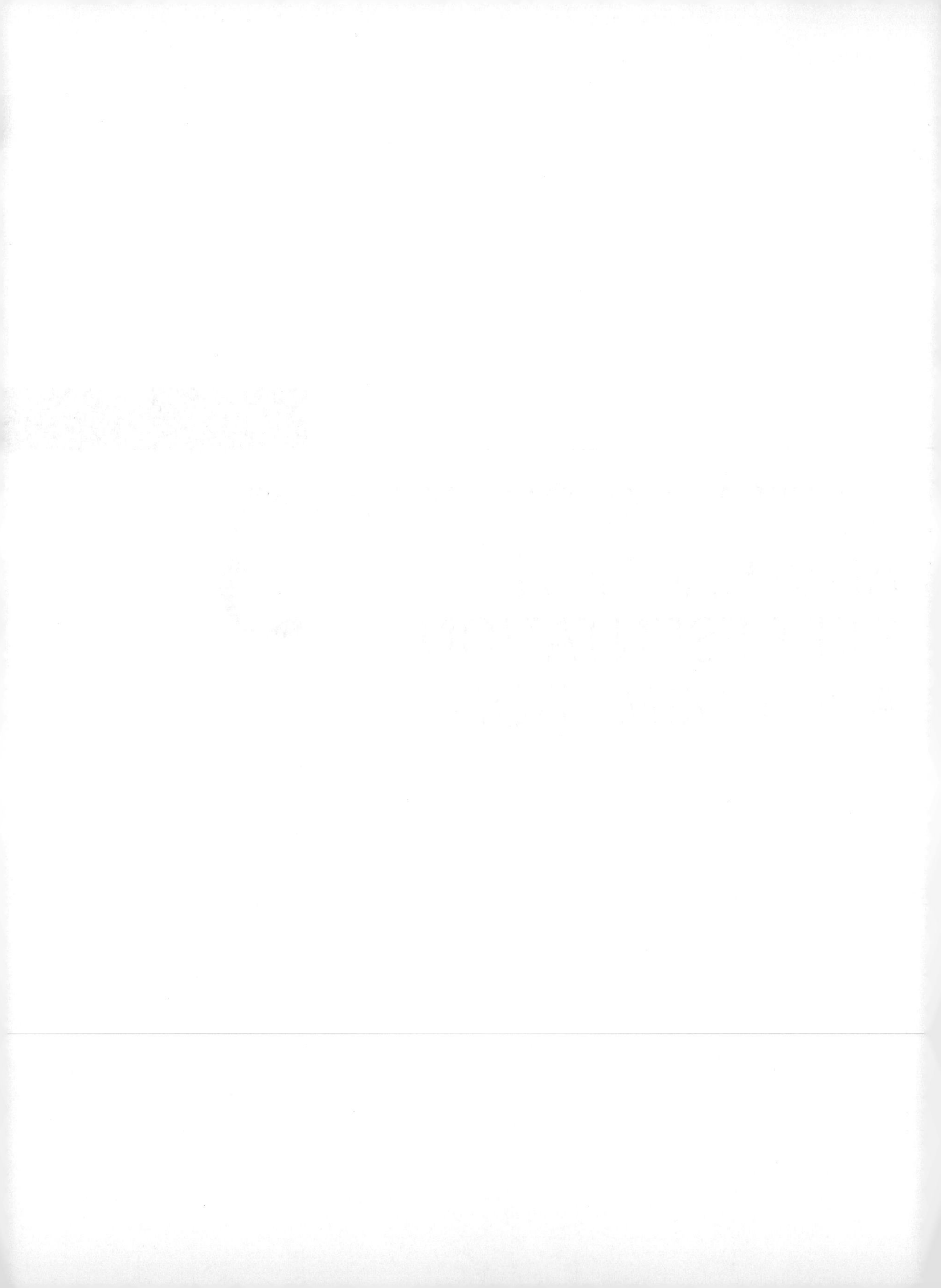

CHAPTER

PARALLEL AND DISTRIBUTED DISCRETE EVENT SIMULATION 14

CONTENTS

JAMES NUTARO WAS THE PRIMARY AUTHOR OF THIS CHAPTER

In Chapter 8 we presented simulators for DEVS models. These simulation algorithms can be called *abstract simulators* since they represent the processing that has to be done to execute DEVS models. As such they can be executed on sequential single processor systems using, for example, event list algorithms to keep track of the times-of-next-event. However, for a number of compelling reasons discrete event simulation execution on multiple machines can be an attractive alternative in various applications:

- Models and their simulators are usually dispersed among many physical locations and there is great potential benefit to sharing such resources. *Distributed Simulation* is the name given to the networking of geographically dispersed simulators of model components to execute a single overall model. The simulators are called federates and the coupled model is called a federation in HLA (Chapter 1).
- The computational demands imposed by large scope, high resolution simulations (Chapter 14), are moving the preferred platforms from single processor systems to multiprocessor architectures. Multiprocessor architectures consist of a number of serial processors interconnected

Theory of Modeling and Simulation. https://doi.org/10.1016/B978-0-12-813370-5.00024-9

through a communications network to enable coordination of their execution and sharing of data. *Parallel simulation* refers to the design of such systems to enable processors to execute the model *concurrently* (or as is often said, in *parallel*).

Whether one uses the term "parallel" or "distributed" (or both) to describe a simulator is largely a matter of objectives in its design. Such objectives include:

- increased speed: if done right, the time required for a simulation run can be reduced in proportion to the number of processors it is allocated
- increased size of models: the combined memory capacity of all processors may be employed
- exploiting the greater data handling or graphics capability provided by specialized nodes
- interoperability and sharing of resources.

A simulation architecture might be called parallel if its main goal is reducing execution time. In contrast, as already indicated, distributed simulation often connotes the goal of interoperating geographically dispersed simulators. While parallel simulation depends on properly distributing model execution among processors, the farmed out work need not be in the form of recognizable model components, or federates, that would characterize distributed simulation.

In this chapter we discuss the basics of *parallel/distributed discrete event simulation* (PDES) algorithms for discrete event models, especially DEVS models, on multiprocessor systems. In our discussion, we'll assume that a model is divided into distinct components and each one executed on a distinct processor. Although real world systems naturally operate in parallel, exploiting this intrinsic parallelism may turn out to be rather difficult. The reasons for this are the strong causal dependencies which exist between the model components. Such dependencies of the model components simulated on distinct processors require careful synchronization to ensure execution correctness.

Traditionally algorithms for parallel/distributed simulation have been categorized as *conservative* or *optimistic* depending on whether *causality violations* are strictly avoided or allowed to occur but detected and repaired. In the second edition we discussed the basics of these approaches to simulation of coupled DEVS models on multiprocessor machines. In this edition, with the benefit of the experience with implementation of Parallel DEVS simulation algorithms, we take a more a general point of view. First we examine, in some depth, the objectives in developing and executing PDES applications, such as increased speed and size of models, and how the approaches taken can be motivated by such objectives. This leads to presentation of a Constructive Cost Model (COCOMO) of PDES software development that takes into account both the objectives and the structures of the models to be simulated. Second, we focus on the critical path of a simulation as the most costly sequence of computations that must be executed to complete a simulation run. This approach throws allows us to better understand the strengths and limitation of traditional PDES algorithms and how the canonical Parallel DEVS simulation protocol to-be-defined provides a vehicle for general robust simulation with near optimal performance.

14.1 THE VALUE OF INFORMATION

Persons new to simulation often see a model as a "world in a box". This perspective is evident in questions such as "do you have a model of X" where X may be a power system, a building, or some other object. Notably absent from this question is an indication of how the model will be used. Implicit in such an utterance is the belief that a model substitutes for the real thing in whatever circumstances are envisioned. If the model is found wanting for our purpose, then we extend it until, eventually, it is almost like the real thing. A related misconception concerns the value of faster execution of a simulation model. This is particularly relevant given the ready availability of multiprocessor workstations and the wide range of algorithms available for parallel execution of discrete event models.

In the context of models used for engineering, a useful perspective on these two issues comes from considering the role simulation plays in constructing a system. In this capacity, the primary purpose of a model is to supply information about the system under consideration. This information may concern design alternatives, performance in a test, limits of operation, or any number of questions that arise when discussing complex technologies and organizations. For our purposes it is useful to measure this information in bits, with the assumption that more bits yield a more accurate, precise, or complete answer to some question.

The value of a bit of information is determined by the costs avoided when it is uncovered in simulation rather than with a real prototype or test. While this is difficult to quantify in practice, let us assume for the sake of illustration a linear relationship between costs avoided a per bit and the number of bits b obtained with a model. The value V of this model is the product

$$V = ab.$$

In practice, the marginal value of the next bit decreases rather than remaining constant as suggested by this equation. You have experienced this in practice when approximating π with 3.14 or acceleration due to gravity with 9.8 $\mathrm{m/s^2}$. Consequently, our linear model of value is optimistic but can still usefully serve to illuminate the economics of simulation.

It is not too much of a stretch to equate bits obtained with a model to software lines of code required to program the simulation. To quantify the cost of doing so, let us equate a bit with a line of code in the basic Constructive Cost Model (COCOMO) (Boehm et al., 1981). This gives a cost C we pay for b bits in the form

$$C = kb^r$$

where $k > 0$ and $r > 1$. For development of a simulation model, it is probably reasonable to use the parameters that COCOMO recommends for "semi-detached" development. This assumes a mix of rigid requirements, which for a simulation are reflected in validation and the need for precise mathematical calculations, and flexible requirements, such as for visualization, output file formats, and so forth. The parameters for semi-detached development are $k = 3$ and $r = 1.12$.

Notice that the cost grows exponentially while the value grows (at best) linearly. Because of this the net value of the model, which is the value of the bits minus the cost of obtaining them, initially rises and then falls steeply. This is illustrated in Fig. 14.1. Our conclusion, based such economics, is that building many small models designed to answer specific questions using a few bits (following the methodology of Chapters 2 and 15) is more cost effective than one or more large models designed to answer a wide range of questions and requiring a correspondingly large number of bits.

14.2 THE VALUE OF PARALLEL MODEL EXECUTION

Parallel computers, a rarity only two decades ago, are now ubiquitous. Your desktop or laptop computer offers substantial opportunity for speeding up simulation tasks by using its multiple computing cores.

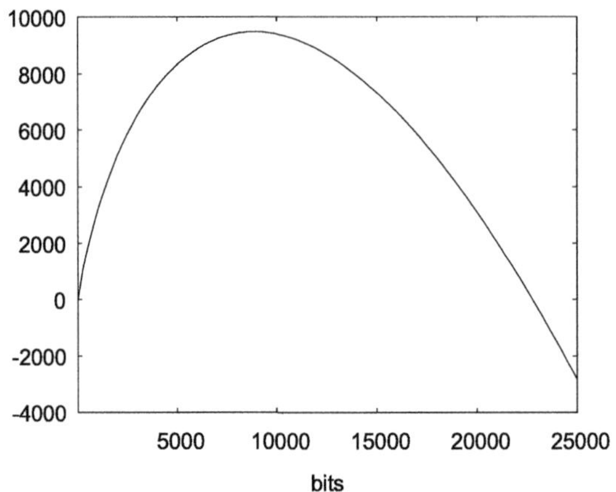

FIGURE 14.1

Net value of b bits using the COCOMO semi-detached model with $a = 10$.

Cloud computers offer an inexpensive way of renting tremendous computing power that can be applied to a simulation task. Several algorithms for exploiting parallel computer hardware have already been introduced in the second edition of this book (and will be repeated later in this chapter). A brief perusal of that material will leave a strong impression that the developmental effort required to use these algorithms, and hence the price of the faster or larger models that they enable, varies substantially. Indeed, this is true in practice and we now consider a basis for selecting a particular algorithm based on the anticipated value of information yielded by the larger, faster, or more detailed simulation study it enables.

Parallel simulation is typically used to address one of the following problems.

1. Sample a larger parameter space than would be feasible with sequential simulation in the available time.
2. Improve the model resolution or detail, which may involve circumventing limits imposed by memory required, execution time, or both (Chapter 15).

In both cases, enlarging the model, its analysis, or both is expected to yield more bits of information. From this perspective, it is convenient to consider only the gain in speedup or problem size obtained by the algorithm with the assumption that this corresponds to an equal gain in bits of information. The cost of obtaining these new bits will depend strongly on the nature of the simulation procedure that is used. The COCOMO model offers less insight in this circumstance because the effects of algorithm choice on k and r have never been quantified.

Nonetheless, we may confidently assert that if the effort expended to develop a parallel simulation is well spent, then that effort must be matched by a corresponding increase in speed, problem size, or both with a commensurate gain in information. Therefore, we proceed to examine how the performance of common algorithms for parallel simulation is intimately related to the characteristics of the model under study. With this insight, algorithms can be selected for greater execution speed or problem size justifying the effort required to obtain it.

14.3 SPEEDUP, SCALING, AND PARALLEL EXECUTION

We begin with a simulation problem of size P that can be solved using a single computer in time T. The size could be measured in computer memory required by the simulation model to obtain a given level of detail, or size could be variations of the model's parameters that we wish to explore. The parallel computer can be used to reduce T or enlarge P, but does so at the cost of an additional workload P' needing time T' to compute. This additional workload comes from the need to exchange data between computers and to coordinate their execution as they work collaboratively on the problem.

We will consider two applications of N computers to a simulation problem, with the aim of using these computers to achieve a scaling factor (or speedup) $S > 1$. In the first application, we wish to solve the same problem of size P in a shorter time $T/S < T$. If P' is negligible, then the speedup achieved will be

$$S = \frac{T}{T/N + T'} \tag{14.1}$$

In the second application, we wish to solve a larger problem $SP > P$ in approximately the same time T. If T' is negligible, then the scale actually achieved will be

$$S = \frac{NP}{P + P'} \tag{14.2}$$

In an ideal world, the overheads T' and P' are both negligible and we may obtain perfect scaling such that $S = N$.

In practice, the overheads depend on N. In a first approximation, we may assume constants α and β so the $P' = \alpha N$ and $T' = \beta N$. Substitution into Eq. (14.1) gives

$$S = \frac{T}{T/N + \beta N} = \frac{NT}{T + \beta N^2} .$$

The N^2 term is sobering. It tells us there is some N that provides the best speedup, and that for larger N our computation slows down! For large enough N, the parallel computation takes more time than is needed with $N = 1$. This behavior is illustrated in Fig. 14.2 for $T = 1$ and $\beta = 0.01$. Further insight into the fact that speedup can actual diminish to zero as N increases comes from a simulation and analytic model of parallel processors communicating over a data network discussed in Chapter 21.

A more encouraging story emerges when we substitute $P' = \alpha N$ into Eq. (14.2). Doing so shows that the expecting scaling is

$$S = \frac{NP}{P + \alpha N} .$$

The N^2 term does not appear and the achievable scale always increases with N. However, each increase yields a diminished return. For a sufficiently large N, the term P in $P + \alpha N$ may be ignored and a limit to the achievable problem size is reached at

$$S \approx \frac{NP}{\alpha N} = \frac{P}{\alpha} .$$

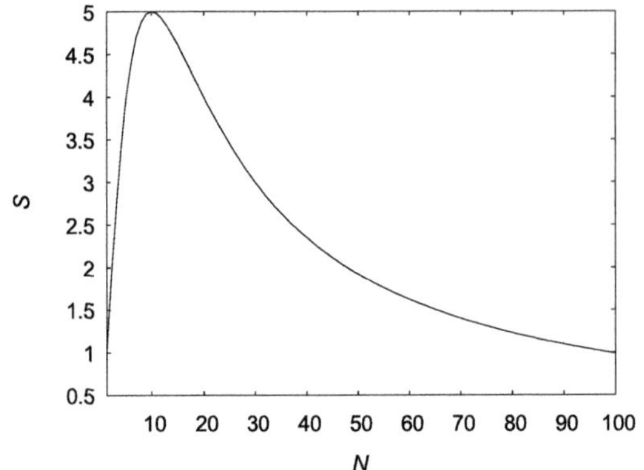

FIGURE 14.2

Speedup and then slow down as the overhead of the parallel simulation algorithm grows with N.

The almost linear speedup and later diminishing return are illustrated in Fig. 14.3 for $P = 1$ and $\alpha = 0.01$.

The above analysis can inform our decisions about the use of a parallel computer for simulation in several ways. Suppose that to obtain some new information from a model, we have the option of reducing T or increasing P. *For instance, we often have the choice of accelerating a single simulation experiment, hence reducing T, or of running several experiments simultaneously, thereby increasing P.* Clearly we should prefer the latter because it makes the best use of our computing resources.

It also frequently occurs that to obtain additional information from a model it is necessary to add details that were initially neglected. This necessitates increasing the problem size to hold new data in the computer memory and increasing the computational time to process these new data. There will

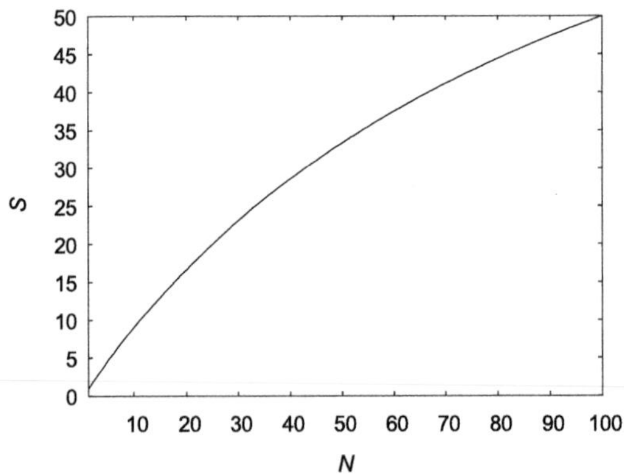

FIGURE 14.3

Diminishing return with growing N.

be several algorithms to choose from when preparing this simulation to exploit a parallel computer, and each algorithm will have a cost associated with its implementation. Generally speaking, the more costly the algorithm is to implement the better it will perform.

For instance, it may be possible to change a simulation model so that it runs on a large, shared memory parallel computer. This effort could involve applying locks to key data structures and enabling for loops to process their data in parallel. With modern compilers, both of these steps can be accomplished by inserting compiler pragmas into the code. The cost of doing so is small, but the resulting α and β will large. On the other hand, we might realize a very effective parallel execution by rewriting large parts of a simulation model. This could involve partitioning the computations in a near optimal way and exploiting speculative computing. Such an effort might take months but result in very small α and β. Which should we choose?

One way to approach this question is to look at the value of the new information imparted by using a parallel computer. While it is difficult to be precise in such a calculation, we can distill the essential elements to a simple, intuitively appealing model. If new detail in a model increases the information gained from it, then in a first approximation let the increase in information be proportional to the increase in P. Similarly, a model that runs more quickly will generate more information in a study of fixed duration, and so the information gained is also proportional to the gain in T. Hence, the total increase in information I is

$$I = \frac{NP}{P + \alpha N} + \frac{NT}{T + \beta N^2} \tag{14.3}$$

If we normalize the information gained by the cost C of implementing the parallel algorithm, then the economic choice maximizes I/C for an appropriate range of N.

This concept is illustrated in Fig. 14.4 for three notional algorithms when $P = T = 1$ so that

$$I = \frac{N}{1 + \alpha N} + \frac{N}{1 + \beta N^2} \tag{14.4}$$

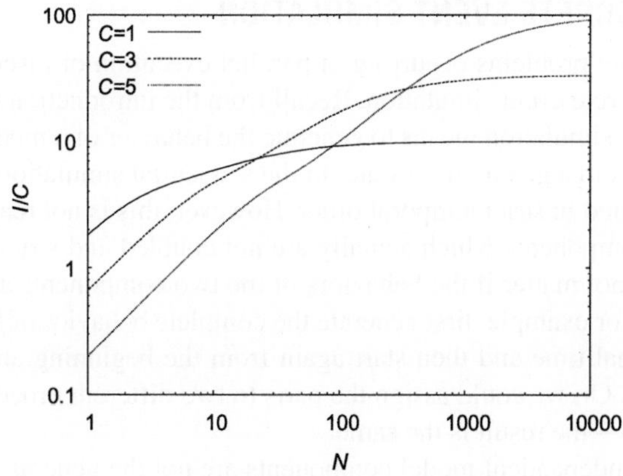

FIGURE 14.4

Comparison of notional I/C curves of three parallel algorithms following Eq. (14.5).

We will consider the ratio I/C for algorithms A with characteristics $(C_A,\ \alpha_A, \beta_A)$

$$\frac{I}{C_A}(N) = \frac{1}{C_A}\left(\frac{N}{1+\alpha_A N} + \frac{N}{1+\beta_A N^2}\right) \tag{14.5}$$

An inexpensive algorithm with $C = 1$ and $\alpha = \beta = 0.1$ is compared with two more sophisticated, more costly algorithms having $C = 3$, $\alpha = \beta = 0.01$ and $C = 5$, $\alpha = \beta = 0.001$ respectively. The choice of algorithm depends on the size of the computer that is expected to be available. For a workstation with small N the cheap algorithm is preferred. The most sophisticated, most expensive algorithm is justified when N reaches supercomputing proportions. Between these extremes the $C = 3$ algorithm is the economic choice.

14.3.1 PARALLEL EXPERIMENTS

The simplest use of a parallel computer is to run several experiments simultaneously. The aim, in terms of our economic analysis, is to increase the size of the simulation task by increasing the number of model configurations that we examine. Because this does not require coordination between the independent simulation runs, the overhead term α can be made very small. With this approach we continue to use a single computer to run each individual experiment, and hence the T terms in Eq. (14.3) may be discarded. It is typical in practice to have $\alpha \approx 0$ and the information gain is

$$I = \frac{NP}{P+\alpha N} \approx \frac{NP}{P} = N\,.$$

We cannot obtain a better gain in information for our effort and conclude that, if parallel experiments will provide sufficient information, then this is the best use of available resources. However, since the cost C of the parallel computer increases with the number of processors, we can expect that there will also be diminishing of returns as for the ratio I/N as well.

14.3.2 PARALLEL DISCRETE EVENT SIMULATION

To get some insight into the problems occurring in parallel execution of discrete event models, let us recall the essentials of discrete event simulation. Recall from the introduction to simulation algorithms in Chapter 8 that computer simulation means to generate the behavior of a model as specified by its dynamic definition starting from a given initial state. In the sequential simulation algorithms we assumed that the behavior is generated in strict temporal order. However, this is not really required. Let us consider a model with two components which actually are not coupled and strictly separated (Fig. 14.5). For such a model, it does not matter if the behaviors of the two components are generated in strict sequential order. We could, for example, first generate the complete behavior of the first component from the starting time to the final time and then start again from the beginning and generate the behavior for the second component. Or, we could assign the parts to two different processors and simulate them (independently) in parallel – the result is the same.

However, completely independent model components are not the general rule. Model components may run independently for a limited time interval, but then show dependencies, eventually to run independent again afterwards. Dependencies in discrete event models occur through events. When an event in one component directly or indirectly influences events in another component, then we say that there

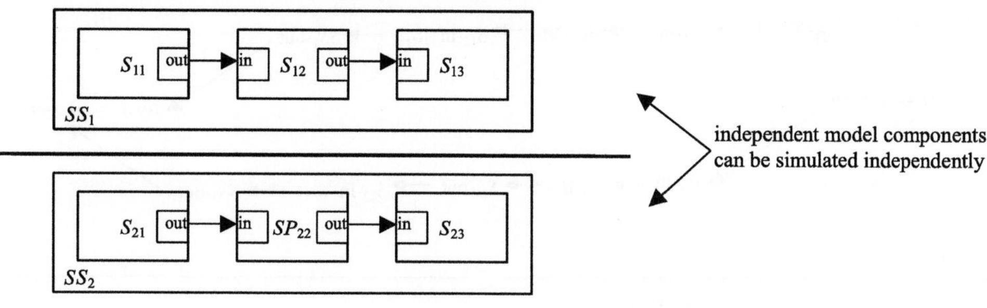

FIGURE 14.5

Two independent models can be simulated in any order.

is a *causal dependency* of events. To guarantee execution correctness, the generation must preserve what is called *causality*, i.e., that the causes have to precede the effects (Fujimoto, 1990). To discuss dependencies between model components, let us restrict the discussion to modular coupled models as presented in Chapter 7. Actually, most parallel discrete event simulation systems adopt a modular modeling approach (model components are usually called *logical processes* in the PDES simulation literature).

In a modular coupled model, dependencies between components are exclusively through input/output couplings. Thus dependencies show as dependencies between output and input events. In the model of Fig. 14.6 the two serial server models are now linked to a model F (fork) at the input side and a component J (join) at the output side. Fig. 14.7 shows the partial ordering of an exemplary sequence of input output events. The ordering shows us that no direct dependencies exist between the two serial servers and, hence, they still can be simulated independently, but events have to be synchronized at the outputs of F and inputs of J.

Sequential simulation algorithms guarantee causality between events by generating, from this partial ordering, a total ordering of events according to the event times (Fujimoto, 1990; Misra, 1986). Parallel discrete event simulation, however, has to exploit the partial ordering to find events that can

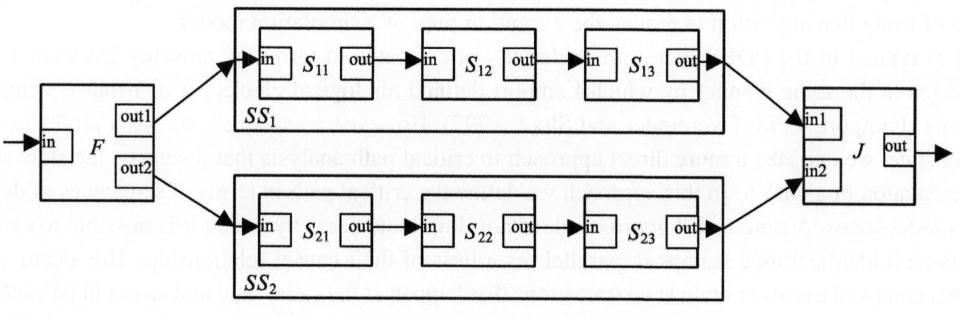

FIGURE 14.6

Two independent model components SS_1 and SS_2 with synchronization points F and J.

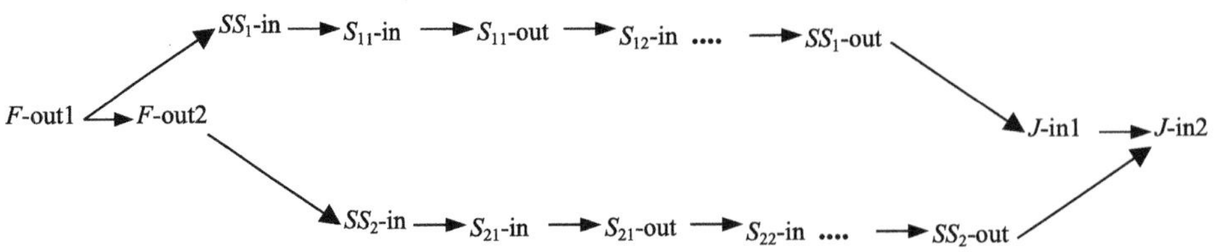

FIGURE 14.7

Dependency relation of input-output events.

be processed in parallel. Fujimoto (1990) defines conditions for correct simulation (without causality violations) as follows:

> *Causality constraint:* A simulation of a model consisting of modular components which exclusively communicate through their modular input/output couplings obeys the causality constraint, if and only if each modular component simulator processes the events in non-decreasing timestamp order.

Parallel discrete event simulation schemes are based on this constraint. While conservative PDES strictly avoids violating the causality constraint, optimistic approaches temporarily violate it but then will detect and repair the violations.

14.3.3 UNDERSTANDING SPEEDUP VIA STATE-BASED CRITICAL PATH ANALYSIS

Critical path analysis is a method for identifying and characterizing chains of events that can be executed in parallel without violating the causality constraint. The critical path in a parallel simulation is the most costly sequence of unavoidably sequential computations that must be executed to obtain a result. If we can identify the critical path in a simulation model then we have determined its shortest possible execution time. This presents an absolute lower limit on the time T needed to execute the model using an ideal algorithm with $\beta = 0$. Given a realizable simulation algorithm, we may use the critical path to estimate the actual speedup achievable for our model, and thereby approximate the value of using that algorithm to reduce the execution time of a simulation model.

It is typical in the PDES literature to define critical paths in terms of causality associated with messages in the same manner by which Lamport defined his logical clocks for distributed computer systems (Lamport, 1978; Overeinder and Sloot, 1995). However, because events are a product of the model state, we can take a more direct approach to critical path analysis that leverages the state space representation of a DEVS. In this approach we define the critical path in terms of sequences of dependent model states. A state based approach does not illuminate cases by which it is possible to execute events with identical time stamps in parallel regardless of their causal relationship. This occur when distinct chains of events contain state transitions that happen at the same time and so could be executed in parallel very easily by any algorithm. Such circumstances are easy to identify by a modeler who is aware of this possibility and if this is kept in mind then nothing is lost by omitting consideration of event times from our analysis method.

In the following, we consider critical path analysis as it applies to Parallel DEVS (PDEVS) coupled models considered as a collection of interacting, atomic DEVS models labeled 1, 2, ... , n. A simulation produces for each DEVS model k a sequence of states $s_{k,1}, s_{k,2}, \ldots, s_{k,m}$.

Definition 14.1. The state $s_{q,j+1} = \delta_q\left(\left(s_{q,j}, e\right), u\right)$ is an immediate successor of state $s_{q,j}$ and all states $s_{p,i}$ such that $\lambda_p\left(s_{p,i}\right) \subset u$.

This definition encompasses two essential aspects of a discrete event simulation. First, to calculate the new state of a model we must know the model's state at the immediately preceding instant; that is, we must know the jth state of the model to calculate its $(j+1)$st state. Second, to calculate the new state we must know the input arriving at the model; that is, we must have the output produced by models that are imminent when the state transition is to occur. Given this information, we may always calculate in parallel the state transitions of imminent models and models receiving input from those imminent models. Indeed, this fact is exploited by the PDEVS abstract simulator to achieve an inexpensive form of parallel computing. In some special cases, other sources of parallelism are available, and the optimistic and conservative simulation algorithms seek to exploit these opportunities.

Definition 14.2. The computational cost of a state s' is the wallclock time required to compute $s' = \delta\left((s, e), u\right)$.

An acyclic, directed graph can be created by drawing arcs from each $s_{k,j}$ to its immediate successors. In this graph, each node is either an initial state with only outgoing edges, a final state with only incoming edges, or has at least one incoming and one outgoing edge. The critical paths through this graph are the most computationally costly that begin at an initial state and finish at a terminal state. This is the shortest time in which a parallel algorithm could execute a simulation of the model.

Fig. 14.8 shows the critical paths of a one dimensional cellular automaton simulation. The structure of the graph clearly reflects the dependence of each state on a cell's previous state and the input from its two neighbors. If for a simulation of t steps we assume that each state transition at each cell requires one unit of time, the critical path has cost t. A sequential simulation of w cells would require tw units of time. Hence the speedup is $wt/t = w$, the number of cells in the model. An algorithm that realizes this performance calculates output functions and state transition functions in parallel at each time step.

In practice, the usefulness of critical path analysis is rather limited. In any simulation model of substantial size, it is infeasible to anticipate the relationships between states. Even if this information can be characterized *posthoc* from simulation output data, the critical paths are likely distinct from simulation run to simulation run, which substantially limits the value of such an analysis for predicting future performance or fine tuning an algorithm.

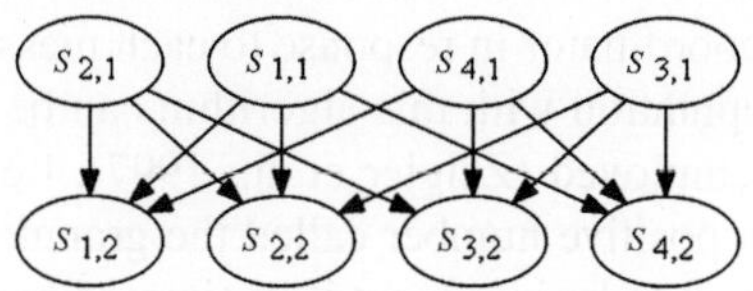

FIGURE 14.8

The critical paths of a cellular automaton simulation.

Nonetheless, critical path analysis provides a theoretical tool for understanding the circumstances under which a particular simulation algorithm performs well or poorly. Given no information about the particular shape of the causality graph, we may wish to choose algorithms that are least sensitive to its structure. On the other hand, given some reasonable anticipation of interactions within the finished model, critical path analysis of those expectations may provide useful insights for algorithm selection.

14.4 PARALLEL DEVS SIMULATOR

Recall that Parallel DEVS was designed so that parallelism intrinsic to DEVS coupled models could be exploited to allow as much parallel processing as possible without engaging in optimistic processing. In each processing cycle, all imminent components can be executed in parallel in an output generation step. In a subsequent input processing step, all imminents and the components they influence can be executed in parallel. Thus the abstract simulator for Parallel DEVS (Algorithms 15 and 16) affords a scheme for PDES that is relatively straightforward to implement, exploits parallelism in moderately to highly active models, and involves no rollback, deadlock or other overhead associated with conservative and optimistic PDES schemes. The latter feature of this algorithm makes it particularly attractive because the modeler may enlarge a model (increase P) or, experience some reduction in execution time (reduce T), with little to no additional development or analysis effort (i.e., cost).

In contrast to classical DEVS, a basic Parallel DEVS model receives a bag of inputs rather than a single input. If the input bag is not empty, processing depends on whether or not the component is imminent. If not, its simulator employs the model's external transition function to process the input. If the model is imminent, the confluent transition function is used instead. An imminent component that does not receive input executes its internal transition function. This simulation procedure can be implemented by exchanging several types of messages between computer systems. These are messages for initialization (i), to compute output ($*$), and to execute a state transition (x). The algorithmic steps taken in response to each type of message are listed below.

The simulator listed above for an atomic model is directed by a coordinator. In contrast to classical DEVS where imminent models are sequentially activated, the coordinator enables concurrent execution of state transitions and output calculations for atomic models. The outputs of these models are collected into a bag called the mail. The mail is analyzed for the part going out because of the external output coupling and the parts to be distributed internally to the components due to internal coupling. The internal transition functions of the imminent models are not executed immediately since they may also receive input at the same simulation time. As with the simulator, the coordinator reacts to i, $*$, and x messages sent by a parent coordinator, and it replies to messages received from a subordinate. At the top of this hierarchy is a root coordinator whose role is to initiate i, $*$, and x messages in each simulation cycle. The steps undertaken by a coordinator in response to each message type are listed below.

Opportunities for parallel computation with this algorithm can be expanded if a less strict test for being imminent than equality is employed (Zeigler et al., 1997). Let t_N be the minimum of all the times-of-next-event and let g be a positive number called the granule size. Then instead of requiring strict equality, we allow a component to be imminent if its time-of-next-event is within g of the global minimum, i.e. we use the relation $\equiv$ defined by

$$t_n \equiv t_N \text{ if } (t_N \leq t \leq t_N + g)$$

Algorithm 15 Simulator for basic Parallel DEVS.

```
Parallel-Devs-simulator
variables :
    parent                                  ▷ parent coordinator
    tl                                      ▷ time of last event
    tn                                      ▷ time of next event
    DEVS                                    ▷ associated model – with total state (s, e)
    y                                       ▷ output message bag
when receive i-message (i, t) at time t
    tl = t − e
    tn = tl + ta(s)
when receive ∗-message (∗, t) at time t
    if t = tn then
        y = λ(s)
        send y-message (y, t) to parent coordinator
    end if
when receive x-message (x, t)
    if x = Φ ∧ t = tn then
        s = δint(s)
    else if x ≠ Φ ∧ t = tn then
        s = δcon(s)
    else if x ≠ Φ ∧ (tl ≤ t ≤ tn) then
        e = t − tl
        s = δext(s, e, x)
    end if
    tl = t
    tn = tl + ta(s)
end Parallel-Devs-Simulator
```

This can greatly increase the number of imminents by including all those that scheduled no later than g after the true global minimum. Of course there is a trade-off here – the larger the granule the greater the concurrency but also the greater the potential for introducing timing error. (Note: $\equiv$ is a "tolerance" not an equivalence relation, since transitivity does not hold.) The concept of granule is consistent with the DEVS extensions that allow the time advance function to return a time window as opposed to a single value, as discussed in Chapter 10 (RTDEVS) of the second edition. The advantage in the latter approach is that, unlike the single global time granule, each component can have a time window that is appropriate to the level of error tolerance needed in the simulation.

14.4.1 CRITICAL PATHS IN PDEVS

The approach of executing simultaneous state transitions in parallel is most effective when such state transitions are frequent. This occurs when the causal graph of the simulation contains very few nodes with a single input and output, and in which most nodes are connected to many others. The causality

Algorithm 16 Coordinator for Parallel DEVS coupled model.

```
Parallel-Devs-coordinator
variables :
    DEVN = (X, Y, D, {M_d}, {I_d}, {Z_{i,d}})        ▷ the associated coupled model
    parent                                           ▷ parent coordinator
    tl                                               ▷ time of last event
    tn                                               ▷ time of next event
    event-list                                       ▷ list of elements (d, tn : d) sorted by tn_d
    IMM                                              ▷ imminent children
    mail                                             ▷ output mail bag
    y_parent                                         ▷ output message bag to parent
    {y_d}                                            ▷ set of output message bags for each child d
when receive i-message (i, t) at time t
    for d ∈ D do
        send i-message to child d
    end for
    sort event-list according to tn_d
    tl = max{tld|d ∈ D}
    tn = min{tnd|d ∈ D}
when receive *-message (*, t)
    if t ≠ tn then
        error: bad synchronization
    end if
    IMM = min(event-list)                            ▷ components with minimum tn
    for r ∈ IMM do
        send *-messages (*, t) to r
    end for
when receive x-message (x, t)
    if not(tl ≤ t ≤ tn) then
        error: bad synchronization    ▷ consult external input coupling to get children influenced by the input
    end if
    receivers = {r|r ∈ children, N ∈ I_r, Z_{N,r}(x) ≠ Φ}
    for r in receivers do
        send x-messages (Z_{N,r}(x), t) with input value Z_{N,r}(x) to r
    end for
    for r ∈ IMM and not in receivers do
        send x-message (Φ, t) to r
    end for
    sort event-list according to tn_d
    tl = t
    tn = mintn_d|d ∈ D
when receive y-message (y_d, t) with output y_d from d
```

```
    if this is not the last d in IMM then
        add (yd, d) to mail
        mark d as reporting
    else if this the last d in IMM then ▷ check external coupling to form sub-bag of parent output
        y_parent = Φ
    end if
    for d ∈ I_N do
        if Z_{d,N}(y_d) ≠ Φ then
            add y_d to y_parent
        end if
    end for
    send y-message (y_parent, t) to parent  ▷ check internal coupling to get children d who receive sub-bag of y
    for child r, xr = Φ do
        for d such that d ∈ I_r do
            if Z_{d,r}(y_d) ≠ Φ then
                add y_d to y_r
            end if
        end for
    end for
    receivers = {r | r ∈ children, y_r ≠ Φ}
    for r ∈ receivers do
        send x-messages (y_r, t) to r
    end for
    for r ∈ IMM and not in receivers do
        send x-messages (Φ, t) to r
    end for
    tl = t
    tn = min tn_d | d ∈ D
    sort event-list according to tn_d
end Parallel-Devs-coordinator
```

graph for cellular automata is an example of this. In general, the best case for PDEVS is when there are m atomic models and each state in the sequence of each model is a predecessor of the next state for all m models. An example of this graph is shown in Fig. 14.9. In this case, the PDEVS algorithm executes all m state transitions in parallel at the corresponding simulation time and the speedup is m.

Sparsity in the causal graph adversely affects speedup. In general, we may expect that only states sharing at least one immediate predecessor will occur simultaneously. As discussed above, these models are guaranteed to be imminent, receiving input, or both at the same simulation time and so will be processed in parallel by the Parallel DEVS algorithm. For example, if a state $s_{i,k}$ immediately precedes $s_{j,m}$, then the output produced just prior to the internal or confluent transition of component i at time t induces a change of state in j to $s_{j,m}$ via an external transition of j. At the same time, the imminent model i changes state to $s_{i,k+1}$. Hence, the new states $s_{j,m}$ and $s_{i,k+1}$ can be computed in parallel. This

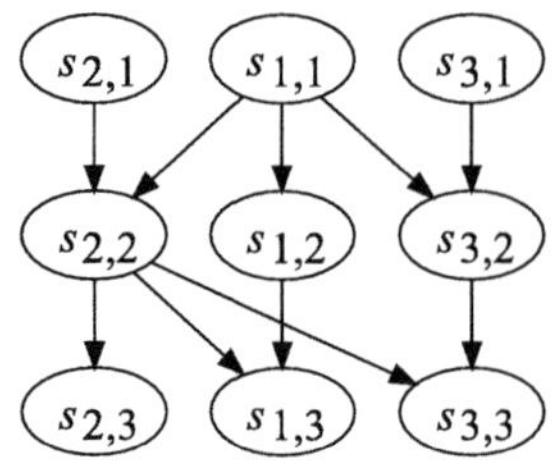

FIGURE 14.9

Best case causal graph for the PDEVS algorithm.

dynamic produces opportunities for parallel execution when dense coupling induces a large fan out for one or more atomic models. An example of a graph with only few such states is shown in Fig. 14.10.

Given a causal graph, we may determine the best possible speedup of the PDEVS algorithm. We begin with an expected computational time of zero. For each set of states that share a predecessor, we select the largest of their computational costs and increment our total expected simulation time by that amount. This accounts for parallel execution of simultaneous events. The remaining states are computed sequentially, and so we add their individual computational costs to the total. These two steps give the best possible execution time for the PDEVS algorithm. The speedup achieved is the sum of the costs of computing all states (i.e., the sequential execution time) divided by the calculated execution time of the parallel algorithm.

To illustrate this procedure, consider the causal graph in Fig. 14.10 and assume that each state transition takes one unit of time to execute. States $s_{2,2}$ and $s_{3,3}$ share predecessor $s_{3,2}$. Also, states $s_{2,2}$ and $s_{1,4}$ share predecessor $s_{2,1}$. Therefore we can compute $s_{2,2}$, $s_{3,3}$, and $s_{1,4}$ concurrently and they collectively contribute one unit of time to the total simulation running time. Similarly, $s_{2,3}$ and $s_{3,4}$ can

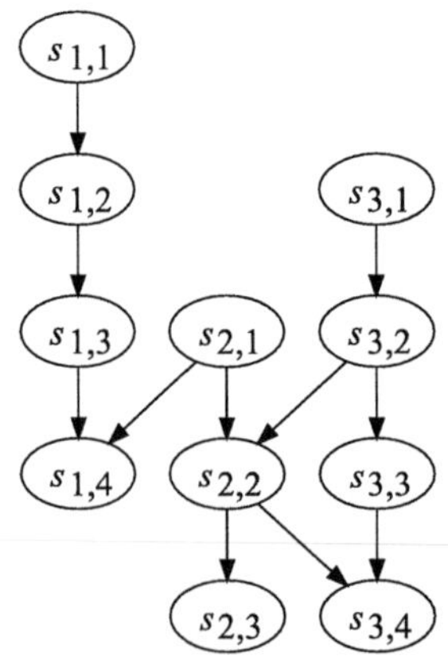

FIGURE 14.10

Causal graph with fan out exploitable by the PDEVS algorithm.

be calculated in parallel. All others must be executed sequentially. Hence, the parallel program needs six units of time to complete and the sequential program needs nine units of time. The best possible speedup in time is $9/6 \approx 1.5$ and the program can use at most three computers effectively. We can fit β to this data by rearranging Eq. (14.1) to obtain

$$\beta = \frac{NT/S - T}{N^2}.$$

Substituting into this expression $T = 9$, $N = 3$, and $S = 9/6$ gives $\beta = 1$.

14.5 OPTIMISTIC AND CONSERVATIVE SIMULATION

A primary feature distinguishing optimistic and conservative simulations from a value-versus-cost perspective is that these algorithms have a greater implementation cost that comes from trading away computational simplicity, modeling power, or both in exchange for a potentially large speedup in time and commensurate increase in information. Optimistic algorithms can simulate arbitrary discrete event systems. However, this comes with a computational overhead for state saving and restoration, potentially substantial development costs to code and test state saving and restoration functions, and further additional effort to debug unanticipated state transitions caused by speculative computation.

Conservative simulations avoid these overheads but require models to have lookahead, and large lookahead is needed to obtain good performance. Lookahead can be defined concisely in terms of a model's input to output functional representation (IOFO; see Chapter 5). Given a state q and input trajectory $x[t, t')$ the input to output function F must be such that, for any trajectories $x_a[t', t' + l)$ and $x_b[t', t' + l)$ the output

$$F\left(q, x\left[t, t'\right) \bullet x_a\left[t', t' + l\right)\right) = F\left(q, x\left[t, t'\right) \bullet x_b\left[t', t' + l\right)\right).$$

Hence, there exists a lookahead $l > 0$ if when given all input up to time t' then all future output to $t' + l$ is completely determined. This is a non-trivial requirement, eliminating from consideration servers with a potential for zero service time (e.g., if service time is drawn from an exponential distribution), most engineering models of physical systems, instantaneous switches, and many other models. When these types of models can be included in a conservative simulation, it is because they are contained within a larger, coupled model that exhibits lookahead or because an approximation has been made to obtain lookahead. An example of the former is a Moore type continuous system with sampled input and output, for which the look ahead is equal to the sample interval. An example of the latter is a server with its service time artificially constrained to be greater than some positive number (e.g., by truncating an exponential distribution). The requirement for positive lookahead also implies additional development costs to identify it and to expose it in a way that the simulator can exploit.

In the second edition of this book, specific algorithms were introduced to enable the optimistic and conservative simulation of DEVS models. This material has been distilled here to provide a brief, self-contained introduction to these algorithms. Following that presentation we focus on the economic value of these simulation approaches, beginning with an examination of speedup based on critical path analysis.

14.5.1 CONSERVATIVE DEVS SIMULATOR

In this section we introduce an approach to conservative parallel simulation of modular DEVS models. It is a scheme which relates to several conservative approaches, most notable the original Chandy-Misra approach with deadlock avoidance (Chandy and Misra, 1979, 1981; Misra, 1986) and the Yaddes algorithm (DeBenedictis et al., 1991). In the definition of the simulator we will make some simplifying assumptions. In particular, we will not explicitly manage simultaneous events and we will formulate the simulator for non-hierarchical models only.

In a parallel simulator for a modular coupled DEVS model the coordinators no longer control the individual simulators which now do their event processing autonomously. The coordinators serve only to distribute the output values as well as to forward next event time estimates as will be described below. Fig. 14.11 shows the structure of a conservative parallel abstract simulator. There are conservative parallel simulators for handling the parallel event processing. Each such parallel simulator has attached a conventional sequential DEVS simulator that is employed to do the event processing under the control of the parallel simulator. The parent coordinator is responsible for distribution of outputs and output time estimates in accordance with the model's coupling structure. The coordinator has to be realized in a distributed form on a distributed computer architecture.

The conservative DEVS simulator maintains a queue Iq to store the received, yet unprocessed input messages. As shown in Algorithm 17, it also maintains variables for the *earliest output time*, EOT, and *earliest input time*, EIT, estimates. The earliest input time estimate $EIT_{i,ip}$ for an input port ip of a component i gives the information that no input will be received at that port at a lower time. It can be computed by considering all earliest output time estimates received from all influencing output ports. So for an input port ip of a component i the earliest input time computes as

$$EIT_{i,ip} = \min\{EOT_{o,op} | ((o, op), (i, ip)) \in IC\}.$$

The minimum of all EITs for a component i determines a lower bound on the time of next input received by i:

$$EIT_i = \min_{ip}\{EIT_{i,ip}\}.$$

This EIT now provides the required information to determine that an event is safe to process. Obviously, any event with a time smaller than EIT can be processed.

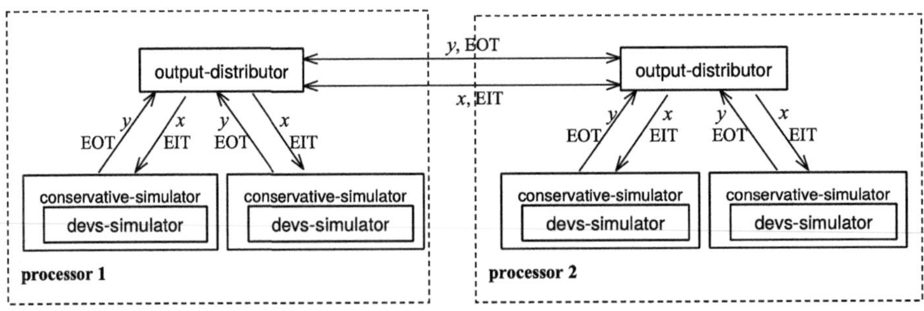

FIGURE 14.11

Structure of the parallel abstract simulator.

Algorithm 17 Conservative parallel DEVS simulator.

```
Conservative-devs-simulator
variables :
    variables:
    parent                   ▷ parent output distributor
    devs-simulator           ▷ associated DEVS simulator or coordinator
    tn                       ▷ time of internal event (maintained by the devs-simulator)
    tl                       ▷ time of last event (maintained by the devs-simulator)
    Iq                       ▷ queue to store the external input messages
    EIT                      ▷ earliest input time estimate
    {EITip}                  ▷ earliest input time estimate for all inputs ip
    EOT                      ▷ earliest output time estimate
    lookahead                ▷ function to compute the lookahead of the model
when receive i-message (i, t) at time t
    send i-message (i, t) to devs-simulator
    EIT = 0, EIT_ip = 0
    compute new EOT = lookahead(q, tl, EIT) and send it back to parent
when receive new EIT_ip estimate for input port ip
    compute EIT = minip{EIT_ip}
    compute new EOT = lookahead(q, tl, EIT) and send it back to parent
when receive x-message (x, t) at time t
    store (x, t) in input-queue Iq with input value x
when receive y-message (y, t) at time t
    send y-message (y, t) to parent from your subordinate devs-simulator
loop:                        ▷ event-loop
    while true do
        (x, tx) is first in Iq and (null, ∞) if Iq is empty
        if tn ≤ EIT ∧ tn ≤ tx then                ▷ process internal event
            send (*, tn) message to your devs-simulator
        else if tx ≤ EIT ∧ tx < tn then           ▷ process external event
            remove first of input queue Iq
            send (x, tx) message to your devs-simulator
        end if
        compute new EOT = lookahead(q, tl, EIT) and send it back to coordinator
    end while
end loop                     ▷ event-loop
end Devs-Simulator
```

The earliest output time estimate is computed based on the state of processing and scheduled events within the simulator, the input time estimates for its input ports and the minimum propagation delay for the model. Let *lookahead* be a function

$$lookahead\colon Q \times T \times T \rightarrow \mathrm{R_0}^{+} \cup \{\infty\}$$

to determine the earliest time of the next output based on the current total state $q = (s, e)$, time of last event *tl* and the current EIT estimate. Hence the EOT_o for the output port *op* of a component *o* is computed by applying the *lookahead* function

$$EOT_{o,op} = lookahead_o\,((s_o, e_o), tl_o, EIT_o).$$

For example, when the model *o* has a minimum propagation delay, D_o, and the next output events are equal to the time of next event, tn_o, the lookahead function is

$$lookahead_o\,((s_o, e_o), tl_o, EIT_o) = \min\{tn_o, EIT_o + D_o\}$$

with $tn_o = tl_o + ta_o(s_o)$ as usual.

To be meaningful, we require that

$$lookahead_o\,(q_o, tl_o, EIT_o) \geq \min\{EIT_o, tn_o\},$$

i.e., the earliest output time estimate computed should not be smaller than the next possible event.

When the parallel simulator receives an initialization message it first forwards it to its DEVS-simulator which initializes the *tl* and *tn* times. Then the parallel simulator initializes its input queue to empty and its input time estimates to 0. Also, it computes its first output time estimate considering the lookahead. This is sent out and eventually will make events in other simulators safe.

When it receives an x-message, the input is not processed immediately but stored in the event list. When it receives new input time estimates, it updates its EOT and sends new estimates back for distribution. Outputs from its DEVS-simulator are forwarded directly to the parent output distributor.

The kernel of the parallel simulator is the infinite event loop, which processes external and internal events as long as it is safe to do so. It sends *- or x-messages to its associated sequential simulator, whichever is next. After each event the EOT estimate is updated and new estimates are sent out.

14.5.2 OPTIMISTIC DEVS SIMULATOR

Optimistic parallel discrete event simulation stands in contrast to the conservative approaches as it take risks that might violate causality. The simulator continues to execute the events even when it cannot guarantee that they are safe to execute. As a consequence, a simulator may get ahead in simulated time of its influencers and, therefore, may receive a message with time smaller than its local simulation time, a so-called *straggler* event. That means a causality violation has occurred – it has to be detected and repaired by "rolling back" in simulated time to take into account the new message. At a rollback a simulator has to

- set back the state before the straggler event,
- annihilate any outputs already sent with simulated time greater than the time of the straggler event,
- proceed forward again by redoing all the input events in correct order.

Obviously to accomplish this, an optimistic simulator must store

- the state information to set back the state,
- the input messages to redo the inputs after rollback,
- the outputs to be able to annihilate the output events later than the rollback time.

In the following we try to illustrate rollback by an example. Fig. 14.12 shows a sample simulation run by means of the input, state, and output trajectories. In the simulation shown in Fig. 14.12A three inputs i_1, i_2 and i_3 are received and processed and outputs o_1 and o_2 are sent until time 9. In Fig. 14.12B a

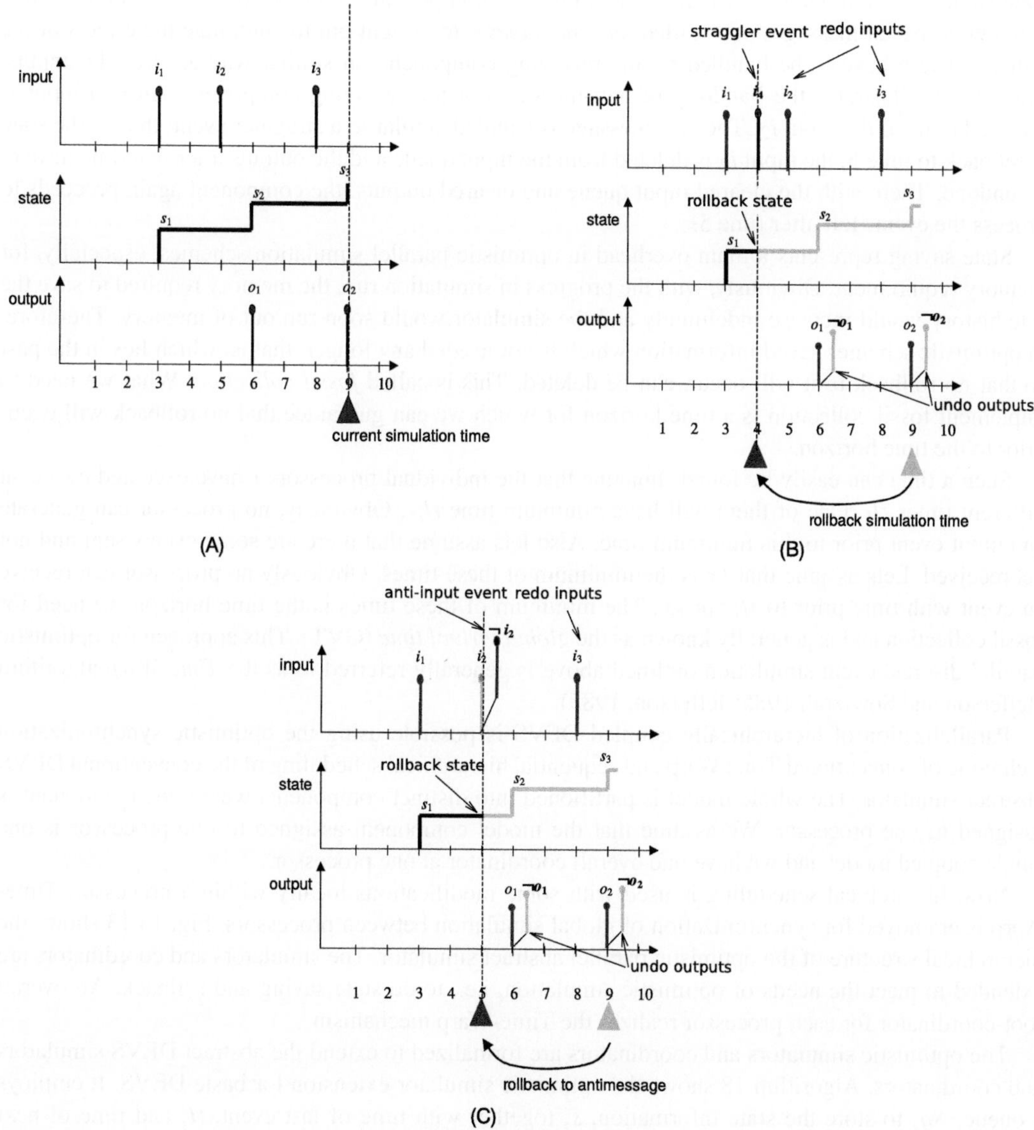

FIGURE 14.12

Straggler event and rollback. (A) Simulation situation with events processed until time 9. (B) Straggler event occurred at time 4. (C) Anti-input $\neg j_2$ received at time 5.

straggler input event i_4 is received at time 4. This necessitates to set the simulation time back to time 4, reset the state at that time, undo the outputs at time 6 and 9, and then step again forward in time, that means, process the straggler input i_4 and redo processing of the inputs i_2 and i_3.

While setting back the state in case of a rollback is straight forward and only requires to store past state values, to undo outputs will affect the influenced components and therefore can be cumbersome. It requires special messages – so-called *anti-messages* – to be sent out to annihilate the effects of the outputs. These have to be handled by the receiving components in similar way as straggler inputs. Fig. 14.12C illustrates the handling of anti-messages at the receiving component. An anti-input is received to undo the input i_2. The anti-message is handled similar to a straggler event, that is, the state is set back to time 5, the input i_2 is deleted from the input queue and the outputs after that time have to be undone. Then, with the cleared input queue and cleared outputs, the component again proceeds to process the events left after time 5.

State saving represents a main overhead in optimistic parallel simulation schemes, especially, for memory requirement. Obviously, with the progress in simulation run, the memory required to save the sate history would increase indefinitely and the simulator would soon run out of memory. Therefore, in optimistic schemes saved information which is not needed any longer, that is, which lies in the past so that no rollback to it will occur, can be deleted. This is called *fossil collection*. What we need to implement fossil collection is a time horizon for which we can guarantee that no rollback will occur prior to the time horizon.

Such a time can easily be found. Imagine that the individual processors i have executed events at different times tl_i. One of them will have minimum time tl_{i*}. Obviously, no processor can generate an output event prior to this minimum time. Also lets assume that there are some events sent and not yet received. Lets assume that tx is the minimum of these times. Obviously no processor can receive an event with time prior to tl_{i*} or tx. The minimum of these times is the time horizon we need for fossil collection and is generally known as the *global virtual time* (GVT). This approach for optimistic parallel discrete event simulation outlined above is generally referred to as the *Time-Warp* algorithm (Jefferson and Sowizral, 1985; Jefferson, 1985).

Parallelization of hierarchically coupled DEVS is possible using the optimistic synchronization technique of conventional Time Warp and sequential hierarchical scheduling of the conventional DEVS abstract simulator. The whole model is partitioned into distinct components where one component is assigned to one processor. We assume that the model component assigned to one processor is one single coupled model and we have one overall coordinator at one processor.

Now, hierarchical scheduling is used with some modifications locally within a processor, Time-Warp is employed for synchronization of global simulation between processors. Fig. 14.13 shows the hierarchical structure of the optimistic parallel abstract simulator. The simulators and coordinators are extended to meet the needs of optimistic simulation, i.e., to do state saving and rollback. An overall root-coordinator for each processor realizes the Time-Warp mechanism.

The optimistic simulators and coordinators are formalized to extend the abstract DEVS simulators and coordinators. Algorithm 18 shows the optimistic simulator extension for basic DEVS. It employs a queue, Sq, to store the state information, s, together with time of last event, tl, and time of next event, tn. At each conventional message, initialization, internal event, and external event, the message is forwarded to the conventional devs-simulator and then state and times are saved. Additionally the optimistic simulator realizes the rollback message which looks up the state queue Sq to restore state s and times tl and tn for the time of the rollback.

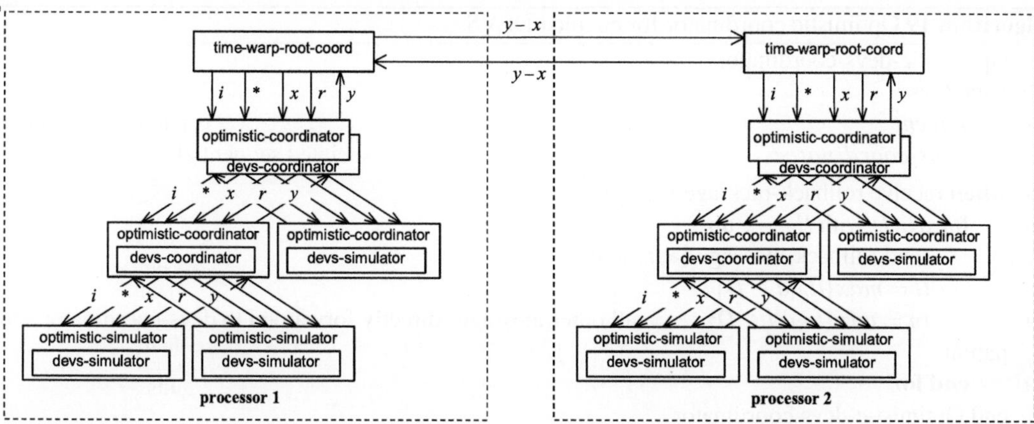

FIGURE 14.13

Structure of the parallel optimistic abstract simulator.

Algorithm 18 Optimistic simulator for basic DEVS.

Optimistic-devs-simulator
variables :
  parent ▷ parent coordinator
  devs-simulator ▷ associated sequential DEVS simulator
  Sq ▷ queue to store the state s plus times tl and tn
when receive i-message (i, t) at time t
  send i-message (i, t) to *devs-simulator*
  initialize *Sq* by inserting initial state and times (s, tl, tn)
when receive x-message (x, t) at time t with input value x
  send x-message (x, t) to *devs-simulator*
  save state and times (s, tl, tn) in Sq
when receive $*$-message $(*, t)$ at time t
  send $*$-message $(*, t)$ to *devs-simulator*
  save state and times (s, tl, tn) in Sq
when receive rollback-message $(*, t)$ at time t
  roll back to state (s, tl, tn) at time t
when receive y-message (y, t) with output y from *devs-simulator*
  send y-message (y, t) to parent
end Optimistic-devs-simulator

The coordinator extension for optimistic simulation (Algorithm 19) forwards the conventional messages unchanged to the subordinates or parent. Additionally, it takes care of rollbacks. When receiving a rollback message for time t, it forwards it to all children d with time of last event tl_d greater than the

Algorithm 19 Optimistic coordinator for coupled DEVS.

```
Optimistic-devs-coordinator
variables :
    parent                                   ▷ parent coordinator
    devs-coordinator                         ▷ associated sequential DEVS simulator
when receive rollback-message (∗, t) at time t
    for dinD with tl_d > t do
        send rollback-message (∗, t) to d
        tl = max{tl_d|d ∈ D}
        tn = min{tn_d|d ∈ D}    ▷ all other messages directly forwarded to devs-coordinator and parent
    end for
end Optimistic-devs-coordinator
```

time t of the rollback. In that way the global state of the associated coupled model is set back to the state at time t. Afterwards the times tl and tn are reinitialized based on the times of the children.

An overall root-coordinator in each processor realizes the Time-Warp mechanism (Algorithm 20). In particular, it stores input and output events from, and to, other processors, takes care of anti-messages, and is involved in global synchronization (GVT and *global-IOq*). In an event loop the root-coordinator sends $*$-messages for next internal events or x-messages for queued input events to its child to process them optimistically. In between, it takes care for global synchronization and fossil collection. Also, whenever it asynchronously receives messages from other processors, it handles them. When an output message is received from the child, the output is stored in the output queue Oq and then sent to the influenced processors immediately. Also it is stored in the global event queue, $pending\text{-}IOq$, to mark it as a sent, but not yet received message. When the output is transmitted to the receiving processor as an input in an x-message, it takes the output event out from the pending global event queue, $pending\text{-}IOq$, and stores the input event in the input queue. Then it sees if it is a straggler event. If so, a rollback message is sent to the child to restore the state at time of the rollback and anti-messages are sent out for all outputs in the output queue with later time. When an anti-message requests cancellation of an input event, the corresponding input message is deleted from the input queue and a rollback is carried out in analogous manner.

Global event synchronization is through the global virtual time, *GVT*. It is computed as the minimum of the last event times tl_i in all processors i and the minimum of the pending events in $pending\text{-}IOq$. Fossil collection can be carried out based on the *GVT*, i.e., all events prior to *GVT* can be are freed (usually this is carried out periodically by traversing all simulators and examining their state queues Sq).

14.5.3 CRITICAL PATHS IN OPTIMISTIC AND CONSERVATIVE SIMULATORS

A state in the causality graph that has more than one predecessor is a synchronization point. At these points, it is necessary to wait while assembling all of the causally preceding information before continuing with the simulation. These points appear, for example, at the forks and joins illustrated in Fig. 14.6 and at each state in Fig. 14.10 that has more than one predecessor. The ideal behavior of a conservative

Algorithm 20 Time-Warp-root-coordinator.

```
Time-Warp-root-coordinator for processor i
variables :
    t                 ▷ current simulation time
    child             ▷ subordinate optimistic simulator or coordinator
    Iq                ▷ queue to store the input events
    Oq                ▷ queue to store the output events
    pending-IOq       ▷ global queue to store the sent, not yet received output events
    tl                ▷ time of last event within the scope of the root-coordinator
loop                                                          ▷ event-loop
    t = t_0
    initialize Iq and Oq empty
    send initialization message (i, t) to child
    tl = tl_child
    while true do
        let (x, tx) be next input event in Iq with smallest time tx
        if tx ≤ tn_child then
            t = tx
            send (x, t) to child
        else
            t = tn_child
            send (*, t) message to child
        end if
        tl = tl_child
        GVT = min{min{tl_i | for all processors i}, min{ty|(y, ty) ∈ global-IOq}}
        do fossil collection for states, inputs, and outputs with time t < GVT
    end while
end loop
when receive y-message (y, t) from child
    save output event (y, t) in Oq
    send request to put (y, t) into global pending-IOq
    send (y, t) to influenced processor
when receive x-message (x, t) from influencing processor
    send request to remove output (y, t) correlating to input event (x, t) from pending-IOq
    save input event (x, t) in Iq
    if t < tnchild then                                       ▷ rollback
        send rollback message (*, t) to child
    end if
    for output events (y, ty) in Oq with ty > t do
        send anti-messages (*y, ty) to influenced processors
        delete (y, ty) from output queue Oq
    end for
when receive anti-message (*x, t) from influencing processor
```

```
    if t < tn_child then                                              ▷ rollback
        send rollback message (*, t) to child
    end if
    for output events (y, ty) in Oq with ty > t do
        send anti-messages (*y, ty) to influenced processors
        delete (y, ty) from output queue Oq
    end for
    delete input event (x, t) corresponding to (*x, t) from Iq
end Time-Warp-root-coordinator
```

or optimistic simulator is to detect these synchronization points while executing sequences between synchronization points in parallel.

To illustrate the benefit of recognizing these synchronization points, consider again the graph shown in Fig. 14.10. As we had determined previously, sequential execution requires nine units of time and the Parallel DEVS algorithm accelerates the simulation such that it requires only six. An ideal optimistic or conservative implementation can execute in parallel all of the simultaneous state transitions, and it can also execute the sequence $s_{1,2}$, $s_{1,3}$ in parallel with the sequence $s_{3,2}$, $s_{3,4}$. Consequently, the total cost of executing these sequences is two rather than four, and the total execution time for the simulation model is four.

The potential benefit of an optimistic or conservative algorithm is even greater for a model with structures like the one shown in Fig. 14.6 and Fig. 14.7. To illustrate the parallel potential in such a model, suppose that the fork generates a single output and then transitions to a state in which it is passive. Similarly, suppose the join undergoes two state transitions induced by input arriving from the models at end of the upper and lower branches. Within each branch, a number a and b of state transitions occur in the upper and lower branches respectively. The causal graph for this model is shown in Fig. 14.14. Again, let us assume that each state transition costs us a single unit of time to execute.

A sequential execution of this model requires $5 + a + b$ units of time: the final state transitions for each branch and the fork (three), the two state transitions of the join (three plus two is five), and the intermediate transitions along each branch. The Parallel DEVS algorithm can execute the state transitions to $s_{U,1}$, $s_{F,f}$, and $s_{L,1}$ in parallel; the state transitions $s_{L,f}$ and $s_{J,1}$ in parallel; and the state transitions $s_{U,f}$ and $s_{J,f}$ in parallel. This reduces the total cost of the simulation by four to make it $1 + a + b$. An ideal conservative or optimistic algorithm can execute the sequences $s_{U,1}, \dots, s_{U,f}$, $s_{L,1}, \dots, s_{L,f}$, and $s_{F,f}$ concurrently. The state $s_{J,1}$ can occur in parallel with $s_{L,f}$, and only $s_{J,f}$ must wait for the upper and lower branches to both be completed. By exploiting this parallelism, the total execution time can be reduced to $1 + \max\{a + 1, b + 1\}$, where the individual one is for $s_{J,f}$ and the other costs are due to the two branches. The total execution times in each case are shown below.

Sequential	$5 + a + b$
Parallel DEVS	$1 + a + b$
Ideal optimistic or conservative	$2 + \max\{a, b\}$

We can see that models with long chains of state transitions have the potential for substantially accelerated simulation. Indeed, it is sometimes possible in practice for the optimistic and conservative

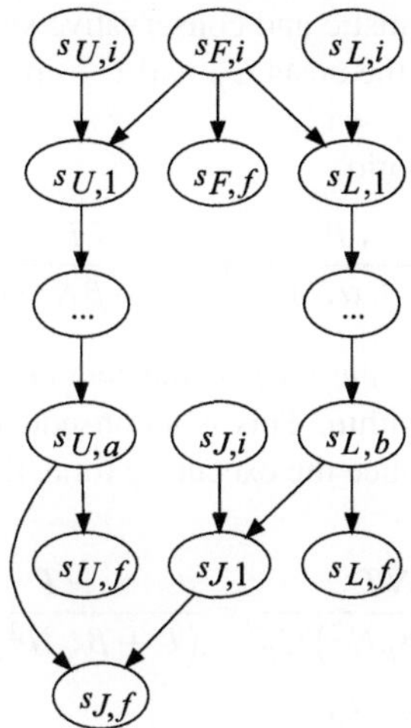

FIGURE 14.14

Causal graph for a model structured as shown in Fig. 14.6.

algorithms to very nearly achieve the best possible execution speed that would be indicated by a critical path analysis. In general, this speedup is achieved by the PDEVS algorithm only in the best case that is illustrated in Fig. 14.9.

This fact explains the tremendous interest in optimistic and conservative simulation algorithms as topic of research in parallel computing. A vast body of literature documents successful efforts to realize this idealized performance for simulation models that exhibit state transition graphs like that shown in Fig. 14.14. Certainly there are large numbers of highly relevant scientific and engineering problems that fall into this category, and the performance of optimistic and conservative algorithms for these models has been studied extensively. In these cases, the additional information obtained by a much faster computation justifies the effort required to realize the rather complicated simulation procedure. At the same time, these important and relevant problems nonetheless constitute a minority of the calculations done in practice to support an engineering activity, and the economics of information may explain the relative absence of advanced optimistic and conservative simulation algorithms in the most widely used, general purpose simulation tools for discrete event models.

To explore this idea, consider the following thought experiment. We have already seen that for all DEVS simulation algorithms, if there are m model components in a simulation then the best possible speedup is m. The actual speedup will be between zero and m where factors such as the level of state-transition activity, the coupling among models, and the causal structure induced by a particular combination of coupling and state transition activity will determine where the speedup falls (Zeigler, 2017; Srinivasan and Reynolds, 1993). This can be reflected in time overhead parameter β_p for the Parallel DEVS algorithm and β_{oc} for the optimistic or conservative algorithm. The actual value of this parameter will depend strongly on the causal graph of the model, but in general we expect $\beta_{oc} < \beta_p$. We might also wish to increase the size of the model by adding new dynamics, details, and other elements requiring more memory. The capacity to do this will depend on the size overheads α_p and α_{oc}. A brief examination of the algorithms for optimistic and conservative simulation supports the conclusion that in general $\alpha_p < \alpha_{oc}$. Finally, there are the costs C_p and C_{oc} in terms of time and effort to realize the parallel simulations where in general $C_p < C_{oc}$.

Now recall the information to cost ratio

$$\frac{I}{C} = \frac{NP}{(P + \alpha N)\,C} + \frac{NT}{(T + \beta N^2)\,C}.$$

If we are interested only in increasing the size of the model, then the information to cost ratio is greatest using the Parallel DEVS algorithm. This is a consequence of the expectation that $\alpha_p < \alpha_{oc}$ and $C_p < C_{oc}$. If we wish chiefly to reduce the execution time, then our choice of algorithm hinges on the equality

$$\frac{NT}{(T + \beta_p N^2)\,C_p} < \frac{NT}{(T + \beta_{oc} N^2)\,C_{oc}}$$

which can be reduced to the ratio $\frac{(T+\beta_p N^2)C_p}{(T+\beta_{oc} N^2)C_{oc}}$.

If this ratio is greater than one, then we will choose an optimistic or conservative algorithm and the Parallel DEVS algorithm otherwise.

We can make several observations immediately. If N^2 is very small relative to T or $C_p \ll C_{oc}$ then the cost terms dominate this ratio and so the Parallel DEVS algorithm will be preferred. Both circumstances are often the case for large engineering models constructed using commercial discrete event simulation tools. These generally begin as small models designed to run on multicore workstations, and the execution time of the model grows steadily as design details and test data accumulate over the course of a project. Costs sunk into the early model development amplify the cost of switching to an optimistic or conservative algorithm, which may require modifications to or further analysis of the model. The simplicity of the Parallel DEVS algorithms tends to separate its cost of use from the size and complexity of the model, allowing the cost of switching to be essentially constant over the course of the project. For the same reason, N is restricted by the initial choice of computer and simulation package.

On the other hand, if N^2 is very large relative to T then the ratio reduces to

$$\frac{(\beta_p)\,C_p}{(\beta_{oc})\,C_{oc}}$$

and the products will determine the decision. Since $C_p < C_{oc}$ is assumed, we must have $\beta_{oc} \ll \beta_p$ to prefer an optimistic or conservative simulation algorithm. This situation is the defining characteristic of supercomputing applications to which these algorithms have been most successfully applied. Supercomputing applications emerge when T becomes very large and reducing it to an acceptable level requires specialized computers with very large N. To make good use of such a computer requires algorithms with sufficiently small β that the N^2 term plays little to no role in determining speedup. Because this criteria is met by the most sophisticated optimistic and conservative parallel discrete event simulation algorithms, we find that $\beta_{oc} \ll \beta_p$. Hence, these types of algorithms, rather than simpler but less costly algorithms, have evolved to become supercomputing tools.

Between these limits, the decision to employ one algorithm or another must be decided on the basis of the modeler's intuition about the patterns of behavior that will emerge in the final simulation

Table 14.1 Comparison of PDES Approaches

Scheme	Approach	Overhead	Pros	Cons
Chandy-Misra Conservative	Process events in strict time stamped order (Causality Preservation) w/Deadlock Avoidance	Null Messages Lookahead Computation	Relatively Simple to Implement	Does not exploit all parallelism and not intrinsically load balancing Relies on lookahead capabilities of the models Only works for heavy message load and uniform message distribution Simultaneous Events are Problematic
DEVS Version	Exploits knowledge of internal events in lookahead			
Time Warp Optimistic	Permits Causality Violation Detects Violations and Remedies using Rollback	State, Message Saving, Fossil Collection Anti-messages Global Virtual Time Computation	Can exploit feedback free couplings	Complex logic is difficult to implement and verify Simultaneous Events are Problematic
Riskfree DEVS Version	Refrain from output until safe	Similar to Time Warp except no Anti-messages and Rollback is local	Relatively Easy to Implement	
Parallel DEVS	Riskfree and Strict Causality Adherence	Global Minimum Time Synchronization Simultaneous Output Collection and Distribution	Easy to Implement Exploits Simultaneous Events Works well for active models with feedback couplings	Does not exploit parallelism in feedback free couplings

program, the perceived need for a large or more rapidly executing simulation model, and the extent to which parallel experiments rather than parallel simulation can satisfy the information requirements of a particular simulation task.

14.5.4 SURVEY OF OPTIMISTIC AND CONSERVATIVE SIMULATION ALGORITHMS

The simulation protocols for parallel discrete event simulation have been a subject of intense study for nearly four decades. Table 14.1 compares several widely known protocols for distributed parallel simulation of discrete event models. As you can see, all have their pros and cons. The selection of a right protocol is heavily dependent on the model in hand as has to be considered carefully. The protocols selected for this table appear in detail in the second edition of Theory and Modeling and Simulation, and their origins can be traced to seminal work on the subject of parallelizing discrete event models.

In some of the field's earliest work, Chandy and Misra introduced basic concepts for conservative distributed simulation (Chandy and Misra, 1979, 1981). They defined causality requirements for correct distributed execution of events and schemes to fulfill them. Several variations and improvements have been developed. Notable impacts came from the Bounded Lag Algorithm of Lubachevsky (1989), the global window synchronization scheme of Nicol and Roy (1991), the Yaddes algorithm (Berthomieu and Diaz, 1991), and the shared memory implementation of Wagner et al. (1989). Lubachevsky's algorithm is synchronous which, with every synchronization cycle, computes input time estimates for each simulator process based on minimum propagation delays and so-called *opaque periods* – lookahead computed dynamically based on the current state. To limit the overhead needed to compute the input time estimates, the bounded lag restriction bounds the difference in the local simulation times of all simulator processes from above by a known finite constant B. The global window synchronization algorithm (Nicol, 1991; Nicol and Roy, 1991) is another synchronous algorithm which tries to exploit static and dynamic lookahead capabilities to define *global time windows* during which event processing is safe. The Yaddes algorithm (DeBenedictis et al., 1991) uses a data flow network to compute input time estimates to guarantee safe event processing. This algorithm in particular is tailored for network models which have a lot of feedback loops, like digital logic models, and therefore are difficult to parallelize using Chandy-Misra or Time Warp.

Optimistic distributed simulation has been introduced by Jefferson (1985) and implemented in the Time-Warp operating system (Jefferson et al., 1987). Ferscha (1995) defines an adaptive optimistic scheme where sending of optimistic interprocessor events is assessed for risk before sending. Riskfree optimistic scheme are first introduced by Dickens et al. (1990) and have been used in the implementations of the Breathing Time Buckets algorithm by Steinman (1992). Excellent reviews of the different approaches can be found in Fujimoto (1990) and Burden and Faires (1989), and recent research on using reversible computations to reduce the cost of recovery from causality errors is described in Perumalla (2013).

Page (1993, 1998) pointed out that research on PDES has often not related itself to the broader context of modeling and simulation methodology in general. Constraints imposed by other parts of the model development life cycle may have greater impact than the local considerations characterizing PDES research. Certainly the economic considerations introduced in this Chapter are among some of the most significant of these constraints in any engineering application. Also, considerations of modeling formalisms can provide useful guidance. For example, classical world views, discussed in Chapter 7, can provide useful guidance for PDES methods (Page, 1998).

The PDS research relating to the DEVS formalism has from the start worked within the modeling and simulation methodology described in this book. Parallel simulators for DEVS modular hierarchical models have been implemented by different researchers (Concepcion and Zeigler, 1988; Zeigler and Zhang, 1990). Reisinger and Praehofer (1995) implemented a combined conservative/riskfree optimistic simulator based on the EOT-EIT network presented in Section 11.2.1 of the second edition of this book. Christensen implemented a Time-Warp DEVS-simulator in Ada using the Time-Warp operating system (Christensen, 1990). Parallel simulation of DEVS models by parallelizing simultaneous external and simultaneous internal events and a mapping of hierarchical DEVS models in a hypercube computer is presented in Seong et al. (1995a,b). Discussion on ordering of simultaneous events in parallel simulation of DEVS models can be found in Kim et al. (1997) with recent solutions based on super dense time appearing in Nutaro (2011). Time Warp parallel simulation of hierarchical DEVS with local hierarchical scheduling instrumented is presented in Kim et al. (1996) with experimental results for a benchmark with various workloads. A riskfree version according to the approach presented in Section 11.3.2 of the second edition has been implemented by Reisinger.

Parallel DEVS and its distributed abstract simulator were formulated by Chow (1996). The DEVS-C++ environment (Zeigler et al., 1997) implements Parallel DEVS using an approach based on closure under coupling (Zeigler and Kim, 1996).

14.5.5 A STATISTICAL APPROACH TO SPEEDUP

As we have noted, it is often impractical to anticipate the structure of a model's causal graph and this problem is particularly acute for simulation tools that are intended for a very broad range of modeling problems. In these circumstances it may be useful to take a statistical approach to understanding the expected value of a particular simulation algorithm given a large sample of existing simulation models. We proceed to develop such an approach here following that of Zeigler et al. (2015), Zeigler (2017).

Let us suppose we have the causal graphs for a large set of simulation runs that are representative of the prior use of a sequentially executing simulation tool. Each component appearing in a model that produced one of these runs generates a path through the causal graph that describes its state transitions. Let the distribution of path lengths appearing in the set of causal graphs be $\boldsymbol{x}$. If we assemble a single model from m components, then the causal graph for a simulation of this model has m samples of $\boldsymbol{x}$. The expected total execution time is

$$Sum\,(m) = E\,\{x_1 + \cdots + x_m\} = mE\{\boldsymbol{x}\}\;.$$

Within this causal graph will be a path of maximum length expressed as the random variable, $max\,\{x_1,\ \ldots,\ x_m\}$. Its expected value is

$$Max\,(m) = E\,\{\max\,\{x_1,\ \ldots,\ x_m\}\}\,.$$

The expected value for the best possible speedup that can be achieved is therefore

$$\frac{Sum(m)}{Max(m)}\,.$$

It is convenient for the sake of illustration to take $\boldsymbol{x}$ to be an exponentially distributed random variable with a mean of one. Doing so trivially yields $Sum\,(m) = m$. Further, if we assume that the samples for the m component paths are independent, then for sufficiently large m we can approximate $Max\,(m) = 0.577 + \ln m$. The expected speedup for a model with m components is approximately

$$\frac{m}{0.577 + \ln m}$$

and the speedup increases as $\mathrm{O}\,(m/\ln m\,)$, which is almost linear in m (since the log increases very slowly for large m).

This approximation depends crucially on independent samples of the path lengths. In practice, this requirement is met with causal graphs like the one shown in Fig. 14.14. These causal graphs will tend to have relatively few simultaneous events in parallel paths. Hence, the Parallel DEVS algorithm will have difficulty using large numbers of processors to speed up this model. This was illustrated with the calculations in Section 14.4. However, this type of causal graph combined with large numbers of processors is what motivated the development of optimistic and conservative algorithms. This suggests that if we can obtain a number of processors close to m and there is good reason to expect infrequent interactions between component models then it may be advisable to offer conservative and optimistic algorithms as part of a simulation package.

On the other hand, we have argued before that for general purpose simulation tools, and particularly when those tools are well established, it is impractical to offer support for optimistic or conservative algorithms, very large numbers of processors, or both. Instead, we will be interested in models with large m, unknown patterns of interaction, and a number of computer processors $N \ll m$. In this circumstance it is instructive to look at the potential speedup achievable by the inexpensive-to-apply Parallel DEVS simulator.

To do so, let us characterize the fraction of state transitions in a causal graph that are simultaneous by a random variable s. For a model with m components, we can expect $mE\{s\}$ such transitions in each iteration of the Parallel DEVS algorithm, and this is also the number of processors that can be effectively utilized by the algorithm. Hence, the expected speedup cannot improve beyond $mE\{s\}$.

Therefore let us consider the case where $N \leq mE\{s\}$, which will occur in any practical model if m becomes sufficiently large or N is sufficiently small. We can maximize our chances in each iteration of using all available processors by assigning m/N components to each processor in a way that maximizes the number of couplings between models residing on separate processors. Specifically, we strive to arrange the models so that the work assigned to each processor in each iteration is close to $mE\{s\}/N$. Because $mE\{s\}$ is the expected effort for each iteration of the serially executing simulator it follows that the best possible speedup will be

$$\frac{mE\{s\}}{mE\{s\}/N} = N .$$

This matches the best possible speedup achievable with the much more expensive conservative and optimistic algorithms. We conclude that the Parallel DEVS algorithm offers, in many cases of practical interest, an economically attractive alternative to optimistic and conservative simulation approaches.

14.6 SUMMARY

This chapter discussed the applications of parallel and distributed discrete event simulation (PDES) algorithms for discrete event models. Although focused on DEVS models, many of the conclusions apply generally due to the universality of DEVS for discrete event systems (Chapter 17). In this edition, based on experience with implementation of Parallel DEVS simulation algorithms, we presented a Constructive Cost Model (COCOMO) of PDES software development that takes into account both the objectives and the structures of the models to be simulated. Also, we focused on the critical path of a simulation as the most costly sequence of computations that must be executed to complete a simulation run. This approach allows us to better understand the strengths and limitations of traditional PDES algorithms and how the canonical Parallel DEVS simulation protocol provides a vehicle for general robust simulation with near optimal performance from the perspective of our cost-based approach to simulation.

Our conclusions are based on the state-based critical path analysis taken which does not account for simultaneous parallel execution of events with identical time stamps regardless of their causal relationship determined by state transition functions. The cost model and critical path approaches started here open a new vista for research into the relative costs and performances of conservative, optimistic, and Parallel DEVS canonical algorithms. One direction is to consider the effects of distinct chains of events containing state transitions that are scheduled to happen at the same time. Often such chains are immediately identifiable, for example, in fixed time step models (Chapter 6) and they may occur in combination with state-produced events as in co-simulation (Chapter 12). The stochastic model of Parallel DEVS simulation presented by Zeigler (2017) takes a step in the direction of comparing the relative performance of various synchronous protocols for Parallel DEVS under combinations of internally and externally caused events. Empirical studies comparing the performance of such protocols are starting to appear at this writing (Cardoen et al., 2016).

REFERENCES

Berthomieu, B., Diaz, M., 1991. Modeling and verification of time dependent systems using time Petri nets. IEEE Transactions on Software Engineering 17 (3).

Boehm, B.W., et al., 1981. Software Engineering Economics, vol. 197. Prentice-Hall, Englewood Cliffs (NJ).

Cardoen, Ben, Manhaeve, Stijn, Tuijn, Tim, et al. Performance analysis of a PDEVS simulator supporting multiple synchronization protocols. In: SpringSim-TMS/DEVS, 2016 April 3–6. Pasadena, CA, USA.

Chandy, K.M., Misra, J., 1979. Distributed simulation: a case study in design and verification of distributed programs. IEEE Transactions on Software Engineering 5, 440–452.

Chandy, K.M., Misra, J., 1981. Asynchronous distributed simulation via a sequence of parallel computations. Communications of the ACM 24 (4), 198–206.

Chow, A.C.H., 1996. Parallel devs: a parallel, hierarchical, modular modeling formalism and its distributed simulator. Transactions of the Society for Computer Simulation 13 (2), 55–102.

Christensen, E.R., 1990. Hierarchical Optimistic Distributed Simulation: Combining Devs and Time Warp.

Concepcion, A.I., Zeigler, B.P., 1988. Devs formalism: a framework for hierarchical model development. IEEE Transactions on Software Engineering 14 (2), 228–241.

DeBenedictis, E., Ghosh, S., Yu, M.-L., 1991. A novel algorithm for discrete-event simulation: asynchronous distributed discrete-event simulation algorithm for cyclic circuits using a dataflow network. Computer 24 (6), 21–33.

Dickens, P.M., Reynolds, P.F., et al., 1990. Srads with Local Rollback.

Ferscha, A., 1995. Probabilistic adaptive direct optimism control in time warp. In: ACM SIGSIM Simulation Digest, vol. 25. IEEE Computer Society, pp. 120–129.

Fujimoto, R.M., 1990. Parallel discrete event simulation. Communications of the ACM 33 (10), 30–53.

Jefferson, D., Beckman, B., Wieland, F., Blume, L., DiLoreto, M., 1987. Time Warp Operating System, vol. 21. ACM.

Jefferson, D.R., 1985. Virtual time. ACM Transactions on Programming Languages and Systems (TOPLAS) 7 (3), 404–425.

Jefferson, D., Sowizral, H., 1985. Fast Concurrent Simulation Using the Time Warp Mechanism. Part I. Local Control, pp. 63–69.

Kim, K.H., Seong, Y.R., Kim, T.G., Park, K.H., 1996. Distributed simulation of hierarchical devs models: hierarchical scheduling locally and time warp globally. Transactions 13 (3).

Kim, K.H., Seong, Y.R., Kim, T.G., Park, K.H., 1997. Ordering of simultaneous events in distributed devs simulation. Simulation Practice and Theory 5 (3), 253–268.

Lamport, L., 1978. Time, clocks, and the ordering of events in a distributed system. Communications of the ACM 21 (7), 558–565.

Lubachevsky, B.D., 1989. Efficient distributed event-driven simulations of multiple-loop networks. Communications of the ACM 32 (1), 111–131.

Misra, J., 1986. Distributed discrete-event simulation. ACM Computing Surveys (CSUR) 18 (1), 39–65.

Nicol, D.M., 1991. Performance bounds on parallel self-initiating discrete-event simulations. ACM Transactions on Modeling and Computer Simulation (TOMACS) 1 (1), 24–50.

Nicol, D.M., Roy, S., 1991. Parallel simulation of timed Petri-nets. In: Simulation Conference. Proceedings. 1991, Winter. IEEE, pp. 574–583.

Nutaro, J.J., 2011. Building Software for Simulation: Theory and Algorithms, with Applications in C++. John Wiley & Sons.

Overeinder, B.J., Sloot, P.M., 1995. Parallel performance evaluation through critical path analysis. In: International Conference on High-Performance Computing and Networking. Springer, pp. 634–639.

Page, E.H., 1993. In defense of discrete-event simulation. ACM Transactions on Modeling and Computer Simulation (TOMACS) 3 (4), 281–283.

Page, E.H., 1998. Zero Lookahead in a Distributed Time-Stepped Simulation.

Perumalla, K.S., 2013. Introduction to Reversible Computing. CRC Press.

Reisinger, G., Praehofer, H., 1995. Object oriented realization of a parallel discrete event simulator. In: EUROSIM, pp. 327–332.

Burden, Richard L., Faires, J. Douglas, 1989. Numerical Analysis, fourth edition. PWS-KENT Publishing Company.

Seong, Y.R., Jung, S.H., Kim, T.G., Park, K.H., 1995a. Parallel simulation of hierarchical modular devs models: a modified time warp approach. International Journal in Computer Simulation 5 (3), 263–285.

Seong, Y.R., Kim, T.G., Park, K.H., 1995b. Mapping hierarchical, modular discrete event models in a hypercube multicomputer. Simulation Practice and Theory 2 (6), 257–275.

Srinivasan, S., Reynolds, P., 1993. On Critical Path Analysis of Parallel Discrete Event Simulations. Computer Science Report No. TR-93-29.

Steinman, J.S., 1992. Speedes: a unified approach to parallel simulation. In: Proceedings of the 6th Workshop on Parallel and Distributed Simulation, vol. 24, pp. 75–83.

Wagner, D., Lazowska, E., Bershad, B., 1989. Techniques for efficient shared-memory parallel simulation. In: Proceedings of the 6th Workshop on Parallel and Distributed Simulation.

Zeigler, B.P., 2017. Using the parallel devs protocol for general robust simulation with near optimal performance. Computing in Science & Engineering 19 (3), 68–77.

Zeigler, B.P., Kim, D., 1996. Design of high level modelling/high performance simulation environments 26 (1), 154–161.

Zeigler, B.P., Moon, Y., Kim, D., Ball, G., 1997. The devs environment for high-performance modeling and simulation. IEEE Computational Science and Engineering 4 (3), 61–71.

Zeigler, B.P., Nutaro, J.J., Seo, C., 2015. What's the best possible speedup achievable in distributed simulation: Amdahl's law reconstructed. In: Proceedings of the Symposium on Theory of Modeling & Simulation: DEVS Integrative M&S Symposium. Society for Computer Simulation International, pp. 189–196.

Zeigler, B.P., Zhang, G., 1990. Mapping hierarchical discrete event models to multiprocessor systems: concepts, algorithm, and simulation. Journal of Parallel and Distributed Computing 9 (3), 271–281.

CHAPTER

HIERARCHY OF SYSTEM MORPHISMS

15

CONTENTS

The essence of modeling lies in establishing relations between pairs of system descriptions. As discussed in Chapters 1 and 2, these relations pertain to a variety of situations:

- the validity of representation of a real system by a model,
- the validity of a lumped model relative to a base model,
- the validity of a system description at one level of specification relative to a system description at a higher or lower level, and
- the correctness of a simulator with respect to a model.

Theory of Modeling and Simulation. https://doi.org/10.1016/B978-0-12-813370-5.00025-0

Based on the arrangement of system levels shown in Fig. 15.1, we distinguish between vertical and horizontal relations. A vertical relation is called an *association mapping*. It takes a system at one level of specification and generates its counterpart at another level of specification. The downward motion in the structure-to-behavior direction formally represents the process by which a simulator generates the behavior of a model. It was presented in Chapter 5. The opposite upward mapping relates a system description at a lower level a higher level one. While the downward association of specifications is straightforward, the upward association is much less so. Recall from Chapter 1, that this is because travel in the upward direction requires us to acquire or specify additional system knowledge. Many structures exhibit the same behavior. Thus recovering a unique structure from a given behavior is not possible except in special circumstances, called *justifying conditions*. Please refer to Chapters 15 and 16 of the first edition of Theory of Modeling and Simulation (Zeigler, 1976) for a detailed discussion of these conditions.

In this chapter we will treat the horizontal relations which relate systems at the same level of specification. Corresponding to each of the various levels at which a system may be known, described, or specified, is a relation appropriate to a pair of systems specified at that level. We call such a relation a *preservation relation* or *system morphism* because it establishes a correspondence between a pair of systems whereby features of the one system are preserved in the other.

For system morphisms, we take the point of view illustrated in Fig. 15.2, where S represents a "big" system and S' a "little" system. For example, S could be a base model and S' a lumped model. Accordingly, the basic orientation is that a part of the behavior of the big system S is mapped on the little system S' and the morphism is taken to be *surjective*. System S' therefore represents part of the system S with a certain degree of accuracy (Fig. 15.2A). On the other hand, we also may employ mappings which go in the other direction, i.e., from the small system to the big one and we have an *injective* mapping. This ensures that all the behavior of small S' is covered by big S (Fig. 15.2B).

We shall develop morphisms appropriate to each level of system specification. These morphisms are such that higher level morphisms imply lower level morphisms. This means that a morphism which

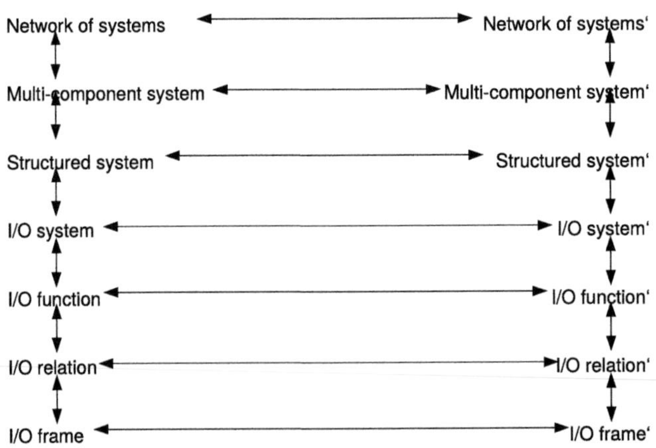

FIGURE 15.1

Hierarchy of system specifications, association mappings (vertical), and morphisms (horizontal).

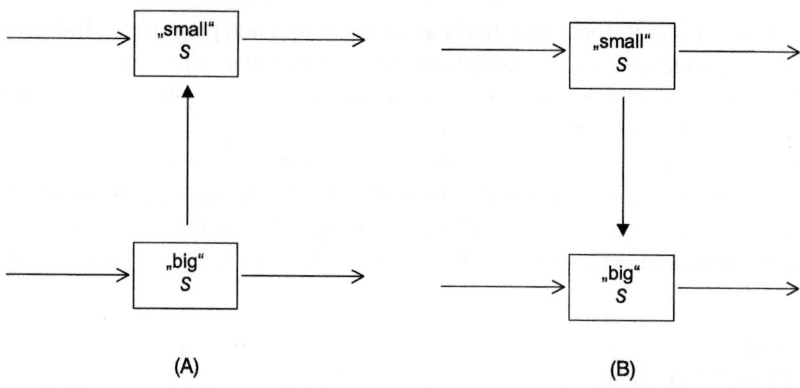

FIGURE 15.2

Morphisms between a big system and a small system. (A) Morphism representing the big by the small system. (B) Morphism embedding the small system in the big system.

preserves the structural features of one system in another system at one level, also preserves its features at all lower levels. We will show this employing the association mappings introduced in Chapter 5. For example, consider a morphism between systems at the I/O System Level in Fig. 15.1. This morphism must be defined so that it induces a morphism at the I/O Function Level between the associated systems at that level and indeed, between the associated systems at the I/O Relation and I/O Frame levels.

15.1 THE I/O FRAME MORPHISM

Recall from Chapter 5 that an I/O frame defines the input and output interface of a system in the form of an input and an output set. We relate two system $IO = (T, X, Y)$ and $IO' = (T', X', Y')$ at the I/O frame level by defining functions to relate the input and outputs interfaces. Let $g : (X', T') \to (X, T)$ be a function to derive a segment over X given a segment over X' and $k : (Y, T) \to^{onto} (Y', T')$ be a function to derive an output segment over output set over Y' of the little system given an output segment over Y of the big system S.

If input sets X and X' are identical, output sets Y and Y' are identical, and functions g and k are identity mappings, then IO and IO' are isomorphic at the I/O frame level. Systems isomorphic at the I/O frame level are said to be *compatible*.

15.2 THE I/O RELATION OBSERVATION MORPHISM

Let S and S' be represented by I/O relation observation (T, X, Ω, Y, R) and $(T', X', \Omega', Y', R')$, respectively. As indicated just now, we want to represent the condition wherein the I/O behavior R' of little S' can be "found in" the I/O behavior R of big S. We take this to mean that if we apply an input segment ω' to S' and observe an output segment ρ', there ought to be an I/O pair $(\omega, \rho) \in R$

corresponding to $(\omega', \rho') \in R'$ (note that this is an example of a mapping from a little system to a big system). This correspondence is to be established by mappings between the segment spaces as follows.

Let $g : \Omega' \to \Omega$. We call g an *encoding map* or *encoder*. If we apply $\omega' \in \Omega'$, it tells us which $\omega \in \Omega$, namely $g(\omega')$, to apply to S.

Let $k : (Y, T) \to^{onto} (Y', T')$. We call k an *decoding map*, or *decoder*. If we observe an output segment $\rho \in (Y, T)$ in the big system, it tells which $\rho' \in (Y', T')$ this represents, namely $k(\rho)$.

The condition that every pair $(\omega', \rho') \in R'$ has a corresponding pair $(\omega, \rho) \in R$ can now be stated as follows: an I/O relation morphism from $(T', X', \Omega', Y', R')$ to (T, X, Ω, Y, R) is a pair (g, k) such that

1. $g : \Omega' \to \Omega$
2. $k : (Y, T) \to^{onto} (Y', T')$
3. For every pair $(\omega', \rho') \in R'$, there exists $(\omega, \rho) \in R$ such that $\omega = g(\omega')$ and $k(\rho) = \rho'$

Let us now consider the other case, where a big system S is mapped to a little one S'. Here we are confronted with a situation where the small system has the represent the I/O behavior of the big system up to a certain accuracy.

Let $g : \Omega \to^{onto} \Omega'$ now be an encoding of the input trajectories from the big system S to the small system S'. If we apply $\omega \in \Omega$, it again tells us to apply $g(\omega)$ to S'. However, through this mapping some information in Ω might get lost. Therefore it is not possible to find an decoder k which uniquely determines an output segment of the big system. Instead we employ a mapping $k : (Y, T) \to^{onto} (Y', T')$ from the big system to the small system which allows us to set the I/O behavior of the big system and the small system in correspondence. We formulate this as follows

1. $g : \Omega \to^{onto} \Omega'$
2. $k : (Y, T) \to^{onto} (Y', T')$
3. for every pair $(\omega, \rho) \in R$, there exists $(\omega', \rho') \in R'$ such that $\omega' = g(\omega)$ and $\rho' = k(\rho)$

Which means that for every I/O relation in the big system S there is a I/O relation in S' so that these relations correspond through mappings g and k. Or in other words, whatever we can observe in the big system we also can observe in the little system, however, through the onto-mapping usually at coarser granularity.

EXAMPLE: SAMPLED DATA REPRESENTATION OF CONTINUOUS I/O SIGNALS

As an example of a I/O behavior morphism let us consider a discrete time encoding of continuous system input and output segments. The system S in Fig. 15.3 has a continuous I/O behavior. We encode the input and output segments of S by discrete time segments of system S', employing coding functions g and k as follows.

$$IORO = (T, X, \Omega, Y, R)$$

with $T = X = Y = \mathbb{R}$ and Ω is the set of bpc segments.

$$IORO' = (T', X', \Omega', Y', R')$$

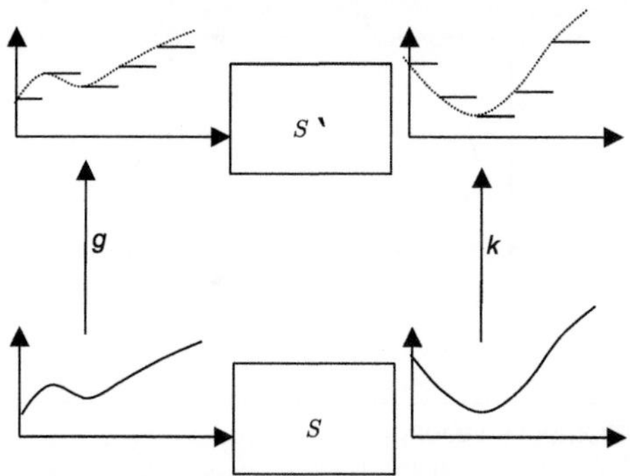

FIGURE 15.3

Digital representation of continuous I/O signals.

with $T' = X' = Y' = \mathbb{R}$ and Ω is the set of piecewise constant segments with constant length s (sample rate). The encoding function $g : \Omega \to \Omega'$ is defined by $g(\omega)(t) = \omega(t')$ with t' is a multiple of sample rate s and just smaller or equal to t. In the same way we define $k(\rho)(t) = \rho(t')$.

Obviously, in sampled representation some information has been lost and the original segments cannot be reconstructed to their full extent (however for bandwidth-constrained signals, there is a sufficiently high sampling rate that permits the original signal to be reconstructed to the desired accuracy).

15.3 THE I/O FUNCTION MORPHISM

Corresponding to the level of system knowledge characterized by the I/O function observation (where the starting state of an experiment is given) is the following preservation concept.

An I/O function morphism from an $IOFO = (T, X, \Omega, Y, F)$ to an $IOFO' = (T', X', \Omega', Y', F')$ is a pair (g, k) where

1. $g : \Omega' \to \Omega$
2. $k : (Y, T) \to^{onto} (Y', T')$
3. For each $f' \in F'$ there is an $f \in F$ such that $f' = k \circ f \circ g$, i.e., for all $\omega' \in \Omega'$, $f'(\omega') = k(f(g(\omega')))$.

We say that f and f' and g and k form a commutative diagram as shown in Fig. 15.4. This is interpreted as follows: suppose the little system S' is in a state whose I/O function is f'. Now an input segment ω' is injected into S' and an output segment $\rho' = f'(\omega')$ is observed – this is the top arrow. On the other hand, suppose the big system S is in a state whose I/O function is f. Then we encode ω' to obtain $\omega = g(\omega')$, inject ω into S, observe the result $\rho = f(\omega)$, and decode ρ to $k(\rho)$ – this is the arrow path labeled with g, f, and k. The result is the same for both processes, namely $f'(\omega') = k(f(g(\omega'))$.

Thus an I/O function morphism runs from S' to S, if and only if, each of the I/O functions of S' can be realized by an I/O function of S using the same pair (g, k) – every "capability" of little S' can be found in big S with uniform encoding and decoding rules.

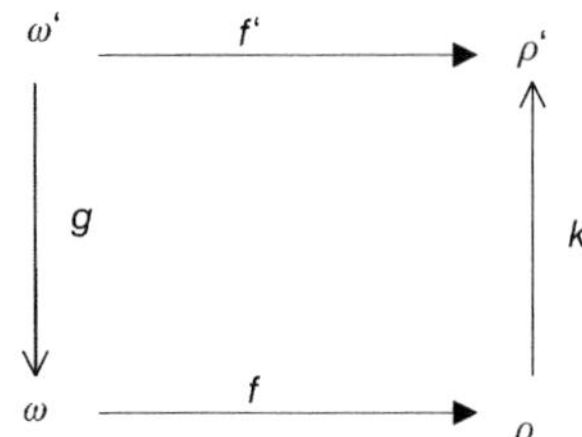

FIGURE 15.4

Commutative diagram of I/O function observation.

EXAMPLE: SCALING A SYSTEM TO DIFFERENT RATES

To relate two systems that are essentially the same except that they operate at different rates, let IOFO be derived from the differential equation

$$\frac{dy}{dt} = x$$

and the IOFO' be derived from the differential equation

$$\frac{dy}{dt} = \tau x$$

Then both systems IOFO and IOFO' are integrators, but IOFO' runs at a rate τ times that of IOFO (consider their responses to a constant input, for example).

Let $g : \Omega' \to \Omega$ be given by

$$g(\omega')(t) = \omega'\left(\frac{t}{\tau}\right)$$

with $l(g(\omega')) = l(\omega') \times \tau$, with $l(\omega)$ and $l(\omega')$ being the length of ω and ω', respectively.

Let $k : (Y, T) \to^{onto} (Y', T')$ be given

$$k(\rho)(t') = \rho(\tau t')$$

with $l(k(\rho)) = l(\rho) \times (1/\tau)$, with $l(\rho)$ and $l(\rho')$ again being the length of ρ and ρ'. Note that g dilates (stretches) a segment to τ times its original width, whereas k does the inverse dilation. (All segments are assumed to begin at zero.)

Exercise 15.1. Show that (g, k) is an I/O function morphism for IOFO and IOFO'.

Exercise 15.2. Show that if $T = T'$, $X = X'$, $\Omega = \Omega'$, and $Y = Y'$, the case $F' \subseteq F$ is given by the I/O function morphism (id, id), with id being the identity mapping.

Exercise 15.3. The above morphism at the IOFO observation level is an example of a morphism from a little system to a big system. Develop, analogous to Fig. 15.4, a corresponding IOFO morphism concept from a big system to a little system.

15.3.1 IOFO SYSTEM MORPHISM IMPLIES IORO SYSTEM MORPHISM

We now show that an I/O function observation morphism also implies the existence of a weaker morphism at the I/O relation observation.

Theorem 15.1. *If (g, k) is an I/O function morphism from $IOFO = (T, X, \Omega, Y, F)$ to $IOFO' = (T', X', \Omega', Y', F')$ then (g, k) is also an I/O relation morphism from the $IORO$ associated with $IOFO$ to the $IORO'$ associated with $IOFO'$.*

Proof. Recall that (T, X, Ω, Y, R) is the $IORO$ associated with (T, X, Ω, Y, F) where $R = \cup_{f \in F} f$.

Let $(\omega', \rho') \in R'$. Then $\rho' = f'(\omega')$ for some $f' \in F'$. Since (g, k) is an I/O function morphism, we have (from line 3 of the definition of the I/O function morphism) that for some $f \in F$

$$\rho' = f'(\omega') = k(f(g(\omega'))).$$

Call $\omega = g(\omega')$ and call $\rho = f(\omega)$ so that $(\omega, \rho) \in R$. Now by the foregoing equation $\rho' = k(\rho)$.

Thus for each $(\omega', \rho') \in R'$, there is a pair $(\omega, \rho) \in R$ with $\omega = g(\omega')$ and equation $\rho' = k(\rho)$ as required by line 3 of the I/O relation morphism definition.

Since (g, k) satisfies lines 1 and 2 of the I/O function and relation morphism directly, we have proved that (g, k) is an I/O relation morphism. □

Exercise 15.4. Show directly that $F' \subseteq F$ implies $R' \subseteq R$ (as would be inferred by applying Theorem 15.1 to (id, id)).

15.4 THE I/O SYSTEM MORPHISM

Moving to the next level, where more about the insides of S and S' is specified, we have, appropriately, a morphism that preserves the specified structural features. As a specialization, this morphism includes the classical notion of homomorphism (Chapter 1). Since the major element introduced at this level is the state space, we must include an additional mapping that establishes a correspondence between the big and little state sets (Fig. 15.5). We refer to the new morphism as a *system morphism* or a *generalized homomorphism*.

A system morphism from big system

$$S = (T, X, \Omega, Y, Q, \Delta, \Lambda)$$

to small system

$$S' = (T', X', \Omega', Y', Q', \Delta', \Lambda') \text{ is a triple } (g, h, k) \text{ such that}$$

1. g: $\Omega' \rightarrow \Omega$
2. h: $\overline{Q} \rightarrow^{onto} Q'$, where $\overline{Q} \subseteq Q$
3. k: $Y \rightarrow^{onto} Y'$

and for all $q \in \overline{Q}$, $\omega' \in \Omega'$

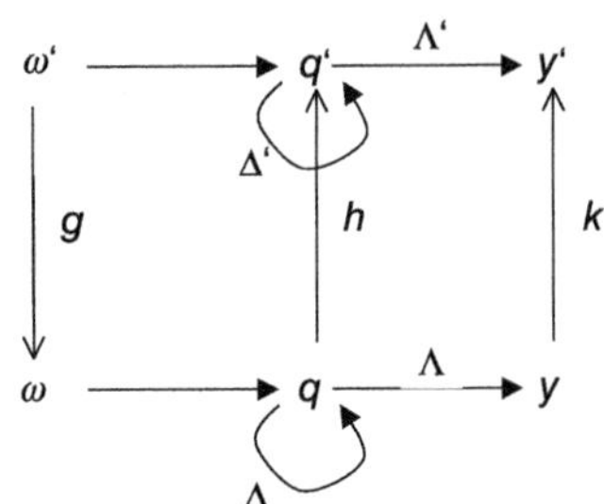

FIGURE 15.5

Commutative diagram showing the I/O system morphism.

4. $h(\Delta(q, g(\omega'))) = \Delta'(h(q), \omega')$ *transition function preservation*
5. $k(\Lambda(q)) = \Lambda'(h(q))$ *output function preservation* [1]

In interpreting line 4, which pertains to the preservation of the state transitions, we start big system in one of the states q in the restricted set $\overline{Q}$. The map h yields a corresponding state in the little system $q' = h(q)$. We then inject a segment ω' into S' and its encoded version $g(\omega')$ into S – sending them to states $\Delta'(q', \omega')$ in S' and $\Delta(q, g(\omega'))$, respectively. These states also correspond under h, i.e., $h(\Delta(q, g(\omega'))) = \Delta'(q', \omega')$.

Line 5 pertaining to the output function preservation, is interpreted as follows. When big system S is in state q and little system S' is in corresponding state $q' = h(q)$, the output values observed in these states are $y = \Lambda(q)$ and $y' = \Lambda'(q')$, respectively. The value y' is also obtained by decoding y with k, i.e., $y' = k(y)$ or $\Lambda'(q') = k(\Lambda(q))$.

Exercise 15.5. Show that for the equation in line 4 to be well defined (both sides defined and equal), $\overline{Q}$ must be closed under $g(\Omega')$, i.e., for all $q \in \overline{Q}$, $\omega' \in \Omega'$, $\Delta(q, g(\omega')) \in \overline{Q}$.

Consider system S and S' described over the same observational base, i.e., $T = T'$, $X = X'$, $Y = Y'$, and $\Omega = \Omega'$. S' is a homomorphic image of S if (g, h, k) is a system morphism, where g and k are identity maps and $\overline{Q} = Q$. The map h is said to be a *homomorphism* in this case. Obviously for compatible systems S and S' the homomorphism conditions 4 and 5 manifest itself as:

4. $h(\Delta(q, \omega)) = \Delta'(h(q), \omega)$
5. $\Lambda(q)) = \Lambda'(h(q))$

If, in addition, h is a one-to-one, it is an isomorphism, and S and S' are said to be *isomorphic*.

When S' is a homomorphic image of S, both systems turn out to have the same I/O function behavior (see), but the difference is that the state space of S' may be much "smaller" than that of S – all the state space Q of S is lumped by h onto that Q' of S'.

When S and S' are isomorphic, there is no essential difference in their structural descriptions at the I/O system level, although they may differ very significantly at higher levels of specification involving more structural detail.

[1] We assume an output function $\Lambda : Q \rightarrow Y$ only dependent of state in the sequel.

EXAMPLE: LINEAR SYSTEM HOMOMORPHISM

Let S and S' be compatible linear systems specified by $\{A, B, C\}$ and $\{A', B', C'\}$, respectively (in discrete time, let a common field F be assumed).

The existence of a homomorphism from S to S' can be reduced to relatively simple relation involving the specifications $\{A, B, C\}$ and $\{A', B', C'\}$ as shown by the following theorem.

Theorem 15.2. *Let $H : Q \to^{onto} Q'$ be a linear mapping. Then H is a homomorphism from S to S' if and only if the following conditions hold:*

1. $A'H = HA$
2. $HB = B'$
3. $C'H = C$

Exercise 15.6. Prove the theorem for the discrete case using the linear machine formalism in Section 5.6.2.

15.4.1 I/O SYSTEM MORPHISM IMPLIES IOFO AND IORO MORPHISM

We now show that if S' is a homomorphic image of S, then S and S' have the same I/O function behavior. This result will be generalized to demonstrate that the holding of a general preservation relation between systems at the I/O system level implies the existence of preservation relations at the lower (behavioral) levels. However, there will be some additional considerations that will have to be brought in to achieve this generalization. So we start with the most straightforward case first.

We say that two states (of the same or different systems) are behaviorally equivalent if they have the same I/O function, i.e.,

$$q \equiv q' \text{ if } \widetilde{\beta}_q = \widetilde{\beta}_{q'}$$

We say that two systems S and S' are behaviorally equivalent if $\widetilde{B}_S = \widetilde{B}_{S'}$, i.e., if for each state $q \in Q$, there is a state $q' \in Q'$ such that $\widetilde{\beta}_q = \widetilde{\beta}_{q'}$ (q and q' are equivalent), and conversely.

Exercise 15.7. Show that $\widetilde{B}_S = \widetilde{B}_{S'} \Rightarrow R_S = R_{S'}$ (S, S' are relationally equivalent).

We can now consider the fundamental theorem on behaviorally equivalence of compatible systems.

Theorem 15.3. *Let h: $Q \to^{onto} Q'$ be a homomorphism from S to S'. Then $\widetilde{B}_S = \widetilde{B}_{S'}$ (S and S' are behaviorally equivalent).*

Proof. We show first that for each $q \in Q$, $h(q)$ is equivalent to q. Consider the last output function $\beta'_{h(q)}$. For $\omega \in \Omega$,

$$\begin{aligned}\beta'_{h(q)}(\omega) &= \Lambda'(\Delta'(h(q), \omega)) \\ &= \Lambda'(h(\Delta(q, \omega))) && \text{(from 4. of homomorphism of compatible systems)} \\ &= \Lambda(\Delta(q, \omega)) && \text{(from 5. of homomorphism of compatible systems)} \\ &= \beta_q(\omega).\end{aligned}$$

Since h is onto Q', every state q' of S' has a representative state $q \in h^{-1}(q')$ such that $h(q) = q'$. Thus $\widetilde{B}_{S'} = \{\widetilde{\beta'}_{q'} | q' \in Q'\} = \{\widetilde{\beta'}_{h(q)} | q \in Q\} \subseteq \widetilde{B}_S$. But also, since h is defined on all of Q, for each $q \in Q$ there is a state $h(q) \in Q'$, thus $B_S \subseteq B_{S'}$. Thus $\widetilde{B}_S = \widetilde{B}'_S$. □

The essential of the above theorem generalizes to broader class of system morphism. First we have

Theorem 15.4. *If (i, h, i) is a system morphism from S to S', then (i, i) is an I/O function morphism from $IOFO_S$ to $IOFO_{S'}$ (hence an I/O relation morphism from $IORO_S$ to $IORO_{S'}$). In other words, if S' is a homomorphic image of an subsystem of S, then $\widetilde{B}_{S'} \subseteq \widetilde{B}_S$ and $R_{S'} \subseteq R_S$.*

Proof. An easy corollary of the above theorem. □

We now would like to show, more generally that, if (g, h, k) is a system morphism from S to S', there is also an I/O function morphism from S to S'. Unfortunately this is not necessarily true for an arbitrary encoding g. To see why, consider the following lemma.

Lemma 15.1. *Let (g, h, k) be a system morphism from S to S'. Then for each $q \in Q$, we have*

$$\beta'_{h(q)} = k \circ \beta_q \circ g$$

Exercise 15.8. Prove the lemma.

This tells us that the last output functions of corresponding states match up, but it does not allow us to conclude that the same is true for the I/O functions, which are obtained by applying the last output functions to successive left segments to obtain segment-to-segment mappings. The correspondence between last outputs must be maintained as this extension is being performed. More specifically, the g mapping must preserve left segmentation. Also, since the k map of the I/O function morphism cannot look at the input segment, the g map must allow such a map to be definable on the basis of k alone.

Thus let us say that $g : \Omega' \to \Omega$ is *invertable* if

1. *g preserves left segmentation:* $\omega'_{t>}$ is a left segment of $\omega' \Rightarrow g(\omega'_{t>})$ is a left segment of $g(\omega')$.
2. *g is one-to-one on length:* $l(\omega'_1) = l(\omega'_2) \Leftrightarrow l(g(\omega'_1)) = l(g(\omega'_2))$.

These properties of g are precisely the ones enabling us to match up instants in an observation interval of the little system with uniquely corresponding instants in the corresponding observation interval of the big system. This is shown by the

Lemma 15.2. *g is an invertable coding if and only if there is a one-to-one (partial) function $MATCH : T_0'^{+} \to T_0^{+}$ such that*

$$MATCH(l(\omega')) = l(g(\omega'))$$
$$g(\omega'_{t>}) = g(\omega')_{MATCH(t)>}$$

With this matching of computation instants $l(\omega')$ and $l(g(\omega'))$, we can define a decoding map $\overline{k} : (Y, T) \to (Y', T')$ as follows.

For any $\rho \in (Y, T)$, $\overline{k}(\rho)$ has length $l(\overline{k}(\rho)) = MATCH^{-1}(l(\rho))$ and

$$\overline{k}(\rho)(t') = k(\rho(MATCH(t')) \text{ for } t' \in dom(\overline{k}(\rho))$$

Exercise 15.9. Check that $\overline{k}$ is well defined in carrying any segment ρ of the big system to a unique segment $\overline{k}(\rho)$ of the little system.

Now using the two previous lemmas we can obtain the strengthened

Lemma 15.3. *Let* (g, h, k) *where* g *is invertable, be a system morphism from* S *to* S'*. Then for each* $q \in Q$,

$$\widetilde{\beta}'_{h(q)} = \overline{k} \circ \widetilde{\beta}_q \circ g$$

where $\overline{k}$ *is defined as before.*

Exercise 15.10. Prove the lemma.

Finally we have the desired

Theorem 15.5. *If* (g, h, k) *where* g *is invertable, is a system morphism from* S *to* S'*, there is an I/O function morphism* (g, k) *from* $IOFO_S$ *to* $IOFO_{S'}$*, hence an I/O relation morphism from* $IORO_S$ *to* $IORO_{S'}$*.*

Proof. The proof follows the lines of the above theorem and is left as an exercise to the reader. □

Exercise 15.11. If h is a homomorphism from S to S', show that for each $q' \in Q'$, the set $h^{-1}(q')$ is a set of equivalent states.

We learn in the next section that the converse of the last statement is also true. That is, if we are given a partition of the sates of S whose blocks consist of equivalent states, we can construct a homomorphic image of S.

15.4.2 THE LATTICE OF PARTITIONS AND THE REDUCED VERSION OF A SYSTEM

Any system has a "reduced version" that represents the most that can be inferred about the structure of S from external observation, i.e., its behavior. Seen from the opposite direction – going from structure to behavior – it turns out that the reduced version is also the system having the "minimal" state set needed to generate the behavior. Thus, the reduced version S_R of a system S has a number of interesting and useful properties which are important in structural inference (Zeigler, 1976) and system representation (Chapter 16).

From here on we restrict our attention to systems whose input segment set Ω contains the null segment Φ. This segment has the useful property that $\Delta(q, \Phi) = q$ for any system state q, and $\Phi \bullet \omega = \omega$ for any segment ω. This restriction is a matter of convenience, since essentially the same results can be obtained without it.

State partitioning is the means to define a reduced version of a system. Let us therefore first review basic concepts and operations of partitions of sets.

Let P_A denote the set of all partitions of a set A. It is well known that P_A is a lattice with underlying partial order $\leq$ given by

$$\pi_1 \leq \pi_2 \Leftrightarrow (\forall a, a' \in A)(a\pi_1 a' \Rightarrow a\pi_2 a').$$

We say that π_1 is a refinement or π_2 or π_1 refines π_2.

Two operations are defined for partitions. The *least upper bound* (lub) and *greatest lower bound* (glb) operators are denoted by $\cap$ and $\cup$, respectively. The glb operator is defined so that $\pi_1 \cap \pi_2$ is the greatest partition with

$$a(\pi_1 \cap \pi_2)a' \Leftrightarrow a(\pi_1)a' \wedge a(\pi_2)a'$$

The lub operator is defined so that $\pi_1 \cup \pi_2$ is the smallest partition with

$$a(\pi_1 \cup \pi_2)a' \Leftrightarrow \exists a = a_1, a_2, ..., a_n = a'$$

such that $a_i(\pi_1)a_{i+1}$ for all odd $i \wedge a_i(\pi_2)a_{i+1}$ for all even i.

These definitions satisfy the required conditions, namely, that $\pi \leq \pi_1 \wedge \pi \leq \pi_2$ implies $\pi \leq \pi_1 \cap \pi_2$ and $\pi_1 \leq \pi \wedge \pi_2 \leq \pi$ implies $\pi_1 \cup \pi_2 \leq \pi$. The finest partition 0 (the partition with all blocks containing a single element) is defined by

$$a0a' \Leftrightarrow a = a'$$

and the coarsest partition I (the partition with one block only containing all elements) is defined by

$$aIa' \Leftrightarrow a, a' \in A.$$

A partition of a state set of a dynamical system is said to be a *congruence* relation $\equiv$ if and only if

$$q \equiv q' \Rightarrow (\forall \omega \in \Omega)\left(\Delta(q, \omega) \equiv \Delta(q', \omega)\right)$$

i.e., if two states are congruent, all next states for all input segments are congruent.

Theorem 15.6. *The set of all congruence relations $\equiv$ of a state set Q of a system with state transition Δ forms a sublattice of all partitions P_Q of state set Q.*

Exercise 15.12. Prove this.

The equivalence relation induces a partition on the set Q consisting of the equivalence classes which we denote by

$$Q/\equiv = \{[q]_\equiv | q \in Q\}.$$

Furthermore, we call a congruence relation *output consistent*, if and only if their states are behavioral equivalent

$$q \equiv q' \Rightarrow \Lambda(q) = \Lambda(q').$$

We are able to apply the behavioral equivalence relation introduced in Section 15.4.1 to states of the same system. Doing so we obtain an output consistent congruence relations.

Theorem 15.7. *A equivalence relation $\equiv$ induced by a homomorphism $h : Q \rightarrow^{onto} Q$ on a state set Q is*

(a) *congruent and*
(b) *output consistent.*

Proof.

Exercise 15.13. Show that $\widetilde{\beta}_q(\omega \bullet \omega') = \widetilde{\beta}_q(\omega) \bullet \widetilde{\beta}_{\Delta(q,\omega)}(\omega')$.

(a)

$$q \equiv q' \Rightarrow \widetilde{\beta}_q = \widetilde{\beta}'_q \Rightarrow (\forall \omega \in \Omega)\left(\widetilde{\beta}_q(\omega) = \widetilde{\beta}'_q(\omega)\right)$$
$$\Rightarrow (\forall \omega \in \Omega)\,(\forall \omega' \in \Omega)\left(\widetilde{\beta}_q(\omega\omega') = \widetilde{\beta}'_q(\omega\omega')\right)$$
$$\Rightarrow (\forall \omega \in \Omega)\,(\forall \omega' \in \Omega)\left(\widetilde{\beta}_q(\omega)\widetilde{\beta}_{\Delta(q,\omega)}(\omega') = \widetilde{\beta}'_q(\omega)\widetilde{\beta}_{\Delta(q',\omega)}(\omega')\right)$$
$$\Rightarrow (\forall \omega \in \Omega)\,(\forall \omega' \in \Omega)\,(\widetilde{\beta}_{\Delta(q,\omega)}(\omega') = \widetilde{\beta}_{\Delta(q',\omega)}(\omega'))$$
$$\Rightarrow (\forall \omega \in \Omega)\,(\Delta(q,\omega) \equiv \Delta(q',\omega))$$

(b)

$$q \equiv q' \Rightarrow \widetilde{\beta}_q(\Phi) = \widetilde{\beta}'_q(\Phi) \Rightarrow \Lambda(\Delta(q,\Phi)) = \Lambda(\Delta(q',\Phi))$$
$$\Rightarrow \Lambda(q) = \Lambda(q') \quad \square$$

The reduced version S_R of a system $S = (T, X, \Omega, Y, Q, \Delta, \Lambda)$ is defined as follows. Let $\equiv_R$ be the *maximal* output consistent congruence relation. Then we define S_R as

$$S_R = (T, X, \Omega, Y, Q/\equiv_R, \Delta/\equiv_R, \Lambda/\equiv_R)$$

where $\Delta/\equiv_R\colon Q/\equiv_R \times \Omega \to Q/\equiv_R$ is defined by

$$\Delta/\equiv_R([q]_{\equiv_R}, \omega) = [\Delta(q,\omega)]\equiv_R$$

and $\Lambda/\equiv_R\colon Q/\equiv_R \to Y$ is defined by

$$\Lambda/\equiv_R([q]_{\equiv_R}) = \Lambda(q).$$

Theorem 15.8. *If S is a system, so is S_R.*

Proof. Use the above lemma to show that $\Delta/\equiv_R$ and $\Lambda/\equiv_R$ are well defined. Then check that $\Delta/\equiv_R$ has the composition property. $\square$

We say that S is reduced if for all pairs $q, q' \in Q$, $\widetilde{\beta}_q = \widetilde{\beta}_{q'} \Rightarrow q = q'$. In other words, each state q has a distinct I/O function $\widetilde{\beta}_q$ associated with it.

Exercise 15.14. Show that S is reduced if and only if for each distinct pair $q, q' \in Q$, there is an input segment ω that will cause S to respond with different output segments when started in states q and q'.

The following theorem presents the important properties of reduced version S_R of the system S.

Theorem 15.9.

(a) *S_R is reduced.*
(b) *S_R is a homomorphic image of S.*
(c) *S_R is behaviorally equivalent to S.*
(d) *For any system S', if S' is behaviorally equivalent to S and S' is reduced, S' is isomorphic to S_R.*

Proof. We show first by a straightforward expansion

$$\widetilde{\beta}_{[q]\equiv_R} = \widetilde{\beta}_q \text{ for all } q \in Q$$

Then (a) and (c) follows immediately for this result. For (b), we consider the mapping $h : Q \rightarrow Q/\equiv_R$ given by $h(q) = [q] \equiv_R$. Then by direct verification h is a homomorphism.

For (d) we establish a mapping $h' : Q' \rightarrow Q/\equiv_R$ of states to their corresponding equivalence classes just as h does. Using the same approach we can show h' is a homomorphism and since S' is reduced, it is also one-one. Thus isomorphism is established. □

The theorem establishes that the reduced version of a system is rather special – it is a homomorphic image of any other behaviorally equivalent system and is thus called *minimal* (in finite state systems it has the smallest number of states). It is also *unique* in the sense that any way of arriving at the reduced version will result in a system that is isomorphic to it.

15.5 SYSTEM MORPHISM FOR ITERATIVELY SPECIFIED SYSTEMS

We have seen in Chapter 5 that it is possible to implicitly specify a system structure by providing an iterative specification, G. For iterative specifications we will define an appropriate system morphism with the natural requirement: if the morphism runs from G to G', then when G is expanded to S_G and G' is expanded to $S_{G'}$, this will imply the existence of a system morphism running from S_G to $S_{G'}$.

To develop the concept we must first construct an input segment encoding map in such a way that a segment may be encoded by composing the encoded versions of the generators in its mls decomposition.

Thus let Ω_G and Ω'_G be admissible set of generators. Let $\overline{g} : \Omega'_G \rightarrow \Omega_G$ be a *generator encoding* map, so that with each generator $\omega' \in \Omega'_G$ (the little input generator set) there is associated a segment $\overline{g}(\omega')$, its coded version (which is not necessarily a generator of the big input set generator set). We extend $\overline{g}$ to an encoder $g : \Omega'^{+}_G \rightarrow \Omega^{+}_G$ by the definition

$$g(\omega') = \overline{g}(\omega'_1)\overline{g}(\omega'_2)...\overline{g}(\omega'_n) \text{ for } \omega' \in \Omega'_G,$$

where $\omega'_1, \omega'_2, ..., \omega'_n$ is the mls decomposition for ω'. Now g is well defined because of the uniqueness of the mls decomposition.

For iterative specifications $G = (T, X, \Omega_G, Y, Q, \delta, \lambda)$ and $G' = (T', X', \Omega'_G, Y', Q', \delta', \lambda')$, a *specification morphism from* G' to G is a triple (g, h, k) such that

1. $\overline{g} : \Omega'_G \rightarrow \Omega^{+}_G$
2. $h : \overline{Q} \rightarrow^{onto} Q'$, where $\overline{Q} \subseteq Q$.
3. $k : Y \rightarrow^{onto} Y'$

   and for all q $\in \overline{Q}, \omega' \in \Omega'_G$
4. $h(\delta^{+}(q, \overline{g}(\omega'))) = \delta'(h(q), \omega')$ *transition function preservation*
5. $k(\lambda(q)) = \lambda'(h(q))$ *output function preservation*

Note that lines 4 and 5 are very similar to lines 4 and 5 of the I/O system morphism definition. The major difference is that, in the present specification morphism, the transition function preservation need only be checked for the generators of the little system Ω'_G.

We say that $\overline{g}$ is *generator preserving* if $\overline{g}(\Omega'_G) \subseteq \Omega_G$. Since in this case, for each generator $\omega' \in \Omega'_G$, $\overline{g}(\omega')$ is also a generator, the function δ need not be extended, and line 4 can read

$$4'.h(\delta(q,\overline{g}(\omega'))) = \delta'(h(q),\omega').$$

This means that the specification morphism can be checked directly by examining the generator transition functions δ and δ'.

15.5.1 ITERATIVE SPECIFICATION MORPHISM IMPLIES I/O SYSTEM MORPHISM

The next theorem establishes the procedure for expanding a specification morphism into a system morphism. The importance of this, as we have indicated, comes in checking whether a system morphism runs from a big to a little system. To establish such a morphism we need to verify only that a specification morphism runs from the *iterative specification* of the one to that of the other. The expansion procedure is then guaranteed by the theorem and need not be carried out.

Theorem 15.10. *If $(\overline{g},h,k)$ is a specification morphism from G' to G, then (g,h,k) is a system morphism from S_G to $S_{G'}$, where g is the extension of $\overline{g}$.*

Proof. We need show only that line 4 holds in the system morphism definition. This is done using induction on the proposition.

$P(n) \equiv [$if $\omega' \in \Omega'^{+}_G$ has the mls decomposition $\omega'_1, \omega'_2, ...\omega'_n$ then for all $q \in \overline{Q} : h(\delta^+(q,g(\omega'))) = \delta'^+(h(q),\omega')]$.

$P(1)$ is just line 4 of the specification morphism definition. Assuming that $P(n-1)$ holds, we can show that $P(n)$ holds. Let ω' have mls decomposition ω'_1, ω'_2,..., ω'_n and let $q \in Q$, then write

$$h(\delta^+(q,g(\omega'_1\omega'_2...\omega'_n))) = h(\delta^+(q,\overline{g}(\omega'_1)g(\omega'_2...\omega'_n))) \text{ (definition of } g)$$
$$= h(\delta^+(\delta(q,\overline{g}(\omega'_1)),g(\omega'_2...\omega'_n))) \text{ (composition property of } \delta^+)$$
$$= \delta'^+(h(\delta(q,\overline{g}(\omega'_1))),\omega'_2...\omega'_n)$$
$$(P(n-1) \text{ using } \omega'_2, ..., \omega'_n \text{ as the mls decomposition of } \omega'_2...\omega'_n)$$
$$= \delta'^+(\delta'(h(q),\omega'_1),\omega'_2...\omega'_n) \ (P(1))$$
$$= \delta'^+(h(q),\omega'_1\omega'_2...\omega'_n) \text{ (composition property of } \delta^{+'})$$

Thus $P(n)$ holds, and by the principle of induction, the theorem, which asserts $P(n)$ for all n, is established. □

Exercise 15.15. Show that if (g,h,k) is a system morphism from S_G to S'_G, then (g,h,k) is a specification morphism from G to G' where $\overline{g} = g|\Omega'_G$. Thus there is a system morphism from S_G to S'_G, if and only if there is a specification morphism from G to G'.

15.5.2 SPECIALIZATION OF MORPHISMS FOR ITERATIVELY SPECIFIED SYSTEMS

Just as we can further specialize the iterative specification schemes to the discrete time, differential equation, and discrete event cases, we can provide appropriate specializations of the specification morphism. The procedure is depicted in Fig. 15.6.

We shall describe the expansion for the discrete time and discrete event cases.

DTSS SPECIFICATION MORPHISMS

Let $M = (X_M, Y_M, Q_M, \delta_M, \lambda_M, h)$ and $M' = (X'_M, Y'_M, Q'_M, \delta'_M, \lambda'_M, h')$ be DTSS. The appropriate morphism is as follows: $(\overline{\overline{g}}, h, k)$ is a discrete time morphism from M to M' if

- $\overline{\overline{g}} : X'_M \to X^+_M$
- $h : \overline{Q}_M \to^{onto} Q'_M$, where $\overline{Q}_M \subseteq Q_M$
- $k : Y_M \to^{onto} Y'M$ and for all $q \in Q, x' \in X'M$
- $h(\delta^+_M(q, \overline{\overline{g}}(x'))) = \delta'_M(h(q), x')$
- $k(\lambda_M(q)) = \lambda'_M(h(q))$

The translation to a specification morphism is very straightforward in this case. The translated morphism is $(\overline{g}, h, k)$ running from $S_{G(M)}$ to $S_{G(M')}$, where $\overline{g} : \Omega_{G(M')} \to \Omega_{G(M)}$ is defined by $\overline{g}(\omega_x) = \omega_{\overline{\overline{g}}(x)}$. The translation to a specification morphism is very straightforward in this case. The generator segments for DTSS are: $\Omega_{G(M')} = \{\omega_x | x \in X, \omega_x : \langle 0, 1\rangle \to \{x\}\}$. The translated morphism is $(\overline{g}, h, k)$ running from $S_{G(M)}$ to $S_{G(M')}$, where $\overline{g} : \Omega_{G(M')} \to \Omega_{G(M)}$ is defined by $\overline{g}(\omega_x) = \omega_{\overline{\overline{g}}(x)}$.

DISCRETE EVENT CASE

We give as an example the construction of a homomorphism for discrete event specifications. Let $M = (X_M, Y_M, S_M, \delta_{\text{ext},M}, \delta_{\text{int},M}, \lambda_M, \text{ta}_M)$ and $M' = (X'_M, Y'_M, S'_M, \delta'_{\text{ext},M}, \delta'_{\text{int},M}, \lambda'_M, \text{ta}'_M)$ be legitimate DEVSs, where $X_M = X'M, Y_M = Y'M$. A DEVS homomorphism from M to M' is a map $\overline{h}$ such that

$$h : S_M \to^{onto} S'_M,$$

and for all $(s, e) \in Q_M, x \in X_M$

1. $\overline{h}(\delta_{\text{ext},M}(s, e, x)) = \delta'_{\text{ext},M}(\overline{h}(s), e, x))$
2. $\lambda'_M(\overline{h}(s)) = \lambda_M(s)$
3. $\overline{h}(\delta_{\text{int},M}(s)) = \delta'_{\text{int},M}(h(s)))$
4. $\text{ta}'_M(\overline{h}(s)) = \text{ta}_M(s)$

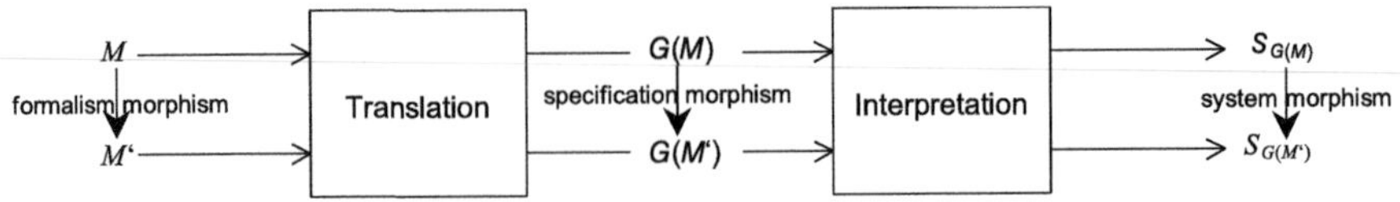

FIGURE 15.6

Specialization of specification morphisms.

The translation of the homomorphism $\overline{h}$ to a specification morphism is given by

Theorem 15.11. *Let $\overline{h}$ be a DEVS homomorphism from M to M'. Let $h : Q_M \rightarrow^{onto} Q'_M$ be a map such that $h(s,e) = (h(s), e)$ for all $(s,e) \in Q_M$. Then h is a (specification) homomorphism from $G(M)$ to $G(M')$.*

Proof. The proof employs the following sequence of assertions, which are left to the reader as an exercise.

- **(a)** h is an onto map.
- **(b)** $\#((s,e), \emptyset_\tau) = \#((\overline{h}(s), e), \emptyset_\tau)$ where $\#(q, \omega)$ is the number of transitions for initial state q and input segment ω (see Section 6.2.3).
- **(c)** $\overline{h}(\delta_{\text{int},M}(s) = \delta'_{\text{int},M}(h(s))$
- **(d)** $h(\delta_G(s,e,\emptyset_\tau)) = \delta'_G(h(s,e), \emptyset_\tau)$
- **(e)** $h(\delta_G(s,e,x_\tau)) = \delta'_G(h(s,e), x_\tau)$
- **(f)** $\lambda'(h(s)) = \lambda(s)$. □

Assertions (d), (e), and (f) show that (i, h, i) is a specification morphism from $G(M)$ to $G(M')$, and since h has domain Q_M, h is a homomorphism from $G(M)$ to $G(M')$.

15.6 THE STRUCTURED SYSTEM MORPHISM

A system at the structured system level differs from an I/O system by its structuring of sets and functions. The sets of a structured system takes the form of a multi-variable set $S = (V, S_1, S_2, S_3, ...)$ identifying an ordered set of variables $V = (v_1, v_2, ..., v_n)$. The functions are multi-variable, which means that they take a multi-variable set as its domain and a multi-variable set as its range. Recall, that a multi-variable function $f : A \rightarrow B$ can be represented as a crossproduct of coordinate functions $f.v_i : A \rightarrow ((v_i), B_i)$ with one function for each variable v_i of the range set B_i. We want a morphism at this level not only to preserve the relations between input, states and output sets of the big and little system as defined at the I/O system level but also to preserve the structural properties as defined by the structuring of sets and transition and output functions.

To introduce a morphism at the structured system level, we construct a system homomorphism from the big to the little system. Let S be the big system and S' the little system, both at the structured system level. There is a structured system homomorphism $h : Q \rightarrow^{onto} Q'$ from S to S' if h is a I/O system homomorphism for the two I/O systems associated with S and S', respectively, and we require that the homomorphism takes a special form as follows.

Let $coord : V \rightarrow^{onto} V'$ be a mapping from the variable set V of Q onto the variable set V' of Q'. Let $V_{vi'} = coord^{-1}(v'_i)$ be the subset of variables of V mapped to v'_i. Then we require that the homomorphism h from Q onto Q' is composed of functions

$$h_{vi'} : Q.V_{vi'} \rightarrow Q'.v_{i'}$$

for each $v_{i'} \in V'$. Thus, a structured systems homomorphism is constructed as a composite of local coordinate maps. Because of its construction the structure morphism can be rephrased to emphasize that the global homomorphism condition reduces to a series of local function preserving conditions

$$h_{vi'}(\Delta(q,\omega).V_{vi'}) = \Delta'(h(q),\omega).v_{i'}.$$

EXAMPLE: STRUCTURE MORPHISM FOR LINEAR SYSTEMS

See TMS2000, pages 312–314.

15.7 MULTI-COMPONENT SYSTEM MORPHISM

A morphism at the multi-component system level should not only preserve the structural properties of the state set but also the coupling structure between pairs of component systems. Therefore a morphism at this level will consist, firstly, of local mappings of blocks of components in the big system to their representatives in the small system and, secondly, of *coupling preservation conditions* which guarantee that the components in the big and the little system are coupled in like manner. We show that under the coupling preservation conditions, the composite of the local preservation conditions define a global morphism at the I/O system level. For simplicity of exposition, we assume that the two multi-component systems to be related are compatible and a homomorphism is established between them.

Let $MC = (T, X, \Omega, Y, D, \{M_d\})$ be a big multi-component system with components $M_d = (Q_d, E_d, I_d, \Delta_d, \Lambda_d)$ and $MC' = (T', X', \Omega', Y', D', \{M'_{d'}\})$ be a little multi-component system with components $M'_{d'} = (Q'_{d'}, E'_{d'}, I'_{d'}, \Delta'_{d'}, \Lambda'_{d'})$. Since MC and MC' are compatible it follows that $X = X'$, $Y = Y'$, $T = T'$, and $\Omega = \Omega'$.

Let $coord : D \rightarrow^{onto} D'$ be a mapping from the set of components D of MC onto the set of components D' of MC'. Let $D_{d'} = coord^{-1}(d')$ be the subset of components of D mapped to d'. We'll call $D_{d'}$ the representing block of d'. Note that since $coord$ is a function the representing blocks $D_{d'}$ form a partition of D. Then we require that the global map

$$h : \times_{d \in D} Q_d \rightarrow \times_{d' \in D'} Q'_{d'}$$

from the state set of MC onto that MC' is composed of $local$ functions

$$h_{d'} : \times_{d \in D_{d'}} Q_d \rightarrow Q'_{d'}$$

for each $d' \in D'$ defined on the states of representing blocks.

To guarantee the preservation of the structural properties we require that the influences between components in the little systems are also, at a finer level, present in the big system (*coupling preservation condition*). Note that this allows some of the influences in the big system to be lost in the small system. However, no new ones can appear in the small system that are not present in the big one.

Formally, for all the influencing components $I'_{d'}$ of a component d' in the small system, there also must be an influence in the big system, i.e.,

$$\forall d' \in D' \forall i' \in I'_{d'} \exists d \in D_{d'} \exists i \in D_{i'} : i \in I_d.$$

And in the same form for the set of influenced components $E'_{d'}$ of components d'

$$\forall d' \in D' \forall e' \in E'_{d'} \exists d \in D_{d'} \exists e \in D_{e'} : e \in E_d.$$

The two conditions guarantee that the set of influencers and influencees are preserved under the coordinate mapping, i.e.,

$$\cup_{i' \in I'_{d'}} D_{i'} \subseteq \cup_{d \in D_{d'}} I_d$$

and

$$\cup_{e' \in E'_{d'}} D_{e'} \subseteq \cup_{d \in D_{d'}} E_d.$$

Exercise 15.16. Prove this.

Fig. 15.7 shows an example multi-components system homomorphism from a big multi-component system MC to a little multi-component system MC'. Subsets of components of MC are mapped to components of MC'. The dependencies between components in MC' also can be observed in MC.

Assuming these coupling preservation conditions, we present the structure morphism as a series of local preservation conditions for the local mapping of each small system component, d' as follows. Let

$$h_{I'_{d'}} : \times_{\substack{i \in \cup D_{i'} \\ i' \in I'_{d'}}} Q_i \to \times_{i' \in I'_{d'}} Q'_{i'}$$

be the mapping from the state sets of the big system influencers of the representatives of d' to the state sets of the small system influencers of d'. This mapping is defined by

$$h_{I'_{d'}}(..., q_i, ...).i' = h_{i'}(..., q_i, ...)$$

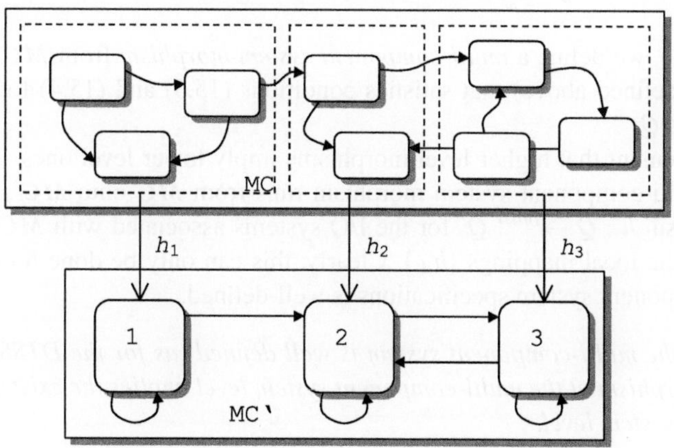

FIGURE 15.7

Multi-component system homomorphism from the “big” system MC onto the “little” system MC’.

Similarly, let

$$h_{E'_{d'}} : \times_{\substack{e \in \cup D_{e'} \\ e' \in E'_{d'}}} Q_e \rightarrow \times_{e' \in E'_{d'}} Q'_{e'}$$

be the mapping from the state sets of the big system influencees of the representatives of d' to the state sets of the small system influencees of d'. This mapping is defined by

$$h_{E'_{d'}}(..., q_j, ...).j' = h_{j'}(..., q_j, ...)$$

Let $q = (..., q_d, ...)$ denote a state of the big system and $q' = (..., q'_{d'}, ...)$ denote a state of the little system. When we apply the transition function of component d' in the little system to the mapped states of all the big system influencers of the representatives of d', we get

$$\Delta'_{d'}(h_{I'_{d'}}(..., q_i, ...), \omega) \tag{15.1}$$

On the other hand when we apply the transition functions Δ_d of all the representatives $d \in D_{d'}$ and then apply the mapping $h_{E'_{d'}}$ to get the states of the influencees of d', we get

$$h_{E'_{d'}}(\times_{d \in D_{d'}} \Delta_d(q.I_d, \omega)) \tag{15.2}$$

We require that computations 1) and 2) yield the same result:

$$\Delta'_{d'}(h_{I'_{d'}}(..., q_i, ...), \omega) = h_{E'_{d'}}(\times_{d \in D_{d'}} \Delta_d(q.I_d, \omega)) \tag{15.3}$$

And we require that outputs of original and lumped components must be identical (since we consider compatible systems here)

$$\Lambda'_{d'}(h_{I'_{d'}}(..., q_i, ...), \omega) = \times_{d \in D_{d'}} \Lambda_d(q.I_d, \omega) \tag{15.4}$$

With this as preamble, we define a *multi-component system morphism* from MC onto MC' as a pair $\langle coord, \{h_{d'}\} >$ (as defined above) that satisfies conditions (15.3) and (15.4) for all input segments, $\omega \in \Omega$, and states $q \in Q$.

Recall our requirement that higher level morphisms imply lower level ones. Thus, we would like to show that if a multi-component system morphism runs from MC onto MC', then there is an I/O system homomorphism $h : Q \rightarrow^{onto} Q'$ for the I/O systems associated with MC and MC', where h is the composite of the local mappings $\{h_{d'}\}$. Clearly, this can only be done for those formalisms in which the multi-component system specifications is well-defined.

Theorem 15.12. *If the multi-component system is well defined (as for the DTSS, DESS and DEVS), the existence of a morphism at the multi-component system level implies the existence of an associated morphism at the I/O system level.*

Proof. The proof follows from the definition of the I/O system associated with a multi-component system as discussed in Chapters 5 and 7. For simplicity reasons in the following we show this for the DTSS case, where $E_d = \{d\}$ for all components d, and for identical input and output interfaces.

For such systems the global state transition function is defined by the crossproduct of the global state transitions functions of the individual components (see Chapters 5 and 7)

$$\Delta.d(q,\omega) = \Delta_d(\times_{i \in I_d} q_i, \omega).$$

To show the preservation of state transitions, we have to show that

$$h(\Delta(q,\omega)) = \Delta'(h(q),\omega)$$

(see section definition of homomorphism in Section 15.4) with h is defined by the crossproduct of local mappings $h_{d'}$, i.e., $h.d'(q,\omega) = h_{d'}(q.D_{d'},\omega)$. With $E_d = \{d\}$ the homomorphism requirement (15.3) from above simplifies to

$$\Delta'_{d'}(h_{I'_{d'}}(\ldots, q_i, \ldots), \omega) = h_{d'}(\times_{d \in D_{d'}} \Delta_d(q.I_d, \omega)).$$

From the definition of h and $h_{I'_{d'}}$ it follows that $h_{I'_{d'}}(\ldots, q_i, \ldots) = h(q).I'_{d'}$. With that observation in hand we conclude that

$$\Delta'_{d'}(h_{I'_{d'}}(\ldots, q_i, \ldots), \omega) = \Delta'.d'(h(q),\omega)$$

and

$$h_{d'}(\times_{d \in D_{d'}} \Delta_d(q.I_d, \omega)) = h_{d'}(\Delta(q,\omega).D_{d'}) = h.d'(\Delta(q,\omega))$$

and finally

$$\Delta'.d'(h(q),\omega) = h.d'(\Delta(q,\omega)$$

for all $d' \in D'$, as required. The output preservation condition can be shown in similar way. □

15.8 THE NETWORK OF SYSTEMS MORPHISM

In a manner similar to that for multi-component systems, we'll define a morphism at the coupled systems level. Besides preserving the dependencies among components, we also require that the interface mappings be preserved under the morphism. Recall that an interface mapping specifies how outputs of influencing components define the inputs of their influencees. Again for simplicity of exposition we assume compatible network systems and restrict our attention to homomorphisms.

Let

$$N = (T, X_N, Y_N, D, \{M_d\}, \{I_d\}, \{Z_d\})$$

be a big network and

$$N' = (T', X'_{N'}, Y'_{N'}, D', \{M'_{d'}\}, \{I'_{d'}\}, \{Z'_{d'}\})$$

be a little network where N and N' are compatible. Let $coord : D \rightarrow^{onto} D'$ be a mapping from the set of components D of N onto the set of components D′ of N′. Then as for multi-component morphisms, we desire that the composite of local maps $h_{d'} : \times_{d \in D_{d'}} Q_d \rightarrow Q'_{d'}$ become a homomorphism

$$h : \times_{d \in D} Q_d \rightarrow \times_{d' \in D'} Q'_{d'}$$

for the I/O systems associated with N and N'.

To guarantee the preservation of the structural coupling properties between components we additionally require that for each component d' in the small system there exists a mapping

$$g_{d'} : \times_{d \in D_{d'}} (X_d, T) \rightarrow (X'_{d'}, T)$$

from the composite set of input segments of the block representing d' of the "big" system to the input segments of d'. In the same way, we require a mapping

$$k_{d'} : \times_{d \in D_{d'}} (Y_d, T) \rightarrow (Y'_{d'}, T)$$

from the composite set of output segments of the block to the output segments of the component d'.

First we require local structure preservation relations for mappings $h_{d'}$ and $g_{d'}$ for state transitions (*state transition preservation conditions*). The preservation condition for state transitions state that state transitions in components $d \in D_{d'}$ and component d' must correspond under mappings $h_{d'}$ and $g_{d'}$

$$h_{d'}(\times_{d \in D_{d'}} \Delta_d(q_d, \omega_d)) = \Delta'_d(h_{d'}(\times_{d \in D_{d'}} q_d), g_{d'}(\times_{d \in D_{d'}} \omega_d)) \tag{15.5}$$

Second, we require that the input/output behavior is preserved under mappings $h_{d'}$, $g_{d'}$, and $k_{d'}$ (*input output behavior preservation condition*). Let ω_d be the inputs in $d \in D_{d'}$ and $\rho_d = \beta_d(q_d, \omega_d)$ be the corresponding output segment for initial state q_d. And let $\omega'_{d'}$ and $\rho'_{d'} = \beta'_{d'}(q'_{d'}, \omega'_{d'})$ denote input/output pairs in the little system. Then we require that input and output pairs in $d \in D_{d'}$ and d' correspond under mappings $h_{d'}$, $g_{d'}$ and $k_{d'}$

$$k_{d'}(\times_{d \in D_{d'}} \beta_d(q_d, \omega_d)) = \beta'_{d'}(h_{d'}(\times_{d \in D_{d'}} q_d), g_{d'}(\times_{d \in D_{d'}} \omega_d)) \tag{15.6}$$

with $d \in D_{d'}$ (Fig. 15.8).

Finally, we require that the coupling relations of the big system is preserved in the little system (*coupling preservation relations*). This manifests itself as a commutative diagram as depicted in Fig. 15.9. When we take the outputs of the influencers $i \in I_d$ of components $d \in D_{d'}$ in the big system and apply the interface mappings Z_d to get the inputs ω_d of $d \in D_{d'}$ and then apply the coding mapping $g_{d'}$ to all ω_d, we obtain the input $\omega'_{d'}$. This value must be the same as if we first applied the coding function $k_{I'}$ to obtain all the outputs $\rho_{i'}$ of components i' influencing d' in the little system and then applied the interface mapping $Z'_{d'}$ to the outputs $\beta'_{i'}$ of $i' \in I'_{d'}$, i.e.,

$$g_{d'}(\times_{d \in D_{d'}} Z_d(\times_{i \in I_d} \rho_i)) = Z'_{d'}(\times_{i' \in I'_{d'}} k_{i'}(\times_{i \in D_{i'}} \rho_i))) \tag{15.7}$$

We define a *network of systems morphism* from N onto N' as a structure $\langle coord, \{k_{d'}\}, \{g_{d'}\}, \{h_{d'}\}\rangle$ (as defined above) such that $\{k_{d'}\}$, $\{g_{d'}\}$ and $\{h_{d'}\}$ satisfy the coupling preservations, state transition preservation and input/output preservation conditions.

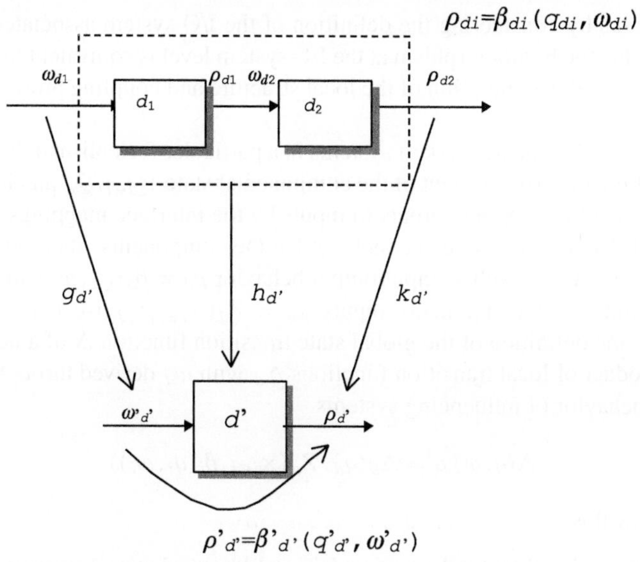

FIGURE 15.8

Input/output pairs preservation relation.

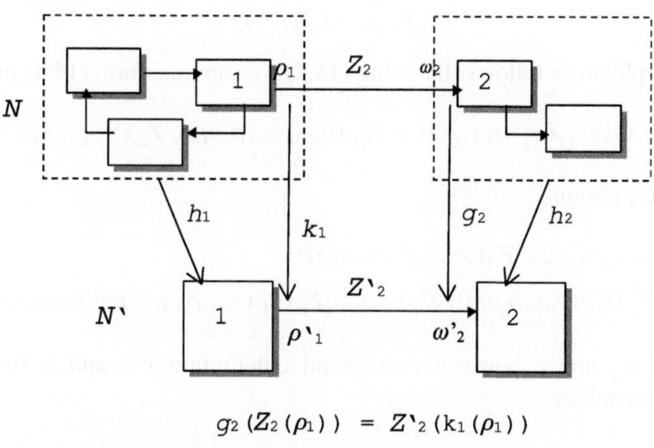

FIGURE 15.9

Network of systems morphisms.

Theorem 15.13. *If the network of systems is well defined, the existence of a network of systems morphism $\langle coord, \{k_{d'}\}, \{g_{d'}\}, \{h_{d'}\}\rangle$ which fulfills the coupling preservations, state transition preservation and input/output preservation conditions implies the existence of an associated morphism at the I/O system level.*

Proof. The proof works by considering the definition of the I/O system associated with a network of system and showing that the homomorphism at the I/O system level is equivalent to the local homomorphism relations $h_{d'}$ under the assumption of the local structure and coupling preservation conditions as defined above.

Recall from Chapter 7 that an network of systems in a particular formalism defines a basic system in that formalism by taking the crossproduct of the components' state $\times_{d \in D} Q_d$ and local state transitions and output functions and by mapping outputs to inputs by the interface mappings Z_d. As components must be well defined I/O systems, we can conclude that the components' state set Q_d, state transition Δ_d and output functions Λ_d, as well as input/output behavior $\rho_d = \beta_d(q_d, \omega_d)$ are well defined.

Also by the definition of components inputs $x_d = Z_d(\times_{i \in I_d} y_i)$ we can conclude that $\omega_d = Z_d(\times_{i \in I_d} \rho_i)$. Hence, the definition of the global state transition function Δ of a network of systems is given by the crossproduct of local transition functions Δ_d with ω_d derived through the interface maps Z_d from the output behavior of influencing systems

$$\Delta(q, \omega).d = \Delta_d(q_d, Z_d(\times_{i \in I_d} \beta_i(q_i, \omega_i))) \tag{15.8}$$

Exercise 15.17. Show this.

With this result in hand and under the prerequisites of the local state transition preservation condition (15.5) the local input/output behavior preservation condition (15.6) and the coupling preservation condition (15.7) we can show that $h(\Delta(q, \omega)) = \Delta'(h(q), \omega)$ with $h : \times_{d \in D} Q_d \to \times_{d' \in D'} Q'_{d'}$ defined by

$$h.d'(q) = h_{d'}(q.D_{d'})$$

is a system homomorphism as follows. By using (15.7) we can transform (15.5) into

$$h_{d'}(\times_{d \in D_{d'}} \Delta_d(q_d, Z_d(\times_{i \in I_d} \rho_i)) = \Delta'_{d'}(h_{d'}(\times_{d \in D_{d'}} q_d), Z'_d(\times_{i' \in I'_{d'}} k_{i'}(\times_{i \in D_{i'}} \rho_i)))$$

and by using (15.6) we obtain

$$\begin{aligned} &h_{d'}(\times_{d \in D_{d'}} \Delta_d(q_d, Z_d(\times_{i \in I_d} \beta(q_i, \omega_i))) \\ &= \Delta'_{d'}(h_{d'}(\times_{d \in D_{d'}} q_d), Z'_{d'}(\times_{i' \in I'_{d'}} \beta'_{i'}(h_{i'}(\times_{i \in D_{i'}} q_i), g_{i'}(\times_{i \in D_{i'}} \omega_i))))). \end{aligned}$$

The forms of Δ_d and $\Delta'_{d'}$ above, however, correspond to definition of Δ and Δ' for the network system in (15.8) and we can conclude

$$h_{d'}(\Delta(q, \omega).D_{d'}) = \Delta'(h_{d'}(q.D_{d'}), \omega')).d'$$

from the definition of h as the crossproduct of $h_{d'}$ we directly conclude

$$h(\Delta(q, \omega)).d' = \Delta'(h(q), \omega')).d'$$

and finally

$$h(\Delta(q, \omega)) = \Delta'(h(q), \omega).$$

The preservation of outputs can be shown in similar form. □

15.9 HOMOMORPHISM AND CASCADE DECOMPOSITIONS

There is close relationship between homomorphism and a particular kind of coupled model, called a *cascade composition*. As illustrated in Fig. 15.10, in such a coupled model there is no feedback from the second, or *Back* component, to the first, or *Front* component. When we are able to construct a homomorphic realization of a system using a cascade composition, we say that the system has a *cascade decomposition*. There is a complete theory of cascade decomposition for finite automata due to Krohn and Rhodes (1963) which used algebraic semigroup theory to breakdown machines into two kinds of primitives called reset and simple group machines – these are memory losing and memory retaining automata, respectively which are not further decomposable. Our approach follows that of Zeiger (1967) in avoiding semigroup theory and rephrases a theorem of Westerdale (1988) which shows directly how any strongly connected finite state system can be decomposed into a memory-losing front component and a memory-retaining back component.

To discuss cascade decomposition into the two kinds of elements, we need to generalize the congruence partition of Section 15.4.2 to allow blocks to overlap. Given a system, $S = (T, X, \Omega, Y, Q, \Delta, \Lambda)$, we define a *cover* to be a family (set) of subsets of Q whose union includes Q. A cover is *maximal* if no subset is included in any other. A partition is a (maximal) cover for which all the sets are disjoint. Just as with *congruence partitions*, a cover has the congruence property if all transitions carry subsets into subsets.

A *cascade composition* is a coupled model containing two component systems, called Front and Back, in which there is no coupling from Back to Front (Fig. 15.10). A cascade composition *homomorphically realizes* a System $S = (T, X, \Omega, Y, Q, \Delta, \Lambda)$ if there is a mapping $h : F \times B \rightarrow Q$ such that h is a homomorphism, where F and B are the state sets of Front and Back, resp. and h must be an onto mapping (covering all of Q) but may be partial (not necessarily defined on all of $F \times B$). When h is one-one we call the realization *isomorphic*.

The basic connection between intrinsic properties of a system (having a congruent cover) and its amenability to cascade decomposition is stated as:

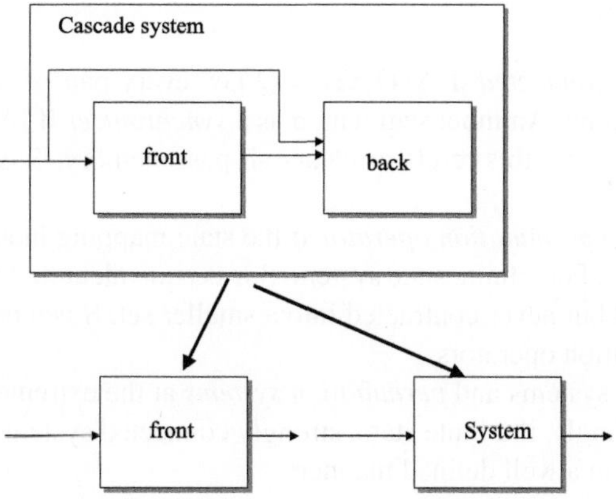

FIGURE 15.10

Cascade Composition with homomorphic mappings.

Theorem 15.14 (Cascade Decomposition). *If a cascade composition homomorphically realizes a System S then there is a maximal cover of Q with the congruence property. Conversely, given a cover of the latter kind we can construct a cascade realization of S.*

Proof. **(i)** Given the mapping $h : F \times B \to Q$ we then identify set of states associated with any front state, f as $Q_f = h(f, B) = \{q[h(f,b)] = q$, for some $b\}$. The sets $\{Q_f | f \in F\}$ form a cover of Q (since h is onto) and is maximal (no Q_f is included in another by definition). This cover has the congruence property due to the fact that there is no feedback from the Back to the Front – the destinations under an input segment of any two states (f, b) and (f, b') with the same f are states that have the same front element (which depends only on f).

(ii) Given a cover with congruence we construct a front component whose states are the subsets of the cover in the same manner as the reduced version of a system. We use the back component to disambiguate the state of S within a cover subset. The transitions are defined as needed. □

The specialization of covers to partitions is stated as:

Corollary 15.1. *The projection of a cascade composition on the front component is a homomorphism and induces a congruence partition. Given a system having a congruence partition, we can construct a cascade composition to* isomorphically *realize the system.*

Exercise 15.18. Prove the corollary.

Note we can define an isomorphic realization because a state is found in exactly one partition block. Thus the front component is guaranteed to have fewer states than the original system. In contrast, since a cover may have overlapping subsets, a system state may have multiple representations, and the front component can be larger than the original system. However, we need the state expansion provided by a cover to identify the memory losing and memory retaining components in a system.

Given $S = (T, X, \Omega, Y, Q, \Delta, \Lambda)$ we can extend Δ to subsets of Q, i.e., $\Delta : 2^Q \times \Omega \to 2^Q$ by defining: $\Delta(Q', \omega) = \{\Delta(q, \omega) | q \in Q'\}$ for any subset Q'.

Exercise 15.19. Show that the extended transition function also satisfies the composition property: $\Delta(Q', \omega v) = \Delta(\Delta(Q', \omega), v)$.

A system, S is *strongly connected* if $\Delta(Q, \Omega) = Q$ i.e., every pair of states has an input segment that takes the first to the second. An input segment, σ is a *synchronizer* if $|\Delta(Q, \sigma)| = 1$ i.e., if it maps every state to a particular state – this resets or clears all past memory. S is *synchronizable* if it has a *synchronizer.*

An input segment, ω is a *permutation operator* if the state mapping induced by ω is a permutation mapping (one-one and onto). For a finite state system, this is equivalent to: $|\Delta(Q, \omega)| = |Q|$, i.e., states are at most shuffled around but never contracted into a smaller set. S is a permutation system if all its input segments are permutation operators.

Clearly, synchronizable systems and *permutation systems* at the extremes of a continuum of memory losing systems. Interestingly, if a finite state, strongly connected system is not one of the extremes, it is a combination of them in a well defined manner:

Theorem 15.15. *Every strongly connected finite state system is either a) a synchronizable system, b) a permutation system, or c) has a cascade decomposition into a synchronizable front system and a permutation back system.*

Proof. Let $S = (T, X, \Omega, Y, Q, \Delta, \Lambda)$ be strongly connected finite state system which is not synchronizable nor a permutation system. Let the *contraction size* of $\omega \in \Omega$ be the size of its range, i.e., $|\Delta(Q, \omega)|$ By definition, the contraction size of a synchronizer is 1. Since S is not synchronizable, call σ a *quasi-synchronizer* if it has the smallest contraction size (> 1) of all $\omega \in \Omega$. Let the range of σ be R, i.e., $\Delta(Q, \sigma) = R$. Since S is not a permutation system, R is a proper subset of Q. Since S is strongly connected, we can show that the family of sets $\{\Delta(R, \omega)|\omega \in \Omega\}$ is a maximal cover of Q with congruence property, (illustrated in Fig. 15.11). Moreover, each of the subsets has the same cardinality.

Exercise 15.20. Use the composition property of the extended transition function to establish the congruence property and to show that each of the subsets of the cover has cardinality $= |R|$, i.e., for all $\omega \in \Omega$. $|\Delta(R, \omega)| = |R|$.

Using the Cascade Decomposition Theorem, we construct a cascade composition whose front component is based on the cover. In this front component, the underlying quasi-synchronizer becomes a true synchronizer. Since all subsets have the same cardinality, the back component has a state set of this cardinality and the induced operations on the back component state set are all permutations. □

Exercise 15.21. Relate the state expansion of the cascade decomposition to the contraction size of the quasi-synchronizer.

The contraction size of a quasi-synchronizer determines the memory loss properties of a finite state, strongly connected system. If the contraction size is unity, there is an input segment that erases all memory. If this segment is a generator, this is like having a button that resets the machine to a ground state. Otherwise, we can synchronize down only to a subset of states whose cardinality is the contraction size. A non-strongly connected finite state system can be decomposed into strongly connected components (Bavel and Muller, 1970) to which the theorem applies.

15.10 CHARACTERIZATION OF REALIZABLE I/O RELATIONS AND FUNCTIONS

Consider the IORO, IOFO, and I/O system level of system description. Given an I/O system S we have seen how to associate the lower level objects IOFO_S and IORO_S with it. Our definition of IORO and IOFO, however, allowed considerable freedom, and we would like to know whether every such

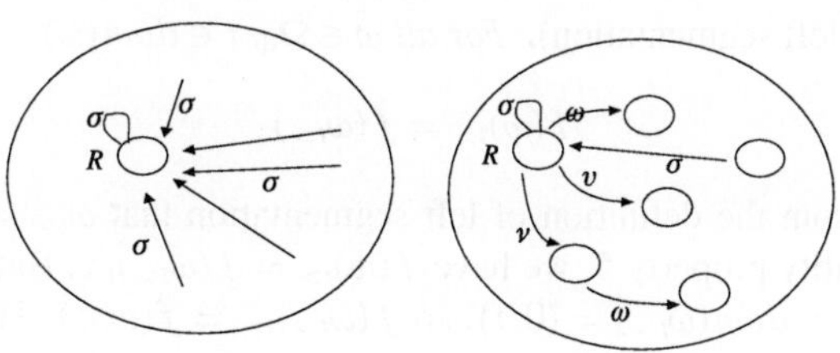

FIGURE 15.11

Illustrating the cover with congruence property induced by a quasi-synchronizer.

object can be associated with some system. The problem can be phrased as follows: an IORO is said to be realizable if there exists an I/O system S such that IORO = IORO$_S$. In other words, an IORO is realizable if we can find a system that has exactly the same IORO associated with it.

Our problem then is to characterize the classes of realizable IOROs and IOFOs by looking for necessary and sufficient conditions that will distinguish a realizable IORO or IOFO from a non-realizable IORO or IOFO. The simpler these conditions are to check, the more useful characterization we will have achieved. After all, one characterization of a realizable IORO is given by the definition of "realizable" itself. But given an IORO, this would require us to search through all possible systems to determine whether there is, or is not, a system S for which IORO = IORO$_S$.

Our characterization are based on the characterization of realizable I/O functions.

A function f is said to be *realizable* if there is an I/O system $S = (T, X, \Omega, Y, Q, \Delta, \Lambda)$ and a state $q \in Q$ such that $f = \widetilde{\beta}_q$.

Theorem 15.16. *A function $f : \Omega \to P$ is realizable if and only if the following conditions hold.*

1. Ω and P are of the right form*: there are sets X, Y, and a time base T such that* $\Omega \subseteq (X, T)$ *and* $P \subseteq (Y, T)$.
2. Ω *is* closed under translation and composition.
3. f is length preserving and in step: *for all* $\omega \in \Omega$, $dom(f(\omega)) = dom(\omega)$.
4. f is time invariant: *for all* $\omega \in \Omega$, $\tau \in T$
$$f(TRANS_\tau(\omega)) = TRANS_\tau(f(\omega))$$
5. f is causal: *for all* ω, $\omega' \in \Omega_0 (= STR(\Omega))$, *and* $t \in dom(\omega) \cap dom(\omega')$
$$\omega_{t>} = \omega'_{t>} \Rightarrow f(\omega)_{t>} = f(\omega')_{t>}.$$
6. f has last values defined: *for all* $\omega \in (X, T)_0$
$$f(\omega)(l(\omega)) \in Y \Leftrightarrow \omega \in \Omega_0.$$

Proof. The necessity of the six listed properties is established by considering the definition of an I/O function $\widetilde{\beta}_q$ of an arbitrary system S. The reader is given the exercise of showing that $\widetilde{\beta}_q$ has these properties.

For establishing sufficiency, we suppose that a function f has the given properties. Based on these properties, we construct a system S_{f*} which realizes f, using the following lemmas. □

Lemma 15.4 (f preserves left segmentation). *For all* $\omega \in \Omega_0$, $t \in dom(\omega)$
$$f(\omega)_{t>} = f(\omega_{t>})$$

Proof. It follows easily from the definition of left segmentation that $\omega_{t>} = (\omega_{t>})_{t>}$. Setting $\omega = \omega$ and $\omega' = \omega_{t>}$ in the causality property 5, we have $f(\omega)_{t>} = f(\omega_{t>})_{t>}$. But by the length-preserving property 3, $dom(f(\omega_{t>})) = dom(\omega_{t>}) = \langle 0, t \rangle$, so $f(\omega_{t>})_{t>} = f(\omega_{t>})$. Thus $f(\omega)_{t>} = f(\omega_{t>})$, as desired. □

Lemma 15.5 (f is specifiable by a last value function $\widetilde{f}$). *There exists a function $\widetilde{f} : \Omega_0 \to Y$ such that for all $\omega \in \Omega_0$, $t \in dom(\omega)$, $f(\omega)(t) = \widetilde{f}(\omega_{t>})$.*

Proof. Define $\tilde{f} : \Omega_0 \to Y$ by $\tilde{f}(\omega) = f(\omega)(l(\omega))$ for $\omega \in \Omega_0$. By property 6, $\tilde{f}$ is defined for every $\omega \in \Omega_0$. Moreover,

$$\begin{aligned} f(\omega)(t) &= f(\omega)_{t>}(t) && \text{(property of left segmentation)} \\ &= f(\omega_{t>})(t) && \text{(Lemma 15.4)} \\ &= \tilde{f}(\omega_{t>}) && \text{(definition of } \tilde{f}\text{, noting } l(\omega_{t>}) = t) \end{aligned}$$

With the help of Lemma 15.5 we construct an object $S_f^* = (T, X, \Omega, Y, Q^*, \Delta^*, \Lambda^*)$, where Ω is the given domain of f and T, X, Y are the sets guaranteed to exist by property 1. In addition, $Q^* = \Omega_0$, $\Delta^* : \Omega_0 \times \Omega \to \Omega_0$ is defined by

$$\Delta^*(\omega, \omega') = \omega \bullet STR(\omega')$$

and $\Lambda^* : \Omega_0 \to Y$ is defined by $\Lambda^* = \tilde{f}$.

We are assuming for simplicity that the null segment $\Phi \in \Omega_0$ so that in particular $\Delta^*(\Phi, \omega') = STR(\omega')$. (If $\Phi \in \Omega_0$, we set $Q^* = \Omega_0 \cup \{\Phi\}$, slightly complicating the definition of Λ^*.)

By Lemma 15.5, Λ^* is well defined and it is easy to check that S_f^* is a system, which moreover is time invariant.

Finally note that for $\omega \in \Omega_0$

$$\beta_\Phi(\omega) = \Lambda^*(\Delta^*(\Phi, \omega)) = \tilde{f}(\omega)$$

so that $\beta_\Phi | \Omega_0 = \tilde{f}$, and it is then easy to check that $\tilde{\beta}_\Phi = f$. Thus S_f^* realizes f as required. □

Having this characterization of realizable functions, we can develop a characterization for realizable IOFOs.

Theorem 15.17. *An* IOFO $= (T, X, \Omega, Y, F)$ *is realizable if and only if the following conditions hold.*

1. *Each function $f \in F$ is realizable.*
2. *The set F is closed under derivatives*

$$f \in F \Rightarrow \forall \omega \in \Omega, \overline{\overline{\tilde{f} L_\omega}} \in F$$

where $\tilde{f}$ is the last output function guaranteed to exist by the realizability of f, $\tilde{f} L_\omega$ is called the ωth derivative of $\tilde{f}$ and is defined by $\tilde{f} L_\omega(\omega') = \tilde{f}(\omega\omega')$ for all $\omega' \in \Omega$, and $\overline{\overline{\tilde{f} L_\omega}}$ is the segment-to-segment mapping specified by $\tilde{f} L_\omega$.

Proof. The necessity of the condition follows by considering an arbitrary system and examining its associated IOFO. Let $S = (T, X, \Omega, Y, Q, \Delta, \Lambda)$ and $\text{IOFO}_s = (T, X, \Omega, Y, \tilde{B}_S)$, where, as we recall, $\tilde{B}_S = \tilde{\beta}_q | q \in Q$. Clearly, each $\tilde{\beta}_q$ is realizable. Moreover, the reader should verify that $\beta_q L_\omega = \beta_{\Delta(q,\omega)}$, i.e., the derivatives of the I/O function of state q are the I/O functions of all the states accessible from q. Thus $\tilde{B}_S$ is closed under derivatives.

To show that the conditions are sufficient, we construct a system S for which IOFO_S is the given IOFO. Since each function f is realizable, we can construct a system S_f^* that realizes it. Now each of these systems has the same state space Ω_0, but it is an easy matter to relabel each of these sets so that they are disjoint.

Naturally, the transition and output functions are likewise altered. For example, let $S^*_f = (T, X, \Omega, Y, \Omega_0 \times \{f\}, \Delta^*_f, \Lambda^*_f)$, where $\Delta^*_f((\omega, f), \omega') = (\omega\omega', f)$, and $\Lambda^*_f(\omega, f) = \widetilde{f}(\omega)$. We can then "lay all the separate systems next to each other" in an operation called the disjoint union. Formally this results in an object $S^*_F = (T, X, \Omega, Y, Q^*_F, \Delta^*_F, \Lambda^*_F)$, where $Q^*_F = \cup_{f \in F}(\Omega_0 \times \{f\})$, $\Delta^*_F|(\Omega_0 \times \{f\}) = \Delta^*_f$ and $\Lambda^*_F|(\Omega_0 \times \{f\}) = \Lambda^*_f$. It is easy to check that S^*_F is a time invariant system. Moreover, we claim that $\text{IOFO}_{\text{SF}*} = \text{IOFO}$. Certainly $F \subseteq \widetilde{B}_{SF*}$, since f is realizable by S^*_f, a part of S^*_F. On the other hand, given $q \in Q^*_F$, it belongs uniquely to some component $\Omega_0 \times \{f\}$, i.e., $q = (\omega, f)$ for some $\omega \in \Omega_0$ and $f \in F$. But since (ω, f) is accessible from $(\emptyset, f)$, we have that $\widetilde{\beta}_{(\omega,f)}$ is a derivative of $\widetilde{\beta}(\Phi, f)$. Noting that $\widetilde{\beta}_{(\Phi,f)} = f$ and that F is closed under derivatives, we have that $\widetilde{\beta}_{(\omega,f)} \in F$. This allows us to conclude that $\widetilde{B}_{SF*} \subseteq F$, hence $\widetilde{B}_{SF*} = F$, as required. □

Finally, we have the

Corollary 15.2 (Characterization of realizable IOROs). *An IORO (T, X, Ω, Y, R) is realizable if and only if there is a realizable IOFO (T, X, Ω, Y, F) such that $R = \cup_{f \in F} f$.*

15.10.1 CANONICAL REALIZATION

Given a realizable I/O function, there is at least one I/O system that realizes it. More than this, from Section 15.4.2, there is a reduced version of this system that has minimality and uniqueness properties. In fact, from the above discussion, we can see that every system that realizes the function has to have states representing the function and its derivatives. Indeed, if the system uses only these states, then we call it the *canonical realization* of the function. It is easy to see that the canonical realization is:

- is *minimal*, defined as being a homomorphic image of a subsystem of any realization of the function
- *unique*, in being isomorphic to any other minimal realization.

Given a realizable IOFO, there is similarly a canonical realization for it. However, given an IORO, there may be several distinct IOFOs that can be associated with it. But, under a condition called system identifiability, a unique IOFO, and canonical realization, exist (Zeigler, 1971).

15.11 SUMMARY

The hierarchy of preservation relations runs parallel to the hierarchy of system specification levels. For each level of system specification there is a corresponding relation that preserved features introduced at this level. Recall that as the level increases, more and more structure is introduced by the system specification. Accordingly, higher level morphisms preserve more structure than do lower level ones. Thus higher level morphisms are stronger in the sense of preserving more structure. Thus it is natural that the existence of a stronger morphism relating a pair of systems should imply the existence of a weaker morphism relating the same pair. In fact, we showed that for each level, a morphism between two system specifications at that level implies a morphism between their associated specifications lower levels.

The utility of a high level morphism, however, lies in the guarantee that a lower level morphism exists, not necessarily in the actual construction of the latter. Indeed, the higher level morphisms allow us to check – by comparison of structures – whether two systems are behaviorally equivalent without having to compare their behavior directly. For example, a homomorphism at the multi-component and network of systems allows us to check several smaller local preservation conditions together with coupling preservation conditions and automatically conclude that a global homomorphism exists. For this reason, higher level morphisms are important in model simplification and in simulator verification.

Although stronger preservation relations imply weaker ones, the reverse is not generally true. This means that as a rule we cannot infer structure uniquely from behavior (Zeigler, 1976). Inductive modeling based on the DEVS formalism and applying non-monotonic logic is discussed in Sarjoughian (1995), Sarjoughian and Zeigler (1996).

15.12 SOURCES

Homomorphisms of general algebras and systems are discussed by Cohn (1974) and Foo (1974). The relationship between I/O function morphism and the system morphism is derived from Krohn and Rhodes (1963) and by Hartmanis and Stearns (1966). Realization of I/O functions is discussed in a number of books, e.g., Kalman et al. (1969), Zemanian (1972). Category theory provides a foundation for abstract properties of morphisms and has been used to characterize canonical realizations (Rattray, 1998). Aggregations in control theory are treated in Aoki (1968). Yamada and Amoroso (1971) discusses behavioral and structural equivalences in cellular spaces.

REFERENCES

Aoki, M., 1968. Control of large-scale dynamic systems by aggregation. IEEE Transactions on Automatic Control 13 (3), 246–253.

Bavel, Z., Muller, D.E., 1970. Connectivity and reversibility in automata. Journal of the ACM (JACM) 17 (2), 231–240.

Cohn, P., 1974. Algebra, vols. 1 and 2. JSTOR.

Foo, N.Y., 1974. Homomorphic Simplification of Systems.

Hartmanis, J., Stearns, R.E., 1966. Algebraic Structure Theory of Sequential Machines, vol. 147. Prentice-Hall, Englewood Cliffs, NJ.

Kalman, R.E., Falb, P.L., Arbib, M.A., 1969. Topics in Mathematical System Theory, vol. 1. McGraw-Hill, New York.

Krohn, K., Rhodes, J., 1963. Algebraic theory of machines. In: Fox, J. (Ed.), Mathematical Theory of Automata, pp. 371–391.

Rattray, C., 1998. Abstractly modelling complex systems. In: Allbrecht (Ed.), Systems: Theory and Practice.

Sarjoughian, H., 1995. Inductive Modeling of Discrete Event Systems: a TMS-Based Non-monotonic Reasoning Approach, p. 256.

Sarjoughian, H.S., Zeigler, B.P., 1996. Inductive modeling: a framework marrying systems theory and non-monotonic reasoning. In: Hybrid Systems.

Westerdale, T.H., 1988. An automaton decomposition for learning system environments. Information and Computation 77 (3), 179–191.

Yamada, H., Amoroso, S., 1971. Structural and behavioral equivalences of tessellation automata. Information and Control 18 (1), 1–31.

Zeiger, H.P., 1967. Cascade synthesis of finite-state machines. Information and Control 10 (4), 419–433.

Zeigler, B.P., 1976. Theory of Modeling and Simulation. Wiley Interscience Co.

Zeigler, B.P., 1971. Canonical realization of general time systems. Information Sciences 12 (2), 179–186.

Zemanian, A.H., 1972. Realizability Theory for Continuous Linear Systems. Courier Corporation.

CHAPTER

ABSTRACTION: CONSTRUCTING MODEL FAMILIES

16

CONTENTS

Theory of Modeling and Simulation. https://doi.org/10.1016/B978-0-12-813370-5.00026-2

Abstraction is key to model construction for simulation. Traditionally, it has been considered to be a black art, a craft without a scientific basis. However, the importance of methodological support for abstraction is becoming increasing evident. Complex distributed simulation systems are being developed that are intended to support interoperation of models at different levels of resolution. Such designs presuppose effective ways to develop and correlate the underlying abstractions. However, discussions of abstraction issues still often proceed through individual anecdotal experiences with little cumulative impact. Although a conceptual and computational framework This chapter presents and concepts and approaches, based on the morphism concepts of Chapter 12, for effectively working with abstraction in model construction.

Abstraction is the process underlying model construction whereby a relatively sparse set of entities and relationships is extracted from a complex reality. We can also define abstraction as *valid simplification*. Thus, to understand and support the abstraction process, we must have concepts for *model validity* and for *model simplification*. A model's complexity is measured in terms of the time and space required to simulate it. Validity is preserved through appropriate morphisms at desired levels of specification. Thus abstraction methods, such as aggregation, will be framed in terms of their ability to reduce the complexity of a model while retaining its validity relative to the modeling objectives and experimental frame (Chapter 2).

In this chapter, we first develop three different measures of complexity. Then we use the M&S framework (Chapter 2) and the morphism specification hierarchy (Chapter 12) to develop a quite general theory of aggregation as well as an approach to developing DEVS abstractions for control. We discuss the importance of maintaining not just one model, but an integrated family of models, to allow flexibility in both calibration and application. The concept of parameter morphism is then introduced to support the organization of models in an integrated family.

16.1 SCOPE/RESOLUTION/INTERACTION PRODUCT

As we have said, *abstraction* refers to a method or algorithm applied to a model to reduce its complexity while preserving its *validity* in an experimental frame. The inevitable resource (time and money) constraints placed on project budgets require working with models at various levels of abstraction. The more detail included in a model, the greater, the greater are the resources required of a simulator to execute it and those required of the development team to build it. We refer the time and memory required to execute a model as its *computational complexity*. The time and cost of development is its *developmental complexity*.

We usually can relate the computational and developmental complexity to the *scope and resolution* of a model. *Scope* refers to how much of the real world is represented, *resolution* refers to the number of variables in the model and their precision or granularity. Scope is a relation between the model and the portion of the real world it is intended to represent. Typically the larger the scope, the greater the number of components in the model. However, this need not be the case since, for example, the model may represent the real world portion of interest as a single component. Consequently, we will use *size*, the number of components in a coupled model, as correlating better with complexity, but most of the time we won't be too wrong if we think of scope as strongly determining size. We will also identify resolution as the number of states in a component and, if we are referring to the model as a whole, then we assume that all components have the same state set (Fig. 16.1). We say that the complexity of

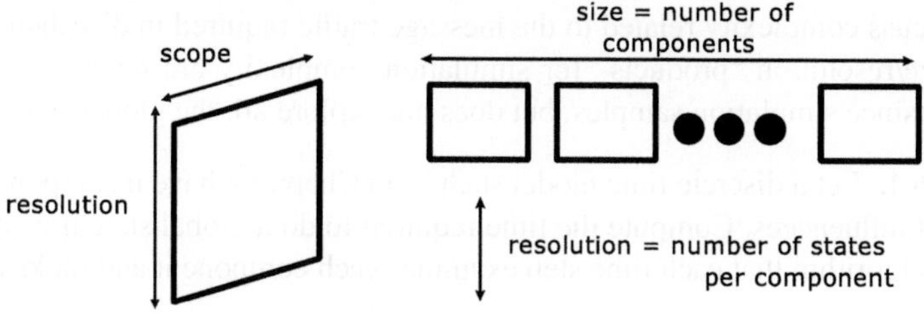

FIGURE 16.1

Size/Resolution Product.

a model depends on the amount of detail in it, which in turn, depends on the *size/resolution* product. "Product" is used symbolically here, i.e., we don't necessarily multiply size and resolution – what we do instead, is discuss several measures in which increasing the size (number of components) and resolution (number of states per component) lead to increasing complexity. Some times we get a better handle on the complexity of a model by including the interactions among components in the product – in this case, we will refer to it as the *size/resolution/interaction* product.

16.1.1 COMPLEXITY

Three kinds of complexity can be understood in terms of the size/resolution/interaction product.

- *analytic complexity* – this is the number of states that have to be considered to analyze, i.e., fully explore, the global state space of the coupled model. The number of states in the coupled model is the cardinality of the cross-product of the local state sets,

  $$|Q| = |Q_1| \times |Q_2| \times \cdots \times |Q_n|$$

  Recall that the coupled model states are called *global* states in contrast to the *local* state sets of the components. Sometimes the increase in states is called "state space explosion" since it involves all the combinations of states and is exponential in the number of components. For example, if each component has m states then the global space of n components has m^n states. Here we see the size/resolution "product" in that increases in both component numbers (size) and component state size (resolution) increase the crossproduct.
- *simulation complexity* – measures the computational resources required to execute the model. There may be many measures that are relevant depending on the resource under focus, time (e.g., execution time) or space (e.g., memory size) and the simulator class considered. Space required is proportional to number of components and the space required for the current state of each – e.g., m states can be coded in log m bits, yielding a size/resolution "product" of $nlogm$ bits. A basic discrete time scanning algorithm may take time proportional to nm^d (where d is the number of influencees, a measure of the interactions in a model) to do a global state transition. A discrete event algorithm can reduce this by a factor related to the model activity. In Chapter 17

we will discuss complexity related to the message traffic required in distributed simulations. In general, size/resolution "products" for simulation complexity are much smaller than analytic complexity since simulation samples, but does not explore all, the global states.

Exercise 16.1. Let a discrete time model such as in Chapter 4 have n components, each with m states and d influencees. Compute the time required to do a global state transition for the basic simulation algorithm that each time step examines each component and looks up a table of next states.

Exercise 16.2. For the same model assume that the fraction of "active" components is a and that a discrete event algorithm (Chapter 3) can limit its scan to only the active components with negligible overhead. What is its time complexity? How does this change when the overhead to manage the list of active components is taken into account?

- *exploratory (search) complexity* – this relates to the resources required to explore a space of models. For example, to evaluate M models requires a time complexity of $M\times$ *size/resolution product*, where the latter depends on the particular analytic or simulation approach used. The size of the model space, M, also depends on the structure of the coupled model and its parameterization. In the worst case, this can grow super-exponentially – the number of transition functions that are possible for an input free discrete time model having n components with m states each is enormous:

$$m^{n^{m^n}}$$

For example, a model with 10 components having 2 states each has 2^{10} or approximately 1000 states and there are approximately $1000^{1000} \sim 10^{3000}$ such models.

Exercise 16.3. Imagine a table with two columns and S rows. The left hand one for current states is filled in, while the right hand column is blank. Each way of filling in the right hand column constitutes a different transition function. Show that the number of ways of filling in the right hand column is S^S.

Clearly real world processes can never explore all the structures in such *Vast* design spaces (see Dennett, 1995), whether such processes be natural (evolution, human intelligent) or man-made (artificially intelligent). We can limit our explorations by imposing parameterized structures on such spaces and then considering only the parameter combinations. For example, there are only approximately 1000 cellular automata having 9 neighbors (including the center cell) with 2 states each, which is independent of the number of cells and is *Vastly* smaller than the unconstrained search space.

Exercise 16.4. Show that there are m^d cellular automata having d neighbors (including the center cell) with m local states. How many linear cellular automata of this type are there? (In a linear transition function, the next state can be specified as weighted sum of the neighbor states.)

In general, the size/resolution product strongly impacts the size of the space of models to explore. Unconstrained, such spaces are *Vast* with cardinalities as high as SR^{SR} (where SR is the size/resolution product). By imposing parameterized structures on them, we can reduce the spaces to more manageable levels – with the risk that something (actually, possibly a *Vast* number of things) of interest maybe left out!

Table 16.1 Some Common Abstractions

Simplification Method	Brief Description	Affects Primarily
Aggregation	combining groups of components into a single component which represents their combined behavior when interacting with other groups	size and resolution
Omission	leaving out • components • variables • interactions	size resolution interactions
Linearization	representing behavior around an operating point as a linear system	interactions
Deterministic/Stochastic Replacement Deterministic ⇒ Stochastic Stochastic ⇒ Deterministic	replacing deterministic descriptions by stochastic ones, can result in reduced complexity when algorithms taking many factors into account are replaced by samples from easy-to-compute distributions. Replacing stochastic descriptions by deterministic ones, e.g., replacing a distribution by its mean	interactions interactions
Formalism Transformation	mapping from one formalism to another, more efficient, one, e.g. mapping differential equation models into discrete event models.	

16.1.2 SIZE/RESOLUTION TRADE-OFF: SIMPLIFICATION METHODS

Since complexity depends on size/resolution product, we can reduce complexity by reducing the size of a model or its resolution (or both). Given a fixed amount of resources (time, space, personnel, etc.), and a model complexity that exceeds this limit, there is a trade-off relation between size and resolution. We may be able to represent some aspects of a system very accurately but then only a few components will be representable. Or we may be able to provide a comprehensive view of the system but only to a relatively low resolution. *Abstraction* is a general process, and includes various simplification approaches such as outlined in Table 16.1.

As indicated, various methods can be used to reduce the complexity of a model as measured by the size/resolution/interaction product. Note that some methods may primarily affect the size, others the resolution, of a model. The reduction in interactions here might be measured by the number of influencees, or more in depth, by the computational complexity of the functions that implement these interactions. Computer Science has characterized this kind of complexity. The challenge in any case, is to reduce complexity while preserving the validity of a model with respect to its counterpart real system in desired experimental frames.

Most abstraction methods work on the structure of a *base* model in an attempt to achieve a valid *lumped* model (hence work at levels 3 and 4 of the specification hierarchy which support state and component descriptions). The homomorphism concept then provides a criterion for valid simplification. Error is introduced when exact homomorphism is not achieved. However, the abstraction may still be valid if the error does not grow so as to exceed the tolerance for goodness of fit (see Chapter 15).

The benefits of abstraction include the ability to perform more rapid analysis and wider ranging exploration at lower cost. This may be crucial in initial attacks at a problem and in gaining an overall understanding of how all the components interact (seeing the forest rather than the trees). Also, abstractions can be more generic and widely applicable then models tuned to specific circumstances. Models based on abstractions can be based on a few general principles in pristine form and hence more reusable than detailed models. Nevertheless, models at various levels of size and resolution are indispensable. More detailed models help in understanding phenomena in the messy real world away from the control of the laboratory. They are necessary for achieving realism for example, in virtual simulations, where new designs (say for a high speed airplane) must be tested under conditions as near as possible to what they will experience in reality.

You might think that one should go about constructing detailed models first – and then derive abstractions from them using simplification methods only when you are compelled to by complexity considerations. Unfortunately, often the data to calibrate and validate high resolution models are not available. On the other hand, data may be available in more aggregate form that enables calibrating lower resolution models. This calls for an approach, called "integrated families of variable resolution models" (Davis, 1995) where models are constructed in ways that support cross-calibration and validation. The parameter morphism concept to be discussed soon facilitates such an approach since it explicitly records how parameters of base and lumped are related.

16.1.3 HOW OBJECTIVES AND EXPERIMENTAL FRAME DETERMINE ABSTRACTION POSSIBILITIES

Modelers must trade off the advantages of model abstraction against the costs. For example, benefits in reduced runtime and memory requirements may be accompanied by an inevitable loss of predictive accuracy. The objectives of the M&S effort should drive the choice of which trade-offs to make.

We return to finite state homomorphisms to demonstrate the importance of objectives and experimental frames to attaining valid abstractions. Recall that homomorphisms can be equivalently rephrased in terms of partitions of the state set. A homomorphism corresponds to a partition that has the congruence property with respect to the transition function and which refines the output partition (Chapter 12). For finite state systems, there is a well-known procedure that generates the desired partition. As illustrated in Fig. 16.2, one starts an iterative process with the output partition. If this does not satisfy the congruence property, then the partition is refined by breaking up blocks that send transitions to different blocks. The resulting partition is tested for congruence and this process continues iteratively until the congruence property is satisfied. (At each step, the partition blocks get smaller, with all states coherently going to the same blocks of the *previous* partition, but not necessarily to the now *smaller* blocks.) The process must terminate since the finest partition – in which each state is in its own block – is the limit. Given a finite state system, the set of congruent partitions is a constructive property of its state transition function. However, which of these partitions can match the output requirements is determined by the output partition – only those that refine the given output partition. And the latter is determined by the experimental frame. The higher the resolution required to compute the desired output, the finer the output partition. The finer the output partition, the finer the congruence required to refine it.

Exercise 16.5. Why doesn't the iteration stop after one step? (Hint: At each step, the partition blocks get smaller, with all states coherently going to the same blocks of the *previous* partition, but not necessarily to the now *smaller* blocks.)

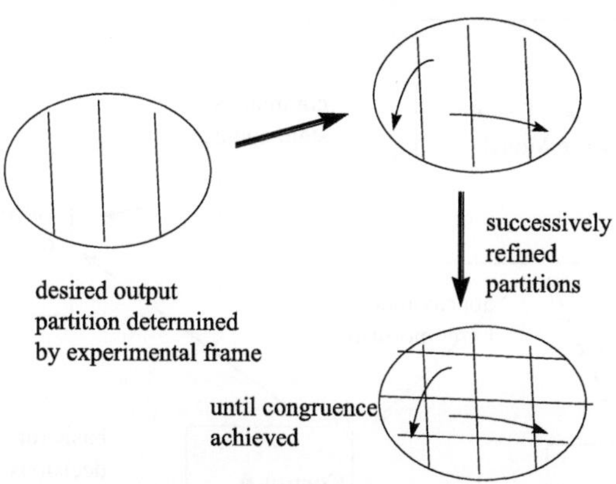

FIGURE 16.2

Homomorphism as Output Respecting Congruence.

Exercise 16.6. Some finite state systems have no non-trivial homomorphisms. Show that a n-counter has congruence partitions that correspond to the factors of n. What can you conclude if n is prime?

16.2 INTEGRATED FAMILIES OF MODELS

To make our discussion concrete, we will introduce an example of an integrated model families.

16.2.1 INTEGRATED MODEL FAMILY EXAMPLE: SPACE TRAVEL

Space ships traveling from earth to other planets and moons will someday be common. We'll develop a base model as a dynamic structure DEVS&DESS. The base model will itself include many simplifications and we'll indicate what simplifications we're introducing from the outset and what general category they illustrate. From the base model, we'll construct two abstractions, one useful for short term control and scheduling of the flights, the other for long term logistical planning. Thus our overall modeling objectives are to construct a space travel planning, scheduling and control system and test it against a realistic base model (Fig. 16.3). The planning model predicts the efficiency of alternative routing policies for given cargo demands (input time trajectory) based on numbers of ships in the various routes. Basically, it is asking will the right number of ships be available at the right places at the right times. The control model will be a DEVS that abstracts the routes of ships and their times of travel, **without** retaining the details of such travel. It can be used to keep track of individual ships and schedule them for picking up and delivering cargo. This abstraction, an example of "transformation of formalism" simplification, supports the design of an event-based controller that can be implemented as a real-time DEVS (Chapter 10). Chapter 18 discusses the general methodology of which this is an example. The logistics model will help us to develop and illustrate the principles underlying simplification by aggregation.

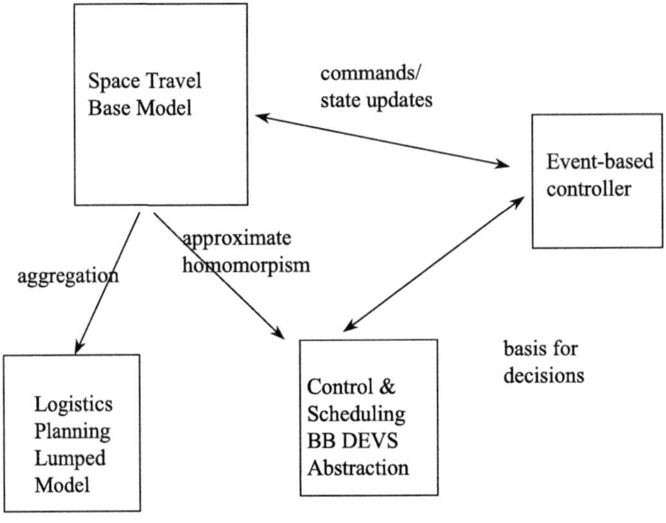

FIGURE 16.3

Space System Integrated Model Family.

16.2.2 SPACE TRAVEL BASE MODEL

Imagine a single space ship starting at earth and destined for the moon and mars. One of the strong influences in its travel is gravity. Recall that gravity is a force exerted on a body by all other bodies in relation to their and their distance away. Theoretically, we might have include the gravitational forces of the earth, moon and mars on the ship (if not all the other planets and objects), but practically these influences are only significant when the ship is near enough to one and then, since the planets are far apart, only this one has any influence. Thus, an example of "omission", we leave out the gravitational influences of all but the most dominant body (if any – in between planets, there is no influence). But now have the problem of deciding when the ship is within the gravitational field of one of the bodies or in free flight.

WHY DEV&DESS?

The motion of the ship is well described by the differential equations of Newtonian mechanics. But the crisp partitioning of space based on gravitational strength is best handled by conditions and state events. Thus DEV&DESS is the natural formalism for the model. We can define a sphere around the earth, say, within which the ship's equations of motion include the influence of earth's gravity. When the ship crosses into, or out of, its equations of motion must be changed accordingly.

Since the sphere cuts out gravitational force that might be there in reality (although, small), there is potential for error in this "omission" approximation. Fortunately, it is a once-only situation and the error should not affect the overall behavior (Chapter 15).

WHY DYNAMIC STRUCTURE?

When the ship enters, or leaves, one of the gravitational spheres, its motion equations change considerably. An elegant way to effect this change is to have the state event trigger a structural change as supported by DS-DEVS (Chapter 10). Fig. 16.4A hows that we can have a component for free flight and for each gravitational region. When the ship crosses a boundary it is sent from the model for the current region to the model for the region it enters.

WHY DISTRIBUTED SIMULATION?

Another way to look at structure change is to assign the components to computers in a distributed simulation (Fig. 16.4B). This makes sense if we are tracking a large number of ships where at any time, there may be many ships orbiting any of the bodies or in free flight between them. We can think of each computer as a server specialized in its particular gravitational conditions and spatial properties. This exploits the parallelism in the model especially if interaction between ships is occurs only between those in close proximity. Then interprocessor message traffic is limited to migrations from one gravitational zone to another.

Fig. 16.5A illustrates how the forces on a space ship are analyzed. Some of the simplifications employed are:

- it is enough to consider the forces acting at the centers of gravity of interacting bodies; indeed, the center of gravity allows us to aggregate particles in a rigid body into a single point that represents their gravitational interaction with any other body.
- we limit ourselves to planar two-dimensional space, thus simplifying 3 dimensional space
- we omit various factors that relate to micro-details of the vehicle such as its orientation, stresses, frictional forces, etc. and these omissions may introduce error.

Arguments about orders of magnitude of error can usually justify such omissions, but sometimes the only way to assess the error effect is to develop a refined model in which the factor is included. *Sen-*

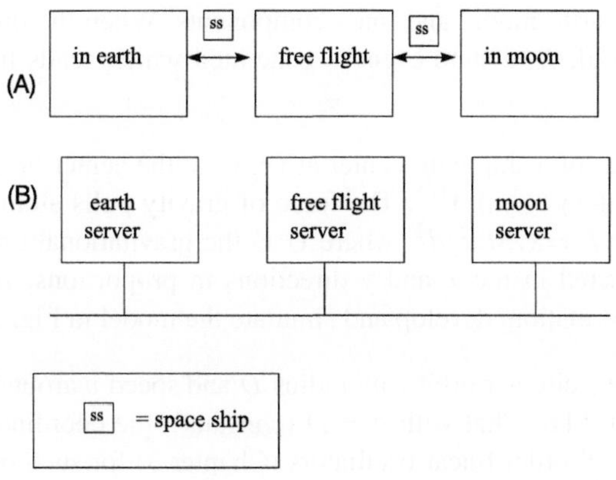

FIGURE 16.4

Dynamic Structure Architecture.

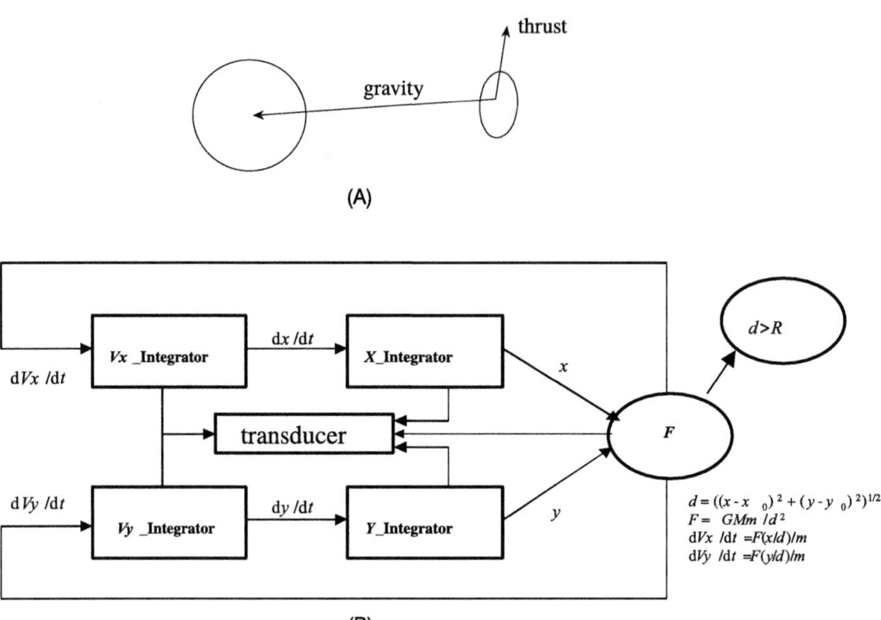

FIGURE 16.5

(A) Forces on a space ship (B) DEV&DESS coupled model for kinetics.

sitivity analysis is the testing of whether these factors affect the ultimate outputs as determined by the experiment frame.

The differential equations we arrive at, based on Newtonian mechanics, are illustrated in the DEV&DESS Fig. 16.5B. The center of gravity coordinates, mass, M, and escape radius, R are parameters that characterize earth, moon, and mars components. When the distance from the center of gravity exceeds the threshold, R, a state event is generated which sends the ship to another model component.

Exercise 16.7. The distance of a ship with center at (x, y) to the center of gravity of a massive body (x_0, y_0) is $d = ((x - x_0)^2 + (y - y_0)^2)^{1/2}$. The force of gravity pulls along the line joining the two centers and has magnitude $F = GMm/d^2$, where G is the gravitational constant, M and m are the masses. The force is projected in the x and y directions in proportions, $p_x = x/d$ and $p_y = y/d$, respectively. Using this information, develop and simulate the model in Fig. 16.5.

Exercise 16.8. To maintain a circular orbit with radius D and speed v around a massive body requires a centripetal force, mv^2/D. Show that with $d = D$ (constant), the coordinate dynamics in Fig. 16.5 separate into independent 2nd order linear oscillators (Chapter 3) for such orbits with frequency $\omega = GM/d^3)^{1/2}$. It is remarkable indeed, that a non-linear 4th order system has such linear 2nd order subsystems. Let $(x_0, y_0) = (0, 0)$. Starting at $x = 0$, $y = D$, what initial velocities v_x and v_y will start the system in a circular orbit with radius D?

The complete model of a space ship is a coupled model containing the kinetic component from Fig. 16.5 connected to its propulsion component which supplies the thrust. Sending such a model on an input port to a server coupled model is illustrated in Fig. 16.6. When the server receives such a model it supplies it the parameters (center coordinates, mass and its radius, escape radius) that characterize its region of space. However, clearly it leave unchanged the position and velocity coordinates and propulsion state variables characterizing the space ship state at the moment it enters the server's region of operation.

16.3 AGGREGATION: HOMOGENEITY/COUPLING INDIFFERENCE PRINCIPLES

As illustrated in Fig. 16.7, *aggregation* is an abstraction method that maps a coupled model into another, less complex, coupled model, with the intent of preserving behavior in some applicable experimental frame. The mapping process is a composition of two steps. In the first, the component set is partitioned into blocks. As in Chapter 12 each block can be described as a coupled model and these block coupled models can be coupled to recapture the full behavior of the original coupled model. Thus this step does not involve any abstraction *per se* but results in a change in structure, called *deepening*, since we have added an extra hierarchical layer it the model. In the second, the block coupled models are abstracted to simpler versions, and are formed into a final coupled model with suitably abstracted coupling.

Fig. 16.8 illustrates how aggregation acts on the base block models and their couplings. The components in each base block model are aggregated into the corresponding lumped block. To enable non-trivial abstraction, the state set of a lumped block must be significantly smaller than its counterpart. We will consider a commonly used class of mappings that ensure that this is true. The mapping for each block is required **to erase the identity** of the components within it.

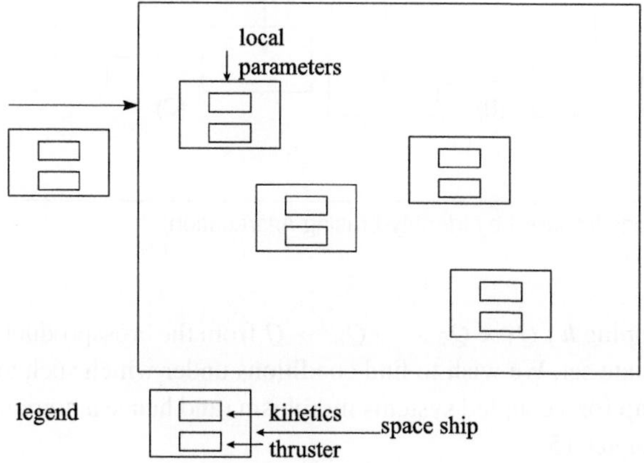

FIGURE 16.6

Server Coupled Model with Space Ship Component Clients.

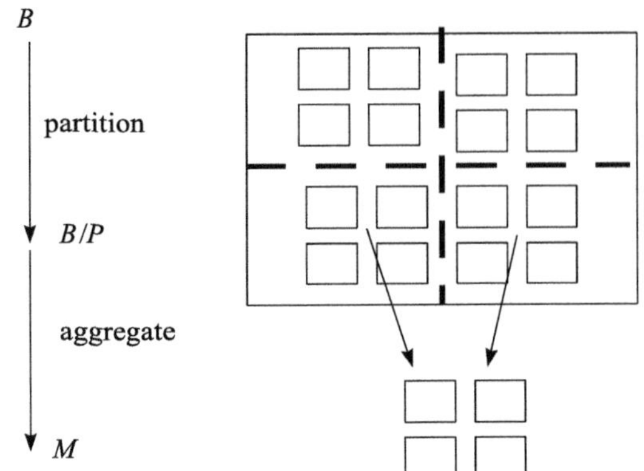

FIGURE 16.7

Aggregation: first Partition then Lump.

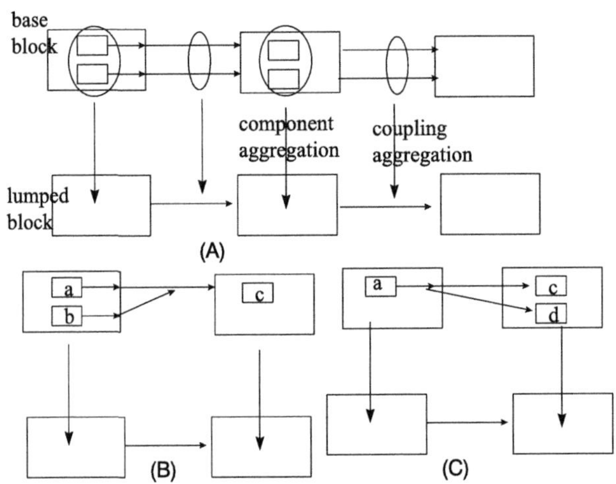

FIGURE 16.8

Almost Necessary Conditions Imposed by Identity-Erasing Aggregation.

Thus, consider a mapping $h : Q_1 \times Q_2 \times \cdots Q_n \rightarrow Q$ from the crossproduct of the component state sets to a lumped block state set. We wish to find conditions under which such an aggregation mapping can serve as the local map for a coupled systems morphism (and hence a homomorphism) as discussed in Theorem 15.13 in Chapter 15.

We first require the mapping to be preserved under *transposition:*

$$h(q_1, \cdots, q_i, \cdots, q_j, \cdots, q_n) = h(q_1, \cdots, q_j, \cdots, q_i, \cdots, q_n)$$

i.e., we can interchange the states of any pair of components without changing the state mapping. Since any *permutation* is a sequence of transpositions, we can just as well require the mapping to be preserved under permutation. The most common mapping used in aggregation is the summation:

$$h(q_1, \cdots, q_i, \cdots, q_n) = \sum q_i,$$

which is clearly preserved under permutation (assuming real valued states and the commutative and associative properties of addition!).

Exercise 16.9. Show that multiplication is also preserved under permutation.

However, summation is often too coarse to achieve valid aggregations. We consider more refined mapping called a *census* in which we count the number of components in each state. Clearly, from a census of states we can compute the summation (if the states are summable) but not conversely. However, a census mapping is *also preserved under permutation* since by interchanging states of components we do not change the number in each state. So a census, while being a more refined aggregation than summation, is still non-trivial in the sense of erasing the identities of components.

Exercise 16.10. Imagine a census taker, C_s asking each component if it is in a particular state s. Show that C_s can be defined recursively by:

$$C_s() = 0$$

$$C_s(q_1, \cdots, q_n) = \begin{cases} C_s(q_2, \cdots, q_n) + 1 & \text{if } q_1 = s \\ C_s(q_2, \cdots, q_n) & \text{otherwise} \end{cases}$$

Then if there are N states are labeled with subscripts $0, 1, \cdots, N-1$, a census-based aggregation mapping takes the form:

$$h(q) = h(q_1, q_2, \cdots, q_n) = (C_0(q), C_1(q), \cdots, C_{N-1}(q))$$

The ability to perform a census of states implies that all components in a block have the same state set. We will go further and require block *homogeneity*, i.e., all components in a block have the same model structure: they are copies of each other. Of course, different blocks may have different characteristic component model structures. These aggregation mappings can be called *identity-erasing* since, for an unconstrained global state space, components lose their identities in the aggregation. We can also say that components become *anonymous* after being mapped by an identity-erasing aggregation.

Exercise 16.11. Show that for the unconstrained global state set of a block (the full crossproduct of its components states) is impossible to recover the identity of a component in a block knowing only the state census. However, it is possible to recover the original state assignments (which component is in which state) if the block state set is suitably constrained. Hint: consider the constraint in which each component can only be in a unique set of states (in effect, the state is used to encode the identity of a component).

Note that although we will be restricting discussion to finite state sets, we don't require linearity of the transition function. At the end of the chapter, we'll turn the special case of linear systems.

16.3.1 COUPLING CONDITIONS IMPOSED BY ANONYMITY

As illustrated in Fig. 16.8, anonymity suggests a restricted form of communication should hold between base blocks. In Fig. 16.8B, consider component c receiving input from component a. Anonymity suggests that any component, b in the same state as a, could have been the one sending the same input. In other words, receiving component c can't distinguish between senders in the same state.

Similarly, in Fig. 16.8C, the sender a should not be able to direct output to particular receiver, any output that can be responded to by a particular component, c in state s can just as well be responded to by any other component, d in the same block, if it is also in s.

This suggests strong requirements on the coupling possibilities in the base model.

We would like to claim that such requirements are necessary, i.e., imposed by the seeking to employ identity-erasing aggregations into homomorphisms. However, simple counterexamples can be given. Indeed, we'll see soon that the uniform coupling requirement imposed by linear lumpability (Chapter 12 is actually weaker than the coupling indifference requirement).

To summarize, along with the identity-erasing aggregations we impose the following prerequisites on a base model partition:

- homogeneity – base model blocks must be homogeneous in structure (components in a block are isomorphic copies of each other),
- inter-block coupling indifference – no receiver can distinguish between senders residing in same block in the same state. Similarly, no sender can direct outputs that distinguish between receivers in same block in the same state.
- output desired by our experimental frame must be compatible with the partition – as in Chapter 12, the partition on the states of the base model defined by the aggregation mapping must be finer than the partition imposed by the desired output mapping. More specifically, we must be able to compute the base model behavior of interest using only the census state counts available from the lumped model.

We can show these are sufficient conditions for the existence of a homomorphism enabling a base model to be abstracted to a lumped model based on identity-erasing block-wise aggregations.

Conjecture: Existence of Identity-erasing Homomorphic Aggregations:

Let there be partition on the components of a base model for which 1) blocks have homogeneous components, 2) the coupling indifference conditions hold for all pairs of blocks and 3) the base model output is computable from block model census counts. Then a lumped model exists for which there is a homomorphism based on an identity-erasing aggregation.

We will only be able to investigate such a conjecture within specific classes of systems. We will restrict our attention to DTSS (discrete time) coupled models as a special but representative case. To proceed will requires us to formalize the coupling indifference conditions Given any pair of blocks B, B' in a partition of components, let a receiving component, $c \in B'$ have all components in B as influencees and an interface map Z_c which is invariant under permutation, i.e.,

$$Z_c(q_1, \cdots, q_i, \cdots, q_j, \cdots, q_n) = Z_c(q_1, \cdots, q_j, \cdots, q_i, \cdots, q_n)$$

for all state assignments to B' and pairs of transpositions. This explicitly requires that receiving component c be indifferent to permutations of the sending block state, hence to the identities of sending components. We also require that each component in B' have the same interface map as c. This guarantees that receivers are indifferent to senders.

A general characterization of such an interface map is that it is a function, z of the state counts, i.e.,

$$Z(q_1, \cdots, q_i, \cdots, q_n) = z(C_0(q), C_1(q), \cdots, C_{N-1}(q)).$$

The underlying function may be non-linear. For example,

$$Z(q_1, \cdots, q_i, \cdots, q_n) = \{1 \text{ if } C_0(q) > C_1(q)0 \text{ otherwise}$$

is a boolean function that compares the number of zeros and ones in the input. In general, although the aggregation mapping is linear, *the lumped models we construct may not be linear*, because for one thing, the coupling between components may be non-linear. Since we are **not trying to linearize** complex behavior, we are more likely to be able to find valid aggregations.

Exercise 16.12. Write out the underlying function z for the above case. Provide other examples of permutation invariant functions.

Fig. 16.9 illustrates a pair of blocks in a base model and the counterpart lumped model components. The coupling between base blocks is mediated by a permutation invariant map Z which sends the same output to each block sub-component. In the lumped model block states are represented by state count vectors. The coupling between blocks is given by z, the underlying block count function of Z.

In the following we investigate how to construct lumped models from a base model having a partition with the conditions stated in Theorem 15.13 of Chapter 12.

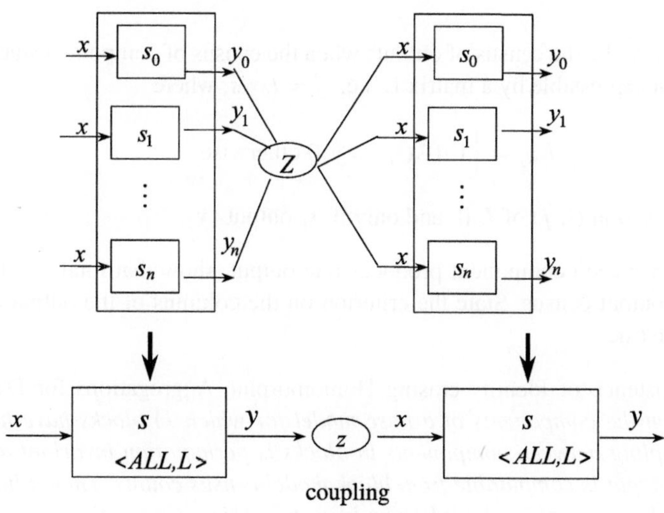

FIGURE 16.9

State Census Mapping.

16.3.2 CONSTRUCTING LUMPED MODELS BASED ON IDENTITY-ERASING AGGREGATION

State Census Redistribution Under Equal Input

As in Fig. 16.9, let $\vec{s} = [s_0, s_1, \cdots, s_N]$ be the vector representing the census of states, i.e., s_i is the number of components in state, i and there are N states are labeled with subscripts $0, 1, \cdots, N$. We call $\vec{s}$ a *census* vector for the block in question. How does the census change after all components in the block have received an input x? (Recall that all components in block receive the same input due the earlier construction.)

The redistribution of states is a linear operation and can easily be computed from the transition function of any component (all being identical). Let $ALL_{i,j}[x]$ be the matrix that represents the *redistribution* under a state transition due to input x applied to all components. Then

$$ALL_{i,j}[x] = \begin{cases} 1 & \text{if } \delta(s_j, x) = s_i \\ 0 & \text{otherwise} \end{cases}$$

i.e., there is a 1 in position (i, j) if, and only if, x sends s_j to s_i.

Let $\vec{s}$ be census at time t, then $\vec{s'} = [n'_0, n'_1, \cdots, n'_N]$, the distribution after each component has received input x, is given by: $\vec{s'} = ALL[x] * \vec{s}$. This is true because:

$$n'_i = \sum ALL_{i,j}[x] * n_j = \sum \{n_j | \delta(s_j, x) = s_i\}$$

Exercise 16.13. Show that total number of components is preserved under a redistribution. State the criterion on the sums of the columns of the state census matrix that guarantees this preservation.

Output Census

Let $\vec{y} = [y_0, y_1, \cdots, y_n]$ be the census of outputs when the census of component states is $\vec{s}$. This is also a linear computation expressible by a matrix L, i.e., $\vec{y} = L * \vec{s}$, where

$$L_{i,j} = \Big\{ 1 \text{ if } \lambda(s_i) = y_j 0 \text{ otherwise}$$

i.e., there is a 1 in position (i, j) of L if, and only if, s_i outputs y_j.

Exercise 16.14. Since each component produces one output, show that total number of components is preserved in the output census. State the criterion on the columns of the output census matrix that guarantees this to be true.

Theorem 16.1 (Existence of Identity-erasing Homomorphic Aggregations for DTSS Models)**.** *Let there be partition on the components of a base model for which 1) blocks have homogeneous components, 2) the coupling between components in blocks is permutation invariant as in Fig. 16.9 and 3) the base model output is computable from block model census counts. Then a lumped model exists for which there is a homomorphism based on an identity-erasing aggregation.*

Proof. We will apply Theorem 15.13 of Chapter 15 which allows to conclude that a homomorphism exists if we can show that a coupled system morphism exists.

Let $N = (T, X_N, Y_N, D, \{M_d\}, \{I_d\}, \{Z_d\})$ be network of DTSS components, $\{M_d\}$ representing the base model. We construct a network $N' = (T, X'_{N'}, Y'_{N'}, D', \{M'_{d'}\}, \{I'_{d'}\}, \{Z'_{d'}\})$ for the lumped model. We define the mapping $coord : D \to onto D'$ as the partition of base model components that places isomorphic components within the same block (which is possible by the homogeneity assumption). The block $coord^{-1}(d')$ is called the block of d'.

The influencers $I'_{d'}$, are those lumped components whose blocks influence the block of d'. By assumption the coupling from such blocks to d' is governed by permutation invariant maps that feed the same input to each component in the block, as illustrated in Fig. 16.9. This allows us to define the lumped interface map $Z'_{d'}$ in terms of the count maps z_d underlying each base interface maps Z_d.

The lumped DTSS model $M'_{d'}$ corresponding to the block of d' is constructed using the census mappings and matrix transformations discussed earlier.

Exercise 16.15. Given a component $M' = (X, Y, S, \delta, \lambda)$ typical of the block $coord^{-1}(d')$ define $M'_{d'} = (X', Y', S', \delta', \lambda')$ appropriately. (Hint: X', Y', and S' are vector spaces of census vectors over sets X, Y, and S respectively, δ' is given by the ALL matrix defined by δ and, λ' is given the L matrix defined by λ.)

To apply Theorem 15.13 of Chapter 15 we need to define the local maps, $h_{d'}$, $g_{d'}$ and $k_{d'}$ for each lumped component d'. We easily identify the local state map $h_{d'}$ as the state census mapping for the block of d'. Recognizing the constraint that all inputs to components in a block are identical, he input map $g_{d'}$ becomes the identity mapping. Finally, the output map $k_{d'}$ is defined as the output census mapping. The theorem then states that a homomorphism will exist if the following conditions are met, where we rephrase the conditions in terms of DTSS networks and components, as follows:

1. *state transition preservation:* this is equivalent to requiring that the state census mapping preserve the state transitions of the components in a block under identical input. By construction using the ALL matrix, it does so.
2. *input output behavior preservation:* this is equivalent to output function preservation by the output census mapping. By construction, the L matrix output function enables the output census to do so.
3. *coupling preservations:* this requires that the commutation of the interface maps in the base and lumped models. By construction, using the count maps $(z_{d'})$ underlying the interface maps (Z_d) does the required preservation.

Exercise 16.16. In Fig. 16.9, show that outputs of the Z and z maps are equal if the block states are related by the census mapping. (Hint: recall that z is the count map that underlies Z.)

Having shown that the conditions required by Theorem 15.13 can be satisfied we can conclude that the state census maps are local homomorphisms that compose together to form a homomorphism at the I/O system level. □

To conclude, we have provided a general way to construct DTSS lumped models from DTSS base models based on identity-erasing aggregations. Base and lumped models constructed in this manner are related by a morphism at the coupled system level and therefore are homomorphic at the I/O system level.

In the next section we provide an example of such aggregated model construction.

Randomized Coupling

In the construction of Theorem 16.1, to guarantee that all components of a receiving block receive the same input from the sending block, we assumed *all-to-all coupling*, in which all pairs of sending and receiving components are coupled. In real world examples this is situation is unlikely to be found, but we can relax the requirement for all-to-all coupling while still expecting that all receiving components received the same, or at least closely matching, inputs. We can think of each receiver as sampling the outputs of the sending block's population (its components). An error-free census emerges from sampling all of the population (which is all-to-all coupling). However, just as in random opinion polls, a large enough finite sample, under the appropriate stochastic conditions, would estimate the census as closely as desired. With this in mind, let the influencers (senders) of each receiver be selected at random and let the interface maps Z_d all have the same underlying count-based function z (Fig. 16.10). If the number of influencees (sample size) is the same for each receiver, their interface maps will receive approximately the same inputs and their outputs will also be approximately the same since the all have the same underlying count-based function. We can even allow the interface maps to have different numbers of inputs provided that they all have the same underlying frequency-based maps, defined in the following exercise.

Exercise 16.17. Let $\vec{y}$ be census vector. By normalizing (dividing each count by the sum of all counts) we obtain a frequency (probability) distribution, $\vec{p}_y$. Say that a count-based function z is *frequency based* if there is a function z' with frequency distribution input such that for all $\vec{y}$, $z(\vec{y}) = z'(\vec{p}_y)$. An interface map Z is *frequency-based* if it has an underlying if its underlying count-based function is frequency-based. In other words, Z is sensitive only to the frequencies, not the actual counts, of the inputs. Two frequency-based maps are *frequency-equivalent* if they have the same underlying frequency-based function. Assuming that all interface maps are frequency-equivalent and sample a

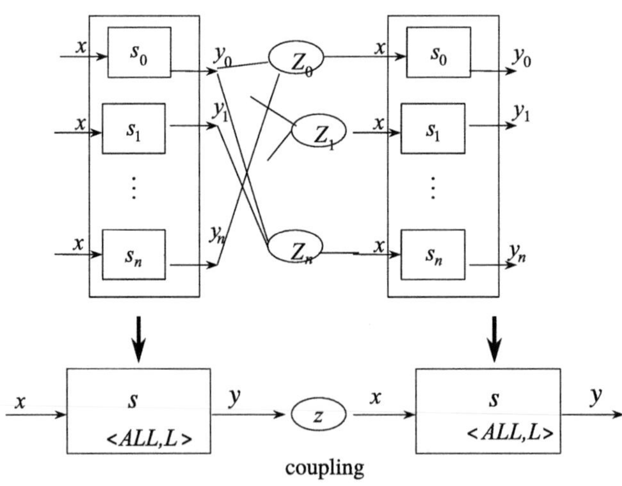

FIGURE 16.10

Randomized Coupling.

sending block's output without error (i.e., obtain the same frequency distribution of inputs) show that their outputs are all the same.

We say that the coupling between components in blocks is randomized if for each pair of blocks the receiving interface maps are all frequency-equivalent.

Corollary 16.1 (Existence of Identity-erasing Approximate Aggregations for DTSS Models)**.** *Let there be partition on the components of a base model for which 1) blocks have homogeneous components, 2) the coupling between components in blocks is randomized and 3) the base model output is computable from block model census counts. Then a lumped model exists for which there is an approximate homomorphism based on an identity-erasing aggregation.*

Exercise 16.18. Prove the corollary using the definitions and exercise just discussed.

The result generalizes the discussion of the first edition of TMS that derived an approximate aggregation of a neural net base model based on firing frequencies. (See TMS76 Chapter 14 for details.)

Exercise 16.19. Model a neuron with refractory period and threshold firing as a finite state DTSS. A base model containing billions of neurons is partitioned into blocks satisfying the homogeneity and inter-block indifference properties through randomized coupling. Construct a lumped model based on such a partition.

16.4 ALL-TO-ONE COUPLING

Fig. 16.11 illustrates another way in which the coupling indifference conditions may be satisfied. We assume that, in addition to the homogeneous set of components assumed earlier, the receiving block has a coordinator that receives all the inputs of the sending block. Based on state feedback from the other components, the coordinator sends an output that causes a non-trivial state change at most one of these others. Receiver coupling indifference requires that the coordinator does not favor any particular component – it can only base its decision on the state of the component not its identity.

Let $ONE[s,x]$ be the matrix that represents sending an input x to any one component in state s, with the restriction that $\delta(s,x) \neq s$ (i.e., x actually does cause a change in state). Then we can represent the change in census state count occasioned by input x to a component in state s by

$$\vec{s'} = \vec{s} + ONE[s,x]$$

where

$$ONE_i[s,x] = \begin{cases} 1 & \text{if } \delta(s,x) = s_i \\ -1 & \text{if } s_i = s \\ 0 & \text{otherwise} \end{cases}$$

Exercise 16.20. State the criterion on $ONE[s,x]$ that preserves the total number of components.

The coordinator, C determines which pair (s,x) to use. By coupling indifference we can assume that C is permutation invariant in both sender outputs and component states of its own block. Thus there

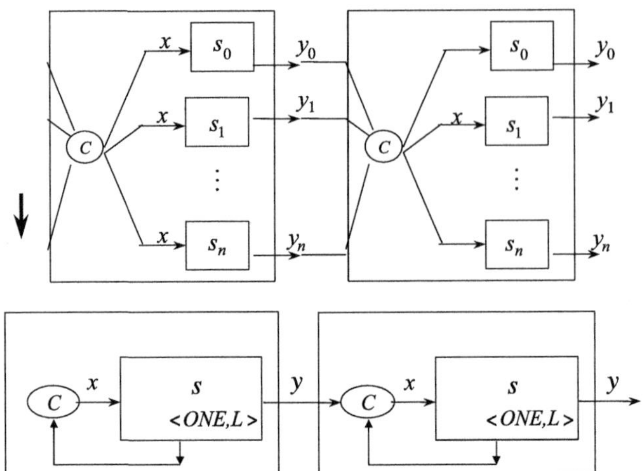

FIGURE 16.11

All-to-one Coupling.

is an underlying census-based function c, which maps output and state censuses $(\vec{y}, \vec{s})$ to a state, input pair (Fig. 16.12). Consequently, the lumped model state transition can be expressed as $ONE[c(\vec{y}, \vec{s})]$.

Corollary 16.2 (Existence of Identity-erasing Approximate Aggregations for DTSS Models). *Let there be partition on the components of a base model for which 1) blocks have homogeneous components with additional coordinator, 2) the coupling between components in a sending block is only to the*

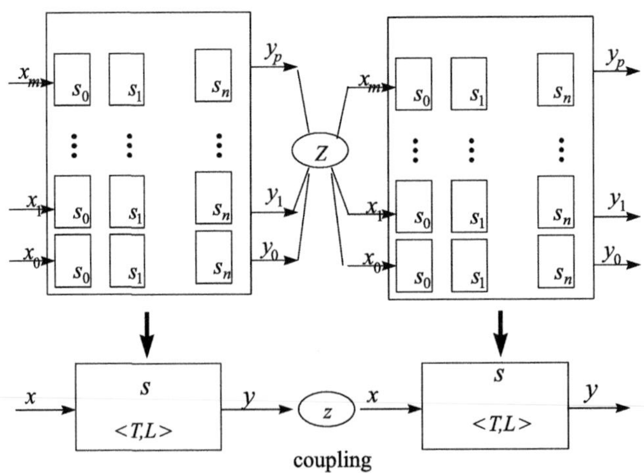

FIGURE 16.12

State Census Mapping.

coordinator of a receiving block and 3) the base model output is computable from block model census counts. Then a lumped model exists for which there is a homomorphism based on an identity-erasing aggregation.

Exercise 16.21. Prove the corollary.

16.4.1 EXAMPLE OF AGGREGATION MODEL CONSTRUCTION: SPACE TRAVEL

Consider the space travel system simplified to an earth-moon shuttle as in Fig. 16.15. Suppose there are 10 ships at any time shuttling between the two stations. Each base model block has 10 components where a component may be empty or be in a non-empty state representing a ship in transit. When a ship reaches its destination it is added to the opposite block for the return journey, being accepted by one, and only, one empty component. Supposing that an output of 1 is emitted when a ship reaches its destination, we send this output to a coordinator that routes to the first empty component in some arbitrary order. Thus we have all-to-one coupling since any component in one block may send a ship to the coordinator in the other block. Moreover, the coupling satisfies the indifference requirement since neither identities of the sending or receiving component determine the coordinator's decision. Of course, only in a specific state can a transaction (ship hand-over) actually occur. To represent a ship's state of travel with a finite state set, we assume a fixed travel time in days and let the number days left represent the current state. For example, if it takes 3 days to travel from the earth to the moon, the state set of a component is $\{0, \cdots, 4\}$ where 0 means "arrival", 4 means "empty", and the rest represent the days left to reach the moon. The transition and output functions are shown in Fig. 16.16. To represent a one day time step, when a component receives a 0, its state reflects the associated reduction in days left. The mapping into census matrices is also shown in Fig. 16.14.

The coordinator function is defined as follows:

$$Coord(y, s) = select(C_0(y), delta(s, 0))$$

where $delta(s, 0)$ is the application of a 0 input to each of the components; $select(1, s)$ is the state resulting from filling in the first empty component in some order and $select(n, s) = select(n - 1, select(1, s))$. In other words, the reduction in days left is done first, then each ship arrival is assigned successively to a (different) empty component.

Exercise 16.22. Show that the expression for Coord implements the informal description. Also, show that it satisfies the permutation invariance requirements.

Theorem 16.2 (Existence of Identity-erasing Homomorphic Aggregations). *Let there be partition on the components of a base model for which 1) blocks have homogeneous components, 2) the coupling indifference conditions hold for all pairs of blocks and 3) the base model output is computable from block model census counts. Then a lumped model exists for which there is a homomorphism based on an identity-erasing aggregation.*

Proof. We will apply Theorem 15.13 of Chapter 15 which allows to conclude that a homomorphism exists if we can show that a coupled system morphism exists.

This will requires us to formalize the coupling indifference conditions. We do so for DTSS coupled models as a special but representative case. Given any pair of blocks B, B' in a partition of components, let a receiving component, $c \in B'$ have all components in B as influencees and an interface map Z_c which is invariant under permutation, i.e.,

$$Z_c(q_1, \cdots, q_i, \cdots, q_j, \cdots, q_n) = Z_c(q_1, \cdots, q_j, \cdots, q_i, \cdots, q_n)$$

for all state assignments to B' and pairs of transpositions. This explicitly requires that receiving component c be indifferent to permutations of the sending block state, hence to the identities of sending components. We also require that each component in B' have the same interface map as c. This guarantees that receivers are indifferent to senders.

A general characterization of such an interface map is that it is a function, z of the state counts, i.e.,

$$Z(q_1, \cdots, q_i, \cdots, q_n) = z(C_0(q), C_1(q), \cdots, C_{N-1}(q)).$$

The underlying function may be non-linear. For example,

$$Z(q_1, \cdots, q_i, \cdots, q_n) = \begin{cases} 1 & \text{if } C_0(q) > C_1(q) \\ 0 & \text{otherwise} \end{cases}$$

is a boolean function that compares the number of zeros and ones in the input. In general, although the aggregation mapping is linear, *the lumped models we construct may not be linear*, because for one thing, the coupling between components may be non-linear. Since we are not **trying to linearize** complex behavior, we are more likely to be able to find valid aggregations.

Exercise 16.23. Write out the underlying function z for the above case. Provide other examples of permutation invariant functions.

Fig. 16.12 illustrates a pair of blocks in a base model and the counterpart lumped model components. The coupling between base blocks is mediated by a permutation invariant map Z which sends the same output to each block sub-component (we can allow Z to map the outputs of sub-components rather than states). In the lumped model block states are represented by state count vectors. The coupling between blocks is given by z, the underlying block count function of Z. We can show that this coupling specification satisfies the internal coupling requirement of Theorem 15.13 in Chapter 15 (recall that this required the interactions to commute with the aggregation). □

In the following we investigate how to construct lumped models from a base model having a partition with the conditions stated in the Theorem.

16.4.2 CONSTRUCTING LUMPED MODELS BASED ON IDENTITY-ERASING AGGREGATION

STATE CENSUS MAPPING

As in Fig. 16.12, let $\vec{s} = [s_0, s_1, \cdots, s_N]$ be the vector representing the census of states, i.e., s_i is the number of components in state, i and there are N states are labeled with subscripts $0, 1, \cdots, N$. We call $\vec{s}$ a *census* vector for the block in question. How does the census change after **all** component have received an input x? The redistribution of states is a linear operation and can easily be computed from the transition function of any component (all being identical). Let $ALL_{i,j}[x]$ be the matrix that represents the *redistribution* under a state transition due to input x applied to all components. Then

$$ALL_{i,j}[x] = \{1 \text{ if } \delta(s_j, x) = s_i 0 \text{ otherwise}$$

i.e., there is a 1 in position (i, j) if, and only if, x sends s_j to s_i.

Let $\vec{s}$ be census at time t, then $\vec{s'} = [n'_0, n'_1, \cdots, n'_N]$, the distribution after each component has received input x, is given by: $\vec{s'} = ALL[x] * \vec{s}$. This is true because:

$$\begin{aligned} n'_i &= \sum ALL_{i,j}[x] * n_j \\ &= \sum \{n_j | \delta(s_j, x) = s_i\} \end{aligned}$$

Exercise 16.24. Show that total number of components is preserved under a redistribution. State the criterion on the sums of the columns of the state census matrix that guarantees this preservation.

At the other extreme, suppose that input x is applied to exactly one of the components. Let $ONE[s, x]$ be the matrix that represents the application of x to any one component in state s, with the restriction that $\delta(s, x) \neq s$ (i.e., x actually does cause a change in state). Then $\vec{s'} = \vec{s} + ONE[s, x]$ where

$$ONE_i[s, x] = \begin{cases} 1 & \text{if } \delta(s, x) = s_i \\ -1 & \text{if } s_i = s \\ 0 & \text{otherwise} \end{cases}$$

Exercise 16.25. State the criterion on $ONE[s, x]$ that preserves the total number of components.

Fig. 16.14 shows an example of a finite DTSS and the redistribution matrices associated with its inputs.

OUTPUT CENSUS

Let $\vec{y} = [y_0, y_1, \cdots, y_o]$ be the census of outputs when the census of component states is $\vec{s}$. This is also a linear computation expressible by a matrix L, i.e., $\vec{y} = L * \vec{s}$, where

$$L_{i,j} = \begin{cases} 1 & \text{if } \lambda(s_i) = y_j \\ 0 & \text{otherwise} \end{cases}$$

i.e., there is a 1 in position (i, j) if, and only if, s_i outputs y_j. The output census mapping is illustrated in Fig. 16.14.

Exercise 16.26. Since each component produces one output, show that total number of components is preserved in the output census. State the criterion on the columns of the output census matrix that guarantees this to be true.

INPUT CENSUS: ALL-TO-ALL AND RANDOMIZED COUPLING

Having the output census for all its influencee blocks gives us part of the information required to compute the input census of a receiving block. The other essential information required is the inter-block coupling. As indicated before such coupling must satisfy the indifference requirement. Two possibilities are illustrated in Fig. 16.13.

All-to-all coupling can be of two kinds. One is the kind we have already characterized in Fig. 16.12. In this case, all sub-components of the receiving block receive the same input $x = Z(y)$ where y is the tuple of outputs of the sending block. In the lumped model, this corresponds to the destination component receiving $x = z(\vec{y})$ where z is the count-based function underlying Z and $\vec{y}$ is the census vector for y.

Exercise 16.27. Show that x has the same value in both cases. This establishes coupling preservation condition in Theorem 15.13 of Chapter 15.

We use the $ALL[x]$ matrix to compute the change in the state counts of the receiving block due to the input x.

Exercise 16.28. Show that with $ALL[x]$ as its transition function, a lumped model can be constructed that satisfies the local structure preservation condition Theorem 15.13 of Chapter 15.

In the second kind of all-to-all coupling, although all the coupling pairs exist, at most one is actually used. In this case, the input $x = Z(y)$ is affects the state of exactly one receiver in some state s. The change in state census is computed using $ONE[s, x]$ where again the same value may be computed in the lumped model as $x = z(\vec{y})$.

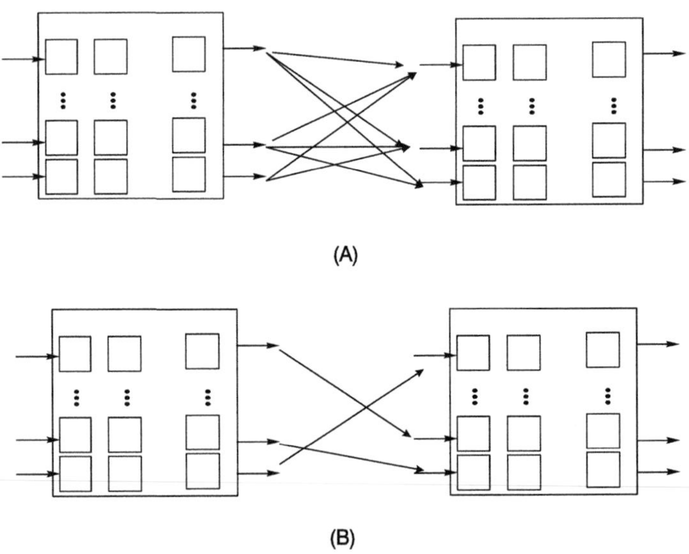

FIGURE 16.13

Two extremes of indifferent block-to-block coupling. (A) All-to-all coupling. (B) Randomized coupling.

In *randomized coupling*, components in a receiving block receive inputs from the sending block in proportion to the probability distribution in which they are generated. Let $\vec{y}$ be output census on the sending block. By normalizing (dividing by the number of outputs) we obtain a probability distribution, $\vec{p}$. For simplicity assume the coupling is the identity mapping so that the input probability distribution is also $\vec{p}$. Randomization requires that for each input, x, the fraction of components receiving input x is p_x., where $\sum_x p_x = 1$. Then the matrix $T[\vec{p}]$ computing the census of states after a transition is described by:

$$T_{i,j}[\vec{p}] = \sum_x p_x * T_{i,j}[x]$$

For example, the matrix $T[p,q,r]$ in Fig. 16.14 describes the census of states s_0, s_1, s_2, s_3 after inputs $0, 1, 2$ are received in proportions $p : q : r$ respectively. In general, let $\vec{s}$ be the census of states in the receiving block and $\vec{s}'$ its census after transitioning under input from a block with census $\vec{s}_B$. Then, $\vec{s}' = T[\vec{p}] * \vec{s}$ where $\vec{p}$ is the normalized distribution of inputs $\vec{x} = Z[L * \vec{s}_B]$.

Exercise 16.29. Show that with $ONE[x]$ as its transition function, a lumped model can be constructed that satisfies the local structure preservation condition Theorem 15.13 of Chapter 15.

Having shown that both the coupling preservation and local structure preservation conditions can be satisfied, we have provided a general way to construct lumped models from base models based on identity-erasing aggregations. Base and lumped models constructed in this manner are related by a morphism at the coupled system level and therefore are homomorphic at the I/O system level.

In the next section we provide a specific example of such aggregated model construction.

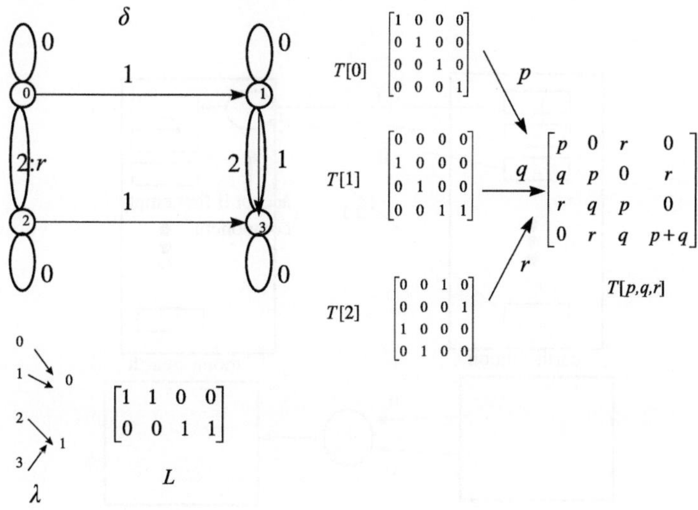

FIGURE 16.14

Example mapping DTSS into stochastic matrix.

16.4.3 EXAMPLE OF AGGREGATION MODEL CONSTRUCTION: SPACE TRAVEL

Consider the space travel system simplified to an earth-moon shuttle as in Fig. 16.15. Suppose there are 10 ships at any time shuttling between the two stations. Each base model block has 10 components where a component may be empty or be in a non-empty state representing a ship in transit. When a ship reaches its destination it is added to the opposite block for the return journey, being accepted by one, and only, one empty component. Thus we have all-to-all coupling since any component in one block may send a ship to any component in the other. Moreover, the coupling is indifferent to identities since neither identities of the sending or receiving component determine the transaction. Of course, only in a specific state can a transaction (ship hand-over) actually occur. In the lumped model, we index the state of a component by $\{0, \cdots, 4\}$ where 0 means "arrival", 4 means "empty", and the rest represent the time (say in days) left to reach the destination. The transition and output functions and their mappings into census matrices are shown in Fig. 16.16.

As an example, let $\vec{s} = [3, 2, 1, 0, 4]$ be the current state of the earth-to-moon block (meaning 3 ships have just arrived at the moon, 2 are 1 day away, and so on, while there are 4 empty slots. Then the output census is $\vec{y} = L * \vec{s} = [7, 3]$ i.e., 7 zero outputs and 3 one outputs (the just-arrived ships).

The coupling causes the 7 zero outputs to be summed to a zero input and applied to all components in the moon-to-earth block. The effect on the state census is $\vec{s'} = ALL[0] * \vec{s}$ If $\vec{s} = [0, 1, 1, 2, 6]$ is the state census of the block, then $\vec{s'} = [1, 1, 2, 0, 6]$ is the state census after the state shift to the left caused by the input 0, corresponding to the reduction in time left of all components except the six that are empty.

The coupling sends the 3 one outputs to the moon-to-earth block. These are applied to three different components so that the effect on the state census is given by $\vec{s'} = \vec{s} + ONE[4, 1] * [3]$ If

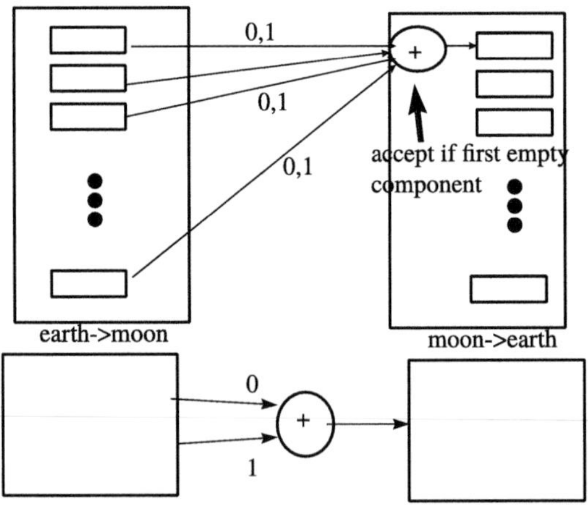

FIGURE 16.15

Earth-Moon Shuttle.

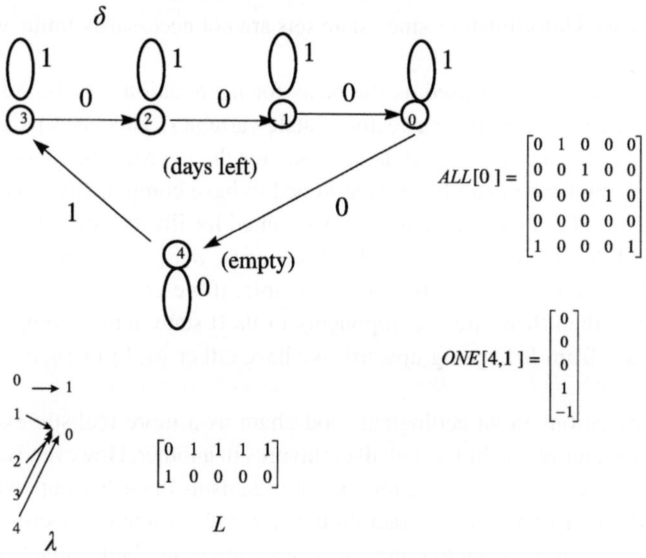

FIGURE 16.16

Census models for Earth-Moon Block.

$\vec{s} = [1, 1, 2, 0, 6]$ is the state census of the block, then after receiving the 3 one inputs, we have

$$\vec{s} = \begin{bmatrix} 1 \\ 1 \\ 2 \\ 0 \\ 6 \end{bmatrix} + \begin{bmatrix} 0 \\ 0 \\ 0 \\ 1 \\ -1 \end{bmatrix} [3] = \begin{bmatrix} 1 \\ 1 \\ 2 \\ 3 \\ 3 \end{bmatrix}$$

Exercise 16.30. Model a neuron with refractory period and threshold firing as a finite state DTSS. A base model containing billions of neurons is partitioned into blocks satisfying the homogeneity and inter-block indifference properties through randomized coupling. Construct a lumped model based on such a partition. (See TMS76 for details.)

16.4.4 CONSTRUCTING AGGREGATIONS THROUGH STATE AND BLOCK REFINEMENT

The homogeneity and coupling indifference requirements seem quite stringent. What if we can't achieve them on our first try at aggregation? As in the finite state reduction case, we can start an iterative process in which we attempt to refine block partitions and/or state sets to increase homogeneity and coupling indifference. Such a process may have to backtrack since refinement of blocks and/or states to increase homogeneity may make it more difficult to achieve coupling indifference, and conversely. This can be understood from the example of the finite machine state reduction process where successive refinements make it easier to control the source of transition arrows but more difficult to control their destinations. Unfortunately, since state sets are not necessarily finite, we cannot guarantee that an iterative process converges.

To understand how the state set used as the basis for aggregation may be refined, we recall that components in the base model may have a vector of state variables. This gives us a lattice of projective spaces, each one of which can serve as a state set basis for the aggregation mapping. For example, in Fig. 16.17, the underlying base model block is assumed to have components, each with 3 binary state variables, labeled a, b, and c. Five components are assumed for illustration each in a different state as illustrated at the top of the figure. The lattice of subsets of $\{a, b, c\}$ is shown in the middle and lowest layers, along with the resulting state census. For example, if we choose only a as the state variable (shown at the bottom), then there are 2 components in its 0 state, and 3 components in its 1 state. Refining the state space from $\{a\}$ (going upward) we have either $\{a, b\}$ or $\{a, c\}$ with their respective counts shown.

Let's look at aggregations in an ecological food chain as a more realistic example. Initially, we lump all herbivores together in one block and all carnivores in another. However, we might find that our lumped model is not good at predicting the number of organisms in each group over time. We surmise that the problem is due to violation of interaction indifference – some carnivores eat other carnivores while some eat herbivores. It is wrong to represent a carnivore as not being able to distinguish between carnivores and herbivores. So we break carnivore block into two: those feeding on carnivores and those feeding on herbivores. After further experiments, we conclude that it won't work to allow all predators in a block to be active all the time. So we add a state variable to an organism's description, representing its state of hunger, in addition to keeping track of its dead/alive status. Only hungry organisms hunt

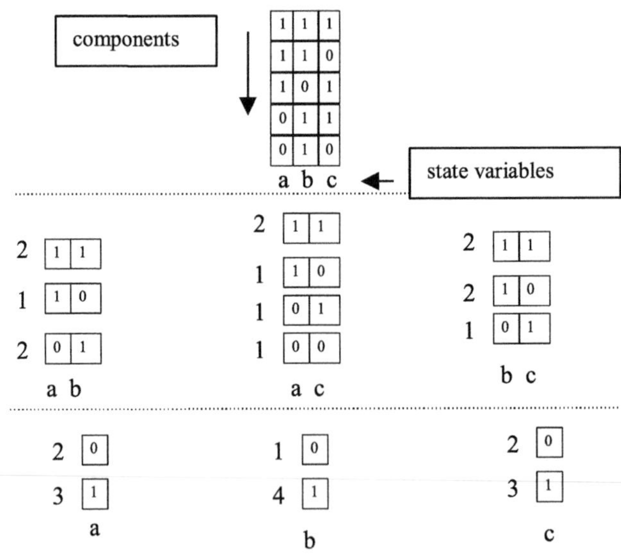

FIGURE 16.17

Illustrating State Set Choice and Resulting State Censuses.

and upon success, leave the hungry state. But now is the delay for re-entering the hungry state uniform over the current blocks (homogeneity)?

So goes the iteration – at each stage, we try refining the blocks and the states employed to develop the lumped model specification. (We may also backtrack and combine blocks or states when a bad choice seems apparent.) The experimental frame – desired output – clearly plays a major role in where we stop. So do the properties of the real system (base model) itself – does it lend itself to component partitions with homogeneity and coupling indifference properties?

Exercise 16.31. Refine the aggregation employed in the space travel logistics model so that it keeps track of the numbers of ships of the different types in addition to their position in space.

16.4.5 UNIVERSALITY/APPLICABILITY OF IDENTITY-ERASING AGGREGATIONS

Many modeling approaches have aggregation as their basis. The complexity of real world systems, such as the foregoing ecosystem example, necessitates such lumping. And in many cases, such as for ideal gases and chemical reactions, such lumped models work quite well. In such cases, we can infer that the homogeneity and coupling indifference requirements are being met. Actually to consider such issues, we need to introduce a more rigorous way of measuring how well a model represents a real system or another model. In the next chapter, we'll discuss error and approximate morphisms. For example, an aggregation where the homogeneity and coupling indifference requirements are not met exactly is an approximate morphism. However, an approximate morphism may still lead to errors that are acceptable in the desired experimental frame.

16.5 ABSTRACTIONS FOR EVENT-BASED CONTROL

Returning to the space travel model family, the second model we consider is an abstraction that supports event-based control of space trips. In such an abstraction, called a boundary-based DEVS, we partition the state space of a base model in a manner that could support a DEVS representation as in Chapter 17. However, instead of constructing this representation, we construct a model that is truly an abstraction by focusing on the boundaries that define the partition. Since the base model is a DESS, and the lumped model is a DEVS, the method in Table 16.1, illustrated by this abstraction is that of "formalism transformation".

The space travel model in Fig. 16.18 furnishes the base model for such a transformation. Here the surfaces of the planetary and lunar surfaces form the boundaries. All trips from one boundary to another are placed into the same equivalence class. For any such class, we retain only information about travel times in the form of a window bracketing the minimum and maximum times. A boundary-based DEVS is thus a variant of the real time DEVS formalism in Chapter 10. Such a DEVS can be used to send control signals to a real system so as to drive it through a desired sequence of boundaries. In our example, this would be to move a ship around the solar system. The time window associated with a boundary-to-boundary trip can be used as information for control. If no word is heard from the space ship after the maximum allowed time has elapsed, a breakdown must have occurred and the controller should initiate corrective action such as a diagnostic procedure. If an arrival signal returns to the controller earlier than the minimum time, then a problem with the sensor communication system

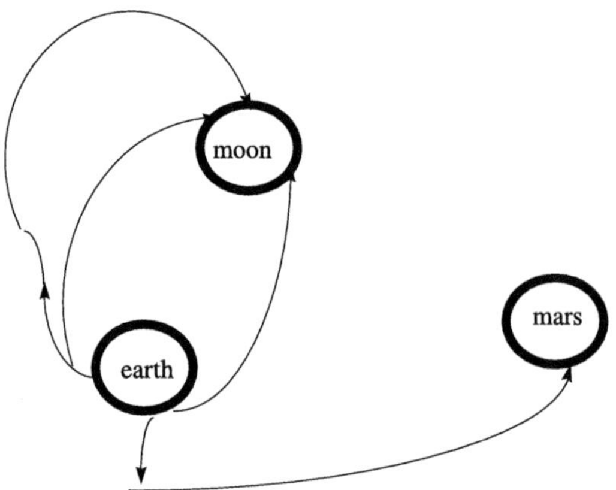

FIGURE 16.18

Space Travel System illustrating Congruence Property.

can be suspected. The formal definition is given next followed by the example application to the space system.

16.5.1 BOUNDARY-BASED DEVS

A boundary-based DEVS is a structure:

$$DEVS = (X, Y, S, \delta_{\text{ext}}, \delta_{\text{int}}, \lambda, \text{ta})$$

where

- $X =$ arbitrary set of symbols representing equivalence classes of input trajectories
- $Y = B$, where B is set of symbols called *boundaries*
- $S \subseteq B \times X$ is a set of symbols called *phases*
- $\delta_{\text{ext}}(b, \phi, 0, x) = (b, x)$ – inputs are accepted only immediately after an internal transition, i.e., when the base model state is on a boundary, then the change is only to store the input. The input is then applied in the subsequent internal transition, as in the following:
- $\delta_{\text{int}}(b, x) = (b', \phi)$ where $b' \in B$ represents the boundary that the base model state reaches after the input trajectory represented by the input. After internal transition input, the store is nullified.
- $\lambda(x, b) = b \in Y$ – the output is the current boundary
- $\text{ta}(b, x) = [t_{min}, t_{max}]$, a time window representing the minimum and maximum times for the internal transition from (b, x) to (b', ϕ) i.e., from one boundary to the next. Also, $\text{ta}(b, \phi) = \infty$.

While we informally described the interpretation of a boundary-based DEVS, its relation as an abstraction of a base model can be formalized:

A system S *abstracts* to a Boundary-based DEVS if

- there is a map of B into the subsets of Q, where the subsets associated with distinct boundaries are disjoint. For simplicity let b stand both for the boundary symbol and the subset it represents.
- there is a map from X to subsets of Ω, where the subsets associated with distinct inputs are disjoint. For simplicity let x stand both for the input symbol and the subset it represents.
- *Congruence property of boundaries:*

$$\forall (b,x) \in S, \forall q, q' \in b, \forall \omega \in x, \exists b' \in B \, such \, that \Delta(q,\omega) \in b' \text{ and } \Delta(q', w) \in b'.$$

  This requires that all states starting on the same boundary are placed on the same destination boundary by all of the input segments belonging to same subset.
- $\forall (b,x) \in S, x \neq \phi, \exists b' \in B$ such that $\delta_{\text{int}}(b,x) = b'$ and $\exists \omega \in x$ such that $\Delta(q,\omega), x) \in b'$. This employs the congruence property to define the next boundary given an input trajectory in the set denoted by the input x (any such input trajectory may be chosen since the destination is the same for all input trajectories applicable to the boundary).
- $\forall (b,x) \in S$, $\text{ta}(b,x) = [min\{l(\omega)|\omega \in x\}, max\{l(\omega)|\omega \in x\}]$ where the minimum and maximum of the segment lengths are taken over all input segments $\omega \in x$. While we assume the congruence property with respect to inputs and states, the times to reach the next boundary may vary, thus resulting in an interval in which the values are spread (see also Chapter 10 on RT-DEVS).

This definition of abstraction can be shown to constitute an appropriately defined homomorphism. The major modification required in the definition of homomorphism in Chapter 12 is to accommodate the time advance window.

Let $S = (T, X, \Omega, Q, \Delta)$ be a system and $M = (X, S, \delta_{\text{ext}}, \delta_{\text{int}}, \text{ta})$ a boundary-based DEVS (for simplicity we omit output considerations – these can be easily added). Fig. 16.19 illustrates the following definition:

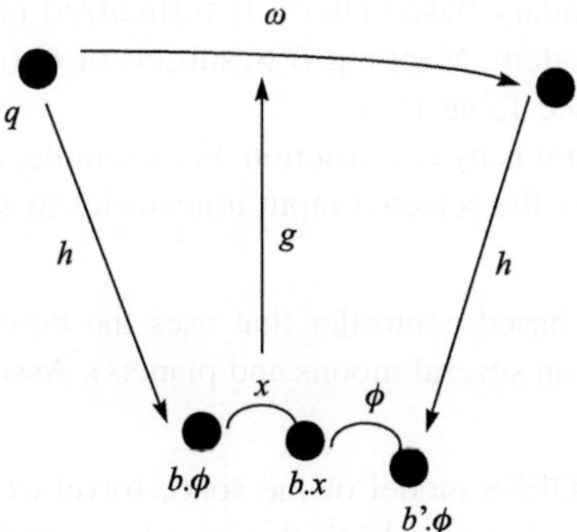

FIGURE 16.19

Illustrating a homomorphism from a model to a boundary-based DEVS.

Consider a mapping pair (g, h), $g : X \to \Omega$ and $h : Q' \to S$ where $h(q) = (b, \phi)$ for some $b \in B$ The pair $(g.h)$ is a *morphism* from S to M if $\forall q \in Q', \forall x \in X$

$$\delta_{\text{int}}(\delta_{\text{ext}}(h(q), 0, x) = h(\Delta(q, g(x))1)$$

and

$$l(g(x)) \in \text{ta}(b, x)2)$$

Due to definition of h, the right hand side of Line 1) requires that starting on a boundary the system winds up on another boundary under the input trajectory input associated with x. The equation therefore requires that the destination boundary be the same as that produced in the DEVS. Line 2) requires that the time taken to make the boundary-to-boundary transition falls within the interval allowed by the time advance function.

Theorem 16.3. *If a system S is abstracted to a boundary-based DEVS, M, then there is a morphism (as defined above) from S to M.*

Exercise 16.32. Provide the proof.

16.5.2 DEVS ABSTRACTION: SPACE TRAVEL EXAMPLE

Let's construct a boundary-based DEVS and show that it is abstracted from the Space System described before. The DEVS will focus on the inter-planetary journeys and will abstract away details of the trajectories of these journeys. Effectively it will place all those with the same source and destination in the same equivalence class, and develop brackets on their travel times (see Fig. 16.20).

Boundary-Based DEVS,

$$M = (X, Y, S, \delta_{\text{ext}}, \delta_{\text{int}}, \lambda, \text{ta})$$

X = {zero,earth-to-moon, earth-to-mars, mars-to-earth,moon-to-earth,
land-on-earth, land-on-moon, land-on-mars}

B = {earth_surface,moon_surface,mars_surface}

The interpretation of the boundary-based DEVS is formalized in the mappings that show it is an abstraction of the Space Travel system. Mapping B to subsets of Q (see Table 16.2).

Mapping X to subsets of Ω (see Table 16.3).

The congruence requirement holds by construction. For example, for input earth-to-moon, all states on the earth's surface are taken by the selected input trajectories to states on the moon's surface. The same is true for the other inputs.

Exercise 16.33. Write an event-based controller that uses the boundary-based DEVS to monitor a multi-leg flight (one that touches on several moons and planets). Assume that the itinerary is given the controller as input.

Exercise 16.34. Write a DEV&DESS model of the space travel example in which the thrust of the space ship can vary within certain bounds. With this as a base model, formulate a procedure to determine the time windows for a boundary-based DEVS. (Hint: using simulation of DEV&DESS base model, search for parameter settings which minimize and maximize the travel times from one planet to another.)

Phase Definition Table

boundary	input	phase
earth_surface	ϕ	on_earth
earth_surface	earth-to-moon	travelling_earth-to-moon
moon_surface	ϕ	on_moon
earth_surface	earth-to-mars	travelling_earth-to-mars
...	...	...

Internal Transition Table

current phase	next phase
on_earth	on earth
travelling earth-to-moon	entering_moon
travelling earth-to-mars	entering_mars
	...

External Transition Table

phase	input	phase
on_earth	earth-to-moon	travelling_earth-to-moon
on_earth	earth-to-mars	travelling_earth-to-mars
on_mars	zero	mars_surface
	...	...

Time Advance Windows

phase	time window
on_earth	∞
travelling_earth-to-moon	[min travel time, max travel time]
on_moon	∞
travelling_earth-to-mars	[min travel time, max travel time]
...	...

FIGURE 16.20

Table definition of Space Travel Example.

Table 16.2 Mapping of boundaries to subsets of Q

boundary	subset of Q
earth_surface	all states whose x, y coordinates lie on the surface of the sphere representing the earth
mars_surface	all states whose x, y coordinates lie on the surface of the sphere representing mars
moon_surface	all states whose x, y coordinates lie on the surface of the sphere representing mars
...	...

Exercise 16.35. The boundary-based DEVS, as defined above, is restricted so that there is always exactly one transition to a next boundary. We can relax this restriction and allow different next boundaries for a initial block and input. This would result in a non-deterministic abstraction, where a set of

Table 16.3 Mapping of X to subsets of Ω

input	subset of Ω
ϕ	$\{0_{t>} \mid t \in T\}$
earth-to-moon	$\{\omega \mid q$ is on earth's surface and $\delta(q, \omega)$ is on moon's surface$\}$
earth-to-mars	$\{\omega \mid q$ is on earth's surface and $\delta(q, \omega)$ is on mar's surface$\}$
...	...

next boundary transitions are possible for states in a block and input. Formalize a non-deterministic boundary-based DEVS to represent such cases. (Hint: Consider the 2-dimensional pot model in Chapter 9, Section 9.4.1, as an example. In the model, the output thresholds are the boundary crossings. For different states in a block either a threshold of the *temp* or the *level* dimension might be reached next. See Praehofer et al., 1993 for details.)

16.6 PARAMETER MORPHISMS

Often we deal with parameterized models and our attempts to build abstractions of them are based on parameter morphisms. As illustrated in Fig. 16.21, a *parameter morphism* maps both the state and parameter spaces of the base model to corresponding lumped model state and parameter spaces. The advantage of a parameter morphism is that it does not just work for a single isolated base model, but for a whole class of such models. Given the parameters of the base model, the parameter mapping uniquely determines the settings of the parameters of the lumped model, for which the state mapping becomes an (approximate) homomorphism. Such unique parameterization allows a "set and forget" approach to

Parameter Morphisms: parameter space mapping yields a lumped model that is a valid abstraction of the base model in some experimental frame

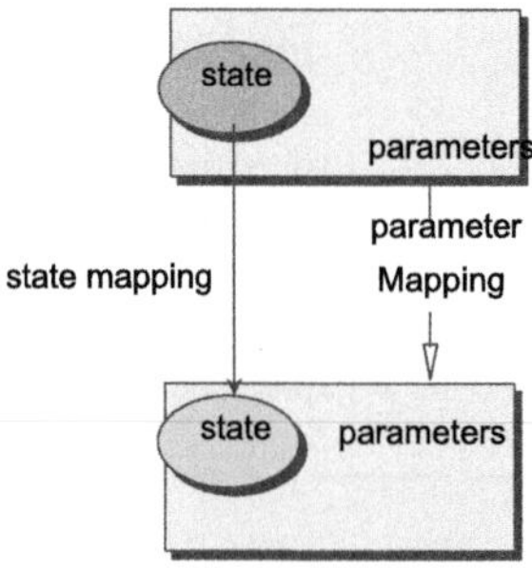

FIGURE 16.21

Parameter Morphism.

lumped model simulation. In contrast, without a parameter mapping we have to find the best setting of the parameters, typically by seeking to minimize the error in the homomorphism. Mapping of parameters is also important for cross model validation (see Chapter 16). Typically, a parameter morphism works only when restricted to a subspace of state and parameter values. Another way of saying this is that there are assumptions underlying the abstraction – often in practice such assumptions are not all made explicit as they should be.

16.6.1 LINEAR SYSTEMS PARAMETER MORPHISMS

Linear systems offer the quintessential parameter morphisms. The state mapping discussed in Chapter 12 is a block-wise summation and thus satisfies the identity-erasing requirements. However, since it does not preserve state counts, it imposes a less demanding form of inter-block coupling indifference requirement then does the state-count based aggregation. For example, if we create blocks $\{1, 2\}$ and $\{3, 4\}$ in a 4th order linear transition matrix, and focus on the influences of 3 and 4 on 1 and 2 we get the sub-matrix:

$$\begin{bmatrix} - & - & g & g_1 \\ - & - & h & h_1 \\ - & - & - & - \\ - & - & - & - \end{bmatrix}$$

Lumpability requires that the sums of the sending block columns restricted to the receiving block rows are equal, namely, that $g + h = g_1 + h_1$. This states that the combined influence of 3 on $\{1, 2\}$ is equal to that of 4 on $\{1, 2\}$. In general, components in a sending block equally influence, in an aggregate manner, any sending block.

Exercise 16.36. Show that the equal influence requirement of lumpability for linear systems is less restrictive than the general coupling indifference requirement. (Hint: show for the above matrix that the coupling indifference requirement is that the sets $\{g, h\}$ and $\{g_1, h_1\}$ are equal (so $g = g_1$ and $h = h_1$ or $g = h_1$ and $h = g_1\}$, which implies (but is not implied by) the lumpability requirement.)

16.6.2 EXAMPLE LUMPABLE: LINEAR DTSS AND PARAMETER MORPHISMS

Fig. 16.22 illustrates an example of a linear DTSS that is lumpable. Note that the influence of the top two delay components on the bottom two is through the same summation of their states. As we have seen, this strong requirement for coupling indifference is one way in which the linear lumpability conditions can be satisfied, but not the only way.

The matrix of the 4th order (4 delay) system and that of its lumped version are:

$$\begin{bmatrix} 0 & 0 & g & g \\ 0 & 0 & h & g \\ k & k & 0 & 0 \\ l & l & 0 & 0 \end{bmatrix} \to\to \begin{bmatrix} 0 & g+h \\ k+l & 0 \end{bmatrix}$$

The aggregation, illustrated in Fig. 16.23, maps coordinates $\{1, 2\}$ to 1 and $\{3, 4\}$ to 2.

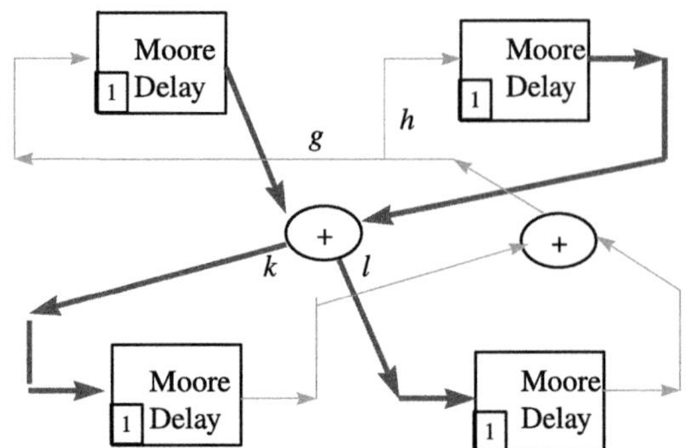

FIGURE 16.22

Linear Lumpable System.

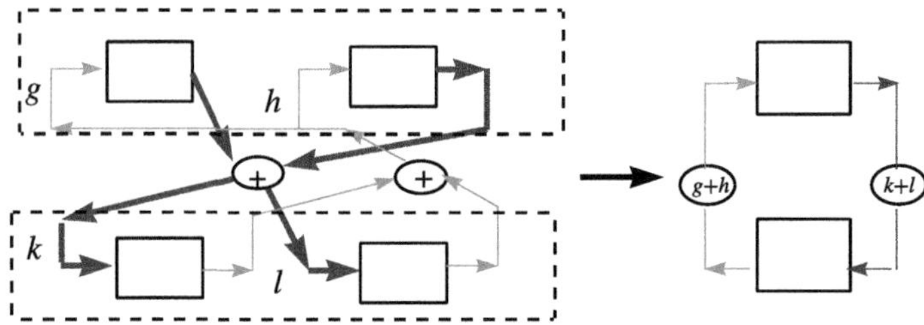

FIGURE 16.23

Aggregation Mapping.

Interpreted as a parameter morphism, the aggregation mapping:

- maps the state space of the base model to that of the lumped model

$$h(q_1, q_2, q_3, q_4) = (q_1 + q_2, q_3 + q_4)$$

- maps the parameter space of the base model to that of the lumped model

$$p(g, h, k, l) = (g + h, k + l)$$

If we take the model in Fig. 16.22 as the parameterized base model, there is no restriction on the state space or parameter space for which the morphism works. However, consider the space of all 4th order models that have the structure:

$$\begin{bmatrix} 0 & 0 & g & g1 \\ 0 & 0 & h & h1 \\ k & k & 0 & 0 \\ l & l & 0 & 0 \end{bmatrix}$$

In this space, the restrictions justifying the morphism are that $g = g_1$ and $h = h_1$. In other words, the lumpability conditions are the assumptions underlying the parameter morphism.

16.6.3 CONSTRAINTS ON PARAMETER AND STATE SPACES: DISAGGREGATION

We conclude from the foregoing discussion that parameter morphisms typically only work when parameters are restricted to a subset of the parameter space. Such constraints on validity should be included with the statement of the morphism. The same can be said for state space constraints. In many cases, lumped models can be constructed under assumptions that constrain the states for which the aggregation mapping is a valid morphism. Equilibrium conditions are often used for this purpose. Typically, when a model is in equilibrium, constraints exist on its parameters that make it easier to construct valid parameter morphisms. Of course, it is then important to include such constraints in the statement of the morphism. For example, it might be stated that the morphism is exact if the base model is operating in equilibrium and that it becomes approximate for other states. As indicated before, we will consider such approximations in the next chapter.

Disaggregation is a process where we wish to associate a base model state with a given lumped model state. Multi-resolution modeling, where we may alternate between models at different levels of resolution, requires such disaggregation capabilities. Unfortunately, by its nature, an aggregation mapping is many-to-one, i.e., many base model states are mapped to the same lumped model state. Thus unique disaggregation can only be done by restricting the base model states to a subset where the aggregation mapping becomes one-to-one. For example, equilibrium conditions may supply such a constrained subset. Indeed, as indicated above, equilibrium conditions, may have justified the mapping as a morphism in the first place.

16.6.4 USING PARAMETER MORPHISMS IN AN INTEGRATED MODEL FAMILY

We have seen that parameter morphisms contain relations between parameter sets of model pairs. These relations can be used to propagate information about parameter values among models in an integrated model family. While greater detail is given in Zeigler (1984), let's see how parameter morphisms look and help propagate information in the space travel example. The base model in Fig. 16.24 contains detailed representations about the ports of call, routes, times, and individual ships in the system. Recall the morphisms we constructed in earlier sections to build the logistics and control models. These should now be extended to parameter morphisms so that they also transmit parameter information from the base model to the abstractions. This means that parameter values employed in the abstractions should be determined by formulas (rules, procedures, tables, etc.) from parameter values in the base model.[1] For example, the time windows in the boundary-based DEVS are derived from the minimum and maximum travel times of equivalent trajectories in the base model. Having a parameter morphism for such a window would not allow it to be frozen forever. For example, if a new flight path were added say made possible by new technology that significantly reduced the minimum time to travel from earth to moon, then the corresponding time window would be automatically updated. This might be done by placing demons in the model base that look for changes in the inputs of parameter formulas and

[1] It might be ideal if **all** such parameters could be set from the base model but this might not be possible since in practice the scope of the base model is itself limited and not all information in an abstraction can be derived from it.

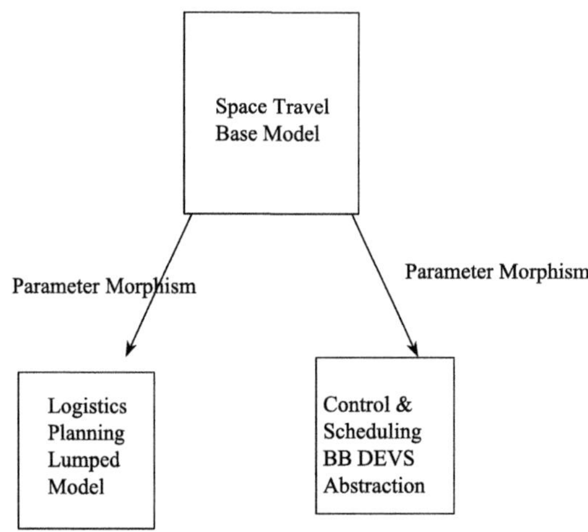

FIGURE 16.24

Parameter Morphisms in Space Travel Example.

invoke a re-computation of the formula when a change occurs. Other base model information that might change such as new ports of call (planets or moons), new routes, and new routing policies would also be propagated automatically by parameter morphisms.

Exercise 16.37. Express the relationship parameters in the logistics model to those of the space travel base model.

16.7 SUMMARY

We have used our M&S framework (Chapter 2) and the morphism specification hierarchy (Chapter 15) for understanding the issues in model abstraction, especially at the coupled model level. While our focus was on aggregation, the same framework applies to the multi-resolution problems that arise in many contexts including distributed interactive simulation. We should not expect that a simple panacea exists that conquers all such problems. Rather the solution lies in a capable tool set that recognizes the influence of modeling objectives and error tolerances, as well as the multi-dimensional choices of bases for aggregation mappings. Formalizing such dimensions of the problem leads to a sophisticated framework which involves concepts such as scope/resolution product, experimental frames, applicability and derivability lattices, conditions for valid simplification, etc. Nothing less than this comprehensive basis would seem adequate to the demands of the problem. The companion book on Model Engineering for Simulation (Zhang et al., 2018) discusses the need for semantic consistency in the use of multiple formalisms, each most suitable to a particular level of abstraction in the same study.

In the next chapter we deal with the error in the abstraction explicitly. This forces us to consider trade-offs between complexity reduction and error accumulation.

16.8 SOURCES

Aggregation has been of interest in economic modeling (Simon and Ando, 1961; Ijiri, 1971) and ecological modeling (Cale, 1995) for some time. Axtell (1992) reviews much of this literature and the issues raised. A more general summary, focusing on engineering and military application, is given by Axtell (1992). A special issue of Transactions of the Society for Computer Simulation on model abstraction contains the papers: Caughlin (1996), Lee and Fishwick (1996), Takahashi (1996), and Zeigler and Moon (1996).

REFERENCES

Axtell, R.L., 1992. Theory of Model Aggregation for Dynamical Systems with Application to Problems of Global Change. Carnegie Mellon University, Pittsburg, PA.

Cale, W.G., 1995. Model aggregation: ecological perspectives. In: Complex Ecology. The Part-Whole Relation in Ecosystems. Prentice Hall, Englewood Cliffs, NJ, pp. 230–241.

Caughlin, D., 1996. Model abstraction via solution of a general inverse problem to define a metamodel. Transactions of the Society for Computer Simulation International 13 (4), 191–216.

Davis, P.K., 1995. An introduction to variable-resolution modeling. Naval Research Logistics (NRL) 42 (2), 151–181.

Dennett, D.C., 1995. Darwin's dangerous idea. The Sciences 35 (3), 34–40.

Ijiri, Y., 1971. Fundamental queries in aggregation theory. Journal of the American Statistical Association 66 (336), 766–782.

Lee, K., Fishwick, P.A., 1996. A methodology for dynamic model abstraction. Transactions of the Society for Computer Simulation International 13 (4), 217–229.

Praehofer, H., Bichler, P., Zeigler, B., 1993. Synthesis of endomorphic models for event based intelligent control. In: Proc. of AI, Simulation and Planning in High-Autonomy Systems. Tucson, AZ, pp. 120–126.

Simon, H.A., Ando, A., 1961. Aggregation of variables in dynamic systems. Econometrica 29 (16), 111–138.

Takahashi, S., 1996. General morphism for modeling relations in multimodeling. Transactions of the Society for Computer Simulation International 13 (4), 169–178.

Zeigler, B.P., 1984. Multifacetted Modelling and Discrete Event Simulation. Academic Press Professional, Inc, London.

Zeigler, B.P., Moon, Y., 1996. DEVS representation and aggregation of spatially distributed systems: speed-versus-error tradeoffs. Transactions of the Society for Computer Simulation International 13 (4), 179–189.

Zhang, L., Zeigler, B., Laili, Y., 2018. Model Engineering for Simulation. Elsevier.

CHAPTER

VERIFICATION, VALIDATION, APPROXIMATE MORPHISMS: LIVING WITH ERROR

17

CONTENTS

This chapter deals with error – *verification* (how to check if a simulator is in error), and *validation* (how to check if a model is in error). The standard for error-free relations between entities in the framework for M&S is the morphism. But in the real world, error is inevitable and we must usually be satisfied with something less than the perfect agreement required by such morphisms. So we have to deal with *approximate* morphisms – relations that are close to, but not one hundred percent, perfect morphisms. While we emphasize morphisms, alternative formulations emphasize statistical techniques (see Balci, 1997).

Theory of Modeling and Simulation. https://doi.org/10.1016/B978-0-12-813370-5.00027-4

17.1 VERIFICATION

Verification is the attempt to establish that the simulation relation (Chapter 2) holds between a simulator and a model. Actually, as we have seen, the simulator may be capable of executing a whole class of models. In this case, we wish to guarantee that it can execute any of these models correctly. There are two general approaches to verification:

- formal proofs of correctness
- extensive testing

The basic concept in establishing correctness is that of **morphism** and we have seen that morphisms can exist at different levels of the system specification hierarchy (Chapter 12). Since both model and simulator are known at the higher levels of structure, it is possible to apply system morphism at such levels to prove correctness. In such a morphism, when a model should go through a state sequence such as the simulator should go through a corresponding state sequence. Typically, a simulator is designed to take a number of *microstate* transitions to simulate a *macrostate* transition of the model. These are computation steps necessary to achieve the desired end result. Moreover, it has a lot of apparatus, represented in its states, necessary to accommodate the whole class of models rather than a single one. The fact that morphisms propagate down the specification hierarchy (Chapter 12) has a practical application if a designer can show that such a morphism holds between the simulator and any model in a class then the simulator correctness will have been established for that class, in other words, any state trajectory of a model in the class will be properly reproduced by the simulator.

Formal proofs employ mathematical and logical formalisms underpinning the concepts in the systems specification hierarchy to rigorously establish the requisite morphism. Unfortunately, such proofs are difficult or impossible to carry out for large, complex systems. Moreover, they may also be prone to error since ultimately humans have to understand the symbols and carry out their manipulations. On the positive side, more automated tools are becoming available to relieve some of the burden.

In the absence of once-and-for-all proofs, extensive testing must be done to assure that all conditions that could arise in simulator operation have been covered by test cases. Time and other resources limit the amount of testing that can be done. However, even though formal proofs are not employed, the morphism concept comes in handy since it offers a framework for laying out the combinations of inputs and states that have to be tested for thorough, yet efficient, testing (Barros, 1996).

17.2 VALIDATION AT THE BEHAVIORAL LEVEL

Validation is the process of testing a model for validity. Fig. 17.1 illustrates how the experimental frame is critical in this process. The frame generates input trajectories to both the source system and the model under test. The corresponding output trajectories of the model and system are fed back into the frame. Validity, whether replicative, predictive or structural (Chapter 2), requires that these trajectories are equal. We will first discuss replicative validation as the basic level of the validation process in which trajectories are compared. Later we will consider predictive and structural validation.

The experimental frame is critical in assessing model validity since it provides the conditions under which both model and system are to be experimented with. Therefore, we'll examine the experimental frame concept in more depth before proceeding further. Recall that a frame is decomposed the into

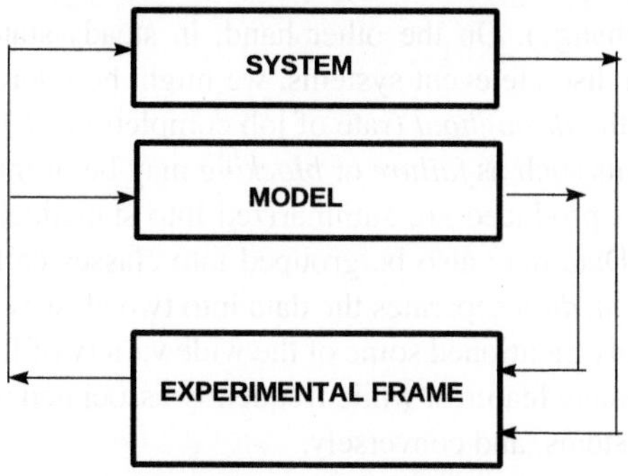

FIGURE 17.1

Validation within Experimental Frame.

Table 17.1 Typical Experimental Frame Components

System Formalism	Typical Generator	Typical Acceptor	Typical Transducer
DESS (Continuous Systems)	periodic: sine wave, square wave	steady state	final value
DTSS (Discrete Time)			
	aperiodic: step, ramp	transient	rise time
DEVS (Discrete Event)			
	periodic arrivals	small queues	throughput
	stochastic arrivals	large queues	turnaround time utilization
	workload characteristics	failure rate	blocking rate

generator, acceptor and transducer components (Chapter 2). Table 17.1 presents some generator, acceptor and transducer behaviors typically associated with continuous, discrete time and discrete event system specifications.

Recall that a *generator* stimulates the system with input trajectories. In continuous systems such trajectories take the form of various kinds of *aperiodic* functions such as steps and ramps, or various *periodic* functions such as sine waves or square waves. In discrete event systems, arrival of events (such as jobs for processing) may be periodic or *stochastic* (random). In the latter case, we need to specify the probability distribution of the inter-arrival times, such as *uniform* or *exponential*. The type of processing required (*workload*) is also part of the generator specification.

We often make a distinction between the *transient* and *steady* state characteristics of system behavior. The *acceptor* is the slot in the experimental frame where conditions limiting the observation of behavior, such as steady state versus transient, can be specified. For example, if interest in transient behavior is specified, then the frame would extract only the parts of the trajectories shown within the corresponding box in Fig. 17.1.

The *transducer* processes the output trajectories, where such post-processing may range from none at all (in which case the trajectories are directly observed) to very coarse summaries where only certain features of interest are extracted. For example, for transient conditions, we may be interested in the *rise-time* (time to adjust to change). On the other hand, in steady-state conditions, *the final value* reached may be of interest. In discrete event systems, we might be interested in the *turnaround times* required to process jobs or in the *throughput* (rate of job completion). *Utilization* of various resources and occurrence of special events such as *failure* or *blocking* may be of interest.

Usually the many numbers produced are summarized into statistical quantities such as the *average*, *maximum*, or *minimum*. Data may also be grouped into classes called *quantiles* with associated breakpoints. For example, the *median* separates the data into two classes of equal sizes.

Please note that we have only mentioned some of the wide variety of features that can be formulated in experimental frames. Also many features, while frequently associated with continuous systems, may also apply to discrete event systems, and conversely.

17.2.1 QUANTITATIVE COMPARISON

Recall that validation requires comparison of model and source system behavior. Please note that the comparison is only performed under the conditions of experimentation specified in the frame; that is, only under the input trajectories generated by the generator, the control conditions checked by the acceptor, and the output summaries performed by the transducer. Such comparison is illustrated in Fig. 17.3 where the input trajectories are the respective output trajectories of the system and model. Each is summarized by a copy of the same transducer and the summaries are compared. In the conventional approach, comparison requires a *metric* and a *tolerance*. The *metric* provides a numerical basis for measuring *goodness-of-fit*. The *tolerance* is a positive number that determines when the fit is good enough.[1] When the fit is outside of tolerance, the model trajectory is judged not to be close enough to that of the system for the model to qualify as valid in the particular frame. On the other hand, if the fit is within tolerance then the model validity is *confirmed* – but **not established** since a possible infinitude of trajectories still need to be compared. Typical metrics are the familiar *distance measure* in *Euclidean space* (square root of sum of squared differences), *largest absolute difference*, and *city block distance* (sum of absolute differences).

When stochastic processes are employed, comparison involves a further consideration: how representative are the observed samples of the underlying spaces. Statistical techniques are then employed to judge significance in terms of the number of samples observed and the variance in their values. However, a word of caution: statistical techniques often make assumptions characterizing the stochastic nature of the data sources, such as normal distributions. These are models themselves and may not be valid!

In doing such comparisons it is easy to forget that the choice of what to observe may have as great, or greater, bearing on validity judgments as the metrics and statistical procedures employed. For example returning to Fig. 17.2 the two curves shown are very close (under most reasonable metrics) in the transient behavior but far apart in the steady state region. Thus a model might be judged valid in a frame which focuses on transient behavior but not in a frame that looks at steady state behavior. Or, a model might agree well in the steady state region but not in the transition to it. Indeed, typically, it is easier to construct models that are accurate with respect to the steady state of a real system but are poor

[1] Tolerance can be more abstractly defined as a relation on states independently of a metric. Insights into error propagation discussed later in the chapter can be obtained with such a formalization (Cin, 1998).

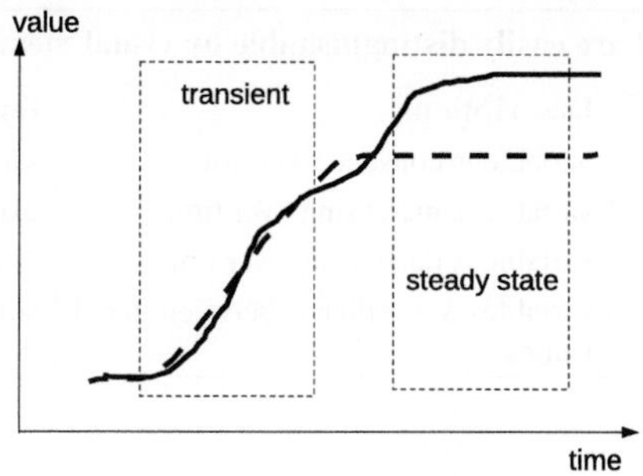

FIGURE 17.2

Experimental Frame Operating Regions: Transient vs. Steady State Conditions.

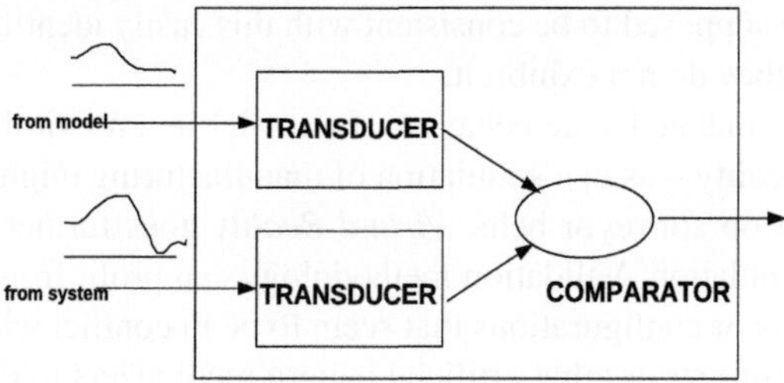

FIGURE 17.3

Comparing Trajectories in an Experimental Frame.

at representing its transient behavior. Unfortunately, especially nowadays, many processes are never in steady state!

17.2.2 QUALITATIVE COMPARISON

While quantitative comparison provides an objective basis for validation, it can miss more qualitative discrepancies or agreements that humans are capable of detecting if provided the right perspectives. Two methods attempting to provide such perspectives are *visualization* and *animation*. *Visualization* attempts to map complex numerical data into graphic structures that humans can more readily understand. Actually, simple graphical displays can suffice for illustrating this idea. Table 17.2 contains some behaviors that were first encountered in linear systems as displayed in figures of Chapter 3. However, the general trends increasing, decreasing, oscillating can be present in general. For example, models of the universe can be steady state (fixed), expanding, or contracting. Graphical display makes it easy for us to immediately distinguish among such behaviors. The importance for validation, is that if a model is supposed to manifest one of these behaviors but actually exhibits another, it is clearly invalid. Quantitative comparison is needed to make finer distinctions between behaviors that agree in their basic form, but qualitative comparison can quickly eliminate models that are not even in the right ballpark. By subjecting models to such qualitative tests early in model development, it may be possible to elimi-

Table 17.2 Some Behaviors that are easily distinguishable by visual means

Behavior	Description	linear system example
fixed	variable is constant over time	same
indefinite increase	variable is increasing over time	exponential explosion
indefinite decrease	variable is decreasing over time	exponential decay
oscillation	variable is oscillates between fixed values	sine wave

nate wrong formulations and focus in promising directions long before serious quantitative comparison can be attempted. For example, recall the space travel example of Chapter 14, where a non-linear 4th order system orbiting system decomposes into two independent second order linear oscillators when initialized to states that establish circular orbits. By initializing in such states, more complex models of orbiting mechanics, that are supposed to be consistent with this easily identifiable oscillatory behavior, can easily be invalidated if they do not exhibit it.

Animation presents the simulated state behavior of a model in a movie-like form that corresponds to the underlying physical reality – as in a simulation of manufacturing might show parts moving from one workstation to the next on conveyor belts. *Virtual Reality* goes further by providing a sensation of actual presence in the simulation. Validation methodology can profit from such presentations since humans can recognize events or configurations that seem to be in conflict with everyday experience. It is also possible to train *neural nets* or other artificial pattern recognizers to do such detection.

17.3 PERFORMANCE/VALIDITY (E.G. SPEED/ACCURACY) TRADE-OFF

Recall (Chapter 14) that scope/resolution product is an indicator of the complexity of a model. Complexity is a measure of the resources required to analyze, simulate, and develop the model. Typically, the performance of a simulator will increase as the complexity of a model decreases. For example, the speed of a simulator (measured say, in the number of model transition executions it can do in a given time) will usually increase as either the number of components or the number of states per component diminishes. We can similarly talk of the performance of model development or model exploration process in terms of how long it takes to develop the model or how many alternatives can be examined in a given time, respectively.

So on one hand, performance can be expected to improve with reduced complexity. On the other hand, reducing the scope/resolution product often entails a loss of validity of the model – which of course, has to be measured within an experimental frame of interest. Thus often we are in a situation where we must sacrifice potential validity for improved performance of a development or simulation process. This performance/validity trade-off is illustrated in Fig. 17.4 where we see, conceptually depicted, performance increasing while validity decreases as the scope/resolution product is decreased. In such a situation, we can set a level of error tolerance which determines the level of error we can live with – any model with more error (less validity) cannot be accepted. The complexity reduction or performance improvement at the error tolerance is the best we can achieve. Of course, the purposes for which the model is being constructed or deployed should determine such threshold limits. We can

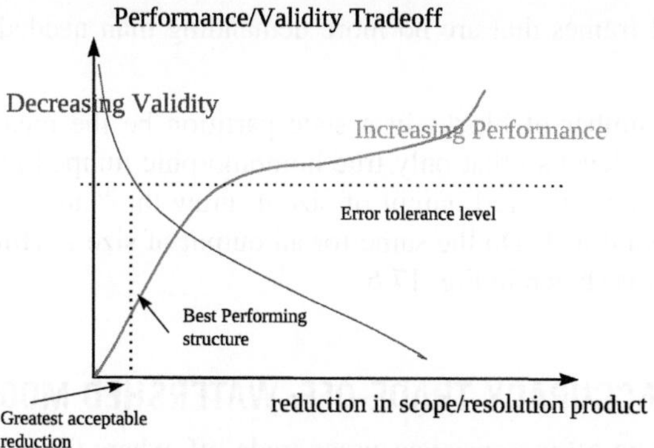

FIGURE 17.4

Performance/Validity Trade-off.

include acceptance criteria within our experimental frame as well, so that the acceptance judgments could be automated.

Before we look at an example, Fig. 17.5 reminds us that the trade-off space we are in is determined by our experimental frame. Thus if we are demanding high resolution in model output then error may increase beyond a tolerable level as the detail included in the model structure decreases. However, it may well be that if we require a lower level of resolution in the model output, then the error may be below tolerance, or even vanish, for the same level of detail included in the model. Indeed, the complexity in the model may be much lower at the threshold of acceptability for the low resolution frame than for its high resolution counterpart. Correspondingly, the fastest simulation might be much faster in the low resolution frame than in the high one. This graphically illustrates the importance

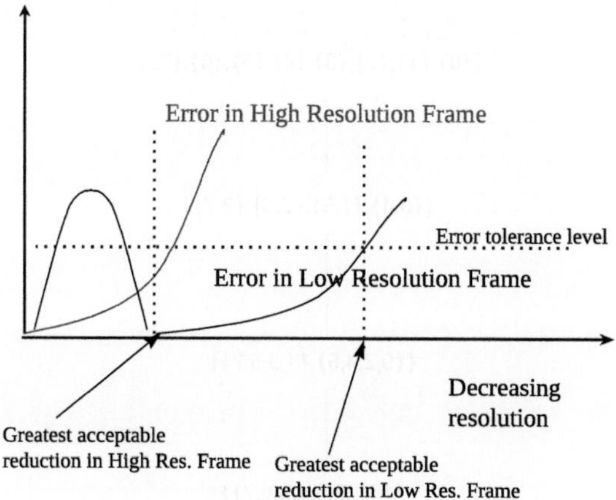

FIGURE 17.5

Breaking the Speed/Accuracy Trade-off with Choice of Experimental Frames.

of choosing experimental frames that are no more demanding than needed for the particular task at hand.

Exercise 17.1. Let the number of blocks in a state partition be the measure of model resolution. Let zero be error tolerance level so that only true homomorphic lumped models are allowed. For an 8-counter (Chapter 14), with a desired output of size 4, draw the "curve" of error versus resolution as the latter decreases from 8 to 1. Do the same for an output of size 2. Hint: the lattice of congruent partitions for an 8-counter is shown in Fig. 17.6.

EXAMPLE OF SPEED/ACCURACY TRADE-OFF: WATERSHED MODELING

One example of such a trade-off is a speed/accuracy trade-off, where for example, we investigate how the speed of a simulation program increases as we successively aggregate model components in more and more coarser partitions.

Consider an experimental frame in which we are interested in surface water accumulation the bottom of a hill as in Fig. 17.7. As rainfall starts the water starts rising and then reaches a maximum level some time after the rain stops. The experimental frame is interested in the this runoff over time. The goodness-of-fit criterion is the maximum of the absolute difference between predicted and actual runoff. The error tolerance is 10% of rainfall input, assumed constant.

DEVS cell space models of surface watershed runoff and their implementation in high performance environments are detailed in Zeigler and Moon (1996). Starting with a base model of 128 cells, a lumped model of 64 cells was created by an aggregation mapping in which adjacent pairs of cells were combined into single cells. A second lumped model was created by using the same process to arrive at a model with 32 cells, and so on. The aggregation map is a parameter morphism (Chapter 14).

The effect of successive aggregation of cells on the accuracy and speed of simulation is shown in Fig. 17.8. As the resolution is reduced, simulation time decreases while error increases. We found that we could obtain a 15 fold speedup with an error tolerance of 10% of input rainfall by decreasing resolution by a factor of eight.

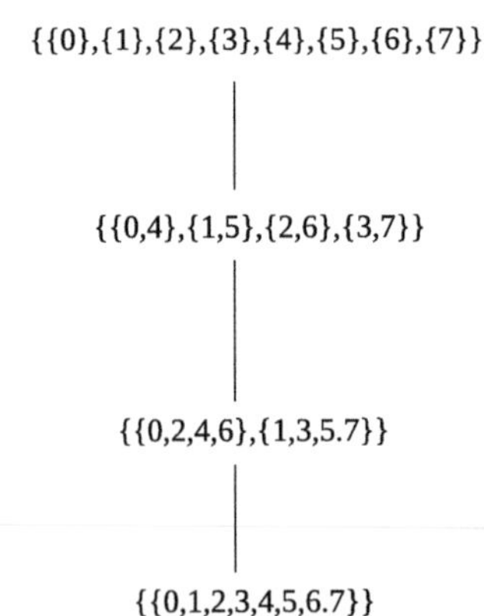

FIGURE 17.6

Lattice of partitions for 8-counter.

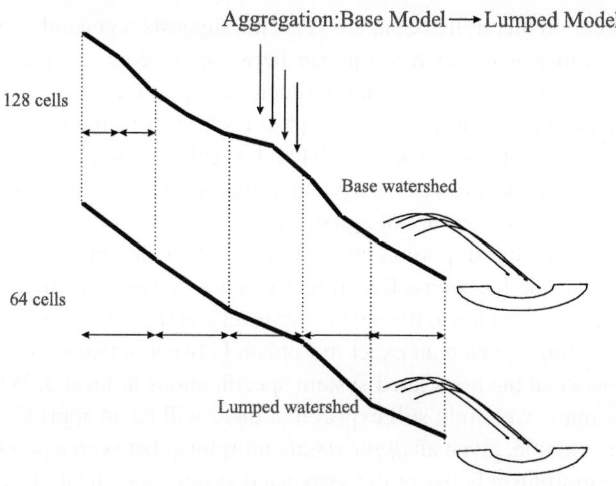

FIGURE 17.7

Aggregation mapping of high-resolution watershed models.

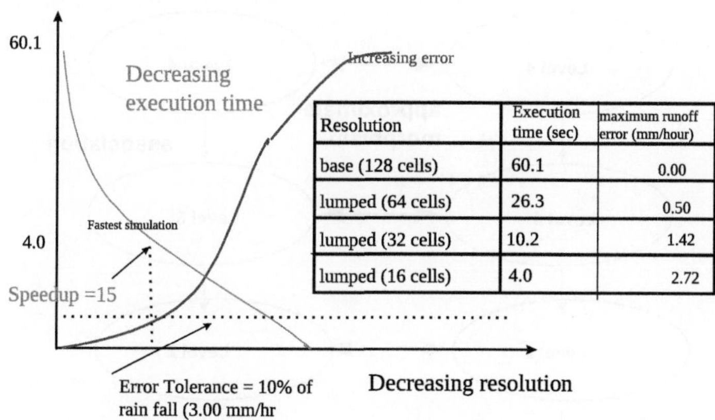

Resolution	Execution time (sec)	maximum runoff error (mm/hour)
base (128 cells)	60.1	0.00
lumped (64 cells)	26.3	0.50
lumped (32 cells)	10.2	1.42
lumped (16 cells)	4.0	2.72

FIGURE 17.8

Speed/accuracy trade-off in aggregation.

17.4 APPROXIMATE MORPHISMS AND ERROR BEHAVIOR

17.4.1 APPROXIMATE MORPHISMS AND THE SPECIFICATION HIERARCHY

So far we have seen how loss of accuracy often arises from reducing the score/resolution product. But we have not discussed why this happens or what can be done about it. Since each level of the specification hierarchy (Chapter 5) offers a means to create a model of a real system, it also presents an opportunity to do so incorrectly. Recall that morphisms afford the means to compare systems at each level of the specification hierarchy (Chapter 12). This suggests a general concept of *approximate morphism* – if all the requirements for a morphism between systems specified at a particular level are not met, then we can call the morphism approximate, as opposed to exact. Thus as illustrated in Fig. 17.9, there is an approximate morphism concept at each level of the specialization hierarchy. If there is an approximate morphism between a model and a real system at some level then we can say that the model is in *error* as an account of the system at that level. As we have seen, just how big this error is, depends on our choice of measure of agreement.

Fig. 17.9 also suggests another important consideration – how do approximate morphisms behave as we descend the levels of the hierarchy. Recall that given a system specification, there is a way to associate with it a system specification at the next lower level and this carries over to morphisms as well (Chapter 12). Thus for example, given an exact morphism between a pair of systems at level 4, there is an exact morphism between the associated system specifications at level 3. When the morphism in question is only approximate, we might still expect that there will be an approximate morphism at the next lower level. So for example, given an *approximate* morphism between a pair of systems at level 4, there is an approximate morphism between the associated system specifications at level 3. Recall that going down the levels corresponds to a simulation process i.e., generating the behavior of the model given its structure. Thus a disagreement in structure between a base model and a lumped model, as

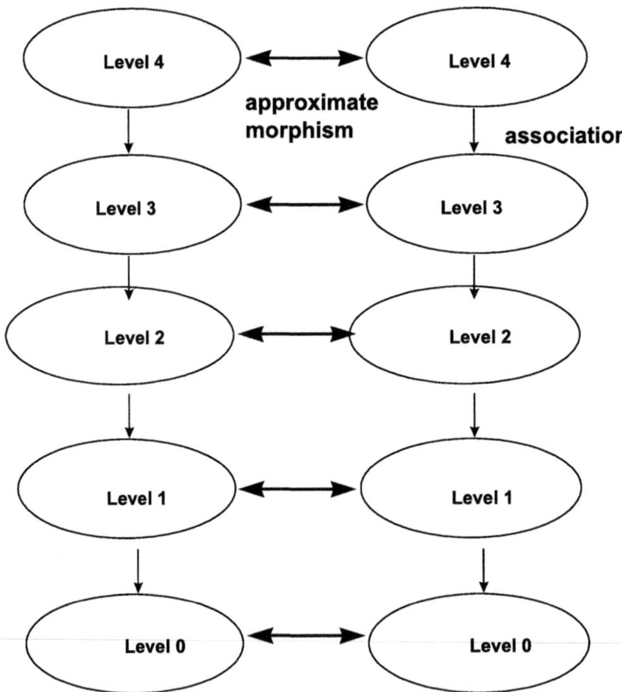

FIGURE 17.9

Approximate Morphisms at System Specification Levels.

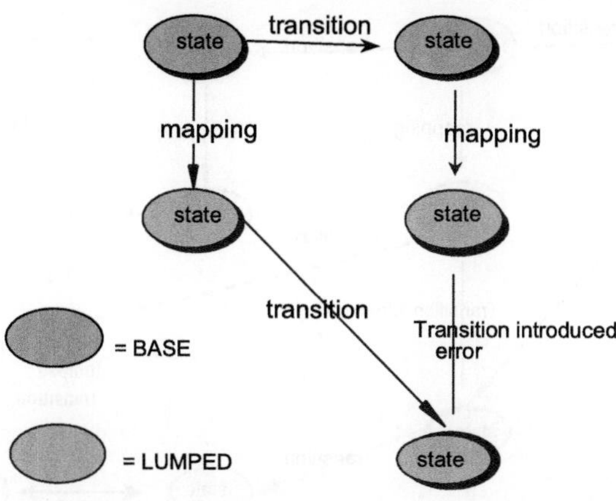

FIGURE 17.10

Error Introduced in Approximate Homomorphism.

encoded in an approximate morphism at level 4, would show up in the mismatch in their behaviors as encoded in an approximate morphism at level 0. Errors in structure may not only be manifested immediately by also at later times as they propagate through the simulation process.

17.4.2 APPROXIMATE HOMOMORPHISMS AND ERROR PROPAGATION

To understand how errors can propagate in the simulation process, we'll focus on approximate morphisms at the state transition level of specification. As illustrated in Fig. 17.10, let there be a mapping from the states of a base model to those of a lumped model. Let the base and lumped models be in states that correspond under this mapping and consider corresponding transitions in these states for both models. Were the state mapping to be an exact homomorphism then the states after the transition would correspond under the mapping. However, in an approximate morphism, the states may differ. More specifically, lumped model may be in a state that differs from the mapped base model state. As shown, this is the error introduced in one transition. Unless we have a metric for the state space of the lumped model, we can say nothing more than that they aren't equal.

Let' continue to track the progress the state trajectories in Fig. 17.11. After two transitions, we compare the mapped base model state to the lumped model state, expecting them to be in error. Similarly, the mapped state trajectory of the base model and the lumped model state trajectory may be compared at every transition. Again, if we have a metric for judging the distance between trajectories, we can say how far apart these trajectories are. More than that, we may sometimes be able to discern how the error introduced at each transition (as in Fig. 17.10) propagates to impact the total error in subsequent transitions.

To see this, look at the box in Fig. 17.11. There we look at a third state in the lumped state space, illustrated at the end of the dashed arrow. This state is obtained by applying the lumped model transition function to the lumped state that was the image of the base model state in the previous transition. In a metric space, it may be "intermediary" between the two states we are interested in – the mapped based model state and the lumped model state. Our goal is to link the error (which is distance between the

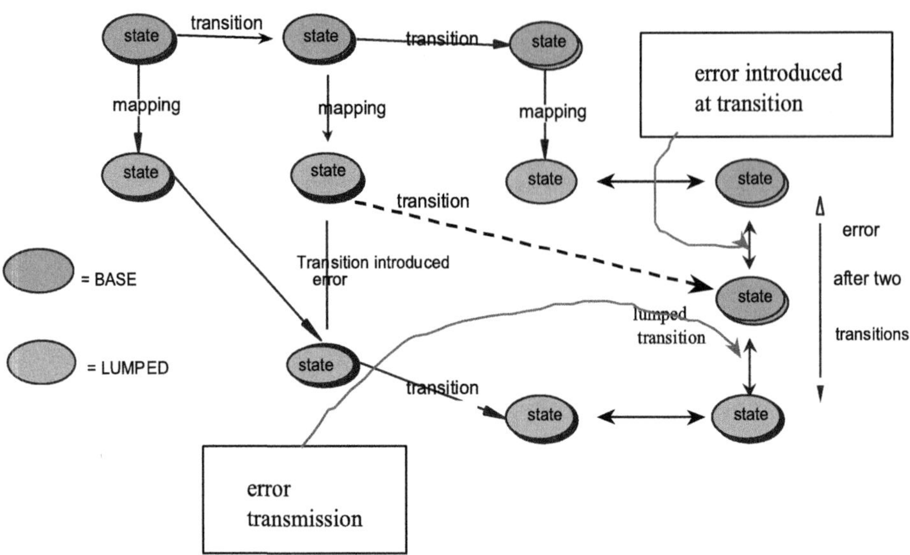

FIGURE 17.11

Error Accumulation Over Time.

latter two states) to their distances to the intermediary state. Why? Because these last two distances can be related to information available in a single transition.

As in Fig. 17.11, let's call the mapped base state, B; the intermediary state, I, and lumped state L. Then notice that the distance between B and I is the error introduced in a single transition. And the distance between I and L is how the lumped model transition function treats two states: does their separation increase, decrease or stay the same after one transition.

Fig. 17.12 depicts how the distances are inter-related using the triangle inequality for distance measures (called metrics), d. Basically, each side is less than the sum of the other two. Looking at the side BL, this tells us directly that

$$d(B, L) \leq d(B, I) + d(I, L). \tag{17.1}$$

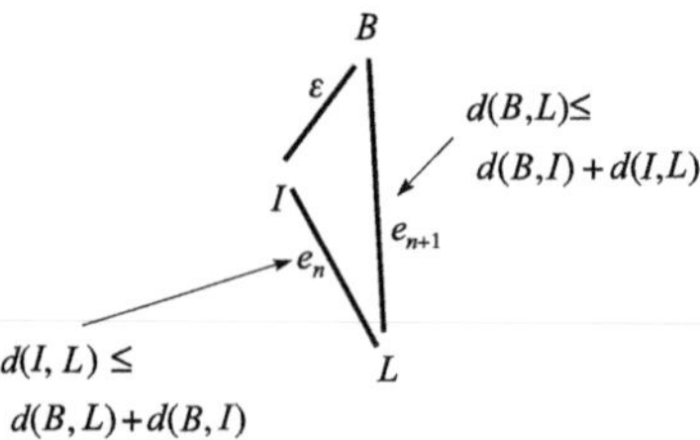

FIGURE 17.12

The Triangle Inequality and the distances between B, I, and L.

This means that if we have *upper bounds* on $d(B, I)$ (the error introduced in a single transition) and $d(I, L)$ (the error propagated from the last transition), then we have an *upper bound* on the error at this transition.

Also, looking at another side of Fig. 17.12, we have that

$$d(B, L) \geq d(I, L) - d(B, I) \tag{17.2}$$

So if we have a *lower bound* on the propagated error and an *upper bound* on the error introduced at this transition, then we have a *lower bound* on the error at this transition.

ERROR PROPAGATION AND ACCUMULATION: BOUNDED AND UNBOUNDED GROWTH

Using the relationships above we can characterize the way error propagates over many transitions in terms of lumped model properties. Let ε_n be the error introduced at transition n and let e_n the total error at transition n where $e_0 = 0$ (since we start in corresponding states). Let Fig. 17.12 represent the situation at transition $n + 1$. This means we can interpret $d(B, I) = \varepsilon_n$, $d(B, L) = e_{n+1}$ and $d(I, L)$ as the error propagated by lumped model from the last transition.

Suppose we know that $d(I, L) \leq a_n \times e_n$, i.e., that the error at a transition grows no bigger than a factor a_n times what it was at the previous transition. Then Eq. (17.1) gives us

$$e_{n+1} \leq a_n \times e_n + \varepsilon_n.$$

If at every transition, $a_n < 1$, we let a be the largest of the a_n and we call a the attenuation factor. In this case, we have bounded growth of the error:

$$e_n \leq \frac{\varepsilon}{1 - a}$$

where ε is the largest of the ε_n.

Exercise 17.2. Using the formula for sums of geometric terms prove the bounded growth relation.

A more subtle bound on growth occurs when the magnification is greater than unity but is counteracted the introduced error. We will encounter this situation in Chapter 17 (DEVS Representation).

Exercise 17.3. Show that if $a > 1$ and ε upper bound the a_n and ε_n, respectively, then

$$e_n \leq \frac{\varepsilon(a^n - 1)}{a - 1}$$

Often, we have $a = 1 + b/n$ where b is small and n is large in which case e_n is bounded by $n\ \varepsilon$ as n goes to infinity. Note that the effect of b disappears.

If on the other hand, we know that $d(I, L) \geq a_n \times e_n$, i.e., that the error grows no smaller than a factor a_n times what it was at the previous transition. Then Eq. (17.2) gives us

$$e_{n+1} \geq a_n \times e_n - \varepsilon_{n+1}$$

If at every transition, $a_n > 1$, we let A be the smallest of the a_n and we call A the amplification factor. Also let ε be the largest of the e_n. And in this case, we have unbounded growth of the error:

$$e_{n+1} \geq A^n \times \varepsilon$$

where ε is the smallest of the ε_n for sufficiently large A.

Exercise 17.4. Using the formula for sums of geometric terms prove the unbounded growth relation.

Finally, if $a_n = 1$ and $\varepsilon_n = \varepsilon$ at every transition, then

$$\varepsilon \leq e_{n+1} \leq n\varepsilon$$

where the growth may be linear at worst. We call this *neutral* growth of error.

Fig. 17.13 shows how to visualize error propagation properties. Attenuation is illustrated in Fig. 17.13A showing how the distance between trajectories may be reduced after a state transition. If errors are introduced in a trajectory bouncing between concave reflectors the trajectory will remain close to where it would be if unperturbed. Contrast this with the error amplification that would arise if the reflectors are convex. In this case, an error introduced in the trajectory may cause it to be deflected in a direction far from the unperturbed case.

17.4.3 EXAMPLE: APPROXIMATE LINEAR SYSTEM HOMOMORPHISMS

Linear systems afford a direct means of illustrating the principles of error propagation. Recall that the requirement for exact homomorphism for linear DTSS systems is expressed by the lumpability condition. What happens if lumpability is not satisfied? We'll see that the concept of approximate homomorphism can be applied and that we can readily predict whether the error propagation will be

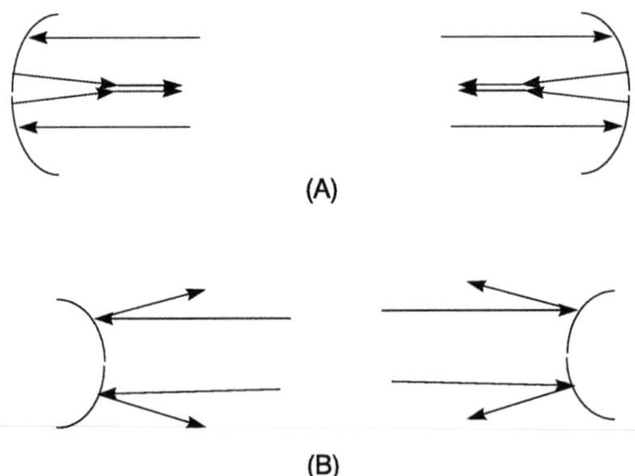

FIGURE 17.13

Illustrating (A) Error attenuation (B) Error Amplification.

bounded, unbounded or neutral. Look at the mapping:

$$\begin{bmatrix} 0 & 0 & g & g1 \\ 0 & 0 & h & h1 \\ k & k & 0 & 0 \\ l & l & 0 & 0 \end{bmatrix} \rightarrow\rightarrow \begin{bmatrix} 0 & (g+g1+h+h1)/2 \\ k+l & 0 \end{bmatrix}$$

If $g = g1$ and $h = h1$ then we have a lumpable system as discussed above. Intuitively, the closer g is to $g1$ and h and is to $h1$, the more likely that a usable lumping is possible. We can formulate a mapping that becomes a homomorphism when the lumpability conditions apply – this is to take the average of the sums within the blocks, as shown in the matrix above on the left. Let's consider a simple special case:

$$\begin{bmatrix} 0 & 0 & g & e \\ 0 & 0 & -e & g \\ g/2 & g/2 & 0 & 0 \\ g/2 & g/2 & 0 & 0 \end{bmatrix} \rightarrow\rightarrow \begin{bmatrix} 0 & g \\ g & 0 \end{bmatrix}$$

Here the average of the upper right hand block is always g and the mapping is illustrated in Fig. 17.14. We expect that the lumped model is a better and better approximation to the base model as e goes to zero.

Table 17.3 shows simulation data that confirms our expectations – as e gets smaller so does the accumulated error at any particular time. More than this, other features of the table are in accordance with the error propagation concepts presented earlier.

To see this, consider the error propagation characteristics of the lumped model. Recall that we need to characterize how the lumped model treats the distances between pairs of states. Since the system is linear, we need only examine how the transition function treats distances relative to the origin. For an arbitrary state (x, y), the transformed state is:

$$\begin{bmatrix} 0 & g \\ g & 0 \end{bmatrix} \begin{bmatrix} x \\ y \end{bmatrix} = \begin{bmatrix} gy \\ gx \end{bmatrix}$$

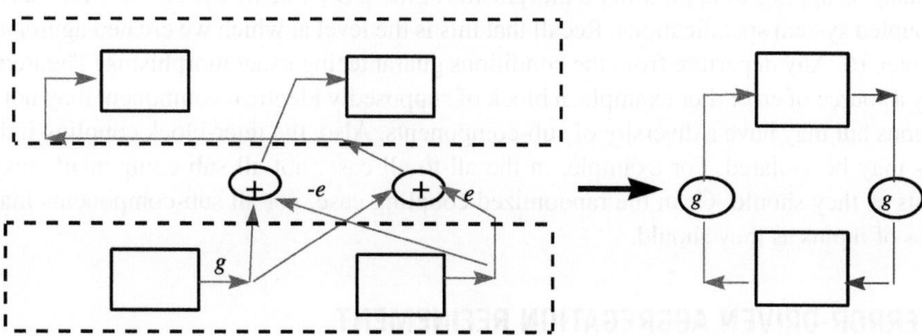

FIGURE 17.14

Approximate Morphism.

Table 17.3 Error in Approximate Morphism

g	e	error at time 5	error behavior
0.8	0.001	−0.002	error decreases over time
0.8	0.01	−0.02	decreases
0.8	0.1	−0.2	decreases
0.8	0.0	0.0	constant
1	0.1	0.4	error increases over time
1.2	0.1	7.0	increases explosively

and the ratio of the distance of point (gx, gy) to the original point (x, y), is:

$$\sqrt{\frac{g^2(x^2+y^2)}{x^2+y^2}} = g$$

Thus, according to the error propagation theory, g plays the role of attenuation or amplification factor depending on its magnitude relative to one. The transition function should cause unbounded error growth if $g > 1$ or neutral growth if $g = 1$. These predictions are borne out in the table above. We see that for the same induced error, the error grows much faster in the unbounded growth condition. For $g < 1$, the theory predicts that error may grow but will reach a limiting value. The table indicates that for the same induced error grows much slower than for the other cases. In fact the error actually starts to diminish with time. This is consistent with the theory that provides *an upper bound envelope* on error growth but is mum on whether it will be attained.

Although it involves more computation, the growth characteristics of any linear system are readily determined. Indeed, they correspond to the usual concepts of stability mentioned in Chapter 2.

17.5 APPROXIMATE MORPHISMS AT THE COUPLED SYSTEM LEVEL

We have seen how error introduced in the transition function of a system can manifest itself in its behavior. This analysis applies to approximate morphisms at the I/O Systems Level. Now let's examine the level of coupled system specifications. Recall that this is the level at which we created aggregated models in Chapter 16. Any departure from the conditions guaranteeing exact morphisms (Theorem 16.1) is potentially a source of error. For example, a block of supposedly identical component may not really be homogeneous but may have a diversity of sub-components. Also, the inter-block coupling indifference conditions may be violated. For example, in the all-to-all case, not all sub-components may get the same inputs as they should. Or in the randomized coupling case, not all sub-components may get the same ratios of inputs as they should.

17.5.1 ERROR-DRIVEN AGGREGATION REFINEMENT

The iterative procedure for creating an aggregation discussed in Chapter 14, can be better formulated by explicitly linking it to error reduction. As illustrated in Fig. 17.15, the procedure tries to reduce the error to below tolerance by reducing its possible sources (Table 17.4).

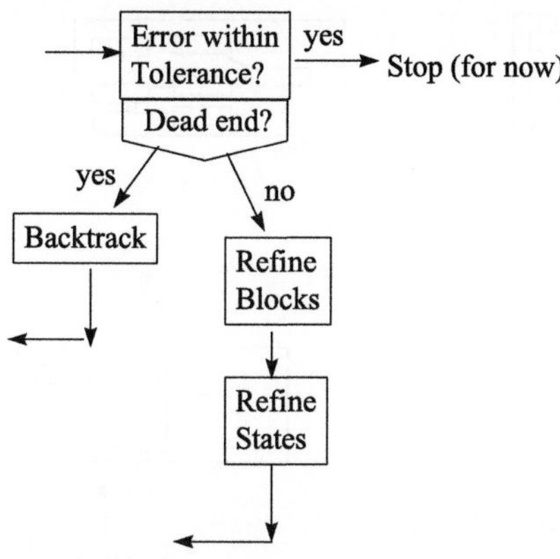

FIGURE 17.15

Aggregation search.

Table 17.4 Sources of Error and Corrective Actions in Aggregation Methods

Source of Error	Corrective Action
inhomogeneity – base model blocks are not homogeneous in structure. The bigger the structural diversity of components in a block, the greater error is likely to be.	Further refine blocks so to make them more homogeneous.
inter-block interaction skewing – preferential interactions between senders and receivers that violate the inter-block indifference requirement.	Further refine blocks and/or further refine the state description of the components (add more state variables) so that interactions occur indifferently between blocks.
output incompatibility – output desired by the experimental frame is too refined for the information preserved by the state count mappings of the base model blocks.	Further refine blocks and/or further refine the state description of the components so that the desired output can be computed from the block state censuses.

Sources of Error and Corrective Actions in Aggregation Methods So far the iterative procedure does not have any guidance as to which blocks or states to try to refine. In the next section we indicate how it might concentrate its efforts on components that are critical sources of error.

IDENTIFYING CRITICAL SOURCES OF ERROR

At the coupled model level, an approximate morphism is like an exact morphism except that it allows local mismatches between blocks of components in the base model and their combined representations in the lumped model. Fig. 17.16 illustrates how a global state transition in the base model may not map exactly to the corresponding global state transition in the lumped model. When we look at the divergence in the mapped base model successor state and the actual lumped model successor states we will see one or more lumped model components where the mismatch occurs. We can distinguish between two kinds of components: critical and non-critical relative to error propagation. A component is *error-critical* if an error in its state propagates in successive transitions to other components and ultimately to the outputs we are interested in through our experimental frame. Accordingly, components

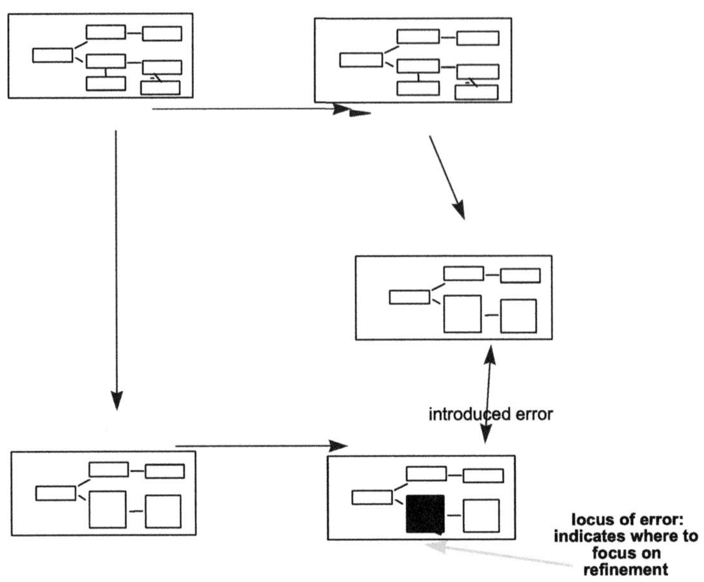

FIGURE 17.16

Localizing the Source of Error in a Coupled Model Approximate Morphism.

whose errors do not ultimately affect the output of interest are non-critical. It is only the critical set of components that we should be concentrating on.

How can we determine if a component is error-critical or not? Two conditions are required for error-criticality:

The component is a source of error growth – we have seen that a test for such growth is whether it has neutral or unbounded error growth characteristics.

There is a path in the influence diagram from the component to the experimental frame specified output – the component can influence the output only if such a path exists even if it is a source of error growth.

These conditions give us some heuristic guidance in constructing lumped models though aggregation. If the error associated with the aggregation is over the tolerance limit, we should isolate the components in the lumped model which are error-critical. Then we should re-examine the base model block that mapped to an error-critical component. If it is not homogeneous, we try splitting it into two or more blocks that are (more) homogeneous. If its coupling indifference conditions are not fully satisfied then may try to refine it and/or its influencer blocks. In general, the conditions offer some guidance in isolating where to concentrate refinements of the current aggregation.

EFFECT OF ERROR ACCUMULATION

The situation with respect to error in aggregation can be viewed from the metaphor of the half empty/half full cup. All modeling can justifiably be viewed as constructing lumped models. Therefore all models are error prone. While much research has been emphasizing the positive – that reasonable conditions justifying abstractions exist in reality (e.g., Cale, 1995) some recent work is more pessimistic (Axtell, 1992). Questions can be raised such as: what if most aggregations generate errors that grow with time? Can't such aggregated models still be useful? How do we actually use models to compensate for their error accumulation? As indicated above, the approach to approximate morphism construction presented here provides some initial answers to such questions. The theory indicates that the dynamics of the lumped model determine how the error is transmitted from one transition to the next, whether it is magnified or attenuated. Thus, we may analyze the lumped model itself to assess its own credibility. One caveat – this is true only provided that the lumped model is close enough so that the bounds on error growth obtained from it are themselves applicable.

Also, the existence of certain lumped features may mitigate uncontrolled error accumulation. For example, error is controlled if the model periodically undergoes a memory erasing reset (all states synchronize to some particular state). Chapter 12 discussed a cascade decomposition of system into a synchronizable front component and a permutation back component. Since it is a permutation system, the back component represents the memory that cannot be erased. So the smaller it is, the more synchronizable the model is. Synchronizable models might be called self-correcting (Westerdale, 1988) but the degree to which this is true will depend on the length of the synchronizer input (relative to the duration of the simulation experiment) and the frequency with which a synchronizer input segment occurs.

17.6 VALIDATION AT STRUCTURAL LEVELS

Recall that there are validity concepts that correspond to the levels of the specification hierarchy: *replicative*, *predictive* and *structural validity*, respectively. The basic mode of validation is at the replicative level since it involves comparison of observable data sets. The problem then is how to validate a model at the predictive and structure levels. Since we are trying to gain additional knowledge beyond that at the data system level, this is the problem of *climbing up the specification hierarchy*. In the case of predictive validation, the issue is whether the proper initial state of a model can be inferred from past system observations. In the case of structural validity, the additional issue is whether the structure of a model can be inferred to be the only one consistent with the observed data. Conditions known that justify such inferences are discussed in Chapter 16 of TMS76. Unfortunately, this is an area where there is a dearth of appreciation for the problem and of computerized tools that might help determine whether the justifying conditions hold. So the issue is ignored at the cost of overconfidence in model prediction. Roughly put, the problem is that if our current model has not been shown to be structurally valid, the real system may in fact be represented by any one of the other model structures consistent with the observed data. Thus, further experimentation may well uncover a significant departure between our model and the real system.

17.6.1 CALIBRATION, PARAMETER IDENTIFICATION, SENSITIVITY

Most models have free parameters whose proper values are unknown. *Calibration* is the process by which parameter values are adjusted so that the model best fits the system data. Unfortunately, as just seen, a model may be replicatively valid but not predictively valid. That is, just because it can be made to fit past data does not guarantee that it will fit future data (i.e., that has not been employed in the calibration process). Indeed, parameters can be viewed as state variables and calibration as state identification. The problem then reduces to that of climbing up the specification hierarchy, as just mentioned. In other words, there is no guarantee that parameter assignments that result in best fits actually lead to credible predictions.

More confidence in the uniqueness of predictions can be obtained by assessing their sensitivity to various perturbations. One approach, that is sometimes feasible, is to try to locate all, or at least many of, parameter assignments that result in good fit. One can then assess how much model predictions vary over the range of these parameter assignments.

Another possibility is to assess the sensitivity of model predictions to variations in the assumptions made in its construction. One particular form of this approach is to assess the impact of adding greater detail on the predictions. This can be viewed as investigating the variation of predictions over a family of higher resolution models that each map to the original lower resolution model.

A related approach is to systematically formulate known sources of error in such a way that the error actually obtained can be used to pin-point the source (Yuan et al., 1998).

17.6.2 LOCAL/GLOBAL, CROSS-MODEL VALIDATION

Relations between high and low resolution models leads us to distinguish between local (applying to the model in focus) and *global* (applying to existing model base) validation. The latter refers to *cross-model* validation, e.g., between models at different levels of resolution. Global cross-validation places greater constraints on model calibration since the parameter values must be consistent with related ones in other models. Parameter morphisms (Chapter 14) facilitate inter-model parameter correlations by providing mappings from base model parameters to lumped model counterparts. For example, suppose that a high resolution model has been validated and there is a high degree of confidence in its parameter values. Then the parameter values in a low resolution model that should be checked for consistency with these values. Indeed, if there is a parameter morphism running from the high-resolution the low-resolution model, the parameters of the latter ("low-res") are computed directly from those of the former ("high-res").

Moreover, the direction of knowledge propagation does not necessarily always go from high to low. For example, if data exists to calibrate a low resolution model, then its parameters should constrain related high resolution models. Indeed in this case, the parameters of the "low-res" model constrain, but do not usually uniquely determine, those of the former "high-res" model.

17.7 HANDLING TIME GRANULARITY TOGETHER WITH ABSTRACTION

As discussed in Chapter 16, an integrated family of variable resolution models organizes models that are constructed in ways that support cross-calibration and validation. However, while we have considered abstraction and resolution the underlying relationships among models, we have not considered how to include time granularity in such relationships. Santucci et al. (2016) proposed an extension of the System Entity Structure that introduced new elements to enable it to support the concepts of abstraction and time granularity in addition to its existing expression of composition and component specialization. The concept they implemented that underlies time granularity relations is illustrated in

Fig. 17.17A where base and lumped models are related by appropriate encoding and decoding maps (Chapter 15). For example a base model of a watershed, as discussed earlier in this chapter, might accept temperature sampled at an hourly rate while a lumped model receives temperature at a daily rate. The input encoding g, from daily to hourly input samples must dis-aggregate a single temperature value every day into a stream of 24 temperature values for a single day. The output decoding k, does the inverse in which every 24 successive values are aggregated to a single corresponding value. Here the base model is operating at a relatively fine level of granularity (hours) while the corresponding lumped model operates at coarser level of granularity.

Fig. 17.17B suggests how the encoding and decoding maps can be implemented by DEVS atomic models called Upward Atomic Model (UAM) and a Downward Atomic Model (DAM) respectively (Santucci et al., 2016).

Exercise 17.5. Write DEVS models that implement the behaviors required by the encoding and decoding maps. Assume that the disaggregation produces a stream of 24 equal values and the aggregation produces the average the last 24 inputs.

Exercise 17.6. Define a morphism pair in which the direction of the input encoding goes from base to lumped model. Compare this with the discussion in Chapter 15 on I/O morphisms.

Exercise 17.7. Give formal definitions for the encoding/decoding maps of the previous exercises.

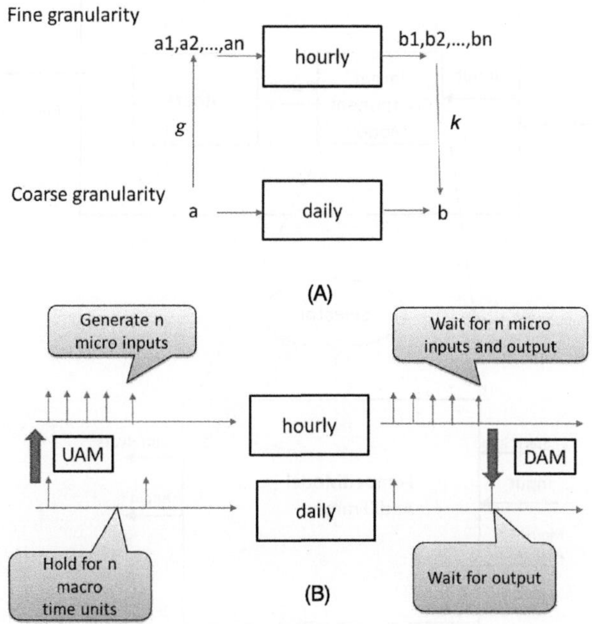

FIGURE 17.17

Illustrating Course/Fine granularity relations.

17.8 MULTI-FIDELITY MODELING AND SIMULATION METHODOLOGY

Another variation on multi-resolution model families is multi-fidelity methodology which is organized around the concept of multi-fidelity models (Choi et al., 2014). The primary goal is to replace a selected component model of a base model by a faster running – possibly less accurate – model to achieve faster execution with acceptable accuracy trade-off. The secondary goal is minimize modification of existing models and the underlying simulator. The multi-fidelity methodology first targets a model component that evidences the possibility of achieving the greatest reduction of execution time which it determines as the one having the greatest contribution to the existing execution time. For such a target, it then selects one or more state variables and a subset of the space spanned by these variables to govern replacement of the target model by a faster "equivalent" model. We formulate this approach as one in which an experimental frame is designed to control the switching between the base component model and the lumped component model. Fig. 17.18 illustrates the situation in which there is a target base component model and an experimental frame in which a lumped model provides a comparable output that allows it to replace the target in the base model under conditions specified by the frame.

As formulated earlier, the experimental frame concept includes control variables and conditions that determined the regime of validity of the lumped model. Fig. 17.19 illustrates a conceptual implementation of the lumped model and its experimental frame in the multi-fidelity context. The frame includes the *acceptor* component which makes the trajectories of the control variables available for testing the

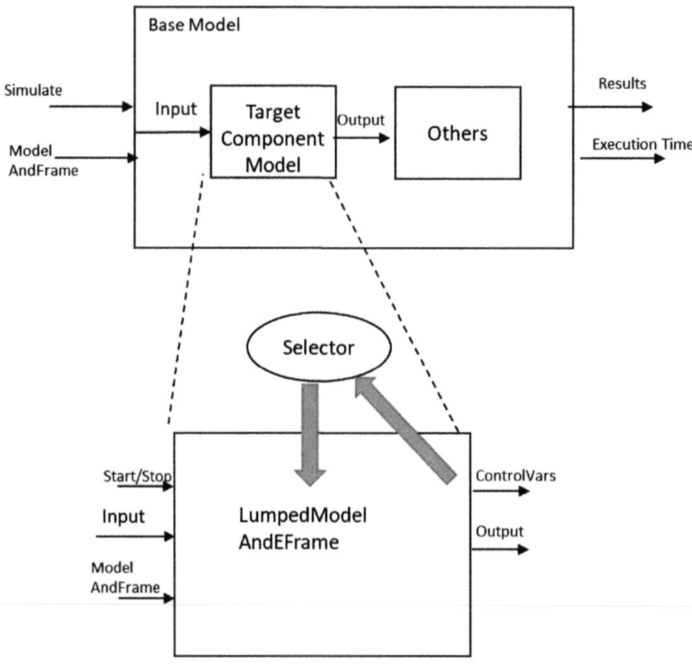

FIGURE 17.18

Target component model replaced by lumped model under specified conditions.

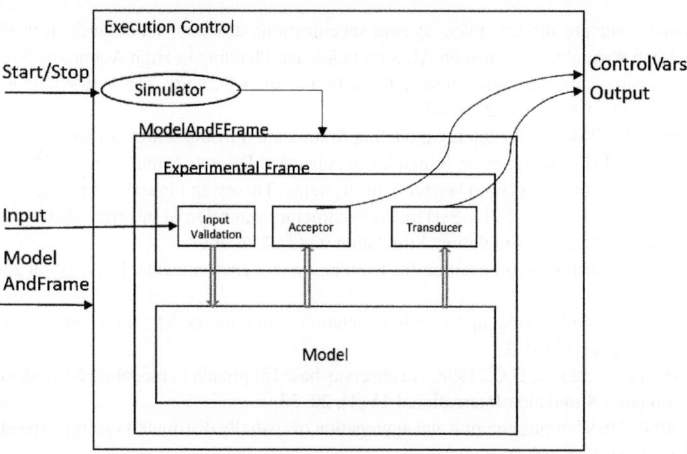

FIGURE 17.19

Execution implementation of multi-fidelity model-frame pair.

validity condition. In this context, this condition determines when the lumped model is considered sufficiently accurate in terms of its input/output relation to replace the target base component model. As shown in Fig. 17.18 a selector added to the original base model makes decisions to switch between low and high fidelity models based on the validity condition applied to the run control variables supplied to it by the frame.

17.9 SUMMARY

This chapter dealt with verification – how to check if a simulator is in error, and validation – how to check if a model is in error. The standard for error-free relations between entities in the framework for M&S is the morphism. Approximate morphisms allow us to deal with more relaxed relations that are close to, but not one hundred percent, perfect morphisms. Morphisms support multi-resolution modeling and multi-fidelity simulation. Yilmaz and Oren (2004) developed a taxonomy of multi-models, a generic agent-based architecture, and single aspect multi-models with exploratory (goal-directed) behavior. This taxonomy offers a framework in which to consider such multi-fidelity methodology.

The companion volume to this edition on model engineering contains a discussion of simulation-based execution of the model-frame pairs and their control of dynamic structure, as required in this context (Zhang et al., 2018).

REFERENCES

Axtell, R.L., 1992. Theory of Model Aggregation for Dynamical Systems with Application to Problems of Global Change. Carnegie Mellon University, Pittsburg, PA.

Balci, O., 1997. Principles of simulation model validation, verification, and testing. Transactions of SCS 14 (1), 3–12.

Barros, F., 1996. Dynamic structure discrete event system specification: structural inheritance in the delta environment. In: Proceedings of the Sixth Annual Conference on AI, Simulation and Planning in High Autonomy Systems, pp. 141–147.

Cale, W.G., 1995. Model aggregation: ecological perspectives. In: Complex Ecology. The Part-Whole Relation in Ecosystems. Prentice Hall, Englewood Cliffs, NJ, pp. 230–241.

Choi, S.H., Lee, S.J., Kim, T.G., 2014. Multi-fidelity modeling & simulation methodology for simulation speed up. In: Proceedings of the 2nd ACM SIGSIM Conference on Principles of Advanced Discrete Simulation. ACM.

Cin, M.D., 1998. Modeling fault-tolerant system behavior. In: Systems: Theory and Practice. Springer, Vienna.

Santucci, J.-F., Capocchi, L., Zeigler, B.P., 2016. System entity structure extension to integrate abstraction hierarchies and time granularity into DEVS modeling and simulation. Simulation 92 (8), 747–769.

Westerdale, T.H., 1988. An automaton decomposition for learning system environments. Information and Computation 77 (3), 179–191.

Yilmaz, L., Oren, T.I., 2004. Dynamic model updating in simulation with multimodels: a taxonomy and a generic agent-based architecture. Simulation Series 36 (4), 3.

Yuan, Z., Vangheluwe, H., Vansteenkiste, G.C., 1998. An observer-based approach to modeling error identification. Transactions of the Society for Computer Simulation International 15 (1), 20–33.

Zeigler, B.P., Moon, Y., 1996. DEVS representation and aggregation of spatially distributed systems: speed versus error tradeoffs. Transactions of SCS 13 (5), 179–190.

Zhang, L., et al., 2018. Model Engineering for Simulation. Elsevier.

CHAPTER

DEVS AND DEVS-LIKE SYSTEMS: UNIVERSALITY AND UNIQUENESS

18

CONTENTS

The DEVS Bus in Chapter 8 provides a means to inter-operate models developed within traditional discrete event world views. But it would be even more reassuring to know that our DEVS Bus would be able to accommodate new modeling formalisms as they emerge. We would like to know how universal the DEVS representation is with respect to the class of all possible discrete event systems. We would also like to know how unique the DEVS formalism is – are there other discrete event formalisms that would support equivalent Bus-like inter-operability? This chapter deals with these issues. Then as a start toward extending the DEVS Bus to a wider class of systems, we will show that the DEVS formalism includes the DTSS (discrete time system specification) formalism in very strong sense. We continue extending the DEVS Bus concept to DESS (differential equation system specification) models in the next chapter.

Theory of Modeling and Simulation. https://doi.org/10.1016/B978-0-12-813370-5.00028-6

Global vs. Local (Component-Wise) Representation

Before proceeding, we need to clarify our concept of representation of one formalism by another. Each of the formalisms DEVS, DTSS, and DESS define a subset of systems. Can the subset defined by DEVS "represent" the other two? The answer depends on the strength of the equivalence, or morphism, that we are interested in. Looking at Fig. 18.1, we say that formalism F can represent a formalism F'' with morphism type M, if for every system specified by F'' there is a system specified by F that simulates it in the sense that there is a morphism of type M that holds between them. For example, we say that the DEVS formalism can represent a formalism at the coupled system level if for every coupled model specifiable in the formalism, there is a DEVS coupled model which simulates it with a coupled system morphism. We'll call this strong, local or component-wise simulation, since we don't have to preprocess the model to get a DEVS simulation. Instead, we convert the components of the model individually to a DEVS representation and then run the resulting DEVS coupled model to generate the original behavior. Contrast this with weak or global representation in which we only know how to create a morphism given the model expressed in the basic formalism but not as a coupled system. In the latter case, any coupled model has to be first converted into its resultant form before being translated to a DEVS equivalent. This is not practical for complex coupled models.

18.1 RELATION BETWEEN CLASSICAL AND PARALLEL DEVS: ARE THERE ONE DEVS OR TWO?

Recall that we have introduced two apparently distinct DEVS formalisms – Classic DEVS and Parallel DEVS. Thus the first question to ask is: are there two DEVS or one? Or indeed many. Is Parallel DEVS represented by Classic DEVS at the coupled systems level? The answer is not known at this writing. However, the answer to the global simulation counterpart can be shown to be "yes". Thus, in principle, Parallel DEVS does not extend the class of systems that can be specified by classic DEVS. This is expressed in

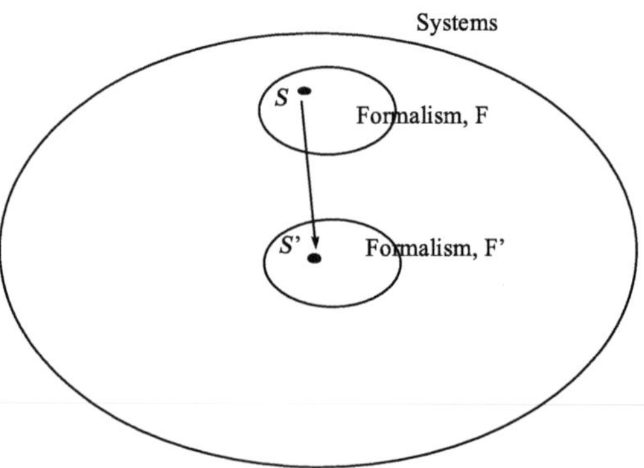

FIGURE 18.1

Formalism Representation.

Theorem 18.1 (Classic DEVS and Parallel DEVS specify the same class of systems). *Given a Parallel DEVS* $MP = (X, Y, S, \delta_{\text{ext}}, \delta_{\text{int}}, \delta_{\text{con}}, \lambda, \text{ta})$, *there is a classic* $\text{DEVS}_{\text{MC}} = (X_{MC}, Y_{MC}, S_{MC}, \delta_{\text{ext},MC}, \delta_{\text{int},MC}, \lambda_{MC}, \text{ta}_{MC})$ *that weakly represents it, and conversely.*

Proof. **(Classic DEVS includes Parallel DEVS)** As suggested just now, we take the input set of the classic DEVS to be set of bags over X, i.e., $X_{MC} = X^b$. Similarly, $Y_{MC} = Y^b$. As a first approach the other components would be transferred over without change: $S_{MC} = S$, $\delta_{\text{ext},MC} = \delta_{\text{ext}}$, etc. However, in the construction of the system, the classic DEVS employs the serialized composition in line 1) while δ_{con} controls tie-breaking in Parallel DEVS. We must show that classic DEVS can achieve the same effect using serialization. To do this, we arrange to let the classic DEVS save the state of the parallel DEVS and apply δ_{con} to this saved state when needed. Thus we set $S_{MC} = S \times S$, where the second component is employed to save the state after an internal transition, i.e., $\delta_{\text{int},MC}(s, s') = (\delta_{\text{int}}(s), s)$. We use the copy component only when a collision occurs, i.e., $\delta_{\text{ext},MC}((s, s'), 0, x) = (\delta_{\text{con}}(s', x), s)$. In mapping MC to the system it specifies, the serial interpretation of collision simulates the parallel interpretation, i.e.,

$$\begin{aligned}\delta_{S,MC}((s, s'), e, x) &= \delta_{\text{ext},MC}(\delta_{\text{int},MC}(s, s'), 0, x)\\ &= \delta_{\text{ext},MC}((\delta_{\text{int}}(s), s), 0, x)\\ &= (\delta_{\text{con}}(s, x), s)\end{aligned}$$ □

(Note that the classic $DEVS_{MC}$ can tell when a collision has occurred because of the zero value for elapsed time.) Thus the state and output of the system specified by the classic DEVS follow those of the Parallel DEVS.

(Parallel DEVS includes Classic DEVS) Given a classic DEVS, we can easily embed it into a Parallel DEVS with the same structure by restricting the inputs of the latter to just bags of size one.

Exercise 18.1. Define a mapping from classic DEVS to Parallel DEVS by projecting pairs of states (s, s') to the state of the left element s. Show that this mapping is indeed a homomorphism.

The theorem shows that the basic formalisms are equivalent in the sense that any basic DEVS in one can be represented by a basic DEVS in the other. Also, by closure under coupling, any coupled model in one can also be represented in the basic formalism of the other. But it is an open question whether or no there is a constructive way to simulate a parallel DEVS coupled model in a classical DEVS coupled model, and conversely.

Exercise 18.2. Investigate the representation of classic DEVS by parallel DEVS as the coupled system level, and the converse as well. This question has some interesting issues in it. For example, can parallel DEVS (which is decentralized) realize the global (centralized) Select function of classic DEVS? Conversely, can classic DEVS realize the parallelism and confluence of parallel DEVS (allowing it several steps to do this)? In each case, if the answer is yes, prove it? if not, provide a counterexample.

Since the class and parallel DEVS are equivalent in terms of the class of systems they specify, which one we employ depends on other considerations. In the following, we will find that parallel DEVS affords convenient properties since it easily handles the simultaneous component transitions inherent in discrete time simulation.

18.2 UNIVERSALITY AND UNIQUENESS OF DEVS

18.2.1 SYSTEMS WITH DEVS INTERFACES

A system with DEVS input and output segments is illustrated in Fig. 18.2. It is not necessarily a DEVS, although it does receive and generate DEVS segments. Recall (Chapter 5) that a DEVS segment consists of an interval of null events punctuated by a finite number of non-null events. The basic generators for such segments are of the form $x_{t>}$ as illustrated in the figure, where x is either an input in X or the null event, ϕ.

We call a system with DEVS interfaces *DEVS-like* for short. We will consider such systems to be class of DEDS (Discrete Event Dynamic Systems) identified by Ho (1989) as containing a variety of discrete event system formalisms, although no subsuming class was formally defined.

Since we want to be sure of taking all considerations into account in characterizing DEVS-like systems, we will start at the lowest level of the system specification hierarchy and work upwards (Fig. 18.3). At the lowest level we have pairs of input/output discrete event segments, the IORO of Chapter 5. Our basic problem is to characterize the I/O systems that realize such segments and see how closely they relate to the class of DEVS systems. We know that DEVS systems have such IOROs associated with them, but we don't know if there are other non-DEVS classes that also do so. In other words, what we would like to know *whether the internal mechanisms of systems with DEVS interfaces are all essentially described by DEVS models.*

First we ask: under what conditions do pairs of discrete event segments actually come from an I/O System? The answer is found in Chapter 12 which characterized realizable IOROs. Essentially the answer is that such pairs must be the union of functions (mapping the input segment sets to the output segment sets) that are realizable, meaning that they can be I/O functions associated with the states of an I/O System. Several properties were shown to be necessary and sufficient for an I/O function to be realizable, the most critical of which was causality – this allows the function to only use past input to determine past output (future input cannot influence past output). The collection of such functions constitutes the IOFO level of specification. For an IOFO to be realizable it has to closed under derivatives. One way to understand this is that if a function, f is associated with a state then each of the states it can access has a function which is derivative of f. Now if we take the functions of an IOFO as labels for states and follow their derivatives, we get a way of constructing an I/O system that realizes the IOFO.

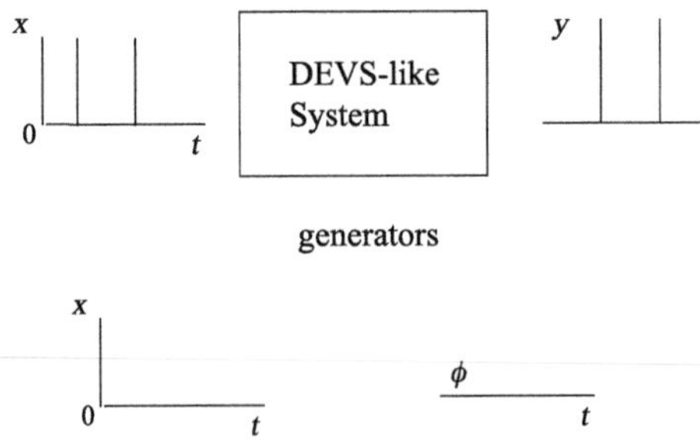

FIGURE 18.2

DEVS-like Systems.

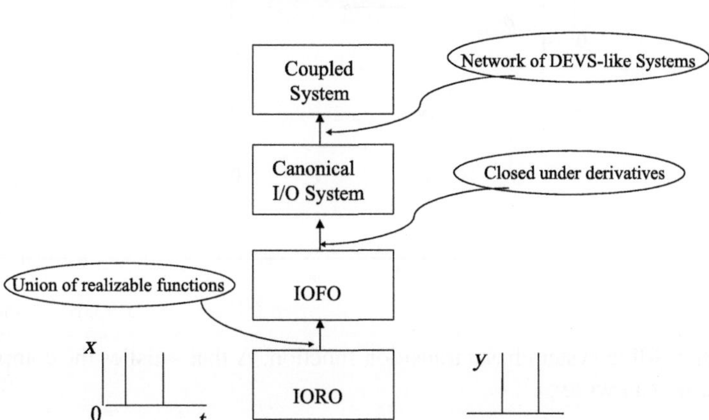

FIGURE 18.3

Hierarchy of DEVS-like System Specifications.

In fact, by the Theorem 15.17 in Chapter 15, such a system is the canonical realization of the IORO – it is reduced, minimal and unique up to isomorphism among all systems that realize the IOFO.

In the rest of this section, we'll assume that we are given as behavior, a realizable I/O function with discrete event input and output segments. We know there is a canonical I/O System that realizes this behavior. What we don't know is whether this canonical realization is in any way related to a DEVS.

Continuing up the system hierarchy we can formulate a way of coupling DEVS-like systems together. Indeed, the coupling rules for DEVS-like systems are identical to those for DEVS. We can also formulate both classic and parallel compositions of DEVS-like systems. We would then like to show that the class of DEVS-like systems is closed under such coupling. Once we have established that the internal structure of such DEVS-like systems is isomorphic to a DEVS, closure will naturally follow as will the fact that all such coupled systems are strongly simulated by DEVS coupled models.

18.2.2 BEHAVIOR OF DEVS-LIKE SYSTEMS

We'll assume that the input segment sets, DEVS(X) have segments with zero length, which become generators, $\{x_0 | x_0 \in X\}$. The other generators are then the non-event segments $\{\phi_{t>}\}$. The composition operation is illustrated in Fig. 18.4. Clearly, $x_{t>} = x_0\phi_{t>}$. For $x_0 y_0$ we represent the result as the generic $(x + y)_0$. In classic DEVS we would set $x + y = x$ to disallow more than one simultaneous external events. In Parallel DEVS, the input set, bags over X, allows the natural interpretation of $+$ as the bag union operation. We obtain a well defined concept of maximal length segmentation by selecting the non-null events and the intervening non-event segments as generator elements.

Exercise 18.3. Show that this definition of mls decomposition satisfies the requirements in Chapter 5.

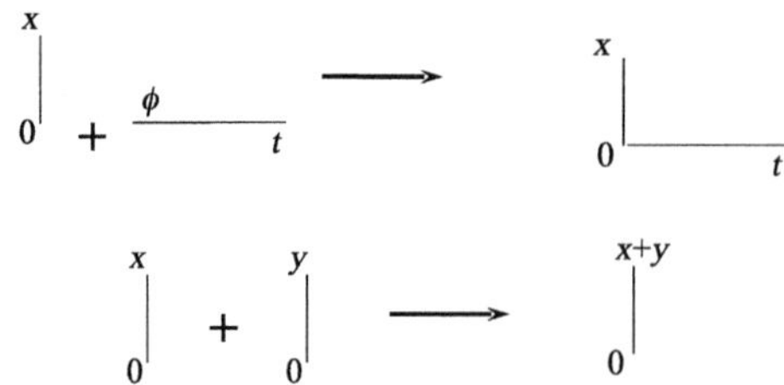

FIGURE 18.4

Composition of Generators.

A realizable DEVS-like system has a transition function, Δ that satisfies the composition property (Chapter 5). Thus, we can write

$$\Delta(q, x_{t>}) = \Delta(q, x_0\phi_{t>}) = \Delta(\Delta(q, x_0), \phi_{t>})$$

Thus there is an instantaneous mapping $\Delta_0 : Q \times X \to Q$, such that $\Delta_0(q, x) = \Delta(q, x_0)$. Then

$$\Delta(q, x_{t>}) = \Delta(\Delta_0(q, x), \phi_{t>}).$$

This shows that an external event has an *instantaneous effect* on the state of the system. We call Δ_0 the *instantaneous input transition function*. In contrast, the effect of an input in a DESS cannot be exerted instantaneously in its state.

Exercise 18.4. Apply a constant input segment to a DESS. Show that as the length of the segment approaches zero, the state transition function approaches the identity. In this sense, the effect of an input does not have instantaneous effect.

Exercise 18.5. Identify the instantaneous input transition function in a system specified by a DEVS.

18.2.3 UNIVERSALITY OF DEVS

We are now ready to show that the behaviors of DEVS-like systems are describable by DEVS.

Theorem 18.2 (Universality Theorem for Systems with DEVS Interfaces). *Every DEVS-like system is a homomorphic image of a DEVS I/O System.*

Proof. Let $S = (T, X, \Omega, Y, Q, \Delta, \Lambda)$ be a DEVS-like system. We will construct a $DEVS, M = (X, Y, S, \delta_{\text{ext}}, \delta_{\text{int}}, \lambda, \text{ta})$ to homomorphically simulate it. We set the DEVS sequential state set, $S = Q$, the state set of the DEVS-like system. The rest of the construction follows:

- The *time advance* from a state q is the time to the next non-null output in S:

$$\text{ta}(q) = min(t|\Lambda(\Delta(q, \phi_{t>})) \neq \phi)$$

- The output of the DEVS at the next internal event is the corresponding system output

$$\lambda(q) = \Lambda(\Delta(q, \phi_{\text{ta}(q)>}))$$

- The internal transition is to the state of the system at the next output:

$$\delta_{\text{int}}(q) = \Delta(q, \phi_{\text{ta}(q)>})$$

- If there is an external input after an elapsed time, e, the DEVS will immediately update its state to the corresponding system state and then take account of the input by applying the instantaneous effect function:

$$\delta_{\text{ext}}(q, e, x) = \Delta_0(\Delta(q, \phi_{e>}), x)$$

The total state set of the DEVS is $Q_{DEVS} = \{(q, e) | q \in Q, 0 \leq e < \text{ta}(q)\}$. We define a mapping $h : Q_{DEVS} \to Q$, by $h(q, e) = \Delta(q, \phi_{e>})$. In other words, when the DEVS system is in total state (q, e), this means it has been resting in state q while the simulated system has been advancing under the null-event input for an elapsed time e. Thus, the state corresponding to (q, e) is indeed $\Delta(q, \phi_{e>})$. □

Exercise 18.6. Show that h is a homomorphism from the system specified by the DEVS onto the DEVS-like system.

The problem now is that h is a homomorphism not necessarily the isomorphism we will need for DEVS uniqueness. To be an isomorphism, h has to be onto and one-one. It is certainly onto but to be one-one requires that for any pair of distinct states (q, e) and (q', e') in Q_{DEVS} the images $h(q, e)$ and $h(q', e')$ must also be distinct. This will not happen unless we define the DEVS appropriately. Let's look at an example to see why.

18.2.4 EXAMPLE: DEVS REALIZATION OF DEVS-LIKE SYSTEM

Fig. 18.5A depicts the accessibility diagram of a DEVS-like system. We'll define this diagram formally in a moment, but for now take it as the state transition diagram of the system with input restricted to null-segments. Consider now how the homomorphism assigns DEVS states to map to the system states. Since state a is an initial state, we represent it in the DEVS by (a, 0). Since d is an output state we need to represent it in the DEVS by a state (d, 0). Now all states with null output along the trajectory between a and d (such as c) are represented by (a, e) which maps to $\Delta(a, \phi_{e>})$ where e ranges between 0 and 2.

However, suppose there is a convergence of states as shown in Fig. 18.5B, where after 1 unit of time, the trajectories from distinct states a and b, merge at c. Then state c is represented by both (a, 1) and (b, 1) in the DEVS, i.e., $\Delta(a, \phi_{1>}) = \Delta(b, \phi_{1>}) = c$. This violates the one-one requirement for isomorphism. To maintain bijectivity (one-oneness) we need to transition to (c, 0) in the DEVS. Thus, the time advance from (a, 0) is 1 (rather than 2, as before) and the internal transition to (c, 0) occurs at the end of this time (if no external events occur). Similarly, the time advance for (b, 0) is also 1. Thus the states transiting out of c are represented only in one way, namely, by (c, e) and we have removed the redundant representation of these states in the DEVS.

18.2.5 UNIQUENESS OF DEVS

The example illustrates the main issue in constructing an isomorphic DEVS realization of a DEVS-like System. To guarantee that the homomorphism mapping is one-one we have to use a sequential state set, S that is a subset of Q, and such that so that every state q has a unique representation in the form

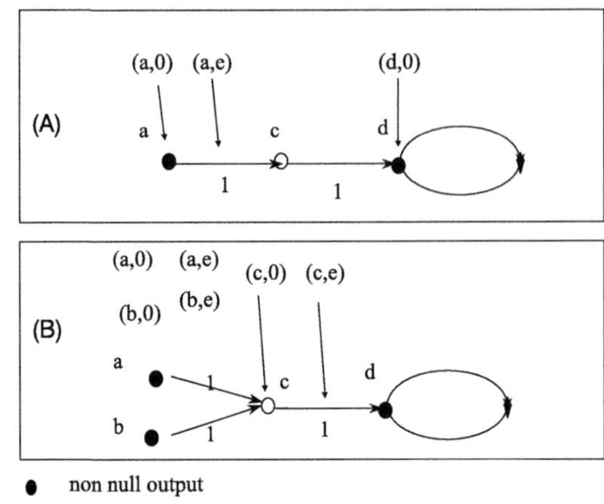

FIGURE 18.5

DEVS-like System Example.

of (s, e). In Fig. 18.5B, $S = \{a, b, c, d\}$. In constructing the simulating DEVS we have to make sure that the DEVS has internal transitions to intersection states in addition to the output generating states already included. The revised definitions will look like the following:

- The *time advance* is the time to the next non-null output or intersection state:

$$\text{ta}(s) = min(t|\Lambda(\Delta(s, \phi_{t>})) \neq \phi \text{ or } \Delta(s, \phi_{t>})) \text{ is an intersection state}\}$$

  For example $\text{ta}(a) = 1$
- The internal transition is to the state of the system at the next internal event:

$$\delta_{\text{int}}(s) = \Delta(q, \phi_{ta(s)>})$$

  For example $\delta_{\text{int}}(a) = c$
- The output of the DEVS at the next internal event is the corresponding system output

$$\lambda(s) = \Lambda(\Delta(s, \phi_{ta(s)>}))$$

  For example $\lambda(a) = \phi$ since c is an intersection state but not an output, state.

We see that in accounting for the intersections we have added a set of *internal events* that are imposed by the underlying transition structure of DEVS-like system. At such an internal event, the DEVS resets its sequential state to the intersection state and the elapsed time clock of the DEVS system is also reset to zero. This guarantees that states following the intersection have a unique representation in the DEVS as required for isomorphism.

- If there is an external input after an elapsed time, e, the DEVS will immediately update its state to the corresponding system state and then take account of the input by applying the instantaneous effect function. However, the resulting state now has to be represented as one of the allowable pairs (s, e). Since the h mapping is now one-one, it has an inverse, we can map any system state into an (s, e) pair using this inverse:

$$\delta_{\text{ext}}(s, e, x') = h^{-1}(\Delta_0(\Delta(s, \phi_{e>}), x))$$

Notice however, that we have to modify the DEVS external transition to set the (s, e) pair since the current function would require us always to set $e = 0$. The modification changes only the external transition function of a DEVS $= (X, Y, S, \delta_{\text{ext}}, \delta_{\text{int}}, \lambda, \text{ta})$. We extend the range of the function so that $\delta_{\text{ext}} : Q \times X \to Q$. In other words, $\delta_{\text{ext}}(s, e, x) = (s', e')$ for s' in S and e' in $[0, \text{ta}(s')]$. The system specified by a modified DEVS is the same as the before the modification except that in response to an external input it resets the total state to the (s, e) pair.

This allows us to state:

Lemma 18.1. *Any reduced DEVS-like system has a DEVS realization that is isomorphic to it.*

The actual proof of the lemma involves taking account of other possibilities where redundant state representation may exist. We rigorously analyze the transition structure of any DEVS-like system and show how to define a set of internal events (including the intersections) that enable us to construct the required isomorphic DEVS realization. The proof is given in the Appendix.

Putting together the Lemma with the earlier discussion we state that the *internal mechanism of any DEVS-like system is in fact exactly described by a modified DEVS:*

Theorem 18.3 (Uniqueness Theorem for Systems with DEVS Interfaces). *Every DEVS-like system has a canonical realization by a DEVS I/O System.*

Proof. The reduced version of the given system always exists and is its canonical realization (Chapter 12). The Lemma shows that this realization is isomorphic to a DEVS. □

18.3 DEVS REPRESENTATION OF DTSS

Our first task in the implementing the DEVS Bus is to provide for DEVS representation of DTSS models. Since DESS models are traditionally simulated in discrete time, this offers an immediate path to representation of DESS models as well. However, we will subsequently look for direct and more efficient means of DEVS representation of DESS models.

It turns out that DEVS can strongly represent the DTSS formalism. In other words, we will show how coupled discrete time systems can be mapped component-wise and simulated in a DEVS environment. Our approach to this demonstration, is to first develop DEVS representatives for each of the component types in DTSS coupled models. Then we show how the DEVS representatives are coupled in a manner preserving their DTSS counterparts to execute the DTSS simulation. Recall from Chapter 7, that there are three types of DTSS components to consider:

- Input Free Moore DTSS – these drive the simulation forward as the ultimate generators of the inputs in a closed coupled model.
- Multi-Ported Memoryless FNSS – these collect outputs from Moore components and transform them into inputs without using any state information.

- Moore DTSS with Input – the outputs are generated for the next cycle based on the current input.
- Mealy DTSS – these include memoryless FNSS. In a well-defined network they form a *directed acyclic graph* (*DAG*) of computations, taking zero time and propagating from Moore outputs back to Moore inputs.

We examine each in turn.

18.3.1 INPUT-FREE MOORE

DEVS representation of an input-free Moore system is straight-forward, since the input-free acts as an un-interruptible generator with a fixed generation period (Chapter 5). Thus, given a $DTSSM = (Y_M, S_M, \delta_M, \lambda_M, h)$ define

$$DEVSh = (Y^+, S, \delta_{\text{int}}, \lambda, \text{ta})$$

where

$$\begin{aligned}
Y &= Y_M \\
S &= S_M \times R^+_\infty \\
\delta_{\text{int}}(s, \sigma) &= (\delta_M(s), h) \\
\lambda(s, \sigma) &= \lambda_M(s) \\
ta(s, \sigma) &= \sigma
\end{aligned}$$

There is a reason we retain the sigma variable even though it is always sets to the predictable step time, h. This relates to the fact that initialization of coupled model DEVS simulations requires setting σ initially to zero (more of this in a moment).

18.3.2 MULTI-PORTED FNSS

DEVS simulation of a memoryless function with a single input is straightforward. When receiving the input value, the model goes into a transitory state and outputs the computed output value. However, if there are multiple input ports, the system must wait for all inputs to be received before going into the output mode. Since inputs may be generated at different points in the processing, we cannot assume that all inputs come in the same message. Thus we formulate a FNSS (Chapter 7) with multiple input ports and define a DEVS to simulate it:

$$\text{Given } FNSSM = (X_M, Y_M, \lambda_M)$$

where

$IPorts = \{in_1, in_2, ..., in_n\}$,

$X_M = \{(p, v) | p \in Iports, v \in V\}$ is the set of input ports and values (for simplicity, we assume all ports accept the same value set),

$Y_M = \{(out, v) | v \in V_{out}\}$ is the single output port and its values, define

$$DEVS = (X^+, Y^+, S, \delta_{\text{ext}}, \delta_{\text{int}}, \delta_{\text{con}}, \lambda, \text{ta})$$

where

$$X = X_M$$
$$Y = Y_M$$
$$S = \{f | f : IPorts \rightarrow V^\phi\} \times R^+_\infty$$

Since inputs need not all arrive together, we must keep track of their arrival status so that we can tell when all have been received. This is the purpose of the function f in the state of the DEVS. Each arriving input is cached (guarding against out-of phase arrivals). The function $Ready$ tells whether all have been received. $Value$ extracts the tuple of values for output by λ. The internal transition function, δ_{int}, then resets the state to all empty input stores. Formally,

$$\delta_{\text{ext}}(f, \infty, e, ((p_1, x_1), (p_2, x_2), ..., (p_n, x_n))) = (f', \sigma)$$

where

$$in_i = p_i \text{ and } f(in_i) = \phi \Rightarrow f'(in_i) = x_i$$
$$\text{otherwise } f'(in) = f(in)$$
$$\text{and } Ready(f') \Rightarrow \sigma = 0$$
$$\text{otherwise } \sigma = \infty$$
$$\delta_{\text{int}}(f, 0) = (f^\phi, \infty)$$
$$\lambda(f, 0) = \begin{cases} \lambda_M(value(f)) & \text{if } Ready(f) \\ \phi & \text{otherwise} \end{cases}$$
$$\text{ta}(f, \sigma) = \sigma$$

Exercise 18.7. Define Ready and f^ϕ.

18.3.3 MOORE DTSS WITH INPUT

The DEVS representation of a Moore DTSS with input combines features of the memoryless and input free implementations. The DEVS waits for all input ports to be heard from before computing the next state of the DTSS and scheduling its next output by setting σ to h. This guarantees an output departure to trigger the next cycle. In the subsequent internal transition the DEVS continues holding in the new DTSS state waiting for the full complement of next inputs.

Exercise 18.8. Implement a Moore DTSS as a DEVS.

18.3.4 MEALY DTSS

A Mealy DTSS is represented as a DEVS is a manner similar to a memoryless function, with the difference that the state is updated when all inputs are received. However, the output cannot be pre-scheduled for the next cycle. (Recall that a Moore system which also updates its state can pre-schedule its output because it depends only on the state (Chapter 7).) So the Mealy DEVS passivates after a state update just as if it were a memoryless element.

Exercise 18.9. Implement a Mealy DTSS as a DEVS.

18.3.5 DEVS STRONG SIMULATION OF DTSS COUPLED MODELS

Now that we have all the component DEVS in place we can couple them together to strongly simulate a DTSS coupled model. First, we assume that the DTSS coupled model is well-defined in that it has no algebraic cycles (cycles of Mealy components that do not include a Moore component, Chapter 7). After this, we will see some interesting correlates between ill-defined DTSS networks and illegitimate DEVS counterparts.

Theorem 18.4. *Given a well-defined DTSS coupled model we can construct a DEVS that can simulate it in a component-wise manner (there is an exact morphism at the coupled model level form the DEVS coupled model to the DTSS coupled model).*

Proof. For each DTSS component, define its DEVS representative in the manner discussed previously. Couple the DEVS representatives to each other in the same manner that their counterparts are coupled in the DTSS network. □

Start each of the DEVS components in the state of their DTSS counterpart. Let the σ's of all the Moore components be set to zero. This means that there will be an output generated from all such components. Now by construction, the DEVS coupled model preserves the topology of DTSS network. So the coupling of DEVS components that represent Mealy (including FNSS) forms a DAG which is isomorphic to the original. Thus, the outputs of the Moore components propagate through the DAG to the inputs of these components in the same manner that they do in the original. When they are arrive, the Moore-representing DEVS components employ them to compute the next states and outputs of their counterparts and schedule these outputs to occur at the next time step. This is the basis for a proof by induction on the number of simulation cycles.

Exercise 18.10. Write the formal version of the proof and complete the proof by induction.

Is the DEVS constructed above legitimate? The next theorem makes a strong connection between DEVS illegitimacy and existence of algebraic cycles in DTSS networks.

Theorem 18.5. *The previous DEVS simulation of a coupled DTSS is legitimate if, and only if, the coupled DTSS has no algebraic cycles.*

Proof. Let the coupled DTSS be well defined. Then the subnetwork of Mealy components does not contain cycles. We show that the DEVS coupled model (actually its associated DEVS at the I/O System level) is legitimate. Since the DEVS coupled model preserves the topology of the original coupling, its representation of the Mealy DAG subnetwork propagates the computation through a series of transitory (zero time) states. Since there are no cycles, eventually activity reaches the inputs to the Moore-representing components. (In parallel DEVS the computations at each level of the DAG all take place in parallel.) The computation of next states and outputs at the Moore-representing components has time advance $h > 0$. If there are no Moore components, then the coupled model has type

Mealy. In this case, the DEVS response to input takes zero time before passivating. In either case, time advances by a fixed amount. Since there are no other sources of internal transitions in the DEVS (by the manner in which it constructed), it is legitimate.

In an ill-defined DTSS coupled model, a cycle exists within the subnetwork of Mealy components. This shows up as a cycle of transitory states in the representing DEVS. This cycle forms a zero time loop of DEVS internal transitions and make the DEVS illegitimate. □

EXAMPLE: DEVS STRONG SIMULATION OF DTSS NETWORK

Fig. 18.6 illustrates a DEVS component-wise simulation of the linear DTSS network discussed in Chapter 15. The DTSS Moore Delay (Chapter 3) is implemented as a DEVS with single input port and the summer as a two-input port memoryless system. The coupling shown preserves that of the simulated network.

Exercise 18.11. Define the DEVS representations of the Moore Delay, the summer, and the DTSS network.

18.4 EFFICIENT DEVS SIMULATION OF DTSS NETWORKS

Recall the event-based simulation of cellular automata in Chapter 3 where the simulation algorithm is based on restricting state updates to only those components that can actually change state. Now that

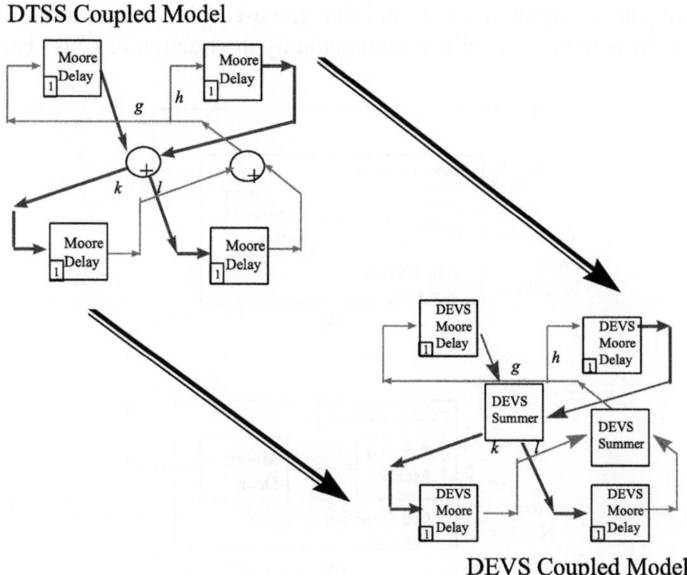

FIGURE 18.6

DEVS Component-wise Simulation DTSS Linear Network.

we have formalized DEVS simulation of DTSS we can investigate how to improve the efficiency of such simulation using the same idea. We can capture the essentials of cellular automata simulation by restricting attention to DTSS networks of Moore Delay components, each fronted by a FNSS. As shown in a), since the memory element is a unit delay, the memoryless element represents the transition function of the cellular automaton – it takes the current states of the neighbors and that of the cell and produces the next state of the cell. Fig. 18.7B, we replace the FNSS by a Mealy system that stores the last values of the neighbor (and self) states. At each cycle, it checks whether any of the inputs has changed. If all are unchanged, it outputs the stored state of the cell. Only if at least one of the inputs has changed does it recompute the next state of the cell using the transition function.

To do a DEVS simulation of such a network, we implement the Mealy and Moore components in the standard fashion with the exception that these components remain passive except for certain conditions. The Moore system remains passive unless, and until, it receives an input from the Mealy system. In this case, it outputs this input after the step time h and then passivates. Thus Moore components only output changed states and do this in unison every cycle. The Mealy system remains passive until it receives changed inputs from its neighbor (and own) Moore systems (since there is only one level components in the Mealy subnetwork, all inputs will arrive in unison). In this case, it computes the transition function and outputs the new cell state, if changed, to the Moore system.

In this way, the coordinating effect of the centralized event-list of Chapter 3 is realized in a distributed fashion. Events are scheduled at only those cells that can change because their neighbors have changed. Only if a cell state changes does it broadcast this news to its influencees.

Exercise 18.12. Write the DTSS change-based Mealy system just described and its DEVS modified-implementation as just discussed.

Exercise 18.13. Compare the straightforward and change-based DEVS simulations of selected cellular automaton. Measure the activity in a cellular automaton by the fraction of cells changing state. Show

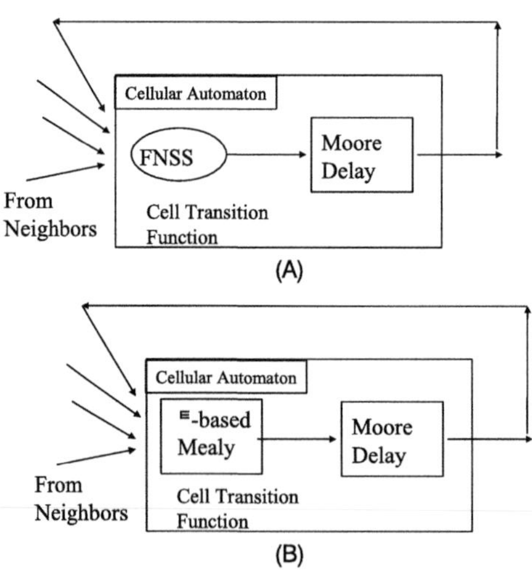

FIGURE 18.7

Cell Automaton Represented as FNSS-fronted Delay Component in DTSS Network.

that the relative efficiency of change-based DEVS simulation over the baseline standard increases as the activity level decreases.

SUMMARY

Are there discrete event systems that are not described by the DEVS formalism? To answer this question we put to work much of the systems theory built up in earlier chapters of the book. We first had to ask whether it matters which formalism, Classical or Parallel DEVS, we are talking about. It turned out that although the formalisms may differ at the coupled system level, they are equivalent at all lower levels. We could then formulate the main issue of DEVS expressibility in terms of discrete event behaviors and climbing up the system specification hierarchy to state space realizations of these behaviors. This approach enabled us to establish both the universality and uniqueness properties of the DEVS formalism. Indeed, not only can every discrete event behavior be realized by a DEVS I/O System but also a DEVS realization be constructed that is isomorphic to the reduced realization of the behavior. Thus, the DEVS formalism (in slightly extended form) can generate all discrete event behaviors (universality), and it can isomorphically represent their canonical internal structures (uniqueness). We went on to establish the ability of DEVS to represent an important subclass of systems the DTSS, or discrete time models such as cellular automata and sampled data systems. This representation is stronger than that which follows from universality since it applies at the coupled system level, not just the I/O systems level. This component-wise mapping enables practical and efficient DEVS simulation of discrete time systems. We continue with exploration of the DEVS Bus concept in the next chapter where we examine its application to simulation of differential equation models.

18.5 SOURCES

Asahi and Zeigler (1993) characterized DEVS-like systems and Sato (1996) raised the issue of uniqueness of DEVS for discrete event systems. Sarjoughian (1995), Sarjoughian and Zeigler (1996) extended the iterative specification approach to provide a means of inferring DEVS structures from DEVS-like behaviors. Ho (1989) defined DEDS (Discrete Event Dynamic Systems) and founded the Journal of DEDS to promote research in the area. Cassandras (1993) provides applications of DEDS to control theory and performance analysis.

APPENDIX 18.A ISOMORPHICALLY REPRESENTING DEVS-LIKE SYSTEMS BY DEVS

We prove the

Lemma 18.2. *Any reduced DEVS-like system has a DEVS realization that is isomorphic to it.*

Proof. Let $S = (T, X, \Omega, Y, Q, \Delta, \Lambda)$ be a reduced DEVS-like system. As indicated in the main text, we need to extract a subset of states that will form the sequential state set S_{DEVS} of the simulating DEVS. Let's call these *milestone* states. Each state q must have a unique representation (s, e) where q

is the state reached from a milestone, s when no external events have been received for e units of time. We have to select a set of milestones that completely cover the state set, Q in this sense.

Consider the autonomous subsystem with non-event input segments, $\phi_{t>}$. The state diagram consists of all state trajectories responding to such inputs. Although for a multi-dimensional state space, its detailed structure would be difficult to capture, the features needed for our purposes can be readily visualized. The composition property of Δ imposes a distinct structure on such state spaces that gives them a structure depicted in Fig. 18.8. The basic relation between states is *accessibility* or reachability:

q' is accessible from q if $(\exists t > 0)\Delta(q, \phi_{t>}) = q'$.

The problem can now be stated as: Identify a set of milestones, S_{DEVS} such that

- each milestone s has an associated *maximum time*, $T(s)$. Let the states accessed from s within time $T(s)$ be its *access* set, $Q(s) = \{\Delta(q, \phi_{e>}) | 0 \leq e < T(s)\}$.
- the access sets are disjoint and cover Q, i.e., form a partition.

To solve this problem we first consider the following types of states:

- *initial state* (e.g., a in Fig. 18.8): states that are not accessible from other states:
  q such that $(\nexists q')$ q accessible from q'
- *final (equilibrium) state* (e.g., b in Fig. 18.8): states that can only access themselves:
  q such that $(\forall t)\Delta(q, \phi_{t>}) = q$
- *state on a finite cycle* (but not final) (e.g., c in Fig. 18.8):
  q such that $(\exists t > 0)\Delta(q, \phi_{t>}) = q$ and $(\exists t' < t)\Delta(q, \phi_{t'>}) \neq q$
- *state on an infinite cycle* (e.g., e in Fig. 18.8):
  q such that there is no initial state that accesses q and there is no final state or cycle of states that q accesses

Fact: the states accessible from any state q include at most one final state or finite cycle of states. □

Exercise 18.14. Prove the fact using the determinism (single valued nature) of the transition function.

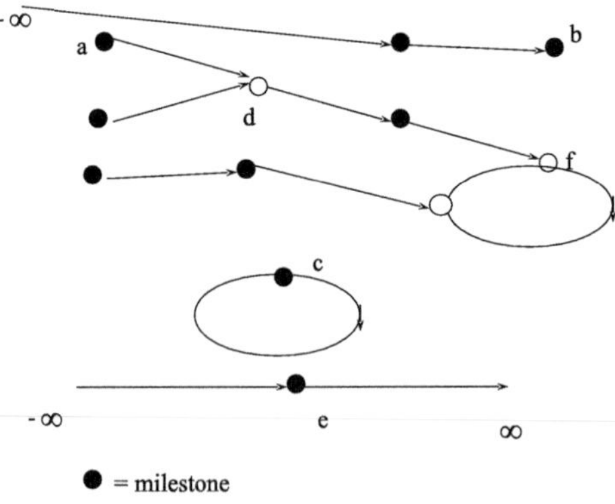

FIGURE 18.8

State Transition Structure.

To identify intersection states we'll need the following definition:

States q_1 and q_2 meet at state q if q is accessible from both q_1 and q_2, i.e.,

$$(\exists t > 0)\Delta(q_1, \phi_{t>}) = q \text{ and } (\exists t' > 0)\Delta(q_2, \phi_{t'>}) = q$$

Fact: states q_1 and q_2 meeting at state q also meet at all states accessible from q.

The proof employs the composition property of the transition function. Any state accessible from q has the form $\Delta(q, \phi_{e>})$. Then replacing q by its two representations, we have

$$\begin{aligned}\Delta(q, \phi_{e>}) &= \Delta(\Delta(q_1, \phi_{t>}), \phi_{e>}) = \Delta(q_1, \phi_{t+e>}) \\ &= \Delta(\Delta(q_2, \phi_{t'>}), \phi_{e>}) = \Delta(q_2, \phi_{t'+e>})\end{aligned}$$

which shows the required accessibility. We can then define an

- *intersection state* as the first meeting point of merging states:
  state q such that q_1 and q_2 meet at $q \wedge q_1$ and q_2 meet at $q' \Rightarrow q'$ is accessible from q

An intersection can take place where two distinct streams meet (e.g., d in Fig. 18.8) or where one stream curls back on itself forming a cycle (e.g., f in Fig. 18.8) or final state (e.g., b in Fig. 18.8).

For our milestones we will include all output states, initial, final and intersection states. However, there may still be states that are not accessible from these states and do not access any of them. Since such states contain no intersections and no initial states they must form self contained cycles that are either finite (such as that containing c in Fig. 18.8) or infinite (such as that containing e in Fig. 18.8). Since the system is reduced every state we can't ignore such states – we have to assume they can access output states using external events. We choose an arbitrary state as a milestone for each such cycle.

There are five types of milestones. A *middle* milestone lies between two other milestones, to the left and right, in the sense that it is accessed by the first and accesses the second. A *right end* milestone only has a milestone on its left. A *left end* milestone only has a milestone on its right. An *isolated* milestone accesses no milestones. The last type is a milestone on an isolated finite cycle, which access itself.

We proceed to develop the access sets, and maximum time for the milestones. We'll also need additional milestones.

As shown in Fig. 18.9, for a middle milestone, s we assign to it all states that it accesses until the next milestone on the right. This its maximum time, $T(s)$ is thus the time to access the next milestone.

Exercise 18.15. Show that the mapping from the interval $[0, T(s))$ to Q given by $f(t) = \Delta(s, \phi_{t>})$ is a one-one correspondence. Hint: if the same state is accessed twice than it forms a cycle contrary to assumption.

For a milestone on an isolated cycle, we use the same approach except that the next milestone is taken to be itself.

For a right end milestone, we define $T(s) = infinity$ and associate its full access set with it.

For a left end milestone, s we use the approach of a middle milestone. However, we have to take care of the infinite half stream on its left, namely, the set of states that access it. For each such state, q, there is a unique time t for it to access to s (t such that $\Delta(q, \phi_{t>}) = s$). The case of an isolated milestone is handled in the same way.

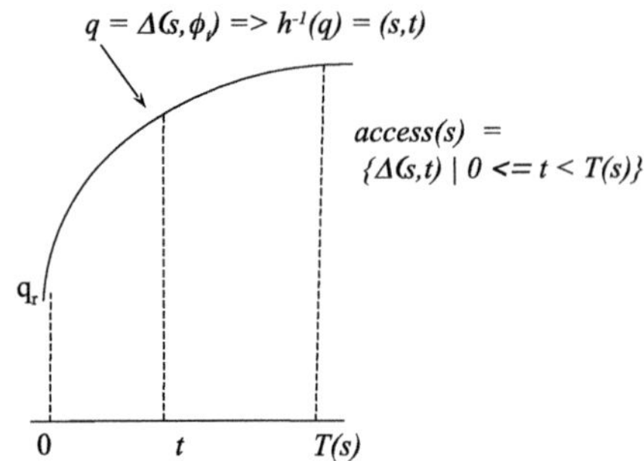

FIGURE 18.9

Access Set Structure.

Exercise 18.16. Show that the uniqueness of access time is true in view of the choice of milestones.

We choose an arbitrary subset of the half stream as a set of middle milestones. This is done, for example, by adding the states $\{..., \Delta^{-1}(s, \phi_{i>}), ..., \Delta^{-1}(s, \phi_{2>}), \Delta^{-1}(s, \phi_{1>})\}$ as milestones. Here $\Delta^{-1}(s, \phi_{i>})$ is the unique state that accesses s in time, i.

Exercise 18.17. Use the composition property of the transition function to show that each additional milestone, $s_i = \Delta^{-1}(s, \phi_{i>})$ has a time $T(s) = 1$, and the mapping $[i, -1) \to Q$ given by $t \to \Delta(s_i, \phi_{t>})$ is one-one, as required.

The chosen set of milestones becomes the sequential state set of the simulating DEVS. The construction of the DEVS follows that in the text and is left as an exercise.

Exercise 18.18. Define a DEVS with the milestones as its sequential states that is isomorphic to the given system, S. (Hint: note that $\text{ta}(s) = T(s)$.)

REFERENCES

Asahi, T., Zeigler, B., 1993. Behavioral characterization of discrete event systems. In: AI, Simulation, and Planning in High Autonomy Systems, 1993. Integrating Virtual Reality and Model-Based Environments. Proceedings: Fourth Annual Conference. IEEE, pp. 127–132.

Cassandras, C., 1993. Discrete Event Systems: Modeling and Performance Analysis. Richard Irwin, New York, NY.

Ho, Y.-C., 1989. Introduction to special issue on dynamics of discrete event systems. Proceedings of the IEEE 77 (1).

Sarjoughian, H.S., 1995. Inductive modeling of discrete-event systems: a TMS-based non-monotonic reasoning approach. In: Electrical and Computer Engineering. University of Arizona, Tucson, p. 256.

Sarjoughian, H.S., Zeigler, B.P., 1996. Abstraction mechanisms in discrete-event inductive modeling. In: Proceedings of the 28th Conference on Winter Simulation. IEEE Computer Society.

Sato, R., 1996. Uniqueness of DEVS formalism. In: AI, Simulation and Planning in High Autonomy Systems. EPD, University of Arizona, San Diego.

CHAPTER

QUANTIZATION-BASED SIMULATION OF CONTINUOUS TIME SYSTEMS

19

CONTENTS

This chapter introduces an alternative approach for the simulation of continuous time systems that replaces the time discretization by the quantization of the state variables. This approach, rather than transforming the original continuous time system into an approximate discrete time system, converts it into an approximate discrete event system that can be easily represented within the DEVS formalism framework.

Theory of Modeling and Simulation. https://doi.org/10.1016/B978-0-12-813370-5.00029-8

The idea behind quantization-based integration is that the simulation steps are performed only when some state experience a significant change, and that each step only affects the state that effectively changes. In that way, calculations are performed only when and where changes occur producing, in some situations, noticeable advantages.

19.1 QUANTIZATION PRINCIPLES

In a previous chapter, a differential equation representation of continuous time systems was introduced. There, we showed that the simulation of those systems requires the usage of numerical integration algorithms that approximate them by discrete time systems. These algorithms are based on answering the following question:

> Given that at time t certain state of a system takes the value $z_i(t)$, which will be the value of $z_i(t+\Delta)$?

for certain parameter $\Delta > 0$. Quantization-based algorithms propose a different question:

> Given that at time t certain state of a system takes the value $z_i(t)$, when will that state take the value $z_i(t) \pm \Delta$?

While the first question implies a time discretization, the second question implies a *quantization* of the state variables since the steps are performed when the states reach certain discrete values that depend on the parameter Δ.

19.1.1 A MOTIVATING EXAMPLE

In order to introduce the basic idea behind this methodology, we shall introduce a simple example consisting on a harmonic oscillator, represented by the set of state equations:

$$\begin{aligned} \dot{z}_1(t) &= z_2(t) \\ \dot{z}_2(t) &= -z_1(t) \end{aligned} \tag{19.1}$$

with initial states $z_1(0) = 4.5$, $z_2(0) = 0.5$. Here, instead of applying a numerical method like Euler's or any similar algorithm, we propose to modify the previous equation as follows:

$$\begin{aligned} \dot{z}_1(t) &= \text{floor}[z_2(t)] \\ \dot{z}_2(t) &= -\text{floor}[z_1(t)] \end{aligned} \tag{19.2}$$

This is, at the right hand side of Eq. (19.1) we replaced each state z_i by its greatest preceding integer floor$[z_i]$. Then, defining

$$q_i(t) \triangleq \text{floor}[z_i(t)] \tag{19.3}$$

as the *quantized states*, we can rewrite the original system of Eq. (19.1) as

$$\begin{aligned} \dot{z}_1(t) &= q_2(t) \\ \dot{z}_2(t) &= -q_1(t) \end{aligned} \tag{19.4}$$

In spite of being non-linear and discontinuous, the *quantized system* of Eq. (19.4) can be easily solved as follows:

- Initially, at time $t = t_0 = 0$ we know that $z_1(0) = 4.5$, $z_2(0) = 0.5$. Then, we have $q_1(0) = 4$, $q_2(0) = 0$. These values for $q_i(t)$ will remain constant until, at certain time t_1, $z_1(t)$ or $z_2(t)$ changes its integer part.
- Evaluating the right hand side of Eq. (19.4), we have $\dot{z}_1(t) = 0$, and $\dot{z}_2(t) = -4$ for all $t \le t_1$.
- Taking into account the last derivative values, the time for the next change in the integer part of $z_1(t)$ is $tn_1 = \infty$, while the time for the next change in the integer part of $z_2(t)$ can be computed as $tn_2 = 0.5/4 = 0.125$.
- At time $t = t_1 = tn_2 = 0.125$ we have $z_1(t_1) = 4.5$, $z_2(t_1) = 0$ with $q_1(t_1) = 4$, and $q_2(t_1) = -1$. These values for $q_i(t)$ will remain constant until, at certain time t_2, $z_1(t)$ or $z_2(t)$ changes its integer part.
- Evaluating the right hand side of Eq. (19.4), we have $\dot{z}_1(t) = -1$, and $\dot{z}_2(t) = -4$ for all $t_1 < t \le t_2$.
- According to these derivative values, the time for the next change in the integer part of $z_1(t)$ is $tn_1 = t_1 + 0.5/1 = 0.625$, while the time for the next change in the integer part of $z_2(t)$ can be computed as $tn_2 = t_1 + 1/4 = 0.375$.
- At time $t = t_2 = tn_2 = 0.375$ we have $z_1(t_2) = 4.5 - 1 \cdot 0.25 = 4.25$, $z_2(t_2) = -1$ with $q_1(t_2) = 4$, and $q_2(t_1) = -2$. These values for $q_i(t)$ will remain constant until, at certain time t_3, $z_1(t)$ or $z_2(t)$ changes its integer part.
- Evaluating the right hand side of Eq. (19.4), we have $\dot{z}_1(t) = -2$, and $\dot{z}_2(t) = -4$ for all $t_2 < t \le t_3$.
- According to these derivative values, the time for the next change in the integer part of $z_1(t)$ is $tn_1 = t_2 + 0.25/2 = 0.5$, while the time for the next change in the integer part of $z_2(t)$ can be computed as $tn_2 = t_2 + 1/4 = 0.625$.
- At time $t = t_3 = tn_1 = 0.5$ we have $z_1(t_3) = 4$, $z_2(t_3) = -1 - 4 \cdot 0.125 = -1.5$ with $q_1(t_2) = 3$, and $q_2(t_1) = -2$. These values for $q_i(t)$ will remain constant until, at certain time t_4, $z_1(t)$ or $z_2(t)$ changes its integer part.
- The calculations continue in the same way until the final simulation time $t = t_f$.

Fig. 19.1 plots the state and quantized state trajectories corresponding to the first six steps of this simulation.

If we advance further the calculations until $t = t_f = 20$, the algorithm performs about 100 more steps and we obtain the trajectories shown in Fig. 19.2. Notice that these trajectories have a close correspondence with what can be expected of a harmonic oscillator like that of Eq. (19.1).

Apparently, replacing $z_i(t)$ by $q_i(t) = \text{floor}[z_i(t)]$ at the right hand side of an ODE we obtain a good approximation of its solution. However, this approximation does not behave in a discrete time way like in classic numerical integration algorithms. Here, we have two types of steps: in times t_1 and t_2 variable $q_2(t)$ changes its value, while $q_1(t)$ remains unchanged. However, at time t_3 we have the opposite situation. Also, the time of the next step is always the minimum between the time of the next change of both quantized states, and, after each step, the time of the next change of each quantized state is rescheduled. This behavior resembles that of a discrete event system.

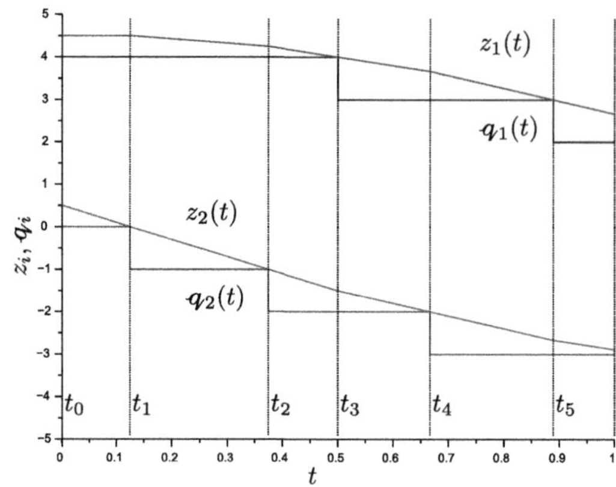

FIGURE 19.1

State and Quantized State Trajectories of Eq. (19.4) (startup detail).

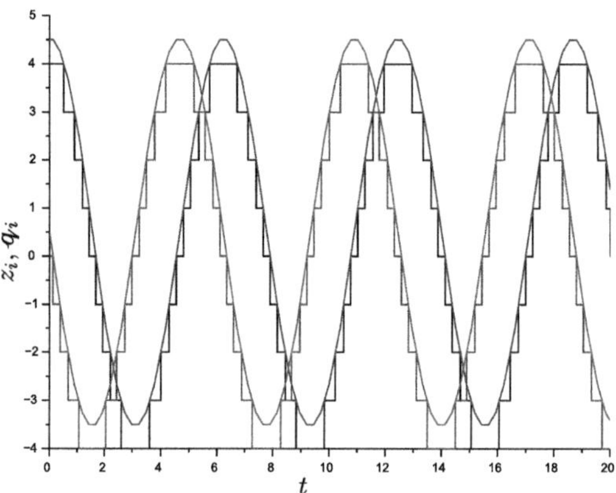

FIGURE 19.2

State and Quantized State Trajectories of Eq. (19.4).

19.1.2 QUANTIZATION AND DEVS REPRESENTATION

In order to find a more direct connection between quantized state approximations and DEVS, we shall first represent Eq. (19.1) and its quantized state *approximation* of Eq. (19.4) as block diagrams by their block diagrams given by Fig. 19.3.

The difference between both block diagrams is the presence of the *quantizer* blocks corresponding to the floor[$z_i(t)$] functions. The output of these blocks are the piecewise constant trajectories $q_i(t)$. Also, the input of the integrators that compute $z_i(t)$ are piecewise constant as they only depend on $q_i(t)$. Thus, we can group together each integrator with the corresponding quantizer to form *quantized integrators* that have piecewise constant input and output trajectories, as shown in Fig. 19.4.

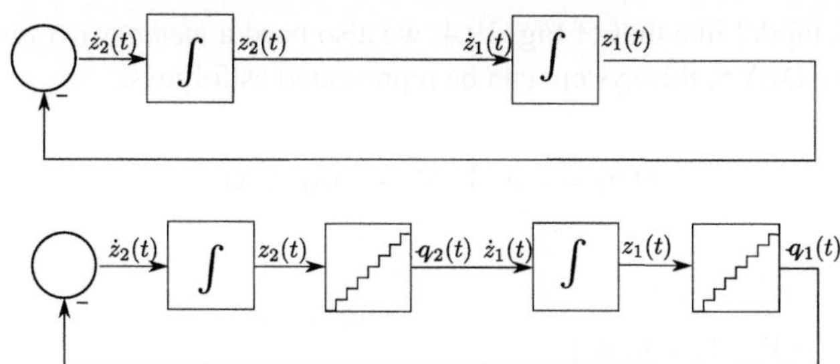

FIGURE 19.3

Block Diagram representation of Eq. (19.1) (top) and Eq. (19.4) (bottom).

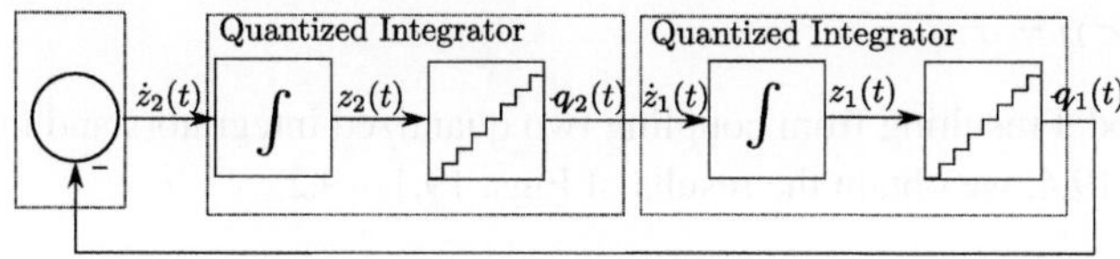

FIGURE 19.4

Quantized Integrators in the representation of Eq. (19.4).

Piecewise constant trajectories can be straightforwardly represented as sequences of events. For that goal, each change in a trajectory can be associated to an event carrying the new value of that trajectory.

Then, taking into account that the input and output trajectories of a quantized integrator can be represented by sequences of events, the quantized integrator itself can be represented by a DEVS model as follows:

$$QI =< X, Y, S, \delta_{\text{int}}, \delta_{\text{ext}}, \lambda, \text{ta} >$$

with

- $X = \mathbb{R} \times \{\text{in}_1\}, = Y = \mathbb{R} \times \{\text{out}_1\}$,
- $S = \mathbb{R} \times \mathbb{R} \times \mathbb{R} \times \mathbb{R}_0^+$,
- $\delta_{\text{int}}(s) = \delta_{\text{int}}((z, q, \dot{z}, \sigma)) = (z + \sigma \cdot \dot{z}, q + \text{sign}(\dot{z}), \dot{z}, |1/\dot{z}|)$,
- $\delta_{\text{ext}}(s, e, x) = \delta_{\text{ext}}((z, q, \dot{z}, \sigma), e, (x_v, \text{in}_1)) = (z + e \cdot \dot{z}, q, x_v, \tilde{\sigma})$,
- $\lambda(s) = \lambda((z, q, \dot{z}, \sigma)) = (q + \text{sign}(\dot{z}), \text{out}_1)$
- $\text{ta}(s) = \text{ta}((z, q, \dot{z}, \sigma)) = \sigma$,

with

$$\tilde{\sigma} = \begin{cases} \dfrac{q - (z + e \cdot \dot{z})}{x_v} & \text{if } x_v < 0 \\ \dfrac{q + 1 - (z + e \cdot \dot{z})}{x_v} & \text{if } x_v > 0 \\ \infty & \text{otherwise.} \end{cases}$$

In order to build a model like that of Fig. 19.4, we also need a *memoryless function* that computes $\dot{z}_2(t) = -q_1(t)$. Using DEVS, this system can be represented as follows:

$$MF_1 = < X, Y, S, \delta_{\text{int}}, \delta_{\text{ext}}, \lambda, \text{ta} >$$

with

- $X = \mathbb{R} \times \{\text{in}_1\}, = Y = \mathbb{R} \times \{\text{out}_1\}$,
- $S = \mathbb{R} \times \mathbb{R}_0^+$,
- $\delta_{\text{int}}(s) = \delta_{\text{int}}((q, \sigma)) = (q, \infty)$,
- $\delta_{\text{ext}}(s, e, x) = \delta_{\text{ext}}((q, \sigma), e, (x_v, \text{in}_1)) = (x_v, 0)$,
- $\lambda(s) = \lambda((q, \sigma)) = (-q, \text{out}_1)$,
- $\text{ta}(s) = \text{ta}((q, \sigma)) = \sigma$.

If we simulate the model resulting from coupling two quantized integrators and the memoryless function as shown in Fig. 19.4, we obtain the results of Figs. 19.1–19.2.

19.1.3 GENERALIZATION OF QUANTIZED SYSTEMS

The idea of replacing the states $z_i(t)$ by their quantized versions $q_i = \text{floor}[z_i(t)]$ on the right hand side of the ODE worked fine in the motivating example and allowed us to simulate the system using a DEVS approximation. The question is then if this idea can be applied to general continuous time systems.

A first observation is that when we reduce the initial states of the oscillator by a factor of 10, the analytical solution consists in harmonic oscillations with amplitude less than 1. Thus, the quantized states $q_i = \text{floor}[z_i]$ will perform too coarse steps to provide a decent solution.

This problem can be easily solved by using a more refined quantization function of the form

$$q_i(t) = \Delta Q_i \cdot \text{floor}\left[\frac{z_i(t)}{\Delta Q_i}\right] \tag{19.5}$$

where the parameter $\Delta Q_i > 0$ will be called *quantum*.

Consider now a general ODE system of the form

$$\begin{aligned}
\dot{z}_1(t) &= f_1(z_1(t), \cdots, z_n(t), u_1(t), \cdots, u_m(t)) \\
\dot{z}_2(t) &= f_2(z_1(t), \cdots, z_n(t), u_1(t), \cdots, u_m(t)) \\
&\vdots \\
\dot{z}_n(t) &= f_n(z_1(t), \cdots, z_n(t), u_1(t), \cdots, u_m(t))
\end{aligned} \tag{19.6}$$

where $z_i(t)$ are the states and $u_i(t)$ are input trajectories. Then, replacing $z_i(t)$ by quantized states $q_i(t)$ in the right hand side of the previous system, and replacing the input trajectories $u_i(t)$ by appropriate piecewise constant approximations $v_i(t)$, we obtain

$$\begin{aligned}
\dot{z}_1(t) &= f_1(q_1(t), \cdots, q_n(t), u_1(t), \cdots, u_m(t)) \\
\dot{z}_2(t) &= f_2(q_1(t), \cdots, q_n(t), u_1(t), \cdots, u_m(t)) \\
&\vdots \\
\dot{z}_n(t) &= f_n(q_1(t), \cdots, q_n(t), u_1(t), \cdots, u_m(t))
\end{aligned} \tag{19.7}$$

that can be represented by the block diagram of Fig. 19.5.

Based on this block diagram, a coupled DEVS representation of Eq. (19.7) can be built using DEVS models of quantized integrators, creating DEVS representations of the different memoryless functions $f_i(\mathbf{q}, \mathbf{u})$ and approximating the input signals $u_i(t)$ by the event trajectories corresponding to $v_i(t)$. That way, in principle, a general ODE like that of Eq. (19.6) can be simulated by a DEVS representation of its quantized approximation given by Eq. (19.7).

Unfortunately, this idea usually fails. Let us consider the following first order model:

$$\dot{z}(t) = -0.5 - z(t) \tag{19.8}$$

with initial state $z(0) = 0$ and consider its quantized approximation

$$\dot{z}(t) = -0.5 - q(t) \tag{19.9}$$

with $q(t) = \text{floor}[z(t)]$ as usual.

This system behaves as follows:

- Initially, at $t = 0$, we have $z(0) = 0$ so $q(0) = 0$ and evaluating the right hand side of Eq. (19.9) it results $\dot{z}(0) = -0.5$.
- As the slope is negative, $z(t)$ immediately becomes negative, so $q(0^+) = -1$ and valuating the right hand side of Eq. (19.9) we have $\dot{z}(0^+) = 0.5$.

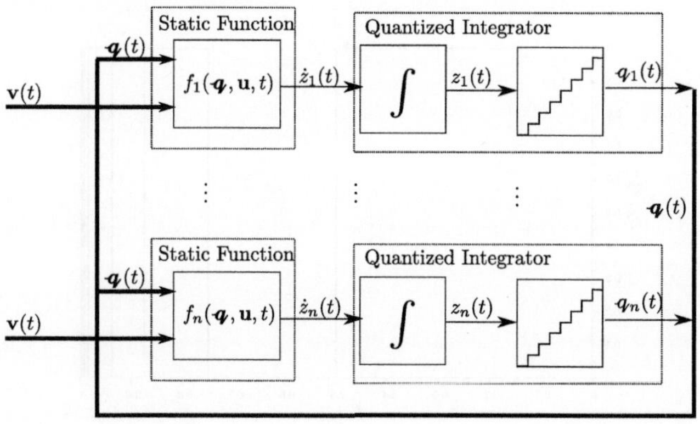

FIGURE 19.5

Block Diagram representation of Eq. (19.7).

- As the slope is now positive, $z(t)$ immediately becomes zero and we are back to the initial situation.

The result of this behavior is an oscillation of $z(t)$ around 0, while $q(t)$ oscillates between 0 and -1. The problem is that the frequency of the oscillation is infinite, so the simulation cannot advance beyond $t = 0$. In fact, if we build a DEVS model coupling a quantized integrator with a memoryless function that computes $\dot{z} = -0.5 - q(t)$, the resulting coupled DEVS model will be illegitimate.

It is clear that in this very particular case the idea of quantizing the states fails. However, does it always fail? Evidently, from what we saw in the introductory example sometimes it works. Anyway, in most cases it fails. For instance, if we replace the value -0.5 in Eq. (19.8) by an arbitrary constant a, the illegitimacy situation will appear for any non-integer value of a. More generally, if we use a quantum ΔQ, the illegitimacy condition appears whenever $a/\Delta Q$ is non-integer.

In higher order systems the problem becomes even worst, as we can also obtain illegitimate trajectories where the distance between events becomes shorter as the time advances, so that the total elapsed time goes to a finite value while the number of events goes to infinite. This is what happens, for instance, in the following second order system:

$$\begin{aligned}\dot{z}_1(t) &= -0.5 \cdot x_1(t) + 1.5 \cdot x_2(t)\\ \dot{z}_2(t) &= -x_1(t) + 1.5\end{aligned} \tag{19.10}$$

Using $q_i(t) = \text{floor}[z_i(t)]$ as quantized states, the solution from $z_1(0) = 1.75$, $z_2(0) = 1$ cannot advance beyond $t = 1$, as shown in Fig. 19.6. Notice there that each interval between changes on q_2 is the half of the preceding interval.

The examples analyzed here show that the quantization strategy with functions like floor$[z_i]$ cannot be used in general. Next, we shall analyze why the idea fails and we shall show an alternative approach that works with general models.

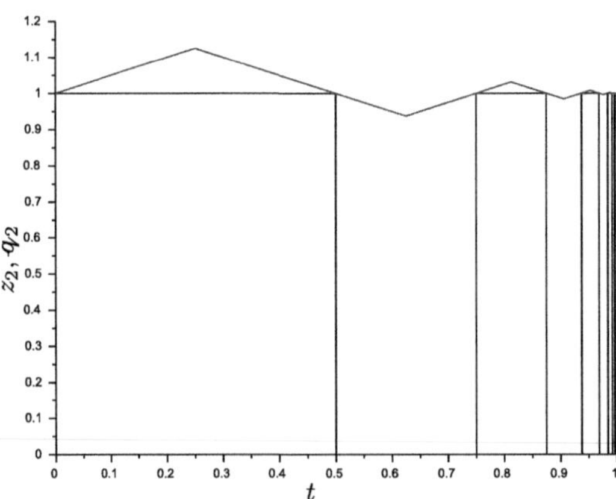

FIGURE 19.6

Illegitimate solution in the system of Eq. (19.10).

19.2 QUANTIZED STATE SYSTEMS

The main drawback of the state quantization idea presented above was the appearance of infinitely fast oscillations in the state and quantized state variables. These oscillations occur when certain state z_i moves around a quantized value. A small change in z_i then provokes a large change in q_i, which in turn can provoke another small change in z_i reaching a cycling behavior.

If we look further to the problem, it results that when $z_i(t)$ is rising and reaches an integer value, it provokes that $q_i(t)$ changes and becomes equal to $z_i(t)$. This is reasonable, as we want that $q_i(t)$ and $z_i(t)$ are close to each other so the solution of the approximate quantized system of Eq. (19.7) is similar to that of the original system of Eq. (19.6). However, when z_i is falling down, once it reaches the integer value $z_i = q_i(t)$, function floor() provokes that $q_i(t) = z_i(t) - 1$. This last change seems unnecessary from an approximation perspective. It looks more reasonable to keep the old value of $q_i(t)$ until $z_i(t)$ is far below $q_i(t)$.

This idea, besides providing a better approximation, also prevents that small changes in z_i provoke consecutive changes in q_i and solves the problem of infinitely fast oscillations. However, using this strategy, the relation between $z_i(t)$ and $q_i(t)$ is no longer memoryless. Given a value for $z_i(t)$, $q_i(t)$ can take two values: floor$[z_i(t)]$ or floor$[z_i(t)] + 1$, according to a *hysteretic* law.

The use of hysteretic quantization, formally defined below, is the basis of the Quantized State Systems method for continuous system simulation.

19.2.1 HYSTERETIC QUANTIZATION

Definition 19.1 (Hysteretic Quantization Function). Given two trajectories $z_i(t)$, and $q_i(t)$, we say that they are related by a hysteretic quantization function provided that

$$q_i(t) = \begin{cases} q_i(t^-) & \text{if } |q_i(t^-) - x_i(t)| < \Delta Q_i \\ z_i(t^-) & \text{otherwise,} \end{cases} \tag{19.11}$$

where the parameter ΔQ_i is called *quantum*.

Fig. 19.7 shows the evolution of a state $z_i(t)$ and its quantized state $q_i(t)$. Notice that until $t \approx 2$ the quantization is identical to that given by the floor() function. However, at that point, when the state falls below q_i there is no change in the quantized state until the difference is equal to the quantum ΔQ_i.

Based on this definition, we introduce next the First Order Quantized State System (QSS1) method.

19.2.2 FIRST ORDER QUANTIZED STATE SYSTEMS METHOD

Definition 19.2 (QSS1 Method). Given the ODE

$$\dot{\mathbf{z}}(t) = \mathbf{f}(\mathbf{z}(t), \mathbf{u}(t)) \tag{19.12}$$

where $\mathbf{z} \in \mathbb{R}^n$, and $\mathbf{u} \in \mathbb{R}^m$ is a vector of known input trajectories, the QSS1 method approximates it by the following *quantized state system*:

$$\dot{\mathbf{z}}(t) = \mathbf{f}(\mathbf{q}(t), \mathbf{v}(t)) \tag{19.13}$$

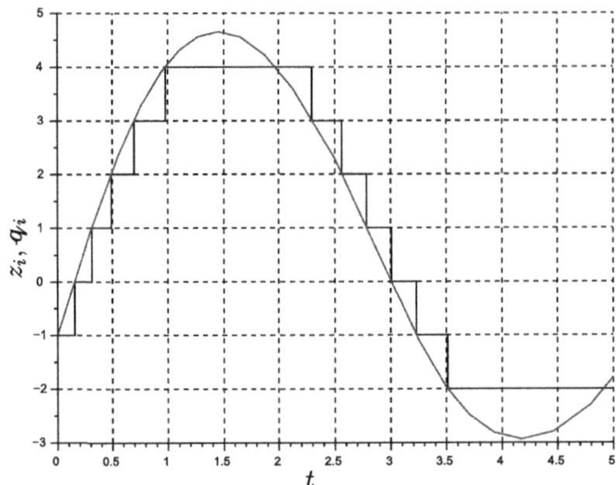

FIGURE 19.7

Hysteretic Quantization with $\Delta Q_i = 1$.

where $q \in \mathbb{R}^n$ is the vector of *quantized states*, whose i-th component q_i is related with the i-th state z_i by a hysteretic quantization function with quantum ΔQ_i, and $\mathbf{v}(t)$ is a piecewise constant approximation of the input vector $\mathbf{u}(t)$.

Notice that the system Eq. (19.12) is just a compact representation of the original system of Eq. (19.6), while the quantized state system of Eq. (19.13) is a compact representation of the quantized system Eq. (19.7).

It can be easily seen that the quantized state trajectories $q_i(t)$ of the QSS1 approximation are piecewise constant, as they are the output of hysteretic quantization functions. Taking into account that the approximated input trajectories $v_i(t)$ are also piecewise constant, it results that the state derivatives $\dot{z}_i(t) = f_i(q(t), \mathbf{v}(t))$ are also piecewise constant. Then, the state variables $z_i(t)$ have piecewise linear trajectories. This particular trajectory forms allows to represent the quantized state systems by simple DEVS models.

19.2.3 DEVS REPRESENTATION OF QSS1

Given a generic ODE like that of Eq. (19.12), the QSS1 approximation of Eq. (19.13) can be represented by a coupled DEVS model composed of *Hysteretic* Quantized Integrators that compute $q_i(t)$ out of $\dot{z}_i(t)$, and Static Functions that compute $\dot{z}_i$ out of $q(t)$ and $\mathbf{v}(t)$, as shown in Fig. 19.8.

Representing the piecewise constant input and output trajectories of each submodel by sequences of events, the hysteretic quantized integrator has the behavior expressed by the following DEVS model:

$$HQI = < X, Y, S, \delta_{\text{int}}, \delta_{\text{ext}}, \lambda, \text{ta} >$$

with

- $X = \mathbb{R} \times \{\text{in}_1\}$, $Y = \mathbb{R} \times \{\text{out}_1\}$,
- $S = \mathbb{R}^2 \times \mathbb{R} \times \mathbb{R}_0^+$,

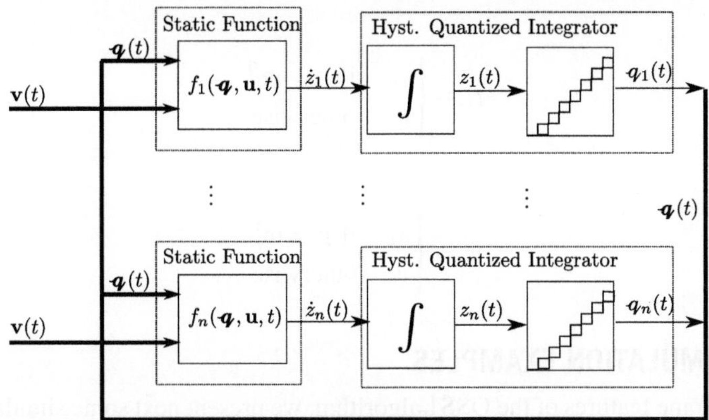

FIGURE 19.8

Block Diagram representation of a QSS1 approximation.

- $\delta_{\text{int}}(s) = \delta_{\text{int}}(((z, \dot{z}), q, \sigma)) = ((z + \sigma \cdot \dot{z}, \dot{z}), z + \sigma \cdot \dot{z}, |\Delta Q/\dot{z}|)$,
- $\delta_{\text{ext}}(s, e, x) = \delta_{\text{ext}}(((z, \dot{z}), q, \sigma), e, (x_v, \text{in}_1)) = ((z + e \cdot \dot{z}, x_v), q, \tilde{\sigma})$,
- $\lambda(s) = \lambda(((z, \dot{z}), q, \sigma)) = (z + \sigma \cdot \dot{z}, \text{out}_1)$
- $\text{ta}(s) = \text{ta}(((z, \dot{z}), q, \sigma)) = \sigma$,

with

$$\tilde{\sigma} = \begin{cases} \dfrac{q - \Delta Q - (z + e \cdot \dot{z})}{x_v} & \text{if } x_v < 0 \\ \dfrac{q + \Delta Q - (z + e \cdot \dot{z})}{x_v} & \text{if } x_v > 0 \\ \infty & \text{otherwise.} \end{cases}$$

In order to build a block diagram like that of Fig. 19.8, we also need memoryless functions that compute $\dot{z}_i(t)$ from $\boldsymbol{q}(t)$ and $\mathbf{v}(t)$. The following DEVS model for a generic memoryless function assumes contains $n + m$ input ports, so that the first n ports receive the n quantized state components $q_i(t)$, while the last m ports receive the input components $v_i(t)$:

$$SF = < X, Y, S, \delta_{\text{int}}, \delta_{\text{ext}}, \lambda, \text{ta} >$$

with

- $X = \mathbb{R} \times \{\text{in}_1^{\boldsymbol{q}}, \ldots, \text{in}_n^{\boldsymbol{q}}, \text{in}_1^{\mathbf{v}}, \ldots, \text{in}_m^{\mathbf{v}}\}$, $Y = \mathbb{R} \times \{\text{out}_1\}$,
- $S = \mathbb{R}^n \times \mathbb{R}^m \times \mathbb{R}_0^+$,
- $\delta_{\text{int}}(s) = \delta_{\text{int}}((\boldsymbol{q}, \mathbf{v}, \sigma)) = (\boldsymbol{q}, \mathbf{v}, \infty)$,
- $\delta_{\text{ext}}(s, e, x) = \delta_{\text{ext}}((\boldsymbol{q}, \mathbf{v}, \sigma), e, (x_v, p)) = ((\tilde{\boldsymbol{q}}, \tilde{\mathbf{v}}), 0)$,
- $\lambda(s) = \lambda((\boldsymbol{q}, \mathbf{v}, \sigma)) = (f_i(\boldsymbol{q}, \mathbf{v}), \text{out}_1)$,
- $\text{ta}(s) = \text{ta}((\boldsymbol{q}, \mathbf{v}, \sigma) = \sigma$,

where

$$\tilde{q}_i = \begin{cases} x_v & \text{if } p = \text{in}_i^{\mathbf{q}} \\ q_i & \text{otherwise} \end{cases}$$

and

$$\tilde{v}_i = \begin{cases} x_v & \text{if } p = \text{in}_i^{\mathbf{v}} \\ v_i & \text{otherwise} \end{cases}$$

19.2.4 QSS1 SIMULATION EXAMPLES

In order to analyze some features of the QSS1 algorithm, we present next some simulation experiments on the following second order model that represents a spring–mass–damper mechanical system:

$$\begin{aligned} \dot{z}_1(t) &= z_2(t) \\ \dot{z}_2(t) &= -\frac{k}{m}z_1(t) - \frac{b}{m}z_2(t) + \frac{F(t)}{m} \end{aligned} \tag{19.14}$$

Here, the variables $x_1(t)$ and $x_2(t)$ are the position and speed, respectively. The parameter m is the mass, b is the friction coefficient, k is the spring coefficient and $F(t)$ is an input trajectory force applied to the mass. For the sake of simplicity, we shall consider that $b = k = m = 1$ and that the input force is a constant $F(t) = 1$.

As the ODE of Eq. (19.14) is linear, it has analytical solution. For initial states $z_1(0) = z_2(0) = 0$, the solution can be computed as:

$$\begin{aligned} z_1(t) &= 1 - \frac{\sqrt{3}}{3}e^{-t/2}\sin\frac{\sqrt{3}}{2}t - e^{-t/2}\cos\frac{\sqrt{3}}{2}t \\ z_2(t) &= \frac{\sqrt{12}}{3}e^{-t/2}\sin\frac{\sqrt{3}}{2}t \end{aligned} \tag{19.15}$$

Fig. 19.9 plots this analytical solution (in dotted lines) together with the one obtained using the QSS1 approximation with quantum $\Delta Q_i = 0.1$ in both state variables. It can be easily noticed that the numerical solution given by QSS1 is not far away from the analytical one, with a maximum difference similar to the quantum ΔQ_i.

Fig. 19.10 then shows the results using a quantum ten times smaller ($\Delta Q_i = 0.01$ in both variables). The results are now much more accurate, and the maximum difference with the analytical solution is again of the order of $\Delta Q_i = 0.01$. However, the price paid for a better approximation is that the number of events is increased by a factor of about 10.

Finally, Fig. 19.11 shows the results using a larger quantum ($\Delta Q_i = 0.25$ in both variables). The results have now a larger error (again of the order of $\Delta Q_i = 0.25$) and finish with oscillations around the equilibrium point of the analytical solution. Anyway, they do not diverge as it usually occur with explicit discrete time numerical algorithms when the step size is too large.

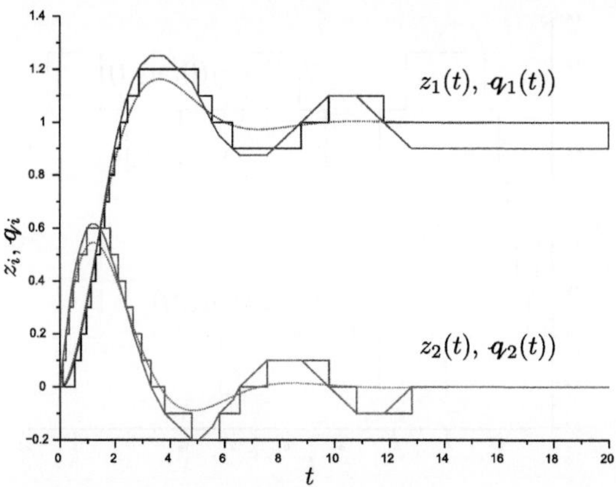

FIGURE 19.9

QSS1 Solution of the system of Eq. (19.14) with $\Delta Q_i = 0.1$. (Analytical solution with dotted lines.)

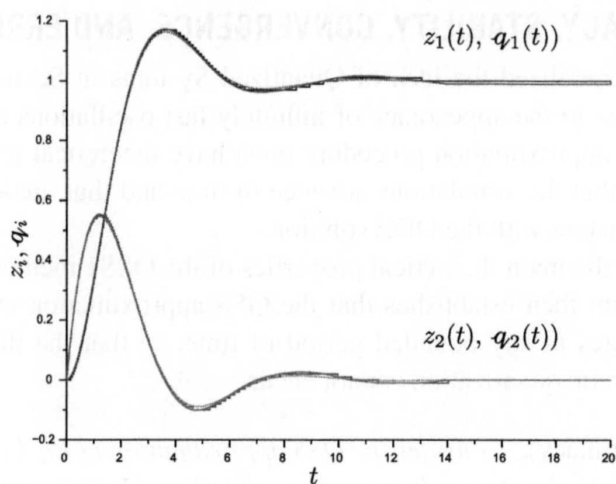

FIGURE 19.10

QSS1 Solution of the system of Eq. (19.14) with $\Delta Q_i = 0.01$. (Analytical solution with dotted lines.)

In these simulation experiments we observed the following facts:

- Infinitely fast oscillations did not appear.
- The numerical results are stable (i.e., they did not diverge) for the different values of ΔQ_i.
- As the parameter ΔQ_i goes to zero, the numerical error apparently goes to zero.
- The numerical error is directly related with ΔQ_i.

We shall see next that all these facts observed for the system of Eq. (19.14) are in fact general properties of the QSS1 algorithm.

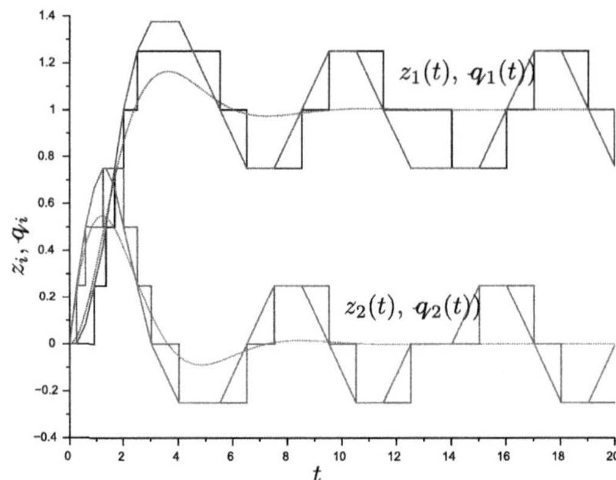

FIGURE 19.11

QSS1 Solution of the system of Eq. (19.14) with $\Delta Q_i = 0.25$. (Analytical solution with dotted lines.)

19.2.5 QSS LEGITIMACY, STABILITY, CONVERGENCE, AND ERROR BOUNDS

When we attempted to generalized the idea of Quantized Systems in Section 19.1.3, we found that the methodology failed due to the appearance of infinitely fast oscillations in most cases. A well established numerical ODE approximation procedure must have theoretical results that ensure that the methodology works, i.e., that the simulations advance in time and that, under certain conditions, the numerical results are consistent with the ODE solution.

Next, we shall present the main theoretical properties of the QSS1 method in the form of different theorems. The first theorem then establishes that the QSS approximation of Eq. (19.13) performs a bounded number of changes in any bounded period of time, so that the illegitimacy condition that appeared using non-hysteretic quantization cannot occur.

Theorem 19.1 (QSS Legitimacy). *Consider the QSS approximation of Eq.* (19.13). *Suppose that function* $\mathbf{f}(\mathbf{z}, \mathbf{u})$ *is bounded while the state* $\mathbf{z}$ *is in certain closed set* $\mathcal{Z} \subset \mathbb{R}^n$ *and the input is in another closed set* $\mathcal{U} \subset \mathbb{R}^m$. *Assume that the quantized state* $\mathbf{q}(t)$ *remains in the set* $\mathcal{Z}$, *and that the input approximation* $\mathbf{v}(t)$ *remains in the set* $\mathcal{U}$ *during the finite interval of time* (t_0, t_f). *Then, the quantized state* $\mathbf{q}(t)$ *can only perform a bounded number of changes in the interval* (t_0, t_f).

Proof. Take any $i \in \{1, \ldots, n\}$ and any $t \in [t_0, t_f]$. The fact that $\mathbf{q}(t) \in \mathcal{Z}$ and $\mathbf{v}(t) \in \mathcal{U}$ implies that $\mathbf{f}(\mathbf{q}(t), \mathbf{v}(t))$ is bounded, and according to Eq. (19.13), $\dot{z}_i = f_i(\mathbf{q}(t), \mathbf{v}(t))$ is also bounded. Then, a constant $M > 0$ exists so that

$$|\dot{z}_i(t)| \leq M \tag{19.16}$$

Suppose that at time $t_1 \in [t_0, t_f]$ the quantized state q_i changes its value. Then, according to the definition of the hysteretic quantization function it results that

$$q_i(t_1^+) = z_i(t_1)$$

The next change in $q_i(t)$ will occur at time t_2, when $z_i(t)$ moves away from $q_i(t)$ so that $|q_i(t) - z_i(t_2)| = \Delta Q_i$, this is, when

$$|z_i(t_2) - q_i(t_1^+)| = |z_i(t_2) - z_i(t_1)| = \Delta Q_i$$

Taking into account Eq. (19.16) and the last equation, it results that

$$\Delta Q_i = |z_i(t_2) - z_i(t_1)| \leq M \cdot (t_2 - t_1)$$

and then,

$$t_2 - t_1 \geq \frac{\Delta Q_i}{M}$$

This last inequality establishes a limit from below to the time between consecutive changes in a quantized state, which in turn establishes an upper bound to the maximum number of changes that can occur in a finite interval and concludes the proof. □

The last theorem establishes that the QSS solutions are legitimate provided that the quantized states do not escape certain set in which function $\mathbf{f}(\cdot, \cdot)$ is bounded. We shall see next that, under certain conditions, some stability properties ensure that the states and quantized state cannot escape from those bounded sets.

In order to establish these stability properties, we need first to provide the following definitions:

Definition 19.3 (Equilibrium Point). Given the ODE

$$\dot{\mathbf{z}}(t) = \mathbf{f}(\mathbf{z}(t), \mathbf{u}(t)) \tag{19.17}$$

with a constant input $\mathbf{u}(t) = \bar{\mathbf{u}}$, we say that $\bar{\mathbf{z}}$ is an equilibrium point of the ODE if

$$\mathbf{f}(\bar{\mathbf{z}}, \bar{\mathbf{u}}) = 0$$

Notice that when $\mathbf{z}(t) = \bar{\mathbf{z}}$, the fact that $\mathbf{f}(\bar{\mathbf{z}}, \bar{\mathbf{u}}) = 0$ implies that $\dot{\mathbf{z}}(t) = 0$ and then the state remains constant at the equilibrium point.

Definition 19.4 (Positive Invariant Set). Given the ODE of Eq. (19.17) with a constant input $\mathbf{u}(t) = \bar{\mathbf{u}}$, we say that the set $\mathcal{I} \subset \mathbb{R}^n$ is positive invariant under the ODE dynamics if the condition $\mathbf{z}(t_0) \in \mathcal{I}$ implies that $\mathbf{z}(t) \in \mathcal{I}$ for all $t \geq t_0$.

Informally, a set is positive invariant if the ODE state trajectories cannot abandon it. Notice that an equilibrium point is a particular case of a positive invariant set.

Definition 19.5 (Stability). Given the ODE of Eq. (19.17) with a constant input $\mathbf{u}(t) = \bar{\mathbf{u}}$, we say that an equilibrium point $\bar{\mathbf{z}}$ is stable if given $\varepsilon > 0$, there exists $\delta > 0$ such that $\|\mathbf{z}(t_0) - \bar{\mathbf{z}}\| < \delta$ implies that $\|\mathbf{z}(t) - \bar{\mathbf{z}}\| < \varepsilon$ for all $t \geq t_0$.

Informally, an equilibrium point is stable provided that the state trajectories that start near that point do not move far away from it.

Definition 19.6 (Asymptotic Stability). Given the ODE of Eq. (19.17) with a constant input $\mathbf{u}(t) = \bar{\mathbf{u}}$, we say that an equilibrium point $\bar{\mathbf{z}}$ is asymptotically stable if it is stable and a compact set $\mathfrak{X} \subset \mathbb{R}^n$ with $\bar{\mathbf{z}} \in \mathfrak{X}$ exists such that $\mathbf{z}(t_0) \in \mathfrak{X}$ implies that $\|\mathbf{z}(t) - \bar{\mathbf{z}}\| \to 0$ as $t \to \infty$.

Informally, an equilibrium point is asymptotically stable when the state trajectories originated in a set around that point converge to that equilibrium. The set $\mathfrak{X}$ from which the trajectories converge to the equilibrium is called *region of attraction*. When the region of attraction is the complete state space $\mathbb{R}^n$, we say that the equilibrium is *globally* asymptotically stable.

Based on these definitions, the following theorem establishes the general stability results for the QSS1 algorithm.

Theorem 19.2 (QSS1 Stability). *Suppose that the point $\bar{\mathbf{z}}$ is an asymptotically stable equilibrium point of the ODE of Eq.* (19.17) *with a constant input $\mathbf{u}(t) = \bar{\mathbf{u}}$ and let $\mathfrak{X}$ be a positive invariant set contained in the region of attraction of that equilibrium point. Assume also that function* **f** *is continuously differentiable. Then, given two positive invariant sets $\mathfrak{X}_1, \mathfrak{X}_2$ such that $\bar{\mathbf{z}} \in \mathfrak{X}_1 \subset \mathfrak{X}_2 \subset \mathfrak{X}$, an appropriate quantization $\Delta Q_i > 0$ for $i = 1, \ldots, n$ can be found such that the solutions of the resulting QSS1 approximation*

$$\dot{\mathbf{z}}(t) = \mathbf{f}(\boldsymbol{q}(t), \mathbf{u}(t)) \tag{19.18}$$

starting in set $\mathfrak{X}_2$ finish inside set $\mathfrak{X}_1$.

The proof of this theorem is based on the fact that the components of $\mathbf{z}(t)$ and $\boldsymbol{q}(t)$ in Eq. (19.18) do not differ from each other in more than ΔQ_i, thus the QSS approximation can be rewritten as

$$\dot{\mathbf{z}}(t) = \mathbf{f}(\mathbf{z}(t) + \Delta\mathbf{z}(t), \mathbf{u}(t)) \tag{19.19}$$

where $\Delta\mathbf{z}(t)$ introduces a bounded perturbation to the original system of Eq. (19.17). Then, under the conditions imposed on function **f**, those bounded perturbation will only provoke bounded changes to the solutions of the perturbed system of Eq. (19.19).

The result does not guarantee that the solutions go to the equilibrium point. However, they ensure that they finish inside a given set $\mathfrak{X}_1$ that can be arbitrarily small. As it can be expected, the smaller this final set is chosen, the smaller the appropriate quantum ΔQ_i will result.

Since the effects of the quantization can be regarded as the effects of perturbations bounded by the quantum, it can be immediately conjectured that when the quantum goes to zero, their effects go to zero and the numerical solution goes to the analytical solution. This conjecture is in fact true, provided that function **f** satisfies the so called *Lipschitz conditions*, defined as follows

Definition 19.7 (Lipschitz Function). A function $\mathbf{f}(\mathbf{z}, \mathbf{u}) : \mathbb{R}^n \times \mathbb{R}^m \to \mathbb{R}^n$ is Lipschitz on $\mathbf{z}$ in a set $\mathfrak{X} \subset \mathbb{R}^n$ if given $\mathbf{u} \in \mathbb{R}^m$, a positive constant $L \in \mathbb{R}^+$ exists so that for any pair of points $\mathbf{z}_a, \mathbf{z}_b \in \mathfrak{X}$ the following inequality holds

$$\|\mathbf{f}(\mathbf{z}_a, \mathbf{u}) - \mathbf{f}(\mathbf{z}_b, \mathbf{u})\| \le L \cdot \|\mathbf{z}_a - \mathbf{z}_b\| \tag{19.20}$$

The previous definition implies that Lipschitz functions are not only continuous, but they also exhibit limited variations in bounded intervals. That way, the Lipschitz condition is stronger than continuity, but it is weaker than differentiability.

Then, the following theorem shows the convergence of the QSS1 solutions to the analytical solutions as the quantum goes to zero.

Theorem 19.3 (QSS1 Convergence). *Let $\mathbf{z}_a(t)$ be the analytical solution of the ODE of Eq.* (19.17) *with initial state $\mathbf{z}_a(0) = \mathbf{z}_0$. Suppose that $\mathbf{z}_a(t)$ remains in a compact set $\mathcal{Z} \subset \mathbb{R}^n$, while $\mathbf{u}(t)$ is confined to a compact set $\mathcal{U} \subset \mathbb{R}^m$. Let $\mathbf{z}(t)$ be the solution of the QSS1 approximation*

$$\dot{\mathbf{z}}(t) = \mathbf{f}(\boldsymbol{q}(t), \mathbf{v}(t)) \tag{19.21}$$

starting from the same initial state $\mathbf{z}(0) = \mathbf{z}_0$, using quantum parameters in all states $\Delta Q_i \le \Delta Q$ for certain $\Delta Q > 0$. Assume that $\mathbf{v}(t)$ is a piecewise constant approximation of $\mathbf{u}(t)$ such that $|v_i(t) - u_i(t) < \Delta Q|$.

Then, provided that function $\mathbf{f}(\mathbf{z}, \mathbf{u})$ is Lipschitz in a set that contains $\mathcal{Z} \times \mathcal{U}$ in its interior, it results that $\mathbf{z}(t) \to \mathbf{z}_a(t)$ as $\Delta Q \to 0$.

The proof of this theorem is also based on writing the QSS1 approximation as a perturbed version of the original ODE. Then, using a similar proof to that of *Continuity of ODE solutions*, the result is almost direct.

The stability and convergence results are *qualitative* properties of QSS1 approximation. They ensure that we can obtain *good* simulation results by using appropriate quantum parameters, but they do not establish a practical relation between the quantum values and the accuracy of the numerical solutions. In order to obtain that *quantitative* information, we need to restrict the systems to *linear time invariant* ODEs of the form:

$$\dot{\mathbf{z}}(t) = A \cdot \mathbf{z}(t) + B \cdot \mathbf{u}(t) \tag{19.22}$$

where the dimension of matrices A and B are $n \times n$ and $n \times m$, respectively.

Before presenting the result that establishes quantitative numerical error bounds, we need to introduce some notation. Given a matrix M (that can be real or complex valued), $|M|$ represents the component-wise modulus of M, i.e., $|M|_{i,j} = |M_{i,j}|$. Similarly, $\mathbb{R}\mathrm{e}(M)$ represents the component-wise real part of M, such that $\mathbb{R}\mathrm{e}(M)_{i,j} = \mathbb{R}\mathrm{e}(M_{i,j})$. Also, given two vectors, $a \in \mathbb{R}^n$, $b \in \mathbb{R}^n$, the expression $a \preceq b$ denotes a component-wise inequality, i.e., $a_i \le b_i$ for $i = 1, \ldots, n$.

Then, the following theorem establishes an upper bound to the maximum error introduced by the QSS approximation:

Theorem 19.4 (QSS1 Error Bound). *Consider the ODE of Eq.* (19.22) *and its QSS1 approximation given by*

$$\dot{\mathbf{z}}(t) = A \cdot \boldsymbol{q}(t) + B \cdot \mathbf{v}(t) \tag{19.23}$$

Assume that matrix A is Hurwitz without repeated eigenvalues. Suppose that the piecewise constant input approximation $\mathbf{v}(t)$ is such that

$$|\mathbf{v}(t) - \mathbf{u}(t)| \preceq \Delta U$$

for certain parameter $\Delta U \in \mathbb{R}^m$. Let $\mathbf{z}_a(t)$ be the analytical solution of the original ODE of Eq. (19.22) *and let $\mathbf{z}(t)$ be the QSS1 solution of Eq.* (19.23) *with $\mathbf{z}_a(t_0) = \mathbf{z}(t_0)$. Then,*

$$|\mathbf{z}_a(t) - \mathbf{z}(t)| \preceq |V| \cdot \left|\mathbb{R}\mathrm{e}(\Lambda^{-1}) \cdot \Lambda\right| \cdot \left(|V^{-1}| \cdot \Delta Q + |\Lambda^{-1} \cdot V^{-1}| \cdot \Delta U\right) \tag{19.24}$$

for all $t \ge t_0$, where $\Lambda = V^{-1} \cdot A \cdot V$ is a matrix of eigenvalues of A and V is a corresponding matrix of eigenvectors.

The proof of this Theorem is based on a change of variables that transforms the original system in a decoupled one, where an error bound can be easily deduced.

Theorem 19.4 shows that the numerical error is always less than a magnitude that is proportional to the quantum ΔQ, a result that we had already observed in the examples we analyzed. When the input $\mathbf{u}(t)$ is not piecewise constant, the error bound contains an additional term proportional to the maximum difference between the input and its piecewise constant approximation.

We can illustrate the use of this result on the example of Eq. (19.14). There, we have

$$A = \begin{bmatrix} 0 & 1 \\ -1 & -1 \end{bmatrix}$$

Matrix A can be decomposed in the following eigenvalues and eigenvector matrices:

$$\Lambda = \begin{bmatrix} -\frac{1}{2} + i \cdot \frac{\sqrt{3}}{2} & 0 \\ 0 & -\frac{1}{2} - i \cdot \frac{\sqrt{3}}{2} \end{bmatrix}; \; V = \begin{bmatrix} 1 & 1 \\ -\frac{1}{2} + i \cdot \frac{\sqrt{3}}{2} & -\frac{1}{2} - i \cdot \frac{\sqrt{3}}{2} \end{bmatrix};$$

and then, Eq. (19.24) gives the following error bound

$$|\mathbf{z}_a(t) - \mathbf{z}(t)| \preceq \begin{bmatrix} \frac{\sqrt{2}}{\sqrt{3}} \cdot (\Delta Q_1 + \Delta Q_2) \\ \frac{\sqrt{2}}{\sqrt{3}} \cdot (\Delta Q_1 + \Delta Q_2) \end{bmatrix}$$

That way, the first simulation we performed using a quantum $\Delta Q_1 = \Delta Q_2 = 0.1$ shown in Fig. 19.9, cannot have an error larger than $\sqrt{2}/\sqrt{3} \cdot 0.2 \approx 0.1633$ in each variable.

The fact that the error bound is proportional with the quantum, implies that obtaining a result ten times more accurate requires the reduction of the quantum by a factor of ten. Unfortunately, reducing the quantum ten times increases the number of simulation steps by the same factor. That way, the number of simulation steps and the computational cost are proportional to the accuracy requested. This relationship between accuracy and computational costs is identical to that of first order accurate classic numerical integration algorithms (Forward and Backward Euler).

As in Euler's methods, the linear relation between error and number of steps is one of the main drawbacks of QSS1. Next, we shall introduce higher order QSS algorithms that solve this problem.

19.3 QSS EXTENSIONS

Classic numerical ODE integration theory provides hundreds of algorithms. They are usually classified by their order (typically from 1 to 5, but methods of over 10th order are sometimes used), and by their explicit or implicit formulation. Higher order algorithms allow to obtain accurate results using large step sizes. Implicit algorithms, in turn, ensure numerical stability for arbitrarily large step sizes allowing the efficient integration of stiff systems.

The QSS1 algorithm introduced above is only first order accurate and it is explicit. We already observed that obtaining accurate results requires the use of a small quantum size ΔQ, which in turn results in small simulation steps. Thus, higher order QSS algorithms are necessary in order to obtain efficient solutions with decent accuracy.

Regarding stiff systems, we cannot expect that an explicit algorithm like QSS1 works. In fact, when we simulate a stable non-stiff system like that of Eq. (19.14) using QSS1, the numerical solutions finish with oscillations like those of Fig. 19.11. Those final oscillations impose a sort of upper limit to the step size, as it occurs in explicit discrete time algorithms like Forward Euler. If the system of Eq. (19.14) were the fast dynamics of a larger stiff system, this upper limit in the step size would provoke that the entire simulation takes too many steps just as it happens in all classic explicit algorithms.

The need of accurate solutions and efficient stiffness treatment motivated the development of the higher order and linearly implicit QSS algorithms that are introduced next.

19.3.1 HIGHER ORDER QSS METHODS

Second order accurate algorithms require to approximate not only the state derivative $\dot{\mathbf{z}}(t)$, but also its second derivative $\ddot{\mathbf{z}}(t)$. Thus, if we want to obtain a second order approximation of the solution of the ODE of Eq. (19.12), then the state derivative $\dot{\mathbf{z}}(t)$ cannot be piecewise constant (because that would imply that $\ddot{\mathbf{z}}(t)$ is zero). In consequence, in a second order accurate method the quantized state $\mathbf{q}(t)$ cannot be piecewise constant.

The conclusion is that we need a quantization function such that $\mathbf{q}(t)$ results piecewise linear. A simple extension of the hysteretic quantization function of Definition 19.1 for QSS1 is given below.

Definition 19.8 (First Order Hysteretic Quantization Function). Given two trajectories $z_i(t)$, and $q_i(t)$, we say that they are related by a first order hysteretic quantization function provided that

$$q_i(t) = z_i(t_k^-) + \dot{z}_i(t_k^-)(t - t_k) \quad t_k \le t < t_{k+1} \tag{19.25}$$

where

$$t_{k+1} = \sup_{t > t_k}(t : |q_i(t) - z(t)| \le \Delta Q_i) \tag{19.26}$$

and the parameter ΔQ_i is called *quantum*.

This definition says that $q_i(t)$ is piecewise linear. Every segment of $q_i(t)$ starts from $z_i(t_k)$ with the same slope of the state ($\dot{z}_i(t_k^-)$, where we use the left state derivative in case $\dot{z}_i(t)$ is discontinuous in t_k). A new segment of $q_i(t)$ starts when the difference between $z_i(t)$ and $q_i(t)$ becomes larger than the quantum ΔQ_i.

Fig. 19.12 shows the evolution of a state $z_i(t)$ and its quantized state $q_i(t)$ that now follows a piecewise linear trajectory. Notice that the fact that every segment of $q_i(t)$ starts with the slope of $z_i(t)$ implies that it takes longer to both trajectories to split from each other. That way, $q_i(t)$ can approximate $z_i(t)$ using fewer segments than those of the hysteretic quantization function used in QSS1.

Making use of the first order hysteretic quantization function, the second order accurate quantized state system method (QSS2) is defined as follows:

Definition 19.9 (QSS2 Method). Given the ODE

$$\dot{\mathbf{z}}(t) = \mathbf{f}(\mathbf{z}(t), \mathbf{u}(t))$$

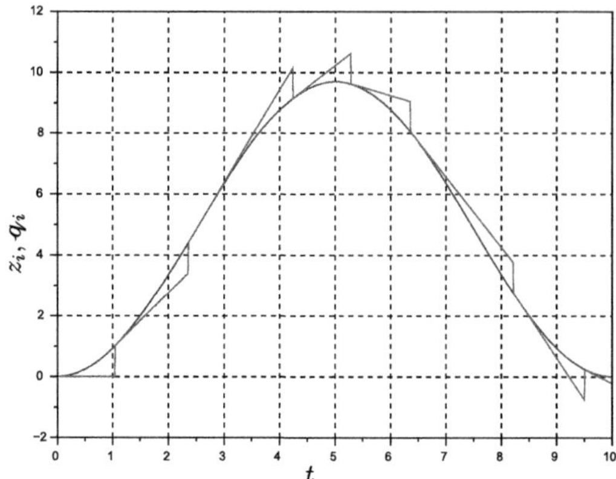

FIGURE 19.12

First Order Hysteretic Quantization with $\Delta Q_i = 1$.

where $\mathbf{z} \in \mathbb{R}^n$, and $\mathbf{u} \in \mathbb{R}^m$ is a vector of known input trajectories, the QSS2 method approximates it by the following *quantized state system*:

$$\dot{\mathbf{z}}(t) = \mathbf{f}(\boldsymbol{q}(t), \mathbf{v}(t))$$

where $\boldsymbol{q} \in \mathbb{R}^n$ is the vector of *quantized states*, whose i-th component q_i is related with the i-th state z_i by a first order hysteretic quantization function with quantum ΔQ_i, and $\mathbf{v}(t)$ is a piecewise linear approximation of the input vector $\mathbf{u}(t)$.

The definition of QSS2 is almost identical to that of QSS1. The only difference is that QSS2 uses piecewise linear quantized state trajectories instead of piecewise constant ones.

Provided that function $\mathbf{f}(\boldsymbol{q}(t), \mathbf{v}(t))$ is linear, it can be easily seen that the state derivatives $\dot{\mathbf{z}}(t)$ follow piecewise linear trajectories and then the state $\mathbf{z}(t)$ follow piecewise parabolic trajectories. In non-linear cases this is not true. However, the state derivatives $\dot{\mathbf{z}}(t)$ can be approximated by piecewise linear trajectories so that the algorithm can be implemented using simple DEVS models.

Assuming that state derivatives are piecewise linear, we can formulate a DEVS model for a QSS2 quantized integrator extending the ideas of QSS1. Taking into account that QSS2 trajectories are piecewise linear, each input or output segment is now characterized by two quantities: the initial segment value and the slope. Thus, the events now carry two real number instead of one.

Then, the following DEVS model represents the behavior of a QSS2 integrator:

$$HQI2 = < X, Y, S, \delta_{\text{int}}, \delta_{\text{ext}}, \lambda, \text{ta} >$$

with

- $X = \mathbb{R}^2 \times \{\text{in}_1\}$, $Y = \mathbb{R}^2 \times \{\text{out}_1\}$,
- $S = \mathbb{R}^3 \times \mathbb{R}^2 \times \mathbb{R}_0^+$,
- $\delta_{\text{int}}(s) = \delta_{\text{int}}(((z, \dot{z}, \ddot{z}), (q, \dot{q}), \sigma)) = ((\tilde{z}, \dot{\tilde{z}}, \ddot{\tilde{z}}), (\tilde{z}, \dot{\tilde{z}}), \tilde{\sigma})$,

- $\delta_{\text{ext}}(s, e, x) = \delta_{\text{ext}}(((z, \dot{z}, \ddot{z}), (q, \dot{q}), \sigma), e, ((x_v, \dot{x}_v), \text{in}_1)) = ((\hat{z}, x_v, \dot{x}_v, (\hat{q}, \dot{q}), \hat{\sigma}),$
- $\lambda(s) = \lambda(((z, \dot{z}, \ddot{z}), (q, \dot{q}), \sigma)) = ((\tilde{z}, \dot{\tilde{z}}), \text{out}_1)$
- $\text{ta}(s) = \text{ta}((z, \dot{z}, \ddot{z}), (q, \dot{q}), \sigma) = \sigma,$

where the state after the internal transition has components

$$\tilde{z} = z + \dot{z} \cdot \sigma + \frac{1}{2} \cdot \ddot{z} \cdot \sigma^2$$
$$\dot{\tilde{z}} = \dot{z} + \ddot{z} \cdot \sigma$$
$$\tilde{\sigma} = \begin{cases} \sqrt{\dfrac{2 \cdot \Delta Q}{\ddot{z}}} & \text{if } \ddot{z} \neq 0 \\ \infty & \text{otherwise,} \end{cases}$$

while the state after the external transition has components

$$\hat{z} = z + \dot{z} \cdot e + \frac{1}{2} \cdot \ddot{z} \cdot e^2$$
$$\hat{q} = q + \dot{q} \cdot e$$
$$\hat{\sigma} = \max\{T : 0 \le \tau \le T \Rightarrow |z + \dot{z} \cdot \tau + \frac{1}{2} \cdot \ddot{z} \cdot \tau^2 - q - \dot{q} \cdot \tau| \le \Delta Q\}$$

where the last condition says that $\hat{\sigma}$ is computed as the maximum time advance such that the state and the quantized state have a difference limited by the quantum ΔQ. This value can be obtained after solving two quadratic equations (one for ΔQ and the other one for $-\Delta Q$), i.e., finding the minimum real positive solution of

$$z + \dot{z} \cdot \hat{\sigma} + \frac{1}{2} \cdot \ddot{z} \cdot \hat{\sigma}^2 - q - \dot{q} \cdot \hat{\sigma} \pm \Delta Q = 0 \tag{19.27}$$

If there is no real positive solution for Eq. (19.27), then $\hat{\sigma} = \infty$ since the state and the quantized state never split from each other farther than ΔQ.

Like in QSS1, the simulation using QSS2 requires building a block diagram like that of Fig. 19.8 in page 497, but replacing the QSS1 integrators by QSS2 integrators. Considering that QSS2 integrators receive and send events containing not also the values but also the slopes of the different trajectories, the memoryless functions that compute $\dot{z}(t)$ from $\boldsymbol{q}$ and $\mathbf{u}$ must also take the slopes into account. Thus, the DEVS models corresponding to QSS2 memoryless functions can be represented as follows:

$$MF_2 = < X, Y, S, \delta_{\text{int}}, \delta_{\text{ext}}, \lambda, \text{ta} >$$

with

- $X = \mathbb{R}^2 \times \{\text{in}_1^{\boldsymbol{q}}, \ldots, \text{in}_n^{\boldsymbol{q}}, \text{in}_1^{\mathbf{v}}, \ldots, \text{in}_m^{\mathbf{v}}\}$, $Y = \mathbb{R}^2 \times \{\text{out}_1\}$,
- $S = \mathbb{R}^n \times \mathbb{R}^n \times \mathbb{R}^m \times \mathbb{R}^m \times \mathbb{R}_0^+$,
- $\delta_{\text{int}}(s) = \delta_{\text{int}}(\boldsymbol{q}, \dot{\boldsymbol{q}}, \mathbf{v}, \dot{\mathbf{v}}, \sigma) = (\boldsymbol{q}, \dot{\boldsymbol{q}}, \mathbf{v}, \dot{\mathbf{v}}, \infty)$,
- $\delta_{\text{ext}}(s, e, x) = \delta_{\text{ext}}((\boldsymbol{q}, \dot{\boldsymbol{q}}, \mathbf{v}, \dot{\mathbf{v}}, \sigma, e, (x_v, \dot{x}_v, p)) = (\tilde{\boldsymbol{q}}, \dot{\tilde{\boldsymbol{q}}}, \tilde{\mathbf{v}}, \dot{\tilde{\mathbf{v}}}, 0)$,

- $\lambda(s) = \lambda(q, \dot{q}, \mathbf{v}, \dot{\mathbf{v}}, \sigma) = (f_i(q, \mathbf{v}), \frac{\partial f_i}{\partial q}(q, \mathbf{v}) \cdot \dot{q} + \frac{\partial f_i}{\partial \mathbf{v}}(q, \mathbf{v}) \cdot \dot{\mathbf{v}}, \text{out}_1)$,
- $\text{ta}(s) = \text{ta}((q, \mathbf{v}, \sigma) = \sigma$,

where

$$(\tilde{q}_i, \tilde{\dot{q}}_i) = \begin{cases} (x_v, \dot{x}_v) & \text{if } p = \text{in}_i^{q} \\ (q_i + \dot{q}_i \cdot e, \dot{q}_i) & \text{otherwise} \end{cases}$$

and

$$(\tilde{v}_i, \tilde{\dot{v}}_i) = \begin{cases} (x_v, \dot{x}_v) & \text{if } p = \text{in}_i^{\mathbf{v}} \\ (v_i + \dot{v}_i \cdot e, \dot{v}_i) & \text{otherwise} \end{cases}$$

Notice that the output function $\lambda(s)$ computes both the value and the time derivative of function $f_i(q, \mathbf{v})$. For that goal, it uses the chain rule:

$$\frac{\mathrm{d} f_i}{\mathrm{d}t}(q, \mathbf{v}) = \frac{\partial f_i}{\partial q}(q, \mathbf{v}) \cdot \dot{q} + \frac{\partial f_i}{\partial \mathbf{v}}(q, \mathbf{v}) \cdot \dot{\mathbf{v}}$$

Fig. 19.13 shows the QSS2 of the system of Eq. (19.14) using a quantum $\Delta Q_i = 0.001$. This solution is very similar to that of QSS1 in Fig. 19.10.

In fact, we can expect that using the same quantum QSS1 and QSS2 obtain results with similar accuracy as Theorem 19.4 also holds in QSS2. The reason is that the theorem is based on the fact that q_i and z_i never differ from each other in more than the quantum ΔQ_i, a fact that is also true in QSS2.

In spite of achieving the same accuracy, QSS1 solution involves 1393 changes in q_1 and 1313 changes in q_2 while QSS2 only involves 73 and 78 changes in q_1 and q_2, respectively. This is due to the second order nature of QSS2.

In order to compare the performance of both algorithms we simulated the system of Eq. (19.14) varying the quantum from $\Delta Q_i = 0.1$ to $\Delta Q_i = 10^{-6}$. The results reported in Table 19.1 show that,

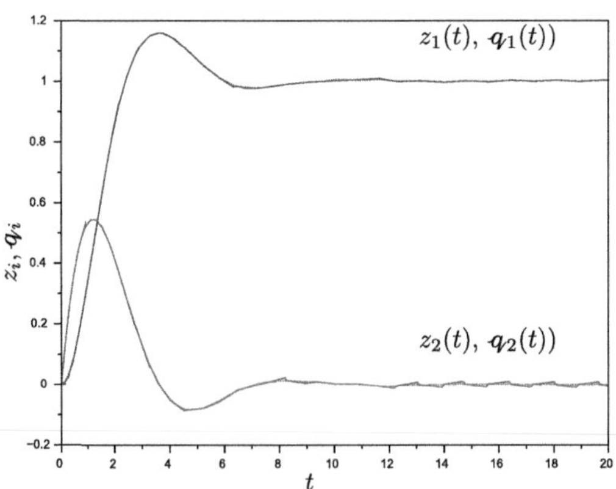

FIGURE 19.13

QSS2 Solution of the system of Eq. (19.14) with $\Delta Q_i = 0.001$. (Analytical solution with dotted lines.)

Table 19.1 Quantum ΔQ_i vs. number of events in QSS1 and QSS2

Quantum ΔQ_i	QSS1 Steps	QSS2 Steps
0.1	47	42
0.01	280	63
0.001	2706	151
0.0001	26,966	457
0.00001	269,500	1457
0.000001	2,694,863	4619

except for the first row, the number of events in QSS1 grow linearly with the accuracy (the inverse of the quantum) while it grows with the square root of the accuracy in QSS2. The relatively larger number of events using a quantum $\Delta Q_1 = 0.1$ shall be explained soon.

While the table shows that QSS2 always performs less steps than QSS1, it is not necessary true that it performs less calculations. A simple comparison of the DEVS models of the quantized integrators and memoryless functions of both algorithms shows that QSS2 events involve more than twice the number of calculations of QSS1 events. Thus, QSS2 performs better when certain minimum accuracy is requested.

Following the idea of QSS2, a third order accurate method called QSS3 was also developed. In this algorithm, the quantized state trajectories follow piecewise parabolic trajectories while the state trajectories follow piecewise cubic trajectories. In QSS3, the number of events grow with the cubic root of the accuracy and it can perform faster than QSS2 provided that the quantum is small enough.

For instance, using QSS3 with quanta $\Delta Q_i = 10^{-6}$ in the system of Eq. (19.14), the simulation takes a total of 547 steps, i.e., about 8 times less events than QSS2. However, QSS3 steps involve about 3 times more calculations than those of QSS2, so the advantages of QSS3 are not so noticeable.

19.3.2 LINEARLY IMPLICIT QSS

A close look at Fig. 19.13 shows that the simulated trajectories have final oscillations around their equilibrium values. While the amplitude of those oscillations has the order of the quantum so the error they introduce is not outside the prescribed tolerance, they do have a negative impact in the simulation performance. The problem is that these *spurious* oscillations are associated to additional simulation steps. Starting from $t = 10$ there is a total of 20 events in both variables, where we would expect that the simulation reaches a final stable point.

These oscillations are the reason why both algorithms (QSS1 and QSS2) do not respect the linear and quadratic relationship between the accuracy and the number of steps for $\Delta Q_i = 0.1$. These final oscillations add events at a fixed rate, so when the overall number of steps is low, the number of events caused by the oscillations is large in proportion.

In the system under analysis the existence of these spurious oscillations does not significantly affect the simulation performance. However, if the frequency of these oscillations were significantly larger, the number of events may become unacceptable.

Let us analyze then the same model of Eq. (19.14) but suppose that now the friction coefficient is $b = 100$, so the equations are

$$\begin{aligned}\dot{z}_1(t) &= z_2(t) \\ \dot{z}_2(t) &= -z_1(t) - 100.01 z_2(t) + 1\end{aligned} \tag{19.28}$$

The solution to initial conditions $x_1(0) = x_2(0) = 0$ is given by

$$\begin{aligned}z_1(t) &= 1 + \frac{1}{9999} e^{-100t} - \frac{10000}{9999} e^{-\frac{t}{100}} \\ z_2(t) &= -\frac{100}{9999} e^{-100t} + \frac{100}{9999} e^{-\frac{t}{100}}\end{aligned} \tag{19.29}$$

Fig. 19.14 shows the state trajectories of these system.

The behavior looks like that of a first order model, governed by the *slow modes* of the form $e^{-\frac{t}{100}}$ in the solution of Eq. (19.29). This can be easily explained by the reason that the *fast modes* of the form e^{-100t} become negligible after a few milliseconds. In fact, when $t = 0.05$ it results that $e^{-100t} \approx 6.7 \times 10^{-3}$.

Fig. 19.15 plots the same solution until $t = 0.05$. There, the *fast* dynamics introduced by the mode e^{-100t} can be appreciated.

Continuous time models exhibiting simultaneous slow and fast dynamics like that of Eq. (19.28) are known as *stiff systems* and they are problematic in the context of numerical integration. Due to stability reasons, the presence of fast dynamics enforces the use of very small step sizes in explicit algorithms like Forward Euler. In the model of Eq. (19.28), any step size $h > 0.02$ would make the simulation with Forward Euler unstable. Taking into account that a reasonable final simulation time of $t_f = 500$ appears to be reasonable according to Fig. 19.14, Forward Euler needs more than $500/0.02 = 25{,}000$ steps to simulate this system.

For these reasons, stiff systems are usually integrated using implicit algorithms like Backward Euler that do not suffer from stability issues and can use larger step sizes. However implicit algorithms are more complicated and their implementation requires the use of iterative routines at each step.

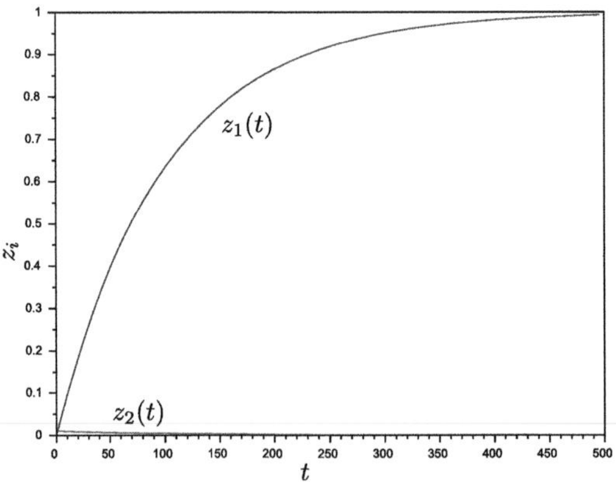

FIGURE 19.14

State Trajectories of Eq. (19.28).

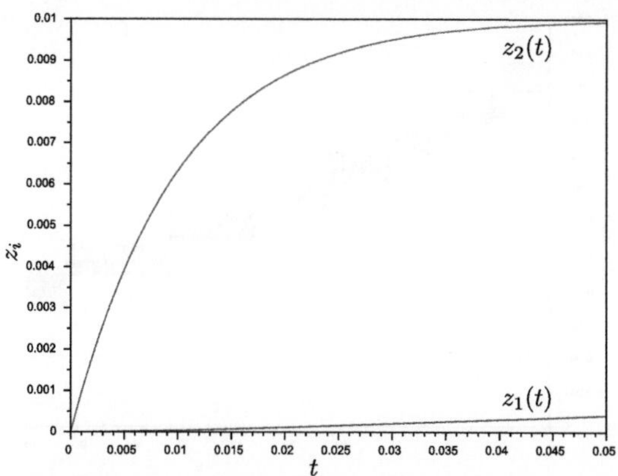

FIGURE 19.15

State Trajectories of Eq. (19.28) (startup detail).

Let us see what happens with QSS1. We first simulated the system using quantum $\Delta Q_1 = 0.1$, $\Delta Q_2 = 0.001$ and obtained the result of Fig. 19.16 after only 30 events (10 changes in q_1 and 20 changes in q_2).

These results may suggest that QSS1 can simulate stiff systems in a very efficient manner. Unfortunately, this is not true. A small change in the quantum so that $\Delta Q_1 = 0.09$ and $\Delta Q_2 = 0.0009$ produces that the simulation has 11 changes in q_1 but 9872 changes in q_2.

Fig. 19.17 shows the trajectories of the state $z_2(t)$ and the corresponding quantized state $q_2(t)$. There, fast spurious oscillations can be appreciated. A more detailed view of the oscillations is shown in Fig. 19.18.

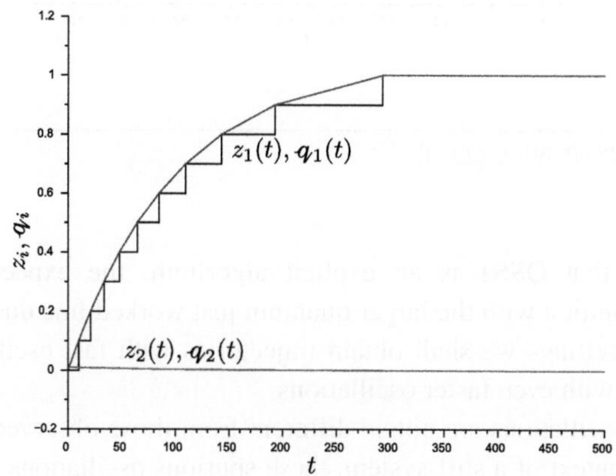

FIGURE 19.16

QSS1 Simulation of Eq. (19.28).

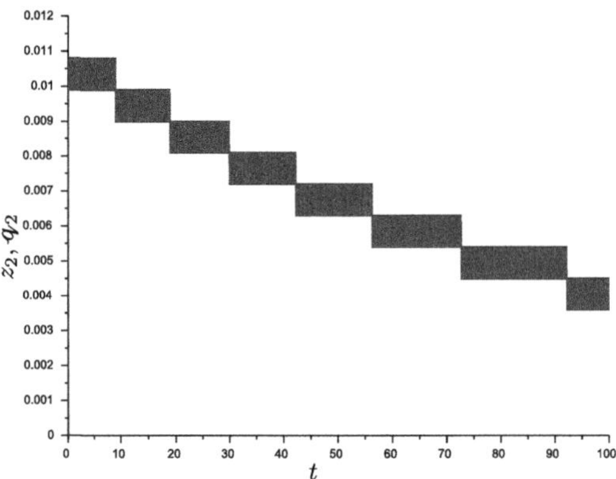

FIGURE 19.17

QSS1 Simulation of Eq. (19.28).

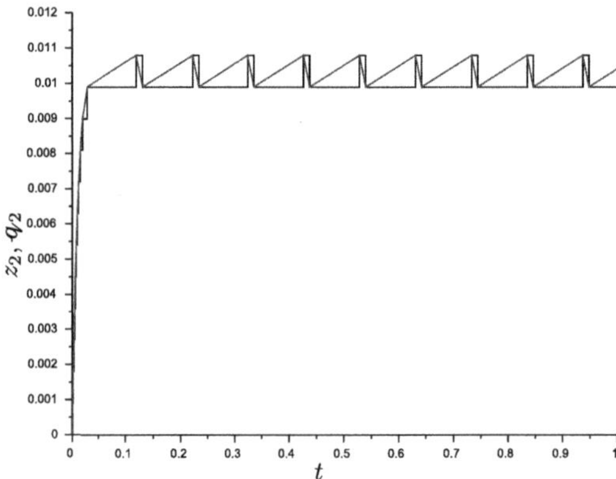

FIGURE 19.18

QSS1 Simulation of Eq. (19.28) (startup detail).

Taking into account that QSS1 is an explicit algorithm, the expected behavior is that of Figs. 19.17–19.18. The solution with the larger quantum just worked fine due to lucky circumstances, but using most quantum settings we shall obtain trajectories with fast oscillations. Using QSS2 the solutions are also similar, with even faster oscillations.

In fact, the observed oscillations are not so different from those observed in Fig. 19.13. The only difference is that in the context of a stiff system, these spurious oscillations are faster with respect to the total simulation time. In consequence, the total number of events becomes unacceptable.

In order to solve this inconvenient, we shall take into account the way in which classic numerical integration algorithms deal with stiff systems. As we mentioned earlier, only implicit algorithms are able to efficiently integrate stiff ODEs. These methods are based on computing future values of the state derivatives.

Taking into account that QSS methods compute the state derivatives using the quantized states $\mathbf{q}$ instead of the states $\mathbf{z}$, the idea to design an implicit quantized state algorithm is that each quantized state q_i contains a *future* value of z_i. Thus, the quantized state q_i must take a value above z_i when the state goes up and a value below q_i when the state goes down.

Thus, in principle, we propose that q_i starts with a value $z_i + \Delta Q_i$ when $\dot{z}_i > 0$ or a value $z_i - \Delta Q_i$ when $\dot{z}_i < 0$ and then the quantized state remains constant until the state z_i reaches q_i. At that point, q_i increases or decreases its value in ΔQ_i according to the sign of $\dot{z}_i$.

The problem with this idea is that the sign of $\dot{z}_i$ may depend on the choice of q_i. We recall that the i-th component of a QSS approximation of an ODE is given by

$$\dot{z}_i(t) = f_i(\mathbf{q}(t), \mathbf{v}(t)) \tag{19.30}$$

so when we increase q_i in ΔQ_i it could happen that $\dot{z}_i$ becomes negative and q_i would not be a future value for z_i. Similarly, if we decrease q_i in ΔQ_i the state derivative may become positive and we would face the same problem. In that situation, we cannot change the quantized state to values $q_i \pm \Delta Q_i$ so that it is a future state value. However, in that case, according to the *Mean Value Theorem*, there is an intermediate value $\tilde{q}_i$ such that $\dot{z}_i = 0$ (i.e., the state z_i reaches an *equilibrium*).

In order to predict the future value of the state derivative $\dot{z}_i$ and to be able to find the intermediate value $\tilde{q}_i$ we first rewrite Eq. (19.30) as follows:

$$\dot{z}_i(t) = J_{i,i} q_i + \nu_i \tag{19.31}$$

where

$$J_{i,i} \triangleq \frac{\partial f_i}{\partial q_i} \tag{19.32}$$

is a *linear approximation* term and

$$\nu_i \triangleq f_i(\mathbf{q}(t), \mathbf{v}(t)) - J_{i,i} q_i \tag{19.33}$$

is an *affine* term. Notice that, provided that function f_i has only linear dependence on q_i, then the *affine* term ν_i does not depend at all on q_i.

Then, computing the *linear term* $J_{i,i}$ and the *affine term* ν_i and neglecting the dependence of ν_i on q_i, we can easily predict the future value for the state derivative $\dot{z}_i$ making use of Eq. (19.31). Moreover, when the future state derivative value changes its sign so that q_i cannot be chosen as a future value for z_i, we can use the linear approximation of Eq. (19.31) to obtain the equilibrium value $\tilde{q}_i$ as

$$\tilde{q}_i = -\frac{\nu_i}{J_{i,i}} \tag{19.34}$$

These ideas are the basis of the first order accurate *Linearly Implicit* Quantized State System (LIQSS1) algorithm. The *linearly implicit* denomination obeys to the fact that the method predicts the future state derivatives making use a linear approximation.

Before presenting a DEVS model for the LIQSS1 quantized integrator, we shall introduce a simple example that shows the behavior of this new algorithm. Consider the first order model:

$$\dot{z}_1(t) = -z_1(t) + 2.5 \tag{19.35}$$

with initial state $z_1(0) = 0$. Let us simulate this system with a LIQSS1 approximation using quantum $\Delta Q_1 = 1$.

The LIQSS1 approximation is expressed as

$$\dot{z}_1(t) = -q_1(t) + 2.5 \tag{19.36}$$

where it can be easily seen that $J_{1,1} = -1$ and $v_1 = 2.5$.

Taking into account that $z_1(0) = 0$ we may set the initial quantized state value $q_1(0) = \pm\Delta Q = \pm 1$. We must choose the positive value $q_1(0) = 1$ because that way the state derivative results $\dot{z}_1(0) = 1.5 > 0$ and the state moves towards the quantized state.

Since $\dot{z}_1 = 1.5$, after $t_1 = 1/1.5$ units of time it results $z_1(t_1) = q_1(t_1) = 1$ and then the quantized state must increase its value so that $q_1(t_1^+) = 2$. Now, it results $\dot{z}_1(t_1^+) = 0.5$ so the state still moves towards the quantized state q_1.

At time $t_2 = t_1 + 1/0.5$ we have $z_1(t_2) = q_1(t_2)$. However, if we increase the quantized state so that $q_1 = 3$, it would result $\dot{z}_1(t_2^+) = -0.5$ and the state does not go towards q_i. Similarly, if we decrease the quantized state so that $q_1 = 1$ we already knew that $\dot{z}_i = 1.5$ and we have the same problem. In this situation, we make $q_i = \tilde{q}_1$ where the equilibrium value is computed from Eq. (19.34) as $\tilde{q}_1 = 2.5$.

Once we make $q_1 = 2.5$ it results that $\dot{x}_1 = 0$ and the simulation ends up at an equilibrium state. The whole simulation required a total of three steps. If we had used QSS1 with the same quantum, the simulation would have finished with spurious oscillations around the equilibrium point $z_1 = 2.5$.

In conclusion, the fact that LIQSS1 uses the quantized states as future values of the states is the reason why spurious oscillations disappear.

In order to formalize the algorithm, we define the LIQSS1 quantization as follows:

Definition 19.10 (Linearly Implicit Quantization Function). Given two trajectories $z_i(t)$, and $q_i(t)$, and two variable parameters $J_{i,i}(t)$ (called linear parameter), and $v_i(t)$ (called affine parameter), we say that the trajectories are related by a linearly implicit quantization function provided that

$$q_i(t) = \begin{cases} q_i(t^-) & \text{if } q_i(t^-) \neq z_i(t) \wedge |q_i(t^-) - z_i(t)| < 2\Delta Q_i \\ \tilde{q}_i(t) & \text{otherwise} \end{cases} \tag{19.37}$$

where

$$\tilde{q}_i(t) = \begin{cases} z_i(t^-) + \Delta Q_i & \text{if } J_{i,i}(z_i(t^-) + \Delta Q_i) + v_i(t) \geq 0 \\ z_i(t^-) - \Delta Q_i & \text{if } J_{i,i}(z_i(t^-) - \Delta Q_i) + v_i(t) \leq 0 \\ -\dfrac{v_i(t)}{J_{i,i}(t)} & \text{otherwise} \end{cases} \tag{19.38}$$

and the parameter ΔQ_i is called *quantum*.

The first condition says that the quantized state remains constant until the state reaches it (as we originally stated) or until the state moves aside beyond twice the quantum size. This last condition prevents that the state moves far away in the opposite direction of the quantized state when it takes the wrong direction due to the presence of non-linear terms or to changes in other quantized states that affect the value of $\dot{z}_i$.

Based on the definition of the linearly implicit quantization function, we can define the LIQSS1 method as follows:

Definition 19.11 (LIQSS1). Given the ODE

$$\dot{\mathbf{z}}(t) = \mathbf{f}(\mathbf{z}(t), \mathbf{u}(t))$$

where $\mathbf{z} \in \mathbb{R}^n$, and $\mathbf{u} \in \mathbb{R}^m$ is a vector of known input trajectories, the LIQSS1 method approximates it by the quantized state system:

$$\dot{\mathbf{z}}(t) = \mathbf{f}(\boldsymbol{q}(t), \mathbf{v}(t))$$

where $\boldsymbol{q} \in \mathbb{R}^n$ is the vector of *quantized states*, whose i-th component q_i is related with the i-th state z_i by a linearly implicit quantization function with quantum ΔQ_i, linear parameter

$$J_{i,i}(t) = \frac{\partial f_i}{\partial q_i}(t)$$

and affine parameter

$$\nu_i(t) = f_i(\boldsymbol{q}(t), \mathbf{v}(t)) - J_{i,i} q_i.$$

$\mathbf{v}(t)$ is a piecewise constant approximation of the input vector $\mathbf{u}(t)$.

The DEVS implementation of LIQSS1 is similar to that of QSS1, except that the LIQSS1 Quantized Integrators are more complex as they must also compute the linear and affine parameters, $J_{i,i}$ and ν_i. LIQSS1 memoryless functions are the same as QSS1 memoryless functions as both algorithms work with piecewise constant approximations.

In order to demonstrate the way LIQSS1 works we simulated again the stiff system of Eq. (19.28) using quanta $\Delta Q_1 = 0.09$ and $\Delta Q_2 = 0.0009$ (the setup that provoked the fast oscillations in QSS1). LIQSS1 completed the 500 seconds of simulated time performing 12 changes in q_1 and 24 changes in q_2, i.e., a total of 36 simulation steps.

Fig. 19.19 shows the trajectories obtained with LIQSS1. They are similar to those of Fig. 19.16, except that the quantized state trajectories are now *future* values of the states. In addition, there are no oscillations so the number of steps performed is very small.

In this example, LIQSS1 solved the problem caused by the system stiffness related to the appearance of fast oscillations in the QSS1 approximation. Unfortunately, the idea behind LIQSS1 sometimes fails.

The quantization function used by LIQSS1 ensures that the quantized state q_i is chosen as a future value of the state z_i. However, when the memoryless function that computes $\dot{z}_i$ receives an event informing a change in some other quantized state q_j, the sign of $\dot{z}_i$ may change. In that case, the state z_i no longer moves in the direction of q_i, and fast spurious oscillations might appear back, depending on the model structure. Anyway, in many practical stiff models LIQSS1 works fine.

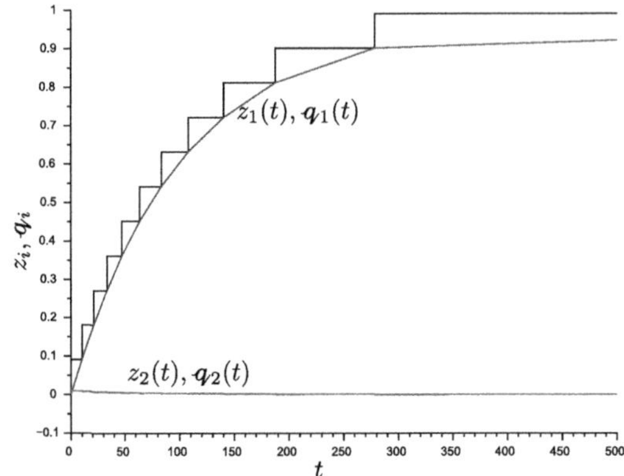

FIGURE 19.19

LIQSS1 Simulation of Eq. (19.28).

Another limitation of LIQSS1 is that, like QSS1, it only performs a first order approximation. Thus, obtaining a decent accuracy implies performing too many steps. Motivated by this fact, a second order accurate LIQSS method called LIQSS2 was also proposed. LIQSS2 combines the ideas of LIQSS1 with those of QSS2: The quantized state trajectories are piecewise linear, but q_i is computed so that a future value of the state z_i and its slope coincide with it.

With the same idea, a third order accurate method called LIQSS3 was also proposed, completing the family of linearly implicit QSS algorithms.

19.4 QSS SIMULATION OF HYBRID SYSTEMS

One of the main advantages of QSS algorithms is that they can simulate hybrid systems in a very efficient fashion.

From a continuous system simulation perspective, hybrid systems contain a continuous time dynamics interacting with discrete time or discrete event subsystems. The problem is that the presence of discrete changes are seen as discontinuities at the continuous time system side, that usually provoke unacceptable errors when those systems are integrated using conventional discrete time algorithms.

In order to safely simulate hybrid systems, conventional algorithms must implement *event handling routines*, where discontinuities are detected and the simulation is restarted after each event occurrence. In presence of frequent discontinuities, these procedures add a huge computational load that significantly slows down the simulations.

Continuous time simulation tools usually represent hybrid systems as follows:

$$\dot{\mathbf{z}}(t) = \mathbf{f}(\mathbf{z}(t), \mathbf{u}(t), \mathbf{d}(t), t) \tag{19.39}$$

where $\mathbf{d}(t)$ is now a vector of *discrete* states that can only change when the condition

$$Zc_i(\mathbf{z}(t), \mathbf{u}(t), \mathbf{d}(t), t) = 0 \tag{19.40}$$

is met for some $i \in \{1, 2, \ldots, Z\}$. Functions $Zc_i(\mathbf{z}(t), \mathbf{u}(t), \mathbf{d}(t), t)$ are called *zero-crossing functions*. Whenever the condition given by Eq. (19.40) becomes true, the discrete states are updated according to the expression:

$$\mathbf{d}(t^+) = Hc_i(\mathbf{z}(t), \mathbf{u}(t), \mathbf{d}(t), t) \tag{19.41}$$

where function $Hc_i(\mathbf{z}(t), \mathbf{u}(t), \mathbf{d}(t), t)$ is the *discontinuity handler* associated to Zc_i.

The event handling routines used by conventional discrete time algorithms usually must iterate on Eq. (19.40) to find the right *event time*, then they must advance the simulation up to that time, execute the handler routine that implements Eq. (19.41), and then restart the simulation from the new condition.

Using QSS algorithms, however, the problem becomes considerably easier. In that case, Eq. (19.39) becomes

$$\dot{\mathbf{z}}(t) = \mathbf{f}(\mathbf{q}(t), \mathbf{v}(t), \mathbf{d}(t), t) \tag{19.42}$$

while zero-crossing conditions are

$$Zc_i(\mathbf{q}(t), \mathbf{v}(t), \mathbf{d}(t), t) = 0 \tag{19.43}$$

Recalling that the quantized state and input trajectories follow piecewise polynomial trajectories, the zero-crossing condition can be straightforwardly detected. Moreover, their times of occurrence can be predicted and scheduled in advance.

In addition, after the value of $\mathbf{d}(t)$ is updated, there is no need to restart the whole simulation. QSS algorithms already perform discrete changes in $\mathbf{q}(t)$ without introducing any problem to the integration of Eq. (19.42).

Let us illustrate these facts in the bouncing ball model we had introduced in Chapter 3:

$$\begin{aligned} \dot{z}_1(t) &= z_2(t) \\ \dot{z}_2(t) &= -g - d(t) \cdot \left(\frac{k}{m} z_1(t) + \frac{b}{m} z_2(t)\right) \end{aligned} \tag{19.44}$$

where

$$d(t) = \begin{cases} 0 & \text{if } z_1(t) > 0 \\ 1 & \text{otherwise} \end{cases} \tag{19.45}$$

We recall that $z_1(t)$ is the ball position and z_2 represents its speed. The discrete state $d(t)$ takes the value 0 when the ball is in the air ($x_1 > 0$) and the value 1 when the ball is in contact with the floor ($x_1 \le 0$).

In order to simulate this system in a correct way, conventional discrete time numerical integration methods needed to be equipped with some routine that detects the condition $z_1(t) = 0$ to change the value of $d(t)$ in the precise moment of the zero crossing. The difficulty is that working at discrete times, the most likely situation is that the crossing occurs in the middle of two steps (i.e., $z_1(t_k) > 0$ and $z_1(t_{k+1}) < 0$). Thus, the algorithm must come backward in time to find the correct value of t_{k+1} so that it coincides with that of the zero crossing. This process usually requires iterations until an accurate value for t_{k+1} is found.

In QSS algorithms the solution is simpler. If first order accurate algorithms like QSS1 or LIQSS1 are used, the change in $d(t)$ can only occur after the quantized state q_1 changes, so the function corresponding to Eq. (19.45) can be implemented in DEVS using a straightforward memoryless function.

In second order QSS algorithms (QSS2, LIQSS2), the change in the sign of z_1 may occur at any instant of time. However, we know that $q_1(t)$ follows a piecewise linear trajectory. Thus, we can easily predict when q_1 is crossing 0. This time, the calculation of $d(t)$ requires using a DEVS model like the following one:

$$HF_2 = < X, Y, S, \delta_{\text{int}}, \delta_{\text{ext}}, \lambda, \text{ta} >$$

with

- $X = \mathbb{R}^2 \times \{\text{in}_1\}$, $Y = \mathbb{R}^2 \times \{\text{out}_1\}$,
- $S = \mathbb{R}^2 \times \mathbb{R}_0^+$,
- $\delta_{\text{int}}(s) = \delta_{\text{int}}(q_1, \dot{q}_1, \sigma) = (q + \dot{q} \cdot \sigma, \dot{q}, \infty)$,
- $\delta_{\text{ext}}(s, e, x) = \delta_{\text{ext}}((q_1, \dot{q}_1, \sigma), e, (x_v, \dot{x}_v, p)) = (x_v, \tilde{x}_v, \tilde{\sigma})$,
- $\lambda(s) = \lambda(q_1, \dot{q}_1, \sigma) = d$,
- $\text{ta}(s) = \text{ta}(q_1, \dot{q}_1, \sigma) = \sigma$,

where

$$\tilde{\sigma} = \begin{cases} -\dfrac{x_v}{\dot{x}_v} & \text{if } x_v \cdot \dot{x}_v < 0 \\ \infty & \text{otherwise} \end{cases}$$

and

$$d = \begin{cases} 1 & \text{if } \dot{x}_v < 0 \\ 0 & \text{otherwise} \end{cases}$$

In a general hybrid system, the QSS approximation of a hybrid system is depicted in Fig. 19.20. Besides the quantized integrators and static functions, the model contains a subsystem with the discrete dynamics that receives the quantized states and input trajectories and computes the discrete state trajectories.

Besides simplifying the event detection and handling, the use of QSS in hybrid systems has the advantage of representing the whole approximated system in a unique formalism (DEVS).

19.5 LOGARITHMIC QUANTIZATION

We already established that the error introduced by a QSS algorithm is proportional to the quantum size ΔQ. Thus, by establishing a quantum $\Delta Q = 0.001$ we might obtain an accuracy of the order of 0.001. However, that accuracy may be adequate when the signals have the absolute value around 1, but completely inadequate when if the signals move around 10^{-6}. Thus, we should choose the quantum in concordance with the signal amplitudes. Unfortunately, in many situations we do not know the signal values before the simulation is run, so we cannot predict a correct value for each quantum ΔQ_i.

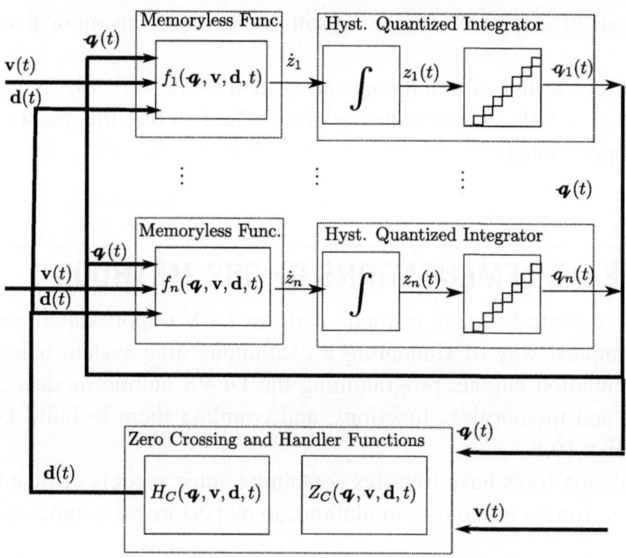

FIGURE 19.20

Block Diagram representation of a Hybrid System approximated by a QSS method.

The idea to overcome this problem is to dynamically change the quantum ΔQ_i so that it is proportional to the signal z_i, i.e.,

$$\Delta Q_i(t) = \Delta Q_{\text{rel}} \cdot |z_i(t)|$$

where the parameter ΔQ_{rel} is called *relative quantum.*

The problem with this approach is that when the signal z_i crosses zero, ΔQ_i also goes to zero and the number of steps may become too large. In order to avoid that problem, the quantization is defined as

$$\Delta Q_i(t) = \max(\Delta Q_{\text{rel}} \cdot |z_i(t)|, \Delta Q_{\text{min}}) \tag{19.46}$$

where the parameter ΔQ_{min} is the *minimum quantum*. Notice that selecting $\Delta Q_{\text{min}} = 0.001$ and $\Delta Q_{\text{rel}} = 0$ we obtain a constant quantization $\Delta Q = 0.001$.

We showed before that using a constant quantum the QSS algorithms intrinsically control the error defined as the difference between the numerical and the analytical solution, i.e.,

$$\mathbf{e}(t) \triangleq \mathbf{z}(t) - \mathbf{z}_a(t)$$

The usage of *logarithmic quantization* as defined in Eq. (19.46), however, produces and intrinsic control of the *relative error*:

$$\mathbf{e}_{\text{rel}}(t) \triangleq \frac{\mathbf{z}(t) - \mathbf{z}_a(t)}{\|\mathbf{z}_a(t)\|}$$

which is the usual goal of step size control algorithms used in conventional numerical integration methods.

Fig. 19.21 shows the results of simulating the system of Eq. (19.14) using a relative quantum $\Delta Q_{\rm rel} = 0.1$ and $\Delta Q_{\rm abs} = 0.01$ in both state variables. The fact that the quantum increases with the signals can be easily appreciated.

19.6 SOFTWARE IMPLEMENTATIONS OF QSS METHODS

The different Quantized State Systems methods perform DEVS approximations of continuous time systems. Thus, the simplest way of simulating a continuous time system using QSS algorithms is by using a DEVS simulation engine, programming the DEVS atomic models corresponding to the quantized integrators and memoryless functions, and coupling them to build the equivalent DEVS model as depicted in Fig. 19.8.

Most DEVS simulation tools have libraries containing some models of quantized integrators and elementary memoryless functions, so the simulation can be performed without actually defining these atomic models.

Among these tools, there is one called PowerDEVS that contains quantized integrators for all QSS methods (QSS1, QSS2, QSS3, LIQSS1, LIQSS2, and LIQSS3). Moreover, PowerDEVS also contains a large collection of DEVS models for memoryless functions so that most continuous time systems can be simulated without creating or modifying any DEVS atomic model.

While DEVS-based tools implement QSS algorithms by coupling quantized integrators and memoryless functions, there is a different approach that builds a single DEVS atomic model containing the whole QSS approximation of a system. This is the idea followed by the *Stand Alone QSS Solver* and the main advantage is that it allows to simulate considerably faster than DEVS-based tools. The reason for this advantage is that state derivatives are computed in single function calls while in DEVS coupled

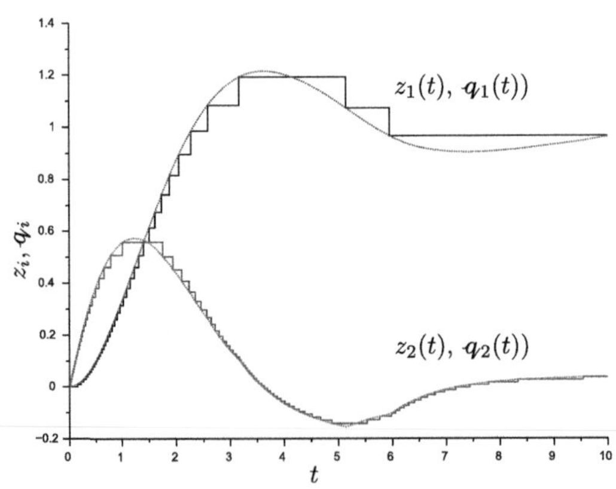

FIGURE 19.21

QSS1 Simulation of Eq. (19.14) with logarithmic quantization.

model they are computed by DEVS models that receive and send events provoking several function calls.

Below, we first introduce the implementation and usage of QSS methods in PowerDEVS and then we describe the Stand Alone QSS Solver.

19.6.1 POWERDEVS

PowerDEVS (Bergero and Kofman, 2011) was originally conceived as a general purpose DEVS simulation tool specialized for QSS simulation of Continuous Time Systems. Thus, it provides a Graphical User Interface similar to those of some popular continuous system simulation tools like Simulink (Matlab) or Xcos (Scilab).

The main PowerDEVS window is shown in Fig. 19.22. The model at the right corresponds to that of Eq. (19.14). The library shown in the left, called *Continuous*, contains the atomic DEVS models corresponding to quantized integrators and different memoryless functions.

Besides the Continuous library, there is a PowerDEVS library called *Sources* containing different atomic DEVS models that produce different input trajectories for the models. One of these *Event Generators* is the *Step Source* in Fig. 19.22 that produces the trajectory for signal $F(t)$ in the model of Eq. (19.14).

Another PowerDEVS library is the one called *Sinks*, with atomic DEVS models that receive events and store them for visualization or post-processing purposes. The *GNU-Plot* block in Fig. 19.22, for instance, was taken from that library and it plots the trajectories corresponding to the received events.

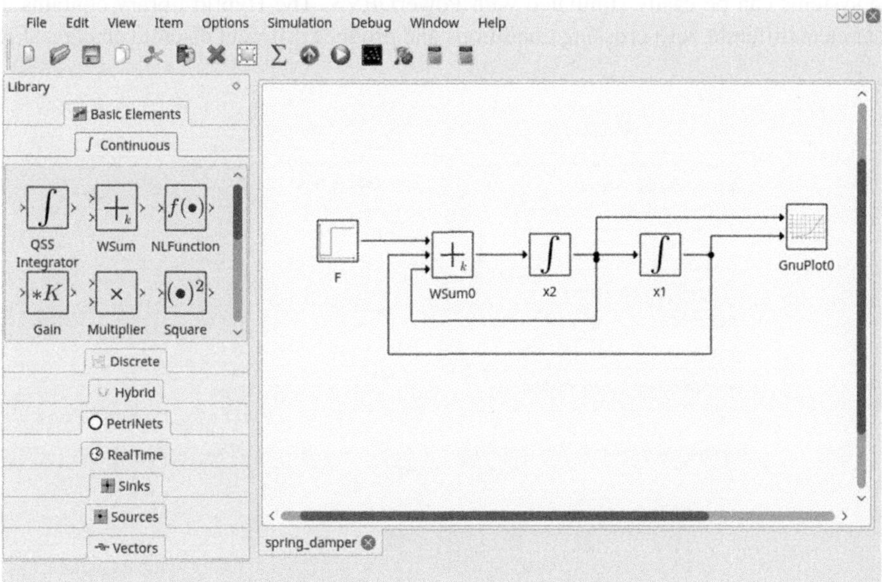

FIGURE 19.22

PowerDEVS main window.

The different atomic models have a set of parameters that can be user-modified from the graphical user interface without changing the DEVS definition. When the user double-clicks on a block, a dialog window like those of Fig. 19.23 is opened allowing the user to change those parameters. In this case, the left parameter windows corresponds to that of a QSS integrator, allowing to choose between different QSS methods, the quantum size and the initial state. The right window is that of a *Weighted Sum* block representing a memoryless function that computes $\dot{z}_2 = F(t)/m - k/m \cdot z_1 - b/m \cdot z_2(t)$.

PowerDEVS simulations are communicated in runtime with Scilab. Thus, atomic PowerDEVS models can read and write Scilab variables. This feature is used in most atomic models to read parameters defined in Scilab Workspace. Notice that the parameters in the right of Fig. 19.23 contain variables m, k, and b, whose values are computed by Scilab.

Once the model is built and parametrized, it can be simulated. Invoking the simulation opens a window like that of Fig. 19.24, that allows to setup the experiment, including options to synchronize the simulation with the wall-clock time, to re-run the simulation several times, to advance only a fixed number of transitions, etc.

In the case of the model of Fig. 19.22, as it contains a GNUPlot block, after running the simulation the results are displayed on a GNUPlot window like that of Fig. 19.25.

The events managed by the different blocks of the Continuous library of PowerDEVS are characterized by an array of values representing the coefficients of each polynomial segment. The quantized integrators are in charge of defining the order of those polynomials according to the chosen algorithm (QSS1, QSS2, etc.). Memoryless functions, in turn, compute their output polynomials with the same order of the polynomials they receive at the input. That way, the same memoryless functions can be used with QSS algorithms of different order.

Hybrid systems can be easily simulated with PowerDEVS. The Hybrid library contains different blocks that detect different zero-crossing conditions and produce different discrete changes. Fig. 19.26

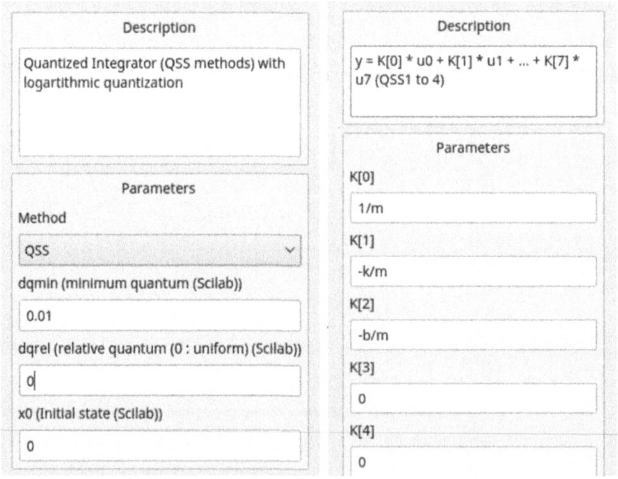

Description

Quantized Integrator (QSS methods) with logartithmic quantization

Parameters

Method

QSS

dqmin (minimum quantum (Scilab))

0.01

dqrel (relative quantum (0 : uniform) (Scilab))

0

x0 (Initial state (Scilab))

0

Description

y = K[0] * u0 + K[1] * u1 + ... + K[7] * u7 (QSS1 to 4)

Parameters

K[0]

1/m

K[1]

-k/m

K[2]

-b/m

K[3]

0

K[4]

0

FIGURE 19.23

Parameter dialog windows.

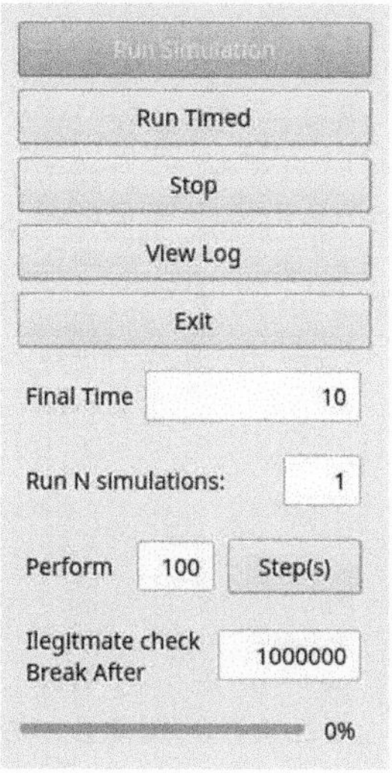

FIGURE 19.24

Simulation dialog window.

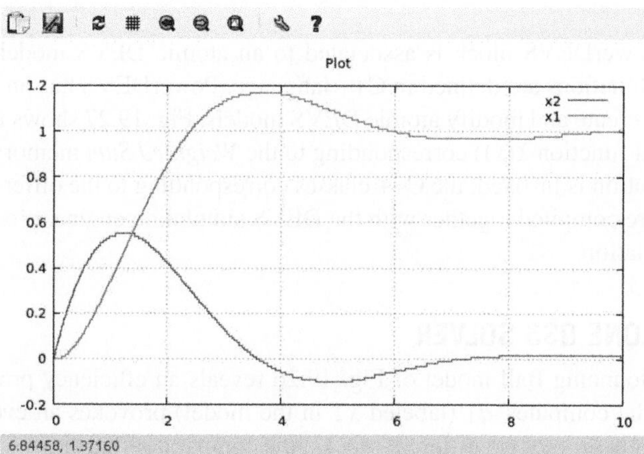

FIGURE 19.25

GNUPlot displaying simulation results.

for instance shows the model of the bouncing ball of Eqs. (19.44)–(19.45). On the left of the model, some blocks corresponding to the Hybrid Library can be seen.

PowerDEVS also contain libraries with discrete time blocks and with a DEVS implementation of Petri Nets. The different blocks use the same type of data in the events than those of QSS approximations, so they can be combined to build multi-formalism models.

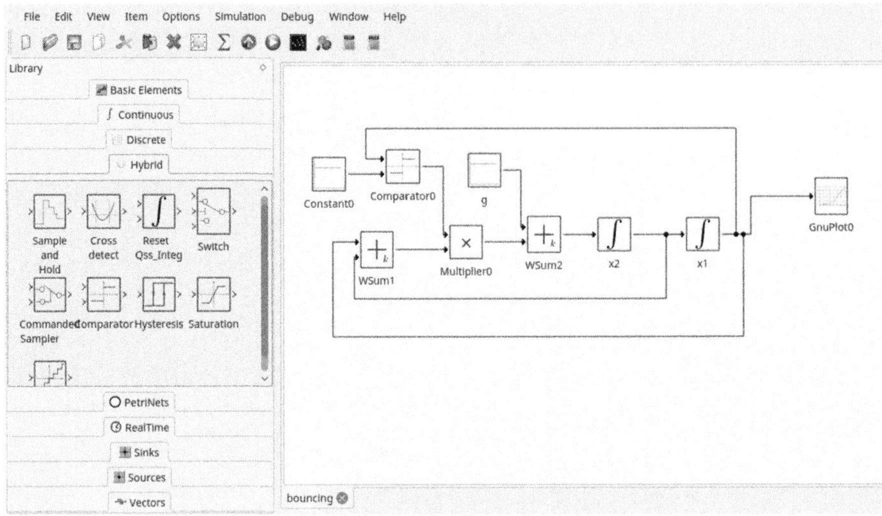

FIGURE 19.26

Bouncing ball model and the PowerDEVS Hybrid Library.

Internally, each PowerDEVS block is associated to an atomic DEVS model where the different transition and output functions are defined in C++ language. PowerDEVS has an atomic DEVS editor that allows the user to create and modify atomic DEVS models. Fig. 19.27 shows this editor with a part of the code (the output function $\lambda(s)$) corresponding to the *Weighted Sum* memoryless function.

Whenever a simulation is invoked, the C++ classes corresponding to the different atomic blocks included in the model are compiled together with the DEVS simulation engine, producing an executable file that runs the simulation.

19.6.2 STAND ALONE QSS SOLVER

A close look to the Bouncing Ball model of Fig. 19.26 reveals an efficiency problem. Whenever the quantized integrator the computes q_1 (labeled $x1$ in the model) provokes an event with a change in that variable, the event is sent to two different blocks: the comparator that checks the condition $q_1 > 0$ and the *WSum1* block that computes the term $\frac{k}{m}q_1 + \frac{b}{m}q_2$. The comparator will compute the future crossing time ad schedule an event for that time. The *WSum1* block, however, sets its time advance to zero, and in the next transition it provokes an event with the value $\frac{k}{m}q_1 + \frac{b}{m}q_2$. That event is sent to the multiplier that computes $d \cdot (\frac{k}{m}q_1 + \frac{b}{m}q_2)$ that sets its time advance to zero so that in the next transition it sends an event with the new output value. This event is received be *WSum2* block that sets its time advance to zero and computes the state derivative $\dot{z}_2 = -g - d \cdot (\frac{k}{m}q_1 + \frac{b}{m}q_2)$. That way, in the next transition it sends this value to the quantized integrator that computes q_2 completing the step.

Thus, the transition that changes the value in $q_1(t)$ triggers a chain of several events that have the unique goal of computing $\dot{z}_2 = -g - d \cdot (\frac{k}{m}q_1 + \frac{b}{m}q_2)$. The computational cost associated to the

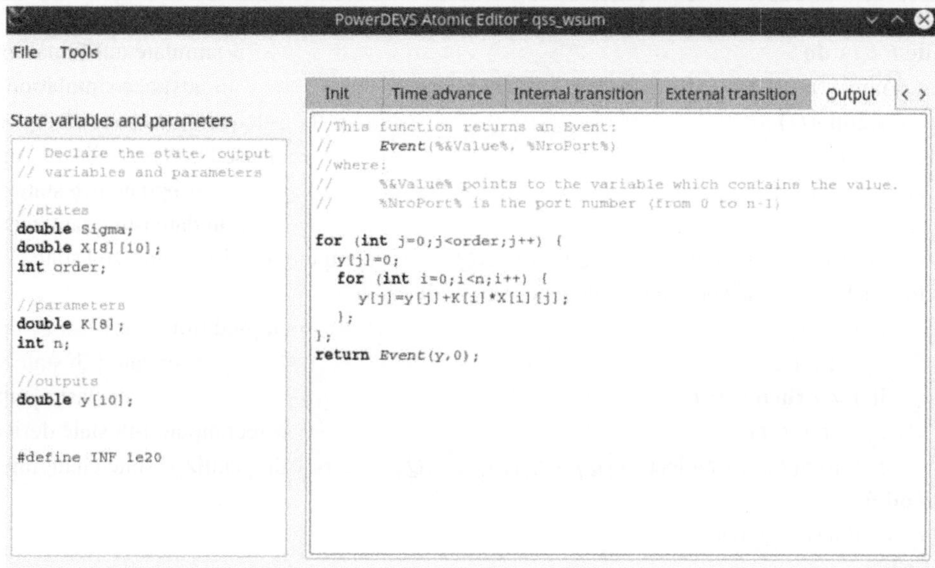

FIGURE 19.27

PowerDEVS Atomic Models Editor.

simulation of that chain of events is considerably higher than what it is required to actually compute the state derivative. In spite of being simple, the implementation of QSS algorithms as DEVS coupled models is inefficient.

A solution to this problem is to build the complete QSS approximation as a single DEVS atomic model. A first approach to do that is to follow the *closure under coupling* property of DEVS building the equivalent atomic model corresponding to the coupled DEVS model. However, the resulting atomic DEVS model would still perform all the internal transitions performed by the coupled DEVS model and the efficiency would not be significantly improved.

A more efficient approach is to translate the logics of the complete QSS simulation into a compact algorithm. For that goal, we consider the QSS approximation of Eq. (19.6) given by

$$\dot{z}_1(t) = f_1(q_1(t), \cdots, q_n(t), v_1(t), \cdots, v_n(t))$$
$$\dot{z}_2(t) = f_2(q_1(t), \cdots, q_n(t), v_1(t), \cdots, v_n(t))$$
$$\vdots$$
$$\dot{z}_n(t) = f_n(q_1(t), \cdots, q_n(t), v_1(t), \cdots, v_n(t))$$

For simplicity, we shall assume that the input trajectories $v_i(t)$ are constant.

Let t_j $(j = 1, \ldots, n)$ denote the next time at which $|q_j(t) - z_j(t)| = \Delta Q_j$, i.e. the time for the next change in q_j. Then, the QSS1 simulation algorithm works as follows:

Algorithm 21 QSS1.

1: **while** $t < t_f$ **do** ▷ simulate until final time tf
2: $t = \min(t_j)$ ▷ advance simulation time
3: $i = \operatorname{argmin}(t_j)$ ▷ the i-th quantized state changes first
4: $e = t - t_i^z$ ▷ elapsed time since last z_i update
5: $z_i = z_i + \dot{z}_i \cdot e$ ▷ update i-th state value
6: $q_i = z_i$ ▷ update i-th quantized state
7: $t_i = \min(\tau > t)$ subject to $|q_i - z_i(\tau)| = \Delta Q_i$ ▷ compute next i-th quantized state change
8: **for** each $j \in [1, n]$ such that $\dot{z}_j$ depends on q_i **do**
9: $e = t - t_j^z$ ▷ elapsed time since last z_j update
10: $z_j = z_j + \dot{z}_j \cdot e$ ▷ update j-th state value
11: **if** $j \neq i$ **then** $t_j^z = t$ ▷ last z_j update
12: $\dot{z}_j = f_j(q, t)$ ▷ recompute j-th state derivative
13: $t_j = \min(\tau > t)$ subject to $\left|q_j - z_j(\tau)\right| = \Delta Q_j$ ▷ j-th quantized state changing time
14: **end for**
15: $t_i^z = t$ // last z_i update
16: **end while**

The algorithm above can be seen as a specialized DEVS simulator for a specific model. Compared with the coupled DEVS model of the QSS approximation in Fig. 19.8, the advantage of the algorithm is that in each step it recomputes all the state derivatives and future event times in a single transition. Moreover, it evaluates each state derivative using a single function call (line 12).

A disadvantage of the new approach is that it requires some extra-knowledge about the model structure: In line 8 there is a condition involving the dependence of state derivatives in state variables. In the coupled DEVS model of Fig. 19.8 that information in intrinsically contained in the coupling structure. In the new algorithm, however, the information must be provided as an *incidence matrix*.

This is the idea followed by the Stand Alone QSS solver, a software tool coded in C language that implements the complete family of QSS methods.

This tool simulates models that can contain discontinuities represented by discrete states, zero-crossing functions and discontinuity handlers, as we did in Eqs. (19.39)–(19.41).

The simulations are performed by three modules interacting at runtime:

1. The **Integrator**, that integrates Eq. (19.42) assuming that the piecewise polynomial quantized state trajectory $\boldsymbol{q}(t)$ is known.
2. The **Quantizer**, that computes $\boldsymbol{q}(t)$ from $\mathbf{z}(t)$ according to the QSS method in use and their tolerance settings (there is a different **Quantizer** for each QSS method). That way, it provides the polynomial coefficients of each quantized state $q_i(t)$ and computes the next time at which a new polynomial section starts (i.e., when the condition $|q_i(t) - z_i(t)| = \Delta Q_i$ is met).
3. The **Model**, that computes the scalar state derivatives $\dot{z}_i = f_i(\boldsymbol{q}, \mathbf{v}, \mathbf{d}, t)$, the zero-crossing functions $Zc_i(\boldsymbol{q}, \mathbf{v}, \mathbf{d}, t)$, and the corresponding event handlers. Besides, it provides the structural information required by the algorithms.

The structure information of the **Model** is automatically extracted at compile time by a **Model Generator** module. This module takes a standard model described in a subset of the Modelica language called μ-Modelica and produces an instance of the **Model** module as required by the QSS solver.

In addition, the Stand Alone QSS Solver also offers a front-end to classic numerical solvers like DASSL, DOPRI, and CVODE.

Fig. 19.28 shows the main window of the Stand-Alone QSS Solver, where the μ-Modelica description of the bouncing ball model of Eqs. (19.44)–(19.45) is under edition. Notice the close correspondence between the equations and the μ-Modelica model description.

Efficiency comparisons between PowerDEVS and the Stand-Alone QSS Solver show that the later is about 30 times faster using first order algorithms (QSS1, LIQSS1), about 12 times faster in second order algorithms (QSS2, LIQSS2) and around 8 times faster in third order methods (QSS3, LIQSS3).

19.7 APPLICATIONS OF QSS METHODS

Compared with conventional ODE solvers, QSS algorithms have the following advantages:

- They can efficiently handle discontinuities: zero-crossing conditions are straightforwardly detected and the occurrence of events only produce calculations in the state derivatives that directly depend on the discrete states that changed.
- In presence of stiff systems with some particular structure (large terms in the main diagonal of the Jacobian matrix), non-iterative LIQSS algorithms can obtain efficient results. This advantage is notorious when the systems are large, as implicit conventional ODE solvers must iterate on large systems of equations.
- In large sparse systems that exhibit localized activity QSS algorithms only performs calculations where the changes occur.

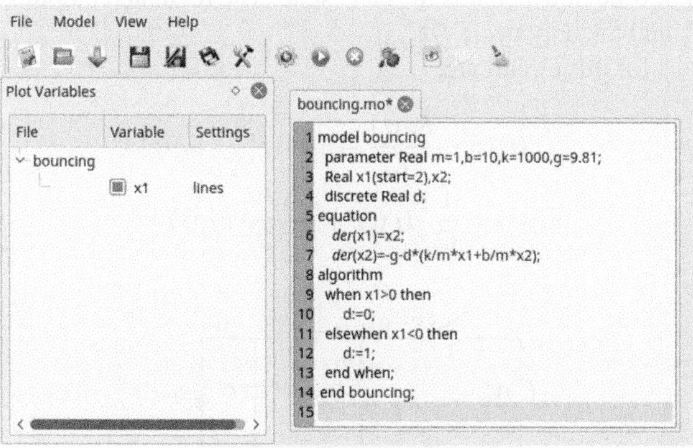

FIGURE 19.28

Stand Alone QSS Solver.

There are several application areas where the continuous time models have some of the features mentioned above:

- Power Electronic Systems: These systems usually contain switches commuting at high frequency so that the models contain frequent discontinuities. In addition, the models are usually stiff.
- Advection–Diffusion–Reaction Systems: These systems are widely used to model the concentration of chemical species (usually contaminants) in fluids (usually rivers). They are represented by Partial Differential Equations that, after being spatially discretized by the Method of Lines, result in a large, sparse, and stiff system of ODEs.
- Building Simulations: Several models in building simulations contain frequent discontinuities (air conditioners turning on and off, for instance).
- Spiking Neural Networks: These models have frequent discontinuities corresponding to the neuron firings.
- Smart Grids: These model usually contain several Power Electronic Converters, so they combine the main advantages of QSS algorithms (discontinuities, stiffness and large scale).

Below, we present some examples and compare the simulation results of QSS algorithms with those of some of the most efficient conventional numerical ODE solvers.

19.7.1 A DC–DC BUCK CONVERTER CIRCUIT

DC–DC power electronic converters circuits that produce an output DC voltage that can be lower or higher than the input DC voltage. The conversion is performed by high frequency commutations of a switching device. Due to their high energy efficiency, these converters are nowadays used in almost every electronic device that requires a regulated DC voltage (DC chargers for mobile phones, laptops, etc.).

A very popular voltage reducer circuit known as *Buck Converter* is depicted in Fig. 19.29. In that circuit, the switch Sw commutates at high frequency (of the order of tens of kilohertz), varying its *duty cycle*, i.e., the fraction of time in which it is in "ON" state. The regulated output voltage is then a result of the input voltage V_s and the duty cycle DC.

The model equations for this circuit are:

$$
\begin{aligned}
\dot{i}_L(t) &= -\frac{R_d(t)}{L} \cdot i_D(t) - \frac{1}{L} \cdot u_C(t) \\
\dot{u}_C(t) &= \frac{1}{C} \cdot i_L(t) - \frac{1}{R \cdot C} \cdot u_C(t)
\end{aligned}
$$

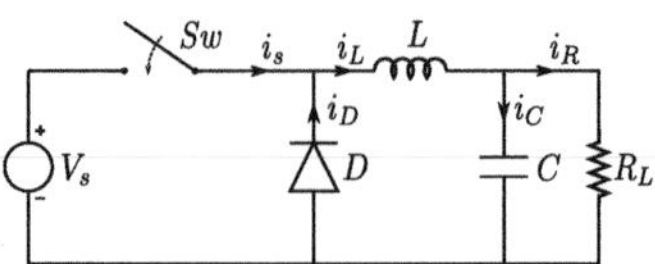

FIGURE 19.29

DC–DC Buck Converter Circuit.

where the current at the diode $i_D(t)$ is computed as

$$i_D(t) = \frac{R_s(t) \cdot i_L(t) - V_s}{R_s(t) + R_d(t)}$$

The switch resistance $R_s(t)$ can take two values:

$$R_s(t) = \begin{cases} R_{\text{ON}} & \text{if the switch is in "ON" state} \\ R_{\text{OFF}} & \text{if the switch is in "OFF" state} \end{cases}$$

Similarly, the diode resistance $R_d(t)$ can take two values

$$R_d(t) = \begin{cases} R_{\text{ON}} & \text{if the diode is in "conduction" state} \\ R_{\text{OFF}} & \text{if the diode is in "cut-off" state} \end{cases}$$

The switch state is driven by the switching strategy, with a frequency f_s and a duty cycle DC. Thus, it is said that the changes in the switch state are driven by *time events*, i.e., by events that are known in advance.

The diode state, in turn, depends on the current and voltage. When the diode current $i_D(t)$ becomes negative, the diode goes to "cut-off" state. Then, when the diode voltage $v_D(t) = R_d(t) \cdot i_D(t)$ becomes positive, the diode enters the "conduction" state. Thus, the diode state is driven by *state events* whose occurrence depend on some variables computed during the simulation.

In order to simulate this system, we adopted parameters $C = 10^{-4}$ F, $L = 10^{-4}$ H, $R = 10\ \Omega$, $V_s = 24$, $f_s = 10^4$ Hz, $DC = 0.5$, $R_{\text{ON}} = 10^{-5}\ \Omega$, $R_{\text{OFF}} = 10^5\ \Omega$.

We first simulated the system using classic solvers. The presence of discontinuities in this case enforces the usage of variable step algorithms equipped with event detection routines. In this case, we used CVODE–BDF, DASSL, and DOPRI solvers, three widely used algorithms for stiff (CVODE–BDF and DASSL) and non-stiff (DOPRI) systems.

We selected the relative and absolute tolerances as $rel_{tol} = abs_{tol} 10^{-3}$, and we set the final time to $t_f = 0.1$ sec. Fig. 19.30 shows the simulated trajectories until $t = 0.003$ sec.

DASSL needed 43,980 steps to complete the simulation. In those steps, it evaluated 140,862 times the whole right hand side of the ODE, and it computed 79,181 times the zero-crossing functions. Running on a AMD Athlon(tm) II X2 270 Processor it took 56.3 milliseconds to complete the simulation.

CVODE–BDF improved DASSL results completing the simulation after 24,004 steps, where the whole right hand side of the ODE was evaluated 49,000 times and the zero-crossing functions were computed 185,712 times. On the same computer, the simulation took 31.4 milliseconds.

In the case of DOPRI, it took 4,729,902 steps to complete the simulation, with 32,400,403 right hand side evaluations and computing 112,524,768 times the zero-crossing functions. On the same computer, the simulation took 7348.77 milliseconds. The reason for the poor performance of this algorithm is that the system becomes stiff when the switch and diode are both in "OFF" state.

We then simulated the system using LIQSS2 algorithm with the same tolerance settings, i.e., $\Delta Q_{\text{rel}} = \Delta Q_{\text{abs}} = 10^{-3}$. This time, the simulation was completed with 6511 changes in variable $u_C(t)$, 13,516 changes in $i_L(t)$, and 4000 steps performed by the event handlers. In consequence, the equation that computes $\dot{u}_c(t)$ was evaluated 40,054 times, the equation that computes $\dot{i}_L(t)$ was evaluated 48,054 times and the zero-crossing functions were evaluated 35,032 times. The simulation, on the

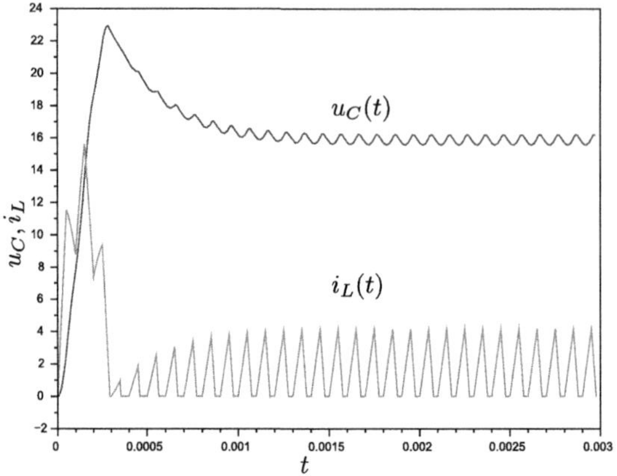

FIGURE 19.30

Buck Converter Simulated Trajectories (startup).

same computer, took 14.1 milliseconds, a significant improvement compared with the conventional algorithms.

The advantages of LIQSS2 observed in this example have two reasons: the algorithm is more efficient to deal with discontinuities and it can integrate stiff system without using implicit formulas.

Regarding accuracy, in this example all the methods provide results within the requested tolerance settings.

19.7.2 A POPULATION OF AIR CONDITIONERS

In several places that experience high temperatures in summer, a significant amount of the power consumed by the population is due to the presence of air conditioners (AC). In order to perform some control on the total power consumption at certain times, some electricity companies are studying the possibility of manipulating in a centralized way the temperature set point of the air conditioner population. Increasing the set point of all air conditioners in a population by 1^0C during some peak consumption period will not affect much the people's comfort but can imply a significant reduction of the power consumption peak. The problem is that the set point change provokes an instantaneous reduction on the power consumption, but then thee is a sort of *rebound* effect. The following model allows to study that phenomenon.

We consider a population consisting in N rooms. The temperature $\theta_i(t)$ of the i-th room is controlled by an air conditioner according to the following law:

$$\dot{\theta}_i(t) = \frac{\theta_a(t) - \theta_i(t)}{R_i \cdot C_i} - \frac{P_i(t)}{C_i};$$

where $\theta_a(t)$ is the ambient temperature, $P_i(t)$ is the power delivered by the i-th air conditioner, C_i is the room thermal capacitance and R_i is the thermal resistance with the ambient.

$$P_i(t^+) = \begin{cases} 0 & \text{if } state = OFF \\ P_{\text{MAX},i} & \text{if } state = ON \end{cases}$$

The i-th air conditioners is turned on when $\theta_i(t) > \theta_{\text{ref}}(t) + 0.5$ and it is turned off when $\theta_i(t) < \theta_{\text{ref}}(t) - 0.5$. Here, $\theta_{\text{ref}}(t)$ is the global reference temperature.

The total power consumption is computed as

$$P_{\text{total}}(t) = \sum_{i}^{N} P_i(t)$$

Notice that the model has N state variables, where N is normally a large number. It also contains discontinuities associated to the air conditioners turning on and off.

Like in the previous case, we simulated this model with conventional and with QSS algorithms.

For that purpose, we simulated a population of $N = 1000$ air conditioners with random (uniformly distributed) parameters in the intervals $C_i \in \{550, 650\}$, $R_i \in \{1.8, 2.2\}$, $P_{\text{MAX},i} \in \{13, 15\}$. The ambient temperature was considered constant $\theta_a(t) = 32^0\text{C}$ and the initial room temperatures were randomly chosen in the interval $\theta_i(0) \in \{18, 22\}$. The reference temperature was initially set at $\theta_{\text{ref}}(t) = 20$, and at time $t = 1000$ it was increased to $\theta_{\text{ref}}(t) = 20.5$.

Fig. 19.31 shows part of the simulated trajectory for $P_{\text{total}}(t)$. There, the undesirable transient due to the change in the reference temperature can be observed.

As before, we first simulated the system using classic solvers. Since the system is non-stiff, DOPRI produced the best results completing the simulation until a final time $t_f = 3000$ after 32,465 steps, 194,792 full function evaluations, and 443,844,631 zero-crossing function computations. That took a CPU time of 11.3 seconds. Regarding stiff solvers, CVODE–BDF performs 37,667 simulation steps taking 29.5 seconds to complete the simulation. DASSL, in turn, performs 81,000 simulation steps that take 1806 seconds of CPU time. The difference between DOPRI and CVODE–BDF can be explained by the fact that CVODE–BDF steps are more expensive than DOPRI steps (BDF are implicit algo-

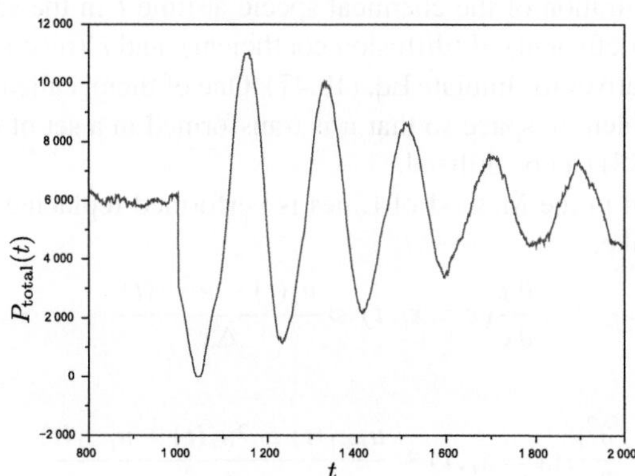

FIGURE 19.31

Total Power Consumption of an Air Conditioner Population.

rithms). In the case of DASSL, the poor performance is the consequence that it cannot exploit the fact that the system is sparse.

QSS2, in turn, performed a total of 103,116 transitions (taking into account changes in quantized states and event handler executions). That took 54 milliseconds of CPU time, a result more than 200 times faster than DOPRI. The stiff QSS solver LIQSS2 performs more transitions (137,733) and takes 84 milliseconds. The problem is that LIQSS2 spends some time trying to find equilibrium points that are never reached as the commutations occur before any equilibrium situation (we recall the system is not stiff). Anyway, the difference between QSS2 and LIQSS2 is not as significant as that of stiff and non-stiff classic solver. In consequence, LIQSS2 works reasonably well in non-stiff models.

If the number of air conditioners is increased to $N = 2000$, for instance, DOPRI needs 46.6 seconds to complete the simulation while QSS2 needs 146 milliseconds (almost 320 times faster). The advantages of QSS algorithms become more significant as the number of AC units is increased. The reason of this is related to the *event density*. When the size of the model grows, the time between successive discontinuities decreases. Thus, classic solvers are enforced to decrease the global step size while the cost of each step grows (because the system is larger). QSS algorithm do not experience this problem because the steps are local to each state.

19.7.3 ADVECTION–DIFFUSION–REACTION EQUATION

Several branches of the science and engineering use continuous models where the variables do not only depend on the time but also on the space coordinates. In these *distributed parameter models*, the variables appear differentiated in time and in space, leading to *partial derivative equations* (PDEs).

The Advection–Diffusion–Reaction (ADR) equation is a PDE that appears in several models corresponding to the transportation of chemical species in water courses. The one dimensional version of the ADR equation is given below:

$$\frac{\partial u(x,t)}{\partial t} + a\frac{\partial u(x,t)}{\partial x} = d\frac{\partial^2 u(x,t)}{\partial x^2} + r(u(x,t)^2 - u(x,t)^3) \tag{19.47}$$

where $u(x,t)$ is the concentration of the chemical specie at time t in the spacial position x. The parameter are: a (advection coefficient), d (diffusion coefficient), and r (reaction coefficient).

There are several alternatives to simulate Eq. (19.47). One of them, called *Method of Lines*, consists in discretizing first the problem in space so that it is transformed in a set of ODEs for which different ODE solvers (classic or QSS) can be utilized.

The space discretization in the Method of Lines is performed replacing the spatial derivatives by finite difference formulas like:

$$\frac{\partial u}{\partial x}(x = x_i, t) \approx \frac{u_i(t) - u_{i-1}(t)}{\Delta x} \tag{19.48}$$

and

$$\frac{\partial^2 u}{\partial x^2}(x = x_i, t) \approx \frac{u_{i+1}(t) - 2u_i(t) + u_{i-1}(t)}{\Delta x^2} \tag{19.49}$$

for $i = 1, \cdots, N$, where N is the number of grid points, Δx is the *grid width*,

$$u_i(t) \approx u(x_i, t) \tag{19.50}$$

is the i-th state variable of the resulting ODE and

$$x_i = i \cdot \Delta x \tag{19.51}$$

is the i-th spatial grid point.

Replacing the partial derivatives in Eq. (19.47) and selecting appropriate *border* conditions (at $i = 0$ and $i = N$), we obtain the following set of ODEs.

$$\dot{u}_i(t) = -a\frac{(u_i(t) - u_{i-1}(t))}{\Delta x} + d\frac{(u_{i+1}(t) - 2u_i(t) + u_{i-1}(t))}{\Delta x^2} + r(u_i^2 - u_i^3) \tag{19.52}$$

for $i = 1, \cdots, N-1$ and

$$\dot{u_N}(t) = -a\frac{(u_N(t) - u_{N-1}(t))}{\Delta x} + d\frac{(2u_{N-1}(t) - 2u_N(t))}{\Delta x^2} + r(u_N(t)^2 - u_N(t)^3) \tag{19.53}$$

In most applications, a large value for N must be used, as it implies using a refined grid that leads to more accurate results. Thus, the system of Eqs. (19.52)–(19.53) is large. Moreover, the term r introduces stiffness. Also, the term d combined with a large value of N also introduces stiffness. For these reasons, ADR equations impose certain difficulties to numerical ODE algorithms.

Setting parameters $N = 1000$, $a = 1$, $d = 0.1$, $r = 1000$, and $\Delta x = 0.01$, and initial conditions:

$$u_i(0) = \begin{cases} 1 & \text{if } i \leq 200 \\ 0 & \text{otherwise} \end{cases}$$

we simulated the system using classic and QSS solvers until a final time $t_f = 2$. The simulated trajectories of some state variables are depicted in Fig. 19.32, showing that the solution is a traveling wave moving in the direction of x.

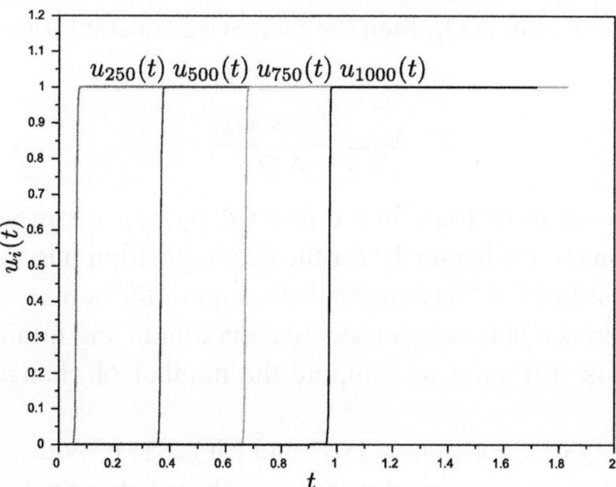

FIGURE 19.32

Advection–Diffusion–Reaction Trajectories.

For classic solvers, the best results were those obtained by CVODE–BDF, that completed the simulation after 1782 steps, where it performed 3101 evaluations of the complete right hand side of Eq. (19.52). Those calculation took 345 milliseconds. DOPRI exhibited a similar performance with 3161 simulation steps and 18,968 evaluations performed in 380 milliseconds (we recall that explicit DOPRI steps are cheaper than implicit CVODE steps). DASSL performed 2340 steps with 38,912 function evaluations that tool 15,760 milliseconds. The poor performance of DASSL is due, again, to the fact that it does not exploit the sparse structure of the problem.

Regarding QSS algorithms, LIQSS2 performed a total of 23,692 individual steps (about 24 steps per state). In consequence, the simulation only took 16 milliseconds (more than 20 times faster than any classic algorithm).

19.8 COMPARISON OF QSS WITH DISCRETE TIME METHODS: ACTIVITY-BASED APPROACH

Before introducing the examples above, we mentioned that QSS algorithms have better performance than classic discrete time methods when the ODEs have certain features (frequent discontinuities, large sparse models, etc.). The concept of *Activity* is useful to formalize this comparison.

Activity was conceived as the measure of the *change* experienced by a trajectory. In its original definition provided in Jammalamadaka (2003), given a trajectory $x_i(t)$, the activity of $z_i(t)$ in the interval (t_0, t_f) was defined as

$$A_{z_i(t_0,t_f)} \triangleq \int_{t_0}^{t_f} |\dot{z}_i(\tau)| \cdot d\tau \tag{19.54}$$

When $z_i(t)$ is a monotonic trajectory, the expression above computes the difference between the initial and the final value of $z_i(t)$ in the corresponding time interval, i.e., it is a measure of the change of the signal.

If we want to approximate $z_i(t)$ using a piecewise constant trajectory $q_i(t)$ so that both signals do not differ from each other more than ΔQ_i, then the piecewise constant trajectory $q_i(t)$ must change at least N_i times where

$$N_i = \frac{A_{z_i(t_0,t_f)}}{\Delta Q_i} \tag{19.55}$$

When the signal $z_i(t)$ is not monotonic in the interval (t_0, t_f), it can be decomposed into a finite number of monotonic sections (see Chapter 10 for the decomposition into monotonic generators). That way, it can be easily seen that Eq. (19.54) computes the sum of the activity in each monotonic section, which is the sum of the difference between consecutive maximum and minimum values of $z_i(t)$. Then, the formula of Eq. (19.55) is still valid to compute the number of changes in a piecewise constant approximation $q_i(t)$.

Recalling that first order QSS algorithms (QSS1 and LIQSS1) provide piecewise constant approximations of the state trajectories, we can expect that the number of changes they perform in the quantized state $q_i(t)$ is at least the value of N_i given by Eq. (19.55). Moreover, if there are not spurious oscillations as those that appeared when we used QSS1 with stiff systems, we can expect that the number of changes in $q_i(t)$ is approximately that of Eq. (19.55).

Thus, in a system of order n, with n state variables $z_1, \cdots, z_n$, we can expect QSS1 or LIQSS1 to perform a total of N steps with

$$N \approx \sum_{i=1}^{n} \frac{A_{z_i(t_0,t_f)}}{\Delta Q_i} \tag{19.56}$$

How about discrete time algorithms?

The first question that comes up is whether to consider both fixed step and variable step methods. Since almost all modern simulation tools working on sequential platforms use variable step algorithms, we will focus the comparison of QSS against them. Moreover, after doing so, we will be able to easily restrict the results to single step counterparts to provide further insight into the QSS comparison.

Let us assume that we use a variable step version of Forward Euler's method (the discrete time counterpart of QSS1). Moreover, let us assume that the step size $h(t)$ is controlled so that the difference between consecutive values in each state $z_i(t)$ is less or equal than ΔQ_i, this is,

$$z_i(t+h(t)) = z_i(t) + h(t)\cdot \dot{z}_i(t) \implies |z_i(t+h) - z_i(t)| = h(t)\cdot|\dot{z}_i(t)| \le \Delta Q_i$$

a condition that is accomplished for all i provided that we choose

$$h(t) = \min_i \frac{\Delta Q_i}{|\dot{z}_i(t)|}$$

Intuitively, this means that the time step required by the highest rate component, i, is the one that must be adopted for all components at any instant in time.

Then, the number of steps per unit of time is

$$\frac{1}{h(t)} = \max_i \frac{|\dot{z}_i(t)|}{\Delta Q_i}$$

Thus the number of steps to compute a global state transition is governed by the component that requires the most steps, and the total number of steps in the interval (t_0, t_f) results:

$$N_d \approx \int_{t_0}^{t_f} \frac{1}{h(\tau)} d\tau = \int_{t_0}^{t_f} \max_i \frac{|\dot{z}_i(t)|}{\Delta Q_i} d\tau \tag{19.57}$$

If we replace the definition of activity of Eq. (19.54) in the expression of Eq. (19.56), we obtain

$$N \approx \sum_{i=1}^{n} \frac{A_{z_i(t_0,t_f)}}{\Delta Q_i} = \sum_{i=1}^{n} \int_{t_0}^{t_f} \frac{|\dot{z}_i(\tau)|}{\Delta Q_i} \cdot d\tau \tag{19.58}$$

A comparison of N_d and N in Eqs. (19.57) and (19.58) immediately tells that the variable time step algorithm performs fewer steps than QSS1 (the maximum is equal or less than the sum). However, the cost of each QSS step can be considerably less than the cost of a discrete time step. Each QSS1 step involves the change of a single quantized state and the re-computation of the state derivatives of those states that directly depend on the quantized state that changes. In contrast, a Forward Euler step involves computing all the state derivatives and updating all component states.

Based on these observations, we compare below the efficiency of QSS and discrete time approximations according to the activity and structural features of the system:

- When a system is *dense*, i.e., when each state derivative depends on most states, the discrete time approach may be more efficient. The reason is that in this case each QSS step becomes almost as expensive as a discrete time step (the change of every quantized state triggers changes in most state derivatives). Thus, taking into account that discrete time algorithms perform fewer steps than quantized state algorithms, QSS methods may be not as efficient.
- When a system has homogeneous activity all time in all state variables, i.e., when the different state variables have simultaneously a similar rate of change, then again discrete time approximations may be better. The reason is now that the activity is similar in all states, so the comparison of Eqs. (19.57) and (19.58) says that $N \approx n \cdot N_d$ and the number of steps performed by QSS1 can be about n times the number of steps of the discrete time algorithm. While QSS1 steps are still cheaper and the total number of state variable updates will be similar, QSS1 must recompute more than one state derivative after each state update, so the total number of computations of the state derivatives results larger in QSS1. In addition, QSS algorithms must recompute the time advance after each step.
- When a system is large and sparse and the activity is mostly localized in a small fraction of the states, QSS algorithms are definitely superior. In large sparse systems QSS steps are significantly cheaper than discrete time steps because those of QSS only involve the calculation of a few state derivatives. In addition, when the activity is localized in few state variables, the difference between N_d and N in Eqs. (19.57) and (19.58) is small. In particular, when the activity is concentrated at a single state variable, both formulas coincide and $N_d = N$.
- When a system is large and sparse and the total activity is homogeneous but it is localized at different components in different periods of time, then QSS algorithms may be superior. This is a similar situation to that of the previous case if it is analyzed period by period. This case is that demonstrated by the Advection–Diffusion–Reaction equation.

While we analyzed here the extremal cases, there are several systems that lie in the middle where QSS or discrete time algorithms may be more efficient according to some other features. We already know that in presence of discontinuities QSS algorithms are superior. Also, in stiff systems the explicit solution provided by LIQSS1 (when it works) lead to much cheaper steps than those of implicit discrete time algorithms.

The original definition of activity limited the analysis to first order accurate algorithms. However, the concept of activity was extended to higher order algorithms in Castro and Kofman (2015), replacing the formula of Eq. (19.54) by

$$A^{(k)}_{z_i(t_0,t_f)} \triangleq \int_{t_0}^{t_f} \left| \frac{\frac{d^k z_i}{dt^k}(\tau)}{k!} \right|^{1/k} d\tau \tag{19.59}$$

where $A^{(k)}$ is called k-th order activity. Notice that when $k = 1$ the formula coincides with that of Eq. (19.54).

Using this definition, it can be proven that when we approximate the signal $z_i(t)$ using a piecewise polynomial trajectory $q_i(t)$ of order $k-1$, the number of steps performed results

$$N_i \approx \frac{A^{(n)}_{z_i(t_0,t_f)}}{\Delta Q_i^{1/n}} \tag{19.60}$$

and we can arrive to similar conclusions regarding the convenience of QSS algorithms or discrete time algorithms of higher order.

Another consequence of Eq. (19.60) is that it formally proves that the number of steps in QSS2 or LIQSS2 grows with the square root of the accuracy while it grows with the cubic root of the accuracy in QSS3 and LIQSS3.

In the foregoing we have focused on variable step discrete time methods. However, by recognizing that a fixed time step method must employ the smallest time step not only overall state variables but throughout the time interval. Thus the comparisons with QSS will be much more favorable for the quantization approach.

Exercise 19.1. Redo the study of Eq. (19.57) for the case of fixed time step methods. Hint: See Chapter 17 of TMS2000.

Exercise 19.2. Redo the study of Eq. (19.58) for the case of logarithmic quantization.

SOURCES AND FURTHER READING

The relationship between state quantization in continuous systems and DEVS was first developed more than two decades ago (Praehofer and Zeigler, 1996; Zeigler and Lee, 1998), with the main results included in the second edition of TMS.

Then, the formalization of Quantized State Systems as a numerical method for ODEs (showing the need of hysteresis), and the proof of its stability and convergence properties were first reported in Kofman and Junco (2001). These results were followed with the second order accurate QSS2 and the proof of the global error bound for linear systems (Kofman, 2002). After that, the use and advantages of QSS algorithms in discontinuous systems were studied in Kofman (2004), and the application of logarithmic quantization, its stability and error bound properties, were first proposed in Kofman (2009). Linearly Implicit QSS methods for stiff systems were developed in Migoni et al. (2013), and recently improved to deal with more complex stiff structures (Di Pietro et al., 2016).

Regarding software implementation of QSS algorithms, the first tool specifically designed to simulate QSS approximations was PowerDEVS (Bergero and Kofman, 2011). Then, the Stand Alone QSS Solver was introduced in Fernández and Kofman (2014).

The uses and advantages of QSS algorithms in different applications were extensively reported in the literature, including spiking neural networks (Grinblat et al., 2012), Power Electronic Converters (Migoni et al., 2015), Advection–Diffusion–Reaction equations (Bergero et al., 2016), Smart Grids (Migoni et al., 2016), Building Simulation (Frances et al., 2014, 2015; Bergero et al., 2018), and particle transport in high energy physics (Santi et al., 2017).

Another application were QSS methods have shown noticeable benefits is that of integrated circuit simulation, where such solvers can run much faster (50 times) with more accurate noise analysis than conventional solvers (Jakobsson et al., 2015)

The problem of parallelization and distributed simulation was also studied in the context of QSS simulation. The asynchronous nature of the algorithms allows obtain good speed up figures in simulations performed on multicore architectures (Bergero et al., 2013; Fernandez et al., 2017). A study of surface water run-off using a high resolution spatial cellular model of a large scale watershed showed a thousand-fold speedup in computation on massively parallel platforms (Zeigler et al., 1997). The cells were modeled with ordinary differential equations and quantization was compared with a standard discrete time numerical method.

EXERCISE FOR FUTURE RESEARCH

Apply the performance models of Chapter 11 to the analysis of Section 19.8 to compare QSS with discrete time methods.

REFERENCES

Bergero, F.M., Casella, F., Kofman, E., Fernández, J., 2018. On the efficiency of quantization-based integration methods for building simulation. Building Simulation 11 (2), 405–418. https://doi.org/10.1007/s12273-017-0400-1.

Bergero, F., Fernández, J., Kofman, E., Portapila, M., 2016. Time discretization versus state quantization in the simulation of a 1D advection–diffusion–reaction equation. Simulation: Transactions of the Society for Modeling and Simulation International 92 (1), 47–61.

Bergero, F., Kofman, E., 2011. PowerDEVS. A tool for hybrid system modeling and real time simulation. Simulation: Transactions of the Society for Modeling and Simulation International 87 (1–2), 113–132.

Bergero, F., Kofman, E., Cellier, F.E., 2013. A novel parallelization technique for DEVS simulation of continuous and hybrid systems. Simulation: Transactions of the Society for Modeling and Simulation International 89 (6), 663–683.

Castro, R., Kofman, E., 2015. Activity of order n in continuous systems. Simulation: Transactions of the Society for Modeling and Simulation International 91 (4), 337–348.

Di Pietro, F., Migoni, G., Kofman, E., 2016. Improving a linearly implicit quantized state system method. In: Proceedings of the 2016 Winter Simulation Conference. Arlington, Virginia, USA.

Fernandez, J., Bergero, F., Kofman, E., 2017. A parallel stand-alone quantized state system solver for continuous system simulation. Journal of Parallel and Distributed Computing 106, 14–30.

Fernández, J., Kofman, E., 2014. A stand-alone quantized state system solver for continuous system simulation. Simulation: Transactions of the Society for Modeling and Simulation International 90 (7), 782–799.

Frances, V.M.S., Escriva, E.J.S., Ojer, J.M.P., 2014. Discrete event heat transfer simulation of a room. International Journal of Thermal Sciences 75, 105–115.

Frances, V.M.S., Escriva, E.J.S., Ojer, J.M.P., 2015. Discrete event heat transfer simulation of a room using a quantized state system of order two, QSS2 integrator. International Journal of Thermal Sciences 97, 82–93.

Grinblat, G., Ahumada, H., Kofman, E., 2012. Quantized state simulation of spiking neural networks. Simulation: Transactions of the Society for Modeling and Simulation International 88 (3), 299–313.

Jakobsson, A., Serban, A., Gong, S., 2015. Implementation of quantized-state system models for a PLL loop filter using Verilog-AMS. IEEE Transactions on Circuits and Systems I: Regular Papers 62 (3), 680–688.

Jammalamadaka, R., 2003. Activity Characterization of Spatial Models: Application to Discrete Event Solution of Partial Differential Equations. Master's thesis. The University of Arizona.

Kofman, E., 2002. A second order approximation for DEVS simulation of continuous systems. Simulation: Transactions of the Society for Modeling and Simulation International 78 (2), 76–89.

Kofman, E., 2004. Discrete event simulation of hybrid systems. SIAM Journal on Scientific Computing 25 (5), 1771–1797.

Kofman, E., 2009. Relative error control in quantization based integration. Latin American Applied Research 39 (3), 231–238.

Kofman, E., Junco, S., 2001. Quantized state systems. A DEVS approach for continuous system simulation. Transactions of SCS 18 (3), 123–132.

Migoni, G., Bergero, F., Kofman, E., Fernández, J., 2015. Quantization-based simulation of switched mode power supplies. Simulation: Transactions of the Society for Modeling and Simulation International 91 (4), 320–336.

Migoni, G., Bortolotto, M., Kofman, E., Cellier, F., 2013. Linearly implicit quantization-based integration methods for stiff ordinary differential equations. Simulation Modelling Practice and Theory 35, 118–136.

Migoni, G., Rullo, P., Bergero, F., Kofman, E., 2016. Efficient simulation of hybrid renewable energy systems. International Journal of Hydrogen Energy 41 (32), 13934–13949.

Praehofer, H., Zeigler, B.P., 1996. On the expressibility of discrete event specified systems. In: Computer Aided Systems Theory – CAST'94. Springer, pp. 65–79.

Santi, L., Ponieman, N., Jun, S.Y., Genser, K., Elvira, D., Castro, R., 2017. Application of state quantization-based methods in hep particle transport simulation. Journal of Physics: Conference Series 898, 042049.

Zeigler, B.P., Lee, J.S., 1998. Theory of quantized systems: formal basis for DEVS/HLA distributed simulation environment. In: SPIE Proceedings, vol. 3369, pp. 49–58.

Zeigler, B.P., Moon, Y., Kim, D., Ball, G., 1997. The DEVS environment for high-performance modeling and simulation. IEEE Computational Science and Engineering 4 (3), 61–71.

CHAPTER

DEVS REPRESENTATION OF ITERATIVELY SPECIFIED SYSTEMS

20

CONTENTS

In this chapter we extend the results of Chapter 19 to distributed simulation of iteratively specified systems. One of the major advantage this affords is to enable sharing of assets across large distances (potentially, connecting earth and space stations, for example). Another advantage is speed of computation when parallelism is exploited particularly in high performance multicore computers. The major difference between distributed simulation and its non-distributed counterpart is that information and data are encoded in messages that travel from one computer to another over a network. This changes our perspective on simulation from one in which only computation matters to one in which communication matters just as much, and sometimes, more. The brunt of simulation analyses in the past (e.g., in the design of numerical methods (Chapters 3 and 8)) had to do with trade-offs between accuracy and computation speed. And while these retain their importance, our new perspective demands that we look at trade-offs between accuracy and communication resources as well. For example, we can ask what is the best simulator architecture to simulate a model distributed among several sites, given limited bandwidth (bits/sec that can be transmitted) or given limited bandwidth and computer memory (the latter taking both computation and communication constraints into account).

Theory of Modeling and Simulation. https://doi.org/10.1016/B978-0-12-813370-5.00030-4

Since messaging requires that continuous quantities be coded into discrete packets or cells and sent discontinuously, discrete event simulation is the most natural means to examine such problems. And since DEVS is the formalism representing system specification in discrete event terms we turn to it as our vehicle for such study. However, the models to be simulated cannot be restricted to discrete event form only since many, especially coming from the physical sciences, are cast in differential equation terms and may come packaged in discrete time simulators, the traditional way of simulating continuous systems.

20.1 DEVS BUS REVISITED

The DEVS Bus provides the concepts we need. Recall (Chapter 8) that we could embed traditional discrete event world views into the DEVS formalism. In such a "wrapping", simulators for models expressed in various event-based formalisms can inter-operate under the control of standard DEVS coordinator. Fig. 20.1 depicts an extended DEVS Bus in which this concept is extended to encompass the other major formalisms, DTSS and DESS. The idea is to provide means to represent DTSS and DESS models in DEVS form thus providing an all-DEVS distributed simulation environment for any modeling domain. For example, we could have rule-based DEVS models representing intelligent agents communicating through a distributed network represented by cellular automata and interacting over terrain represented by spatially continuous (partial) differential equation models.

20.1.1 APPROACHES TO DEVS REPRESENTATION OF CONTINUOUS SYSTEMS

To make the extended DEVS Bus a reality requires that we have workable representations of the DTSS and DESS formalisms in DEVS. We already have shown how DEVS can simulate discrete time models at both the global and local levels. This provides a start towards practical means of wrapping the traditional kinds of differential equation models into DEVS. This chapter discusses approaches to strongly simulate DESS systems in DEVS. Since necessarily these are approximations, we need to

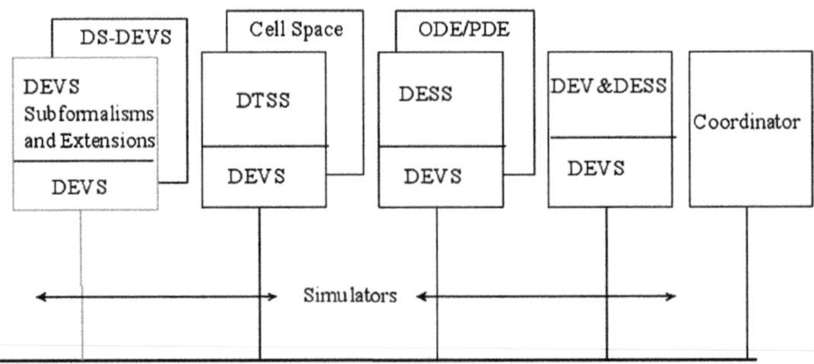

FIGURE 20.1

DEVS Bus Extended to Non-Discrete Event Formalisms.

establish the conditions under which the error can be made as small as desired. Backing off from such limiting case theorems, we need to study how the DEVS representation can provide more efficient simulation than the traditional discrete time simulation of differential equation systems.

Having established the means to simulate discrete time coupled models in component-wise DEVS manner, we are in a position to develop one approach to DEVS simulation of differential equations systems. This approach, the DEVS equivalent of traditional numerical integration, corresponds to the top pathway in Fig. 20.2. The major step in this approach is to discretize time to fixed time steps and employ standard numerical methods to obtain DTSS representations of the DESS components (Chapter 3). Since as shown in Chapter 8, this can be done in a component-wise manner, we have a component-wise simulation by DTSS of DESS coupled models. Now applying the DEVS strong-simulation of DTSS (Chapter 18), we can directly construct a DEVS coupled model to simulate the DTSS representation, and hence, the original DESS coupled in the required component-wise manner. Of course, as indicated in the figure, error may be introduced in the DTSS approximation of the DESS model and this will carry-over to the DEVS simulation. However, since the DEVS simulation of the DTSS stage is error-free, the DEVS simulation will not suffer more error than the DTSS simulation.

However, in going second hand through a DTSS simulation we are not taking advantage of reduced computation and message passing that a direct DEVS representation might provide. Thus, for our second approach to DEVS representation of DESS (the left hand pathway in Fig. 20.2) we recall that DEVS offers an alternative way of handling continuous time. Instead of advancing time in discrete steps, we can advance time based on discrete events, where an event is a significant change in an input, state or output variable. This is the approach followed in Chapter 19.

The key to handling the continuous variables of a DESS is to determine when a significant event occurs in a component. Such an event can then be reported to other components. We will introduce a significant event detector, called a quantizer, which monitors its input and uses a logical condition to decide when a significant change, such as crossing a threshold has occurred. The concept of quantization will formalize the operation of significant event detectors. A quantum is measure of how big a change must be to be considered significant.

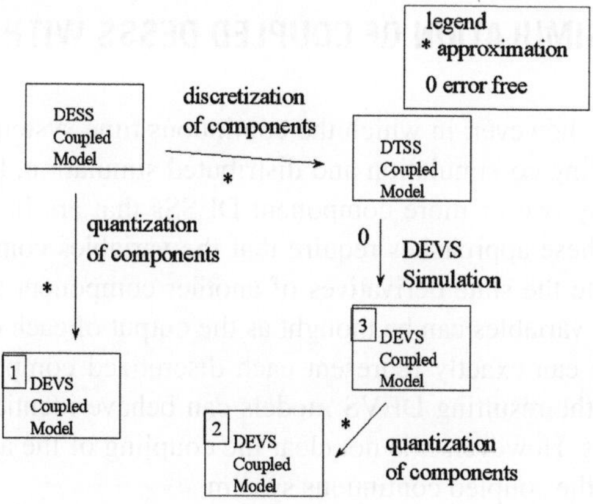

FIGURE 20.2

Approaches to DEVS Simulation of DESS Coupled Models.

To reduce local computation and message passing we would like to have the quantum size be large. However, the decision of what is a significant change can't just be made locally since it also involves other components. We are really asking how big a change must be before we need to tell others about it. The answer to this kind of question is through error analysis of coupled models whose components are quantized. Put another way, we need to look at closed loop behavior, i.e., how any error propagates through a network where feedback loops exist. Such analysis is well known for numerical integration of differential equation systems, so to prepare the way for our quantization approach, we first turn to such classical analysis.

In the next few sections, we first look at discrete time and quantized simulation of ODEs with arbitrarily small error. We will show that DEVS can simulate DESS through approximation via DTSS and via Quantization.

20.2 DEVS SIMULATION OF DESS WITH ARBITRARILY SMALL ERROR

20.2.1 DISCRETIZED SIMULATION OF A DESS WITH ARBITRARILY SMALL ERROR

By simulation with an arbitrarily small error we mean that there is an approximate morphism (Chapter 16) where there is a parameter such that error accumulation in a finite time interval can be made as small as desired with a choice of parameter value.

If a single continuous time ODE underlying a DESS is approximated using any conventional numerical algorithm (Euler, Runge–Kutta, etc.) we know from Chapter 3, and from numerical ODE literature (see Hairer et al., 1993, for instance) that the numerical solution of an ODE converges to the analytical solution as the step size h goes to 0.

Taking into account that the DTSS approximation provided by any of these methods can be exactly represented by a DEVS model, then the DEVS representation also verifies this convergence property and can simulate the DESS with arbitrarily small error.

20.2.2 DISCRETIZED SIMULATION OF COUPLED DESSS WITH ARBITRARILY SMALL ERROR

There are several situations, however, in which the continuous time system is not discretized at once. In some applications including co-simulation and distributed simulation, for instance, the continuous time system is composed by two or more component DESSs that are individually discretized using some numerical method. These approaches require that the variables computed by some component that are required to calculate the state derivatives of another component are communicated between them. These communicated variables can be thought as the output of each component system.

Here, we know that we can exactly represent each discretized component system using DEVS. Moreover, the coupling of the resulting DEVS models can behave identically to the coupling of the approximated DTSS models. However, it is not clear the coupling of the approximated DTSS models is a good approximation to the coupled continuous system.

Intuitively, it seems rather obvious that if the underlying numerical methods used in each component converge, and provided that the communication step size goes to zero with the integration step size h, then the overall solution should converge. Note that typically a communication is a state update that occurs only after an integration step. Thus the communication step size is a multiple of the integration step as just stated.

There are indeed several results that confirm our intuition (Trčka et al., 2009; Moshagen, 2017). We present below a simplified version of these results.

We shall consider an ODE composed by two component systems:

$$\begin{aligned}\dot{\mathbf{z}}_1(t) &= \mathbf{f}_1(\mathbf{z}_1(t), \mathbf{z}_2(t))\\ \dot{\mathbf{z}}_2(t) &= \mathbf{f}_2(\mathbf{z}_1(t), \mathbf{z}_2(t))\end{aligned} \tag{20.1}$$

where $\mathbf{z}_1(t)$ is the state vector computed by the first component and $\mathbf{z}_2(t)$ is the state vector computed by the second component.

We shall consider that a one-step method of order p is used on both subsystems with a step size h. Also, the values of $\mathbf{z}_i(t)$ are communicated every H units of time between both subsystems.

Then, the discretized ODE has the following representation

$$\begin{aligned}\mathbf{z}_1(t_{k+1}) &= \mathbf{z}_1(t_k) + h \cdot \mathbf{f}_1(\mathbf{z}_1(t_k), \mathbf{z}_2(t_j)) + \mathbf{d}_1 \cdot h^2\\ \mathbf{z}_2(t_{k+1}) &= \mathbf{z}_1(t_k) + h \cdot \mathbf{f}_2(\mathbf{z}_1(t_k), \mathbf{z}_2(t_j)) + \mathbf{d}_2 \cdot h^2\end{aligned} \tag{20.2}$$

where the expression term $\mathbf{d}_i \cdot h^2$ represents the remaining terms of the Taylor series expansion of the algorithm in use. Here, t_k are the time steps of the numerical algorithm with step size h while t_j are the time communication steps with step size H.

If Forward Euler is used, the terms $\mathbf{d}_i \cdot h^2$ does not exist. Moreover, if $h = H$ then $t_k = t_j$ and Eq. (20.2) coincides with the Euler approximation of the system of Eq. (20.1). Thus, we arrive at the conclusion that splitting a simulation with Forward Euler using a step size equal to the communication step size is equivalent to simulating the entire coupled system with Forward Euler.

Let us suppose now that the communication step size satisfies $H = c \cdot h$ with $c > 0$ (in most practical cases we will have $c \geq 1$ as it does not make much sense to have a communication interval less than the step size).

We can rewrite the term

$$\mathbf{f}_1(\mathbf{z}_1(t_k), \mathbf{z}_2(t_j)) = \mathbf{f}_1(\mathbf{z}_1(t_k), \mathbf{z}_2(t_k)) + \mathbf{f}_1(\mathbf{z}_1(t_k), \mathbf{z}_2(t_j)) - \mathbf{f}_1(\mathbf{z}_1(t_k), \mathbf{z}_2(t_k)) \tag{20.3}$$

Assuming that $\mathbf{f}_1$ is Lipschitz with constant L, this results in

$$\|\mathbf{f}_1(\mathbf{z}_1(t_k), \mathbf{z}_2(t_j)) - \mathbf{f}_1(\mathbf{z}_1(t_k), \mathbf{z}_2(t_k))\| \leq L \cdot \|\mathbf{z}_2(t_k) - \mathbf{z}_2(t_j)\| \tag{20.4}$$

The fact that $\mathbf{f}_1$ is Lipschitz implies that it is also bounded by some constant M, so $\dot{\mathbf{z}}_1(t)$ is bounded by the same constant and the difference

$$\|\mathbf{z}_2(t_k) - \mathbf{z}_2(t_j)\| \leq M \cdot |t_j - t_k| \leq M \cdot H = M \cdot c \cdot h$$

where we note that the difference between t_j and t_k is less or equal than the communication step size H.

Then, plugging the last inequality in Eq. (20.4), and defining

$$\Delta\mathbf{f}_1(t_k) \triangleq \mathbf{f}_1(\mathbf{z}_1(t_k), \mathbf{z}_2(t_j)) - \mathbf{f}_1(\mathbf{z}_1(t_k), \mathbf{z}_2(t_k))$$

we can rewrite the first component of Eq. (20.2) as

$$\mathbf{z}_1(t_{k+1}) = \mathbf{z}_1(t_k) + h \cdot (\mathbf{f}_1(\mathbf{z}_1(t_k), \mathbf{z}_2(t_k)) + \Delta\mathbf{f}_1(t_k)) + \mathbf{d}_1 \cdot h^2 \tag{20.5}$$

Taking into account that

$$\|\Delta\mathbf{f}_1(t_k)\| \leq M \cdot c \cdot h$$

we have that $\|h \cdot \Delta\mathbf{f}_1(t_k)\| \leq M \cdot c \cdot h^2$ and then we can merge the term with those of the order of h^2, that is,

$$\mathbf{z}_1(t_{k+1}) = \mathbf{z}_1(t_k) + h \cdot \mathbf{f}_1(\mathbf{z}_1(t_k), \mathbf{z}_2(t_k)) + \tilde{\mathbf{d}}_1 \cdot h^2 \tag{20.6}$$

If we follow an identical procedure with $\mathbf{z}_2(t)$, we finally arrive to

$$\begin{aligned} \mathbf{z}_1(t_{k+1}) &= \mathbf{z}_1(t_k) + h \cdot \mathbf{f}_1(\mathbf{z}_1(t_k), \mathbf{z}_2(t_k)) + \tilde{\mathbf{d}}_1 \cdot h^2 \\ \mathbf{z}_2(t_{k+1}) &= \mathbf{z}_1(t_k) + h \cdot \mathbf{f}_2(\mathbf{z}_1(t_k), \mathbf{z}_2(t_k)) + \tilde{\mathbf{d}}_2 \cdot h^2 \end{aligned} \tag{20.7}$$

This last expression, which does not depend on t_j, has some consequences:

- If we are using Forward Euler's method, the resulting approximation is still correct up to the term of h (i.e., it is first order accurate). Thus, the convergence result of Theorem 19.3 holds.
- If we are using a higher order one step method, then the resulting approximation is only first order accurate. The reason is that when we placed the term $h \cdot \Delta\mathbf{f}_1(t_k)$ with those of the order of h^2, we modified those terms and they will no longer coincide with that of the analytical solution. In consequence, the whole approximation will be only first order accurate, so, in principle, it does not make much sense using those higher order algorithms (we will see soon a way that this can be solved).
  In any case, Theorem 19.3 holds and once again we have convergence of the coupled model simulation.

The conclusion is then that, for any one-step method, when both the communication and the integration step size h go to zero, the simulation error goes to zero.

Regarding the approximation order, the work of Moshagen (2017) showed that when the values $\mathbf{z}_i$ are communicated using interpolation polynomials of the same order as the method (i.e., order p) then the numerical solution converges to the analytical solution with h^p, preserving the approximation order. We already saw in Chapter 19 that we can represent piecewise polynomial trajectories using DEVS (we used them to implement high order QSS algorithms). Thus, the approach of Moshagen (2017) that preserves the approximation order can be easily represented by DEVS.

Exercise 20.1. Outline a proof for the case of order 1 (line) for a pair of communicating component systems.

While this analysis only proved convergence for the coupling of two components, it can be straightforwardly extended for the coupling of an arbitrary number of them.

Exercise 20.2. Outline a proof of convergence of the simulation for an arbitrary number of communicating component systems.

Finally, we assumed that the components communicate the state values $\mathbf{z}_i$. In many situations, they communicate an output value $\mathbf{y}_i(t) = g_i(\mathbf{z}_i(t))$. This analysis can be easily extended to that case also.

20.2.3 SIMULATION OF A DESS BY QSS WITH ARBITRARILY SMALL ERROR

We have seen in Chapter 19 that ODEs can be simulated using QSS algorithms that produce an approximate model (called Quantized State System) that has an exact representation as a DEVS model.

We already saw that QSS algorithms have a convergence property given by Theorem 19.3. Thus, using this approach, a single ODE can be simulated with arbitrarily small error provided that a parameter that bounds the quantum in all integrators goes to zero.

20.2.4 QSS SIMULATION OF COUPLED DESS WITH ARBITRARILY SMALL ERROR

As with the case of conventional discrete time approximations, there are situations in which we may be interested in coupling the DESS component systems after they are approximated by the QSS algorithms. In such case, we must prove that the coupling of the QSS approximations converges to the true solution when the quantum goes to zero.

Regarding the coupling, to preserve the discrete event nature of QSS approximations, we will consider that the variables are transmitted between subsystems in an asynchronous fashion when they experience noticeable changes.

In our approach, we shall not transmit the state variables $z_i(t)$, but the quantized states $q_i(t)$. Each quantized state $q_i(t)$ will be transmitted whenever the difference with the last transmitted value is greater than certain parameter $\Delta\tilde{Q}_i$ (that might not be equal to the quantum ΔQ_i used by the QSS algorithm). This strategy implies using a hysteretic quantization (like that of QSS) that ensures the legitimacy of the coupled model.

We shall suppose that there exists a maximum quantum $\Delta Q_{\max}$ such that

$$\Delta Q_i \le \Delta Q_{\max}, \quad \Delta\tilde{Q}_i \le \Delta Q_{\max} \quad \forall i$$

Thus, given the coupling of two subsystems

$$\begin{aligned}\dot{\mathbf{z}}_1(t) &= \mathbf{f}_1(\mathbf{z}_1(t), \mathbf{z}_2(t))\\ \dot{\mathbf{z}}_2(t) &= \mathbf{f}_2(\mathbf{z}_1(t), \mathbf{z}_2(t))\end{aligned} \tag{20.8}$$

the use of QSS in each subsystem and the communication of $\mathbf{q}_i(t)$ to the other subsystem leads to the following equations:

$$\begin{aligned}\dot{\mathbf{z}}_1(t) &= \mathbf{f}_1(\mathbf{q}_1(t), \tilde{\mathbf{q}}_2(t))\\ \dot{\mathbf{z}}_2(t) &= \mathbf{f}_2(\tilde{\mathbf{q}}_1(t), \mathbf{q}_2(t))\end{aligned} \tag{20.9}$$

where $\mathbf{q}_1(t)$ is the quantized state vector of the first subsystem and $\tilde{\mathbf{q}}_2(t)$ is the transmitted value of the quantized state vector of the second subsystem (with analogous definitions for $\mathbf{q}_2$ and $\tilde{\mathbf{q}}_1$. This last equation can be rewritten as:

$$\begin{aligned}\dot{\mathbf{z}}_1(t) &= \mathbf{f}_1(\mathbf{q}_1(t), \mathbf{q}_2(t) + \Delta\mathbf{q}_2(t))\\ \dot{\mathbf{z}}_2(t) &= \mathbf{f}_2(\tilde{\mathbf{q}}_1(t) + \Delta\mathbf{q}_1(t), \mathbf{q}_2(t))\end{aligned}$$

where $\Delta \mathbf{q}_i(t) \triangleq \tilde{\mathbf{q}}_i(t) - \mathbf{q}_i(t)$. Defining also $\Delta \mathbf{z}_i(t) = \mathbf{q}_i(t) - \mathbf{z}_i(t)$, we obtain

$$\begin{aligned}\dot{\mathbf{z}}_1(t) &= \mathbf{f}_1(\mathbf{z}_1(t) + \Delta\mathbf{z}_1(t), \mathbf{z}_2(t) + \Delta\mathbf{z}_2(t) + \Delta\mathbf{q}_2(t))\\ \dot{\mathbf{z}}_2(t) &= \mathbf{f}_2(\tilde{\mathbf{z}}_1(t) + \Delta\mathbf{z}_1(t) + \Delta\mathbf{q}_1(t), \mathbf{z}_2(t) + \Delta\mathbf{z}_2(t))\end{aligned} \tag{20.10}$$

which is a perturbed version of the original ODE of Eq. (20.8). The fact that each component $\tilde{q}_i(t)$ is transmitted whenever it differs from q_i in a quantity $\Delta\tilde{Q}_i$ implies that

$$|\Delta q_i(t)| = |\tilde{q}_i(t) - q_i(t)| \le \Delta\tilde{Q}_i \le \Delta Q_{\max} \tag{20.11}$$

We also know that

$$|\Delta z_i(t)| = |q_i(t) - z_i(t)| \le \Delta Q_i \le \Delta Q_{\max} \tag{20.12}$$

Thus all the perturbation terms in Eq. (20.10) are bounded by $\Delta Q_{\max}$. Then, the convergence result of Theorem 19.3 in Chapter 19 holds and the solution of the coupled QSS goes to the true solution as the parameter $\Delta Q_{\max}$ goes to zero.

As in the discrete time case, we can make some similar remarks:

- If we use QSS1 and use the same quantum in the algorithm and the transmission ($\Delta Q_i = \Delta\tilde{Q}_i$), then we have $\tilde{\mathbf{q}}_i(t) = \mathbf{q}_i(t)$ (the quantized states are exactly transmitted). Thus Eq. (20.9) coincides with the QSS1 approximation of the coupled model which implies that the coupling of the QSS approximation is equivalent to the QSS approximation of the coupled model.
  Notice that this result is analogous to that of coupling Forward Euler simulations with $h = H$.
- If we use a higher order approximation (QSS2, etc.) then we still have convergence. However, if we send the values of q_i in a piecewise constant fashion, there will be much more values transmitted of $q_i(t)$ than steps on the corresponding variable. Thus, for efficiency reasons, a polynomial of the same order than the algorithm must be used to transmit $q_i(t)$. This was the same conclusion we arrived to regarding conventional DTSS approximations.

This analysis showed that we can also simulate coupled QSS approximations preserving convergence.

20.2.5 CONVERGENCE OF COUPLING OF QSS AND DTSS

Taking into account that DEVS can represent both, QSS and DTSS approximations of continuous systems, we raise the question of whether it is possible to couple a QSS approximation with a DTSS approximation. This is, given the coupling of two systems,

$$\begin{aligned}\dot{\mathbf{z}}_1(t) &= \mathbf{f}_1(\mathbf{z}_1(t), \mathbf{z}_2(t))\\ \dot{\mathbf{z}}_2(t) &= \mathbf{f}_2(\mathbf{z}_1(t), \mathbf{z}_2(t))\end{aligned} \tag{20.13}$$

we propose the following approximation

$$\begin{aligned}\mathbf{z}_1(t_{k+1}) &= \mathbf{z}_1(t_{k+1}) + h \cdot \mathbf{f}_1(\mathbf{z}_1(t_k), \tilde{\mathbf{q}}_2(t_k))\\ \dot{\mathbf{z}}_2(t) &= \mathbf{f}_2(\mathbf{z}_1(t_j), \mathbf{q}_2(t))\end{aligned} \tag{20.14}$$

where we are using a conventional Forward Euler approximation for $\mathbf{z}_1$ and a QSS1 approximation for $\mathbf{z}_2$ with the communication strategies used in both cases (a communication step size of H for $\mathbf{z}_1$ and a quantum $\Delta\tilde{Q}_i \leq \Delta Q_{\max}$ for $\tilde{q}_i(t)$.

While the DEVS representation of this approximation is straightforward, it is not clear again that the results converge to the true solution.

In any event, we conjecture that when h and $\Delta Q_{\max}$ *go simultaneously to zero, the solution of Eq.* (20.14) *converge to the true solution of Eq.* (20.13).

20.3 DEVS COMPONENT-WISE SIMULATION OF ITERATIVELY SPECIFIED COUPLED SYSTEMS

Chapter 12 showed that iterative specifications are closed under coupling and that DEVS can simulate iterative specifications in both direct and component-wise forms. However, the simulation is question was not realistic in that it requires that input segments be encoded directly as values of DEVS events. Practical simulation requires that restrictions be placed on the encoding of input segments into DEVS events. Likewise messages must be restricted in the information they transmit in component-wise simulations of iterative specifications. These restrictions can be approached from the perspective of component-wise simulation of iteratively specified systems with arbitrarily small error. Much of the groundwork for such strong simulation has already be laid. TMS2000 showed (and this edition confirmed in Chapter 19) that DEVS can simulate DESS through approximation via DTSS and via Quantization. Recalling that DESS are in fact iterative specifications this amounts to DEVS strong simulation of a subclass of systems with arbitrarily small error. However, the challenge remains to extend this result to the whole class of iteratively specified systems. This would include the DEV&DESS subclass which characterizes what can be called co-simulation of event-based component systems with ordinary differential equation component systems (Camus et al., 2018; Zeigler and Sarjoughian, 2017). Additionally, it would include new iterative specification families introduced in Chapter 12 of this edition.

Recapping work in this chapter, we first extended earlier results to DEVS component-wise simulation of DESS coupled models. We did this first by component-wise discretization of DESS components and allowing the step size, h in Fig. 20.3A, for component-self state updating to be different from that (H, in Fig. 20.3A) for updating other components with one's own state sent through the DEVS Bus. We then obtained a similar result by replacing discretization by quantization in Fig. 20.3B in which case, the quantum sizes for self-updating and informing others of updates are allowed to differ. This is followed by consideration of a DEVS coupled model of discretized and quantized DESS components representing a form of co-simulation.

With such very practical cases in hand, we will now go on to consider the general case of DEVS component-wise simulation of coupled iterative specifications with arbitrary small error. The approach will be to represent generator segments as concatenations of generators that are simple in the sense that they can be "cheaply" encoded into values of DEVS events. For example, a set of piecewise constant generators that is closed under segmentation qualifies because a DEVS encoding event need carry only a pair consisting of the constant and the length of the segment. Thus simulating an iterative specification by another one that employs such constant generators with arbitrarily small error is an example of the approach we will be taking (see Fig. 20.4).

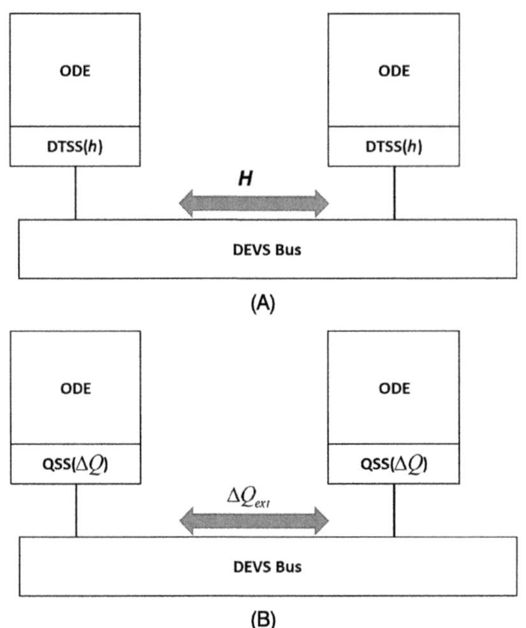

FIGURE 20.3

DEVS component-wise simulation of DESS coupled models.

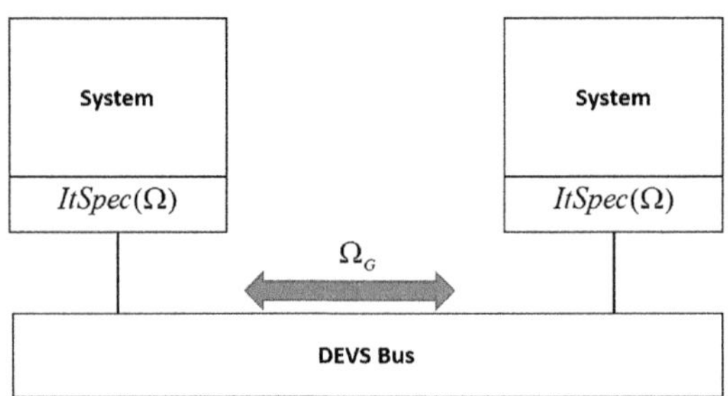

FIGURE 20.4

DEVS component-wise simulation of iteratively specified coupled systems.

In Chapter 12, an admissible coupled iterative specification was defined and a theorem stated that such a structure specifies a well-defined Iterative Specification at the I/O System level. Recall that this amounts to closure under coupling of the set of Iterative System Specifications.

Such closure allows us to consider a basic iterative specified system with feedback from output to input as the representative of such coupled iterative specifications as illustrated in Fig. 20.5A. Our focus will be on a single generator fed from output to input so that it satisfies the fixed point equation: $\beta_q(\omega_G) = \omega_G$. Our approach is to show that ω_G can be expressed as a concatenation of smaller gen-

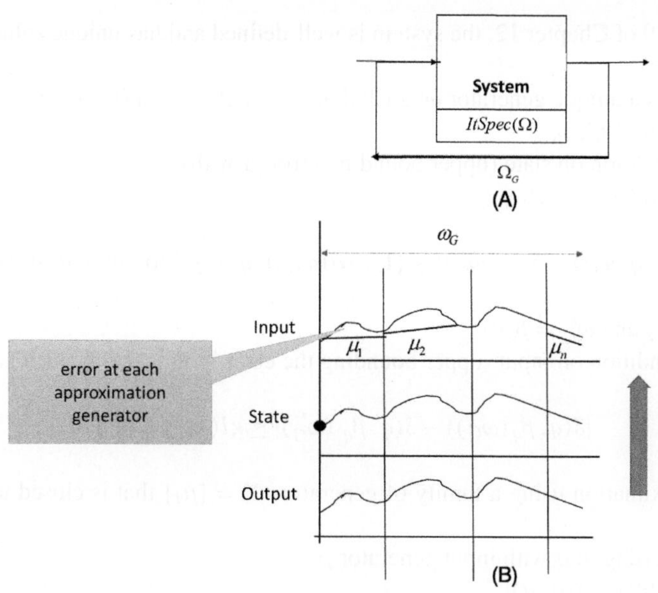

FIGURE 20.5

Iteratively Specified System with Feedback Loop.

erators such that we can get an approximation of the system behavior with a small error as desired. This will extend to any input segment of the system, i.e., any concatenation of generators such as ω_G, since we can make the error vanishingly small for such a segment by making the errors for each of its subsegments as small as needed. Furthermore the subsegments can be considered as the generating set for a (more refined) system that can simulate the original one (Fig. 20.5B).

Theorem 20.1. *Let an iterSpec $G = <X, Q, Y, \Omega_G, \delta_G, \lambda>$ specify a system $S = <X, Q, Y, \Omega_G^+, \delta, \lambda>$. Let $\Omega_{G'}$ be a generator set which generates Ω_G i.e., $\Omega_{G'}^+ = \Omega_G$. Then, an iterative system specification G' can be defined using $\Omega_{G'}$ as its generating set which specifies a system S' isomorphic to S.*

Exercise 20.3. Prove the theorem. Hint: use the composition property to show that a single step of the transition function of S can be simulated by multiple steps of the transition function of G'.

We are ready to state and prove the main theorem.

Theorem 20.2 (Component-wise Simulation of Iteratively Specified Coupled Systems with arbitrarily small error). *Given a coupled model of iteratively specified systems satisfying conditions given below, there is a coupled model of DEVS components that can strongly simulate it with arbitrarily small error.*

The conditions that are required are Lipschitz conditions as well as an assumption concerning error contraction that is stated after some initial development.

Proof. From Chapter 11, recall an iterSpec, $G =< X, Q, Y, \Omega_G, \delta_G, \lambda >$ specifies a system $S = < X, Q, Y, \Omega_G^+, \delta, \lambda >$ with I/O Functions associated with S are given by $\beta_q : \Omega_G^+ \rightarrow (Y, T)$ for $q \in Q$ with $\beta_q(\omega_G) = \lambda(q, \omega_G)$ and $\beta_q(\omega_G\omega) = \beta_q(\omega_G)\beta_{\delta(q,\omega_G)}(\omega)$ for $\omega_G \in, \Omega_G, \omega \in \Omega_G^+$.

By Theorem 12.9 of Chapter 12, the system is well-defined and has unique solutions to fixed point feedback equation:

For $t > 0$ there is a unique generator $\bar{\omega}^t$ such that $\bar{\omega}^t = \beta_q(\bar{\omega}^t)$ and $l(\bar{\omega}^t) = t$.

For any generator of G, ω_G

1. Lipschitz condition on state (upper bounding error growth):

For small enough $l(\omega_G) = l(\omega'_G)$

$$|\delta(q, \omega_G) - \delta(q', \omega'_G)| \leq (1 + \alpha l(\omega_G))| q - q' | \text{ for all } q, q' \text{ in } Q$$

where $\omega_G = \beta_q(\omega_G)$ and $\omega'_G = \beta_{q'}(\omega'_G)$.

2. Lipschitz condition on input (upper bounding the error introduced by each generator):

$$|\delta(q, \beta_q(\omega_G)) - \delta(q, \beta_{q'}(\omega'_G)| \leq \kappa l(\omega_G)| q - q' |$$

We consider approximation using a family of generators $\Sigma = \{\mu_i\}$ that is closed under left and right segmentation.

$e(\mu, q) =$ error starting in q with input generator μ

$e(\mu, q) = |\delta(q, \bar{\omega}^{l(\mu)}) - \delta(q, \mu)|$

$\delta(q, \bar{\omega}^{l(\mu)})$ is the unique correct solution for the length of μ

$\delta(q, \mu)$ is the possibly different state that μ brings the system to.

$e(\mu) = max\ e(\mu, q) =$ the largest error over states q in a set containing the subset visited.

The amplification of the state error in the system is determined by:

$$\begin{aligned}
&|\delta(q, \omega_G) - \delta(q', \omega'_G)| \text{ (where } \omega_G = \beta_q(\omega_G) \text{ and } \omega'_G = \beta_{q'}(\omega'_G))\\
&= |\delta(q, \omega_G) - \delta(q, \omega'_G) + \delta(q, \omega'_G) - \delta(q', \omega'_G)|\\
&\leq |\delta(q, \omega_G) - \delta(q, \omega'_G)| + |\delta(q, \omega'_G) - \delta(q', \omega'_G)|\\
&\leq \kappa l(\omega_G)| q - q' | + (1 + \alpha l(\omega_G))| q - q' |\\
&\leq (1 + \kappa \alpha l(\omega_G))| q - q' |
\end{aligned}$$

For $\tau = l(\omega_G)$, choose $\{\mu_i | i = 1...n\}$. Then $n \leq \frac{\tau}{min(l(\mu_i))}$.

Also choose the longest generator μ_* so that $l(\mu_*) \leq f\frac{\tau}{n}$ where $f \geq 1$.

The bound on amplification at each step

$$a_{max} = max\{\frac{|\delta(q_i, \mu_i) - \delta(q'_i, \mu_i)|}{| q_i - q'_i |}\} = (1 + \kappa\alpha * max\ l(\mu_i)) = (1 + \kappa\alpha * f\frac{\tau}{n})$$

Using the error propagation approach of Section 17.4 (for the case where $a = 1 + b/n$), the error after n steps, for large enough n is

$$e_n \leq n\ max\{e(\mu_i)\} = \frac{\tau max\{e(\mu_i)\}}{min(l(\mu_i))}$$

which goes to zero for by the following assumption.

Assumption: We can make $\frac{max\{e(\mu_i)\}}{min(l(\mu_i))}$ as small as desired by choice of ω_G e.g., reducing the time step or the quantum size.

The case $f = 1$ requires that all approximation generators are the same length in which case the assumption becomes

We can make $max\{e(\mu_i)\}$ as small as desired by choice of approximation generators all of length $\frac{\tau}{n}$. So the error must decrease as n increases. This is the case of DTSS component-wise simulation with time step that decreases in inverse proportion to n. (Section 16.2, TMS2000.)

The case f>1 requires that the longest generator continues to become smaller as the number of generators increases – this is a natural requirement since otherwise we are not using all generators efficiently. This the same kind of requirement for uniform segmentability in Section 16.3.2 in TMS2000 which requires predictability of quantization levels with decreasing quantum size. Moreover, the assumption now requires that maximum error decrease as least as fast as the shortest generator which is similar to the requirement for quantization.

So we see that the component-wise simulation of systems by DTSS and QSS are subsumed as special cases by the more general theorem on component-wise simulation of iteratively specified systems established here.

Notice that the approximation becomes exact for an iteratively specified system whose generating set is closed under segmentation and for which sufficiently short generators can be exactly represented by DEVS events. □

Exercise 20.4. Consider an admissible iteratively specified coupled system with a generator set that is closed under segmentation. Consider a set consisting of all left and right subsegments of the original generator set. Show that this set is also a generating set and that considered as an approximating set of generators, it satisfies the assumption made in the theorem.

20.4 SIMULATION STUDY OF MESSAGE REDUCTION UNDER QUANTIZATION

In this section, we take a retrospective view of heuristic arguments and conjectures made in TMS2000 on the relative performance of quantized distributed simulation/co-simulation as represented in the DEVS Bus. Our discussion is based on the updated presentation of QSS in Chapter 19 and its comparison with conventional numerical methods. The extra dimension introduced here is that context of distributed/co-simulation and its message traffic demands.

Recall that the motivation for developing the quantization approach to system representation was to support the use of the DEVS Bus for both event-based and time-driven simulation. Using quantization, we are primarily looking for a way to reduce message traffic between components, however reduction in computation and error trade-off are also important considerations. First we review a simulation study that provided some promising results and helps to focus on the issues more intensely. The results on significant reduction of message overhead were verified in simulation studies in the DEVS/HLA environment (Lee and Zeigler, 2002; Zeigler et al., 2002).

Fig. 20.6 reproduces the two approaches to quantization of DESS that were contrasted. The first (Fig. 20.6A) employed conventional time-stepped numerical methods to do the integration of the differential equations of each component in a coupled model (hence, a DTSS representation). We then attach a quantizer to the output, thus sending messages to other components only when large enough changes occur. We call this the *Quantized DTSS* approach. In the second approach (Fig. 20.6B), instead of using numerical analysis to do the integration, we directly quantize each integrator and represent it with a DEVS simulator. We call this the *Quantized DEVS* approach.

20.4.1 SOME INDICATIVE SIMULATION RESULTS

The Quantized DTSS and Quantized DEVS approaches were studied with an example of well-understood behavior that enabled comparing both message reduction and error properties. Fig. 20.7 shows the example is a DESS model of a Newtonian body orbiting around a fixed point (e.g. earth around the sun).

The results of the orbiting body to be presented in Table 20.1 were obtaining using a trapezoidal method for both DEVS and DTSS integration. The results to be presented in Table 20.3 were obtained using the simple Euler method for both. We note that greatly improved numerical methods have been developed for such marginally-stable cases (see Chapter 3) so that the times for simulation have been greatly reduced. However, the context under consideration is that of distributed simulation in which message traffic among federates is under study so that the efficiency of component simulators is not the main focus. The context is different from the non-distributed case, in that although a numerical method on a component can take steps larger than the quantum size it still has to send messages out when it crosses quantum levels between steps.

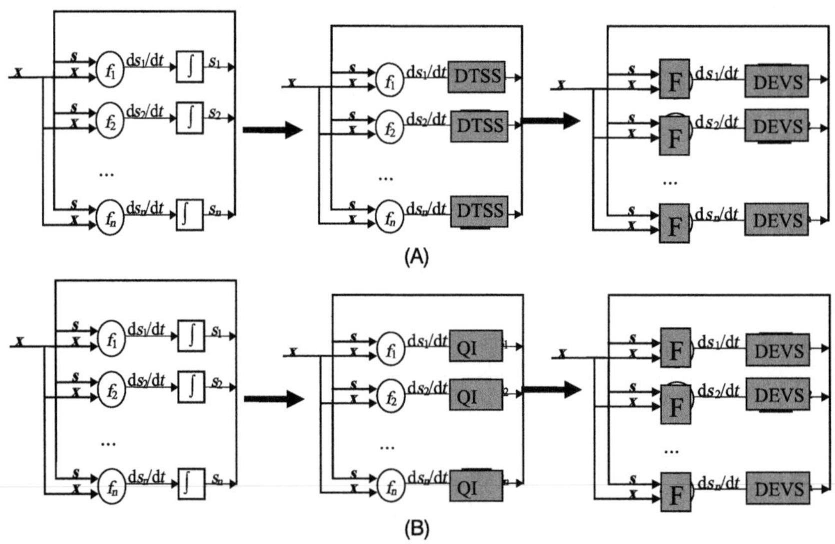

FIGURE 20.6

Quantized DTSS and Quantized DEVS Simulation Approaches.

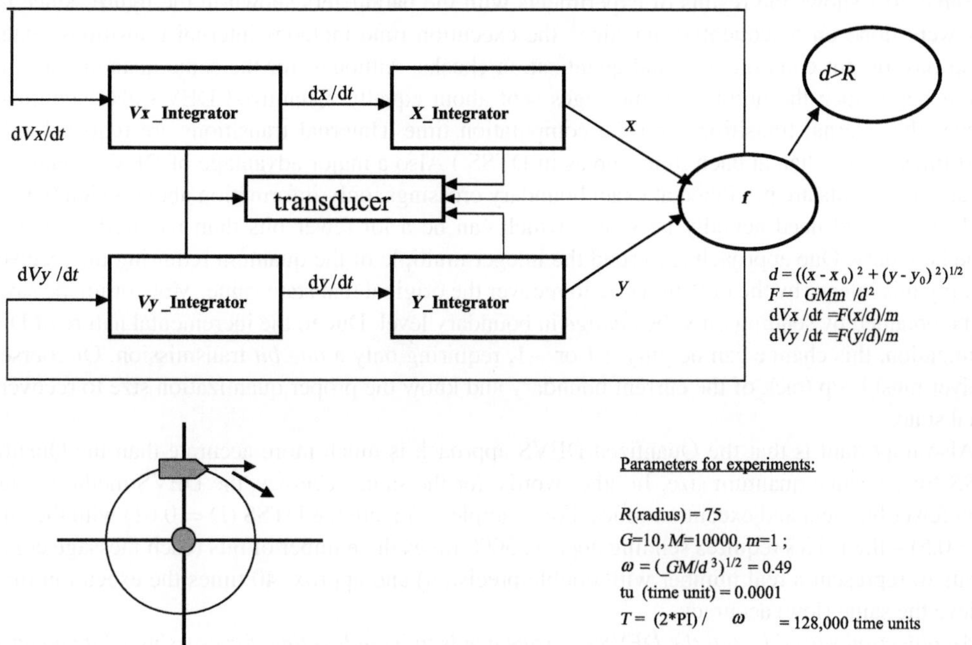

FIGURE 20.7

Test Example.

Table 20.1 Comparison of Quantization Approaches (messages shown as the sum of those exiting the integrators and the memoryless function)

Approach	Quantum size	Performance Measure				
		Number of messages sent	Number of bits transmitted	Number of internal transitions	Execution time (seconds)	Error (avg. deviation of radius from constant)
DTSS (h = 0.1 tu)		4,000,000 + 200,000	384,000,000	4,000,000	118,774	0.0099
DTSS (h = 1 tu)		400,000 + 200,000	38,400,000	400,000	12,776	0.015
Quantized DTSS (h = 1 tu)	$D = 0.5$	1,215 + 1,664	184,256	400,000	9,250	3.73
	$D = 0.1$	6,256 + 8,330	933,504	400,000	9,243	1.47
	$D = 0.01$	64,954 + 74,552	8,928,384	400,000	9,456	0.441
Quantized DEVS (with ±1)	$D = 0.5$	1,272 + 1,710	2,982	1,272	219	0.441
	$D = 0.1$	6,360 + 8,550	14,910	6,360	1,207	0.073
	$D = 0.01$	63,602 + 85,510	149,112	63,602	13,130	0.007

Table 20.1 shows the results of experiments with the parameters shown in the figure. Since these tests were done on a sequential machine[1] the execution time includes internal transitions, external transitions, output computations and quantization checks. Although for the same quantum size, both approaches reduce the number of messages sent about equally, Quantized DEVS also significantly reduces the internal transitions and the computation time. (Internal transitions are required only at event times rather than at each time step as in DTSS.) Also a major advantage of DEVS quantization is that since events are predicted at exact boundary crossings, only information about which boundary has been crossed need actually be sent – which can be a lot fewer bits than required for the state at the boundary. One approach is to send the integer multiple of the quantum requiring the receiver to multiply this integer by the quantum size to recover the original real state value. More dramatic savings can be obtained by sending only the *change* in boundary level. Due to the incremental nature of DEVS quantization, this change can be only $+1$ or -1, requiring only a *one bit* transmission. Of course the receiver must keep track of the current boundary and know the proper quantization size to recover the actual state.

Also important is that the Quantized DEVS approach is much more accurate than the Quantized DTSS for the same quantum size. In other words, for the same accuracy, the DEVS method requires many fewer bits sent and execution times. For example, compare the DTSS ($D = 0.01$) with the DEVS ($D = 0.5$) – the DTSS requires sending approx. 3000 times the number of bits (each message contains 64 bits to represent a real number with double precision) and approx. 40 times the execution time to achieve the same (low) accuracy.

An objection raised is that the DEVS message needs to include a time for crossing while a constant time step does not. While extra overhead plays a role in a single processor context, in the usual distributed/co-simulation context under consideration every message is time stamped so that the overall global time is correctly maintained.

Caveat: These tests were done with first order methods (see Chapter 19) and may not represent the results that would be obtained with higher order and/or more efficient modern methods.

We also compared Quantized DEVS with "pure" DTSS, by which we mean using the DTSS without quantization. Effectively, this is Quantized DTSS with a quantum size matching the precision of the computer word size. The remarkable results is that the DEVS achieves a better accuracy than the pure DTSS ($h = 1$) with approx. the same execution time but with a reduction of approx. 250 times the number of bits transmitted. When we run the DTSS with $h = 0.1$, the results are even more dramatic. Now both have about the same accuracy but the DEVS is 10 times faster and has a 2500 times reduction in bits sent.

20.4.2 COMPARING QUANTIZED DEVS WITH DTSS IN DISTRIBUTED SIMULATION OF DESS

How can we explain the remarkable gains in distributed simulation performance by Quantized DEVS displayed in this example? Is this example just a fluke or is there something fundamentally different about the DEVS approach? In TMS2000 we raised question and addressed with several conjectures and discussions of them. Here we revisit this section with the additional knowledge gained in the interim.

[1] We used the DEVSJAVA environment using double precision for real number representation.

Conjecture 20.1. *For any exactly quantizable DESS coupled model, a DEVS component-wise simulation is never less, and often much more efficient, than a counterpart variable or fixed step simulation with the same final accuracy.*

We note that the conjecture is preconditioned on the *exactly quantizable* property. This requires that quantization does not introduce errors into the system representation as defined in TMS2000 Chapter 18. This idealization obviates the introduction of quantization-specific errors such as spurious oscillations (Section 19.3.2). The focus is on bringing out the underlying mechanism at work for efficiency gain in both execution and messaging, recognizing the other factors might reduce or nullify this benefit.

Consider a DEVS simulation of a DESS network of integrators as in Fig. 20.6 in comparison to conventional numerical solver for the same network.

We now present a heuristic argument why this conjecture should be true. The argument is limited to *DEVS simulation based on first order QSS and first order time stepped methods.*

By assumption for any given final error there is a quantum size D that enables the DEVS simulation to stay within that error. Let us choose the largest D that satisfies the tolerance requirement – so that crossing a boundary later than required by this D will result in a final error that is above the tolerance limit.

Let's look at a DEVS simulation cycle as illustrated in Fig. 20.8A. Each DEVS integrator, i schedules itself for the next boundary crossing at $\tau_i = D/|x_i|$, where is the current input derivative to integrator, i. The time advance to the next event is then $\sigma = \min\{\tau_i\}$ and the imminent integrators are those for which $\tau_i = \sigma$. In Fig. 20.8A, the imminents are $1, 4, 6$. The coupling then determines which integrators receive outputs from these imminents. Imminents $1, 6$ send to $2, 4, 7$ in the figure (shown as heavy vertical arrows). The imminents and the influencees $1, 2, 4, 6, 7$ then execute their transition functions – $1, 6$ the internal, and $2, 7$, the external function and 4 the confluent function (which in the case of an integrator is the same as the external transition function. This results in new time advances for imminents and their influences (shown as dotted arrows for the latter). None of the

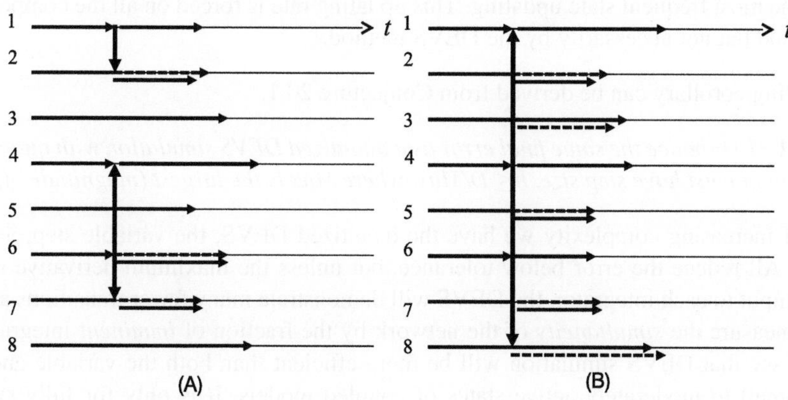

FIGURE 20.8

Comparing DEVS and DTSS simulation of DESS.

remaining integrators (e.g., 3, 5, 8) execute computations and their next event times are unchanged. We are now ready for the next DEVS cycle.

To measure the complexity of such a cycle we see that we need to obtain measurements from the imminents and their influencees. But we don't necessarily have to identify them at each cycle to do this. This is because the complexity of a run is determined by accumulating the measured elements in the individual cycles. We instrument each component to count its internal transitions, external transitions and confluent transitions. This will catch all the executions without requiring us to identify the imminents and their influencees. We also have all outputs sent to a transducer thus capturing all the messages sent from the imminents to influencees, again without necessarily identifying who did what when.

Now consider Fig. 20.8B where we illustrate a synchronized variable step simulation strategy. The global time advance is again $\sigma = \min\{\tau_i\}$. Instead of activating only the imminents however, we activate all the integrators at this time. They exchange messages as prescribed by the coupling and execute their integration step (external transition function). Thus communication and computation occur at each event. We can call this a variable step integration method since at each cycle, σ may be different. Note that some integrators such as $\{3.5, 8\}$ are now updated that were not it in the original DEVS simulation. Since we have chosen the quantum size for error tolerance, any increased accuracy that might result from such updating does not matter to us. (Had we wanted this improved accuracy we should have said so in the beginning – we can always go back and do so!)

In our final case, at each cycle we set $h = D/Max$ where *Max* is the largest (magnitude of) derivative of all the inputs to the integrators. Here the time advance is fixed for the whole run, so we have a DTSS simulation with time step *h*. A bigger time step than *D/Max* runs the risk of increasing error since we would be too late for the boundary crossing of the integrator(s) that receives this largest derivative input (and D was chosen as liberally as possible within the error tolerance). Thus, *D/Max* is the largest acceptable (constant) time step. Note that the updating in any cycle in which the maximum derivative is not present, is superfluous, since the earliest required updating occurs later.

Remark 20.1. The claim is that the fastest component (the one with the largest derivative) is the one that requires the most frequent state updating. This updating rate is forced on all the components in the time step method but not necessarily by the DEVS method.

The following corollary can be derived from Conjecture 20.1:

Corollary 20.1. *To produce the same final error as a quantized DEVS simulation with quantum size, D, a DTSS simulation must have step size, $h \leq D/Max$, where Max is the largest (magnitude of) derivative.*

In order of increasing complexity we have the quantized DEVS, the variable step, and the fixed step methods. All reduce the error below tolerance, but unless the maximum derivative is present at each cycle as input to each integrator, the DEVS will demonstrate many fewer transitions and message traffic. If we measure the *simultaneity* of the network by the fraction of *imminent* integrators at each cycle, we can see that DEVS simulation will be more efficient than both the variable and fixed step methods for small to moderately active states of coupled models. It is only for fully simultaneous states that persist for all of the run that a variable step method will equal the DEVS in performance. In addition, the maximum derivative has to be input to at least one integrator at each time cycle for the fixed time step method to match the DEVS performance.

Remark 20.2. We need to clarify the definition of network *simultaneity*. It is measured by the fraction of quantized integrators that are imminent in each cycle. In the DEVS case the integrators have the largest derivative in the global state at that cycle (hence the smallest time advance.) We distinguish this property from network connectivity which determines the number of influencees of an imminent when it sends an output.

Conjecture 20.2. *Highly Simultaneous DESS Networks underlying distributed/co-simulation are rare.*

In the context of distributed/co-simulation a number of arguments suggest that highly simultaneous networks are rare – the first consideration relates to the heterogeneity of simulation composed of diverse DESS components. Internally they might well be homogeneous, i.e., governed by simultaneous dynamics (all contained integrators having the same time steps). However, at the coupled level, their time steps might well be different, hence low in simultaneity as defined above.

Second, it might be instructive to take a stochastic characterization of the selection of derivatives, if we assume each integrator chooses its derivative independently, then the chance that all integrators receive the same input derivative, let alone the same maximum value, decreases exponentially with the number of components. Of course, the derivative process may be highly correlated, since the underlying model is deterministic. Nevertheless, the exponential dependence on number of integrators might well hold in any sequence of models with any sort of interesting behavior.

Third, we might get some insight from examining common types of curves. For example, in a second order linear oscillator, the integrators see sine wave inputs from each other that are 90 degrees out of phase. Indeed, when one integrator is getting the maximum value (amplitude times frequency) the other is getting the minimum, 0. Only every eighth cycle do the integrators see the same input magnitude and as we shall see the DEVS has approx. 30% fewer internal transitions. The 4th order orbiting oscillator above shows a much greater reduction for DEVS integration in internal transitions over the DTSS simulation with the same accuracy.

20.4.3 INSIGHT FROM 2nd ORDER LINEAR OSCILLATOR

The linear oscillator shown in Fig. 20.9A provides a revealing example of the efficiency advantages of DEVS. It is well known that the exact behavior is characterized by sinusoidal oscillations with the integrators 90 degrees out of phase with each other.

Thus, while one is generating a sine wave the other is generating a cosine wave. Indeed, from the input perspective of each one, we can replace the other with the corresponding sinusoidal generator in open loop fashion as in Fig. 20.9B. The derivative input trajectories are shown in Fig. 20.9C. Taking into account that the step size is inversely related to the magnitude of the derivative, the same pattern of relative step sizes is repeated every eighth of a cycle. In every such period, one integrator always sees higher derivatives than the other, and they alternate among these fast and slow behaviors. Think of it this way: which ever is currently the slow integrator needs the fast integrator to speed it up; conversely, the fast integrator needs the other to slow it down.

Now in the DEVS simulation, in every eighth of cycle, the time advances of the fast integrator are larger than those of the slow integrator. It is only at the end of this period where they are equal. As shown in Table 20.2, we can estimate the average time step by following the steps of one integrator through a quarter cycle containing its slow and fast phases. We predict that DEVS cuts down on the number internal transitions and output messages by about 30% over the DTSS with the same error. In contrast, the variable step method is predicted to produce only a 10% reduction.

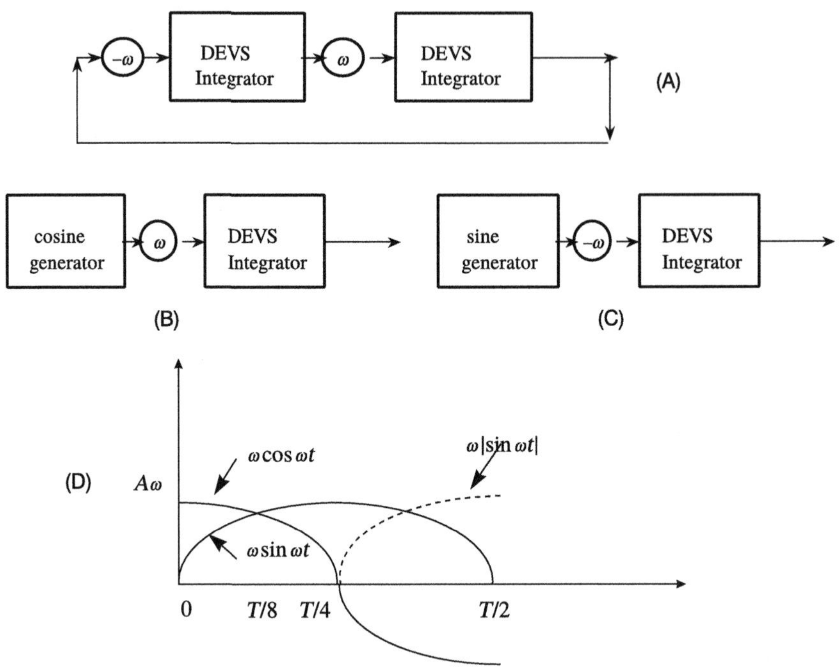

FIGURE 20.9

Second Order Linear Oscillator.

Table 20.2 Predicted Comparison of DEVS and Variable Time Step Relative Baseline DTSS

Method	Means to Estimate	Internal transitions (output messages) relative to DTSS as D gets small
fixed time step DTSS	D (set max. derivative = 1)	1
variable time step	Average of minimum time step of both integrators over one eighth cycle	0.9
Quantized DEVS	Average of time step of one integrator over one quarter cycle	*2/3.14 = 0.64*

Exercise 20.5. Write the integral for the average time step of one integrator over a quarter cycle. Look up the integral in a table. In general, the average time step is not equal to the quantum divided by the average derivative. Compute the latter and compare.

Exercise 20.6. Write a short program to compute, given a quantum size, successive time step sizes in the DEVS and variable time step methods. As the quantum gets smaller, do your averages converge to those in the table?

Table 20.3 Simulation Comparison of DEVS Quantization Approach with Pure DTSS: $\omega = 0.0001$, $A = 75$, run time = 62,800 (One cycle)

	Avg. Error	Number of messages passed	Number of data bits passed	Number of internal transitions	Execution time (seconds)
DTSS (h = 66.7)	0.79	1882	120,448	1882	125
(h = 13.3)	0.16	9418	602,752	9418	550
(h = 1.33)	0.016	94,198	6,028,672	94,198	10,140
Quantized DEVS (D = 0.5)	0.96	1213	1213	1213	153
(D = 0.1)	0.2	6019	6019	6019	657
(D = 0.01)	0.02	59,999	59,999	59,997	6583

Exercise 20.7. Show that the requirement that $h \leq D/Max$ implies that $h/T \leq D/A$ for the oscillator, i.e., h/T and D/A are closely correlated. Explain why this should be so.

Simulation of a test case showed that the predictions of relative performance of quantized DEVS versus pure DTSS were remarkably close. We compared the Quantized DEVS with quantum sizes, D shown in Table 20.3 against the DTSS with step sizes, $h = D/max$ where the maximum derivative is the amplitude multiplied by the frequency with values shown in the table. Note that the ratio of messages sent (and internal transitions) is the same as predicted to 2 decimal places! The errors incurred in quantized DEVS are however, slightly higher than the corresponding DTSS values, although of the same order of magnitude. The data bits passed (using the DEVS binary output technique) are 100 times fewer for DEVS than for DTSS. The execution time for DEVS, although initially slightly larger, improves to about 40% that of DTSS as the quantum size decreases.

20.4.4 CONJECTURES FOR FURTHER RESEARCH

Caveat: Effect of Integration Method

The results presented provide strong evidence for the claim that quantized DEVS can reduce the message traffic burden in distributed simulation *visa vis* quantized DTSS for the same accuracy. Moreover they support the stronger conjecture that quantized DEVS does better than pure DTSS in both message passing and total execution time, even for small systems such as the 2nd order oscillator. However, TMS2000 stated the caveat that *the effect of integration method has to be considered.* There are many varieties of integration methods for differential equation systems with arrays of parameter settings and accuracies. Thus we have to be cautious with respect to the stronger conjecture concerning DEVS as a better integration approach than conventional integration methods in conventional uniprocessor simulation. This issue was considered in depth with respect to QSS in Chapter 19. It may well be that conventional methods are preferred for small differential equation systems on isolated sequential processors or even multiprocessors with shared memory. However, as the size of the system increases and also the number of processors with message passing required for interprocessor communication, it appears that DEVS quantization has distinct message reduction advantages for the same accuracy.

The developments so far suggest that we can refine our conjecture into a restricted and strong from:

Conjecture 20.3 (Restricted). *The message passing efficiency of a quantized DEVS simulation of differential equation systems is never less than that of counterpart quantized DTSS to achieve the same accuracy.*

Conjecture 20.4 (Strong). *The efficiency (both message passing and execution time) of a quantized DEVS simulation of differential equation systems is never less than that of counterpart conventional integration methods to achieve the same accuracy.*

The efficiency of quantized DEVS relative to the alternative approaches increases with the:

* order of the differential equation system, i.e., with increasing numbers of integrators the frequency of occurrence of simultaneity is diminished.
* temporal heterogeneity of the network, i.e., the dispersion in the derivative values (interestingly, a class of such systems are called stiff systems in the literature and are hard to solve with conventional methods. TMS2000 asked the question: are such systems easier to solve with DEVS? Chapter 19 of this edition addresses this question with respect to the QSS approach.
* connectivity of the network, i.e., with the fraction of other integrators influenced by an imminent integrator. In the extreme of a fully disconnected network (a not very interesting case!), the only cost (internal transitions) of conventional approaches is that due to less than full simultaneity. As the connectivity of the network, increases there is a greater penalty for the communication required for unneeded state updating.

Further research is needed to establish the correctness of the conjectures. Especially, as indicated, the effect of integration method on the comparison must be investigated for the strong conjecture.

Exercise 20.8. Consider quantization-based DEVS simulation of the 2nd order oscillator with positive and negative damping respectively. Predict which model will have the largest efficiency.

20.5 SUMMARY

This chapter argued that distributed simulation requires expansion of the concept of simulation complexity to include measures of communication among nodes. We showed that discrete event, rather than traditional approaches, are better at reducing such complexity while enabling accurate representations of continuous systems. To do this we developed the theoretical foundation for the quantized systems, and used this foundation to show how the DEVS representation can provide more efficient simulation at lower error cost than the traditional integration methods of differential equation systems, especially in a distributed simulation context. Based on this foundation and empirical run data, we make several conjectures on the conditions where DEVS-based quantization provides more efficient simulation than quantization or discretization of continuous systems. This forms the basis for the DEVS Bus, a distributed simulation framework for multi-formalism modeling.

20.6 SOURCES

DEVS representation of piecewise constant systems was first described in Zeigler (1984) and further developed in Zeigler (1989), Praehofer and Zeigler (1996), Zeigler et al. (1996), Zeigler and Moon

(1996), Kim and Zeigler (1997). Discrete time methods for real-time distributed simulation were developed by Howe (1988, 1990) and Laffitte and Howe (1997). Dead Reckoning plays a major role in distributed interactive simulation (Bassiouni et al., 1997; Goel and Morris, 1992; Lin and Schab, 1994).

REFERENCES

Bassiouni, M.A., Chiu, M.-H., Loper, M., Garnsey, M., Williams, J., 1997. Performance and reliability analysis of relevance filtering for scalable distributed interactive simulation. ACM Transactions on Modeling and Computer Simulation (TOMACS) 7 (3), 293–331.

Camus, B., et al., 2018. Co-simulation of cyber-physical systems using a DEVS wrapping strategy in the MECSYCO middleware. Simulation Journal. https://doi.org/10.1177/0037549717749014. Published online.

Goel, S., Morris, K., 1992. Dead reckoning for aircraft in dis. In: Proceedings of AIAA Flight Simulation Technologies Conference. South Carolina.

Hairer, E., Nørsett, S., Wanner, G., 1993. Solving Ordinary Differential Equations. I, Nonstiff Problems. Springer-Verlag, Berlin.

Howe, R., 1988. Dynamic accuracy of the state transition method in simulating linear systems. Transactions of the Society for Computer Simulation International 1, 27–41.

Howe, R., 1990. The use of mixed integration algorithms in state space. Transactions of the Society for Computer Simulation International 7 (1), 45–66.

Kim, D., Zeigler, B.P., 1997. Orders of magnitude speed up with DEVS representation and high performance simulation. Enabling Technology for Simulation Science, SPIE Aerosense 97.

Laffitte, J., Howe, R.M., 1997. Interfacing fast and slow subsystems in the real-time simulation of dynamic systems. Transactions of SCS 14 (3), 115–126.

Lee, J.S., Zeigler, B.P., 2002. Space-based communication data management in scalable distributed simulation. Journal of Parallel and Distributed Computing 62 (3), 336–365.

Lin, K.-C., Schab, D.E., 1994. The performance assessment of the dead reckoning algorithms in dis. Simulation 63 (5), 318–325.

Moshagen, T., 2017. Convergence of explicitly coupled simulation tools (cosimulations). arXiv preprint arXiv:1704.06931.

Praehofer, H., Zeigler, B.P., 1996. On the expressibility of discrete event specified systems. In: Computer Aided Systems Theory – CAST'94. Springer, pp. 65–79.

Trčka, M., Hensen, J.L., Wetter, M., 2009. Co-simulation of innovative integrated HVAC systems in buildings. Journal of Building Performance Simulation 2 (3), 209–230.

Zeigler, B.P., 1984. Multifacetted Modelling and Discrete Event Simulation. Academic Press Professional, Inc, London.

Zeigler, B.P., 1989. DEVS representation of dynamical systems: event-based intelligent control. Proceedings of the IEEE 77 (1), 72–80.

Zeigler, B.P., Cho, H.J., Kim, J.G., Sarjoughian, H.S., Lee, J.S., 2002. Quantization-based filtering in distributed discrete event simulation. Journal of Parallel and Distributed Computing 62 (11), 1629–1647.

Zeigler, B.P., Moon, Y., 1996. DEVS representation and aggregation of spatially distributed systems: speed-versus-error tradeoffs. Transactions of the Society for Computer Simulation International 13 (4), 179–189.

Zeigler, B.P., Moon, Y., Lopes, V.L., Kim, J., 1996. DEVS approximation of infiltration using genetic algorithm optimization of a fuzzy system. Mathematical and Computer Modelling 23 (11–12), 215–228.

Zeigler, B.P., Sarjoughian, H.S., 2017. Guide to Modeling and Simulation of Systems of Systems, 2nd edition. Springer.

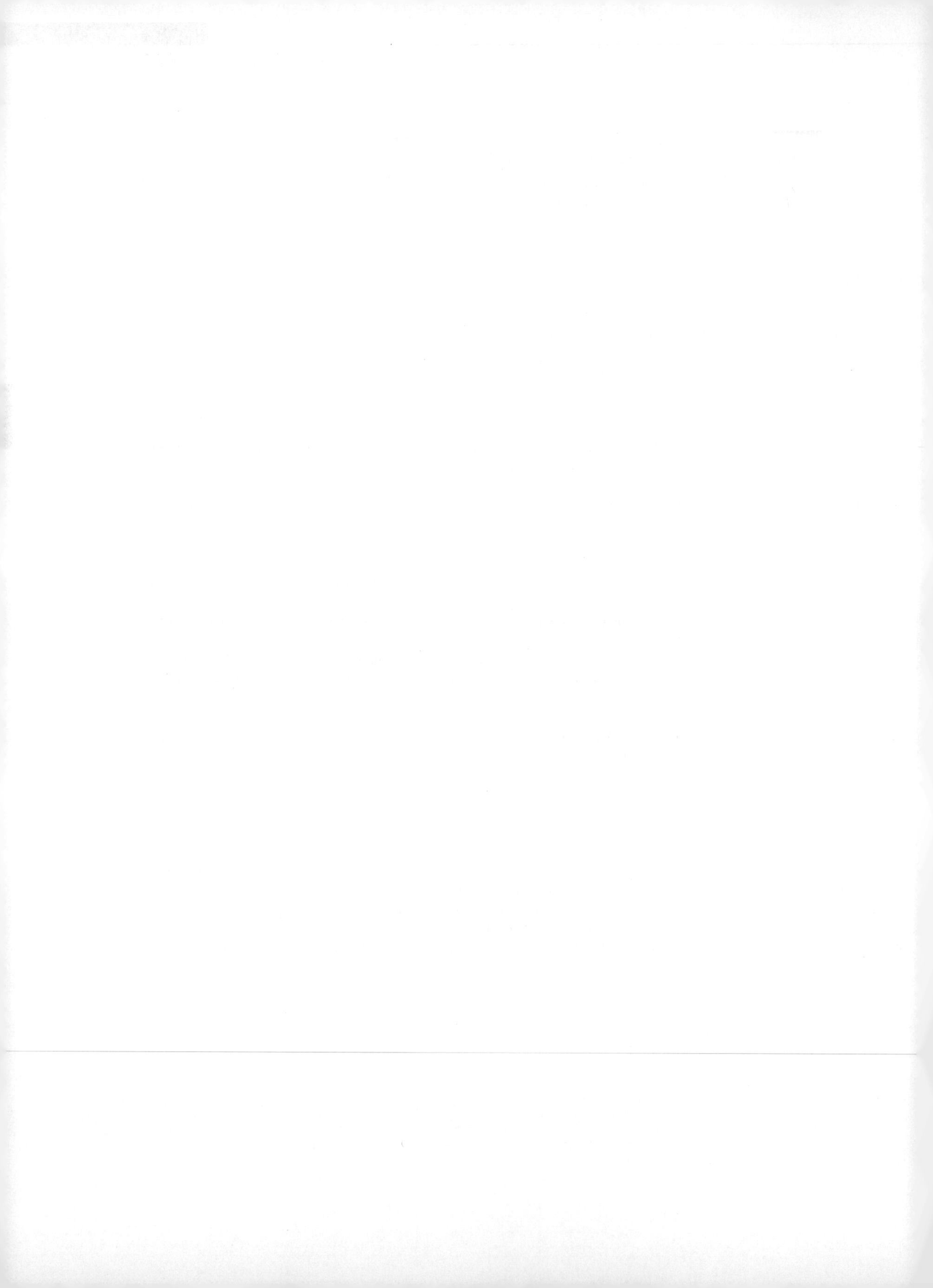

PART

ENHANCED DEVS FORMALISMS

4

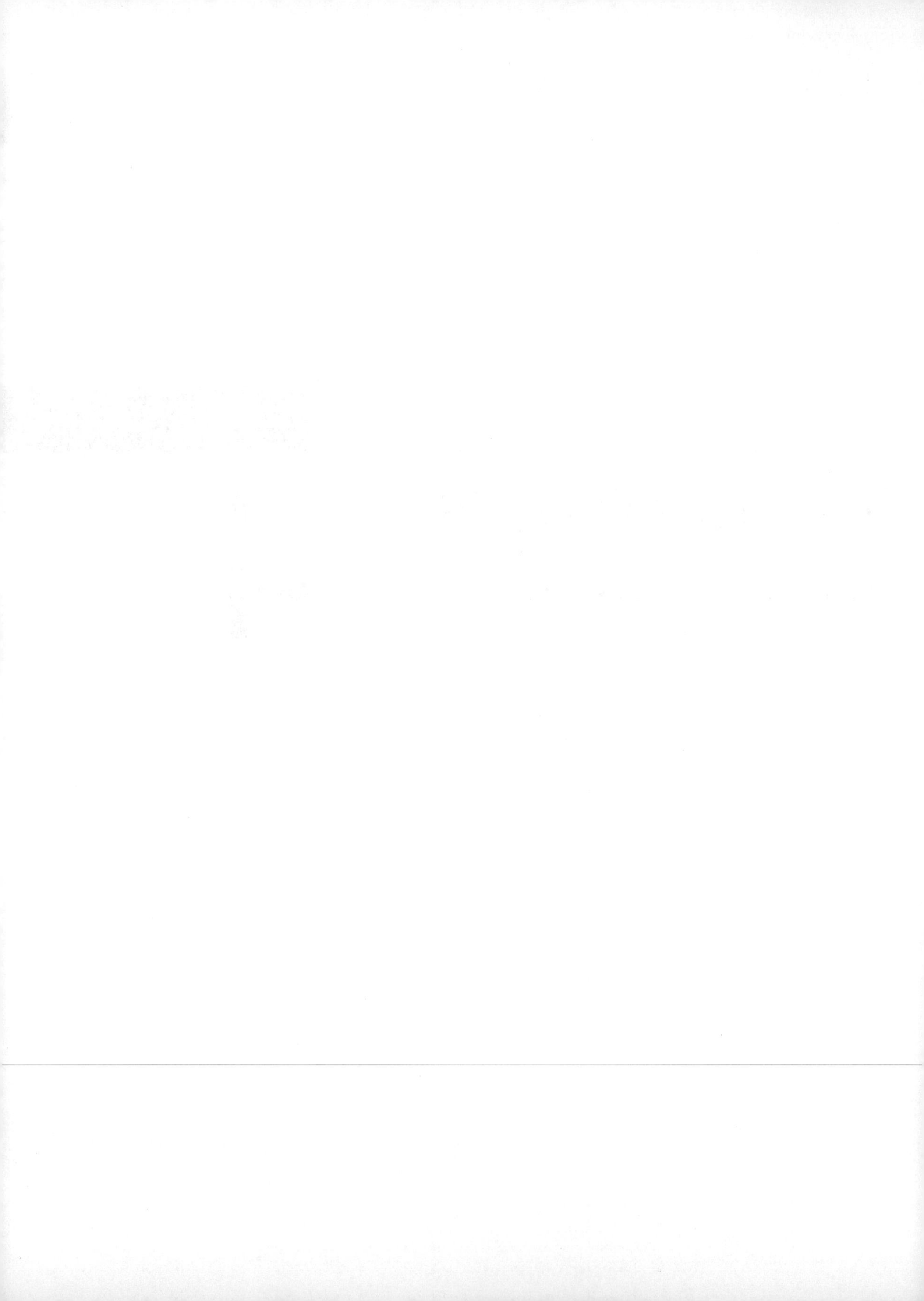

CHAPTER

DEVS MARKOV MODELING AND SIMULATION

21

CONTENTS

DEVS provides a unified framework for analysis, design, modeling and simulation of systems, both deterministic and stochastic. However, at the time of publication of TMS2000, it lacked a com-

Theory of Modeling and Simulation. https://doi.org/10.1016/B978-0-12-813370-5.00032-8

pletely well-defined foundation in probability theory. This gap was filled by the general framework for Stochastic DEVS developed by Castro et al. (2010). We note however that TMS76 employed ideal random generators to characterize stochastic simulation in DEVS, a concrete realization employed by Stochastic DEVS as well. Here we build on the foundation with specification of Markov models as DEVS model classes. Markov Modeling is among the most commonly used forms of model expression. Indeed, Markov concepts of states and state transitions are fully compatible with the DEVS characterization of discrete event systems. These concepts are a natural basis for the extended and integrated Markov modeling facility developed within M&S environments (e.g., MS4 Me). DEVS Markov models, which are full-fledged DEVS models, and able to be integrated with other DEVS models just like other DEVS models. From this point of view, a Markov modeling facility makes it much easier to develop probabilistic/stochastic DEVS models and to develop families of DEVS models for cutting edge challenging areas such as Systems of Systems, agent-directed systems, and DEVS-based development of Web/Internet of Things. Besides their general usefulness, the Markov concepts of stochastic modeling are implicitly at the heart of most forms of discrete event simulation.

DEVS Markov modeling differs fundamentally from existing packages such as PRISM (Kwiatkowska et al., 2011) and SAN (Deavours and Sanders, 1997) in fundamental respects based on its origins in general mathematical systems theory (Seo et al., 2018). In brief, this means that the total state of a system which includes its sequential state and elapsed time in that state, is employed rather than the sequential state only as is the case for other simulation model formalisms. Implications of this fact for Markov modeling are discussed below. Moreover, the ability to compose in a hierarchical manner with well-defined coupling specification is a second fundamental consequence of the systems theory basis.

Due to their explicit transition and time advance structure, DEVS Markov models can be individualized with specific transition probabilities and transition times/rates which can be changed during model execution for dynamic structural change.

21.1 MARKOV MODELING

DEVS Markov Models can represent complex systems at the level of individual subsystems and actors. In this guise, systems and actors can be represented as components with states and transitions as well as inputs and outputs that enable them to interact as atomic models within coupled models using coupling in the usual way. Briefly stated, these atomic and coupled models are useful because:

- The DEVS simulator provides a Monte Carlo layer that generates stochastic sample space behavior.
- DEVS Markov models can express probabilistic agent-type alternative decisions and consequences, generate and analyze both transient and steady state behavior.
- Together with experimental frames, DEVS Markov models support queuing-like performance metrics (queue sizes, waiting times, throughput) as well as a wide range of others (win/loss ratios, termination times).

Markov Finite Chain State-transition Models are computationally much faster because they employ deterministic computation of probabilities interpreted as frequencies of state occupation of the corresponding DEVS Markov Models (DMM). Such models are very useful because:

- They yield probabilities for ergodic DMMs in steady state.
- They yield probabilities for DMMs that reach absorbing states.
- They support computation of state-to-state traversal times for models where time consumption is of essential interest.
- They provide simplifications of DMMs that are accurate for answering certain questions and can be composed to yield good approximations to compositions of DMMs.

21.2 BACKGROUND

Appendix 21.A gives some background for this chapter on exponential distributions, Poisson processes, and Markov basics. A chapter in the second edition of Zeigler and Sarjoughian (2017) shows how such concepts are fully compatible with the DEVS characterization of discrete event models and a natural basis for the extended and integrated Markov modeling facility developed within the MS4 Me environment. Well known Markov model classes were defined by Feller and others (Kemeny et al., 1960; Feller, 2008), with both discrete and continuous time bases: Continuous Time Markov Model (CTM), Discrete Time Markov (DTM) and the Markov chain class. Recently attention has been drawn to Semi-Markov models and Hidden Markov models (Barbu and Limnios, 2009). DEVS simulation of a formalism called General Semi-Markov Processes (GSMP) GSMP (Glynn, 1989) was discussed by Rachelson et al. (2008) to take account of time as a crucial variable in planning which introduces specific structure along with additional complexity, especially in the case of decision under uncertainty.

21.3 MAPPING DEVS MARKOV MODELS

Integrating Markov modeling into DEVS opens up a wide variety of model types that can be incorporated within the same framework. It helps to organize such models into classes that relate both to the traditional ones encountered in the mathematics and applications literature as well as to the structural features that characterize all DEVS models as specifications of input/output dynamic systems.

21.3.1 GENERAL FRAMEWORK FOR STOCHASTIC DEVS

Before proceeding, we recall the general framework for Stochastic DEVS developed by Castro et al. (2010), where a Stochastic DEVS has the structure:

$$M_{ST} = < X, Y, S, \mathcal{G}_{int}, \mathcal{G}_{ext}, P_{int}, P_{ext}, \lambda, ta > \tag{21.1}$$

where X, Y, S, λ, ta have the usual definitions (Chapter 5).

Here $\mathcal{G}_{int} : S \to 2^S$ is a function that assigns a collection of sets $\mathcal{G}_{int}(s) \subseteq 2^S$ to every state s. Given a state s, the collection $\mathcal{G}_{int}(s)$ contains all the subsets of S that the future state might belong to with a known probability, determined by a function $P_{int} : S \times 2^S \to [0, 1]$. When the system is in state s the probability that the internal transition carries it to a set $G \in \mathcal{G}_{int}(s)$ is computed by $P_{int}(s, G)$.

In a similar way, $\mathcal{G}_{ext} : S \times \Re_0^+ \times X \to 2^S$, is a function that assigns a collection of sets $\mathcal{G}_{ext}(s, e, x) \subseteq 2^S$ to each triplet (s, e, x). Given a state s and an elapsed time e, if an event with value x arrives, $\mathcal{G}_{ext}(s, e, x)$ contains all the subsets of S that the future state can belong to, with a known probability calculated by $P_{ext} : S \times \Re_0^+ \times X \times 2^S \rightarrow [0, 1]$.

Coupled models with Stochastic DEVS components are taken to be defined in the usual DEVS form (Chapter 7) and in this sense Stochastic DEVS is closed under coupling (Castro et al., 2010). Proof of such closure shows that Stochastic DEVS is defined in a way that is consistent with the underlying fundamental probability theory since the resultant of coupling has to be shown to have same probability structure underlying the basic models.

21.3.2 STOCHASTIC DEVS WITH FINITE SETS

To simplify the approach, assuming that S is finite, we let

$$P_{int}(s, G) = \sum_{s' \in G} Pr(s, s') \tag{21.2}$$

where $Pr(s, s')$ is the probability of transitioning from s to s'. This is possible when S is finite and in other circumstances. The fact that the general formulation has been shown to be closed under coupling allows us to work with the simpler formulation. Similarly, we set

$$P_{ext}(s, e, x, G) = \sum_{s' \in G} P_{s,e,x}(s') \tag{21.3}$$

where $P_{s,e,x}(s')$ is the probability of transitioning from s to s' having received event with value x after elapsed time e. Note that this requires not only the input but the elapsed time at which it occurs to be taken into account. When we impose the restricted condition that the elapsed time does not matter, we will have the special subclass that corresponds to Continuous Time Markov (CTM) models with external input.

Stochastic DEVS can be more simply presented as

$$M_{ST} =< X, Y, S, Pr, \{P_{s,e,x}\}, \lambda, ta > \tag{21.4}$$

where $Pr : S \times S \rightarrow [0, 1]$ and $P_{s,e,x} : S \rightarrow [0, 1]$ for all $x \in X$ and $e \in [0, \mathrm{ta}(s)]$ determine the probabilities of internal and external transitions, respectively.

Since Stochastic DEVS is an extension of DEVS that is based on fundamental probability theory it provides a sound foundation to develop Markov modeling classes which provide direct practical tools for stochastic simulation.

21.3.3 SES FOR DEVS MARKOV MODELS

In the following, we employ a System Entity Structure (SES) to organize the model classes of interest. The SES sees a DEVS Markov model specification as composed of a time base, phase set, external event set and transition function mapping with specializations for each component. Classes and subclasses of such models then can be formed by choice of elements within some of specializations, perhaps leaving others unspecified. The broader the class, the fewer elements of the SES are fixed. Conversely, the narrower the class, the greater are the number of selections made.

Exercise 21.1. Show that in an SES, a class of models specified by one pruning is a subclass generated by another pruning if every selection of the second is also made by the first.

For simplicity of exposition we restrict attention to the Classic DEVS family.

The specializations in our SES for DEVS Markov models are, as shown in Fig. 21.1:

Time Base – can be discrete or continuous. Most of our concentration will be on continuous time but simplifying properties of discrete time make it useful at times.

Phase Set – this is the state set typically referred to in math expositions. The reason we refer to it as the phase set since the state of a DEVS will include sigma, the explicit representation of the time advance. In addition, the global state of DEVS includes the elapsed time. We focus mostly on finite phase (Markov state) sets, however, much of the approach will extend to infinite state sets.

External Event Set – this is the external interface including the input and output sets. Sometimes, it is convenient to consider only the transition portion of the DEVS structure, which omits the external event set related elements, and will be called the core of the structure. On the other hand, the external interface is needed for including model types such as hidden Markov and Markov Decision Processes that interact with the environment and other agents.

Also not shown,

External Transition Function – models can disallow the elapsed time since the last event to influence the effect of an input. This restriction is in effect for DTM and CTM as will be shown.

Transition Function Mapping – this will help us to understand the differences in the varieties of Markov models in the literature, e.g., semi-Markov, GSMP, hidden Markov, etc.

The kernel of the Markov implementation in DEVS is the Transition Function Mapping, illustrated in Fig. 21.2. The phase and sigma play the roles of Markov state and transition time, respectively. At every internal transition the current phase and sigma together with a random number select a next phase and sigma.

Subclasses of Markov models can be characterized by pruning operations on the overall SES.

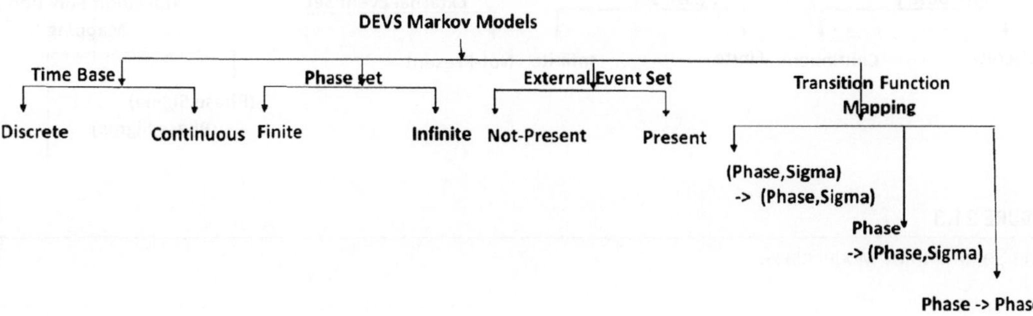

FIGURE 21.1

System Entity Structure for DEVS Markov Models.

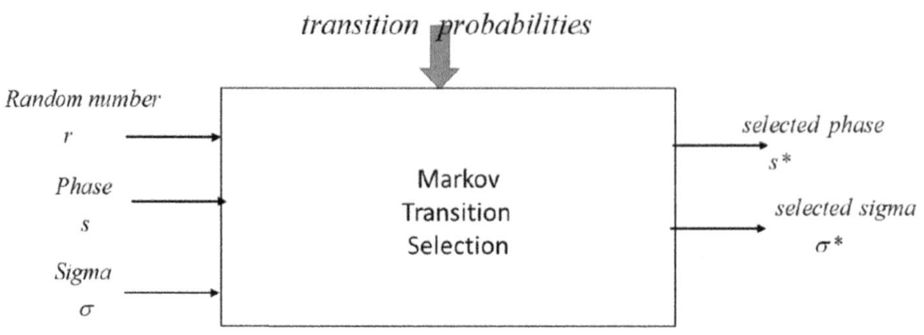

FIGURE 21.2

The Select function that produces a state to which to transition and the time it takes.

The DEVS Semi-Markov model (Fig. 21.3) employs the most flexible transition mechanism which uses the current phase and sigma to select the next phase and sigma. When composed into coupled models such a models include the GSMP (Glynn, 1989; Rachelson et al., 2008).[1]

For example, the Discrete Time Markov (DTM) model class is generated Fig. 21.4 where the time base is discrete with finite state set. The transition specification employs transitions from phases to phases which take a fixed time step also called a cycle length.

In contrast the Continuous Time Markov (CTM, Fig. 21.5) employs a continuous time base and a transition specification that employs only the current phase to determine both the next phase and the transition time (Fig. 21.6). It can do so because the transition probabilities are interpreted as rates in exponential distributions for transition times. That is, all transition times are based on the same type of distribution (exponential) while in the more general case, a time distribution can be associated with each non-self-transition pair (phase, phase'). We'll go into more detail later (Section 21.6.2).

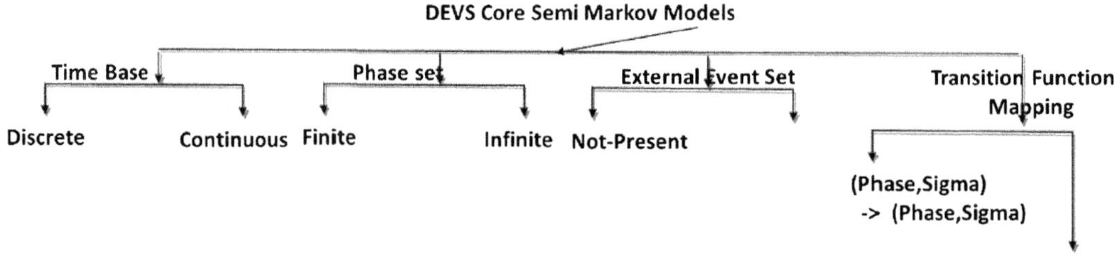

FIGURE 21.3

The Semi-Markov Model class.

[1]The use of the adjective "Semi" for Markov models in the literature stems from the fact that they loosen the requirement that inter-event times be exponentially distributed (i.e., fully Markov). From the present perspective this relaxation extends the class of models significantly while still retaining essential analytic and simulation properties (Barbu and Limnios, 2009).

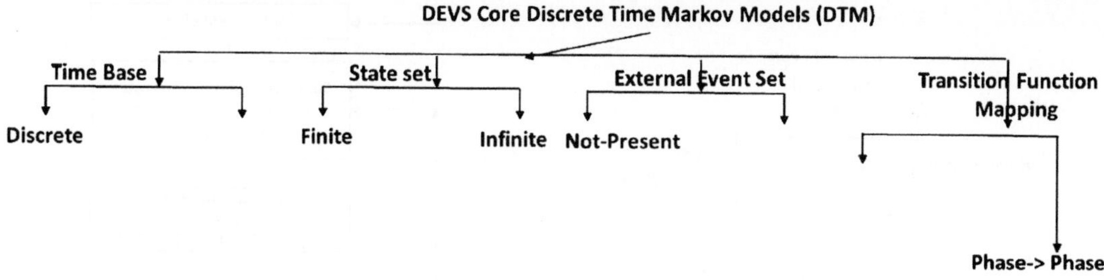

FIGURE 21.4

The Discrete Time Markov Model class.

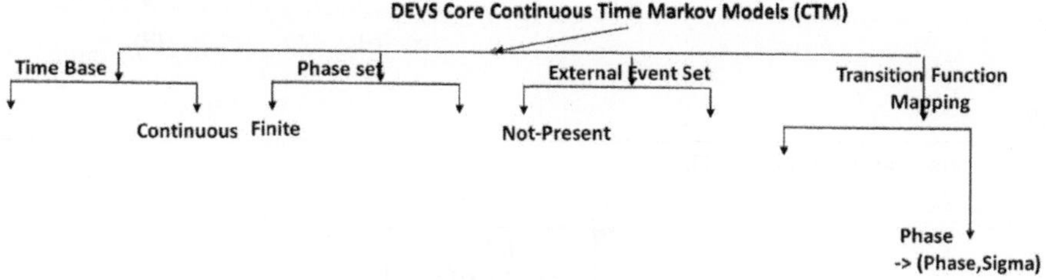

FIGURE 21.5

The Continuous Time Markov Model class.

21.3.4 UNCOUPLING DECISION PROBABILITIES FROM TRANSITION TIMES

The continuous time approach links probabilities of choices for jumping to other states with rates of transition. Since times for transition are directly specified by rates this forces the modeler to try capture both decision probabilities and times of transition in one degree of freedom. The full scale Markov model class provides two degrees of freedom to specify decision probabilities and transition times separately and independently. The ability to do so is has its origins in the semi-Markov models formulation which allows distributions other than exponential and geometric even though these are not necessarily memoryless (Barbu and Limnios, 2009).

As illustrated by the example in Fig. 21.7, consider fielder trying to field a baseball in the air hit by a batter. If she can catch the ball cleanly she can throw it immediately back to the infield. But if she does not control the ball, shown as missing, then the time to get back to controlling the ball, is largely independent of the probability of catching it cleanly.

The ratio of p1 to p2 will be specified based on the catching skill of the fielder but independently of the time it takes to proceed to the next state whether to throw directly, t1, or chase the ball down and then throw, t2. Furthermore, since a missed ball is likely to continue on farther afield, the time it takes to reach the infield after being thrown by the fielder is most often greater than the time it would have taken had it been caught. By separately controlling probabilities and times, the modeler is free to enforce such dependencies as well (see Table 21.1). However, the down side of this freedom is that if applied to all pairs of transitions, it greatly increases the number of parameters that have to be supplied. One solution is to employ the CTM convention as default and allow the modeler to override it as needed.

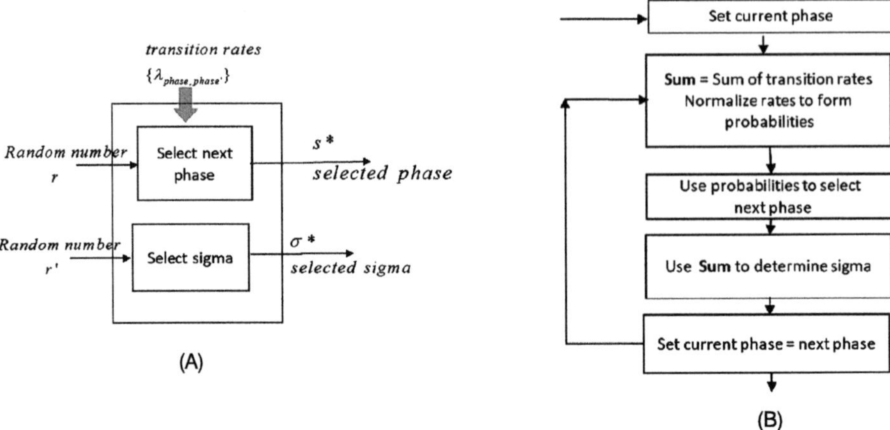

FIGURE 21.6

Selection of next phase and sigma for CTM.

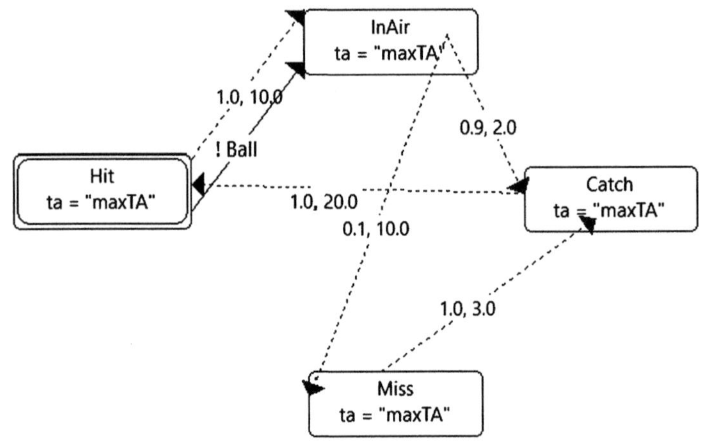

FIGURE 21.7

Markov Model Example: Baseball scenario.

Exercise 21.2.

1. Give an example where the time for decision is the same for all choices but the probabilities for outgoing transitions are all different.
2. Give an example where the time for decision is dependent on the choices to be made.

Exercise 21.3. Show that a DEVS Markov model reduces to a DEVS CTM if all distributions are exponential and all the transitions emerging from any state have the same average time.

Table 21.1 Uncoupled probability and time transition specifications

Transition from *InAir* to	Probability	Mean Time	Subsequent time to Throw
Catch	p1	t1	t3
Miss	p2	t2	t4 > t3

Table 21.2 Model types with external event set present

	Input set	Output set	Atomic or Coupled
Finite State Controllers	Control inputs	Readouts	Atomic
Server Queue Network	Customer IDs	Customer IDs	Coupled
Evaluation of Patient Treatment Protocols	Interventions	Evaluation variables	Mostly atomic
Stochastic Agents	Perceptions	Actions	Mostly atomic
Markov Decision Process	Environment response, Reward	Action	Atomic
Hidden Markov	Input may be absent	observable state variables	Coupled

21.3.4.1 DEVS Markov Model Classes

Model classes based on full-fledged DEVS include the input, output, and associated external, confluent, and output functions. Some of the varieties are shown in Table 21.2.

21.4 HIDDEN MARKOV MODELS

Mathematically speaking, a hidden Markov process is actually a combination of two processes, one that represents the dynamics of a system and the other representing its observable output (Barbu and Limnios, 2009). In simple cases such a process might be handled by a single atomic model with the output representing the observable process. However, for the case where the underlying structure is undergoing dynamic change, a full representation in DEVS may require a coupled model of several components as illustrated in Fig. 21.8A. As an example, we implement the hidden Semi-Markov model of Barbu and Limnios (2009) in which switching between a fair and loaded (unfair) dice constitute the invisible dynamics and the visible results of throwing of dice. The throwing of a dice is a Finite State Probability DEVS (FPDEVS) which is a Markov model for which the transition times are fixed. In this case, while active, the model stays in state Throw for one time unit, and selects one of six faces with uniform probability (fair version). In contrast, in the loaded version, face 6 is favored (probability .5) with the others splitting up the remaining probability. Formally, this is a Semi-Markov model since the sojourn times are fixed values (1 for Throw and 0 for the others). A face state returns to the Throw state after outputting its value. The model can be activated, deactivated and reactivated through its input port. The underlying structure change is handled by the HiddenControl model which is a DEVS CTM, switching between states Fair and Unfair with rates 0.5 and 0.1 respectively (so spending much more time in the Unfair state). Outputs from the controller switch between the fair and loaded versions of the dice according to which state is being entered. Such switch is accomplished by simultaneously activating and deactivating the fair and loaded models. Outputs from the dice show up at the output of the coupled model depending on which fair/loaded version is currently active.

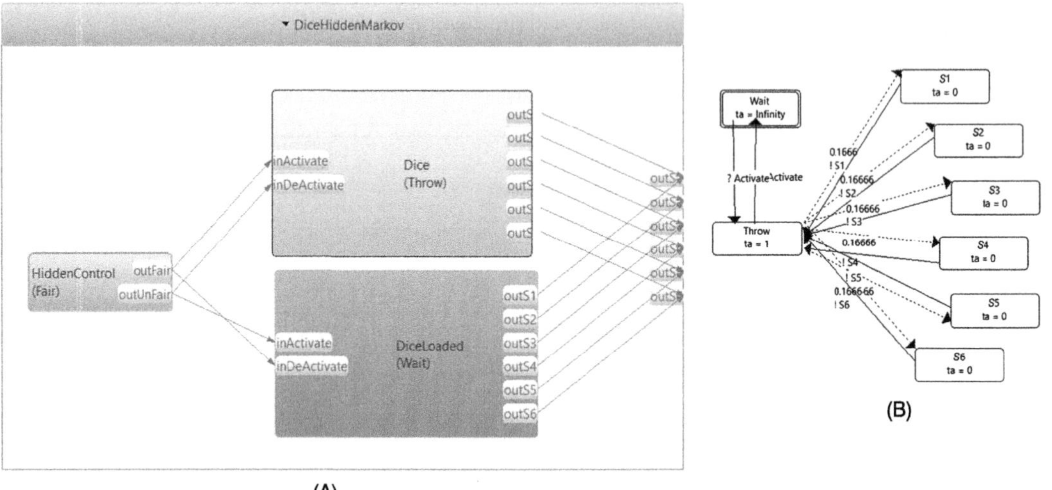

FIGURE 21.8

Hidden DEVS Markov Model.

This example illustrates a number of interesting points:

- The output trajectory is a semi-Markov process in the sense that the underlying time step is one unit with variable duration of the hidden state (fair or loaded dice) as measured by such a time unit.
- All subtypes of Markov models are employed: DEVS CTM, Fixed time Semi-Markov (FPDEVS), and hidden Markov (the coupled model itself).
- All features of DEVS coupled models are illustrated: input ports, output ports, couplings. However external events are only of the form where there is no elapsed time dependence.
- The output trajectory differs markedly from that of a fair dice in that there are periods of high frequency of the "6" face that are highly unlikely and that set up the basis for discovering the underlying dynamics of the control (Barbu and Limnios, 2009).

We return to formal characterization of DEVS Hidden Markov models later.

Exercise 21.4. Write a DEVS atomic model with dynamic structure that represents the fair and loaded versions of the dice through change in the underlying transition probabilities.

Exercise 21.5. Obtain the steady state probabilities of the fair and loaded states of the HiddenControl (a CTM model) and use them to obtain the long term frequencies of the six face values.

21.5 PREVIEW: CLOSURE UNDER COUPLING OF DEVS MARKOV MODELS

It will help to get a view of the inclusion of Markov classes within one another before undertaking more detailed exposition. Fig. 21.9 shows such inclusion as a basis to discuss closure under coupling of such

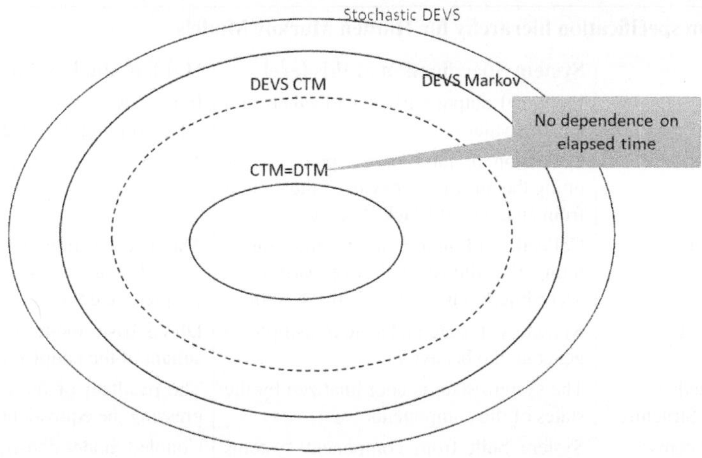

FIGURE 21.9

Classes that are closed (solid line) and not closed (dashed line) under coupling.

classes. (Recall that the importance of proving closure under coupling was explained in Chapter 10.) As mentioned above, the Stochastic DEVS framework provides the outer DEVS class and is closed under coupling as we have seen. We show later that the DEVS Markov class is closed under coupling. Also the smaller class DEVS CTM (and DTM which is equivalent) is closed under coupling where the external transition function does not depend on its elapsed time argument. However, when dependence on elapsed time is allowed, the DEVS CTM class is not closed under coupling (shown by the dotted lines) and expands to the full DEVS Markov class under coupling, i.e., the latter is the smallest closed class contained in the former.

21.5.1 SYSTEM SPECIFICATION HIERARCHY FOR HIDDEN MARKOV MODELS

The hierarchy of system specifications (Table 21.3) provides another way to organize DEVS Markov models along the behavior/structure dimensions.

In the following we provide some of the essential formal constructions underlying the descriptions of the table. Some of the remaining assertions are left as an exercise. See the development in Section 4.6 for the Turing machine hierarchy of specifications example.

The hierarchy of system specifications also sets up an associated hierarchy of system morphisms (Chapter 15). We explore some of the important ones here.

21.6 FORMALIZATION OF DEVS MARKOV MODELS

21.6.1 PROBABILITY CORE DEVS

We introduce two probabilistic structures that will provide the basis for specifying the DEVS Markov formalism and that will formalize the Transition Function Mapping of Fig. 21.1. First, we define a

Table 21.3 System specification hierarchy for Hidden Markov Models

Level	Name	System Specification at this level	Hidden Markov Model Example
0	I/O Frame	Input and output variables together with allowed values.	Input may be absent, output variables are those forming the observable process.
1	I/O Behavior	Collection of input/output pairs constituting the allowed behavior of the system from an external Black Box view.	The union of the I/O functions in level 2.
2	I/O Function	Collection of input/output pairs constituting the allowed behavior partitioned according to initial state of the system.	The set consisting of functions computed as the I/O functions associated with each coupled model state.
3	I/O System Structure	System with state and state transitions to generate the behavior.	DEVS basic model representing the resultant of the coupled model at level 5.
4	Coordinatized I/O System Structure	The system state is coordinatized by the states of the components.	The resultant of the coupled model expressing the equivalent basic model.
5	Coupled Systems	System built from component systems with coupling recipe.	Coupled model constructed from DEVS Markov models of various types. The coupling may involve models that disallow influence of elapsed time as in CTM and DTM

Probability Transition Structure (PTS) in set-theoretic form to be given in a moment. It often takes on the familiar form of a matrix of probability values. For example, the matrix

$$\begin{bmatrix} p_{00} & p_{01} \\ p_{10} & p_{11} \end{bmatrix}$$

Formally, a Probability Transition Structure is a structure
$PTS = < S, Pr >$
where
$Pr : S \times S \rightarrow [0, 1]$
$Pr(s, s') = v, 0 \leq v \leq 1$
As a relation Pr contains triples of the form (s, s', v)
which stands for state s transitions to state s' with probability v.
For each $s \in S$, define the restriction of Pr to s,
$Pr|s : S \rightarrow [0, 1]$
defined by $Pr|s(s') = Pr(s, s')$.
Then Pr is subject to the constraint that
it is fully defined (every transition has a probability)
and the probabilities of transitions out of every state sum to 1.
That is, for each $s \in S$,
$\sum_{s' \in S} Pr|s(s') = 1$

Exercise 21.6. Show that for a matrix representation such as the one above, the constraint means that the sum of elements along each row equals 1.

Remark. This set-theoretic representation of the usual matrix form render it more convenient to create and manage derived DEVS models because it supports necessary manipulations equivalent to data structure operations.

21.6.2 MARKOV CHAIN

The basic interpretation of a PTS is of a Markov chain, i.e., a set of states that generate sequences determined by the probability structure. The probabilities of the ensemble of all such state sequences can be described by a vector representing probabilities of being in the states and iterative application of the associated matrix. In a very abbreviated summary of Markov chain theory (see e.g., Kemeny et al., 1960; Feller, 2008), we have

At each step $n = 0, 1, \ldots$
$p(n+1) = p(n) \bullet P$
where p *is a row vector and* P *is the Markov matrix.*
The state vector in equilibrium reached at step $n*$; *is defined by*
$p(n+1) = p(n)$, *for all* $n > n*$
which implies that
$p(n*) \bullet P = p(n*)$
and $p* = p(n*)$ *is the equilibrium vector where a unique solution exists.*

The second structure to be introduced allows us to work with the times of the transitions. These can be referred to variously as sojourn times, transition times, time advances, elapsed times, or residence times depending on the context.

Time Transition Structure
$TTS = < S, \tau >$
where
$\tau : S \times S \rightarrow ProbabilityDensityFunctions(pdf)$
such that the time for transition from s *to* s' *is selected from a pdf* $\tau(s, s') : R^{+}_{0,\infty} \rightarrow [0, 1]$

For example,

$$\begin{bmatrix} \tau_{00} \; \tau_{01} \\ \tau_{10} \; \tau_{11} \end{bmatrix}$$

is represented by the structure: TTS $= < \{0, 1\}, \tau >$ where
$\tau : \{0, 1\} \times \{0, 1\} \rightarrow [0, 1]$ such that $\tau(i, j) = \tau_{ij}$ and
τ_{ij} *is a pdf (probability density function)* $\tau_{ij} : R^{+}_{0,\infty} \rightarrow [0, 1]$.
For example, $\tau_{ij}(t) = e^{-t}$ *represents the exponential pdf for selecting a time for transition from* i *to* j.

We use the pair of probability and time structures to specify the internal transition and time advance functions of a DEVS model as follows:

Probability Transition Structure
$PTS = < S, Pr >$
and
Time Transition Structure
$TTS = < S, \tau >$
gives rise to a *DEVS Markov Core*
$M_{DEVS} = < S_{DEVS}, \delta_{int}, ta >$
where $S_{DEVS} = S \times [0, 1]^S \times [0, 1]^S$
with typical element (s, γ_1, γ_2) with $\gamma_i : S \to [0, 1], i = 1, 2$
where
$\delta_{int} : S_{DEVS} \to S_{DEVS}$ is given by:
$s' = \delta_{int}(s, \gamma_1, \gamma_2) = (SelectPhase_{PTS}(s, \gamma_1), \gamma_1', \gamma_2')$
and $ta : S_{DEVS} \to R^{+}_{0,\infty}$ is given by:
$ta(s, \gamma_1, \gamma_2) = SelectSigma_{TTS}(s, s', \gamma_2)$
and $\gamma_i' = \Gamma(\gamma_i), i = 1, 2$

The random selection of phase and sigma employed here corresponds to that illustrated in Fig. 21.2. As in TMS76, the selections employ ideal random number generators. Note that the selection of transition time is performed after the next phase has been picked.

Exercise 21.7. Define the Stochastic DEVS that corresponds to the pair (PTS, TTS) where $PTS = < S, Pr >$ and $TTS = < S, \tau >$. Hint: Use the normal form in which the time advance is kept explicitly in the state, i.e., let $M_{ST} = < S', Pr', ta >$ where $S' = S \times R_0^+$, and ta$(s, \sigma) = \sigma$. Define $Pr : S' \times S' \to [0, 1]$ appropriately.

21.6.2.1 Example of DEVS Markov Core

Consider the situation depicted in Fig. 21.7 in which a fielder has a probability of catching or fumbling a ball. In contrast to the CTM approach depicted in the figure, using the DEVS Markov formulation, we can separately account for the probability of one or the other eventuality as well as the times taken in each case. Here is an example

Let the phases be coded by integers as follows:
$Hit = 0$
$InAir = 1$
$Catch = 2$
$Miss = 3$
Then the PTS and TTS can be represented by matrices:

$$\begin{bmatrix} p_{00} & p_{01} & p_{02} & p_{03} \\ p_{10} & p_{11} & p_{12} & p_{13} \\ p_{20} & p_{21} & p_{22} & p_{23} \\ p_{30} & p_{31} & p_{32} & p_{33} \end{bmatrix} \quad \begin{bmatrix} \tau_{00} & \tau_{01} & \tau_{02} & \tau_{03} \\ \tau_{10} & \tau_{11} & \tau_{12} & \tau_{13} \\ \tau_{20} & \tau_{21} & \tau_{22} & \tau_{23} \\ \tau_{30} & \tau_{31} & \tau_{32} & \tau_{33} \end{bmatrix}$$

The non-zero probability elements are:
$p_{01} = 1,\ p_{12} = 0.1, p_{13} = 0.9, p_{20} = 1, p_{32} = 1.$

Note that a transition that is the only one emerging from a phase gets a probability of 1. This contrasts to the case of the CTM shown in Fig. 21.7. Now let the elements of the *TTS* be density functions for transition times with mean values as follows:

$$\tau_{01}\ mean\ = 1, \tau_{12}\ mean\ = 2, \tau_{13}\ mean\ = 10,\ \tau_{20}\ mean\ = 20, \tau_{32}\ mean\ = 3$$

With time units as seconds, these values assert that the ball is in the air for one sec., and if it is caught, then it takes 2 sec. to return it to the catcher. However, if it is missed it takes 10 sec. to recover and be in a position to return it to the catcher. Using exponential pdfs we need only specify these values for the means while other distributions might require specifying more parameters.

The conversion of the pair *(PTS, TTS)* into a DEVS Markov Core model is the formal equivalent of the flow chart in Fig. 21.16. We see that to make a transition, first we randomly select a next phase based on PTS which specifies a transition pair. We then randomly select a sigma for the time of this transition. The random selections are performed by the functions *SelectPhase* and *SelectSigma*, respectively.

The sequential dependence of transition time selection dependence on probability selection and its evolution over time are illustrated in Fig. 21.10A and B respectively. A state trajectory can be visualized as starting in an initial phase such as S2, and advancing with variable time steps determined by the successive states. For example, the transition from S2 to S4 is selected based on transition probabilities out of S2 and then the transition time $T_{S2->S4}$ is selected from its distribution. This sequential dependence allows us to use the representation in Fig. 21.10A where the successive selection of probabilities is considered as a Markov chain as front component with the time advance mapping as back component. Thus if we are only interested in the probabilities of states at transitions (and not in the timing of the transitions) then we can consult the Markov chain interpretation of the *PTS*. In particular for an ergodic chain, the steady state probabilities can be computed using the Markov

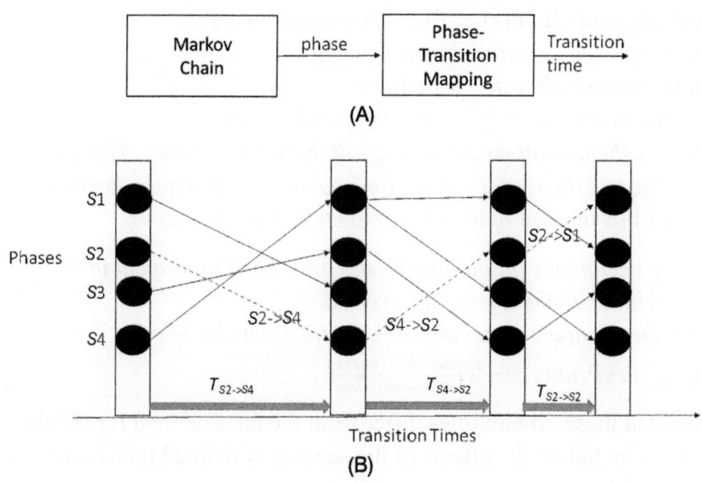

FIGURE 21.10

State trajectories for DEVS Markov core model and sequential decomposition.

Matrix model. Having these probabilities we can then compute the expected sojourn times using the *TTS* distributions.

Exercise 21.8. Write a coupled model that implements the cascade composition in Fig. 21.10.

Thus this cascade decomposition leads us to take the following approach to compute the expected sojourn times for a DEVS Markov Core model specified by a pair *(PTS, TTS)*:

Probability Transition Structure
$PTS = < S, Pr >$
where
$Pr : S \times S \to [0, 1]$
$Pr(s, s') = v, 0 \leq v \leq 1$

Time Transition Structure
$TTS = < S, \tau >$
where
$\tau : S \times S \to ProbabilityDensityFunctions$
such that the time for transition from s to s' *is selected from* $\tau(s, s') : R^{+}_{0,\infty} \to [0, 1]$
The expected transition time $T(s, s') = \int tf(t)dt$ *where* $f = \tau(s, s')$
Assuming the model stays in s until it completes its transition to other states,
the $SojournTime(s) = \sum_{s' \in S} Pr(s, s')T(s, s')$

Assuming the model is ergodic, let P be the steady state distribution.
Note that here P(s) is the frequency of entry into state s
and it stays in s for an average of SojournTime(s) after each entry;
so $ExpSojournTime(s) = P(s) * SojournTime(s)$
is the expected duration of residence in s.
Summing over all states, s we have
$ExpSojournTime = \sum_{s \in S} ExpSojournTime(s)$
which is the overall model average of the sojourn times. This can be viewed as the model's basic time advance step: on the average, it takes this amount of time to complete a state transition.

Let the occupancy probability be the fraction of time spent in a state relative to the time spent over all.
Then the occupancy probabilities are for each s in S
$OccupancyProb(s) = \frac{ExpSojournTime(s)}{ExpSojournTime}$.

Note that these probabilities differ from the ones defined for steady state by including the effects of the separately defined transition time distributions.

For the baseball example we have
$SoujournTime(0) = 1,$

$$
\begin{aligned}
&SoujournTime(1) = p_{12}\tau_{12} + p_{13}\tau_{13} = 2*0.1+0.9*10 = 9.2\\
&SoujournTime(2) = 20\\
&SoujournTime(3) = 3\\
&\textit{It turns out that the steady state probabilities are as follows:}\\
&P(0) = 0.256\\
&P(1) = 0.256\\
&P(2) = 0.256\\
&P(3) = 0.232
\end{aligned}
$$

See Exercise for explanation.
So the durations in each state are

$$
\begin{aligned}
&ExpSojournTime(0) = P(0) * SojournTime(0) = P(0) * 1 = 0.256\\
&ExpSojournTime(1) = P(1) * SojournTime(1) = P(1) * 9.2 = 2.3\\
&ExpSojournTime(2) = P(2) * SojournTime(2) = P(2) * 20 = 5\\
&ExpSojournTime(3) = P(3) * SojournTime(3) = P(3) * 3 = .7\\
&So\ ExpSojournTime = \textstyle\sum_{s \in S} ExpSojournTime(s) = 8.256
\end{aligned}
$$

and the model changes state approx. every 8 seconds on the average.

Now the

$$
\begin{aligned}
&OccupancyProb(s) = \tfrac{ExpSojournTime(s)}{ExpSojournTime}\\
&i.e.,\\
&OccupancyProb(0) = .04\\
&OccupancyProb(1) = .28\\
&OccupancyProb(2) = .60\\
&OccupancyProb(3) = .08
\end{aligned}
$$

That is, seven percent of the time is spent in fumbling with the ball.

Exercise 21.9. For the example of Fig. 21.7, confirm that the equilibrium equations are:

$$
\begin{aligned}
&P(0) = P(2)\\
&P(1) = P(0)\\
&P(2) = .1 * P(1) + P(3)\\
&P(3) = .9 * P(1)
\end{aligned}
$$

$$
\begin{aligned}
&Let\ x = P(0) = P(1) = P(2)\\
&Then\ P(3) = .9x\ and\ also\ P(3) = 1 - 3x\\
&with\ solution\ P(0) = P(1) = P(2) = x = \tfrac{1}{3.9}\ and\ P(3) = .232
\end{aligned}
$$

Use the same approach to get the equilibrium solution for $p_{13} = f$, an arbitrary value between 0 and 1.

Discuss how the occupation probabilities vary as f increases from 0 to 1.

Exercise 21.10. Show that if all mean transition times for a state are equal to 1 and the distributions are exponential then the occupancy probabilities agree with those predicted by the equivalent CTM. (Hint: the occupancy probabilities are equal to the steady state distribution values.)

Exercise 21.11. Consider the case where the common value is not necessarily 1.

Exercise 21.12. Consider the case where all transitions have the same probability. (Hint: show that the steady state probabilities are equal, and derive the implications for occupancy probabilities.)

Exercise 21.13. Compute the expected sojourn times for a DEVS Markov Core model for which the PTS specifies an ergodic Markov chain.

21.6.3 TRANSIENT BEHAVIOR

In the following we restrict our attention to models that can be fairly easily analyzed for transient behavior. Given a starting state and an ending state we assume that there are no cycles in the state transition diagram and that the ending state is the only absorbing state. This means that as we start from the starting state and continue from one transition to the next, we will eventually arrive at the ending state without going in circles or reaching a dead end.

More formally, we assume that the underlying digraph is a directed acyclic graph (DAG). The underlying digraph is the directed graph whose nodes are the states and there is an edge from s to s' just in case $Pr(s, s') > 0$. This digraph is acyclic if there are no cycles in it (Chapter 7). The access relation is defined where s accesses t if there is a path in the digraph from s to t. We then require that the model contains only states that are accessible from the starting state and that can access the ending state.

Now let M be a DEVS Markov model satisfying the requisite conditions. Then for any pair (s, t) in $S \times S$, the average traversal time from s to t is:

$$AvgTime(s,t) = \begin{cases} meanT(s,t) & if\, Pr(s,t) > 0 \\ \sum_{s' \in S} Pr(s,s')(meanT(s,s') + AvgTime(s',t)) & otherwise \end{cases}$$

where $meanT(s, t)$ is the expectation of the pdf $\tau(s, t)$.

This is a recursive definition that depends on the assumed acyclic property of the model to be well-defined.

Exercise 21.14. In the baseball example of Fig. 21.7, if we are interested in the average time from hitting (state 0) to catching (state 2) we can disregard the throwback, i.e., remove the transition from 2 to 0. This yields an acyclic digraph with the properties needed for transient computation. The non-zero probability elements are:

$$p_{01} = 1,\; p_{12} = 0.1,\; p_{13} = 0.9,\; p_{32} = 1.$$

and

$$\tau_{01}\; mean = 1,\; \tau_{12}\; mean = 2,\; \tau_{13}\; mean = 10,\; \tau_{32}\; mean = 3$$

Compute the average time from hitting to catching the ball. Hint: consider the traversal time from state 0 to state 2.

21.6.4 EXTENSION WITH INPUT/OUTPUT SETS AND FUNCTIONS

To specify a fully fledged DEVS model we add input and output sets, X *and* Y as well as external transition and output functions, δ_{ext} and $\tilde{\lambda}$, resp.

$\delta_{ext} : Q_{DEVS} \times X \to S_{DEVS}$ specifies the effect of the input,
where $Q_{DEVS} = \{(s, e)|s \in S_{DEVS}, e \in [0, ta(s)]\}$
and
$\tilde{\lambda} : S_{DEVS} \times S_{DEVS} \to Y$
specifies outputs for transitions pairs – we will see why in a moment.

The extended structure specifies a DEVS Markov
$M_{DEVS} =< X, S_{DEVS}, Y, \delta_{int}, \delta_{ext}, \lambda, ta >$

The output function
$\lambda : S_{DEVS} \to Y$
with
$\lambda(s, \gamma_1, \gamma_2) = \tilde{\lambda}(s, s*)$ *where SelectPhase*$(s, \gamma_1) = s*$

To explain the output effect we note that $s*$, is the phase selected randomly so that the transition $s \to s*$, is taken. It follows that the output is the one specified by the pair $(s, s*)$.

Exercise 21.15. As presented, the δ_{ext} function is a deterministic mapping.
Define a Stochastic DEVS, $M_{ST} = < X, Y, S', Pr', \{P_{s,e,x}\}, \lambda', ta >$
where $Pr : S' \times S' \to [0, 1]$ *and* $P_{s,e,x} : S' \to [0, 1]$ *for all* $x \in X$ *and* $e \in ta(s)$
that allows the input event and elapsed time to probabilistically specify the next state. Hint: let S' be the normal form of state set as in the previous exercise.

Exercise 21.16. Show how to implement the effect of *probabilistic input by* expanding the state set S' to include *more random variables.*
Note: For a model whose external transition function does not depend on elapsed time we have:

$$\delta_{ext}(s, e, x) = \widehat{\delta}\ (s, x)$$

The manner in which $SelectPhase_{PTS}$ and $SelectSigma_{TTS}$ perform random selections is reviewed in Appendix 21.B where we also compare this approach to the one described for continuous time Markov models in the literature.

21.6.5 INPUT/OUTPUT BEHAVIOR OF DEVS MARKOV MODELS

The I/O behavior of a DEVS Markov model is defined in the manner generally described in Chapter 2. Here the ideal random number generator provides an essential element to enable a deterministic specification of at the I/O Function level. When we remove the generator and account for the transition and time probability structures we get the state trajectories and their probability distributions discussed in Section 21.7. With the added input and output specifications, these state trajectories become the basis for computing the distributions for the I/O behaviors. However, we can skip this step if our interest is in relating the rate of events in the output trajectory to the rate of events in an input trajectory in steady state.

In Fig. 21.11A assume we apply an input trajectory with known rate of external events and that steady state has been reached. We want to relate the rate of output events ideally as a time-invariant function of the input rate. For example, consider a neuron that is quiescent unless activated by sufficient input stimulus. Then the transfer function may take a functional form as in Fig. 21.11B where the rate of firing output pulses increases with the input pulse rate until a saturation level is reached. The Rate IO Transfer Function can be used to predict the smallest size of neural net with given (usually sparse) connectivity that can exhibit sustained firing (Chapter 23).

Let Ti be the average time between successive external events so that 1/Ti is the input rate. To represent the effect of inputs on the DEVS Markov model we replace the external transitions by suitable internal transitions. As a special case where this is straightforward consider the case where external events only act on states that are passive in their absence. Thus for each such state we replace the external event by an internal transition with probability 1 and mean time = Ti. Given this refashioned model, which contains only internal transition and output specifications, we can compute the expected sojourn times in steady state as illustrated above. Moreover, the expected overall sojourn time can be computed and can be considered as the average time between transitions illustrated in Fig. 21.11A as To. Furthermore the probability that an output occurs in such transitions can also be computed, shown as P in Fig. 21.11A.

Exercise 21.17. Prove that output rate is P/To. Hint: consider that N successive transitions occur in a time interval NTo and produce NP outputs on the average.

Exercise 21.18. Show that the probability P does not depend on the input inter-event time Ti. Hint: examine the dependence of steady state probabilities. Hence, the function in Fig. 21.11B can be expressed as P*f(input rate).

The IO Rate Transfer Function can be measured in a simulation using an appropriate experimental frame discussed in an exercise that will appear after we consider coupled models of DEVS Markov components.

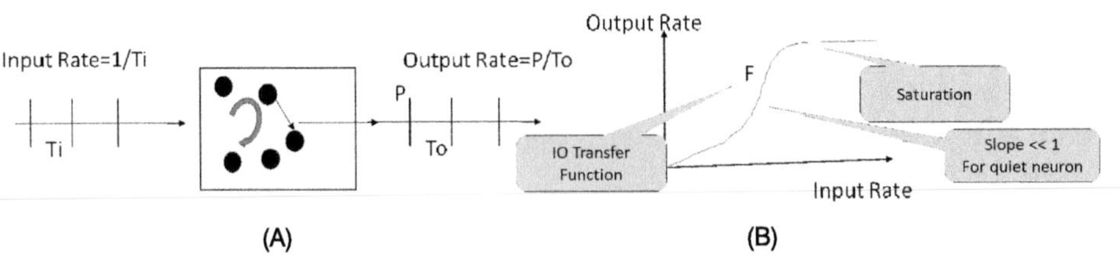

FIGURE 21.11

Relating Input and Output Rates.

21.6.6 COUPLED MODELS – DEVS NETWORKS OF MARKOV COMPONENTS

Based on the standard specification of networks of DEVS components in Chapter 7 we can define coupled models of fully fledged DEVS Markov models. For simplicity we restrict attention to the Classic DEVS family and to input-free resultant, leaving further extensions to exercises.

$N =< D, \{M_d\}, \{I_d\}, \{Z_{i,d}\}, Selectfn >$
with
D a set of component references
For each $d \in D$,
M_d is a DEVS Markov model.
For each $d \in D$,
I_d is the influencer set of d.
and for each $i \in D$,
$Z_{i,d}$ is a function, the "i-to-d" output translation with
$Z_{i,d} : Y_i \to X_d$,
$Selectfn : 2^D - \{\phi\} \to D$

The resultant, $DEVSN =< S, \delta_{int}, ta >$
where
$S = \times_{d \in D} Q_d$
where
$Q_d = \{(s_d, \vec{\gamma}_d, e_d) | s_d \in S_{PTS}, 0 \le e_d \le ta_d(s)\}$
where
$ta : S \to R^+_{0,\infty}$
with
$ta(s) = \min\{\sigma_d | d \in D\}$ *and* $\sigma_d = ta_d(s) - e_d$
where the imminents
$IMM(s) = \{d | d \in D \wedge \sigma_d = ta(s)\}$

The selected imminent is:
$d* = Selectfn(IMM(s))$
To define $\delta_{int} : S \to S$
with $\delta_{int}(s) = s'$
Let $s = (..., (s_d, \vec{\gamma}_d, e_d), ...)$
and let $\delta_{int}(..., (s_d, \vec{\gamma}_d, e_d), ...) = (..., (s'_d, \vec{\gamma}'_d, e'_d), ...)$ *where*

$$(s'_d, \vec{\gamma}'_d, e'_d) = \begin{cases} (\delta_{int,d}(s_d, \vec{\gamma}_d), 0) & \text{if } d = d* \\ (\delta_{ext,d}(s_d, \vec{\gamma}_d, e_d + ta(s), x_d), 0) & \text{if } d \in I_{d*} \wedge x_d \neq \phi \\ (s_d, \vec{\gamma}_d, e_d + ta(s)) & \text{otherwise} \end{cases}$$

with $x_d = Z_{d*,d}(\lambda_{d*}(s_d, \vec{\gamma}_d))$

Exercise 21.19. Design a coupled model containing a DEVS Markov model and an experimental frame that measures its IO Rate Transfer Function. The frame contains a generator of external events with a mean inter-arrival time that can be varied and a transducer that obtains the rate of output events in steady state in response to a given input rate. From this data it computes the desired ratio of output to input rate.

21.6.7 PROOF OF DEVS MARKOV CLASS CLOSURE UNDER COUPLING

The above definition of the resultant leads us to proof of closure under coupling as follows:

Theorem: *The DEVS Markov Class is Closed Under Coupling*

Proof: *We show how the resultant DEVSN has the underlying form of the basic DEVS Markov model.*

First assume that $d*$ *influences all other components so that they all recompute their next states and time advances.*
With $S = \times_{d \in D} Q_d$ *we need to define the Probability Transition Structure*
$PTS = < S, Pr >$
where $Pr : S \times S \rightarrow [0, 1]$
But, since each component transition is selected independently, the PTS of $DEVN$ *is the crossproduct of the individual component PTSs.*
For the Time Transition Structure
with $TTS = < S, \tau >$
$\tau : S \times S \rightarrow ProbabilityDensityFunctions$
with the time for transition from s to s' *selected from* $\tau(s, s') : \mathbb{R}^+_{0,\infty} \rightarrow [0, 1]$
Given that global state s' *has been selected independently using the SelectSigma functions from the global state* s, *the*
probability density function $\tau(s, s')$ *can be expressed as the minimize operation applied to the crossproduct of the individual density functions.*

The assumption that the imminent's influencees include all components allows us to express the probability and time transition structures of the resultant as crossproduct of the individual probability structures and the minimize operation applies to the crossproduct of the density functions, respectively.

Now if $d*$*'s influencers don't include all of D then the above discussion applies to* $d*$ *and its influencers. Each element of the complementary set does not change state except that its time left is reduced by the global elapsed time (i.e.,* $ta(s)$*). These components are easily identified because their elapsed times are* not *zero (while the others are set to zero).*

The final time advance is a minimization of the minimum of the SelectSigma results for $d*$'s influencees and the time advances of the other components.

Exercise 21.20. Provide a formal proof a the above sketch.

21.7 CONTINUOUS AND DISCRETE TIME SUBCLASSES OF MARKOV MODELS

We now formalize the traditional DEVS Markov Continuous Time and Discrete Time models (CTM and DTM, respectively) as special cases of DEVS Markov models. We will show that they are equivalent subclasses of DEVS Markov models. As mentioned before (Fig. 21.2), these subclasses derive both the probabilities and transition times from the same set of parameters, the rates of transition. The follows makes this restriction explicit:

Definition 21.1 (DEVS CTM).

Given a Probability Transition Structure
$PTS =< S, Pr >$, *where for all* $s \in S$, $Pr(s, s) = 0$
Note there is no "self-transition" – only the other states are eligible as destinations.
PTS induces a Time Transition Structure
$TTS =< S, \tau >$
where for all $s, s' \in S$,
$\tau(s, s')$ *is the exponential distribution with parameter* $\sum_{s' \neq s} Pr(s, s')$.
The pair (PTS, TTS) gives rise to a DEVS Continuous Time Markov (CTM) Core which is a subclass of the overall DEVS Markov Model class.

Exercise 21.21. Compare the specification just given with the traditional one given in Appendix 21.B. Discuss the implications of using each one to implement a simulator for the DEVS CTM class.

Definition 21.2 (DEVS DTM).

A Probability Transition Structure
$PTS =< S, Pr >$, *and constant time step*, h
induces a

$\widehat{PTS}_h =< S, \widehat{Pr}_h >$, *where*
where $\widehat{Pr}_h$ *is the original* Pr *with each probability multiplied by* h
i.e., $\widehat{Pr}_h\,(s, s') = h\ Pr(s, s')$

PTS also induces a Time Transition Structure
$TTS =< S, \tau >$
where for all $s, s' \in S$, $\tau(s, s') = PDF(h)$ *(the pdf that assigns* 1 *to value* h, *and* 0 *otherwise)*

The pair (PTS, TTS) gives rise to a DEVS Discrete Time Markov (DTM) Core which is a subclass of the overall DEVS Markov Model class.

Note: for large h, a better approximation is given by using $p' = 1 - \exp(-hp)$ as the value assigned to the new transition probability instead of h.

Exercise 21.22. Show how the DEVS DTM just formalized specifies a DEVS model by specializing the definition in Section 21.7.

The DEVS DTM just defined follows the traditional formulation in which its models take a fixed time step to advance time. We can show the equivalence of DEVS CTM and DTM by considering the following multi-step extension of the DTM:

Definition 21.3 (Multi-step DEVS DTM).

> *Given a Probability Transition Structure*
> $PTS =< S, Pr >$, *and constant time step*, h
> *it induces a*
>
> $\widehat{PTS}_h =< S, \widehat{Pr}_h >$, *where*
> *where* $\widehat{Pr}_h$ *is the original Pr with each probability multiplied by h*
> *i.e.*, $\widehat{Pr}_h\,(s, s') = h * Pr(s, s')$
> *(or transformed to* $p' = 1 - \exp(-hp)$ *for large h).*
> *PTS also induces a Time Transition Structure*
> $TTS =< S, \tau >$
> *where for all* $s, s' \in S$, $\tau(s, s')$ *is the geometric distribution*
> *with parameter* $\sum_{s' \neq s} \widehat{Pr}_h\,(s, s') = h \sum_{s' \neq s} Pr(s, s')$
>
> The pair (PTS, TTS) gives rise to a multi-step DEVS Discrete Time Markov (DTM) Core which is a subclass of the overall DEVS Markov Model class.

Informally, a multi-step DTM takes the number of steps that a DTM takes to leave its current state as its sojourn time (when multiplied by h). Note that the probability of the DTM leaving a state on a single step is the sum of the outgoing probabilities. This can be taken as the probability of success in each trial of a sequence of trials. The geometric distribution represents the distribution of number of trials until success.

We can now show formally that the DEVS CTM and DTM are equivalent classes.

Theorem 21.1. *DEVS CTM and DEVS DTM derived from the same PTS define the same DEVS Markov subclass in the limit as h goes to 0.*

Proof. The proof proceeds in two steps. In the first, the multi-step DTM is shown to be the correct multi-step extension of the single step DTM. In the second step, the geometric distribution and exponential distribution are noted to be the same in the limit as the step size h goes to zero. □

Exercise 21.23. Carry out the two steps of the proof.

21.8 RELATIONS BETWEEN DEVS CTM AND DTM

From fundamental Markov theory, CTM is equivalent to DTM in the limit of small time step. The DEVS extension of each type introduces an external transition function which has elapsed time in state as an argument. However, only CTM can provide elapsed time that represents sojourn time in state. This is because the time advance function randomly selects the sojourn time using the correct computation. On the other hand, the external transition function in the DTM case can only supply the

Table 21.4 Behavioral comparison of CTM and DTM with/without auxiliary timing.

DEVS DTM\DEVS CTM	As defined	No elapsed time dependence
As defined	DEVS CTM > DEVS DTM	DEVS CTM = DEVS DTM
With stopwatch	DEVS CTM = DEVS DTM	DEVS CTM = DEVS DTM

time step value being employed. We can get the actual sojourn time by adding a stopwatch which starts when a state is first entered and stops when the state is changed.

Exercise 21.24. Show how to add stopwatch functionality to a DEVS DTM by adding a state variable that accumulates the transition count while holding in a state.

Thus, as in Table 21.4, with both DEVS CTM and DEVS DTM as defined, the DEVS CTM has more capability than the DEVS DTM. Adding the stopwatch to the DEVS DTM gives it the same capability as DEVS CTM. In contrast, if we restrict the external transition function to not depend on elapsed time, then both types are equivalent.

21.8.1 EXAMPLE: DEVS MARKOV COUPLED MODEL

The following example employs DEVS Markov models as components and investigates the different types of models that characterize the resultants and their behaviors. The example is a coupled model of a DEVS CTM with a DEVS Markov (fixed transition time) model representing the usual timed situation in reality. Also the latter is replaced by a DEVS CTM where the timer is allowed to be a random element and the resulting types of models and behaviors are investigated.

Mobile Worker

At any time, a worker is located in a locality with variability in the available jobs. Having just arrived to a locality, the worker begins a search for employment. Finding a job, the worker stays in the job until losing it. At that point she remains unemployed until finding a new one. The worker allows a fixed amount of time to find such a job and leaves the area in search of new employment if a new job does not materialize within that time. In Fig. 21.12 we represent such a worker with states inJob, JobLess, and JobSearch together with a timer that alerts the worker to start a search after a fixed time. Also this timer is replaced by one which has an exponentially distributed time duration.

- **a)** Coupled model for the timed mobile worker containing worker and timer components where the timer is the CTM alternative
- **b)** CTM of mobile worker
- **c)** Fixed duration alternative for timer (set to 3 days)
- **d)** CTM alternative for timer (exponential duration with mean of 3)
- **e)** Part of the state diagram of the coupled model – starting with (inJob, Expired), the loss of job with output StartTimer! leads to (JobLess, Timing); then, depending on which happens first, finding job or timer timeout, transition back to (inJob, Expired) or (JobSearch, Expired).

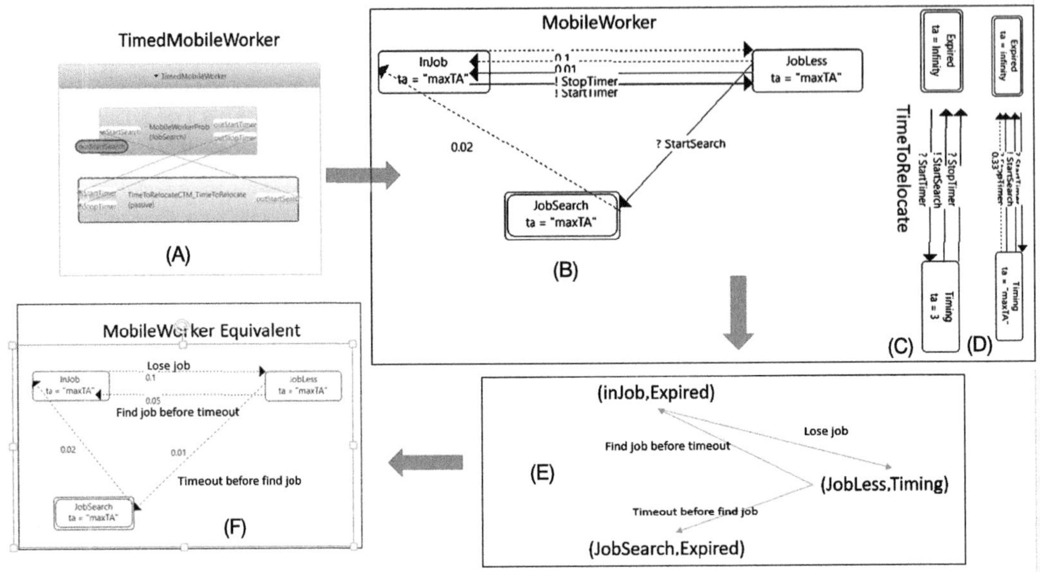

FIGURE 21.12

Mobile Worker.

21.8.2 DEVS HIDDEN MARKOV MODELS

We return to characterize the hidden Markov model in Fig. 21.8 noting that it is of the form of a Cascade Composition presented in Section 15.9. The structure of the resultant model can be characterized as in Fig. 21.13. Using the terminology of Section 15.9, the front component is correlated with a partition on the state set of the resultant defined by the equivalence classes of the states mapped to the same front component state. State trajectories remain within an equivalent class and are governed by a back component which is a DEVS Markov model associated with the equivalent class. Output equivalence classes are orthogonal to the blocks of the state mapping so that outputs can't reveal the state of the front block. Switching between such classes is governed by the front component which is the hidden Markov model.

The mapping can be characterized as a multi-step homomorphism of stochastic timed models (Chapter 9).

Exercises.

1. Write a CTM DEVS model for the mobile worker and a DEVS model for the fixed duration timer.
2. Write a coupled model of with components in the previous exercise.
3. Write a CTM DEVS model for the variable duration timer and substitute it for the timer in the coupled model.

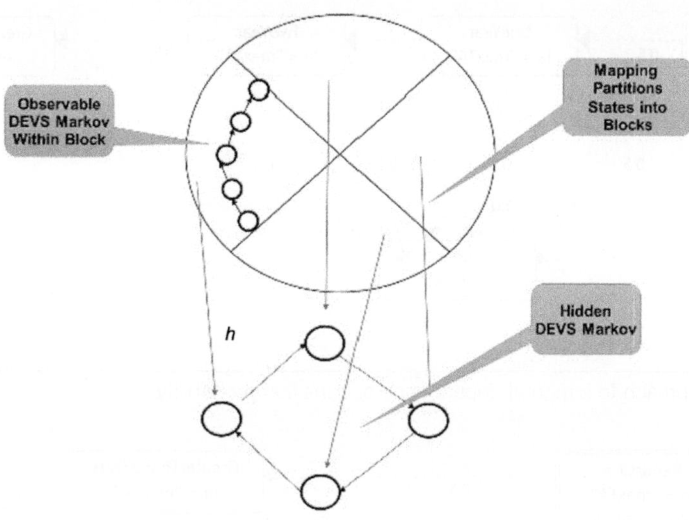

FIGURE 21.13

Hidden Markov Model.

4. Show that the Probability Transition Structure underlying the coupled model state diagram in E can be shown to be isomorphic to the atomic DEVS CTM in F where the correspondence merely projects out the state of the worker from the paired state of the worker and timer.
5. Argue that the coupled model with fixed duration timer is not equivalent to a CTM.
6. If instead of immediately jumping to the JobSearch state when the timer expires, the worker waits for the time remaining and jumps to JobSearch instead. Show that this is an example of a DEVS Markov model where the elapsed time is being used and prove (or disprove) that the resultant is not equivalent to a DEVS CTM.

21.8.3 EXAMPLE: DYNAMIC STRUCTURE VIA VARIABLE TRANSITION PROBABILITY

The dynamic structure capability of DEVS (Chapter 1) can be applied to DEVS Markov models to vary the underlying PTS and TTS structures. This allows more granular control of probabilities and transition times than is possible within fixed structure models. For example, the DEVS CTM model in Fig. 21.14 uses a "tunnel" sequence to lower the probability of death with longer remission. The DEVS Markov model in Fig. 21.15 employs the sojourn time of the initial state to lower the probability in a more granular manner.

Exercise 21.25. Show how to implement the change in PTS in Fig. 21.15 based on the sojourn time of the initial state. Hint: obtain the sojourn time as the time advance of the initial state and define a function to map the values obtained to the probability of transition of GreaterThan2Years to Death in the PTS.

The flow chart in Fig. 21.16 illustrates the control flow for the internal transitions for an atomic DEVS Markov model based on a pair of PTS and TTS.

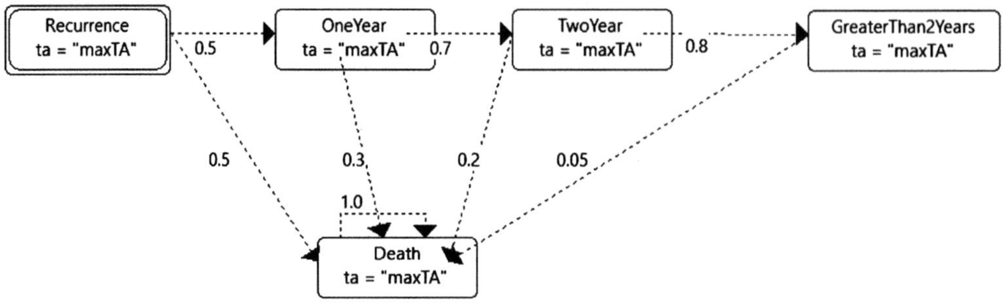

FIGURE 21.14

Tunnel sequence approach to temporal dependence of transition probability.

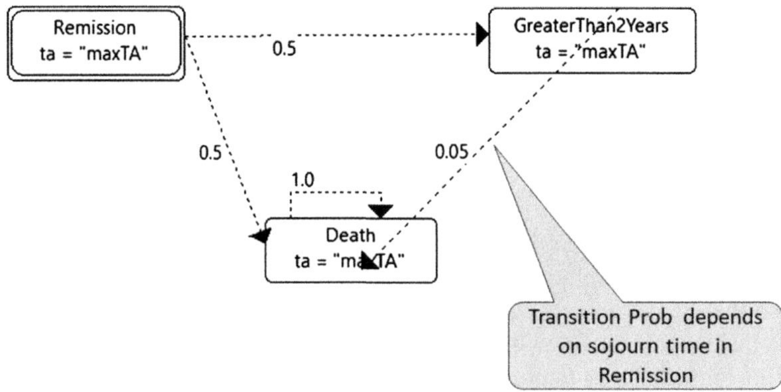

FIGURE 21.15

Sojourn time approach to temporal dependence of transition probability.

To illustrate the implementation of dynamic structure change, Fig. 21.16 considers that the PTS or TTS structures can depend on some index such as elapsed time or input value. Moreover, the current value of the index is incorporated into the state as a state variable. Thus, when the index is changed to a new value, the value is stored to help determine the values of the PTS and/or TTS. This might happen due to internal or external transitions. Dependence, of this kind, on external input values has been associated with Markov Decision Processes and Markov Agent Models. The DEVS ability for an external event to supply the elapsed time as index appears to be novel but can be considered to be the natural feature that exploits the coupling capability of coupled models having DEVS Markov model components. That is, if we allow such models to be influenced by elapsed time in external transitions then the resultants of coupled models will also depend on the global time advance (or residence time in the last global state) and hence will be Semi-Markov in nature. Furthermore, any change in the PTS or TTS structures amounts to dynamic structure change since it changes the basic properties of the model's transition definitions.

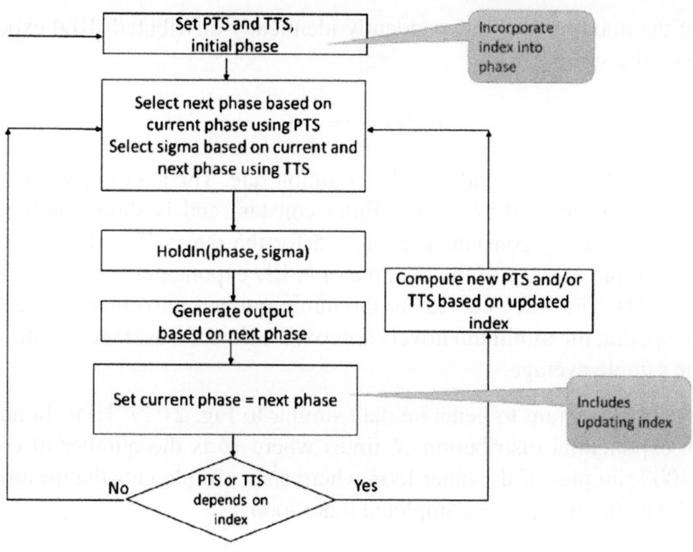

FIGURE 21.16

Flow chart to illustrate DEVS Markov atomic model.

APPENDIX 21.A EXPONENTIAL DISTRIBUTION AND MARKOV MODEL BASICS

21.A.1 EXPONENTIAL DISTRIBUTION PROPERTIES

$pdf,\ p(t) = \lambda e^{-\lambda t}$
$cdf,\ F(\tau < t) = 1 - e^{-\lambda t}$
$mean,\ E(t) = 1/\lambda$
λ *is* called the *rate*
$T_{avg} = 1/\lambda$ *is* called the *average time*
Either T_{avg} or λ can specify the parameter, they are inversely related.

Random variables which are defined by operations on sets of random variables such as the average, minimum, and maximum, come up in discussion of DEVS Markov models. Here is some background for that discussion.

1. The minimum of independently distributed exponential variables is exponentially distributed with rate equal to the sum of the rate parameters.
2. The mean of a sum of independently distributed exponential variables equals the sum of their means (but the distribution is not exponential),

3. The mean of the maximum of independently identically distributed (IID) exponential variables is described by the sum

$$\frac{1}{\lambda}(1+\frac{1}{2}+\frac{1}{3}+...+\frac{1}{N})$$

where N is the number of variables and λ is the common rate. The sum in parentheses is an approximation of $0.577 + ln(N)$ where 0.577 is the Euler constant and is shown as the smooth curve in Fig. 21.17. Note that it is nearly constant at approx. 5 for the range 75 to 100. The curve just below the sum curve is the sample average of the maxima of N IID exponential variables each sampled 1000 times for $N = 1, \ldots, 100$. The wavy curves are the minimum and maximum of the sampled averages for each N. We can see that the minimum hovers approx. 2 below the average and the maximum hovers approx. 7 above the sample average.

Exercise 21.26. Write a program to generate data similar to Fig. 21.17. Hint: In an inner loop write code to sample an exponential distribution N times where N is the number of components. In an outer loop obtain 1000 samples of the inner loop where each sample gets the minimum, average, and maximum of the N data points of each completed inner loop.

21.A.2 ZERO-MEMORY PROPERTY

Suppose X is a continuous random variable whose values lie in the non-negative real numbers. The probability distribution of X is memoryless if for any non-negative real numbers t and s, we have

$$Pr\,(X > t + s|\, X > t) \;= Pr\,(X > s\,)$$

For the exponential

$$Pr\,(X > t + s|\, X > t) \;=\; \frac{exp(-\lambda(t+s))}{exp(-\lambda t)}$$
$$= \frac{\exp(-\lambda t - \lambda s)}{exp(-\lambda t)} = \exp(-\lambda s) = Pr\,(X > s\,)$$

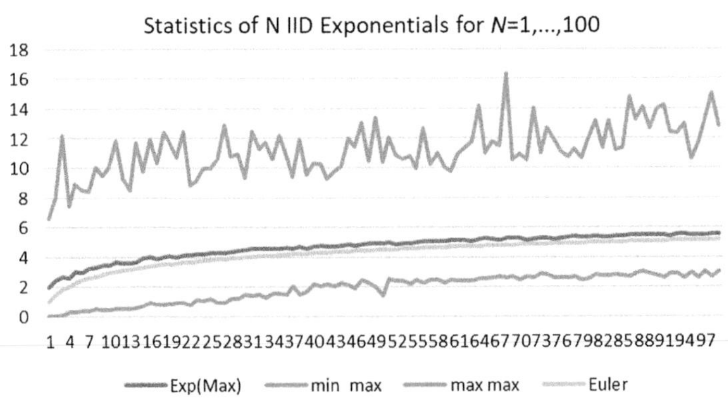

FIGURE 21.17

Plots of statistics for the Maximum of N IID exponential variables for $N = 1, \ldots, 100$.

Exercise 21.27. Draw the exponential curve and illustrate the property graphically. Try other distributions such as uniform and normal, showing graphically why they don't have the memoryless property.

21.A.3 POISSON PROCESS

A Poisson process is a sequence of event instants, $t_1\ t_2 \ldots\ i_i \ldots$ such that the inter-event times are independently and identically exponentially distributed:

$$p(t_{i+1}/t_i) = \lambda e^{-\lambda t_{i+1}}$$

The rate and mean are the same as the underlying exponential distribution.

21.A.4 MARKOV MODELING

Referring to the model in Fig. 21.18,

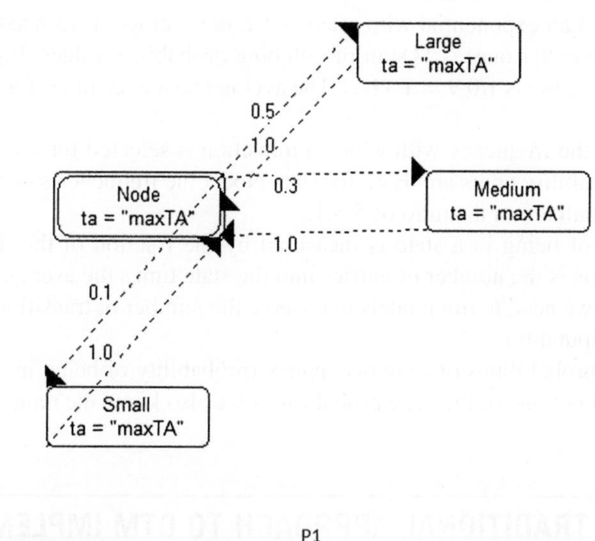

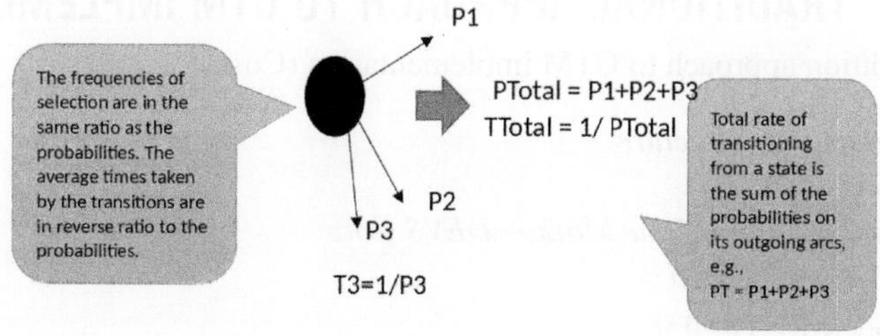

FIGURE 21.18

The ExponTest model for summarizing Markov modeling.

1. A probability on a dotted arrow (state transition) is interpreted as the probability of transitioning to the target state (or the rate of making such a transition in time), e.g., 0.5 is the probability of transition from Node to Large.
2. Each of the probabilities emerging from a state must >=0 and must sum to <=1. The remaining 1 – sum is the probability of remaining in the state, e.g., the sum of the probabilities out of Node is .9, leaving .1 as the probability of remaining in Node.
3. Given a state, the simulator independently samples an exponential distribution with each probability value as its rate parameter and takes the minimum to compute the resting time in the state before transitioning to the target state (one with the minimum time).

E.g., for state Large, there is only one sample taken from the exponential with parameter 1.0; for state Node, samples are taken from distributions for 0.5, 0.3, and 0.1 with the smallest sampled value determining the result.

1. The total rate of transitioning out of a state is the sum of the rates along its arcs. Mathematically, the distribution of the minimum of exponential distributions is exponential with the sum of the rate parameters as its rate parameter, e.g., the rate parameter for the minimum of the 3 distributions for state Node is 0.9.
2. Since the mean of an exponential with rate r is 1/r, the average residence time in a state (before transitioning to another one) is 1/ sum of outgoing probability values. E.g., for state Node, the average residence time is $1/0.9 = 1.11\ldots$ The average residence times for the other states are all the same $= 1$.
3. Mathematically, the frequency with which a transition is selected for a state by the simulator is equal to the probability on its arc. E.g., for state Node, the frequencies of transitioning to Large, Medium, and Small are in the ratio of 5:3:1.
4. The probability of being in a state is measured by the fraction of the time spent in the state during a run – this is the number of entries into the state times the average residence time in the state. In general we need to run models to observe the number of transitions into a state (it is not analytically computable).
5. The sum of the probabilities of state occupancy (probability of being in a state) equals 1. So if we computed all but one of the state probabilities we also know the remaining one.

APPENDIX 21.B TRADITIONAL APPROACH TO CTM IMPLEMENTATION

We formulate the tradition approach to CTM implementation (Costa)

Probability Transition Structure

$PTS =< S_{PTS}, Pr >$

gives rise to a Continuous Time Markov DEVS Core

$M_{DEVS} =< S_{DEVS}, \delta_{int}, ta >$

$S_{DEVS} = S_{PTS} \times [0, 1]^{|S_{PTS}|}$

with typical element (s_{PTS}, γ) *with* $\gamma : S_{PTS} \to [0, 1]$

If S_{PTS} *is represented as a vector* $(s1, s2, \ldots)$ *then* γ *will become a vector of random numbers* $(\gamma 1, \gamma 2, \ldots)$.

To represent the random selection of the next state and residence time in that state:
Define $Select_{PTS} : S_{PTS} \times [0, 1]^{|S_{PTS}|} \to S_{PTS} \times R^{+}_{0,\infty}$
by $Select_{PTS}(s, \gamma) = (s*, \sigma*)$
where for each $s' \in S_{PTS}$
$\sigma(s', \gamma) = -\left(1/Pr(s_{PTS}, s')\right) \ln\left(\gamma(s')\right)$
and $s*$ *is such that* $\sigma(s*, \gamma) = \sigma*$
where $\sigma* = \min\{\sigma(s',\ \gamma)|s' \in S_{PTS} - \{s_{PTS}\}\}$
We define the projections on the state and sigma:
Define $Select_{PTS}(s, \gamma) = (SelectState_{PTS}(s, \gamma), SelectSigma_{PTS}(s, \gamma)) = (s*, \sigma*)$
$ta : S_{DEVS} \to R^{+}_{0,\infty}$
$ta(s_{PTS}, \gamma) = SelectSigma_{PTS}(s_{PTS}, \gamma)$
Note there is no self-transition-only the other states are eligible as destinations.
$\delta_{int} : S_{DEVS} \to S_{DEVS}$ *is given by:*
$\delta_{int}(s_{PTS}, \gamma) = (SelectState_{PTS}(s_{PTS}, \gamma), \gamma')$
and $\gamma' = \Gamma(\gamma)$ *where* Γ *is the ideal random number generator.*

Note that if there is only one transition out of state s
$Select_{PTS}(s, \gamma) = SelectState_{PTS}(s, \gamma)$
and $ta(s) = SelectSigma_{PTS}(s, \gamma)$

REFERENCES

Barbu, V.S., Limnios, N., 2009. Semi-Markov Chains and Hidden Semi-Markov Models Toward Applications: Their Use in Reliability and DNA Analysis, vol. 191. Springer Science & Business Media.

Castro, R., Kofman, E., Wainer, G., 2010. A formal framework for stochastic DEVS modeling and simulation. Simulation: Transactions of the Society for Modeling and Simulation International 86 (10), 587–611.

Deavours, D.D., Sanders, W.H., 1997. "On-the-fly" solution techniques for stochastic Petri nets and extensions. In: Proceedings of the Seventh International Workshop on Petri Nets and Performance Models.

Feller, W., 2008. An Introduction to Probability Theory and Its Applications, vol. 2. John Wiley & Sons.

Glynn, P.W., 1989. A GSMP formalism for discrete event systems. Proceedings of the IEEE 77 (1), 14–23.

Kemeny, J.G., Snell, J.L., et al., 1960. Finite Markov Chains, vol. 356. van Nostrand, Princeton, NJ.

Kwiatkowska, M., Gethin, N., Parker, David, 2011. PRISM 4.0: verification of probabilistic real-time systems. In: Proc. 23rd International Conference on Computer Aided Verification. CAV'11. In: LNCS, vol. 6806. Springer, pp. 585–591.

Rachelson, E., Quesnel, G., Garcia, F., Fabiani, P., 2008. A simulation-based approach for solving generalized semi-Markov decision processes. In: ECAI, pp. 583–587.

Seo, C., Zeigler, B.P., Kim, D.H., 2018. DEVS Markov modeling and simulation: formal definition and implementation. In: TMS/DEVS Symposium. SpringSim Conference.

Zeigler, B.P., Sarjoughian, H.S., 2017. Guide to Modeling and Simulation of Systems of Systems, 2nd edition. Springer.

CHAPTER

DEVS MARKOV MODEL LUMPING

22

CONTENTS

This chapter builds on the background of Chapters 15 and 21 by applying lumping concepts to DEVS Markov models. We will find that the main conditions such as uniformity of structure in blocks and identity-independence in block-to-block interaction also apply to DEVS Markov models. Moreover, these conditions support powerful morphisms and approximate morphisms for such models.

From this vantage point, we match up with Chapter 2 of "Model Engineering for Simulation" (Zhang et al., 2018). There we show that the computational approach to evaluation of approximate morphism error makes it possible to quantify and test the extent to which aggregations and simplifications of simulations may be usefully employed.

Theory of Modeling and Simulation. https://doi.org/10.1016/B978-0-12-813370-5.00033-X

22.1 OVERALL APPROACH

Our overall approach is the creation of base-lumped model pairs as depicted in Fig. 22.1 in an experimental frame (not shown for convenience). As in Fig. 22.1A we partition the components of the coupled model into blocks and create a coupled model of lumped versions of the components. The underlying theory is to create base and lumped models with a homomorphism holding in the given experimental frame. We ask questions like: 1) how likely is it to be able to construct lumped models that satisfy the strict conditions for exact morphism? 2) How far can such conditions be relaxed while still obtaining "reasonable" outputs from lumped-model simulations? After transferring the general theory of systems morphisms, discussed earlier in Chapter 15, to DEVS Markov models, we extend the theory to deal with approximate morphisms. As illustrated in Fig. 22.1B, we will develop a methodology to relate the departure from strict "lumpability" conditions to the error in behavior mismatch that results.

22.2 HOMOMORPHISM OF TIMED NON-DETERMINISTIC MODELS

Our first step is to extend the morphism concepts developed in Chapter 15 and further elaborated in Chapter 16 to timed proto-systems introduced in Chapter 12. Recall that a timed non-deterministic model is defined by:

$$M = < S, \delta, ta >,$$
where $\delta \subseteq S \times S$ *is the non-deterministic transition relation,*
and $ta : \delta \rightarrow R_0^{\infty}$ *is the time advance function that assigns to each state pair in* δ *a time to make that transition.*

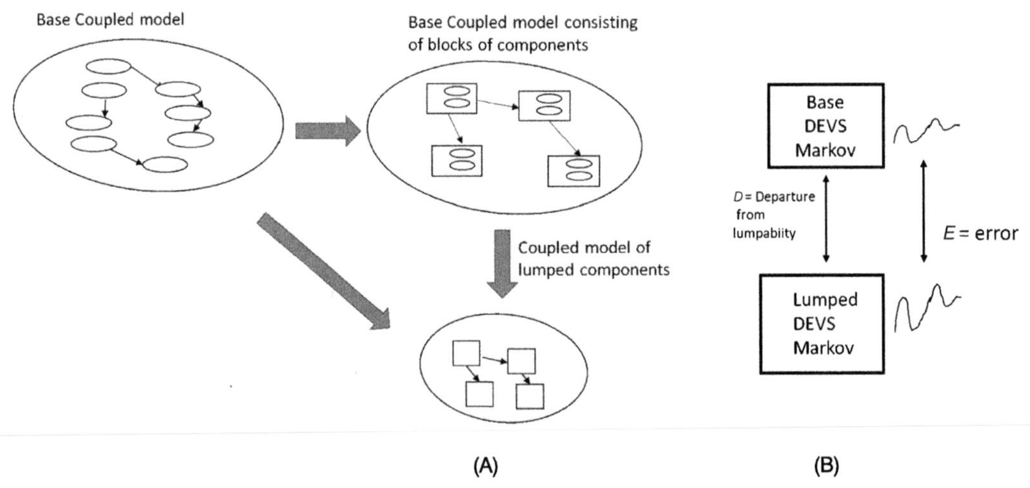

FIGURE 22.1

Overall approach to creation of base-lumped model pairs.

For such a model, we define the state trajectory associated with the sequence $s_1, s_2, ..., s_n$ as: $STRAJ_{s_1,s_2,...,s_n} :< 0, \tau > \rightarrow S$ where $(s_j, s_{j+1}) \in \delta,\ j = 1, 2, ..., n-1$ i.e., the trajectory follows a path in the state graph of the model.

The timing associated with the trajectory can be defined by imagining the following: at time $T_1 = 0$, the model enters s_1 and takes $ta(s_1, s_2)$ to reach s_2. At time $T_2 = 0 + ta(s_1, s_2)$ the model enters s_2. At time $T_3 = 0 + ta(s_1, s_2) + \text{ta}(s_2, s_3)$ the model enters s_3, and so on until the final state of the sequence is reached. In other words, we have $T_1 = 0$, and for $i = 1, ..., n-1$, $T_{i+1} = T_i + ta(s_i, s_{i+1})$.

Then $STRAJ_{s_1,s_2,...,s_n}(t) = s_i$ if $t \in < T_i, T_{i+1} >$. In this way, $STRAJ_{s1,s2,...,sn} :< 0, \tau > \rightarrow S$ is defined by the sequence of states $s_1, s_2, ..., s_n$ and the associated sequence of times at which these states are entered: $0, T_1, T_2, ..., T_n$.

We can now define the behavior of M as the set of all state trajectories as just defined.

On the way to a theorem concerning homomorphisms of timed models we start with

Proposition 22.1. *Consider two timed models:*

$$M =< S, \delta, ta >, \text{ where } \delta \subseteq S \times S, \text{ and } ta : S \times S \rightarrow R_0^\infty$$
$$M' =< S', \delta', ta' >, \text{ where } \delta' \subseteq S' \times S', \text{ and } ta' : S' \times S' \rightarrow R_0^\infty.$$

Consider a mapping $h : S \rightarrow S'$ which is a one-to-one correspondence such that:

For all pairs $(s_1, s_2) \in S \times S$,
1. $(s_1, s_2) \in \delta \Leftrightarrow (h(s_1), h(s_2)) \in \delta'$
2. $ta'(h(s_1), h(s_2)) = ta(s_1, s_2)$

Then M and M' have the same behavior.

Proof sketch : Due to 1) every trajectory in M is also in M' and conversely. Due to 2) they have the same timing.

Proposition 22.1 can be extended to a homomorphism of models with outputs.

Proposition 22.2. *Consider two timed models with output:*

$$M =< S, Y, \delta, \lambda, ta >, \text{ where } \delta \subseteq S \times S,\ \lambda : S \times S \rightarrow Y, \text{ and } ta : S \times S \rightarrow R_0^\infty$$
$$M' =< S', Y, \delta', \lambda', ta' >, \text{ where } \delta' \subseteq S' \times S',\ \lambda' : S' \times S' \rightarrow Y, \text{ and } ta' : S' \times S' \rightarrow R_0^\infty.$$

Consider a mapping $h : S \rightarrow S'$ (onto)
which is such that:
For all pairs, $(s_1, s_2) \in S \times S$,
1. $(s_1, s_2) \in \delta \Leftrightarrow (h(s_1), h(s_2)) \in \delta'$
2. $ta'(h(s_1), h(s_2)) = ta(s_1, s_2)$
3. $\lambda'(h(s_1), h(s_2)) = \lambda(s_1, s_2)$

Then M and M' have the same output behavior where the output behavior of a model can be defined by applying the output function to every transition in a state trajectory and maintaining the associated timing.

Proof sketch: Due to 1) every trajectory in M is also in M' and conversely. Due to 2) they have the same timing. Due to 3) they have the same output behavior.

Exercise 22.1. Define the output behavior of a timed model. Hint: consider any state trajectory with its state and time sequences. Apply the output function to the state sequence while keeping the time sequence.

22.3 HOMOMORPHISM OF DEVS MARKOV MODELS

DEVS Markov models can be viewed as extensions of the timed non-deterministic models that have probabilities assigned to outgoing transitions potentially making some transitions more likely than others. When iterated over multiple transitions, this means that there are also probabilities associated with state trajectories and also with the observable behaviors associated with models. In extending the homomorphism concept to DEVS Markov models our main task is to make sure to match up the single step elements in the right way to assure that the multi-step consequences also match up.

Fig. 22.2 illustrates our approach to defining homomorphisms for DEVS Markov models. Given a map h from the phases of M onto those of M' we must start by assuring that the first square in the commutative diagram commutes. In the deterministic case this means that the selected phases correspond in both models. Then, going to the second square, for corresponding phases it must be that the selected sigmas are equal to assure that the transitions have the same timing. In the probabilistic case however these depend on probabilistic choices so we must check that the *underlying probability distributions are the same*. Consequently, for selecting phases we must have that the probabilities of transition from the image phase, $h(ph)$ in M' must be the same as those from the pre-image, ph in M. Then given corresponding selected phases, we must have that the probability distributions for the transition times must be the same. This requires that the probability transition structures, *PTS* and *PTS'*, and the transition time structures, *TTS* and *TTS'*, must become equal under the state correspondence.

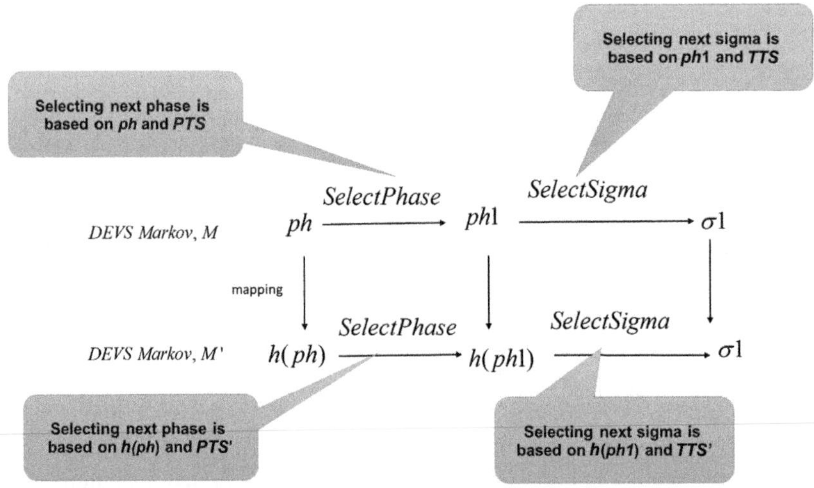

FIGURE 22.2

Approach to homomorphism of DEVS Markov models.

Having done this, we can go on to show that the probabilistic behaviors of the models are the same. Again, in contrast to the deterministic case the trajectories have probabilities associated with them and we want corresponding state trajectories to have the same probabilities in order to declare that the models have equivalent behavior.

For DEVS Markov model, M, let $STRAJ_{s1,s2,...,sn} :< 0, \tau > \rightarrow S$ be a state trajectory defined by the sequence of states $s_1, s_2, ..., s_n$ and the associated sequence of times at which these states are entered, $0, T_1, T_2, ..., T_n$. The probability of this trajectory is $P(s_1, s_2, ..., s_n) * P(T_1, T_2, ..., T_n/s_1, s_2, ..., s_n)$ where $P(s_1, s_2, ..., s_n) = Pr(s_1, s_2)Pr(s_2, s_3)...Pr(s_{n-1}, s_n)$ and $P(T_1, T_2, ..., T_n/s_1, s_2, ..., s_n) = P(T_2 - T_1/\tau(s_1, s_2)))P(T_3 - T_2/\tau(s_2, s_3)))...P(T_n - T_{n-1}/\tau(s_{n-1}, s_n)))$ where $P(T_{i+1} - T_i)/\tau(s_i, s_{i+1})))$ denotes the probability assigned to the inter-event time $T_{i+1} - T_i$ by $\tau(s_i, s_{i+1})$, the pdf associated with the transition from s_i to s_{i+1}. Strictly speaking, the probability is assigned to small intervals around the given times for continuous valued times. The Behavior of M is the set of state trajectories with their associated probabilities.

Note that as in Fig. 21.10, since the state transitions occur independently of their timing, their sequencing can be computed in Markovian fashion, i.e., by multiplying the conditional transition probabilities. Then the probabilities of the associated inter-event times can be computed from the corresponding state transitions by referring to the pdfs associated with the transitions.

Theorem 22.1. *Let*

$M =< S, \delta, ta >$ and $M' =< S', \delta', ta' >$ specified by structures PTS, TTS and PTS', TTS' respectively.

Consider a mapping $h : S \rightarrow S'$ (onto)
such that:
For all pairs $(s_1, s_2) \in S \times S$,
1. $Pr'(h(s_1), h(s_2)) = Pr(s_1, s_2)$
2. $\tau'(h(s_1), h(s_2)) = \tau(s_1, s_2)$

Then M and M' have the same probabilistic state behavior.

Proof. (Sketch) Consider any state trajectory of M, $STRAJ_{s1,s2,...,sn}$. *Due to* 1), *M and M' assign the same probabilities to the state sequence $s_1, s_2, ..., s_n$ as discussed above. Due to* 2), *M and M' assign the same probabilities to the associated inter-event times, $T_1, T_2 - T_1, ..., T_n - T_n - 1$. Therefore, $M$and M' assign the same probabilities to the trajectory.* □

Note that while the requirements for homomorphism look innocent enough, they actually may be hard to satisfy exactly in the case that the mapping h is not one-one. In that case, the conditions require that all states in the same partition block (that map to the same state) have the same transition probabilities and timing distributions.

Fig. 22.3 explains the strong requirements in detail. Note that both $s1$ and $s2$ map into the same state, s′ under h. Similarly, for $t1$ and $t2$ with respect to t'. Now, p the probability of transition from s′ to t′ must be reflected back into every pair of transitions from the block of s′ to that of t′. This means that the probabilities of transitions $s1$ to $t1$, $s1$ to $t2$, $s2$ to $t1$, and $s2$ to $t2$ must all equal p. A similar requirement holds for the time transition structure. This essentially requires that lumped model be replicated in the base model multiple times.

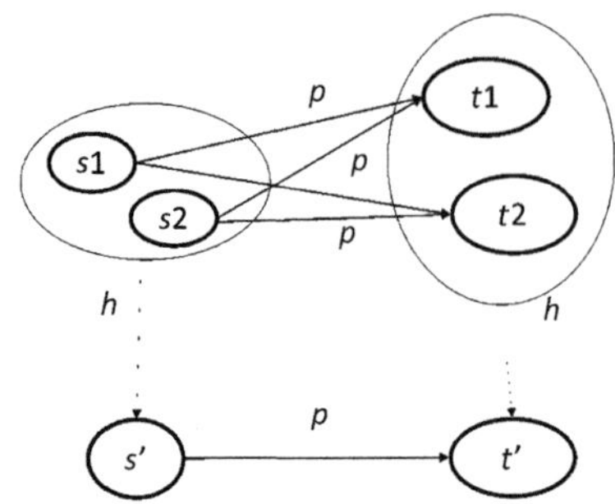

FIGURE 22.3

Explaining the strong requirements for homomorphism.

A weaker condition, usually called lumpability (cf. Chapter 16) for Markov chains, is that the sum of the transition probabilities from any state in a block to all states in another block be the same for all such states. This requires for example, that the probability of transition from $s1$ to $t1$ or $t2$ be the same as from $s2$ to $t1$ or $t2$. In this case, the homomorphic image gets the sum of these transitions as its transition probability.

In Section 22.5, we back off these strong requirements and consider approximate morphisms where the conditions are not strictly satisfied.

The statistics associated with state trajectories are also extended from the non-deterministic timed model case with the following definitions:

Statistics associated with a state trajectory with path
$s_1, s_2, ..., s_n$ and the associated sequence of times $T_1, T_2, ..., T_n$:
Estimated probability of occurrence of state $s = \frac{\#visits\ to\ s}{length\ of\ path}$
Sojourn time in $s = \sum_{(s,s') \in SuccPairs(s_1,s_2,...,s_n)} ta(s,s')$
where $SuccPairs(s_1, s_2, ..., s_n) = \{(s_i, s_{i+1}) | i = 1, ..., n-1\}$.
Recall that for $i = 1, ..., n-1$, $T_{i+1} - T_i = ta(s_i, s_{i+1})$.
Finally we have: Estimated probability of s in state trajectory = Estimated prob. of occurrence of s ∗ Sojourn time in s

Remark 22.1. We can extend the homomorphism defined in Theorem 22.1 to preserve the input/output behavior of fully-fledged DEVS models by requiring it to satisfy the usual requirements for external and output function preservation.

Before proceeding to consider approximate morphisms, we present examples of homomorphisms of DEVS Markov models that illustrate their use in the methodology underlying Fig. 22.1.

22.3.1 RANDOM PHASE REGIME FOR COUPLED DEVS MARKOV MODELS

Recall that due to closure under coupling, coupled DEVS Markov models are equivalent to atomic DEVS Markov models. Recall Theorem 21.6.7 there are a *PTS* and *TTS* that characterize the resultant

DEVSN as a DEVS Markov model. Sometimes such models admit a solution to their steady state behavior called "random phase" which we characterize as follows: DEVSN enters a steady state in which the sequence of inter-event times is distributed over a fixed probability distribution. More particularly, consider the sequence of internal event times as generated by the sequence of selected imminents t_1, $t_2, \ldots t_n, \ldots$.

Then the inter-event times, $T1, T2, \ldots, TN, \ldots$, where $Ti = t_i - t_{i-1}$ are selected from the given distribution. For example, consider a coupled model of components, each of which has the same exponential distribution for its time advance. Then the next event time of the first coupled model transition is determined by selecting from an exponential distribution which has the rate parameter of any one of them multiplied by the number of components. This distribution is not necessarily the final steady state distribution since the non-imminent components continue to hold in their states until their chosen time advances becomes due. However, after sufficient iteration, this exponential distribution does indeed become the distribution for inter-event times of the coupled model. (In this particular case the inter-event times become a Poisson process.)

22.4 EXAMPLE: COMBAT ATTRITION MODELING

We revisit a pair of models where Continuous Time Markov (CTM) modeling in DEVS was used to express base and lumped models with an explicit morphism relation between them (Zeigler, 2018). Two opposing forces of equal size and unit fire power are arrayed against each other (Fig. 22.4). The base model has red and blue groups consisting of individual atomic DEVS Markov models representing shooter objects, e.g., tanks. Following the pattern of Fig. 22.1, the lumped model is formed by considering these groups as blocks and employing a single atomic model component to represent each block. In the model that we consider here (which is slightly different than that of Zeigler, 2018) the state of a lumped model component is the number of base model shooters still alive in the group it represents. In the *attrition* experimental frame, the outcome (one side reaching zero shooters) and the time it takes are random variables to be observed through simulation.

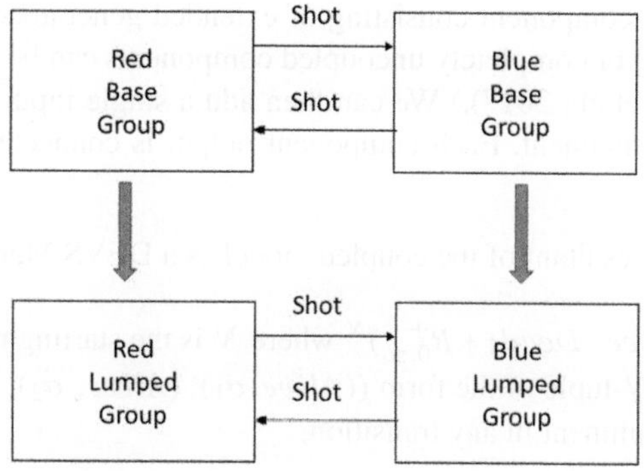

FIGURE 22.4

Opposing tank forces with attrition morphism.

22.4.1 BASE MODEL COMPONENT: REALIZING A POISSON PROCESS WITH DEVS MARKOV MODEL

The basic component of the base model is a tank that shoots a stream of bullets. Traditionally, this stream has been represented by a Poisson process which means that the inter-event durations are exponentially distributed with the same rate parameter. This behavior can be realized by the simplest DEVS Markov model consisting of one state with self-transition and output.

Exercise 22.2. Write a one-state DEVS Markov model with output that realizes a Poisson process. Hint: The *PTS* and *TTS* have only one value each.

Because it does not allow self-transitions, a DEVS CTM requires at least two states to realize a Poisson process. So to realize this behavior, we have we shuttle back and forth between two identically looking states at the desired rate while outputting at each transition.

Exercise 22.3. Write a *PTS* and a DEVS CTM with output that realizes a Poisson process.

Exercise 22.4. Express the two state DEVS CTM as a two state DEVS Markov Model and apply Theorem 22.2 to show that there is a homomorphism from the two state model to the single state model.

To model a vulnerable tank, we extend the single-state Poisson generator by adding an input that sends the model to a dead state with probability, v.

Exercise 22.5. Add a second state, *Dead*, to the generator and an input port called Kill. When an input is received on this port, the model transitions to the Dead state with probability v. Otherwise, the model continues with its shot generation. Extend your *PTS*, *TTS* and DEVS model structures accordingly.

The tank model realized by the extended Poisson generator is the basic component of the base model.

22.4.2 BASE COUPLED MODEL

Now we have a base model component consisting of extended generators with no internal coupling. (A coupled model consisting of completely uncoupled components can be represented as a crossproduct of components (Zeigler et al., 2017).) We can then add a single input port that sends a kill input to the input port of each component. Each component output is connected via an output port to the coupled model output port.

Exercise 22.6. Express the resultant of the coupled model as a DEVS Markov atomic model.

Hint: the state set is $(\{Alive,\ Dead\} * R^{+}_{0,\infty})^N$ where N is the starting number of tanks and the initial state is an N-tuple of the form $((Alive, \sigma_1), (Alive,\ \sigma_2), .., (Alive, \sigma_N))$. Assume only one component is imminent in any transition.

The base model is made of two such component coupled models, called Red and Blue, coupled to each other with kill inputs and outputs as illustrated in Fig. 22.4. Thus, for example, an output from one of the red tanks is sent to each tank in the blue component model.

Lumped Model Component

We can represent a Red or Blue coupled model by an atomic model with a state space consisting of the crossproduct $\{Alive, Dead\} \times N$ (where N denotes the natural numbers). A state $(Alive, n)$ then represents the current number of alive generators where $n>0$; and $(Dead, 0)$ is the state in which all components have been killed. Based on the random phase steady state assumption, the rate of self-transition while in the *Alive* phase is the individual rate times n (see Section 22.3.1). Each transition generates a kill output so that the lumped model generates such outputs in the same manner that the coupled model does.

Exercise 22.7. Express each force (Red or Blue) group lumped model as a DEVS Markov atomic model. Hint: the state set is now compressed to the set of elements of the form (ph, n) with n the number of alive elements. Now, as long as ph is the *Alive* phase, we have self-transitions occurring at the same rate as in the base model. A kill input sets $n = n - 1$ with probability $1 - (1 - v)^n$, which is the probability of at least one component being killed (and is approximated by $n*v$ for small enough $n*v$). When n becomes 0 this represents the state where there are no remaining alive components and the phase is set to Dead.

Proposition 22.3. *The base and lumped models of a force group have the same probabilistic behavior.*

Proof. We show that Theorem 22.1 holds for the DEVS Markov models.

Consider any state $s = (ph_1, \sigma_1), (ph_2, \sigma_2), .., (ph_N, \sigma_N)$ *of the base model*
Let's say s *is alive if* $census_{Alive}(s) > 0$ *i.e., if at least one tank is alive.*
Then define a map h *from the base state set to the lumped state set:*

$$h(s) = \begin{cases} (Alive, census_{Alive}(s)) & \text{if } s \text{ is alive} \\ (Dead, 0) & \text{otherwise} \end{cases}$$

By the random phase assumption, in the alive state, the probability is unity that the base model will
transition to a state st *with* $census_{Alive}(st) = census_{Alive}(s) - 1$
and the mean time is $\frac{T}{census_{Alive}(s)}$
where T *is the mean time for an individual component.*
Note that because of uniformity all base components have the same probability of being killed which justifies the above statement.

In the lumped model we have the transition $(Alive, n)$ *to* $(Alive, n - 1)$ *for* $n > 1$.
Here we also have unity probability of occurrence and the mean time is $\frac{T}{n}$ *by design.*
Thus since h *sets* n *to* $census_{Alive}(s)$, *we have* $Pr'(h(s), h(st)) = 1 = Pr(s, st)$
and $\tau'(h(s), h(st)) = Exp(\frac{n}{T}) = \tau(s, st)$.
Thus for alive states h *satisfies the requirements of Theorem 22.1 for homomorphism.* □

Exercise 22.8. Show that the same is true for the dead state.

Exercise 22.9. Show that h satisfies the conditions for homomorphism when taking the input and output into account.

22.4.3 LUMPED COUPLED MODEL

As indicated, the base model represents a battle encounter of Red and Blue forces as a pair of the DEVS Markov coupled models each sending Kill outputs to the other. If the Kill outputs are distributed uniformly to the internal combat units and the vulnerabilities within each group are the same then the coupled models can each be aggregated to lumped models of the form of the atomic model just given. These lumped models are then cross coupled by Kill output to input ports in simple fashion to form the lumped coupled model.

Exercise 22.10. Define the lumped model of the battle encounter as a coupled model of a pair of the above mentioned DEVS Markov atomic models representing Blue and Red forces in combat.

22.4.4 HOMOMORPHISM FROM BASE TO LUMPED MODELS

The proof of homomorphism from base to lumped model follows the approach of Section 15.9 where we employ the individual group homomorphism just established in combination to define the mapping of the overall base state to the lumped state in Fig. 22.4. Thus the lumped model maintains correspondence with the base model by maintaining the correct counts of alive tanks in each group as time progresses. In the exact morphism, the groups generate the same rate of fire and suffer the same probability of casualties over time until one or the other has no tanks left.

Exercise 22.11. Follow the approach of Section 15.9 to prove that a homomorphism exists from the force-on-force base model to the lumped model developed above.

22.5 APPROXIMATE MORPHISMS

As indicated in the introduction, and fortified by an instance in which exact morphism is seen to be quite stringent, our interest now proceeds to examine the approximate version of the morphism just established. We want to determine how far the uniformity condition can be relaxed while still obtaining a reasonably good lumped model. This requires defining appropriate measures for departure from uniformity and error.

The assumptions generally required for exact homomorphism in Chapter 15 relate to homogeneity (here called uniformity) of both structure of components and of coupling among components within and between blocks. Having developed the class of DEVS Markov models we are in a position to examine these conditions to justify constructed homomorphisms in the new context. In the stochastic case such conditions may be called the *lumpability* conditions.

22.5.1 INTRODUCING NON-UNIFORMITY INTO THE BASE MODEL

In the attrition base model uniformity of firing conditions can be formulated as the following requirements:

1. *Uniformity of structure*: all components have the same rate of fire and the same vulnerability
2. *Uniformity of coupling*: uniformity of emitted and received fire distribution

(a) *Emitted* fire uniformity: Every unit distributes the same total fire volley to the opposing group;

(b) *Received* fire uniformity: Every unit receives the same fire volley from the opposing group.

First we examine the uniformity of the base model by recognizing that the vulnerability parameters, v_i may not be the same for all units i within a force group. To consider such heterogeneous distributions of values, we set up a mapping from base model parameters to lumped model parameters that can be defined as follows:

For each block of components (blue and red force groups) the mapping of vulnerability parameters is an aggregation which sends the individual values to their average (mean) value:

$$F_{Block} : \textit{Base Parameters} \rightarrow \textit{Lumped Parameters}$$
$$F_{Block}(\{v_i \mid i = 1, .., N_{Block}\}) = Mean_{Block}(\{v_i \mid i = 0, .., N_{Block}\})$$

where v_i are the individual vulnerability values in a block and v_{Block} is their mean value assigned to the lumped block.

This is a particular instance of a *parameter morphism*, where the morphism not only sets up a correspondence between states but also maps base model parameter values to lumped values for which the correspondence is expected to hold (Chapter 17).

Since the morphism is expected to hold exactly only when exact uniformity within bocks holds, we measure the departure from uniformity by employing the standard deviation relative to the mean:

$$LumpedSTD = STD(\{v_i \mid i = 0, .., N_{Block}\})$$

Note that strict uniformity holds (all block vulnerability parameter values are equal) if, and only if, the standard deviation vanishes as required for such a measure.

The relaxed morphism concept, enabled by allowing departure from uniformity in parameter mappings, is called an *approximate morphism* (Chapter 17). To operationalize this concept, we need to characterize the error we can expect from a lumped model which assumes exact uniformity. In other words, an approximate morphism must also be accompanied by information that indicates the departure from exact homomorphism in relation to the departure from uniformity. This information can be compiled into a *lumpability zone* statistic that characterizes the neighborhood of *LumpedSTD* near zero in which acceptable predictions of the lumped model may be obtained.

A useful methodology for integrated families of models at various abstraction levels would allow multiplicity in calibration and cross-calibration within diverse experimental frames. DEVS modeling has shown instances of a well-justified process of abstraction from traditional differential equation models to DEVS representation and spatial aggregation to assure relative validity and realism within feasible computational constraints.

As in Fig. 22.1, consider base and lumped models with a homomorphism holding in an experimental frame (not shown for convenience). We ask questions like: 1) how likely is it to be able to construct lumped models that satisfy the strict conditions for exact morphism. 2) How far can such conditions be relaxed while still obtaining "reasonable" outputs from lumped-model simulations? In outline the process to be exemplified is:

1. Define a class of base models.
2. Specify a lumping process to construct lumped models from base models.

3. Define a metric for measuring the departure of a lumped model from the required condition for exact lumpability.
4. Define an experimental frame and metric for error between predictions of a base/lumped model pair.
5. Perform an experiment in which a representative sample of base models from the class of interest is generated and plot the results. In the stochastic case, typically numerous samples are generated in pursuit of statistical significance.

Fig. 22.5 depicts a notional (E, D)-space which is populated by pairs of prediction error, E and departure from lumpability, D (defined in steps 3 and 4) and sampled in experiment 5). As illustrated on the left for each generated base model and derived lumped model, the error extracted from results of simulation runs is plotted against the lumpability measurement. The sloped and plateaued curve shown is an idealization of the noise-laden results expected in actual examples. At the origin are images of (B, L) pairs satisfying exact lumpability. The slope represents (B, L) pairs for which the lumpability conditions nearly hold and the error incurred is relatively small. The sloped line eventually dissolves into a horizontal line, the background level of error. This represents the average difference in predictions expected when sampling pairs of models at random without employing the lumping process. The outer edge of the lumpability space is the largest value of the lumpability metric observed. Within it, the lumpability zone is demarcated by the smallest value of D for which E equals the background level. The ratio of the lumpability zone to the lumpability space is a measure of the pervasiveness of lumpability. The average error within the lumpability zone can be taken as a measure of the effectiveness of lumpability; the ratio of this average to the background level indicates the influence of the lumping process in reducing error below the level that could otherwise be expected. The ratio of average error in the lumpability zone to average error overall indicates the effectiveness of the lumping approach – the smaller this ratio, the greater the utility of the lumping process.

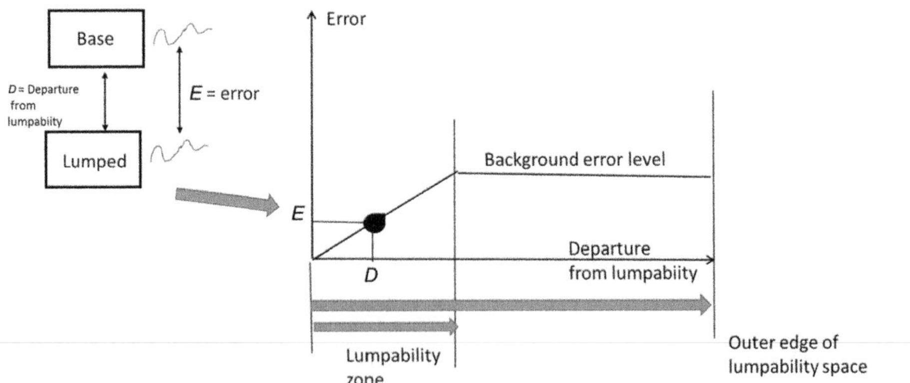

FIGURE 22.5

Scaling the zone of useful lumpability.

22.5.2 LUMPABILITY ZONE EVALUATION FOR DEPARTURE IN UNIFORMITY OF STRUCTURE

To illustrate the lumpability zone information about an approximate morphism that might be carried along with its definition, we performed an empirical investigation of the attrition model in Fig. 22.5 using a methodology described in the companion volume, Model Engineering for Simulation (Zhang et al., 2018).

Recalling that accuracy is relative to an experimental frame, we focus on two model outputs of interest, the *win/loss outcome* and the *time to reach such an outcome*. In the first case, a natural error metric is the distance between the base and lumped model Blue winning percentages; in the second, the distance between the termination times of the models. In this section we focus on termination times in relation to departure of uniformity in vulnerability.

The parameters and their values were:

- *Force Ratio:* 2:1 (Number of Blue tanks: 100, Number of Red tanks: [40, 60])
- *Fire rate*: uniform across at 0.1 shots/unit time and the same for both sides
- *Output to input couplings*: uniformly all-to-all.
- *Vulnerability*: sampled from a Beta distribution with mean 0.5 and standard deviation ranging from 0.05 to 0.25 in steps of 0.05 to create increasingly large *LumpedSTD*

At $LumpedSTD = 0$ the error between lumped model base model termination times was close to zero as predicted. The error actually decreased slightly until $LumpSTD = 0.15$ and then rose sharply. Thus the lumpability zone $[0, 0, 0.15]$ is 60% of the full range (0.15/0.25) indicating that the lumped model can be used in the termination frame for relatively large departures from uniformity in vulnerability for force ratio 2:1.

Lumpability Dependence on Force Ratio

However, the size of the lumpability zone depends on force ratio. The size shrinks when the forces are equal in strength (i.e., a ratio of 1:1) to 10% of full range. This indicates increased sensitivity to non-uniformity when firing rates and vulnerabilities are equally balanced. On the other hand, the size expands to 100% at ratios approaching infinity. When one side completely dominates the other, then sensitivity to small variations becomes only a second order effect. This shows more generally that *lumpability zones for some parameters can depend on settings of other parameters*.

Lumpability Dependence on Force Sizes

Lumped model error decreases with increasing force sizes (at force ratio of 2:1). Indeed, agreement in mean termination time is very close for sizes larger than 50, while the approximation is less good for smaller numbers. This is consistent with the convergence to the mean for large sample sizes expressed in the central limit theorem.

Interestingly, the simulation run times for the base model grow much more rapidly for larger sizes than those of the lumped model. Thus in the lumpability zone [100, 1000] we have high accuracy with short executions of the lumped model – ***indicating a zone where accuracy and execution time do not have to be traded off.***

Departure From Uniformity in Coupling

Turning to the effect of departure from uniformity in coupling, we leave untouched the "Emitted fire distribution" requirement which normalizes the fire distribution emitted by each attacker. Focusing on the "Received fire distribution," uniformity requires that every target receives the same total fire.

The distribution of fire from blue to red can be represented by an n x m matrix, called the Fire matrix, where the blue and red forces are of size n and m respectively:

$$P_{blue \rightarrow red} = [p_{i,j}]$$

Here $p_{i,j}$ represents the rate of fire from blue unit i to red unit j. The Emitted fire uniformity condition requires that the influence of any blue tank on all the red tanks sums to the same value. We normalize this value to unity giving the matrix entries the flavor of probabilities. In other words,

$$\forall_{j,k} (\sum_{i=0}^{n-1} p_{i,j} = \sum_{i=0}^{n-1} p_{i,k} = 1)$$

which states that the columns each sum to 1.

Received fire uniformity requires that the influence of all the blue units on any red unit is the same:

$$\forall_i (\sum_{j=0}^{n-1} p_{j,i} = \sum_{k=0}^{n-1} p_{k,i})$$

This requires that the rows each sum to the same value.

Relaxing uniformity of received fire means allowing such row sums to vary and suggests that we employ a measure of their variation as a metric of departure from uniformity. To implement this concept, we treat matrix elements as sampled values, calculating the mean of row sums and choosing the metric to be the standard deviation from the mean. As with vulnerability, calling the metric *LumpSTD*, we have that exact uniformity is then equivalent to $LumpSTD = 0$. Recalling that accuracy is relative to an experimental frame, we consider the win/loss outcome as well as the time to reach such an outcome.

Lumpability Dependence on Distribution of Fire

We followed the procedure outlined earlier and performed an empirical investigation by generating 10,000 pairs of randomized stochastic matrices. The square matrices were sampled from a uniform distribution of sizes m from 5 to 45 in steps of 5. Matrices were generated as sequences of m x m entries sampled with Binomial distribution with probability increased in 10 steps from 0.1 to 1 each with 1000 trials (with normalization to assure column summation to unity). The objective in more refined terms became to assess how rare are matrices satisfying the exact uniformity requirement, and assuming this to be very small, how big a neighborhood can be expected in which approximate uniformity can be effective.

We follow the outlines of Fig. 22.5 with prediction error E defined as absolute difference between base and lumped output values of interest (blue success ratio and finish time) and lumpability departure metric defined by *LumpSTD*. The plots corresponding to Fig. 22.5 are shown in Fig. 22.6 for blue success and finish time, respectively on the y-axis. *LumpSTD* points are shown in a continuous curve

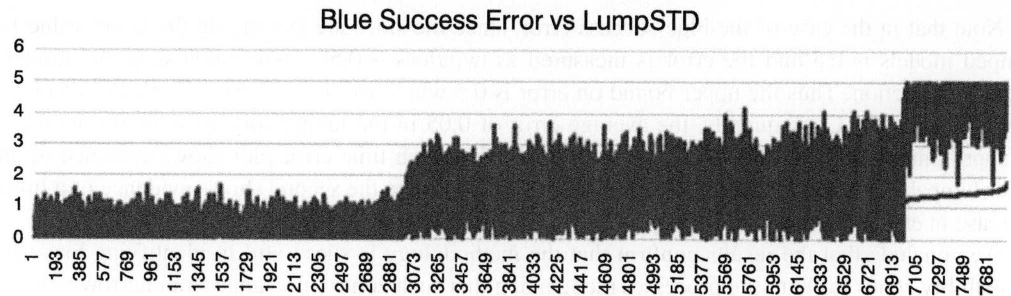

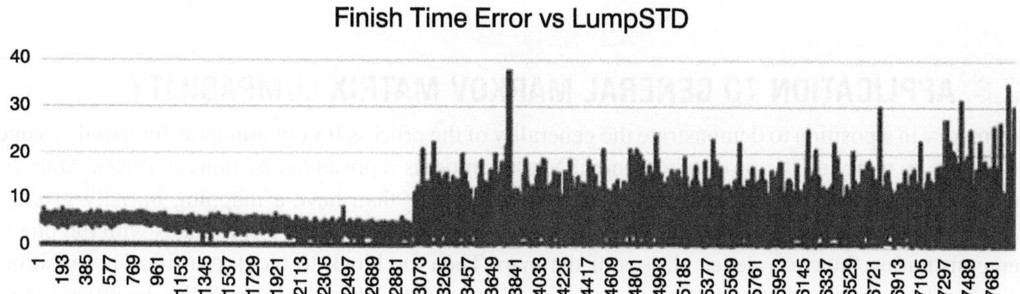

FIGURE 22.6

Plots of blue success error and finish time error instantiations of Fig. 22.5.

Table 22.1 Characteristic Values for Combat Model Lumpability Evaluation

Element	Value for Blue Success Ratio	Value for Finish Time
LumpSTD boundary for lumpability zone	0.6	0.6
LumpSTD outer limit	1.85	1.85
LumpSTD zone/outer limit	33%	33%
Avg error within lumpability zone	0.05	5
Avg error overall	0.5	5.7
Avg error in zone/ avg error overall	10%	88%

with increasing order with paired error values comprising the noisy curves (blue success ratio error is shown scaled X10 for easier viewing).

Table 22.1 presents results for both outputs of interest. The lumpability zones occupy approximately 33% of the full range suggesting that near uniformity is quite pervasive in randomly generated lumped models. In the case of blue success error, the ratio of average error in the lumpability zone to the overall average error (10%) suggests a strong influence of the lumping process in reducing error below the level that could otherwise be expected. This is less true in the case of finish time where the average error ratio is 88%. Another sign of lower effectiveness for finish time prediction is the large variance in its error that is evident from the bottom plot of Fig. 22.6.

Note that in the case of the blue success error, since the sides are symmetric the target value for lumped models is 0.5 and the error is measured as |win/loss – 0.5| where win/loss is that model's success prediction. Thus the upper bound on error is 0.5 which we see realized in high *LumpSTD* region of the space. Consequently, the average error of 0.05 in the lumpability zone shows the utility of approximate lumping. Also, it is interesting that the finish time error plot shows evidence of another lumpability zone besides the one already identified where the second shows evidence of a linear increase in error with increasing *LumpSTD*.

We conclude that the results confirm that the methodology is successful in identifying low-error lumpability zones for lumping processes recognizing that such zones may range from narrow to broad revealing the sensitivity to non-uniformity of aggregation conditions.

22.6 APPLICATION TO GENERAL MARKOV MATRIX LUMPABILITY

We are now in a position to demonstrate the generality of the process for obtaining the lumpability zone. We do this by applying the process to general Markov models represented by finite matrices. Matrices are generated similarly to the ones discussed above. Rather than have a mapping initially given, a matrix is partitioned into blocks based on a randomly generated partition of the states with the blocks then constituting the equivalence classes of the map. Then the block-to-block transition probabilities are defined using averaging with the standard deviation used to define the lumpability departure metric in a manner similar to that described for the fire distribution study. Fig. 22.7 illustrates how this is done. The blocks are shown as a column of ovals on the left and replicated on the right. Transitions

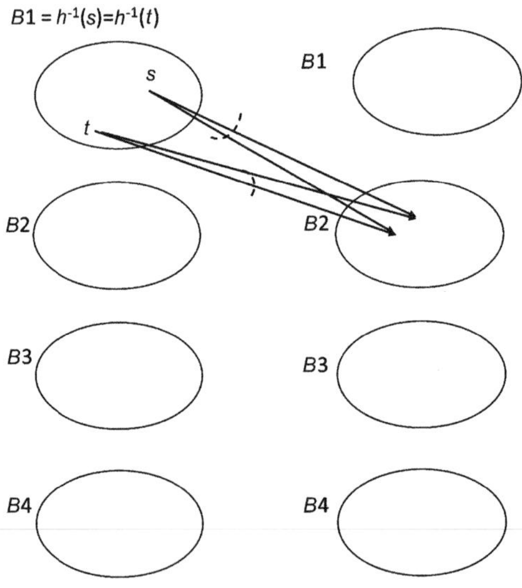

FIGURE 22.7

Illustrating the computation of lumped model transitions.

from individual states in blocks are indicated by arrows. Exact lumpability requires that the sums of transitions from states, e.g., s and t, to any block are equal (suggested by the dotted arcs).

To relax this requirement, we use the average of the sums as the probability of block-to-block transition. In more detail:

As in Chapter 21, let M be a CTM based model defined PTS, P. We will construct a lumped model M' with PTS, P' to be defined as follows:
Let π be a partition on the states of S. Let $B(i)$ denote the partition block containing i. Let $h : S \rightarrow \pi$ be defined by $h(i) = B(i)$, i.e., h maps each state into the block containing it.
For each Pair of Blocks, B and B' and for each state $i \in B$, let
$m_i = \sum_{j \in B'} p(i, j)$
be the sum of the probabilities from i to states of B'.
We define the probability of transition from B to B'
$Pr(B, B') = Avg_B(m_i)$, *the average of the sums, m_i.*
with the standard deviation
$std_{B,B'} = STD(m_i)$.
Then let the departure from lumpability be the maximum of these values over all pairs of blocks:
$LumpSTD = \max_{B,B'}(std_{B,B'})$.

The measure of quality for the lumping is taken as the error in steady state distribution of the lumped model relative to that of the base model. More detailed description is given in Zeigler et al. (2017).

First we restate the lumpability theorem in the exact case:

Theorem 22.2. *An exact homomorphism from M to M' exists if $LumpSTD = \max_{B,B'}(std_{B,B'}) = 0$.*

Proof. (Sketch) We show that $Pr'(h(s), h(t)) = Pr'(B(s), B(t)) = Pr(s, t)$. □

Exercise 22.12. Under the given condition, show that the traditional exact lumpability just stated holds. Hint: Note that if $m_i = m$ for all $i \in B$, then $\sum_{i \in B} \sum_{j \in B'} p(i, j)p(i) = m \sum_{i \in B} p(i) = mp'(B) = p_{B,B'}p'(B)$.

Corollary 22.1. *The probabilistic behaviors of M and M' are the same modulo π.*

Exercise 22.13. Prove the corollary. Hint: consider state trajectories as defined in Section 22.2.

Exercise 22.14. For a deterministic model, an equilibrium (also called steady state) is a state which is invariant under the transition function. Prove that if $s*$ is an equilibrium state of M and h is a homomorphism then $h(s*)$ is an equilibrium state of the homomorphic image.

Corollary 22.2. *An equilibrium state of DEVS CTM M maps to an equilibrium state of a lumped model M' of M.*

Exercise 22.15. Prove the corollary. Hint: treat the matrix transformation of M as a deterministic model.

Table 22.2 Characteristic Values for General Markov Matrix Lumpability Evaluation

Element	Steady State of Lumped Model
LumpSTD boundary for lumpability zone	0.03
LumpSTD outer limit	1.2
LumpSTD zone/outer limit	2.5%
Avg error within lumpability zone	0.06
Avg error overall	0.15
Avg error in zone/ avg error overall	40%

Exercise 22.16. Based on the result of Exercise 22.14 show that a measure of the accuracy of a homomorphism is how far the image of an equilibrium state, $h(s*)$ moves in the homomorphic image in a specified finite time interval.

In experimentation mentioned above, we showed that the steady state error depended on the lumpability metric as expected. The obtained values are shown in Table 22.2. The maximum value of *LumpSTD* across all samples was found to be 1.2. The lumpability zone was found to be bounded by $LumpSTD = 0.03$ containing approximately 2.5% (= 0.03/1.2) of randomly generated partitions. The average overall error was 0.1 with the average error within the lumpability zone as 0.06.

So we see that lumpability is effectively able to reduce the error to 40% of that which can be expected without using the standard lumping method. Therefore it appears that approximate lumpability for general Markov matrices is more difficult to satisfy than exhibited by the examples explored in combat modeling.

22.7 LUMPING OF DEVS MARKOV MODELS TO STUDY SPEEDUP IN MULTIPROCESSOR COMPUTATION

22.7.1 BASE MODEL OF MULTIPROCESSOR

In this example we illustrate how lumping of DEVS Markov models can be used to understand some strengths and limitations of using networked clusters of processors to speedup computations. We outline a base model that takes into account communication among processors by including a physical medium that constrains messaging to a single channel with waiting and access protocol. The protocol could take on various forms such as a first in/first out queue, random selection of waiting processors or others. For reference, the simulation study in Chapter 19 of Zeigler and Sarjoughian (2017) describes the model family and reports simulation results. *Here our emphasis is to show how to employ relatively simple and analyzable lumped models to throw light on the behavior of more complex, simulation-oriented base models.* Together the models render an understanding of how the base model's performance depends on its parameter values. In particular, the base-lumped model pair allows us to get a handle on the communication time dependence on number of processors, N and its effect on the possible increase in computation speed occasioned by a multiprocessor cluster.

Each processor in Fig. 22.8 cycles between two states, *active* and *wait*. In *active* it takes a time with mean *CompTime* to compute its current part of a job and upon completion issues a request for more

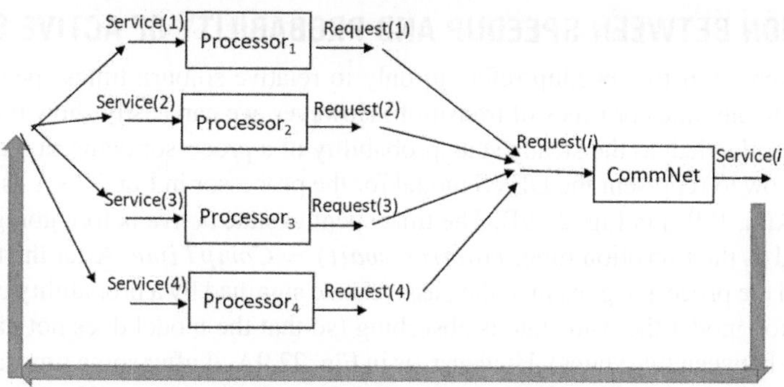

FIGURE 22.8

Base model: DEVS coupled Model of Multiprocessor.

communication services and waits to receive an acknowledgment to proceed. Upon getting a service response, it transitions to ***active*** and the cycle continues until all processors have finished their parts of the job. We assume that a request from a processor to the controller takes zero time but that once issued, a service (sending of messages in the distributed case or access to common memory in the shared case) takes *CommServTime*.

The base model is a coupled model of DEVS Markov components and by closure under coupling (Section 21.5) is representable as a DEVS Markov model. *This is important since we want to study the effect of variance as an independent variable in the computation times of the processors (Zeigler et al., 2015) and this exceeds the capabilities of DEVS CTM models (Chapter 21).*

Since each of N processors gets an equal part the job to do and takes mean time *CompTime* to complete it, N***CompTime* is the expected time taken in sequential processing. If all operate independently the expected time taken for parallel processing is *CompTime*. Thus

$$\begin{aligned}Speedup &= sequential\ time/parallel\ time\\ &= \frac{N*CompTime}{CompTime}\\ &= N\end{aligned}$$

Now taking the communication service into account, we see in Fig. 22.8 that a processor must wait its turn in the queue while other waiting processors are served before proceeding to its active state. Thus we must add *CommTime*, the **total** service time, to the parallel processing time attributed to each processor, so that

$$\begin{aligned}Speedup &= sequential\ time/parallel\ time\\ &= \frac{N*CompTime}{CompTime+CommTime}\end{aligned}$$

So the speedup relative to the maximum possible is:

$$\begin{aligned}RelativeSpeedup &= \frac{Speedup}{N}\\ &= \frac{CompTime}{CompTime+CommTime}\end{aligned}$$

22.7.2 RELATION BETWEEN SPEEDUP AND PROBABILITY OF ACTIVE STATE

Notice that the expression for speedup refers to only to relative sojourn times spent in working and waiting states not to any rates or times of transition. However, we can easily show that in steady state, *RelativeSpeedup* is identical to the steady state probability of a processor being in the *Active* state.

First let's see how to represent the DEVS model for the processor in Fig. 22.9A as a DEVS Markov model (Ahn and Kim, 1993) in Fig. 22.9B. The time spent in state *active* before going to the wait state can be represented as the transition time, $\tau(active, wait) = CompTime$. After the transition the Request is issued and the processor goes in to the passive state signified by a probability of transition equal to one. As a Markov model the wait state is absorbing (so that the model does not have a steady state behavior shuttling between two states). However, as in Fig. 22.9A, if after some time, a service input arrives, and then the model transits to the active state and the cycle can repeat. We can represent this time as a random variable and identify it as the transit time $\tau(wait, active) = CommTime$ in Fig. 22.9B.

Note that we don't know at this point whether this random variable is exponentially distributed as would be required for a CTM but we still can proceed with the model as a DEVS Markov model. To do so we assign unity probabilities to the two transitions. Then following the procedure in Section 22.8, we have the steady state equations as:

$$P(active) = 1 * P(wait)$$
$$P(wait) = 1 * P(active)$$

With the obvious solution that the two probabilities are equal to 1/2. Furthermore, accounting for sojourn times we can show that the relative speedup equals the probability of being active:

$$\begin{aligned} P_{Active} &= \frac{\frac{1}{2}CompTime}{\frac{1}{2}CompTime + \frac{1}{2}CommTime} \\ &= \frac{CompTime}{CompTime + CommTime} \\ &= RelativeSpeedup \end{aligned}$$

Exercise 22.17. Show that you can get the same result by applying CTM theory. However, you have to assume the exponential distribution for the service transition and you have convert the transition times

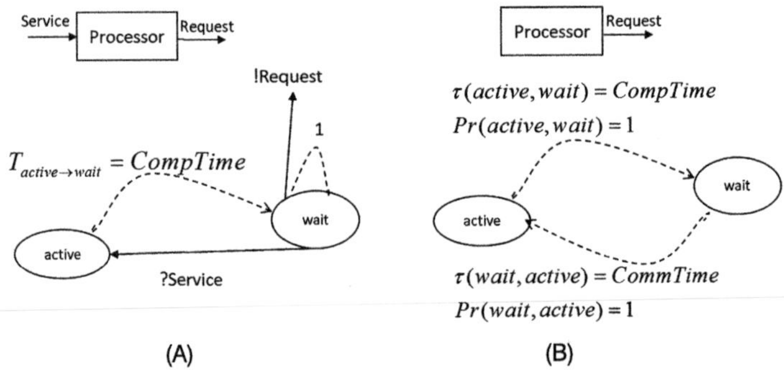

FIGURE 22.9

DEVS Markov Model of Processor.

into rates as well. Hint: show that you get:

$$P_{Active} = \frac{CommRate}{CommRate+CompRate}$$
where
$$CommRate = \frac{1}{CommTime}, \; CompRate = \frac{1}{CompTime}$$
Thus
$$\begin{aligned} P_{Active} &= \frac{1/CommTime}{1/CommTime+1/CompTime} \\ &= \frac{CompTime}{CompTime+CommTime} \\ &= RelativeSpeedup \end{aligned}$$

22.7.3 LUMPED MODEL OF MULTIPROCESSOR

It turns out that an informative path to development of a lumped model follows from partitioning the states of the base model into two regimes as illustrated in Fig. 22.10. In the Normal regime, the communication medium is fast enough (i.e., *CommServTime* << *CompTime*) to keep the queue almost empty and almost all processors active. In this case *CommTime* is close to *CommServTime* and speed up is close to *N*.

In the Congested regime, communication is slow and the queue fills up and can only service a finite number of processors, N_{crit} at any time. We work out N_{crit} in a moment. Since only a finite number of processors are active at any time, a processor must contend with for $N - N_{crit}$ others that are also waiting so that $CommTime = (N - N_{crit})*CommServTime$ increases with *N* so that relative speed up goes to zero as *N* increases. The switching point, N_{crit}, occurs when the arrival rate of requests to the queue just exceeds the service rate – which is where standard queueing theory (itself Markov-based) predicts that the queue size grows without bound.

The components of the lumped model are Markov Matrix models of the processors and of the communication network (Fig. 22.11). The lumped models of the processors each have a probability of being active, P_{active} computed in steady state. The major assumption in constructing the lumped

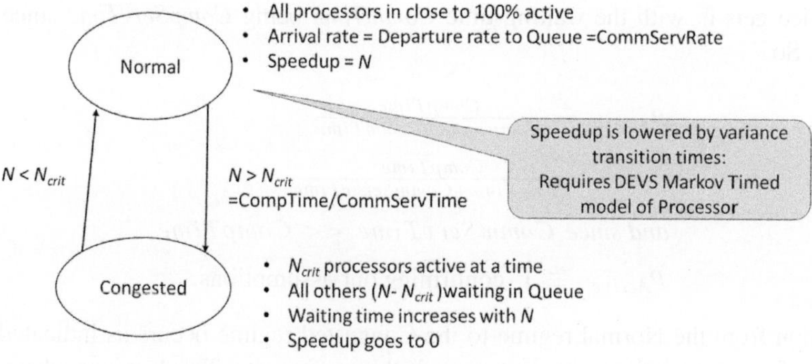

FIGURE 22.10

Multi-regime representation derived from lumped model.

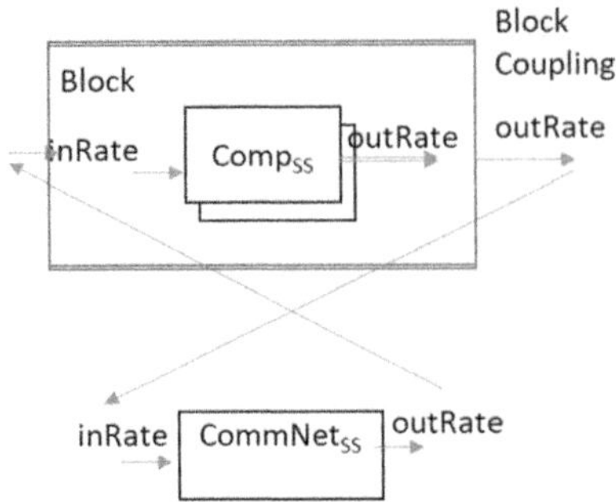

FIGURE 22.11

Grouping of components in the lumped model.

model is that in steady state, the number of active processors is:

$$N_{active} = N * P_{active}$$

To justify this assertion we require 1) uniformity of structure, i.e., all processors have the same structure, and 2) that the transitions of the components are sufficiently mixed that there are no permanent deviations from the current probability value of any one component (see Chapter 16).

Now each active processor outputs a service request so that the arrival rate to the *CommNet* component is $N_{active} * CompRate$.

Now we assume uniformity of distribution of the CommNet output to the processors. That is, there is no priority for processors in the underlying queueing discipline (but see the Appendix where priorities are considered).

In the ***Normal*** regime, we assume all processors are active and we will show that this is a consistent solution under the conditions that *CommServTime << CompTime.*

Let $P_{active} \cong 1$ *and the arrival rate is* $N * CompRate$. By uniformity of output, each processor expecting service gets it, with the waiting time *CommTime* being *CompServTime* since there are no others waiting. So

$$\begin{aligned} P_{Active} &= \frac{CompTime}{CompTime+CommTime} \\ &= \frac{CompTime}{CompTime+CommServeTime} \end{aligned}$$

and since $CommServTime << CompTime$

$P_{Active} \cong 1$ confirming our assumptions.

The transition from the Normal regime to the Congested regime occurs as indicated above when the arrival rate of requests to the queue just exceeds the service rate. This happens where

$$N * CompRate \geq CommServRate$$

In terms of times,

$$\frac{N}{CompTime} \geq \frac{1}{CommServTime}$$

$$N * CommServTime \geq CompTime,$$

$$i.e., \ N_{crit} = \frac{CompTime}{CommServTime}$$

In the ***Congested*** regime, we assume that N_{crit} processors are active at any time. In this case, we can show that the probability of the active state goes to zero:

$$P_{Active} \rightarrow 0 \, with \, N \rightarrow \infty.$$

Indeed, after going to the wait state, an active processor must wait for the other $N - N_{crit}$ processors to be served and its *CommTime* is $(N - N_{cri})$ **CompServeTime*. Thus

$$P_{Active} = \frac{CompTime}{CompTime+CommTime}$$

$$= \frac{CompTime}{CompTime+(N-N_{crit})*CommServeTime}$$

$$\rightarrow 0 \text{ with } N \text{ increasing.}$$

Recall that relative speedup is given by the probability of a processor being in the active state in steady state. The above analysis predicts that as the number of processors increases from one to the critical value, speed up increases linearly (a statement of Amdahl's law (Zeigler et al., 2015)). Beyond the critical value, the speed up decreases to zero. Note that realistically the computation cycle speed (1/CompTime) has to be much smaller than the network bandwidth (1/CommServTime) and therefore the critical point for network saturation (N_{crit}) should be very large. Simulation results described in Zeigler and Sarjoughian (2017) Chapter 19 verifies these predictions and gives information about the region around the saturation value.

22.7.4 TAKING COMPONENT IDENTITY INTO ACCOUNT

An interesting elaboration of the formulation just given is the case where communication bandwidth given to processors may vary. In this case the components of the base model do not satisfy the identity-independence requirement for interacting with the network so that we can't directly apply strict lumping theory. We have seen that one way to handle departure from such uniformity is through the approximation morphism which assesses the effect of such structural deviation on the lumped model's behavior in relation to that of the base model. However, in the case of processor prioritization the departure from uniformity is systematic and not amenable to higher order approximations. Nevertheless, in this case, as the Appendix shows, there is a natural way to develop the lumped model that makes it a good predictor of base model behavior.

22.8 SUMMARY

Building on the background of Chapter 17 and 21 we applied lumping concepts to DEVS Markov models. The main conditions such as uniformity of structure in blocks and identity-independence in block-to-block interaction were shown also apply to DEVS Markov models. Moreover, these conditions were shown to support powerful morphisms and approximate morphisms for such models. This theory lays the basis for application of the computational approach to evaluating approximate morphism error dependence on deviation from strict lumpability requirements. This makes it possible to quantify and test the extent to which aggregations and simplifications of simulations may be usefully employed as discussed in Chapter 2 of "Model Engineering for Simulation" (Zhang et al., 2018). The chapter closed with an example that showed how to employ relatively simple and analyzable lumped models to throw light on the behavior of more complex, simulation-oriented base models. We employed a base-lumped model pair to characterize how the possible increase in computation speed occasioned by a multiprocessor cluster depends on its number of processors. This example also illustrates how it may be possible to construct usable lumped models in cases where the departure from uniformity in base model structure is systematic and not amenable to higher order approximations. This provides a single instance but begs the question of how the theory can be extended more generally to new kinds of base-lumped model constructions.

APPENDIX 22.A PRIORITIZED COMMUNICATION IN MULTIPROCESSOR CLUSTERS

Here we consider the case where communication bandwidth given to processors may vary. For example, let processors be indexed in an order, $1, 2, 3, \ldots$, and consider that service time, *CommServTime* increases with increasing index. Recall that completing the job requires that all processor complete their parts, the slowest communication determines the speedup as in:

$$\begin{aligned} Speedup &= sequential\ time/parallel\ time \\ &= \frac{N*CompTime}{\max_{n\in[1,N]}\{CompTime(n)+CommTime(n)\}} \end{aligned}$$

where n is the index of the processor in the order. In view of Fig. 22.12, $CommTime$ increases with n but $CompTime$ does not. So

$$RelativeSpeedup = \frac{Speedup}{N} = \frac{CompTime}{CompTime + CommTime(N)} = \frac{1}{1 + \frac{CommTime(N)}{CompTime}}$$

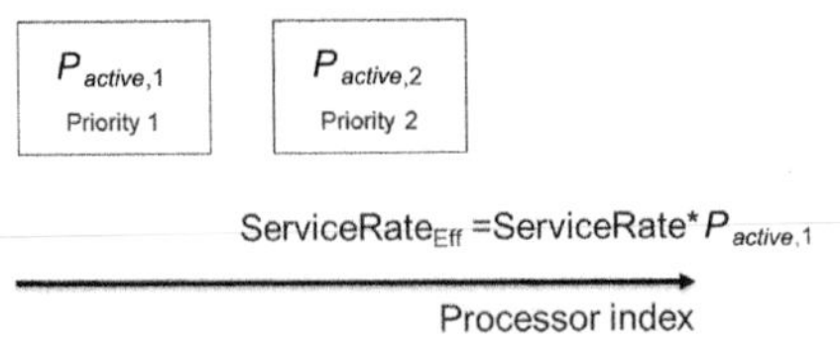

FIGURE 22.12

Effective service rate reduction with increasing index.

Assuming that communication time is much smaller than computing time. $RelativeSpeedup \cong 1 - \frac{CompTime}{CommTime(N)}$ i.e., speed up is now determined by the communication service delay that the processor with smallest priority experiences.

We can extend the lumped model to compute the communication delays experienced by the processors in iterative fashion following Bobbio et al. (2016). Since a processor can only employ the communication medium if all processors with higher priority are busy with their computation the interaction of components can be handled in a serial manner. Fig. 22.12 illustrates the principle: Suppose the first processor has been shown to have probability P_{Active} of being in state active so is busy computing and not trying to send messages. This leaves the medium free for second processor. Then the latter's service rate is effectively reduced by P_{Active} because it only gets to use the communication medium with that probability. Using this effective rate we can employ the Markov steady state result to compute the probability of being in state active for the second processor. By iteration we proceed to compute the probabilities of being active and effective service rates of the processors in order of high to low priority.

Exercise 22.18. Show how to carry out the computation process sketched just now.

Suppose that there are groups of processors of size N_i for each priority index, i. Then the group with highest priority never waits for others, what is its speedup?

Under what circumstances might it make sense to prioritize processors to achieve speedup?

REFERENCES

Ahn, M.S., Kim, T.G., 1993. Analysis on steady state behavior of DEVS models. In: AI, Simulation, and Planning in High Autonomy Systems Conference. IEEE.

Bobbio, Andrea, Cerotti, Davide, Gribaudo, Marco, Iacono, Mauro, Manini, Daniele, 2016. Markovian agent models: a dynamic population of interdependent Markovian agents. In: Seminal Contributions to Modeling and Simulation. Springer.

Zeigler, B.P., 2018. Simulation-Based Evaluation of Morphisms for Model Library Organization. Chapter 2 in Zhang et al. (2018). Elsevier.

Zeigler, B.P., Sarjoughian, H.S., 2017. Guide to Modeling and Simulation of Systems of Systems. Springer-Verlag, London.

Zeigler, Bernard P., Nutaro, James J., Seo, C., 2015. What's the best possible speedup achievable in distributed simulation: Amdahl's law reconstructed. In: DEVS TMS. SpringSim.

Zeigler, B.P., Nutaro, J.J., Seo, C., 2017. Combining DEVS and model-checking: concepts and tools for integrating simulation and analysis. International Journal of Simulation and Process Modelling 12 (1), 2–15.

Zhang, L., Zeigler, B., Laili, Y., 2018. Model Engineering for Simulation. Elsevier.

CHAPTER

SPIKING NEURON MODELING – ITERATIVE SPECIFICATION 23

CONTENTS

This chapter provides an example of developing a general system model for a real system using abstraction and iterative system specification. A neuron is a highly complex electro-chemical system that can be abstracted to a DEVS that represents the input/output behavior that, in turn, allows it to interact with other neurons in a brain network. The iterative specification captures bursting and packets of spikes that may be the way neurons synchronize to accomplish cooperative behavior (Muzy et al., 2017).

23.1 A BIOLOGICAL NEURON AS A DYNAMIC SYSTEM

Fig. 23.1 depicts a single biological neuron. Most commonly, inputs from other neurons are received on *dendrites*, at the level of *synapses*. The circulation of neuronal activity (electric potentials) is due to the exchange through the neuron *membrane* of different kinds of ions. Dendrites integrate locally the variations of electric potentials, either excitatory or inhibitory, and transmit them to the *cell body*. There, the genetic material is located into the *nucleus*. A new pulse of activity (an *action potential*) is generated if the local electric potential reaches a certain threshold at the level of the *axon* hillock, the small zone between the cell body and the very beginning of the axon. If emitted, action potentials continue their way through the axon in order to be transmitted to other neurons. Action potentials, once emitted, are all or nothing phenomena: 0, 1. The propagation speed of action potentials can be increased by the presence of a *myelin* sheath, produced by *Schwann cells*. This insulating sheath is not continuous along the axon. There is no myelin at the level of the *nodes of Ranvier*, where ionic exchanges can still occur. When action potentials reach the tip of the axon, they spread over all *terminals* with the same amplitude, up to synapses. The neuron can then communicate with other following neurons. Notice that a focus on electrical signals (without dealing with chemical signals) is taken here.

Theory of Modeling and Simulation. https://doi.org/10.1016/B978-0-12-813370-5.00034-1

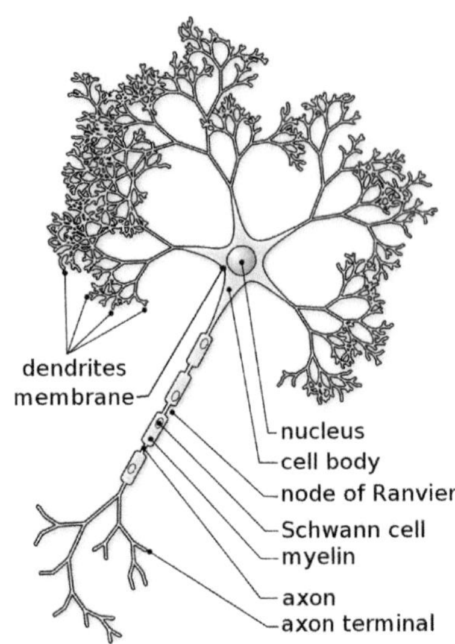

FIGURE 23.1

Sketch of a biological neuron (adapted from http://fr.wikipedia.org/wiki/Neurone).

Fig. 23.2 describes the continuous potential propagation through the dendrite (input), the soma (locus of potential integration) and the axon (output) of a neuron. A neuron can be described as an input-output system with dendrite PostSynaptic Potential (PSP) as input, soma membrane potential as state and axon potential as output. Segmentation is represented by vertical lines.

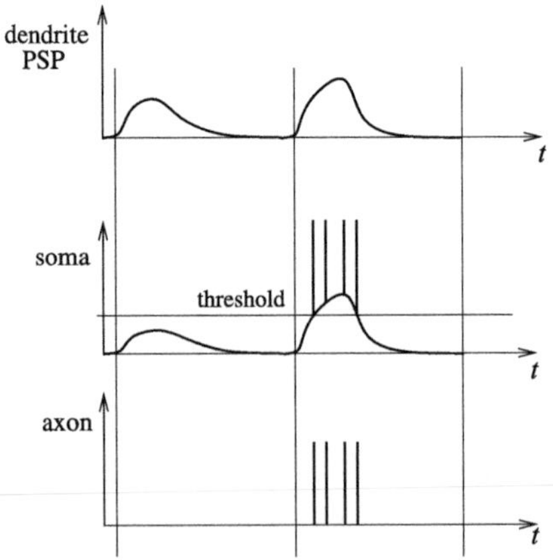

FIGURE 23.2

Continuous potential propagation through the dendrite, the soma and the axon of a neuron.

The use of *dynamical systems* in neurosciences (Izhikevich, 2007) concerns mainly continuous generations of spikes (Izhikevich, 2003; Maass, 1997; Grinblat et al., 2012; Mayrhofer et al., 2002). The iterative specification of systems to be presented here aims at being more general and applicable to any discrete or continuous (neuronal) system or network.

23.2 DISCRETE EVENT MODELING OF A LEAKY INTEGRATE AND FIRE NEURON

Discrete event spiking neurons have been widely implemented in several software environments (Brette, 2007). From the neuronal nets perspective, the Discrete Event System Specification (DEVS) formalism has been used mainly for proposals of novel neuron models (Zeigler, 2005), the specification of dynamic structure neurons (Vahie, 2001), the abstraction of neural nets (Zeigler, 1975) and for the specification of continuous spike models (Mayrhofer et al., 2002, Grinblat et al., 2012). In addition to DEVS representations, discrete events have been used successfully in neuronal nets for modeling (Tonnelier et al., 2007, Brette, 2007) and simulation (Hines and Carnevale, 2004, Tang et al., 2013, Brette, 2007, Mouraud et al., 2005).

The goal of developing a discrete event model hereafter is to propose a model, as simple as possible, capturing the essence of discrete event characteristics in a leaky integrate and fire (LIF) neuron application. Spikes here do not explicitly model variations in potential. Rather they are abstracted into event segments, i.e. eventually generators for an iterative specification. A LIF neuron consists of a memory-based model storing the potentials previously received in the membrane (especially here in the soma, thus simply referring to the "soma membrane potential" as the "membrane potential" hereafter). An interesting aspect is that the remaining potentials decrease in time due to leakage leading to nonlinearities in neural behaviors. Corresponding simple discrete event models discussed here provide a basis that can be easily extended to deal with further details (such as multiple inputs, synaptic weights, etc.). Fig. 23.3 depicts the dynamics of a basic discrete event model of a leaky integrate and fire neuron.

A discrete event input is a couple (x_i, t_i), with $x_i = 1$ and $t_i \in \mathbb{R}^+_\infty$ respectively the value and the time stamp of the discrete event. Discrete event inputs occur in a sequence $(x_0, t_0), (x_1, t_1), \ldots, (x_i, t_i), \ldots, (x_n, t_n)$. The same characterization holds for each discrete event output (y_i, t_i).

Based on the current value of membrane potential m, the new value m' consists of:

$$m' = \begin{cases} r^e m + 1 & if\, m < \theta \\ 0 & otherwise \end{cases} \tag{23.1}$$

with $e \in \mathbb{R}^+_\infty$ the *elapsed time since the last transition, initial membrane potential* is $m = 0$, *new membrane potential* m' depending on: *remaining potential* $r^e m$, with $r \in [0, 1]$ the *remaining coefficient*,[1] and $\theta \in \mathbb{R}^+$ the *firing threshold.*

Discrete event neuron models consider only membrane potential transitions when receiving an external event (an input spike) or when having scheduled an internal event (the membrane potential going

[1]When *remaining coefficient* value $r = 1$, there is *no leak*, all the potential received remains in the membrane. When $r = 0$, all the potential received at time $t - 1$ is lost (the model is then equivalent to McCulloch & Pitts' model).

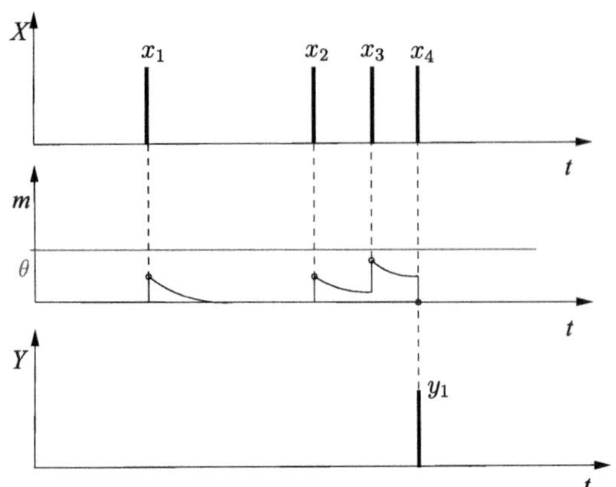

FIGURE 23.3

Discrete event model of a leaky integrate and fire neuron.

to zero because of the potential leak). Also, notice that both change of potential membrane value and firing are immediate (without the commonly employed "artificial" discrete time delay of one unit). For membrane potential, this means that membrane potential is not updated (considering it is not changing significantly) between two discrete events.

Spike value emission y depends on *threshold* $\theta \in \mathbb{R}^+$:

$$y = \begin{cases} 1 & if\, m \geq \theta \\ \phi & otherwise \end{cases} \tag{23.2}$$

23.3 MULTI-LEVEL ITERATIVE SPECIFICATION

Our purpose is to model spiking neurons (cf. Fig. 23.4) using the iterative specification formalism. "Spikes" are concatenated together to form high-level "packet" generators.

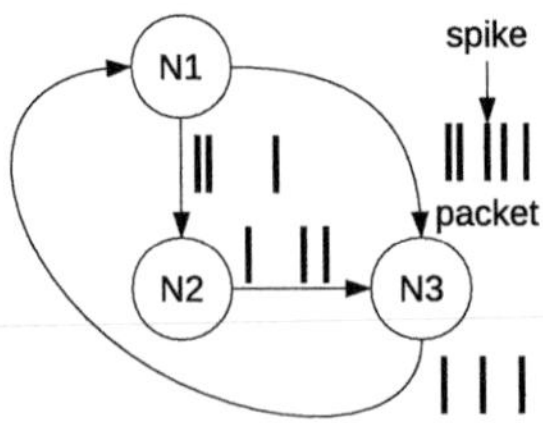

FIGURE 23.4

Neurons N1, N2 and N3 exchanging spikes and packets.

Fig. 23.5 presents the corresponding multi-level state transition diagram of an iterative specification of a biological neuron. Fig. 23.6 presents the corresponding trajectories. A neuron is modeled here as exchanging packets (of spikes) with other neurons. When the neuron is exchanging packets it is assumed to be "synchronizing" with other neurons.

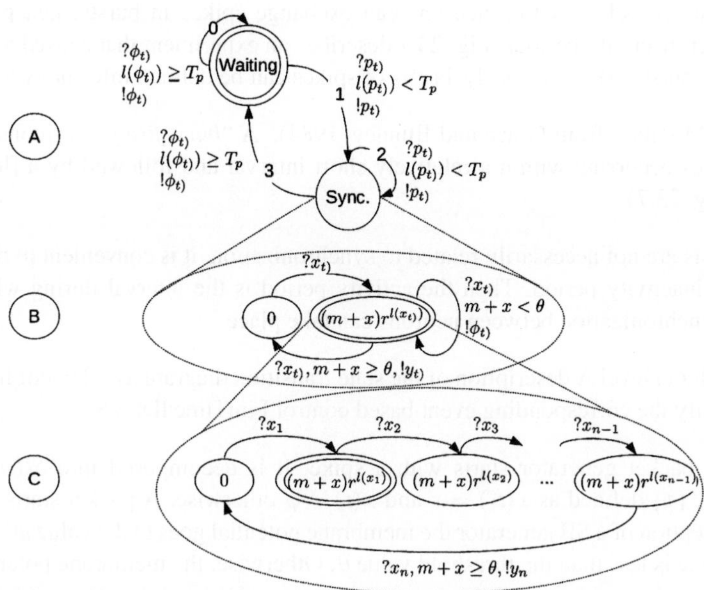

FIGURE 23.5

Multi-level state transition diagram of iterative specification of LIF neurons with: Level A the composition of packet generators $p_{t)}$, level B: the "compressed" spike initiated interval (SII) generators $x_{t)}$, and level C "decompressed" SII generators occurring in a sequence $x_1, x_2, ..., x_k, ..., x_n$.

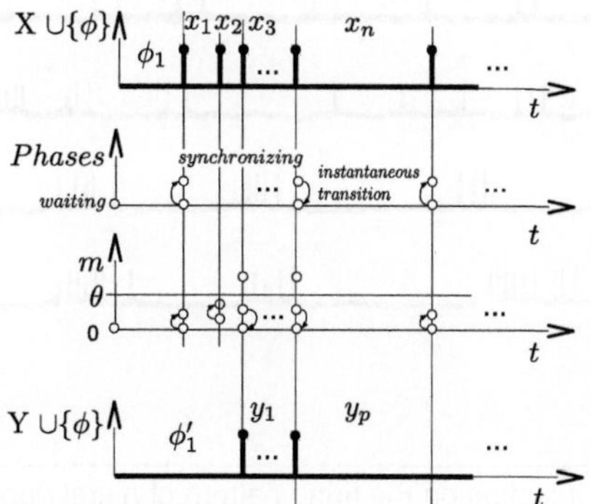

FIGURE 23.6

Trajectories of synchronizing LIF neurons.

In Fig. 23.5, at level A, when the neuron receives no spikes (i.e., null event generators $\phi_{t\rangle}$ of length greater than T_P) it remains in phase "Waiting". When the neuron receives a first generator $p_{t\rangle}$ of length less than T_P, it goes to the "Synchronizing" phase. The neuron receives and sends then a packet generator. At this level, as we will see, a packet generator is conceived as a composition of non-instantaneous generators.

It is a well known behavior that neurons can exchange spikes in bursts, i.e., packets of spikes followed by longer inactivity periods. Fig. 23.7 describes an experiment that caused a single neuron to generate spikes in bursts. More precisely, bursts of spikes can be defined informally as follows.

Definition 23.1 (Modified from Grace and Bunney, 1984)**.** A "*burst firing* (...) [consists] of trains of two or more spikes occurring within a relatively short interval and followed by a [longer] period of inactivity" (cf. Fig. 23.7).

Although bursts are not necessarily related to synchronization, it is convenient to think of the waiting phase as the inactivity period. Then the activity period is the interval during which packets are exchanged and synchronization between neurons can take place.

Exercise 23.1. Model level A description of the state transition diagram as a TimedFinIterSpec. Based on the latter, specify the corresponding event based control FiniTimedDEVS.

At level B, a packet generator starts with a spike. It is decomposed into *SII input generators* $x_{t\rangle} : [t_1, t) \rightarrow X \cup \{\phi\}$ defined as $x(t_1) = x$ and $x(t) = \phi$ otherwise. A packet starts and ends with a spike. At each reception of a SII generator the membrane potential goes to the value $m' = (m + x)r^{l(x_{t\rangle})}$, as long as this value is less than the threshold value θ. Otherwise, the membrane potential is reset to 0. At level C, each *non-instantaneous generator* consists of $x_k : [l(x_{k-1}), l(x_k)) \rightarrow X \cup \{\phi\}$ occurring in a sequence $x_1, x_2, ..., x_k, ..., x_n$. When the neuron receives a non-instantaneous generator x_k it achieves an instantaneous transition followed by a non-instantaneous transition. The different transition functions are detailed in the text below.

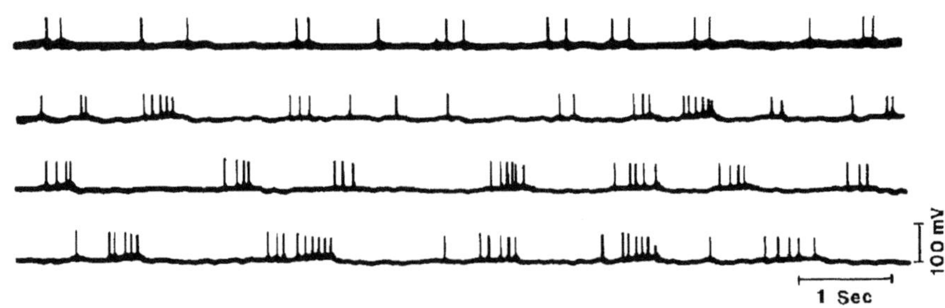

FIGURE 23.7

Burst firing: "Effect of intracellular injection on the firing pattern of nigral dopamine (DA) cells. In the first few minutes following impalement with a calcium-containing electrode, the stabilized DA cell demonstrates its typical slow, single spike firing pattern (top trace). As calcium leaks from the electrode into the cell, the pattern slowly changes over the next 10 to 20 min into a burst-firing pattern (second through fourth trace)" (from Grace and Bunney, 1984).

23.4 ITERATIVE SPECIFICATION MODELING OF SPIKY NEURONS

The previous discrete event model provides an introduction to the discrete event dynamics of a LIF neuron. The model is considered here from an iterative system specification perspective, ensuring the global system properties as defined in Section 12.2. Fig. 23.8 depicts the segment-based description. Notice that due to exponential decay, a SII segment lasts indefinitely until a next SII segment.

Definition 23.2. The translation of a LIF model M_S into a *spike iterative specification* $G(M_S)$ of a system (Level C and B of Fig. 23.5) consists of:

$$G(M_S) = (\delta_S, \lambda_S)$$

where for each generator $\omega_i \in \Omega_G^S$, $dom(\omega_i) = [0, t_i) \cap \mathbb{R}$ (generators start at time zero[2]), $X_S = Y_S = \{1, \phi\}$, $Q_S = \{m \mid \mathbb{R}^+\}$, $\Omega_G^S = \Omega_G^X \cup \Omega_G^\Phi$ is the *set of input generators* where $\Omega_G^X = \{x_i \mid x_i(0) = 1 \wedge x_i(t) = \phi \; for \; t \in (0, t_i]\}$ is the *set of SII generators*, $\Omega_G^\Phi = \{\phi_i \mid \phi_i : [0, t_i) \to \{\phi\}\}$ is the *set of non event inputs*.

For each input $x_i \in \Omega_G^S$, the *spiky transition function* consists of

$$\delta_S(m, x_i) = \begin{cases} 0 & if \; m + x \geq \theta \\ (m + x)r^{t_i} & otherwise \end{cases},$$

where the top part of the equation corresponds to a spike emission and the bottom part corresponds to an update of membrane potential with input spike addition. For each null value input $\phi_i \in \Omega_G^S$,

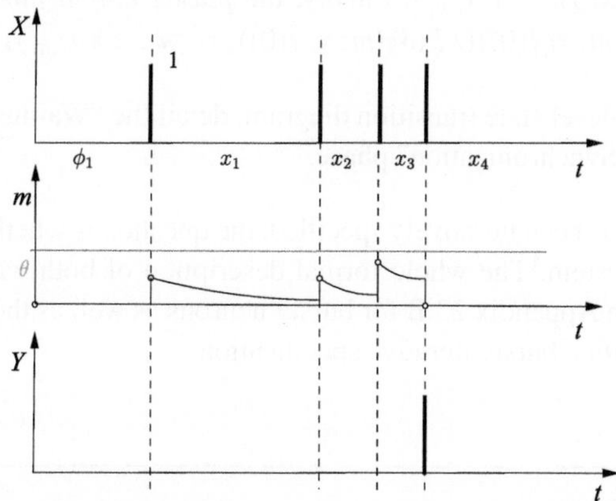

FIGURE 23.8

Discrete event model of a leaky integrate and fire neuron. Notice that state computations occur only at the beginning and at the end of segments.

[2]Remember from Section 5.6.2 that generators can be translated to start from time zero.

the *non-instantaneous transition function* consists of $\delta_S(m, \phi_i) = mr^{t_i}$ which corresponds to a non-instantaneous membrane potential decrease when no input spike is received.

Finally, the *spiky output function* $\lambda_S : Q \times \Omega_G^X \to Y$ consists of $\lambda_S(m, x_i) = \lambda_S(\delta_S(m, x_i(0)), \phi_i))$ defined applying first the instantaneous transition $\delta_S(m, x_0) = m'$ and after

$$\lambda_S(m', \phi_i) = \begin{cases} 1 & if\, m' \geq \theta \\ \phi & otherwise \end{cases}.$$

The proof that the SII iterative specification can be associated with a system is provided in Appendix 23.A.

Definition 23.3. The translation of a LIF model M_S into an *packet iterative specification* $G(M_P)$ of a system (Level A of Fig. 23.5) consists of:

$$G(M_P) = (\delta_P, \lambda_P)$$

where for each generator $\omega_i \in \Omega_G^P$, $dom(\omega_i) = [0, t_i) \cap \mathbb{R}$, $X_P = Y_P = \{1, \phi\}$, $Q_P = \{m \mid \mathbb{R}^+\}$, $\Omega_G^P = \Omega_G^X \cup \Omega_G^\Phi$ is the *set of input generators* where $\Omega_G^P = \{p_i \mid p_i(t_j) = 1\, for\, all\, t_j \in [0, t_i) \wedge p_i(t) = \phi\, elsewhere\}$ is the *set of packets*, $\Omega_G^\Phi = \{\phi_i \mid \phi_i : (0, t_i] \to \{\phi\}\}$ is the *set of non event segments*. A packet $p_i \in \Omega_G^P$ starts and ends with spikes and in between it is a concatenation of non-instantaneous generators: $p_i = x_{i,1} \bullet x_{i,2} \bullet \ldots \bullet x_{i,n_i}$. The packet transition function consists of the sum of extended spiky transitions $\delta_S^+(m, x_{i,1} \bullet x_{i,2} \bullet \ldots \bullet x_{i,n_i})$ such that $\delta_P(m, p_i) = \delta_S^+(m, x_{i,1} \bullet x_{i,2} \bullet \ldots \bullet x_{i,n_i}) = \delta_S(\delta_S(\ldots \delta_S(\delta_S(m, x_{i,1}(0)), x_{i,2}), \ldots), x_{i,n_i})$. Finally, the *packet output function* $\lambda_p : Q \times \Omega_G^P \to Y$ consists of $\lambda_P(m, p_i) = \lambda_S(m, x_{i,1}(0))(\lambda_S^+(\delta_S(m, x_{i,1}(0)), x_{i,2} \bullet \ldots \bullet x_{i,n_i}))$.

Exercise 23.2. In the multi-level state transition diagram, detail the "Waiting" phase at levels B and C as done previously for the "Synchronization" phase.

Now that a LIF neuron has been iteratively specified, the question is whether this specification leads to a well-defined general system. The whole formal description of both waiting phase and synchronization phase is provided in Appendix 23.B for bursty neurons as well as the proof that a well-defined system can be associated with a bursty iterative specification.

23.5 SUMMARY

This chapter presented a multi-level approach to describe high-level generators (bursts) based on basic generators (spikes). The graphical state transition representation allows following the sequence of both state transitions and input generators. Using such an approach dynamical modules can be composed and combined to study formally the combination of basic iterative specifications at structural level. For example, bursty neurons can be combined to study the synchronization of populations of neurons to achieve biological functions in the brain (Aviel et al., 2003).

APPENDIX 23.A ITERATIVE SYSTEM SPECIFICATION OF A SPIKING NEURON

Theorem 23.1. *A spiky iterative specification $G(M_S) = (\delta_S, \lambda_S)$ can be associated to a system $S_{G(M_S)} = (\Delta_S, \Lambda_S)$ through a concatenation process $COPRO_{G(M_S)} = (\delta_S^+, \lambda_S^+)$.*

Proof. Sufficient conditions for iterative specification (cf. Theorem 12.3) can be checked as:

1. Considering SII input segments as a subset of all discrete event segments, i.e., $\Omega_G^S \subset \Omega_G^E$, where each input segment starts with initial value zero or one, the set of generated input segments Ω_G^{S+} is *strongly proper* being based on condition $c(\omega_{t)}) = \begin{cases} true & if\, n_E \le 1 \\ false & otherwise \end{cases}$, for left segment $\omega_{t)} = x_i(t)$, with the set of input generators $\Omega_G^S = \{x_i \mid x_i(0) = 1 \wedge x_i(t) = \phi \;\; for \;\; 0 < t < t_i\}$.
2. *Consistency of composition*: The *single segment state transition function* for a SII input segment $x_i \in \Omega_G^S$ is defined as $\delta_S(m, x_i) = \begin{cases} 0 & if\, m + x \ge \theta \\ (m+x)r^{t_i} & otherwise \end{cases}$, where the top part of the equation corresponds to a spike emission and the bottom part corresponds to an update of membrane potential with input spike addition. The output function consists of $\lambda_S(m, x_i) = \begin{cases} 1 & if\, m + x \ge \theta \\ \phi_i & otherwise \end{cases}$. Now consider that each SII generator $x_i \in \Omega_G^S$ can be decomposed into basic discrete event generators: (i) an event input generator $z_i \in \Omega_G^Z$ such that $\Omega_G^Z = \{z_i \mid z_i(0) = 1\}$, and (ii) a *non event input generator* $\phi_i \in \Omega_G^\Phi$ *such that* $\Omega_G^\Phi = \{\phi_i \mid \phi_i : [0, t_i) \to \{\phi\}\}$. It can be shown that these atomic generators can be *concatenated* to define recursively the single segment state transition function δ_S of the spiky iterative specification $G(M_S) = (\delta_S, \lambda_S)$, and then both extended state transition function δ_S^+ and extended output function λ_S^+ of the concatenation process $COPRO_{G(M_S)} = (\delta_S^+, \lambda_S^+)$:
   (a) For a null event input generated segment $\omega \in \Omega_S^+$, e.g.:

   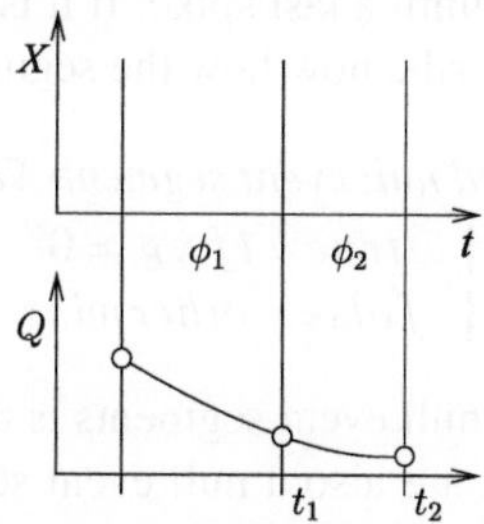

   considering an exponential remaining coefficient $r^t = exp(-\alpha t)$, with α a constant, null event input generators, $\phi_1, \phi_2 \in \Omega_G^\Phi$ can be concatenated while membrane potential computation remains true, i.e., $\delta_S^+(m, \phi_1 \bullet \phi_2) = \delta_S(\delta_S(m, \phi_1), \phi_2)$ or through membrane potential $m' = \delta_S^+(m, exp(-\alpha(t_1 + t_2))) = \delta_S(\delta_S(m, exp(-\alpha t_1)), exp(-\alpha t_2))$.
   (b) For each SII segment $x_i \in \Omega_G^S$, $\delta_S^+(m, x_1 \bullet x_2) = \delta_S(\delta_S(m, x_1), x_2)$, as the single segment state transition function δ_S is recursively defined for a SII input event segment as

$\delta_S^+(m, x_i) = \begin{cases} \delta_S(0, \phi_i) & if\, m + x \geq \theta \\ \delta_S(m + x, \phi_i) & otherwise \end{cases}$, where the top part of the equation corresponds to a spike emission and the bottom part corresponds to an update of membrane potential with input spike addition.

(c) Finally, the extended output function is defined following the general approach given earlier. □

APPENDIX 23.B ITERATIVE SPECIFICATION MODELING OF BURSTY NEURONS

Based on bursts and on the previous definition of basic discrete event generators, we aim to define *bursty LIF neurons* as having the usual LIF properties while being able to exchange, through input-output interfaces, and to compute, bursty segments. A bursty segment consists of packet segments followed by longer null segments. Until now only *packets* (and their duration) have been used formally. Before being able to segment bursty segments we need to precisely define what is a bursty segment in order to be able to identify each subsegment.

Definition 23.4. A *bursty segment* $\omega \in \Omega_B$, can be segmented alternatively into an active segment (with events) followed by an inactive segment (with no events), followed by an active segment,..., leading to a generated segment $\omega^+ \in \Omega_G^{B+}$ such that $\omega^+ = p_1 \bullet \phi_1 \bullet p_2 \bullet \phi_2 \bullet ...$, where ϕ_i *are null event segments* and $p_j = x_{j1} \bullet \ldots$ *are packet segments*: a concatenation of *SII segments* $x_{jk} \in \Omega_G^S$. Bursts follow also two *requirements*: (i) each packet duration $l(p_j)$ is less than a length T_p, i.e., $l(p_j) \leq T_p$, and (ii) each inactivity duration $l(\phi_i)$ is greater than a packet duration, i.e., $l(\phi_i) > T_p$. Notice that active segments are called *packets* to reflect the discrete nature of spikes.

Fig. 23.9 presents an example of bursty segmentation in a bursty neuron. The algorithm starts the segmentation either with a burst segment or an empty segment. If it is a packet segment, the algorithm will concatenate all SII segments until a last spike. If it is an empty segment, the algorithm will search for the first next spike. Let us describe now how the segmentation can be properly achieved.

Lemma 23.1. *The set of generated null event segments* $\Omega_G^{\Phi+}$ *is proper and based on generators* $\Omega_G^{\Phi} = \{\omega | c_\phi(\omega) = true\}$, *with* $c_\phi(\omega) = \begin{cases} true & if\, n_E = 0 \\ false & otherwise \end{cases}$, *indeed* $\Omega_G^{\Phi+} = \Omega_G^{\Phi}$.

Proof. The concatenation of two null event segments is another null event segment. The left and right segments of a null event segment are also a null event segments. Therefore the set of null event generated segments, $\Omega_G^{\Phi+}$, is *strongly proper*. Then the set of null event generators Ω_G^{Φ} is closed under concatenation. This means that the concatenation of a finite number of null event segments is also a null event segment. Therefore the MLS always results in termination with the input segment $\omega \in \Omega$ as the result. □

Now that null event segments can be detected and concatenated, let us detect packets. To do so, the density of segments is defined. This density can be *low* (being lower than a threshold D) or *high* (being above a threshold D). More precisely:

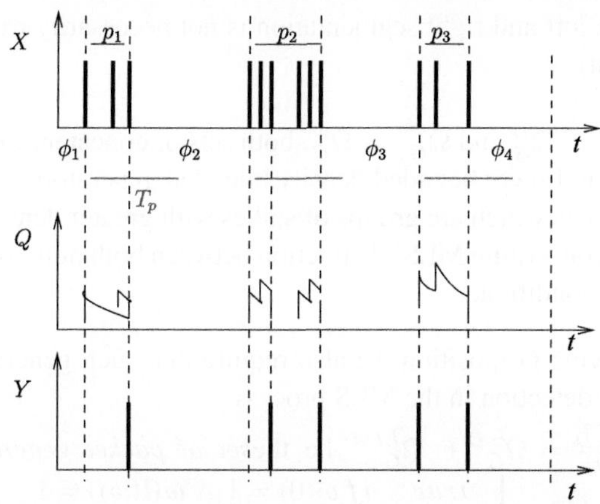

FIGURE 23.9

Bursty segmentation at input, state and output levels.

Definition 23.5. *Density* of a segment $\omega \in \Omega$ consists of $Density(\omega) = \frac{n_E(\omega)}{l(\omega)}$ with $n_E(\omega)$ defined as the number of events in the segment. *Generator set* $\Omega_G^{>D} = \{\omega | c_{>D}(\omega) = true\}$, with

$$c_{>D}(\omega) = \begin{cases} true & if\ Density(\omega) > D \\ false & otherwise \end{cases},$$

is *lower bounded. Generator set* $\Omega_G^{<D} = \{\omega | c_{<D}(\omega) = true\}$, with

$$c_{<D}(\omega) = \begin{cases} true & if\ Density(\omega) < D \\ false & otherwise \end{cases},$$

is *upper bounded.* Generator set Ω_G^D is said to be *density bounded* if it is either upper or lower bounded.

As for the set of null event generators, the set of density bounded generators is closed under concatenation.

Remark 23.1. Density bounded generator sets are *closed under concatenation.* This can be easily shown considering all pairs of generators $\omega_1, \omega_2 \in \Omega_G^D$, such that $n_E(\omega_1) < (or >) Dl(\omega_1))$ and $n_E(\omega_2) < (or >) Dl(\omega_2)$, leading to $n_E(\omega_1) + n_E(\omega_2) < (or >) Dl(\omega_1) + Dl(\omega_2)$. Density bounded generator sets are closed under concatenation.

The set of null event generators, Ω_G^Φ, was strongly proper. Let us see now the propriety of the set of density bounded generators Ω_G^D.

Lemma 23.2. *The set of generated density bounded segments* Ω_G^{D+} *is* weakly proper, *indeed* $\Omega_G^{D+} = \Omega_G^D$.

Proof. By closure under concatenation, the concatenation of a finite number of density bounded segments is also density bounded. Therefore the MLS always results in termination with the input segment as the result. Closure under left and right segmentation is not necessarily true (depending on the uniformity of spike distribution). □

Having shown that $\Omega_G^{D+} = \Omega_G^D$ and $\Omega_G^{\Phi+} = \Omega_G^{\Phi}$, both sets of concatenated generators are finite and thus respects MLS definition. Lower bounded density bounded generators can be employed as the obvious means to represent bursts which are groups of spikes with greater density than their surrounding. Now, the question remains concerning MLS' distinction between both null event and packet generators. This requires adding a new condition:

Remark 23.2. In the following proposition we also require that such generators start and end with a spike to facilitate boundary detection in the MLS process.

More formally, let $\widetilde{\Omega_G^{>D}} = \Omega_G^{>D} \cap \Omega_G^{LIM}$ be the*set of packet generators* with $\Omega_G^{LIM} = \{\omega \mid c_{LIM}(\omega) = true\}$ and $c_{LIM}(\omega) = \begin{cases} true & if\,\omega(0) = 1 \wedge \omega(l(\omega)) = 1 \\ false & otherwise \end{cases}$, then

Proposition 23.1. $\widetilde{\Omega_G^{>D}} \bigcup \Omega_G^{\Phi}$ *is a* weakly proper *generator set.*

Proof. In view of the Lemmas 23.1 and 23.2, it is enough to consider heterogeneous segments of the form, $\omega = \phi_1\omega_1 \bullet \phi_2\omega_2 \bullet ... \bullet \phi_n \bullet \omega_n$. MLS will produce a left segment that ends at the first spike of the first density-bounded segment. The remaining right segment will then have the density-bounded segment as its MLS left segment. The subsequent remaining segment is of the same form subject to repeated MLS. The process terminates because the remaining segment always gets shorter. Note that the requirement for boundary spikes greatly simplifies the MLS segmentation. Without this requirement MLS can stop in the middle of a generator, between two spikes. □

Notice that Ω_G^{LIM} is consistently *weakly proper* (starting with spike is closed under left-segmentation, similarly ending is...).

The proposition gives us the means to generate and uniquely decompose segments containing high density subsegments. However it does not assure that high density segments are separated by long null event segments. One solution to do so is to synthesize generators that have encapsulated high density segments and null event segments of equal lengths. Let $\Omega_G^B = \Omega_G^{\Phi}\widetilde{\Omega_G^{>D}} = \{\omega \mid \omega = \phi \bullet \omega', \phi \in \Omega_G^{\Phi}, \omega' \in \widetilde{\Omega_G^{>D}}, c_L(\phi, \omega') = true\}$ be the set of bursty generators, with $c_L(\phi, \omega') = \begin{cases} true & if\,l(\phi) = l(\omega') \\ false & otherwise \end{cases}$, a generator set of paired concatenated null event and spike-enclosed density lower-bounded segments (or *packets*).

Notice that $\omega = \phi \bullet \omega'$ which is a higher level generator composed of two concatenated generators. Also having both null event and packet segments of equal length has been set for sake of simplicity. This assumption can be easily extended using both length conditions $c_L(\phi) = \begin{cases} true & if\,l(\phi) > T_p \\ false & otherwise \end{cases}$ and $c_{L'}(\omega') = \begin{cases} true & if\,l(\omega') \leq T_p \\ false & otherwise \end{cases}$ with T_p an arbitrary packet length.

Lemma 23.3. Ω_G^B *is a* weakly proper *generator set.*

Proof. The MLS process will detect the first null event segment and continue to the following density-bounded segment, stopping at its end, which marks the end of the first paired null event-density-bounded segment. The remaining right segment, if not a null event, is of the same form as the original (thus leading to a weak proper set). The process terminates because remaining segment always gets shorter. □

With all the previous apparatus, it is now easy to detail the structure of a bursty neuron model as (cf. Level B of Fig. 23.5)

Definition 23.6. The translation of a bursty model M_B into an *iterative specification* $G(M_B)$ of a system consists of:

$$G(M_B) = (\delta_B, \lambda_B)$$

where $dom(\omega_{X_B})$, X_B, Y_B, Q_B are defined as for the spike model, Ω_B^G is the *set of input segment bursty generators* defined with both *single segment state transition* δ_B and *output function* λ_B in following theorem.

Again we need to prove now that the bursty specification leads to a well-defined system.

Theorem 23.2. *A bursty iterative specification* $G(M_B) = (\delta, \lambda)$ *can be associated to a system* $S_{G(M_B)} = (\Delta_B, \lambda_B)$ *through a concatenation process* $COPRO_{G(M_B)} = (\delta_B^+, \lambda_B^+)$.

Proof. Sufficient conditions for iterative specification (cf. Lemma 23.3) can be checked as:

1. Considering bursty input segments as a subset of all discrete event segments, i.e., $\Omega_G^B \subset \Omega_G^E$, they can be properly segmented based on Ω_G^B the *weakly proper generator set* as previously defined (cf. previous lemmas and proposition).
2. *Consistency of composition*: Considering a generated bursty input segment $\omega \in \Omega_G^{B+}$ as a concatenation of spiky segments such that $\delta_B{}^+(m, \omega) = \delta_B{}^+(m, x_1 \bullet x_2 \bullet ... \bullet x_n) = \delta_S^+(\delta_S^+(\ldots \delta_S^+(\delta_S^+(m, x_1), x_2), \ldots), x_n)$. As spiky transitions proved to be composable (cf. Theorem 23.1), i.e., $\delta_S^+(m, x_i) = \delta_S(m + z_i, \phi_i)$, the final membrane potential value is equal to $\delta_B{}^+(m, \omega) = \delta_S^+(m, x_i)$, with generator spiky transition $\delta_S^+(m, x_i)$ described in Theorem 23.1. Finally, the *extended bursty output function* λ_B^+ can also be defined using both the concatenation of extended spiky transition functions and the definition of extended spiky output function λ_S^+. □

REFERENCES

Aviel, Y., Horn, D., Abeles, M., 2003. Synfire waves in small balanced networks. Neurocomputing 58, 123–127.

Brette, R., 2007. Simulation of networks of spiking neurons: a review of tools and strategies. Journal of Computational Neuroscience 23 (3), 349–398. https://doi.org/10.1007/s10827-007-0038-6.

Grace, A., Bunney, B., 1984. The control of firing pattern in nigral dopamine neurons: burst firing. The Journal of Neuroscience 4 (11), 2877–2890.

Grinblat, G.L., Ahumada, H., Kofman, E., 2012. Quantized state simulation of spiking neural networks. Simulation 88 (3), 299–313.

Hines, M., Carnevale, N., 2004. Discrete event simulation in the NEURON environment. Neurocomputing 58–60, 1117–1122. http://www.sciencedirect.com/science/article/pii/S0925231204001808.

Izhikevich, E.M., 2003. Simple model of spiking neurons. IEEE Transactions on Neural Networks 14 (6), 1569–1572.
Izhikevich, E.M., 2007. Dynamical Systems in Neuroscience. MIT Press.
Maass, W., 1997. Networks of spiking neurons: the third generation of neural network models. Neural Networks 10 (9), 1659–1671.
Mayrhofer, R., Affenzeller, M., Prähofer, H., Höfer, G., Fried, A., Fried, E., 2002. DEVS simulation of spiking neural networks. In: Cybernetics and Systems: Proceedings EMCSR 2002, vol. 2, pp. 573–578.
Mouraud, A., Puzenat, D., Paugam-Moisy, H., 2005. DAMNED: a distributed and multithreaded neural event-driven simulation framework. Computing Research Repository abs/cs/051.
Muzy, A., Zeigler, B.P., Grammont, F., 2017. Iterative specification as a modeling and simulation formalism for I/O general systems. IEEE Systems Journal 11 (4), 1916–1927. https://doi.org/10.1109/JSYST.2017.2728861.
Tang, Y., Zhang, B., Wu, J., Hu, T., Zhou, J., Liu, F., 2013. Parallel architecture and optimization for discrete-event simulation of spike neural networks. Science China Technological Sciences 56 (2), 509–517. https://doi.org/10.1007/s11431-012-5084-2.
Tonnelier, A., Belmabrouk, H., Martinez, D., 2007. Event-driven simulations of nonlinear integrate-and-fire neurons. Neural Computation 19 (12), 3226–3238.
Vahie, S., 2001. Dynamic neuronal ensembles: neurobiologically inspired discrete event neural networks. In: Discrete Event Modeling and Simulation Technologies: A Tapestry of Systems and AI-Based Theories and Methodologies. Springer-Verlag.
Zeigler, B., 2005. Discrete event abstraction: an emerging paradigm for modeling complex adaptive system. In: Adaptation and Evolution (Festschrift for John H. Holland).
Zeigler, B.P., 1975. Statistical simplification of neural nets. International Journal of Man-Machine Studies 7 (3), 371–393. https://doi.org/10.1016/S0020-7373(75)80018-6.

CHAPTER

OPEN RESEARCH PROBLEMS: SYSTEMS DYNAMICS, COMPLEX SYSTEMS

24

CONTENTS

RODRIGO CASTRO WAS THE PRIMARY AUTHOR OF THIS CHAPTER

In this final chapter we use the developments in Systems Dynamics, a well-known M&S methodology, to expose challenges to the theory presented in this book posed by global systems modeling. We first lay the groundwork by relating concepts of Systems Dynamics to those of systems theory and DEVS. Our aim is to give readers of the book a window into the universe of open questions that the theory M&S of may help to address[1]

[1] This chapter was contributed by Dr. Rodrigo Castro.

Theory of Modeling and Simulation. https://doi.org/10.1016/B978-0-12-813370-5.00035-3

24.1 SYSTEMS DYNAMICS, DEVS, AND CHALLENGES FOR M&S OF COMPLEX SYSTEMS

24.1.1 INTRODUCTION

In TMS76 the DYNAMO simulation language was introduced to illustrate continuous system modeling within general systems theory. DYNAMO was for many years the standard language to encode the System Dynamics formalism and its models. Since then it has formed a broad community of System Dynamics users, mainly in the social sciences. In this edition we return to the relationship between the System Dynamics modeling paradigm and system-theoretic based modeling and simulation. Since the concepts, tools, and models of both communities have greatly matured, the time is ripe to offer a taste of the synergy that could arise by bringing the two methodologies together. The spirit remains the same, though: to understand how to convert complex networks of interconnected models into a form compatible with the prototypical simulation procedures developed in this book. The following aims to briefly overview some promising directions for future research.

24.2 SYSTEM DYNAMICS

SD BASICS

System Dynamics (SD) is a modeling methodology developed since the late 1950's at the MIT School of Industrial Management by the engineer Jay Wright Forrester (Forrester, 1961).

SD was developed as a strategy for inductive modeling, i.e. to model systems where the behavioral meta-laws governing the constituents of the system are not known. Typical systems amenable to be described with SD are those for which complexity hinders the overall understanding by relying exclusively on domain-based first principles (e.g. demographic, biological, economic, environmental, health, and managerial systems, to name a few). Thus, the structure and behavior of a system are built up essentially from empirical observations.

System Dynamics is rendered also as a low-level modeling paradigm. Dynamics are mostly described by basic components of type "level" (or "stock") representing state variables, and "rate" representing first order differential equations. SD operates on a continuous-time, continuous-state scheme (see Chapters 3 and 19). A typical SD model is built as an interconnection of levels and rates that influence each other, yielding a (usually non-linear) set of ordinary differential equations of order n for n stocks. In 2017 the SD discipline celebrated 60 years of continued growth, with tens of published books and volumes of specialized journals,[2] spanning dozens of application domains. A key aspect contributing to the wide spread adoption of SD is its visual appeal and intuitive "bathtub" metaphor, depicted in Fig. 24.1.

The level of water in the bathtub is increased at a rate controlled by an intake valve, and decreased at a rate controlled by an outtake valve. The source providing the flow entering a stock (and not coming from another), and the sink of material leaving a stock (and not entering into another) are abstracted away with "cloud" elements. These are "unmodeled aspects" of the system (i.e., boundaries of the model).

[2]System Dynamics Society https://www.systemdynamics.org/.

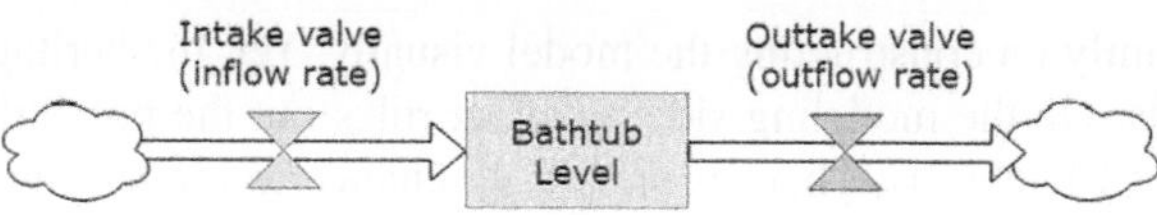

FIGURE 24.1

The System Dynamics bathtub metaphor.

The modeling strategy pursued by SD is to identify first the state variables for the model (those that define the *stocks*) and then identify a *list of influencing variables* that act upon each *rate* for every stock. Each list can include parameters, table functions, or state variables, which can be combined into arbitrary functions to define the law governing that rate (Sterman, 2000).

After levels, rates and functions are placed in a model they are interconnected to transport information across the system. The resulting graphical layout helps the modeler with identifying "causal loops" in the structure. Loops can be either "reinforcing" (positive) or "balancing" (negative).

Although its graphical visualization is appealing, ultimately any SD model represents a set of (possibly non-linear) Ordinary Differential Equations (ODE). The mathematical beast created behind the scenes brings along the full repertoire of difficulties known for centuries: stability, sensitivity to parameters and initial conditions, chaotic behavior, stiffness, etc.

24.2.1 SD STRENGTHS AND LIMITATIONS

In this section we quickly outline some features that have made SD an attractive methodology for systems modeling and simulation as well as some difficulties that limit its robust application.

SD's visual insights proved extremely successful at lowering the entry barrier into the arena of continuous systems modeling and simulation for scholars in soft sciences without a background on mathematical calculus. The concept of visually organizing competing causal loops (reinforcing and balancing) are effective to elicit mental models (even with groups of stakeholders, rather than modelers) and to communicate complex dynamic concepts among modelers coming from varied disciplines and with disparate backgrounds.

But every rose has its thorn. Let us split the discussion into two aspects along the vein of a key concept put forward in this book: the strict separation between modeling and simulation (see Chapter 2).

SD took off under the umbrella of the DYNAMO language, which tightly coupled together model specification with simulation technique. The modeler himself "programs" both his domain-specific knowledge along with the ODE solver (1st order Euler method). The whole model is then conceived around the concept of a fixed time step *DT* (also *h* in some literature), a central variable throughout the code. We could even argue that DYNAMO actually encodes a DTSS rather than a DESS (see Chapter 6).

This limitation was (only partially) overcome throughout the decades. Current environments supporting SD (e.g. Stella,[3] Vensim,[4] AnyLogic[5]) provide a palette of global solvers to choose from, so

[3] https://www.iseesystems.com/store/products/stella-simulator.aspx.

[4] https://www.anylogic.com/.

[5] http://vensim.com/.

the modeler focuses mainly on constructing the model visually. Yet, the heritage of fixed DT lurks in the shadows of SD tools. On the modeling side, this fact rules out the possibility to express generalized discrete events (see Chapter 4). There is simply no chance to model time events or state events that take place at timestamps in between integer multiples of DT. This hinders progress towards sound hybrid systems modeling and simulation (see Chapter 12). On the simulation side, it prevents the use of adaptive step solvers[6] which are key to allow the scaling up of large continuous/hybrid models (at reasonable computational costs). Along the same line it is not possible to profit from attractive scalability properties offered by asynchronous solvers such as those discussed in Chapter 19. We will provide ways to circumvent this limitations in the sections below.

Returning to modeling aspects, the SD bathtub metaphor (along with the reinforcing/balancing loops) rely on a very strong assumption: the rates never assume negative values. If they do, new bidirectional flows are needed, and the interpretation of the causality of loops can get dubious as they can invert roles. We shall revisit this issue shortly. Finally, a modeling strategy pervasive in SD practice is to control rates by means of multiplicative factors. In a nutshell, if a given rate of change is an unknown complex function of n variables $dx(t)/dt = f(x1, \ldots, xn)$ then the approach simplifies it down to modeling $dx(t)/dt = x\_op * f1(x1) * \ldots * fn(xn)$, shifting the problem to identifying n "simpler" functions. In non-linear systems this is known as a "small signal" linearization that produces models valid only within small intervals around a selected operational point $x = x\_op$. This can become a risk when adopted as a generalized recipe (see a discussion of this issue in Cellier (2008) in the context of large socio-ecological models, a topic that motivates further discussions towards the end of the chapter).

24.3 MAPPING SD INTO DEVS

We will show a possible path for retaining the positive aspects of SD while circumventing some limitations. It was suggested in this book (Chapter 17) that DEVS can be seen as a universal common modeling formalism for simulation purposes. The idea is then to bring SD to a common ground that can open the door both to improve some weak aspects and also to make it interoperable with a broader scope of formalisms.

This could be performed in different ways. A possibility is to resort to a library-based approach, with reusable DEVS components that match SD components in a one-to-one fashion. The user is then left in charge of interpreting an SD system and build the equivalent DEVS model interconnecting components from the new library.[7]

Another possibility is an automatic translation-based approach. It also relies on a set of predefined components but it removes the user from the workflow, performing the parsing and building stages automatically. We shall adopt this second approach. Though, automatic algorithmic translation demands

[6]Some SD toolkits such as Vensim (Eberlein and Peterson, 1992) offer adaptive step size Runge Kutta methods. Yet, these are amendments not robust, as some SD modeling primitives (e.g. step, pulse, ramp) will change behavior depending on underlying step adaptation techniques not accessible by the modeler (see https://www.vensim.com/documentation/index.html?rungekutta.htm).

[7]This path was adopted by others to provide SD capabilities in the context of the Modelica language (see https://github.com/modelica-3rdparty/SystemDynamics).

unambiguous source and target formats. We shall rely on two relatively recent technologies to support our endeavor: the XMILE standard for SD (Eberlein and Chichakly, 2013) and the DEVSML standard for DEVS (Mittal et al., 2007; Mittal and Douglass, 2012).

24.3.1 XMILE

XMILE emerged as an alternative to provide SD with an interchange language that would enable conversion of models between different software packages. The goal is to store and share SD simulation models in unambiguous manner across tools, using a common lexicon.[8] In this chapter we will rely on the XMILE format as exported by Stella Architect v1.5.1.

24.3.2 DEVSML

DEVSML emerged to address the problem of model interoperability using DEVSML as the transformation medium towards composability and dynamic scenario construction. The composed models are then validated using universal atomic and coupled standard definitions. Using DEVSML as the communication medium gives the modelers independence to concentrate on the behavior of components in their preferred programming language (e.g. C++ and Java). DEVSML also gives modelers the capability to share and integrate models with other remote models and get integrated validated models back in their own language. In this chapter we will adopt a DEVSML-like format.

24.3.3 CORE MAPPING IDEA

At its essence, the key structure of SD consists of a stock and its input/output controlled rates, which can be seen as a **generalized stock-and-flow submodel**. A reasonable first step is to find a basic unit of structure and behavior in DEVS that is equivalent to the stock-and-flow submodel. Said structure should be replicated as many times as needed, in principle producing as many stock-and-flow submodels as state variables (differential equations) exist in the system.

In theoretical terms, this is equivalent to establishing an SD-to-DEVS morphism at the Coupled Component level (Level 4 in the Morphism Relations Between Systems, see Chapter 14). It describes how a system is composed where components of the systems can be placed into correspondence so that corresponding components are morphic and also the couplings among corresponding components are equivalent.

A possible morphic DEVS model for the stock-and-flow submodel can be seen in Fig. 24.2. We can see four DEVS atomic models: one dynamic **integrator** (e.g., Quantized System Specification, see Chapter 19) and three **memoryless functions,** namely a derivative function (F_der), an increment rate function (F_incr) and a decrement rate function (F_decr).

The correspondence between components of the SD and DEVS models is summarized in Table 24.1.

[8]We could say that XMILE plays the modern role similar to that played by DYNAMO for more than 25 years until the Stella product was introduced in the mid 1980's and then several other GUI-based toolkits emerged, each using custom formats to store SD models.

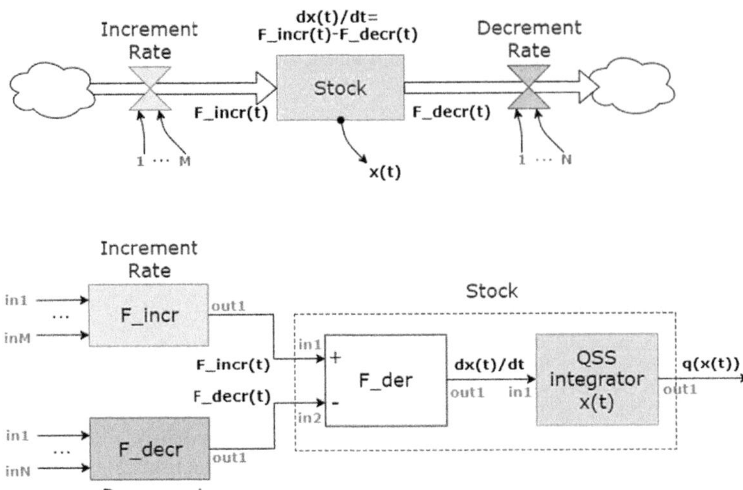

FIGURE 24.2

The basic stock-and-flow structure. SD (above) and a DEVS equivalent (below).

Table 24.1 Correspondence between components of the SD and DEVS models

SD component	DEVS component (atomic model)	Basic parameters	Role	Expression (port based)	Number of ports	Required
Stock	QSS	Initial condition (x0), accuracy (DeltaQ)	State quantization-based integration	out1=QSS(in1)	1 in, 1 out	Yes
	F_der		Calculate the derivative of the state variable	out1=in1-in2	2 in, 1 out	Yes
Increment Rate	F_incr		Calculate the additive term for F_der	out1 = F_incr(in1,. . . ,inM)	M in, 1 out	No
Decrement Rate	F_decr		Calculate the subtractive term for F_der	out1 = F_decr(in1,. . . ,inN)	N in, 1 out	No

QSS integrator (QSS) The initial condition x0 is the same as that of the Stock in SD. The fact that the output of QSS is q(t) while the output of the SD Stock is x(t) does not pose an inconsistency:

q(t) is the quantized version of x(t) as explained in Chapter 19, observing that the quantization error e(t)=x(t)-q(t) is globally bounded in linear systems by the *state accuracy parameter* DeltaQ. Regarding the x(t) output in the SD Stock, it also includes the inevitable integration error produced by *time step parameter* (Delta Time, or "DT" in SD jargon). In the QSS model x(t) is available as an internal state variable.

Derivative function (F_der) This is a static function (i.e., it updates its output at the same timestamp a new event arrives at any input port). It calculates the value of derivative of x(t) as a combination of an additive term (in1) and a subtractive term (in2). This function is performed implicitly in SD, because the visual "bathtub paradigm" assumes that the net rate of change of the stock is the combined action of an increment and a decrement rate. This approach relies on the assumption that rates never assume negative values. It is obvious that a **negative increment** rate will not contribute incrementally to the stock, and the same reasoning holds for a **negative decrement** rate (which would actually contribute incrementally). This is an awkward feature of the SD paradigm. In fact, a special "biflow" (bi-directional flow) has been added to SD tools to cope with this limitation, allowing for a rate valve to control both input and output flows. At this point, the bathtub metaphor gets weakened and loses part of its visually intuitive appeal.
In the DEVS approach, F_der will always compute $dx(t)/dt = F\_incr(t) - F\_decr(t)$ without any particular consideration about the signs that its input signals might bear.

Increment Rate function (F_incr) This is a memoryless function that calculates the increment rate, i.e. the additive portion of F_der. It implements a generalized function of the input signals 1 to M.

Decrement Rate function (F_decr) Analogously, this is a memoryless function that calculates the decrement rate, i.e. the subtractive portion of F_der. It implements a generalized function of the input signals 1 to N.

The arrows entering F_incr and F_decr carry information coming from any other parts of the model. In SD, generalized input-output functions can be interconnected through their input and output ports similarly to DEVS models. In SD, input-output functions are encoded into "Converters" and their pictorial representation are small circles with arrows inputting and outputting signals[9] (see the use this functions in the example in the next section).

Finally, different SD tools offer varied "inline generators". This is, the mathematical expression defining a function in an SD Converter can include generator primitives such as PULSE, STEP or RAMP, and also random number generation such as UNIFORM, WEIBULL, etc. A translation of these primitives into DEVS is straightforward, using specialized atomic models to provide each function. We leave this issue open for an advanced discussion, and concentrate here on more basic aspects.

[9] Actually, F_incr and F_decr could be regarded as SD Converters. The reason why we decide not to match them with a Converter but only with a Rate, is that Rates themselves are a very particular kind of converters: those which can modulate Flows for Stocks. Thus, for the sake of clarity and concordance between the SD and DEVS approaches, F_incr and F_decr are kept as special cases, while generalized SD converters can be simply mapped into generalized DEVS functions performing the equivalent instantaneous calculations.

24.3.4 EXAMPLE: PREY-PREDATOR MODEL

A classical domain for SD models is population dynamics. A reference ecological model in this domain is the Predator-Prey type of system, with its simplest form being the Lotka-Volterra model (Lotka, 1956; Volterra, 1928).

A population of predators Pred and a population of preys Prey interact in the following way. Predators hunt for preys and, when encounters happen, a portion 0<k_eff<1 of the energy of preys is absorbed by the predators (defining their birth rate). The system is as follows:

$$d\mathbf{Prey}(t)/dt = \text{-}prey\_death\_r * \mathbf{Pred} * \mathbf{Prey} + prey\_birth\_r * \mathbf{Prey}$$
$$d\mathbf{Pred}(t)/dt = \text{-}pred\_death\_r * \mathbf{Pred} + prey\_death\_r * \mathbf{Pred} * \mathbf{Prey} * k\_eff$$

where:

1. $prey\_birth\_r$ is the natural growth rate of the prey population feeding on natural resources available in abundance
2. $prey\_death\_r$ is the death rate of the prey population due to encounters with predators
3. $pred\_death\_r$ is the natural death rate of the predator population
4. k_eff is the energy efficiency in growing predators from preys

Fig. 24.3 shows the schematic representations of the Lotka-Volterra model both in SD and DEVS. The DEVS model was obtained automatically using a translator from an XMILE source file into a DEVSML target file, applying the morphism relations proposed in the previous section.

Below are illustrative examples of excerpts taken from the XMILE and DEVSML specifications.

The XMILE specification for the Prey <stock> component is shown below.

```
<xmile>
...
 <header>
   <name>Lotka-Volterra</name>
   ...
 </header>
 <model>
         <variables>
           ...
           <stock name="Prey">
                  <eqn>100</eqn>
                  <inflow>BirthPrey</inflow>
                  <outflow>DeathPrey</outflow>
                </stock>
           ...
         </variables>
         ...
 </model>
</xmile>
```

The <eqn>100</eqn> tag sets the initial condition of the state variable to 100 units. The <inflow> and <outflow> tags define the name for the increment and decrement rates, namely BirthPrey and DeathPrey, respectively.

The DEVSML specification for the Prey stock component is the following:

```
<devs>
 <scenario>
   <coupled name="Lotka-Volterra" modelType="Coupled">
                ...
```

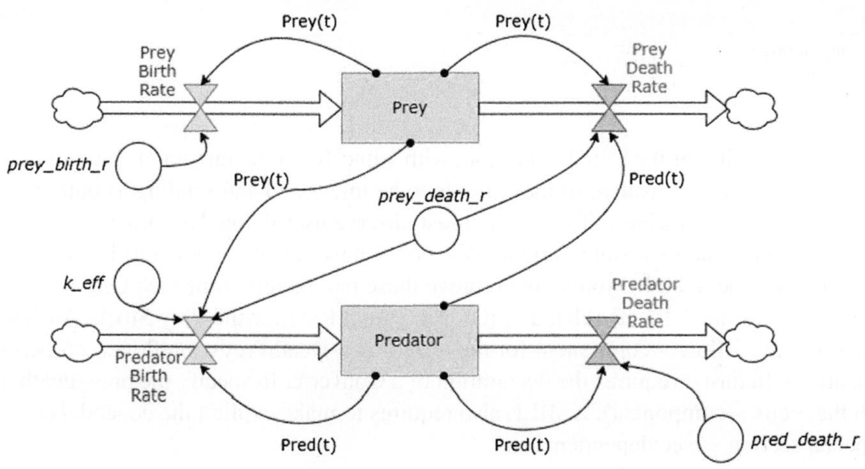

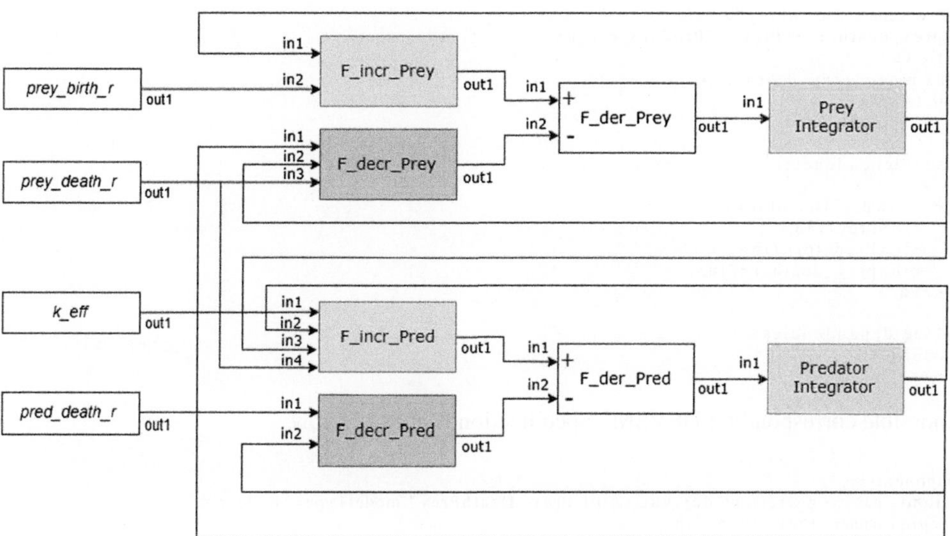

FIGURE 24.3

Model of the Lotka-Volterra system. System Dynamics schematic (top) and DEVS coupled model (bottom).

```
<components>
  <atomic name="Prey population" modelType="QSS">
          <port name="derivative of Prey" type="in" />
          <port name="quantized Prey" type="out" />
    <parameter name="initial state" value="100" />
    <parameter name="relative quantum size" value="1e-2" />
    <parameter name="minimum quantum size" value="1e-4" />
  </atomic>
```

```
      ...
      <components>
    </coupled>
  </scenario>
</devs>
```

We see the mapping of the initial condition, with value 100. One input and one output port are created, taking as input the derivative of the variable to be integrated and sending as output the quantized version of the integrated variable. This model must observe user-defined accuracy requirements for the integration procedure at each state variable. We could have left this choice to be set globally for all integrators (as is done in all SD tools) and remove these parameters from each QSS atomic model.[10]

Below we focus on elements related to the Decrement Rate for the Prey Stock. An XMILE declaration involves the <flow> component for the <outflow>DeathPrey</outflow> property defined in the stock above. In turn it requires the declaration of a Converter to specify the prey_death_r parameter (set with the <aux> component). XMILE also requires to make explicit the dependencies of variables in the special section <isee:dependencies>.

```
...
 <variables>
          ...
   <flow name="DeathPrey">
<eqn>prey_death_r * Prey * Predator</eqn>
   </flow>
   <aux name="prey_death_r">
<eqn>0.1</eqn>
          </aux>
          ...
   <isee:dependencies>
     ...
     <var name="DeathPrey">
       <in>Prey</in>
       <in>Predator</in>
       <in>prey_death_r</in>
     </var>
     ...
   </isee:dependencies>
 </variables>
...
```

A possible corresponding DEVSML specification is shown below.

```
...
 <components>
   <atomic name="decrement derivative of Prey: DeathPrey" modelType="F_decr">
     <port name="Prey" type="in"/>
     <port name="Predator" type="in"/>
     <port name="prey_death_r" type="in"/>
     <port name="DeathPrey_Rate" type="out"/>
     <parameter name="function" value="prey_death_r * Prey * Predator"/>
   </atomic>
   <atomic name="parameter: natural prey death rate" modelType="ConstGen">
```

[10]Yet, there is a strong argument for leaving the accuracy as a per-integrator setting. Namely, among the most salient features of QSS (and DEVS in general) are the asynchronous and distributed simulation capabilities (see the discussion in Chapters 13 and 19). Thus, we purposely preserve the relative accuracy (relative quantum size) and maximum accuracy (minimum quantum size) as per-atomic settings. Finally, regarding practical usability aspects, a user could just ignore these parameters – leave them not set – and rely on the DEVS simulation tool to request the user for a default QSS accuracy at a global level before simulating the model. (See Context-dependent simulation accuracy adjustment below.)

```
      <port name="prey_death_r" type="out"/>
      <parameter name="prey_death_r" value="0.1"/>
    </atomic>
    ...
  </components>
  <internal_connections>
      ...
    <connection model_from="Prey population" port_from="quantized Prey"
                model_to="decrement derivative of Prey: DeathPrey" port_to="Prey"/>
    <connection model_from="Predator population" port_from="quantized Predator"
                model_to="decrement derivative of Prey: DeathPrey" port_to="Predator"/>
    <connection model_from="parameter: natural prey death rate" port_from="prey_death_r"
                model_to="decrement derivative of Prey: DeathPrey" port_to="prey_death_r"/>
      ...
  </internal_connections>
...
```

The listing above presents DEVS atomic models of type F_decr and ConstGen. We also make explicit the Internal Couplings to wire the input ports of the F_decr model in the context of the overall DEVS coupled model. The full specification will of course require also the QSS model type for the Predator population, the F_der_Prey model to combine the decrement rate with the increment rate, and so on.

The atomic model ConstGen for the constant value generation could be avoided by hardcoding that value directly into the "function" of the F_decr model. Yet, the atomic model option enables desirable flexibilities such as easy replacement by more complex generators (e.g. a slowly varying modulated parameter, yielding a new variable) or easy rewiring to other submodels of the environment within which the population is evolving (e.g. a death rate could be a function of average temperature, in turn related to climate change effects)

24.3.5 EXPERIMENTAL RESULTS

To illustrate the correctness of the approach, we present a simple experiment that shows that the DEVS simulation closely agrees with the SD standard. We used Stella for simulating the SD model, and the CD++ toolkit (Wainer, 2002) for the DEVS model. Stella exported the XMILE specification for the Lotka-Volterra system.

We developed a translation framework that produces the corresponding morphic DEVSML specification, as well as a language-specific model (in this case a DEVSML-to-CD++ translation).

The simulation results are shown in Fig. 24.4 for model parameters prey_birth_r=0.02, prey_death_r=0.1, pred_death_r=0.3, k_eff=0.1. The final virtual simulation time is set to 500 (time unit is arbitrary, and could typically represent days, months, years, etc.).

The integration control[11] for SD was set (at simulation tool level) to second order Runge-Kutta (RK2) with fixed time step DT=1/4. The accuracy control for DEVS was set (at model specification

[11] A detailed comparative discussion about integration performance and error control is beyond the purpose of this section. Suffice it to say that we tuned the RK2 and QSS1 parameters to yield visually indistinguishable results. Due to the very different nature of the two solvers, it can be seen by the naked eye (dots in the curves of Fig. 24.4) that the QSS1 integrators adapt their step sizes independently, and often they require fewer calculations (dots) of the derivative functions than those required by the discrete time counterpart. Note that QSS1 is only a 1st order accurate method, while RK2 is 2nd order accurate. We opted for RK2 instead of 1st order Euler (also available in Stella) because the latter required a DT as small as 1/1100 for comparable qualitative results, making the simulation noticeably slower.

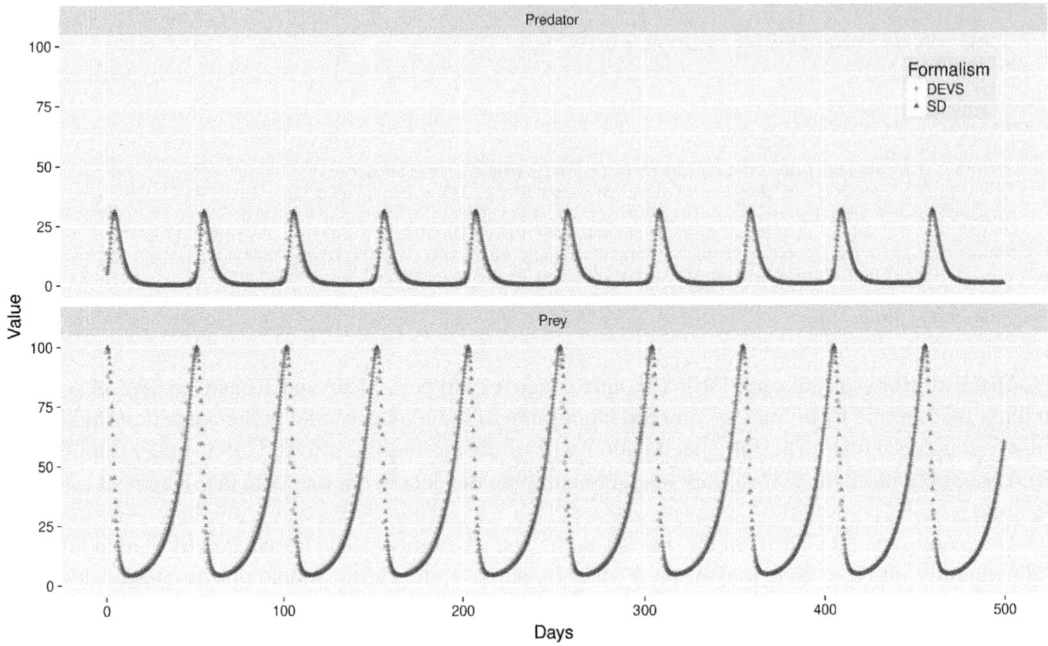

FIGURE 24.4

Simulation results for the Lotka-Volterra system.

level, for each integrator) to first order QSS (QSS1) with adaptive quantum size deltaQrel=1% and maximum accuracy deltaQmin=deltaQrel/100.

These results, obtained with a proof of concept XMILE-to-DEVSML translation tool, suggest that the line of research is worth pursuing, aiming at a comprehensive SD-to-DEVS translation framework, which is a work in progress.

24.4 CHALLENGES FOR SOUND INTERDISCIPLINARY M&S OF LARGE COMPLEX SYSTEMS: THE CASE OF SOCIO-ECOLOGICAL GLOBAL SUSTAINABILITY

The emergence of SD defined a new thread in the evolution of studies in complexity and systems thinking, harvesting both adopters and detractors along the way. We would like to look at this evolution through the lens of large socio-ecological models known as Global Models, an interdisciplinary modeling practice that was born in the realm of SD and then took off as a broad discipline in itself. In this setting, we shall identify some major challenges and offer some important research topics in light of concepts presented this book.

24.4.1 BRIEF HISTORY OF GLOBAL MODELING

First we briefly review some history highlighting aspects relevant to the evolution of the state of M&S practice. In TMS76 the exposition of SD (and its seminal DYNAMO language) was exemplified with World2, a model presented in Forrester (1971) concerned with global socio-ecological sustainability (namely, interactions between global pollution, natural resources depletion, capital and agriculture activities, and population size).

Forrester's book was the third in his series on SD: Industrial Dynamics, Urban Dynamics and World Dynamics. This shows a clear upsurge in the level of systems' scale and complexity faced by the methodology. Predictably, claims made about a potential global population collapse in 50 years backed by a computer simulation of a 5th order ODE can spur lively reactions.

Can we possibly grasp the complexity of the whole world and lump it into such a simplified model? If simulation-based projections operate in timespans of decades, how can we give accreditation to such a model (without waiting for decades to validate it)? No one had gone that far before.

The successor was Meadow's World3, which tried to address criticisms by augmenting some degrees of detail in the model. The corresponding book The Limits to Growth (Meadows et al., 1972) became a best-seller, arguably the first ever backed by computer-based M&S.[12] SD became also known worldwide and the book had two updates (in 1992 and 2004, with 12+ million copies sold). More disaggregated dynamics were introduced for domains such as cohort dynamics, human fertility, ecological footprint, human welfare index, labor utilization, food production, land fertility, service sector, etc. We can interpret the World2-to-World3 evolution as part of a workflow that created a family of models at two different resolutions, by involving different abstractions, approximations, and trade-offs. (Chapters 15, 16, and 21.)

The discipline of global modeling was born, and more than 2 decades followed of intense debate and simulation modeling, with a dozen of competing Global Models produced by research groups internationally (Jacovkis and Castro, 2015). SD played a role in most of them, while others rejected it completely (Herrera et al., 1976). What was obtained is a family of models, all trying to tackle very similar problems but with very different approaches, such as projective vs. normative, spatially aggregated vs. disaggregated (e.g. a "single world" model vs. a per-continent/per-country splitting), etc. The book Groping in the Dark-The First Decade of Global Modeling (Meadows et al., 1982) is an outstanding example of efforts put in systematic cross-comparison of a set of known Global Models, including structured questionnaires to the models' authors.

The M&S of large scale socio-ecological systems witnessed then very interesting developments, supported by interdisciplinary congresses and publications supported by organizations such as The Club of Rome and IIASA.[13]

During the 1990's and 2000's global attention was captured by other kind of models, environmental models, which have many common traits with the traditional Global Models, although more focused

[12]Interestingly, the Limits to Growth book did not include a single line of SD code nor dynamic equations, but only some SD's high level "causal loop diagrams". The underlying computational specification and analysis for World3 was Dynamics of Growth in a Finite World in 1974, a 600 pages thick technical book that sold only a few hundred copies. Thanks to this book the SD specification of World3 became openly accessed, analyzed and criticized, with all of its pros and cons openly exposed. A Stella version was later made available, and currently an XMILE specification can be readily exported.

[13]The Club of Rome https://www.clubofrome.org/, The International Institute for Applied Systems Analysis (IIASA) http://www.iiasa.ac.at/.

in environmental problems (notably climate change) than in social development problems. Perhaps pushed by the global financial crises in the late 2000's and the evident and growing social inequality worldwide, there is now a resurgent interest in global modeling.

24.4.2 CURRENT M&S CHALLENGES IN GLOBAL MODELS

The interplay between human society and nature is again a major concern as our hyperexponential demographic growth (e.g. Varfolomeyev and Gurevich, 2001) imposes a critical ecological footprint (e.g. Kitzes and Wackernagel, 2009). Societies rely on complex processes to sustain or improve their living standards, often at rates exceeding nature's capacity to renew its resources (Rockström et al., 2009). Said processes create social tension due to the asymmetric distribution of wealth, an issue receiving increasing attention from the M&S field (e.g. Castro et al., 2014).

Human activities and ecological processes interact on many scales, resulting first in continuous gradual changes that may cause eventually discrete-event-like disturbances. The latter are often perceived as surprises, since the behavior of environmental systems result from hard to understand complex, hierarchical non-linear interactions of many heterogeneous subsystems, operating at disparate spatio-temporal scales, rates and intensities.

Understanding this complexity calls for M&S techniques able to handle complex systems. Sound and rigorous, yet flexible and efficient M&S methodologies tailored to such uses are needed. But socio-ecological M& S faces today many challenges. Domain-specific models have reached a degree of complexity that makes them more and more difficult to understand in their strengths and weaknesses. Some reasons for this are:

1. As models evolve they tend to accumulate knowledge from various disciplines, which too often roots in basic assumptions of varied reliability. Inasmuch as such assumptions affect the overall results they are generally difficult to analyze, since intertwined with the rest of the model they can't be easily isolated to understand their influence and relevance.
2. Many M&S approaches are not based on a well understood theoretical framework and therefore risk lack of rigor (e.g. state-of-the-art ecological models are considered too complex to be published in a standard mathematical form). They exist only in implemented forms (code that entangles simulation and model) sometimes written in programming languages that hinder a clear distilling of the underlying equations and the understanding of the involved mathematics.
3. The lack of a theoretical framework has several disadvantages: a) Mathematical analysis (e.g. analytical parameter sensitivity) is basically not possible, b) Heuristic coding impedes effective numerical experimentation (e.g., numerics depending on language implementations that may not be generally available), c) Too often it is impossible to exchange submodels to test competing hypotheses encapsulated in them (requiring entire rebuilds from scratch, tedious reverification and retesting of models), d) The integration of models that stem from disparate expertise is highly complicated, resource demanding, and a risky process; as a consequence collaborative research may suffer, becoming too costly and time consuming.
4. Many M&S efforts attempting to integrate subsystems fall short in checking the effects from interconnecting submodels developed independently. For instance, there is a need to automatically validate physical consistency and constraints, such as whether the new integrated system still conserves mass and energy.

Table 24.2 Theory-based Research Challenges in Socio-Economic Modeling

Research Needed in ...	Relevant Chapters in this book	Other related Sources of Theory
More and better formalism translation strategies	This Chapter (SD-to-DEVS) Chapters 10, 11, 12 (Iterative System Specification)	Guide to M&S of Systems of Systems (Chapter 4, 12)
Evolvable multi-resolution modeling	Chapter 15 (Integrated Families of Models), Aggregation (Uniformity and Indifference conditions, Parameter Morphisms) Chapter 20 (Homomorphisms of DEVS Markov Models)	Guide to M&S of Systems of Systems (Chapter 3–8, 16)
Quality and accuracy of model simplifications	Chapter 16 (Approximate Morphisms) Chapter 21 (Approximate Morphisms of DEVS Markov Models)	Model Engineering For Simulation Chapter on Simulation-based Evaluation of Morphisms
Automatic checking of coupling constraints	Chapter 7 (Basic Formalisms: Coupled Multi-Component Systems)	(Castro et al., 2015)
Context-dependent simulation accuracy adjustment	Chapter 19 (QSS, activity measurement of simulation)	Model Engineering For Simulation Chapter on Model complexity analysis
Automated M&S interoperability	Chapters 10, 11, 12 (Iterative System Specification)	Model Engineering For Simulation Chapters on HLA-compliant DEVS Modeling/Simulation and Model Management
Model self-documentation capabilities	Chapter 2 (M&S Framework)	Model Engineering For Simulation Chapter on Model Management ...DEVS.. (Sesartić et al., 2016; Grimm et al., 2010, 2017; Zeigler and Hammonds, 2007)

What has resulted from these difficulties may be seen as an "M&S crisis" in the context of socio-ecological complex systems: Too many "islands of knowledge" rigidly encapsulated within specific models and tools have emerged that are difficult to generalize or reuse. These islands risk pushing M&S in the exactly opposite direction than is required for tackling the challenges of global modeling in an integrative and transdisciplinary manner.

24.5 THEORY-BASED RESEARCH NEEDED

With this background, we present lines along which more theory-based research in M&S is needed in Table 24.2. These are first briefly described:

***More and better formalism translation strategies*:** (exemplified by the SD-to-DEVS translation in this chapter) that are independent of the application domain.

***Evolvable multi-resolution modeling*:** Computational methods to progressively lump/un-lump models, with assistance to choose (and declare explicitly) the level of spatio-temporal granularity at which the model operates and to tie this granularity to the resolution at which the motivating questions can be answered. Models at municipality, city, region, country, continent or world need to be obtained, carrying along the sectors represented (food, energy, education, health, industry, agriculture, natural resources, etc.). In turn each sector should offer different levels of aggregation (e.g. population as a whole or as a more detailed aggregate of cohorts with particular dynamics). Parameters at one level can be seen either as: an emergent property determined by faster dynamics at "lower levels" or a boundary condition determined by slower dynamics at "upper levels" (where top level is the most aggregated level, one and bottom level is the opposite, following the framework of bottom-up/top-down modeling).

***Quality and accuracy of model simplifications*:** Automatically assess measures of credibility according to the level of simplification. For instance, a ratio such as Food per Capita vs. Life Expectancy at Birth in a continent can be an emergent result of complex underlying dynamics at varied sectors. Yet, for some questions to be addressed, this relation could also be encoded as a memoryless input/output function, based on a regression made on statistical sources describing said ratio for each country in the continent. It should be possible to (perhaps automatically) assess the quality and accuracy of such a regression to determine its impact on the model behavior.

***Automatic checking of coupling constraints*:** When coupling models together some "consistency" aspects need to be considered. For instance, when a model of an industrial sector is coupled with a model of the natural resources from which the industries consume raw material, the rates of consumption should not exceed the availability or rate of renewal of the resources (or at least it should provide the modeler with means to detect and handle explicitly such scenarios). Research is needed to translate models automatically into adequate formalisms capable of systematically checking for physical consistency and constraints.

***Context-dependent simulation accuracy adjustment*:** When model interconnection yields large-sized complex systems (e.g. thousands of state variables) the numerical integration performance can become a serious limitation. Therefore accuracy requirements should be kept at a minimum acceptable. In the case of asynchronous integration techniques each integrator can have its own accuracy requirements. Said requirements could then be made context-dependent on a per-submodel/per-state variable basis, i.e. accuracy can be a function of the role played by each state variable in a larger composite system, so as to minimize the simulation activity in that portion of the system. What are the criteria and possible methods to define context-dependent accuracy requirements?

***Automated M&S interoperability*:** Interoperability involves 1) data exchange compatibility – federates in a distributed simulation need to understand each other's messages which involves syntactic, semantic, and pragmatic agreements, 2) time management compatibility – a correct simulation requires that all federates adhere to the same global time and their transitions and message exchanges are timed accordingly. Research is needed to enable automated construction of mediation to harmonize data exchange and time management among federates which is not supported by today's interoperation standards and environments.

***Model self-documentation capabilities*:** Storing informative meta-data together with model and real system data supports sustainable digital preservation of such data for model reuse over decades and can help advance scientific progress through effective reuse of validated models and replication of simulation results. Separation of model data from the mathematical structures allows the same data to be used for alternative model variants and extensions as well as to apply different sets of parameters to the same mathematical model. Protocols exist to structure the description of Individual- and Agent-based simulation models, mostly tested in the domain of ecological and social systems. These are user-driven, natural language-based approaches. Can we design generalized, automated documentation protocols that are decoupled from the underlying modeling and simulation techniques?

These points are summarized in Table 24.2.

REFERENCES

Castro, R.D., Cellier, F.E., Fischlin, A., 2015. Sustainability analysis of complex dynamic systems using embodied energy flows: the eco-bond graphs modeling and simulation framework. Journal of Computational Science 10, 108–125.

Castro, R., Fritzson, P., Cellier, F., Motesharrei, S., Rivas, J., 2014. Human-nature interaction in world modeling with Modelica. In: Proceedings of the 10th International Modelica Conference, number 96. March 10–12, 2014, Lund, Sweden. Linköping University Electronic Press, pp. 477–488.

Cellier, F.E., 2008. World3 in Modelica: creating system dynamics models in the Modelica framework. In: Proc. 6th International Modelica Conference, vol. 2. Bielefeld, Germany, pp. 393–400.

Eberlein, R.L., Chichakly, K.J., 2013. XMILE: a new standard for system dynamics. System Dynamics Review 29 (3), 188–195.

Eberlein, R.L., Peterson, D.W., 1992. Understanding models with Vensim. European Journal of Operational Research 59 (1), 216–219.

Forrester, J.W., 1961. Industrial Dynamics. Pegasus Communications, Waltham, MA.

Forrester, J.W., 1971. World Dynamics. Wright-Allen Press.

Grimm, V., Berger, U., DeAngelis, D.L., Polhill, J.G., Giske, J., Railsback, S.F., 2010. The odd protocol: a review and first update. Ecological Modelling 221 (23), 2760–2768.

Grimm, V., Polhill, G., Touza, J., 2017. Documenting social simulation models: the odd protocol as a standard. In: Simulating Social Complexity. Springer, pp. 349–365.

Herrera, A.O., Scolnik, H.D., Chichilnisky, G., Gallopin, G.C., Hardoy, J.E., 1976. Catastrophe or New Society?: A Latin American World Model. IDRC, Ottawa, ON, CA.

Jacovkis, P., Castro, R., 2015. Computer-based global models: from early experiences to complex systems. Journal of Artificial Societies and Social Simulation 18 (1), 13.

Kitzes, J., Wackernagel, M., 2009. Answers to common questions in ecological footprint accounting. Ecological Indicators 9 (4), 812–817.

Lotka, A.J., 1956. Elements of Mathematical Biology. Dover Publications, New York.

Meadows, D.H., Meadows, D.L., Randers, J., Behrens, W.W., 1972. The Limits to Growth. Universe Books, New York.

Meadows, D., Richardson, J., Bruckmann, G., 1982. Groping in the Dark: The First Decade of Global Modelling. John Wiley & Sons.

Mittal, S., Douglass, S.A., 2012. DEVSML 2.0: the language and the stack. In: Proceedings of the 2012 Symposium on Theory of Modeling and Simulation-DEVS Integrative M&S Symposium. Society for Computer Simulation International, p. 17.

Mittal, S., Risco-Martín, J.L., Zeigler, B.P., 2007. DEVSML: automating devs execution over SOA towards transparent simulators. In: Proceedings of the 2007 Spring Simulation Multiconference, vol. 2. Society for Computer Simulation International, pp. 287–295.

Rockström, J., Steffen, W., Noone, K., Persson, Å., Chapin III, F.S., Lambin, E.F., Lenton, T.M., Scheffer, M., Folke, C., Schellnhuber, H.J., et al., 2009. A safe operating space for humanity. Nature 461 (7263), 472.

Sesartić, A., Fischlin, A., Töwe, M., 2016. Towards narrowing the curation gap—theoretical considerations and lessons learned from decades of practice. ISPRS International Journal of Geo-Information 5 (6), 91.

Sterman, J.D., 2000. Business Dynamics: Systems Thinking and Modeling for a Complex World, number HD30. 2 S7835 2000.

Varfolomeyev, S., Gurevich, K., 2001. The hyperexponential growth of the human population on a macrohistorical scale. Journal of Theoretical Biology 212 (3), 367–372.

Volterra, V., 1928. Variations and fluctuations of the number of individuals in animal species living together. ICES Journal of Marine Science 3 (1), 3–51.
Wainer, G., 2002. CD++: a toolkit to develop devs models. Software: Practice and Experience 32 (13), 1261–1306.
Zeigler, B.P., Hammonds, P.E., 2007. Modeling and Simulation-Based Data Engineering: Introducing Pragmatics Into Ontologies for Net-Centric Information Exchange. Academic Press.

Index

D

R

S

V

X

Z

Printed and bound by CPI Group (UK) Ltd, Croydon, CR0 4YY
08/05/2025
01864926-0001